ENVIRONMENTAL CHEMISTRY

ENVIRONMENTAL CHEMISTRY

Edited by

J. O'M. Bockris
Flinders University of South Australia
Bedford Park, Australia

PLENUM PRESS • NEW YORK AND LONDON

628.5
B665

Library of Congress Cataloging in Publication Data

Main entry under title:

Environmental chemistry.

    Includes bibliographical references and index.
    1. Environmental chemistry. I. Bockris, John O'M.
QD31.2.E58   628.5   76-21081
ISBN 0-306-30869-X

First Printing — January 1977
Second Printing — October 1978

© 1977 Plenum Press, New York
A Division of Plenum Publishing Corporation
227 West 17th Street, New York, N.Y. 10011

Printed in the United States of America

Contributors

B. G. BAKER, School of Physical Sciences, Flinders University of South Australia, Bedford Park, South Australia

J. O'M. BOCKRIS, Department of Chemistry, School of Physical Sciences, Flinders University of South Australia, Bedford Park, South Australia

J. BRICARD, Laboratoire de Physique des Aerosols, Université de Paris VI, 11 Quai Saint-Bernard, Paris, France

R. R. BROOKS, Department of Chemistry, Massey University, Palmerston North, New Zealand

CHARLES L. COONEY, Department of Nutrition and Food Science, Massachusetts Institute of Technology, Cambridge, Massachusetts

J. CREANGE, Sterling-Winthrop Research Institute, Rensselaer, New York

R. M. E. DIAMANT, Department of Applied Chemistry, University of Salford, Salford, England

E. D. HOWE, Department of Mechanical Engineering, University of California, Berkeley, California

BURTON H. KLEIN, Department of Economics, Division of Humanities and Social Sciences, California Institute of Technology, Pasadena, California

A. W. MANN, Minerals Research Laboratories, Division of Mineralogy, C.S.I.R.O., Private Bag, Wembley, Western Australia

T. MULLINS, New South Wales Institute of Technology, Broadway, New South Wales, Australia

R. W. RUSSELL, Vice-Chancellor, Flinders University of South Australia, Bedford Park, South Australia

SPENCER L. SEAGER, Department of Chemistry, Weber State College, Ogden, Utah

K. SEKIHARA, On leave of the Meteorological Research Institute, Tokyo, Aerological Observatory, Tateno, Ibaraki-ken, Japan

L. E. SMYTHE, Department of Analytical Chemistry, University of New South Wales, Sydney, New South Wales, Australia

D. J. SPEDDING, Department of Chemistry, University of Auckland, Auckland, New Zealand

E. J. STERNGLASS, Department of Radiology, University of Pittsburgh, Pittsburgh, Pennsylvania

H. STEPHEN STOKER, Department of Chemistry, Weber State College, Ogden, Utah

W. STRAUSS, Department of Industrial Science, University of Melbourne, Parkville, Victoria, Australia

Preface

There is no need in the 1970s to explain the writing of a book on "Environmental Chemistry." The despoliation of the environment by man's activities has long been clear to chemists. However, it has been the subject of public debate for a short time—since the late 1960s.

Curiously, there has been little reaction in the textbook literature to reflect this concern. Apart from some brief and sketchy paperbacks for schools, there has not yet been published a substantial review of environmental chemistry. One reason for this is the breadth of the chemistry involved: it could scarcely be covered by one or two authors, for it is as wide as chemistry itself.

The ideal way to write such a book would be to gather a couple of dozen authors in one place and keep them together for 6 months of discussions and writing. This not being very practical, it was decided to do the next best thing and to attempt to network a number of men together in mutual correspondence and interaction, which would lead to a book that had the advantages of the expertise of a large number of persons, and lacked many of the usual disadvantages of the multiauthor book. Thus, synopses of the various articles were sent to each author, and they were encouraged to interact with each other in attempting to avoid repetition and in keeping their symbols uniform and their presentation style coordinated.

In respect to the choice of authors, they have been sought independently of geography wherever good reputation had been heard of in a given scientific area. An exceptionally large number of authors are from the antipodes. Since the present editor has relocated his activities to this area of the world, he has been impressed by the fact that, in countries so relatively unsullied by air and water pollution, such a (relatively) large number of people are working vigorously in areas of environmental chemistry. Perhaps Australian and New Zealand scientists have taken the hint from their frequent visits to the Northern Hemisphere and are inspired to prevent similar things from happening as their countries grow more industrialized.

The next question which had to be decided was the audience. There are two extreme audiences for which a book on environmental chemistry could be written: the scientific public—the readers of the *Scientific American*—or the graduate students who are working in subjects relevant to environmental chemistry and the professionals who are employed to work in the field. We have tried to appeal to both ends of the spectrum by writing a text that we believe will appeal primarily to the undergraduate university student, as the main user, and will also include some persons at both ends of the spectrum.

What of the choice of topics? This is the editor's greatest responsibility. It is a difficult one to make in an area which is so broad. The subjects in the present book were chosen in collaboration with the author's colleagues, particularly those at Flinders University, but also in correspondence with workers in many parts of the world, and with the publishers. Some subjects may be criticized as borderlines to chemistry, e.g., discussions of electrochemical transportation. However, there is no doubt whatsoever that the change of motive power in transportation would be the most important part of a change toward a less-polluting environment, and it is the *chemistry* of batteries and fuel cells, and their possible advances, that is the rate-determining step in this area.

Lastly, a brief word concerning the evolution of chemistry. There has been pessimism in many quarters during the last decade concerning the future of chemistry. The thin front edge in fundamental and theoretical chemistry is full of operators, themselves used by physicists thinly disguised as theoretical chemists. The fat back end of the subject in organic chemistry is being increasingly pulled into biology. And as for that increasingly thin wraith called "inorganic chemistry," who will say nowadays what *that* means?

But it does not seem that there is a case for pessimism when one considers the future of chemistry, as long as one makes a slight change and calls it chemical science. Thus, the realm which fans out in front of chemically oriented scientists and chemical engineers is the enormous one of making man's desire for a comfortable world continue to be realizable without the consequences of pollution and the exhaustion of materials which it is indeed bringing. In short, the chemist is in for very active years; he again must become the center of the science and technology picture, for the essential task now is to make every process on which we now run our works into a nonpolluting recyclic process with zero material loss. There is enough chemistry in that to employ most chemists for (nearly) all time.

J. O'M. Bockris

Flinders University of South Australia
Bedford Park, South Australia

Contents

1

Environmental Chemistry

J. O'M. Bockris

1. The General Cultural Background of the Last Century

Seventeenth-century man lived in a world which, in respect to the amount of energy per capita, differed little from that in which he had lived since he became a recognizable species. In the eighteenth, Newcomen invented a steam engine which pumped water from coal mines; Watt made the steam engine a practical transducer of heat to mechanical work and took out a patent on it in 1796.

The nineteenth century was a time of great optimism. It bore one of the greatest achievements of fundamental science, the electromagnetic theory, from which is derived much of today's technology. Rationalism, coupled with science and engineering, was thought capable of bringing enlightenment and comfort to all and was thought to be wholly good. The novels of Jules Verne[1] and the early novels of H. G. Wells[2,3] reflected this period of optimism in man's ability to fend for himself.

The first quarter of the twentieth century was also a period of major upheaval and progress in fundamental science: it brought the theory of

J. O'M. Bockris • Department of Chemistry, School of Physical Sciences, Flinders University of South Australia, Bedford Park, South Australia relativity, with the realization of the equivalence of mass and energy; and it brought the strange quantum mechanics, in which was widely realized the extremely disturbing concept that different laws govern those worlds which our senses do not perceive. Dirac wrote that in the quantum mechanical equations lay all the possibilities for the theoretical determination of material events.

In the 1920s there was already a strong reaction against rational and scientific thinking in the Weimar Republic of Germany, where man's spirit bubbled up in multicolored display; and the deep plunge into the blood baths of the wars of 1914 and 1939 brought a similar reaction.

Then, in 1939 and 1942, two energy-oriented events of great importance occurred. In the first, Hahn and Strassmann[4] observed nuclear fission. In the second, a team under Fermi made the first energy-producing atomic reactor at the University of Chicago.[5] During the twenty years from 1945 to 1965 there occurred the climax of a period of optimism concerning science and engineering. Scientists, and particularly atomic physicists, were highly esteemed and were looked to as progenitors of a rising living standard. The end of physical work and the automated paradise was seriously considered by many to be only a few decades away for those in technologically advanced countries. The era is ambivalently characterized in Stent's book, *The End of Science and the Coming of the Golden Age.*[6] Stent proposed that the Uncertainty Principle had a macro-equivalent: that realizations of further principles in science were unlikely. Science had run not only into one of its plateaus of progress, but into a final asymptote.* Little more could be conceived by man, since knowledge had become too complex for further progress. He also proposed that the basis for the materialistic Utopia was here, unrecognized only because of inhibitions due to social and economic factors. A limitless supply of energy, obtainable principally from the atom, he thought, would give man the power earlier attributed to gods.

Synthesis is followed by antithesis, and that by thesis. The antithesis was *Future Shock.*[7] The bright and brittle 1960s in the United States brought a diminution of American confidence and hegemony. Instead of glorying in the further results of applied science, Americans—materially the most advanced of men—began to revolt against the pace and some of the results of technology, and in particular against pollution. A loss of spirit could be discerned in the mid-1960s, and became a clear downturn in 1968, a year of inflections in America. In 1970, the United States Congress voted to let other countries build the first supersonic passenger aircraft. A malaise came upon American students, precipitated by Leary and Alpert's experiments at Harvard on perception-broadening drugs. Those who thought they had perceived the whole picture to a greater degree wanted no more to work hard at attaining more material goods. By 1969, those few who made long term projections began to see many icebergs in the path of the further growth of population and the spread of affluence to the two-thirds of the world which lacked it (Fig. 1).[8]

---

* Of course, Stent is not referring to engineering.

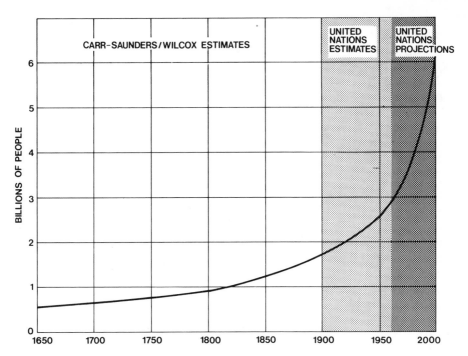

Figure 1. World population since 1650 has been growing exponentially at an increasing rate. Present world population growth rate is about 2.1% per year, corresponding to a doubling time of 33 years.[8]

Some of these icebergs were identified in 1972 in a seminal work by Meadows *et al.*[9] Here, what simple logic could have—but did not seem to have—predicted qualitatively was set forth quantitatively: industry and prosperity gave rise to air and water pollution (Fig. 2).[10] Previous estimates of the number of years in which resources (e.g., metals) would be available had been made with linear extrapolations of growth, neglecting the fact that growth of demand becomes exponential with time. The chill which came with the Meadows *et al.* book was the realization that exhaustion times came much sooner than expected: in respect to metals, many would be exhausted within 1–3 decades (Fig. 3).[9] Fossil fuels, the basis of affluence, were in sight of ending before atomic energy production could be developed to replace them. The race between the materials-depletion technologies and the abundant energy and recycling technologies was being won overwhelmingly by the former. Mankind's "progress," which in 1969 looked hardly worthwhile to a growing number of university students, now seemed threatened with extinction even to the majority who wanted it. Stent's Golden Age had not lasted half a decade before it was seen that a possibility of its doom existed.

But something of the lusty nineteenth-century heart—the Hertzian oscillator and its electromagnetic radiation—still beat. Man's lonesomeness was

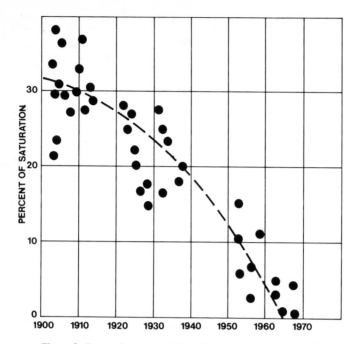

Figure 2. Increasing accumulation of organic wastes in the Baltic Sea, where water circulation is minimal, has resulted in a steadily decreasing oxygen concentration in the water. In some areas, especially in deeper waters, oxygen concentration is zero and almost no forms of aquatic life can be supported.[10]

less than in earlier crises. Electronics made communication simultaneous throughout the world. Small village communities now contained conservation societies; their members followed the declining situation through the media and through the easily available books and sent petitions to central governments.

Childhood's End, Clark's novel about the evolution of a world consciousness, seemed to demand attention.[11] One hundred years after Maxwell and Gibbs man began to reach out with a reaction characteristic of the new time. Quantum mechanics showed that "sense data" deceive: experimental laws of the behavior of things were conglomerate laws, and laws governing the tiny particles which make up macro-objects were entirely different. Time occasionally seemed to reverse. Smith, a former employee of the scientific staff at the Boeing Scientific Laboratory, examined the ability of persons to know which light would next be lit in a random event generator and found a number of persons who could predict these events.[12]

This book is being written in 1976, a time which again presages great changes in fundamental physics and much promise in the relation between science and society. There hangs over this period the great threat of the rising

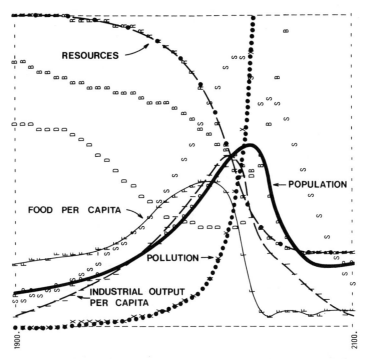

RESOURCES

POPULATION

FOOD PER CAPITA

POLLUTION

INDUSTRIAL OUTPUT
PER CAPITA

1900.   2100.

Figure 3. To test the model assumption about available resources, the resource reserves in 1900 were doubled, keeping all other assumptions identical to those in the standard run. Industrialization can now reach a higher level since resources are not so quickly depleted. The larger industrial plant releases pollution at such a rate, however, that the environmental pollution absorption mechanisms become saturated. Pollution rises very rapidly, causing an immediate increase in the death rate and a decline in food production. At the end of the run, resources are severely depleted in spite of the doubled amount initially available.[9]

price of energy, i.e., its exhaustion, and the parallel threat of air pollution from fossil, and probably from atomic, sources. Concepts which may reverse the situation demand an abundance of clean energy and an extremely high degree of recycling of the products of its use. If population growth can be stopped below some ten billions, and if man has the will to organize the necessary research and development efforts in time, Stent's Golden Age is still to come (Fig. 4).

2. Too Many People?

The concept is quite new of a branch of chemistry devoted to the untoward happenings which follow the injection of a number of industrial side products

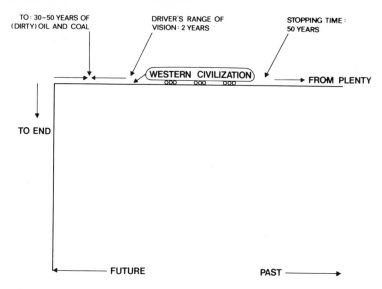

Figure 4. In Western political systems, an elected representative on the average is about 2 years away from the time at which he must prove popular in terms of the present, or else give up his chance of influencing events in the future.

into the air and the water. In 1965, it would have been thought esoteric, as we now regard, say, lunar chemistry.

Man is the original and basic pollutant. Life on this planet has existed for billions of years, animals for hundreds of millions of years, and man for a few million. During all this very long time there has been harmonious ecological development. The individual life, as it evolved, has been subject to the whole, and it has fitted in with the chemical happenings going on around it. The balanced situation was not affected significantly by man until this century. There have been two causes of the increasing disturbance which man must now face. On the one hand, the rate at which chemical and other industries are now pumping materials into the waters and the atmosphere is too much for the natural processes of recycling to deal with. Natural processes, mostly bacterial, do exist for dealing with the ingress of new materials into the waters, but not at the rate at which this has been happening in many of our rivers or in large bodies of water, e.g., the Baltic Sea. The result is visible pollution, and a whole series of changes which come upon the loss of balance in a part of the natural and long-lasting—and therefore balanced—system. Likewise, for the atmosphere, there are photochemical processes which recycle artificial impurities and reattain balance. However, from the beginning of the twentieth century, the amount of unnatural materials which man has placed into the atmosphere has increasingly beaten nature's methods of reestablishing balance. It is the rate of contamination which is crucial, and this depends entirely upon the number of people present in the world and upon their living standards. This then leads us to the background for Chapter 2, which concerns the biochemistry of contraception.

3. *Controlling Pregnancy*

Of all the things which have to be done if we are to attain a balanced world, without being subject to the horrible previous methods of reestablishing balance after accelerated aggregation in numbers, i.e., famine and mass starvation, the most important is to attain rational control of the rate of inception of new individuals. In earlier times, the number of pregnancies per woman must have been approximately the same as those which now occur in communities in which contraception is not practiced. The difference is that, in earlier times, a large number of these pregnancies were terminated naturally before birth, or resulted in children who died at an early age. The spread of the use of antiseptics and of pre- and postnatal medical treatment for pregnant women has, however, been very thorough, even in nonaffluent societies, such as the Indian. What has not kept pace, however, with this population-increasing speed of medical care, has been the corresponding necessary reduction in the number of successful pregnancies per female. The net result must be a vast increase in the rate of population growth in the less educated, and therefore less affluent, parts of the world, those being the economies and resources which are least able to adjust to a large increase in population.

Adequate care of pregnant females must be accompanied by contraception, except perhaps in those very scarce parts of the world which could bear a population increase. But artificial contraception has been well known, available, and practiced for many decades in affluent countries (Don Juan used the condom). There is now little difficulty with population growth in such communities. The difficulty occurs in those communities in which people are less disciplined in handling the artificial aspects of natural processes. The successful practice of contraception demands remembering and bothering, at a time when the libido is streaming toward its natural goal. What is needed is a contraceptive method which would offer a way of reversibly turning off the procreative function. For developments of this kind to become likely, however, it is necessary for chemists to learn something about the basic biochemistry of the processes which lie behind the chemistry of fertilization.

In these considerations, it must not be overlooked that even an ideally simple and successful method of individual contraception needs the volition of one partner in the sexual act. Thus, in India, the social system is such that the only support people can expect in old age is from their children. A maximization of successful pregnancies per female is the desired result. Here, therefore, the social arrangements must change before contraception is desired. Famine and disaster are also less feared when they occur rhythmically, and, to the illiterate population, seemingly unavoidably. Contraceptive chemistry can only be effective at a sufficient level of recognition.

4. *Do Pollutants Affect Our Minds?*

Pollution of the environment is directly related to overpopulation, and one of the feedback effects of some of these pollutants may be upon the brain. This area of knowledge is as yet a tenuous one. Sheer poisons (e.g., CO) and their physiological actions are well known, but there exist other situations, in which parts of biological functioning, including the brain, are very slightly affected below the toxic level by an element such as lead, giving rise to changes in mood and in psychic energy. One may envisage, therefore, the danger of a wide-spread downturn in activity of those in affluent surroundings and in their determination to achieve, which could arise as a result of new materials in the atmosphere at low levels. The effects would creep in gradually over a number of years, and their objective establishment (and separation from parallel effects of affluence) would be difficult.

Such considerations are conjectural. They certainly could happen—and with some pollutants which we know are in the atmosphere. But our knowledge as to whether they do happen, what the mechanism is, and what we can do about it, is in a rudimentary state.

5. *Can We Get Food Chemically?*

While the diminution of the population must be the first goal in a rational environmental policy, there are limits to the degree to which men will agree to kill each other. Hence, any intended rational diminution of the population can only occur over several generations.* During the time in which the present overly large number of people continues to live upon this planet, our natures (identification and empathy) are such that we shall try to feed them. This is becoming increasingly difficult, largely because the fertility of land in many parts of the world is becoming exhausted due to over-use. There are chemical approaches for overcoming this difficulty, such as the utilization of new fertilizers, or of new agriculture in respect to the development of new crop strains (e.g., wheat). However, there is a growing appreciation of the concept that men do occasionally improve upon nature,† and the concept that we could

---

* This is true of the democratically based wish. However, government leaders can be less humane. The further away a people is from sources of masses of food (e.g., in the North American countries and Australia), the easier it is *not* to identify with the populace. When the giving of food tends to deprive their own population, even if overdeveloped, the far-off country and its starving citizens is rapidly forgotten, and the older, more irrational mechanisms for population balancing (i.e., famine and mass starvation) reassert themselves.

† In view of the general area dealt with in this book, and the difficulties caused by man's unsuccessful disturbance of nature, the validity of this statement may be doubted. However, it is obvious that, in some respects, man has improved upon nature, at least in small areas, e.g., computing machines do perform arithmetic more quickly than man. And cars do transport him to more places than those to which he could walk.

make food, instead of having it made by natural photochemical and biochemical processes, is an attractive one and above all a timely one. There are some things which man still has in abundance, in spite of the awesome problem of allowing more people to grow up on the planet than he can easily maintain. Oxygen, nitrogen, CO_2, salt water, carbon, iron, aluminum, and light are all materials that are available in great abundance. If CO_2, water, and energy from light are utilized to combine with nitrogen and carbon, food could, in principle, be synthesized. More than normal chemistry would be needed, it may be thought, for man to be able to create the complex chemicals which form, say, the fleshy section of the legs of young sheep. Perhaps, however, materials acceptable to the digestive system can be devised which are less complex than proteins, fats, and starches. More probably, man may employ the enzymes, the naturally occurring very selective super-catalysts, to do the job. There is a commencement of this attitude in the combining of CO_2 with hydrogen to form formaldehyde, and the union of this substance with air nitrogen, in the presence of enzymes, to form proteins. The key factor is the enzyme. But the increased food supply which could (with sufficient research) arise from this situation must be used carefully, and with ecological considerations in mind. Thus, suppose one were to imagine a hypothetical chemical food factory, manufacturing edible materials from solar sources for energy, with enzymes as catalysts, and CO_2, nitrogen, water, and oxygen as raw materials. If the resulting food were given indiscriminately to all peoples, without demanding of them an effort which would have evolutionary significance, the principal result would be that their health would increase, and therefore their rate of reproduction. The possibilities of a massive production of artificial food are, therefore, to be taken with caution, and only in conjunction with the presence of a wish to use contraception, and with the applied biochemistry necessary to make this wish sufficiently easy and cheap to be achievable by the extremely poor illiterates who are the main source of the population increase.

6. *The Necessary Aftermath of Eating*

The other end of the food process, i.e., dealing with the degraded materials which are feces and urine, must be faced. Here is a difficulty directly resulting from too many people. Earlier species deposited their excreta on the earth, and the earth's chemicals used them in a cycle, one which evolved billions of years in the past. However, the excreta of 10 million people living within a few hundred square miles of concrete and asphalt is a burden too great for the natural processes of recycling. The present method of dealing with sewage from cities represents difficulties which could be alleviated by the development of methods of recycling such materials and upgrading them, perhaps in the house, to reusable substances.

7. *Will There Be the Materials Needed by the People?*

The Limits to Growth (Meadows *et al.*, 1972)[9] will probably be regarded in time as a book of importance, comparable to other great seminal works which have changed the course of man's thinking, e.g., Darwin's *Evolution of Species.* In their book, Meadows *et al.* showed that the rate at which resources are being used up will cause most of them to pass through a maximum rate of production within a short time, less than a century, and many of them in a few decades.* What would be the result of the removal of (i.e., a great increase in the price of), for example, copper, or lead, from our lives? If not dealt with, it would cause a sudden diminution of living standards.

There seem to be two kinds of answers to the problem of resource exhaustion, but only one of them is chemical. First, there may be advances in mining technology. Thus, when it is said, "the metal X will be exhausted in Y years," there are a number of tacit assumptions behind the statement. It is assumed that the price of the item, in constant dollars, will not be greatly exceeded (e.g., by four times), and this then feeds back upon the technology which can be used. However, the development of new technologies (e.g., in communications and transport) sometimes gives rise to a diminution in cost (e.g., as a consequence of the introduction of transistors in radios). Thus it may be that developments in technology will reveal to us greater resources of materials at depths too great for the present means of extraction to be economical. The underground gasification of coal is one such technology, although it does not at present seem to have a great future. Atomic mining with the use of atomic explosives has been discussed, although its applications may be difficult to make safely.

However, in the end, the problem of resources is analogous to the problem of overpopulation and the theory of Malthus. If one *does* have enough food to feed people, they breed more people. But the resultant increase in need for food is multiplicative, not additive; as food increases linearly with the number of people able to make it, the increase in the number of people (all wanting more food) is more than linear, so that the battle for a successful balance of food production and number of mouths to feed is lost by using food to feed an

---

* The ever-present interjection of the free enterprise ethic (with its typical mixture of good and bad contributions to net progress) must not be forgotten. Owners of lead mines, for example, have no wish that the full extent of their resources be known. They desire shortages, or the image of them, to be abroad across the land. The more the customer fears shortage, the more he will be willing to pay for purchasing the resource. New methods in mining would only come with research. Research has to be funded. There, in the decision of what is to be research-funded, is the turnkey to social growth. The funding of research into radically new methods of mining (to get at resources more deeply buried) will be less likely in free enterprise countries, where the corporate leaders will clearly discourage other research on resources. The corresponding tragedy is that research in non-free enterprise countries will be less likely to be organized for successful application and exploitation, because of the absence of the activating factor of profit and the resulting lack of exuberant entrepreneurship.

excess population. Similarly, if more resources are found at a greater depth, and thus more people are able to exist on the earth, more resources will be needed, and there will come a time at which, whatever the technology, no more resources can be found. In the end, better mining would give larger scrap heaps, and it would be better to interrupt this noncyclical process and replace it by a cyclical one as soon as possible. Of course, this is already done to a small extent (e.g., some steel from scrapped cars is recycled). What we need is a maximum degree of cyclicity, the same materials being run through the system again and again, as the fresh water supply rained down upon us from the skies is used for various purposes, rejected to wastes and the sea, and then reevaporated by solar energy so that it is pure again, and so on. Finally, what number of people can exist upon this planet, in some degree of comfort, will depend more upon the completeness of the recycling process for important materials than upon the availability of energy (although the possibilities of recycling clearly depend, along with much else, upon the cost of energy).

8. *Dirty Air*

We must now hasten away from these areas which concern the coming decades and get down to immediate matters, without which this book would not have been written. The first of these is air pollution. Citizens of most cities will have already experienced this phenomenon. In abnormal cases, persons have fainted due to local perturbation (in heavy traffic) of carbon monoxide. A degree of wretchedness is experienced across the centers of our cities as the result of air pollution accompanied by dirt. Most of the considerations within this area of pollution are highly chemical. It is simple to devise, at a freshman chemical level, cures for much of the trouble, and if freshman chemistry—indeed chemistry at any level—were the only consideration, much could be rectified in a few years. The sulfur in coal, emitted in the form of SO_2 from factory chimneys, could be prevented from entering the atmosphere by filtering smokestack effluent to produce sulfuric acid, which could be made in this way instead of by the present synthetic processes. The NO produced in combustive processes could produce commercial nitric acid. The simplicity of these ideas does not prevent them from having merit, though there are sometimes difficulties. Some stacks do not smoke so well if the effluent is being captured and made to pass through filtering devices.* But dealing with the sulfur and the nitric acid emitted from factories would be simple compared with the complex situation arising from automobile exhausts. Here, the chemistry student must

---

* The difficulty is more the balance between economics and incentives. The surrounding populace may need the cleanup, but the owner of the chimney will reduce his profit if he fits the chimney with cleanup equipment. The only thing which will make him fit it is a law. And laws can be fought, distorted, diluted, and disobeyed.

come up against a problem which he may at first find a bit dismaying. A chemist may look into the chemistry of a pollutive problem, e.g., the conversion of sulfur dioxide to produce sulfuric acid, but such a consideration, though essential, is usually not enough. If ecology says that everything must be balanced into a harmonious whole from the point of view of long-term materials balance, there is a corresponding law, in terms of the social, economic, and therefore political, balance. Being able to write a chemical equation, to show that the change in the free energy is negative, i.e., that the process can take place, and finally to show that the kinetics are acceptable so that the process will take place, and with no significant competing reactions, are all necessary processes in the practice of environmental chemistry. However, they are insufficient. If the economics are not acceptable to the corporate groups who determine the economic progress of the community, the process which the chemist works out will not be bought. What is good for the community is a force in the so-called democratic situation (e.g., the injunction brought out by community groups against the building of reactors), but it is only one force, often the weakest, because it is the least coherent and least financially concentrated. If the antipollution advance disturbs the advance of the capital of the corporate groups, it will have little chance of being accepted, however good it may be for the community. Suppose that a change in transportation damages the corporations which makes the present types of cars, to the extent that they must fire a large number of their employees because they cannot afford the continuation of their production, and also because they must purchase new rigs, joists, tools, etc., to make a totally new kind of car, say, the electric car. What then about the net good to the community of nonpolluting electric cars? The same applies to the costs of dealing with air pollution. One must take into account the corporate viewpoint in all these things, as long as it is the controlling one.* One year ahead is the normal time for the thinking in a corporation, and 10 years is regarded as an absurd period in which to make serious economic planning, since the situation cannot be foreseen with precision over this long a time. The chemist must accept these extraneous limitations on his activity. Just as the industrialist thinks it poor that he can no longer eject waste products from his factory into the nearby river and has to buy a plant for recycling them, so the research chemist must accept the fact that he cannot disturb the capital structure of the corporations and be accepted, i.e., continue to receive research grants. Of course, these statements are meant as observations of trends and tendencies rather than as absolutes. They have to be taken into account in developments which may be considered practical, just as the slow rate constant and the expense of a catalyst have to be taken into account to modify the simple initial thermodynamic considerations of an actual chemical situation.

---

* The analogous situation in a Communist country would be taking into account the view of the Party executive. They, too, will guide things for their own advantage, which, of course, will be described euphemistically, like media massage in capitalism.

9. *Car Exhausts*

Of aspects of the general environmental chemical picture which have the greatest impact, and which draw the anxious public's attention, the noxious exhaust pipe is the most well known. There are several radical solutions to the car exhaust pollution problem. These start by getting rid of it, i.e., replacing the internal combustion engine with some other motive force. But the concept of a kind of ecology in social and economic matters is applicable here. It is gradualism which must reign. It may be "unacceptable" at this time to make electric cars, not at all because they perform less well than internal combustion driven cars; that is true and is loudly pointed out, but is not the reason that they are unacceptable, for they would have so many counter-advantages in avoiding noise, and eliminating air pollution, that the public would most likely make the compromise with a lesser top speed if left alone and not molded in their opinions by the corporate use of the media. However, it may not now be acceptable to make them for reasons of disturbing the economy by diminishing the profits of corporate groups and suffering their resulting inability to give a salary to the same number of persons. Hence, the sensible chemist will keep within the balance, just as he will ask himself what he can do within the acceptable situation to reduce the effects of the exhaust pipe. It turns out that he can do a great deal. Many of the unsaturated compounds can be oxidized to CO_2 and, of course, the question is to what degree, and at what cost to the corporation which decides?

10. *A Benign Disaster*

Supposing for a moment that we could get rid of the noxious, unsaturated molecules in exhausts, there is still one substance which it is impossible to remove, because we are trying to make everything go toward it. This is CO_2. At first, this seems to be a very benign pollutant. Indeed, it is a less obvious and damaging one than, say, nitric oxide, carbon monoxide (these are fatal if ingested in sufficient amounts), or even the unpleasant pollutants, the unsaturates, which form the depressing smog. But the benign pollutant can creep up upon us and cause effects which could be most damaging. We are referring here to the greenhouse effect, which has long been discussed and in which the aggregation of CO_2 into the atmosphere causes the fraction of light which is normally reflected away from the earth and out into space to be decreased, thus heating up the earth's atmosphere. Calculations upon this effect are not reliable at the moment, but if one takes into account what has been calculated, it seems as though, after the end of the century, we would be in for unpleasant temperature rises, which could eventually decrease the habitable areas of the earth in the century when world population will be reaching its maximum, particularly in the warmer countries.

11. Feedback?

Some scientists who have agreed with the greenhouse effect theory have been puzzled to learn that a drop in world temperatures has occurred in the last 15–20 years. This contradiction may be resolvable, however, by the observation that the drop in temperature that has taken place since the middle 1950s is associated with the rise of the number of solid particles in the atmosphere, mostly in the form of colloidal particles, e.g., residue from smog. The stability of aerosols in our present atmosphere is one subject in which many considerations are not chemical, and this brings us to an important aspect of the situation in environmental chemistry. Environmental chemistry can seldom exist as a subject by itself. More than most fields, it must have its connections well wired up with surrounding fields. "A chemist who is not also a physicist is nothing at all," is supposed to have been a statement of the great early German chemist, Bunsen. Chemists, in fact, would have decreasing impact if they tried to exist solely as chemists and took no notice of surrounding areas of research. The leading characteristic of chemistry is that it is a central science. Physics is the basic science, but many of the real and actual things, those which concern the environment, are most effectively discussed in terms of chemistry, for chemistry deals with systems at a level too complex for the physicists' approach. So a chemist who wants to do environmental chemistry, and wants to know something of the stability of colloidal particles and aerosols, i.e., pure surface chemistry, should also know about light and reflectivity, meteorological effects, cloud physics, and cloud reflectance. He can then apply his chemistry effectively and be a useful scientist.

12. Dirty Water

Another starting point in environmental chemistry, apart from air pollution, was water pollution. Headlines pass through the mind as one thinks about the parts of this field which have come to public attention during the last decade. Mercury found in fish is a common topic. Parts of the great Rhine river in Germany are said to be "dead," i.e., to be so polluted that they lack the oxygen necessary to drive their bacteria and give them life, so that they cease to act on chemical pollutants in the river. Thor Heyerdahl has stated that he found pollutant materials even 1000 mi from land in a second trip across the Pacific Ocean in 1969. Here, once more, there is a tendency to apply a simple freshman chemistry, namely the bubbling of high-pressure oxygen through tubes through which the lake is made to flow, and thus to revivify it. Such processes may indeed be practical. It is usually a question of the economics of given processes when compared with alternatives.

13. *Small Amounts, but Vital*

One special problem of aqueous pollution has been given much attention in the last few years, and that is pollution by trace elements. The effects of lead on the additives to gasoline do appear to have a special danger, just as trace elements in our diet, e.g., calcium, are essential to our health. There is a physiological ecology, too, i.e., the need for balance, and it turns out that tiny concentrations of certain things in natural waters may be exceedingly damaging, just as the lack of traces of other substances could be dangerous.

14. *Breeder Reactor Fears*

One special type of ionic pollutant is radioactive materials which are produced as a result of reactor use. The existence of such difficulties has been given much publicity. The difficulties do not appear to be restricted to the breakdown of vessels which hold the products of reactor use, but, disturbingly, extend to air pollution, because the processing plant frequently malfunctions, ejecting pollutants in the air.

15. *DDT, etc.*

Much more usual and visible to us, and exceedingly unpleasant, is the organic pollutant. DDT has been the most obvious of those feedback aspects of the application of technology to everyday life, in which everything seemed at first so good but turned out to be finally a cause for worry. Thus, plastics could be the hoped-for replacement in many situations involving the corrosion of metal parts, but what about their disposal? Detergents may make laundering easier, but what of white-flecked lakes?

16. *Fuel Cells and Energy Policy*

Do questions of energy supply really involve chemistry? There is certainly not much chemistry in wind farms, aerogenerators, or the prospective fusion reactors. But there is a good deal of chemistry in what we can do to get ourselves a more efficient (and hence cheaper) use of the remaining fossil fuels, and one of the more important subjects here is the electrochemical fuel cell. This machine, treated with reverence during the space age research of the 1960s, is the only way in which we can obtain more than 50% efficient conversion of our chemical fuels. The general situation in energy supply is that what we have done until now has always involved the fossil fuels (coal, oil, and natural gas), and these are now within a few decades of running out. The trouble is that the exhaustion of these fuels would bring us, and particularly those of us who live in colder climates, to a kind of desperation. Shivering mothers of families will not be concerned with the finer distinction of percentage pollution increases. The Zeus of the modern world is the oil corporation and it owns the residual fossil fuels. One can see here a direct and dramatic connection between the corporate ruling influences on our lives and the possibility of practicing environmental chemistry. The big message here is: The fossil fuels have done damage to us in our atmosphere, but it is no use changing quickly to the atomic fuels, for their pollution of the atmosphere may be worse than that which we have with fossil fuels, although less visible. The controversy as to how clean we can make the atomic reactors is still active, but the worst part of the situation is that many of the answers will not be known for some decades. Some answers, and particularly those concerned with genetic effects, the worst possible damage of all, will not be known for three generations, or 100 years. Thus, many scientists now think that the commitment of so much of our financial weight for the future of energy to the atomic methods is a poorly chosen path, which we should try to retreat from while the going is still (fairly) good. Solar energy is plentiful, and solar energy could be developed to take over all the burdens which we hoped that atomic energy would do with complete cleanness; but, once more, the system is there. Its inertia makes it go on running down the atomic (rather than the solar) research path, and there is only a limited amount which the individual can do about it, unless he aims to become a professional politician or a member of the board in a very large corporation.

17. *Hydrogen?*

One of the most interesting and attractive areas in energy chemistry is the replacement of the power source for cars. Both batteries and fuel cells have been mentioned as possible replacements of the power source; they would produce electric power, and the pollution problem would be solved completely. One of the intermediate possibilities here would be to go through the phase of an intermediate period with hydrogen as a fuel, perhaps liquefied to resemble gasoline. Many advantages would arise from the use of hydrogen in place of electricity. For one thing, hydrogen is able to transport energy cheaply over very long distances, and it may be of use to us to take our energy over long distances in the foreseeable future, e.g., to avoid the dangers of the atomic reactor by utilizing the solar energy plentifully available in the Southern Hemisphere and transferring it over 3000–4000 mi to the northern climates.

Another advantage, and a powerful one, would be that the same rigs could be used by car manufacturers to make hydrogen-fueled cars as are used to make gasoline fueled cars. The change to the less-polluting alternative may actually occur.

18. *Fresh Water*

Finally, there are areas in environmental chemistry which do entirely concern chemistry. One is the desalination of water, the importance of which in extending the availability of certain parts of the earth is clear. It is a subject in which there are many techniques and possibilities. The main thing in all of them is what energy must be used to achieve the final result, the unit of fresh water. The effects of success here (1000 gal of fresh water for, say, 25 cents) would be immense, and might increase by as much as 20% the amount of the world habitable by man.

19. *Analysis*

A topic which is basic to the achievement of many environmental chemical goals is analytical chemistry itself, but analytical chemistry often has to function within environmental chemistry in a fairly special way. In particular, very small amounts must usually be measured. However, the estimates nearly always must be made in mixtures, so that there may be interference between one substance and another, or the detection of tiny traces of one substance in overwhelmingly large amounts of another—in air, waters, food, and sewage—has to take place.

20. *The Systems*

Last of all, an entirely nonchemical topic has been added to our list. This is the relationship between the economic system and the practice of environmental chemistry. This is often referred to within our considerations, and there is no avoiding it. However, it is good to look at it in a rational and quiet way, and to perceive it, rather like the necessary rain, as a not easily changeable part of the whole. For to change, one must know: to what better system? A discussion of that, per se, is more than that of environmental chemistry.

References

1. J. Verne, *Twenty Thousand Leagues Under the Sea*, Paris, 1869.
2. H. G. Wells, *The Invisible Man*, C. A. Pearson & Sons, London, 1897.
3. H. G. Wells, *The Time Machine*, Heinemann, London, 1895.
4. O. Hahn and F. Strassman, *Naturwiss.*, **27** : 11 (1939).
5. E. Fermi, *Collected Papers*, Vol. II, University of Chicago Press, 1965.
6. G. Stent, *The Coming of the Golden Age, A View of the End of Progress*, The American Museum of Natural History, New York, 1969.
7. A. Tofler, *Future Shock*, Random House, New York, 1970.
8. D. J. Bogue, *Principles of Demography*, John Wiley and Sons, New York, 1969.

9. D. H. Meadows, D. L. Meadows, J. Randers, and W. W. Behrens III, *The Limits to Growth*, Potomac Associates, Washington, 1972.
10. S. H. Fonselius, "Stagnant Sea," *Environment*, July/August (1970).
11. A. C. Clarke, *Childhood's End*, Sidgwick and Jackson, London, 1970.
12. H. Smith, *J. Appl. Phys.*, **41** : 462 (1970).

The Limitation of Population

Whatever the future of our environment and whatever the contributions which chemical scientists may make to it toward modifying it, the basic problem is that of population growth. In a sense, therefore, limiting the total world population to a given number, and probably decreasing it, is the most important practical act which can be accomplished in solving our pollution problems. It is clear that chemistry plays a vital part in this act, for it is universally agreed that contraception is a better method for reducing the excessive numbers of people on the planet than the more brutal methods of periodic starvation and disease that have kept population down in the past.

But contraception has been known and available in technologically advanced countries for many years. The present methods are clearly not working in populations with a lesser degree of technological advancement. What is needed, therefore, is an enhanced knowledge of the chemical aspects of the fertilization process, so that research into and development of a practical method of contraception—one that needs only a simple act of initiation and would have effect over a year or more—can be begun.

2

Biochemical Control of Human Fertility

J. Creange

1. Introduction

Throughout history, a number of population control methods have been known and used. Condoms were known in the time of Casanova, intrauterine devices have a long history, and abortion has always been a widely used technique. Biochemical control of conception by the use of menses-inducing drugs has been known for over 2000 years,[1] although truly effective methods have been available on a large scale only in the last quarter-century.[2,3]

Despite all efforts to date, it is clear that population growth worldwide is still going on at a potentially disastrous rate. Indeed, of all the global problems facing mankind today, perhaps none is more central than that of sheer numbers of people. Pollution, utilization of resources, energy production, starvation, the list of the woes of mankind which are exacerbated by population pressures is endless. For this reason research to develop effective and acceptable methods

---

J. Creange •Sterling-Winthrop Research Institute, Rensselaer, New York 12144

for control of human fertility is of the utmost urgency. The problem is not a simple one at either the biochemical or the sociological level.

2. Reproductive Processes

2.1. Male

2.1.1. General

The male reproductive system is designed for the production of vast numbers of motile gametes, or spermatozoa, and their transport and delivery to the female, where fertilization and subsequent fetal development take place. The spermatozoa are formed in the testes, in structures known as seminiferous tubules. These tubules are lined with a specialized layer of cells called the germinal epithelium, which contains the germ cells, or spermatogonia. The spermatogonia in turn are the ultimate source of spermatozoa. They undergo specialized sequences of cell divisions which serve both to replenish the supply of spermatogonia and to form the spermatozoa. The process of sperm production, called spermatogenesis, takes about 74 days from start to finish.[4] Following their release into the tubular lumen, the spermatozoa travel through the tubules and are stored in the ducts of the epididymis for varying amounts of time. At ejaculation they are sent through the vas deferens to the penis. During this time, secretions from sex accessory structures, such as the seminal vesicles, the prostate, and the Cowpers and urethral glands, are added to the semen.

2.1.2. Biochemical Parameters

Control of the reproductive processes in males resides in an integrated system comprising the testes and the central nervous system, in particular the hypothalamus and the pituitary. The system is diagrammed in Fig. 1.

2.1.2a. *Testosterone.* Testosterone, the principal male hormone (androgen), is secreted by the interstitial, or Leydig, cells of the testes. These cells reside outside the seminiferous tubules. The androgen is involved in maintenance of the spermatogenic process; it is responsible for the growth and development of the male sex accessory structures and masculine physical characteristics, such as body conformation, hair distribution, voice, and body chemistry. To some extent, testosterone is also involved in male sexual behavior, especially potency and libido.

2.1.2b. *Pituitary and Hypothalamic Hormones.* Secretion of testosterone is stimulated by a pituitary hormone called luteinizing hormone (LH). This hormone is also sometimes called interstitial cell stimulating hormone (ICSH) in specific reference to its function in the male. Production of LH in turn is governed by a releasing factor, LH–RF, secreted by the hypothalamus. Testosterone interacts with the hypothalamus in a negative feedback loop to control production of LH–RF.[5] LH itself feeds back upon the hypothalamus to

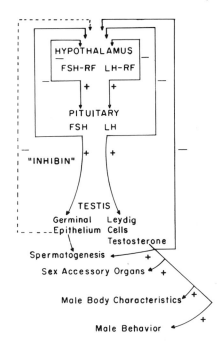

Figure 1. Hormonal control of male reproduction. The + signs refer to stimulatory influences, the − signs to inhibitory influences.

regulate LH–RF production in a "short-loop" system, and there is also evidence of an "ultra-short-loop" system in which LH–RF controls its own production.[6]

A second pituitary hormone, follicle stimulating hormone (FSH), is also involved in the maintenance of spermatogenesis, although its role is not well understood. Secretion of FSH is controlled by a hypothalamic releasing factor, FSH–RF. Feedback control of the FSH system is not clear; testosterone does not appear to be directly involved.[7,8] An unknown factor, "inhibin,"[9] secreted by the seminiferous tubules, has been postulated to be the factor controlling FSH secretion.[10]

2.1.3. *Male Contraception*

2.1.3a. *General.* At this time there is no acceptable method available for the biochemical control of male fertility. Research in this area has lagged behind that for females. One reason for this is that more of the reproductive process occurs in the female, hence, the bulk of clinical and research effort has been there. Another reason is that the process of spermatogenesis is continuous rather than cyclical. That normal ovulation and menstrual cycles could be interrupted by pregnancy, lactation, or a variety of other treatments is an easily observed phenomenon and has been known over the years in both humans and domestic animals. The production of an egg in the female can be inferred from achievement of pregnancy and from a variety of indirect clues, such as circulating hormone levels, body temperature, vaginal mucus characteristics, etc. No such readily identifiable clues are available as indices of spermatogenesis. Similarly, determination of male fertility, the actual end point in question, is not easily done in experimental animals, much less in humans. In a screening procedure involving measurement of fertility of a population of animals, a loss of fertility resulting from a blockade of spermatogenesis might not be manifest for about 2 months in rats, and longer than that in humans. This is due to the length of time required for the production of spermatozoa. The time period is further obscured by the fact that sperm can be stored in the epididymis for extended periods of time.

This inability to measure male fertility rates easily or reliably has forced investigators to use indirect indices for the evaluation of male contraceptive agents. These indices include sperm count, motility, and a variety of biochemical properties of semen. Thus, the end point of screening programs is often not the desired effect, namely reduced fertility, but some other parameter which is then related to fertility rate. In this regard, it has been pointed out that there are no objective methods for the determination of the fertility of semen samples.[11] Fertility can only be established by probability methods.

Another complicating factor is the question of the appropriateness of the animal models available. In females, animal studies have correlated well with clinical experience, even to the level of dosage. This has not been true with males. For example, it is possible to maintain spermatogenesis in animals lacking pituitaries or those with suppressed pituitary activity by administration of testosterone. To date this has not been achieved in man. There is at least a quantitative, if not a qualitative difference between experimental animals and man in this respect.[12] Also, the effect of the pituitary hormones FSH and LH on spermatogenesis is not clearly understood. Although combination of FSH and LH appears to be necessary, interpretation of the findings is obscured by the difficulty in obtaining pure hormones. Interpretations are complicated in humans by the effects of underlying pathological conditions causing the subjects to be given pituitary hormones. Finally, alterations in spermatogenesis are not always correlated with actual measurements of fertility rate.

2.1.3b. *Current Approaches.* Despite the difficulties, there have been a number of attempts to regulate male fertility by biochemical means. These have followed two major lines of approach. On one hand, there have been attempts using hormonally active agents, an endocrine approach similar to that used with females. A second line has been to induce reversible sterility by the use of drugs which are not hormonal in action. A number of fairly potent compounds have been used, including alkylating agents, antimetabolites, nitrofurans, and even carcinogenic hydrocarbons, such as dimethylbenzanthracene.[13] Generally speaking, the toxicity of these nonsteroidal compounds has precluded their use in man.

One class of compounds, the *bis*-(dichloroacetyl) diamines, blocked spermatogenesis in man without showing overt toxicity. Unfortunately, they induced severe alcohol intolerance, a property which eliminated them as useful contraceptive agents.[14]

Another class of compounds, α-chlorohydrin (3-chloropropane-1,2-diol) and related compounds,[15] chemically related to glycerol, induces sterility in males. Toxicity has been a problem in this group, but recently an analogue has been developed which is much less toxic in experimental animals.[16] These compounds are not blockers of spermatogenesis, but induce infertility by some other mechanism. This class of compounds so far appears to be the most promising lead among nonsteroidal drugs.

Using the endocrine approach, it has been known for years that the female sex hormones, estrogens, will suppress testicular function largely by virtue of their potency as inhibitors of pituitary hormone production.[17] However, the side effects of estrogen treatment, including impotence and breast development, render them unacceptable as male contraceptive agents.

Administration of testosterone or other androgenic hormones to humans results in decreased spermatogenesis, also apparently due to suppression of the pituitary.[18] Upon cessation of androgen treatment, a "rebound" takes place in which increased spermatogenesis and in some cases improved fertility is the result. Long-term suppression of testicular activity in humans by androgen administration is not acceptable, however, due to the incidence of side effects including hypermetabolism, hypercalcemia, elevated hematocrit, edema, altered liver function, and plasma lipids.

Progestins are hormones which are not normally present in any great quantity in males. Their function in females relates to the maintenance of pregnancy. These agents have been used to reduce male fertility. They also act at the level of the central nervous system, causing a fall in LH and thereby a fall in testosterone production. There is also a concomitant decrease in libido and loss of potency, causing these compounds to be unacceptable as male antifertility agents.[19,20]

Much the same can be said about antiandrogenic compounds. These agents act by blocking the action of androgens at their sites of action.[21,22] Antiandrogens, which, it should be noted, also often show progestational activity, will reduce libido and potency, and also cause a reduction in size of the sex accessories, and are thereby probably not acceptable as male contraceptives.[23]

2.1.3c. *Prospects.* Overall, it must be stated that progress in the biochemical control of male fertility has been discouraging. There are a number of factors working against success in this area. As mentioned above, there are problems associated with testing and screening a noncyclic and drawn-out phenomenon, such as spermatogenesis. Side effects are another serious problem. In general the nonhormonal agents used have been too toxic, and the hormonal agents cause loss of libido and potency.[24] Males appear highly suceptible to these latter side effects.[25] In this regard, it should be noted that oral contraceptives as commonly used in females essentially mimic the normal physiological state of pregnancy. The biochemical interruption of fertile, ovulatory menstrual cycles is therefore merely an extension of normal physiological processes. There is no comparable state, other than senescence or disease, in which spermatogenesis is blocked in men. Other side effects which occur, such as reduction in gonad size, are not noticeable in women, but are highly objectionable in men. Sexual potency is another uniquely male problem, linking both biochemical and psychological parameters.

One of the most interesting hormonal approaches has been the concurrent administration of androgens and a new steroid drug, Danazol. Danazol is an agent which blocks pituitary production of FSH and LH, but does not behave like an estrogen or progestin.[26] The blocking agent, by lowering FSH and LH production, reduces spermatogenesis; the low doses of androgen are added to compensate for decreased libido. Skoglund and Paulsen[27] found that the androgen exerted a synergistic effect, resulting in a much greater reduction of spermatogenesis with the combination than would have been seen with either drug alone. Similar studies, using combined androgen and progestin therapy, have recently been tried in rats[28] and in man.[29,30,31]. This is an interesting and promising approach which warrants further attention.

2.2. Female

2.2.1. Physiology and Biochemistry

2.2.1a. *General.* The control of female reproductive processes, like that of the male, involves an integrated gonadal, hypothalamic and pituitary system, diagrammed in Fig. 2. It is a complex system designed for the regular, cyclic production and fertilization of a single egg and its subsequent transport, implantation, and fetal development in a uterus which has been suitably prepared for the purpose.

2.2.1b. *Ovum Maturation and Ovulation.* Unlike the male, the human female starts her reproductive years with a fixed and limited number of cells which can develop into fertilizable eggs.[32] During each menstrual cycle, a

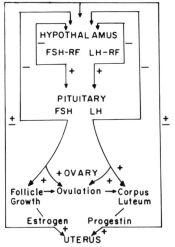

Figure 2. Hormonal control of female reproduction.

dozen or so of these so-called primordial follicles start to grow; as the cycle progresses only one follicle eventually reaches maturity. The remainder undergo a degenerative process called atresia. During this, the follicular phase of the cycle, growth of the follicles appears to require FSH. As its maturation progresses, the follicle itself secretes increasing amounts of estrogen and some progestin. As these hormones reach a certain level, they abruptly exert a positive feedback on the pituitary, causing a sudden, dramatic outpouring of both FSH and LH.[33,34] About a day after this surge, ovulation occurs.

At ovulation, the mature follicle, which has developed into a fluid-filled ball lined with cells and containing the single mature ovum, ruptures and releases the ovum into the fallopian tubes. It then takes about three days for the egg to travel down the fallopian tubes to the uterine cavity.

2.2.1c. *Uterine Growth.* During the follicular phase of the menstrual cycle, the uterine lining, known as the endometrium, responds to increasing steroid hormone levels by growing in thickness and vascularity. In this way, maturation of the ovum is linked in time to preparation of the endometrium for possible implantation of the fertilized ovum, or conceptus.

2.2.1d. *Cervical Mucus.* Also linked to the hormonal mileu are the chemical and physical properties of the cervical mucus. At the time of the FSH and LH surge and ovulation, the cervical mucus is at its lowest viscosity. Changes in the protein and salt composition are also seen, the net result being that sperm penetration and transport are facilitated.

2.2.1e. *Corpus Luteum.* After the follicle ruptures and releases the ovum, the cells remaining behind coalesce into a glandular structure called the corpus luteum. Under the influence of LH, this structure persists for about 2 weeks and secretes large amounts of progestins. This portion of the menstrual cycle is known as the luteal phase. During this time, progestins from the corpus luteum support the uterine endometrium in its thickened, highly vascular state in preparation for implantation. Should fertilization fail to occur, the corpus luteum degenerates, progestin levels drop, the uterine lining is sloughed, and menstruation takes place. The complex hormonal changes that are seen during a normal menstrual cycle are diagrammed in Fig. 3.

Should fertilization occur, followed by implantation, the endometrium is maintained throughout the duration of pregnancy, initially by progestins from the corpus luteum. Subsequently, steroid hormones secreted by the placenta supplant those from the corpus luteum in maintaining the pregnancy. For the duration of pregnancy, cyclic pituitary activity ceases, as does further follicle maturation and ovulation.

2.2.1f. *Sperm Transport.* It is generally believed that the ovum is short lived and can be fertilized for only about 12 hr after ovulation. The life span of human sperm is not known with certainty. Estimates usually are in the neighborhood of 2–3 days; however, motile sperm have been found up to 7 days post insemination.[35] Sperm transport is facilitated at midcycle, near the time of ovulation, not only through the cervical mucus, as noted above, but also up the uterus and fallopian tubes. Sperm have been found in the tubes as little

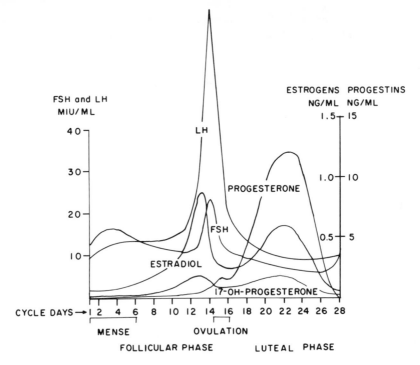

FSH and LH
MIU/ML

ESTROGENS PROGESTINS
NG/ML   NG/ML

Figure 3. Hormonal patterns seen during a normal menstrual cycle.

as 30 min after coitus.[36] Conversely, at other times during the cycle, e.g., during the luteal phase when progestin levels are high, the cervical mucus is thick, its proteins are highly cross-linked, and sperm penetration is markedly inhibited.

    2.2.1g. *Fertilization.* The union of sperm and egg usually takes place in the fallopian tubes about a day or so after ovulation. Very little is known about the mechanism of this phenomenon in humans. Studies in experimental animals have shown that sperm undergo a "maturation" process called capacitation before they can fertilize an egg.[37] This phenomenon, which takes place in the female genital tract, has not conclusively been shown to occur in humans, although there is some presumptive evidence that it may.

    2.2.1h. *Implantation.* This is the process by which the conceptus, having emerged from the fallopian tubes, attaches to the endometrium in order to complete further development. Little is known about the biochemical parameters involved. It is necessary that the endometrium be suitably prepared by prior action of estrogens and progestins; alterations in the hormonal pretreatment of the uterus can prevent implantation. Other agents, such as metallic copper, also interfere with implantation.[38]

2.2.2. *Biochemical Fertility Control*

    There are a number of sites where drugs might act to disrupt or to prevent pregnancy. Follicle maturation and ovulation could be blocked. Fertilization could be prevented by blockading or killing the sperm. The uterine endometrium could be rendered inhospitable to the conceptus, thus preventing implantation or subsequent fetal development. Another method could be disruption of the timing of the ordered sequence of events in the menstrual cycle, so that pregnancy does not occur. To some degree, the presently available biochemical methods of contraception affect all these parameters.

2.2.2a. *Oral Contraceptives.* The biochemical contraceptive methods in widespread use today, with the exception of the copper-containing intrauterine devices (IUD), all rely on the activity of drugs which have either estrogenic or progestational properties. The action of these agents is essentially a distortion of the normal hormonal patterns of the female, resulting in a disruption of the events in the menstrual cycle. Perhaps the best known and most widely used of the "classical" oral contraceptives is the combination of estrogen and progestin, administered daily. Continued dosage with estrogen and progestin causes growth of the endometrium, as would be seen in a normal cycle. However, pituitary activity is suppressed such that FSH and LH are low throughout the cycle; there is no surge of those hormones at midcycle.[39,40]. The alteration in hormonal levels results in the inhibition of follicular growth and maturation. The end result is that ovulation does not occur. When medication is stopped, the endometrium sloughs and menstruation takes place.

A variant of this is the sequential dose regimen, in which estrogen is given alone for about 2 weeks, followed by a combination of estrogen and progestin. The estrogen administration causes endometrial growth and may even cause a surge of LH production. However, LH surges produced by estrogen alone often are not accompanied by FSH, as would be the case in a normal cycle. Other LH surges also take place during the medication period.[41] Thus, ovulation may be prevented by the eccentric and erratic timing of the LH peaks and by the relative lack of FSH.

While the presently available oral contraceptives are highly effective, there are disadvantages. One drawback is that they require faithful adherence to a daily medication schedule. Missed medications can result in ovulation and subsequent unwanted pregnancy. There is also concern because with oral contraceptives healthy people are being given fairly powerful drugs for extended periods of time. The oral contraceptives, as any other drugs, do exhibit some undesirable side effects. Among these are a higher incidence of thromboembolisms, cerebrovascular diseases, hypertension, altered blood sugar metabolism, diminished lactation, and nausea. The literature in this area is extensive and often confusing.[42] Some questions have been raised concerning the possibility of carcinogenicity with the oral contraceptives. This emotional issue remains clouded, partly due to difficulties in proving the lack of carcinogenicity, and to controversies over appropriate animal models for testing this phenomenon. As yet, there has been no clear evidence that the oral contraceptives presently used increase the incidence of cancer in humans. Generally speaking, most of the side effects seen with the oral contraceptives relate to the hormonal activity of the drugs and are similar to the side effects of pregnancy.

2.2.2b. *Minipills.* Progestational agents have been known since 1937 to block ovulation.[43] Low doses of progestins have been shown to be effective antifertility agents in humans.[44] The "minipills" are progestin-only pills, with the active ingredients being in the smallest possible doses. They exert a variety of effects. They alter the timing of events in the cycle; they appear to disrupt the midcycle surge of FSH and LH; and they cause changes in the cervical mucus similar to those seen during the luteal phase of the cycle, when sperm penetration and survival is minimal.

Side effects are still seen with the minipills. The major clinical drawbacks include menstrual irregularities, spotting, and breakthrough bleeding.

2.2.2c. *Drug-Impregnated Vaginal Rings and Silastic Implants.* These are devised for the long-term administration of progestins, either locally, with the rings, or systemically, with the implants. The idea is to eliminate the need for daily medication. The side effects seen are a reflection of the hormonal nature of the drug, with the additional problem of possible effects of the plastic.[45]

2.2.2d. *Long-Acting Preparations.* There are a number of drugs, mostly progestins, which can be injected and which will act over a period of months. These compounds either are absorbed into body fats, from which they bleed out slowly, or else they form a depot at the site of injection.

One drawback of this type of medication is that it is not readily reversible. Also, variation of the active period may occur between individuals or from time to time in the same individual.

2.2.2e. *Intrauterine Devices.* Although IUDs were introduced early in this century for clinical use, they have only recently become a popular means of contraception.[46] Today there are three types of devices. Some are simply inert plastic, variously shaped like loops or coils. The mechanism of action of these types is not well understood, although some shapes appear more effective than others. Some IUDs have copper wire wound around them, since metallic copper *in utero* has been shown to be an effective contraceptive.[47] An IUD which is impregnated with progestins has recently been introduced. The attempt here is to administer the drug locally and thus minimize side effects. The progestin-impregnated IUD may act in part by blocking capacitation of sperm in the uterus.[48]

IUDs are not as effective as the "classical" oral contraceptives, and there is a high incidence of uterine bleeding associated with their use.[49] As noted by Moyer and Shaw[50] this may be critical where people are living on borderline or deficient diets, precisely where effective contraceptive therapy is most needed.

2.2.2f. *Post-coital Antifertility Agents.* In this area there are two main courses of therapy. Abortion can be induced by infusion of prostaglandins; high doses of estrogens given prior to implantation also will prevent pregnancy.

Prostaglandins are natural substances related to long-chain fatty acids and they are found in a wide variety of tissues. Unlike sex hormones, whose primary functions relate to reproduction, prostaglandins are involved in many physiological processes. Their use to induce abortion causes many unpleasant side effects. Further, they must be infused; no orally active prostaglandins have as yet been developed. It will be a major research problem to increase their specificity of action and develop orally active agents.

The estrogens used in this type of therapy must be given in high doses. The side effects seen are correspondingly worse. Therefore, neither treatment would appear to be useful on a scale commensurate with world need.

2.2.3. Prospects

Despite the existence of effective methods for control of fertility in humans, world population growth is still excessive. Part of this problem is political in nature, since use of contraception is not encouraged, nor are methods made available in many areas where the need is greatest. Political and social attitudes, discussion of which is beyond the scope of this chapter, will have to change before there is a worldwide acceptance of any techniques available. For a stimulating discussion of this subject, see the recent articles by Potts[51] and by Venning.[52] Perhaps most illuminating is a quote from the latter article: "Mankind's response to mankind's problem of population growth seems at present only marginally more intelligent than the response of the lemmings."

The biochemical and biological problems involved are formidable enough. There is a continuing effort to develop agents which will be safer and more effective than those now available. Considerable effort is being expended with nonsteroidal agents such as the prostaglandins. With steroids, a significant breakthrough has been the development of Danazol, which, as mentioned above, blocks FSH and LH production, but it not estrogenic or progestational. The blockading of cyclic pituitary activity in females would be expected to result in inhibition of ovulation and disruption of other cyclic events, such as endometrial development, cervical mucus changes, etc. The lack of hormonal activity of this drug suggests that side effects may be less than those now seen with presently available estrogens and progestins.

Another area of potential usefulness has been opened by isolation and characterization of the hypothalamic releasing factors involved in FSH and LH production. Although it is too soon to tell whether this avenue will prove fruitful, analogues of these factors have been synthesized and are being tested.

Whether a truly worldwide effective program for fertility control can be developed in a reasonable length of time cannot be predicted. What is required is a massive synthetic program for both steroids and nonsteroidal agents, coupled with an extensive screening and toxicology program. The screens in females should be set up so as to detect activity both prior to and after fertilization and implantation. Basic research should be encouraged, along with applied clinical research. Finally, and perhaps most importantly, the fruits of all this work must be applied. Meaningful population control programs must be instituted in those areas of the world where population growth threatens the well-being of the people.

References

1. W. Jöchle, *Contraception,* **10** : 425 (1974).
2. J. W. Goldzieher and H. W. Rudel, *J.A.M.A.,* **230** : 421 (1974).
3. G. Pincus, *The Control of Fertility,* Academic Press, New York, N.Y., 1965.
4. C. G. Heller and Y. Clermont, *Rec. Progr. Horm. Res.,* **20** : 545 (1964).
5. J. M. Davidson, in *Frontiers in Neuroendocrinology* W. F. Ganong & L. Martini, eds., Oxford Univ. Press, New York, N.Y., 1969, p. 343.
6. O. Vilar, in *Human Reproduction* E. S. E. Hafez and T. N. Evands, eds., Harper & Row, Hagerstown, Md., 1973, p. 12.
7. R. S. Swerdloff, P. K. Grover, H. S. Jacobs, and J. Bain, *Steroids,* **21** : 703 (1973).
8. C. A. Paulsen, in *Gonadotrophins, 1968* E. Rosemberg, ed., Geron-X Inc., Los Altos, Calif., 1968, p. 163.
9. H. F. Klinefelter, Jr., E. C. Reifenstein, Jr., and F. Albright, *J. Clin. Endocr.,* **2** : 615 (1942).
10. C. A. Paulsen, in *Textbook of Endocrinology,* 4th ed. R. H. Williams, ed., W. B. Saunders Co., Philadelphia, Pa., 1968, p. 405.
11. R. Eliasson, in *Human Reproduction* E. S. E. Hafez and T. N. Evans, eds., Harper & Row, Hagerstown, Md., 1973, p. 39.
12. E. Steinberger, *Physiol. Rev.,* **51** : 1 (1971).
13. B. W. Fox and M. Fox, *Pharmacol. Rev.,* **19** : 21 (1967).
14. C. G. Heller, M. F. Lall, and M. J. Rowley, in *Proc. 3rd Internat. Pharmacol. Meeting 1966,* Vol. 2, E. Diczfalusy, ed., Pergamon Press, London, 1968, p. 61.
15. J. A. Coppola, *Life Sciences,* **8** : 43 (1969).
16. J. A. Coppola and R. J. Saldarini, *Contraception,* **9** : 459 (1974).
17. A. Albert, in *Sex and Internal Secretions,* 3rd ed. W. C. Young, ed., Williams & Wilkins, Baltimore, Md., 1961, p. 105.
18. C. G. Heller, W. O. Nelson, and A. A. Roth, *J. Clin. Endocr.,* **3** : 573 (1943).
19. C. G. Heller, W. M. Laidlaw, H. T. Harvey, and W. O. Nelson, *Ann. N.Y. Acad. Sci.,* **71** : 649 (1958).
20. C. G. Heller, D. J. Moore, C. A. Paulsen, W. O. Nelson, and W. M. Laidlaw, *Fed. Proc.,* **18** : 1057 (1959).
21. F. A. Kincl, M. Maqueo, and R. I. Dorfman, *Acta Endocr.,* **39** : 223 (1965).
22. F. Neumann and R. V. Berswordt-Wallrabe, *Acta Endocr. Suppl.,* **100** : 42 (1965).
23. F. A. Kincl, M. Maqueo, and R. I. Dorfman, *Acta Endocr.,* **39** : 145 (1965).
24. A. J. Bateman and H. Jackson, *Acta Endocr. Suppl.,* **185** : 224 (1974).
25. D. M. DeKretser, *Contraception,* **9** : 561 (1974).
26. W. P. Dmowski, H. F. L. Scholer, V. B. Mahesh, and R. B. Greenblatt, *Fertil. Steril.,* **22** : 9 (1971).
27. R. D. Skoglund and C. A. Paulsen, *Contraception,* **7** : 357 (1973).
28. C. Terner and J. MacLaughlin, *J. Reprod. Fertil.,* **32** : 453 (1973).
29. J. Frick, *Contraception,* **8** : 191 (1973).
30. E. M. Countinho and J. F. Melo, *Contraception,* **8** : 207 (1973).
31. E. D. B. Johansson and K. G. Nygren, *Contraception,* **8** : 219 (1973).
32. E. Block, *Acta Anat.,* **14** : 108 (1952).
33. W. D. Odell and R. S. Swerdloff, *P.N.A.S.,* **61** : 529 (1968).
34. W. D. Odell and D. L. Moyer, *Physiology of Reproduction,* C. V. Mosby, St. Louis, 1971, p. 64.
35. W. H. Perloff and E. Steinberger, *Am. J. Obstet. Gynec.,* **88** : 439 (1964).
36. B. B. Rubenstein, H. Strauss, M. L. Lazarus, and H. Hankin, *Fertil. Steril.,* **2** : 15 (1951).
37. C. R. Austin, *Aust. J. Sci. Res.,* **4** : 581 (1951).
38. J. P. Polidoro, R. M. Culver, S. Thomas, and D. W. Hahn, *Contraception,* **10** : 481 (1974).
39. G. T. Ross, W. D. Odell, and P. L. Rayford, *Lancet,* **2** : 1255 (1966).
40. C. M. Cargille, G. T. Ross, L. A. Howland, and P. L. Rayford, *Clin. Res.,* **16** : 33 (1968).
41. R. S. Swerdloff and W. D. Odell, *J. Clin. Endocr. Metab.,* **29** : 157 (1969).
42. W. D. Odell and M. E. Molitch, *Ann. Rev. Pharmacol.,* **14** : 413 (1974).

43. A. W. Makepeace, G. L. Weinstein, and M. H. Friedman, *Am. J. Physiol.*, **119** : 512 (1937).

44. J. Martinez-Manautou, V. Cortez, J. Giner, R. Aznar, J. Casazola, and H. Rudel, *Fertil. Steril.*, **17** : 49 (1966).

45. F. Bischoff and G. Bryson, *Acta Endocr. Suppl.*, **185** : 296 (1974).

46. E. Richter, *Deutsch. Med. Wschr.*, **35** : 1525 (1909).

47. J. Zipper, M. Medel, and R. Prager, *Am. J. Obstet. Gynec.*, **105** : 529 (1969).

48. A. Rosato, J. J. Hicks, R. Aznar, and E. Mercado, *Contraception*, **9** : 39 (1974).

49. G. Tietze and S. Lewitt, *Ninth Progress Report of the Cooperative Statistical Program, Studies in Family Planning*, Pop. Council, 1970, #55.

50. D. L. Moyer and S. T. Shaw, Jr., in *Human Reproduction*, E. S. E. Hafez and T. N. Evans, eds., Harper & Row, Hagerstown, Md., 1973, p. 309.

51. M. Potts, *Clinics in Endocr. and Metab.*, **2** : 577 (1973).

52. G. R. Venning, *Clinics in Endocr. and Metab.*, **2** : 589 (1973).

Pollutants can have many deleterious effects upon bodily health, but the most insidious effects are those in which the inhaling of pollutant materials affects the brain and results in changes in normal behavior. We are still ignorant about the long-range effects of air pollution, but the effects of lead in the atmosphere have now been given extensive consideration. More attention will need to be given to this problem immediately, so that the effects of the noxious materials that our present energy systems put into the atmosphere on our capability to adjust to our ecosystem can become known, understood, and controllable.

3

The Psychochemistry of Pollutants

R. W. Russell

1. Introduction

For many centuries man has possessed the knowledge that his behavior can be altered—in a sense, "controlled"—by the chemical nature of his external environment. Holding a prominent place among more recent developments of this knowledge has been an empirical understanding of how the selective use of chemicals constitutes a major breakthrough in the treatment of mental illness. Drugs of other chemical configurations produce distortions of normal behavior, some resulting in addiction, with its characteristic tolerance development, physical dependence, and withdrawal symptoms. But, extending even beyond drugs taken for medical or other purposes is the vast ambient chemical environment upon which, through biochemical evolution, man has come to depend, and in which our technologically sophisticated society is setting "ecological traps" by altering its quality. Regardless of the routes by which

R. W. Russell • School of Biological Sciences, Flinders University of South Australia, Bedford Park, South Australia 5042

chemical agents enter the body, the fact that they may affect behavior by interacting with biochemical events is the basic set of interrelations which defines the term "psychochemistry."[1]

According to this view, two major interactions are involved: first, the change in biochemical processes introduced by the entry of chemicals into the body; and second, a consequent alteration of normal relations between biochemical processes and behavior, which appears as changes in the latter. It is a very important principle in pharmacology that the effects of chemicals entering the body by whatever route—orally, by respiration, by absorption, or by injection—are not to introduce anything new, but, rather, to alter the dynamics of biochemical processes that are already there. Changes in these processes may, in turn, be reflected in changes of behavior. Indeed, there is extensive evidence that subtle changes in behavior may appear well before effects are readily detectable in other ways.

It is important for the newcomer to psychochemistry to appreciate that much can be known about the effects of exogenous chemicals on behavior without understanding the mechanisms underlying the effects. Indeed, it has been the rule rather than the exception that drugs used in the treatment of mental illness have been discovered and validated on essentially empirical grounds: the input of chemical substance X produces systematic changes in behavior Y. Such information can be put to use, even when events occurring between X and Y are not known. Because it has received such little attention in the past, the state of knowledge about the "psychochemistry of pollution" is now in the early stages of such empirical study. Its development can, however, benefit from the increasingly sophisticated body of information which has been accumulated by investigations in the fields of neuropharmacology and of psychopharmacology.[2,3] This is the approach of the present discussion.

2. Psychochemistry

The external and internal chemical environments have a common ancestry in the primitive atmosphere which existed one to five billion years ago. One of the earlier basic alterations during the chemical evolution was the change from the reducing to the oxidizing atmosphere, which made possible development of the kinds of biological systems which characterize life on earth today.[4] The economy within this evolution is strikingly illustrated in ". . . an underlying *unity* which implies a *biochemical evolution* much more elaborate and much earlier than the *biological evolution* that gave us all the various forms, performances, and behaviors of the plants and animals today."[5] Constraints of the external chemical environment determined the nature of organisms which could survive.

The appearance during evolution of the capability for an organism to reproduce itself was another landmark in survival. The decoding of DNA, the basic genetic material passed from generation to generation, has shown that development of an individual organism is controlled at a molecular level and proceeds by a wide variety of biochemical reactions. The concept which emerges from these considerations is one in which all other properties of living organisms—including anatomical, electrophysiological, and behavioral—are, in a sense, dependent upon biochemical events. This fact gives such events a kind of priority not only in the creation of new individuals, but also in the processes by which individuals adjust to the dynamics of their external environments.

One important feature of this evolutionary process has been the appearance of the capability of an organism to react as an integrated whole to changes in the external environment, a property which has become known as *behavior.* From its earliest days, the study of living organisms has suggested that certain behavior patterns, often referred to as "instinctive," are genetically determined. These are behaviors which appear without an opportunity for learning through experience: they are chemically "precoded," or "programmed," in the genetic material. Many other behavior patterns are acquired, i.e., learned during an organism's lifetime, by processes which are also limited by genetically determined potentialities and involve a great variety of biochemical events.

The intimacy with which the external and internal chemical environments and the behavioral property of living organisms have been related during the lengthy process of evolution suggests that these interrelationships have been important to survival. It also suggests that changes in any one of the three may be reflected in changes in the others. In fact, the three merge into each other at their interfaces, and one of the basic questions about relations between them concerns the processes by which the influences of one are transduced into effects on the others.

Basic to effects on behavior are neurochemical events which take place within the nervous system, particularly within the central nervous system (CNS). The sheer complexity of the CNS provides opportunities for a multitude of such events to occur. The human brain contains some 100 billion cells, or "neurons," separated from each other by gaps, or "synapses," in the order of hundred-millionths of a centimeter. The cleft between two neurons is shown in Fig. 1. Activity between cells results from the flow of neurochemical transmitter substances across the gaps in amounts as little as billionths of a gram. The activity can be recorded as changes in electrical potential in the magnitudes of millionths of a volt.

Figure 2 summarizes the step reactions involved in the synthesis and, later, inactivation of a typical neurohumoral transmitter substance, acetylcholine (ACh):

$$CH_3-N-CH_2 \cdot CH_2 \cdot O \cdot COCH_3$$

The precursor, choline, is readily acetylated with energy available from

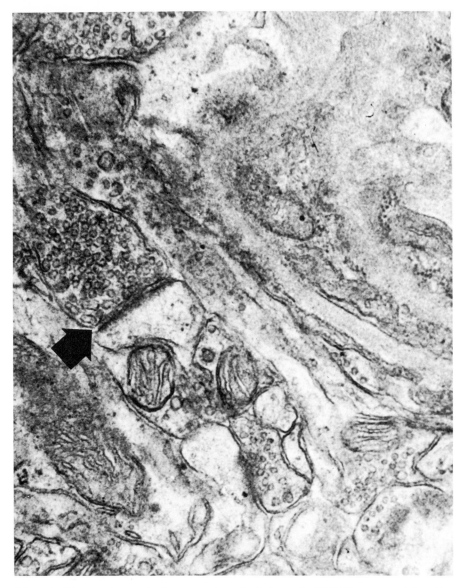

Figure 1. Electronmicrograph of a synapse in rat brain (cortex) × 4800. The arrow points to the synaptic cleft (Cotman).

adenosine triphosphate, coenzyme-A, and the specific enzyme, choline acetyl-ase. ACh is then stored in "bound" form in morphological units, "synaptic vesicles," which appear in Fig. 1 as the small circles located presynaptically. On stimulation of the neuron, "free" ACh is released and diffuses across the synaptic cleft, combining with free postsynaptic receptors to form a transmitter–receptor complex, and initiating activity in the postsynaptic

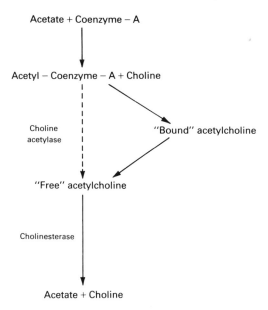

Figure 2. Summary of the step reactions involved in the synthesis and later inactivation of acetylcholine (ACh).

neuron. The inactivating enzyme, acetylcholinesterase (AChE), present in blood and tissues, catalyzes the dissociation of the complex to the free receptor and the metabolic products, acetate + choline. Normal activity in certain areas of the nervous system depends upon these events.

In other areas, different neurohumoral transmitters serve similar roles. Figure 3 diagrams one of these, in which the major element is norepinephrine (NE). Again, effects of exogenous chemicals on this system can best be understood in terms of the events involved in the synthesis, storage, release, and inactivation of the neurotransmitter. The biosynthesis of NE involves the sequential action of three enzymes: tyrosine hydroxylase converts tyrosine to 3,4-dihydroxyphenylalanine (Dopa); decarboxylation of Dopa to 3,4-dihydroxyphenylethylamine (Dopamine) by Dopa decarboxylase occurs rapidly; and dopamine hydroxylase is involved in conversion of Dopamine to NE, which is stored in presynaptic vesicles. Release of the transmitter into the synaptic cleft leads to its action on receptor sites of the postsynaptic membrane. This action ceases when reuptake of NE into the presynaptic membrane occurs, or when NE is metabolized either by catechol-O-methyl-transferase or by monoamine oxidase (MAO).

Chemicals entering the body from the external environment may induce changes in such neurochemical processes by affecting the amount of substrate available for any particular step in the chain, by altering the level of activity of an enzyme essential for catalyzing one of the step reactions, or by interfering with release or reuptake processes. A standard experimental approach in

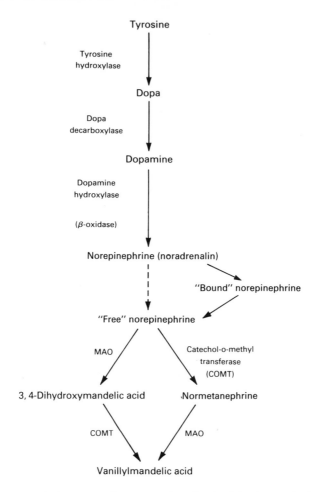

Figure 3. Summary of step reactions involved in the synthesis and later inactivation of norepinephrine (NE).

psychochemistry for studying interactions between biochemical events and behavior has been to use drugs as "tools" to vary the former while measuring concomitant changes in the latter.[6]

There are multiple routes by which chemical substances may enter the body, e.g., oral ingestion, inhalation, absorption by mucous membranes or by skin, injection into particular sites, and pellet implantation. Initial entry can therefore be into a variety of specific internal environments from which a chemical may have to travel some distance before reaching its locus of action within the body. It may have to cross a number of hurdles, which include the blood–brain barrier, boundaries of various tissue cells, and even intracellular barriers. If it reaches its site of action in sufficient concentration, the duration of its effects depends upon such factors as the degree of its localization at the site, its biotransformation in reacting with other substances which it contacts during its journey to the site, and the interrelation of its absorption and excretion.

The general model described in the preceding paragraphs views the behavioral output of living organisms as being dynamically related to biochemical events occurring internally, which are in turn affected by changes in the external chemical environment. *Input* from the latter is reflected in the behavioral *output* of the system. In its search for the *internal biochemical mechanisms* which intervene between input and output, psychochemistry recognizes the key role of the nervous system, while appreciating that sense organs are involved in the input and motor elements, i.e., muscles and glands, at the output ends of the overall system. It is also appreciated that the chief organs of metabolism and excretion, the liver and the kidneys, are inevitably involved in the metabolic fate of most chemical substances entering the body.

The very significant extent to which the behavioral adjustment of an organism can be affected by a specific biochemical abnormality is illustrated in a disorder, phenylketonuria (PKU), resulting from an inherited failure of the body to carry out one particular step in a normal series of biochemical events because of a congenital absence of the enzyme required for that step. In PKU, the fault lies with a defect in the breakdown of the amino acid, phenylalanine, characterized by failure of its normal conversion to tyrosine (see Fig. 4). When the enzyme is absent, phenylalanine is converted into phenyl ketones which have a toxic effect on parts of the CNS. It is generally recognized that the metabolic abnormality is due to a single defective gene and that its site of occurrence is in the liver. Individuals diagnosed as phenylketonurics have some intellectual impairment, often to a severe degree. Accentuation of reflexes, spasticity, and seizures may occur. In contrast are the effects of two other genetic faults, alkaptonuria and tyrosinosis, which are biochemically closely related to PKU, but which are not associated with significant behavioral effects of any kind.

Psychochemistry is concerned with the search for relations between biochemical events within the body and the capacity of an organism to react as an integrated whole to the ecosystem in which it lives, its behavioral adjustment. Whatever the reason may be—genetic fault, malnutrition, or the introduction of exogenous chemicals—changes in normal relations between these two properties of living organisms increase the risks of producing maladjustments and threatening survival. It is within this frame of reference that the psychochemistry of pollutants takes on a special importance.

A major current concern is to what extent organisms which have evolved under one set of environmental conditions can continue to function effectively when the quality of those conditions is significantly modified. With the explosion of population and the giant strides in technological advance which characterize our times have come significant changes in both the composition and the variability of the external chemical environment. In most large cities today, more waste products are being discharged into the atmosphere than can be dissipated. Many of these pollutants are capable of inducing changes in

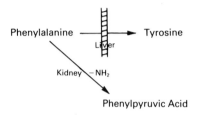

1. Normal

Phenylalanine $\xrightarrow[\text{Liver}]{\text{+ OH}}$ Tyrosine

2. Phenylketonuria

Phenylalanine ---||--→ Tyrosine
Liver

Kidney — NH_2

Phenylpyruvic Acid

Condition	Serum (mg. %)		CSF		Urine (mg. 1 day)
	PhA[a]	PhP[b]	PhA	PhP	PhP
Normal	1.0	0	Trace	0	0
Phenylketonuria	30.0	0–3	7.0	Trace	1000
Phenylketonuria and Diet	1.0	0	0	0	0

[a] Phenylalanine.
[b] Phenylpyruvic acid.

Figure 4. Phenylalanine metabolism (1) normally, and (2) in phenylketonuria (PKU). The table at the bottom of the figure summarizes the relative amounts of phenylalanine and phenylpyruvic acid in serum, cerebrospinal fluid, and urine under the three conditions given at the left.

endogenous biochemical events which, in turn, result in behavioral maladjustments.

To appreciate the significance of this possibility—and certainly to research it—requires an understanding of the *sensitivity* of behavior to changes in the external chemical environment and of the *plasticity* of behavioral reactions to such changes.

3. Sensitivity of Behavior to Changes in the External Chemical Environment

Sensitivity of behavior to changes in the two chemical environments may be defined in terms of *limits* or *thresholds of change* within which the behavior can vary without significant distortion, either temporary or permanent. A sometimes useful analogy is with Hook's law regarding dynamic relations between stress and strain in physical systems, where a system becomes distorted beyond an "elastic limit," or disintegrates from its original structure when a "yield point" is exceeded. In psychochemical terms, the "*basal threshold"* is the minimum concentration or dose of a chemical which induces a significant change in the behavioral parameter measured. Doses below the minimum concentration produce no effects. The other extreme of the sensitivity range is defined by a *"terminal threshold,"* the maximum dose beyond which anomalies of behavior occur, or some other response is elicited that interferes with the behavior being measured.

An example of the quantitatively precise relations between chemical stimulation and an organism's behavioral output will illustrate the meaning of "sensitivity" in the present context. Essential to the survival of more highly evolved organisms, including man, is the maintenance of proper fluid balance in body tissues. Research has shown that, at least in some species, water intake—drinking behavior—is controlled by a functional circuit in the brain which responds to changes in concentration of the neurohumoral transmitter, ACh, described above.[7,8] Drinking behavior can be elicited experimentally even in satiated animals by direct stimulation of the circuit with the proper chemical substance. Experimental manipulation of the concentration of that substance produces changes in drinking, shown in Fig. 5.[1] The S-shaped curve is defined by the two thresholds discussed earlier. Sensitivity is limited at low dosages by a "noise" level analogous to those which characterize communications circuits generally; drinking is not elicited until a critical dose level is reached. It is also limited at high dose levels by the capacity of the neural circuit. Between these thresholds, behavioral output is a linear function of the dose of the chemical stimulation expressed logarithmically, indicating that for each increase in chemical stimulation there occurs a corresponding increase in the behavioral response.

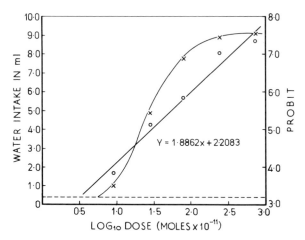

Figure 5. The sigmoid-shaped curve shows the relation between concentration of the cholinomimetic substance, carbachol, injected into the lateral hypothalamus and water intake. The straight line is a log-probit plot of the same data.[1]

Psychochemistry has traditionally analyzed the effects of interactions between behavior and the two chemical environments in terms of the dose level of an external chemical substance to which organisms are exposed, or in terms of some parameter of the internal biochemical processes involved, e.g., the activity level of an enzyme which catalyzes a process or the content of a neurohumoral transmitter substance which is essential to the behavior. Figure 6 summarizes the kinds of relations commonly found when the sensitivity of behavior to acute changes in the chemical environments is studied. At the top is represented the S-shaped curve which is typical of behavioral effects as a function of dose level, being schematically similar to the plot for empirical data shown in Fig. 5 (the basal threshold at *a* and the terminal threshold at *b*). The lower curve illustrates the general nature of relations between behavioral effects and the level of activity of endogenous enzymes. It is typical that effects are not apparent in quantitative measures of behavior until the decrease in enzyme activity reaches a critical level (at *c* in Fig. 6) after which further changes are often associated with exponential changes in behavior as the enzyme loses control of its substrate.

4. Behavioral Plasticity

4.1. Characteristics of Plasticity

Despite their sensitivity to acute changes in the chemical environments, living organisms show remarkable *behavioral plasticity* when such changes are chronic. Systematic adjustments in a wide variety of behavioral patterns have

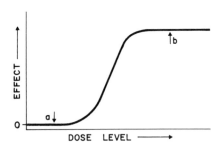

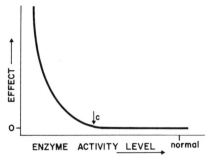

Figure 6. The curve at the top shows the relation between dose level of an exogenous chemical introduced into the body and the consequent behavioral effects. The lower curve illustrates the general relation between level of endogenous enzyme activity and behavioral effects.

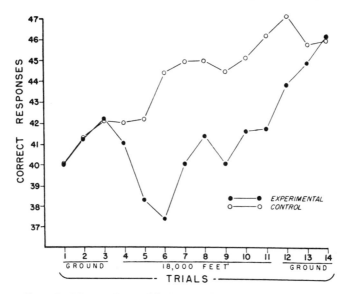

Figure 7. Effects of lowered O_2 level at simulated 18,000-ft altitude on a numerical task. Control subjects remained at ground level.[10]

been reported when man and other living organisms have been subjected to continuing or repeating exposures to unusual chemical environments.[9] The sequence of this adjustment follows a general pattern: chronic exposure is evidenced initially as a change, decrement or increment, in the behavior observed; this acute change is followed by return of behavior to its pre-exposure state, although there may be some oscillation before it reaches that level; withdrawal from the unusual environment may be followed by new behavioral changes, often in a direction opposite to that initially induced by exposure to that environment. This describes the phenomena of "behavioral tolerance" and the so-called "withdrawal syndrome." Exposure to unusual conditions of the natural environment provides classical confirmation of the basic features of this phenomenon of adjustment. Figure 7 shows the effects, on a task requiring simple numerical addition, of lowering the oxygen level in respired air to simulate that at an altitude of 18,000 ft. It is clear that control subjects, exposed to exactly the same experimental conditions as the anoxic subjects except for the decrease in oxygen level, showed improvement in performance at a decelerating rate throughout the experimental period. By contrast, performance of the anoxic subjects was adversely affected initially, but recovered with prolonged exposure at a rate that equaled that of the controls within a very few trials following the return to normal environmental conditions.

Such behavioral tolerance may also develop during contact with polluted environments, as urban populations are exposed to such atmospheric pollutants as carbon monoxide and sulfur oxides which induce significant changes in the body's internal chemical milieu. Some very important illustrations of tolerance development are to be found in persons exposed to special industrial and agricultural environments. For example, studies of individuals with varying degrees of contact with organophosphorus pesticides have shown biochemical effects in the form of significant reductions in the activities of enzymes important to the normal functioning of the nervous system. Clinical studies report that clearly recognizable improvement in symptoms occurs with continued contact, and the only fatal or near-fatal cases have been in instances characterized by acute, massive exposure.[11,12]

Figure 8 summarizes results of an experimental study of the effects of one such compound on the behavior of animal subjects who had learned to press a lever in order to obtain water when thirsty. Having established a stable level of performance under standard conditions of water deprivation, the subjects were injected intramuscularly every third day with diisopropylfluorophosphate (DFP) at a dose sufficient to reduce the level of activity of the brain enzyme, cholinesterase (ChE), to approximately 30% of normal and to maintain it at that level throughout the remainder of the experiment. The initial effect of the treatment was to cause very significant decreases in the behavior. However, this effect diminished with subsequent injections until it disappeared entirely, despite the continuing low level of enzyme activity.

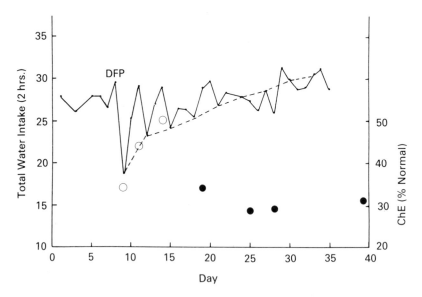

Figure 8. The process of tolerance development in a simple operant response is shown by the dashed curve. The solid line shows the water intake. DFP was administered every third day; its effects decreased with repeated injections. The level of brain ChE activity is shown by the circles (acute DFP by open circles, chronic DFP by solid circles), remaining at about 30% of normal while behavioral tolerance developed.[13]

To these examples of accidental and of experimental exposure to chemical substances which are not normally present in the external environment may be added those instances in which chemical agents are taken deliberately by clinical prescription or for "kicks." There is a broad range of circumstances in which interactions between behavior and chemical environments show that the plasticity of tolerance development is, indeed, an effective process in the adjustment and, often, in the survival of living organisms.

4.2. *Limits of Behavioral Plasticity*

A very important problem in understanding the nature of plasticity in the relations between behavior and chemical environments is to *define its limits.* It is not surprising that effects on the magnitude of acute changes in behavior, and on the time for tolerance to develop during chronic treatment, are systematically related to the dose level, or severity, of an exposure.[14] This general relation is illustrated in Fig. 9. Curve *a* represents the baseline responses of control subjects. Subjects treated with low, subthreshold doses do not differ from controls in their overt behavior, yet there may be very important differences, which will be described in the next paragraph. Curve *b* is schematic of reactions to supraliminal doses: an initial effect disappears with chronic exposure. Curve *c* represents the effect produced by high doses, when behavioral changes occur rapidly and gross physical pathology results in incapacitation. The limits of behavioral adjustments are bounded on one side by no observable change and, at the other extreme, by death.

There is another and more subtle kind of "limit" to adjustment. Even when behavioral tolerance has fully developed, and there are no observable differences between the behaviors of tolerant subjects still on chronic treatment and nonexposed subjects, other differences still exist. These appear most

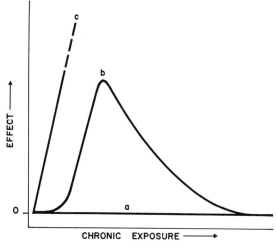

Figure 9. The limits of behavioral plasticity are shown schematically: *a* represents effects of a low dosage which produces no behavioral change; *b* represents the typical process of behavioral tolerance development; and *c* represents effects of a high dosage, which leads eventually to a damaged organism and no behavioral recovery.

dramatically when the two types of subjects encounter other changes in their chemical environments. In our laboratory, tolerant subjects have in some instances proved to be much more sensitive than normal animals to challenges by new chemicals and, in other instances, much more resistant.[15] Which of these states occurred depended on the chemical characteristics of the substances involved. However, the major point here is that adjustment to certain changes in the chemical environments may place limits on behavioral adjustment to chemical changes of other kinds.

The existence of the limits, plus the fact that within them both the magnitude of acute changes in behavior and the time for tolerance to develop are systematically related to the dose level, or severity of the exposure, suggests that at least two biochemical processes are involved in adjustment: the first underlying the acute changes in behavior, and the second involved in the development of tolerance. In Fig. 10, the former is schematically represented by process A and the latter by process B. The typical pattern of behavioral tolerance development is observed when process B counteracts the initial effects of process A. With high levels of exposure to a chemical agent, process A may occur too rapidly for process B to compensate, as represented in A'. In this case, behavior becomes completely distorted and the exposure may prove fatal.

5. Ecological "Traps"

When one understands the nature of the interactions discussed above, it is not difficult to appreciate that, by his own actions, man has been and continues to create ecological traps in which he himself becomes ensnared. The psychochemistry of pollutants is concerned with the effects on behavioral adjustment of an external chemical environment whose quality is being

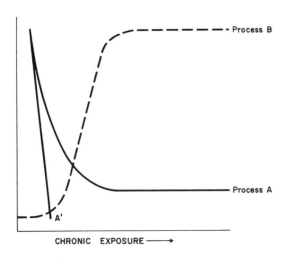

CHRONIC EXPOSURE ⟶

Figure 10. Theoretical processes underlying behavioral plasticity: process A is induced directly by the effects of an exogenous chemical on internal biochemical events underlying behavioral change; process B is hypothesized to begin shortly after process A and to counteract effects of A on behavior. A' shows the case at high dosages, when biochemical changes occur so rapidly that process B does not operate fast enought to counteract effects of A'.

degraded in many of man's habitats.[16] Other characteristics of these traps are discussed elsewhere in this text. Views from the several vantage points must eventually be pooled in order to provide the integrated body of working knowledge required to solve such ecological problems confidently. Our knowledge about the processes by which changes in the external chemical environment are transduced into changes in biochemical events within the body, and these, in turn, into alterations of behavior, is growing. At the same time, there are great pressures to provide solutions for very practical and often immediate problems of environmental management in which behavior is one of the major features. Some examples may make the reasons for these pressures more understandable. It is important to remember that "psychochemistry" has a special interest in behavior, and therefore examples will be chosen in which the behavioral component is prominent, even though other effects of pollutants, e.g., pathophysiologic,[17] are often involved.

5.1. *Heavy Metals*

One of the most dramatic instances of the effects of water pollution on man began to unfold in May, 1956, when a physician reported to the Minamata Health Center of the Kumamota Prefecture in Japan on the hospitalization of a patient who had developed neurological symptoms of then unknown origin.[18] The causal agent for what has become known as "Minamata disease" was methyl mercury that had drained into the sea with waste water from a chemical plant and had polluted marine products eaten by the local inhabitants. The disease, which is considered to be incurable, includes particular involvement of the CNS and has extensive behavioral as well as bodily symptoms: severe disturbances of speech and of gait, loss of normal hearing, narrowing of the visual field and other sensory disturbances, and, occasionally, hallucinations.[19] These effects produce impairment of normal behavior either through distortion of sensory input about the external environment or through deficiencies in the capability to carry out responses required for adjustment to that environment. Experimental studies with animal subjects are beginning to add information to these clinical data about how subtle the consequences of exposure to methyl mercury may be, even into the next generation. For example, the offspring of mice exposed during early pregnancy are apparently unaffected during postnatal development: brain weight, protein, ChE, etc., are not altered; however, significant differences in general behavioral activity and in motor coordination between experimental and control subjects can be measured later in life.[20]

5.2. *Gases*

The ecological traps man is setting for himself are not always as overtly apparent as the Minamata incident. Adverse pathological and clinical effects have been associated with the exposure of human beings and of animals to those atmospheric pollutants contained in automobile exhaust products, and to those, e.g., sulfur oxides and associated small particles, contained in industrial and household stack effluents. Such pollutants may also affect behavior. For example, exposure of urban populations to carbon monoxide (CO) becomes a matter of immediate concern when we consider that there are certain urban areas where episodes have occurred of community contact with CO sufficient to cause very significant increases of its content in body fluids. CO is not a component of normal dry air. Although colorless, odorless, and nonirritating, it can be very toxic. The maximum allowable concentration for health workers in industry, for an 8-hr working day, is 50 parts per million (ppm) in the respired air. A concentration of 1000 ppm can produce unconsciousness in 1 hr and death in 4 hr; effects on behavior occur at concentrations considerably lower than those which produce clinically pathological signs in the usual medical sense. In the body, CO combines with the hemoglobin of the blood, forming carboxyhemoglobin (COHb), which interferes with the blood's normal function in the transport of oxygen. The CNS is one of the tissues most sensitive to a deficiency of oxygen. Research has shown that, by such action, exposure to CO, even at relatively low levels which produce 2–5% COHb, may impair behavior in ways so subtle as to go unobserved by the person involved.[21] The effects of such relatively small changes in the internal chemical environment are first observable in more complex skills such as the perception of time, producing no apparent impairment in simpler behaviors until the chemical changes are of greater magnitude.[22] Under such conditions, impairment is likely to occur when attempts are made to maintain a high level of attention to a monotonous, routine task, e.g., prolonged turnpike driving or heavy machine operation. Epidemiological appraisals suggest that significant positive correlations can be found between automobile accidents and the presence of other pollutants, e.g., photochemical pollutants, in the atmosphere.[23–25]

5.3. *Pesticides*

Other potential "traps" can be found among the chemicals man has synthesized as aids in his agricultural efforts. One class of pesticides are the organophosphorus compounds, of which di-isopropylfluorophosphate (DFP) is an example:

$$(CH_2)_2\overset{\overset{\displaystyle H}{|}}{C}-O-\underset{\underset{\displaystyle O}{\|}}{\overset{\overset{\displaystyle F}{|}}{P}}-O-\overset{\overset{\displaystyle H}{|}}{C}(CH_3)_2$$

These agents are highly specific, producing virtually irreversible inactivation of the enzyme acetylcholinesterase (see Fig. 2) and certain other esters by alkyl phosphorylation. Their high lipid solubility results in penetration into the CNS when entering the body peripherally.[26] Dose–response analyses showed that there exists a critical level, at about 40% normal ChE activity, at which several basic changes in functions occur: the enzyme begins to lose control of its substrate, the transmitter substance ACh;[27] a sharp drop in nerve conductance

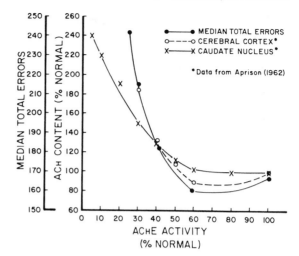

Figure 11. The curve for median total errors in a serial problem-solving situation shows a highly significant trend toward increasing error scores, with decreasing AChE activity below a critical level between 40% and 50% of normal. The other curves show that at this level AChE loses control of its substrate ACh.[32]

begins to appear;[28] and a crossover takes place from potentiation to a decline in a respiratory reflex.[29] Of special significance for psychochemistry is the observation that a wide variety of behavioral patterns begin to be significantly disrupted at this same critical level, including motor coordination, reflexes, and innate and learned behavioral patterns.[30–32] In Fig. 11, effects are composed of changes in AChE activity, in ACh content in two CNS areas and in errors made in complex problem-solving behavior.[32] Analogies of the kind just described between the two sets of relations are clearly apparent.

Due to the limitations of present knowledge, it is currently impossible to establish a concentration of any pollutant—other than zero—below which it can be asserted with full confidence that behavior will not be adversely affected. Factors of safety built into public health standards will, of course, differ significantly, depending upon what criterion of adverse effect is used. If avoidance of significant behavioral changes is the criterion, standards must often be set at lower concentrations than would be the case were the criterion an avoidance of adverse physiological change, and both would differ from standards based upon lethality.

Psychochemistry's concern about man's ecological traps must extend even beyond such considerations as these. There is emerging evidence which suggests that behavior may be more sensitive to mixtures of pollutants than to primary pollutants alone. There also is evidence that a person's physical characteristics, e.g., age and general health, may affect his sensitivity to pollutants and may do so differentially, some characteristics increasing sensitivity and others not.

6. *Environmental Management*

During the course of biological evolution, man survived in part because he was peculiarly adapted to his chemical environments. Later, he also developed skills in controlling those environments. Demands upon such skills have increased rapidly as man's technological knowledge has led him to introduce more and more changes into his ecosystems, until, today, there are those who predict biological disaster in the foreseeable future[33] unless greatly improved methods of environmental management are soon introduced.

There are others who, quite rightly, include in their analyses of such management other variables which, although only indirectly related to psychochemistry as the term has been defined here, are of great significance in affecting the *quality* of life, e.g., architectural style, nature of the workplace, personal safety, administration of justice, opportunities for time and leisure, control over goods and services, and man's place in society. These kinds of variables have been referred to as "social cost–benefits."

Environmental management becomes a process of "trade-offs" between the *biological limits* within which man can adjust to his ecosystem, and the *social cost–benefits* which he perceives as contributing to the quality of his life. The community near an abattoir may consider that the disadvantages of odors generated are outweighed by the work opportunities provided by the plant. The urban dweller may accept the dangers of atmospheric pollution in exchange for the way of life which the city provides. There is still controversy about the noise pollution created by supersonic transport and the advantages of more rapid modes of travel. The rules for trade-offs are not laid down by ecologists. They are determined by societies. Fortunately, societies are becoming increasingly aware of their need for the assistance of experts in making such decisions.[34] Models for this kind of environmental management process are being created and analyzed.[35]

Although the major flexibility in such decision making resides among the social cost–benefit variables, there can be no doubt about the significant contributions psychochemistry can make when standards are being set for the biological limits of pollution. It is only within the safety of such limits that trade-offs with social cost–benefits can be considered. Attention must be paid, in the setting of limits, to the effects of pollutants on the behavior which plays so central a role in an organism's adjustment to its environment.

Today's general cry is for "relevance," and it should be heard here as elsewhere. Indeed, the new "psychochemistry," as the term is used here, is a child of relevance. Two decades ago, psychochemistry was directed toward a frontier, primarily toward empirical search for chemical agents of value in the treatment of mental illness. Discovery, mainly by serendipity, of a major or minor "tranquilizer," or of an "energizer" which could withstand the test of mass clinical trials, was a very significant contribution to human health and happiness. Since then, public attention has been drawn to other social issues: to problems of drug addiction and to problems of environmental quality and man's survival.

The chemotherapy of behavior disorders is still relevant, but the new psychochemistry has been extended to encompass other issues of social concern. In so doing, it is maturing from the limitations of its earlier period, constrained by a dependence upon empiricism alone, to a much more comprehensive and systematic body of knowledge and a fuller repertoire of technical sophistication. This advance is enabling psychochemistry to broaden its relevance in areas it has thus far chosen to study (e.g., mental health, environmental quality), and to increase the excitement and challenge it has to offer those whose imagination sees equal opportunities in the study of such other basic matters as the effects of early environments on later development, the effects of population density on behavioral adjustment, and even the nature of learning and memory.

References

1. R. W. Russell, *Frontiers in Physiological Psychology*, Academic Press, New York, 1966.
2. R. W. Russell, *Ann. Rev. Psychol.*, **15** : 87 (1964).
3. R. Kumar, I. P. Stolerman, and H. Steinberg, *Ann. Rev. Psychol.*, **21** : 595 (1970).
4. S. L. Miller and H. C. Urey, *Science*, **130** : 245 (1959).
5. J. D. Bernal, in *Horizons in Biochemistry*, M. Kasha and B. Pullman, eds., Academic Press, New York, 1962.
6. R. W. Russell, in *Drugs and Behavior*, L. Uhr and J. G. Miller, eds., Wiley, New York, 1960.
7. A. E. Fisher and J. N. Coury, *Science*, **138** : 691 (1962).
8. S. P. Grossman, *Amer. J. Physiol.*, **202** : 872 (1962).
9. E. M. Killich, *J. Physiol.*, **107** : 27 (1948).
10. R. W. Russell, *J. Exp. Psychol.*, **38** : 178 (1948).
11. R. W. Russell, D. M. Warburton, and D. S. Segal, *Commun. Beh. Biol.*, **4** : 121 (1969).
12. W. T. Summerford, W. J. Hayes, J. M. Johnston, K. Walter, and J. Spillane, *A.M.A. Arch. Ind. Hyg.*, **7** : 383 (1953).
13. R. W. Russell, *Psychologia*, **14** : 136 (1971).
14. T. J. Chippendale, G. A. Zawolkow, R. W. Russell, and D. H. Overstreet, *Psychopharmacologia*, **26** : 127 (1972).
15. R. W. Russell, B. J. Vasquez, D. H. Overstreet, and F. W. Dalglish, *Psychopharmacologia*, **20** : 32 (1971).
16. J. R. Goldsmith, *Amer. J. Public Health*, **57** : 1532 (1967).
17. T. T. Higgins and J. R. McCarroll, in *Air Quality Standards*, A. Atkinson and R. S. Gaines, eds., C. E. Merrill, Columbus, Ohio, 1970.
18. Environment Agency, Government of Japan, *Pollution related to diseases and relief measures in Japan*, Environment Agency, Tokyo, 1972.
19. S. C. Harvey, in *The Pharmacological Basis of Therapeutics*, L. S. Goodman and A. Gilman, eds., Macmillan, New York, 1970.
20. J. M. Spyker, S. B. Sparber, and A. M. Goldberg, *Science*, **177** : 621 (1972).
21. V. G. Laties, R. R. Beard, B. B. Dinman, and J. N. Schulte, in: Committee on Effects of Atmospheric Contaminants on Human Health and Welfare, *Effect of Chronic Exposure to Low Levels of Carbon Monoxide on Human Health, Behavior, and Performance*, National Academy of Sciences, Washington, D.C., 1969.
22. R. R. Beard and G. A. Wertheim, *Amer. J. Public Health*, **57** : 2012 (1967).
23. A. J. Haagen-Smit, *Arch. Environ. Health*, **12** : 548 (1966).
24. J. Lewis, A. O. Baddeley, K. G. Banham, and D. Lovitt, *Nature*, **225** : 95 (1969).
25. H. K. Ury, *Arch. Environ. Health*, **17** : 334 (1968).
26. G. B. Koelle, in *The Pharmacological Basis of Therapeutics*, L. S. Goodman and A. Gilman, eds., Macmillan, New York, 1970.
27. M. H. Aprison, *Recent Advanc. Biol. Psychiat.*, **4** : 133 (1962).

28. I. B. Wilson and M. Cohen, *Biochim. Biophys. Acta,* **11** : 147 (1953).
29. B. Metz, *Amer. J. Physiol.,* **192** : 101 (1958).
30. R. W. Russell, R. H. J. Watson, and M. Frankenhaeuser, *Scand. J. Psychol.,* **2** : 21 (1961).
31. P. H. Glow and S. Rose, *Nature,* **206** : 475 (1965).
32. A. Banks and R. W. Russell, *J. Comp. Physiol. Psychol.,* **64** : 262 (1967).
33. D. H. Meadows, D. L. Meadows, J. Randers, and W. W. Behrens, *The Limits to Growth,* Earth Island Ltd., London, 1972.
34. I. de Groot, *J. Air Pollution Control Assoc.,* October : 681 (1967).
35. S. Schwartz and G. B. Siegel, in *Air Quality Standards,* A. Atkinson and R. S. Gaines, eds., C. E. Merrill, Columbus, Ohio, 1970.

Once we have succeeded in controlling the overgrowth of the population and reducing the effects of unwanted materials in the atmosphere on our lives, it will be necessary to assure that we have sufficient food of the type we need. To go out and kill animals to provide the urban dweller with protein smacks of a holdover from an earlier time. It is clear, both from the point of view of a rational approach to the problem and for the safety of food supplies in the future, that chemical methods of producing food have much to recommend them. The subject is at an early stage of development, but the present position suggests that development might make possible the cyclical production of food from chemical means, involving basic materials such as carbon dioxide, nitrogen, hydrogen, and water.

4

Chemical Sources of Food: An Approach to Novel Food Sources

Charles L. Cooney

1. Introduction

1.1. Scope

The question of feast or famine has been a central issue for mankind from the very beginning. Technology has developed throughout the centuries; man has increased his ability to produce more food and then expand the population not only to meet the available food supply but to demand even more of it. While the basic question of food supply and demand has not changed with time, the intricacies and sophistication of the problem have increased. Prehistoric man was faced with the need for more efficient animal traps and new varieties of plants, while today we are faced with the need to develop alternative foods and more efficient ways to produce them. This dilemma is further compounded by uneven distribution of food and other natural resources as well as economic

Charles L. Cooney • Department of Nutrition and Food Science, Massachusetts Institute of Technology, Cambridge, Massachusetts, 02139

and political factors which, depending on how they are used, can lessen or increase the problem of food supply and demand.

This chapter deals with one alternative to the development of new food sources: the bioconversion of chemicals to foods. The discussion will deal primarily with protein as the end objective in this conversion, although some consideration will be devoted to the production of food additives. After initially looking at the need for alternative food sources, and establishing the rationale for the use of chemicals as a source of food, we will examine in some detail the basic principles underlying the conversion of chemicals to food. For reasons which will become apparent, the conversion of chemicals to food is a bioconversion process, and because of the need to minimize cost and maximize volumetric productivity the bioconversion is carried out by microorganisms. In this light, we will focus on the capabilities, problems, and limitations of fermentation as a food-producing process. This will be done from the point of view of the microbiologist, the biochemist, and the chemical engineer; where possible, a quantitative treatment will be employed to define the bioconversion process. Once fermentation process concepts have been established, some consideration will be given to economics. Finally, a review of the state-of-the-art and a look into future prospects will be presented in an attempt to provide and overview of where we are today and where we may go tomorrow.

1.2. *Alternative Food Sources*

The alternative sources of food available to an individual, a community, or a country depend very much on location, climatic variations, physical geography, state of technical development, and availability of basic natural resources. Thus, it is not surprising that the number and quantity of alternative foods is to some extent proportional to the level of technological and social development and to the availability of resources in a country. A country with a long seacoast has great potential food resources from the sea; a country in a temperate environment with vast plains has potential food resources from cereal and legume crops; and a country with large open grass plains has potential food resources from cattle. However, in all of these cases, unless the technological development and capital equipment are available to exploit and distribute these potential resources, they will remain only potential and will not be available as alternatives, or the degree of exploitation will be less than optimum. To further compound the availability of alternative food sources, the productivity of the land is a strong function of climatic and environmental conditions. As a consequence, food productivity will fluctuate with changes in climate, availability of fertilizer and animal feeds, and other environmental parameters. It is in this light that there is a strong desire to seek out methods of food production which are independent of climatic and other environmental fluctuations. There is a strong driving force to seek methods for deriving food from better-defined chemical sources.

The ability of novel food sources to compete with traditional and alternative sources depends on several factors, but clearly there must be some incentive or driving force that favors the novel food source if it is to have an impact. The primary traditional foods are derived from animals such as beef, poultry, and swine, and from plants such as cereal and legume crops. Man has actively adapted these foods for his own use (i.e., as cultivated crops) for several thousand years. Chemical sources of food must undergo the equivalent "several thousand years of evolution" in a decade or less. As a consequence, many novel food sources will find their initial application as supplemental sources of protein, carbohydrate, fat, etc., and only in the long term will they become accepted as foods by themselves. In this context, we will focus on foods from chemical sources and will pay particular attention to single-cell protein or, as it is more simply called, SCP.

1.3. Single-Cell Protein (SCP)

Single-cell protein is a generic term for crude or refined sources of protein derived from unicellular or multicellular microorganisms, e.g., bacteria, yeast, molds, and algae. During the past 2 decades much has been written about SCP, but two references which provide a good overall perspective on both the nutritional and technological aspects are the volumes edited by Mateles and Tannenbaum[1] and Tannenbaum and Wang.[2]

The need for single-cell protein (SCP) as a supplement to the world protein source is well established; research and development on its production has been intense for over a decade. As a consequence, there are today a number of SCP plants in operation in Europe, the United States, and elsewhere, and more are under construction. Thus in many ways, we are no longer talking about a protein source of the future, but rather about a protein source in use today, even though the total amount is still small. Research and development today is addressing the problems of future second- and third-generation processes for SCP production for both animal feed and human food. These problems are associated with low-cost production and novel applications.

For a number of reasons some form of SCP can ultimately be expected to play a major role in meeting world protein needs. The population of the earth is bound to reach a point at which agriculture becomes insufficient or uneconomic as the sole source of food. We can only guess when this will occur. Malthus, in 1798, proposed convincing arguments for a widening gap between food production and population, in his essay on the law of diminishing returns. We can see, in retrospect, that his prophecy on starvation has been delayed since he could not predict rapid technological advances. However, if we are to continue to prevent the food–population gap from becoming larger, then additional technological advances must be developed and applied. The production of food from chemical sources has this potential. It has been calculated by Humphrey[3] that a 10% supplement to the world's food supply could be provided by a fermentation plant occupying less than one square mile of the earth's surface. Thus, there is potential for large volumetric productivity. This may be placed in perspective by the following analogy. In Table 1, the mass-doubling times of

Table 1. Mass-Doubling Times[a]

Organism	Time for one doubling in mass
Bacteria and Yeast	20–120 min
Mold and Algae	2–6 hr
Grass and Some Plants	1–2 weeks
Chickens	2–4 weeks
Hogs	4–6 weeks
Cattle	1–2 months
People	1/4–1/2 year

[a]Maximum rates are being compared.

selected organisms are compared. From this analogy it may be seen that sources of SCP offer a major rate advantage compared with traditional protein sources. Furthermore, SCP is the only potential food which has absolutely no dependency on agricultural inputs or other harvesting operations. That is, it can be a truly synthetic yet complete source of food whose composition we can control. It is important to stress at this point that SCP contains numerous nutrients in addition to protein, e.g., carbohydrates, vitamins, minerals, etc., and has even been used in the past as an approach to intensive production of fat for human consumption.

It is also important to recognize that some forms of SCP have been used as human food for millenia. The idea of using microorganisms as a direct food source has been attributed[4] to Delbrück's work in 1910. This, however, is not really the case, for there are records dating back several centuries that suggest that microorganisms have been a direct source of food for considerably longer. Foods such as cheese, bread, soy sauce, and alcoholic beverages have supplemented the human diet for at least 5000 years. While ingesting these foods, man has consumed the microorganisms which were responsible for its production or modification, and he certainly does not appear to have suffered from this unknowing consumption of microorganisms; in fact, the supplementary proteins and vitamins were a benefit. In either case, however, this cannot be considered as a direct utilization of microorganisms as a food source.

Accounts referring back to the sixteenth century provide evidence that a form of algae was consumed by Aztec Indians living in Tenochtitlan, presently Mexico City. Accounts of this form of hydroponics were brought through the Spanish conquistadors who entered what is now Mexico in the mid-sixteenth century. Two illustrative historical examples of SCP are referred to by W. V. Farrar,[5] writing on Aztec food technology:

Not contented with feeding on living things, they ate also a certain muddy substance that floats on the water of the lake, which they dried in the sun and preserved, to make use of it as cheese, which it resembled in flavour and taste. They gave to this substance the name of Tecuitlatl, or excrement of stones.

and:

... then there were fishmongers, and others who used to sell little loaves which they make out of a sort of slime which they gather from the great lake, which they thicken and make loaves of it, having a flavour like cheese.

The former is by F. S. Clavigero, in his book on the conquest of Mexico, and the latter is by Bernal Diaz, who was one of the first Spaniards to enter Tenochtitlan.

In other accounts quoted by Farrar, descriptions are provided as to how the "slime" was gathered, prepared, and eaten. It was evidently deemed palatable by both the Spaniards and the natives, and was described as having a cheese-like taste. The important point to be noted is that people living in an overpopulated area (the city was supposedly larger than contemporary European cities) actually cultivated and consumed microorganisms. The practice of eating what appears to be an algae was undoubtedly in effect before the Spaniards entered Mexico, and could conceivably have been a common practice for centuries.

There is also a recent account by a Belgian Saharan expedition that Africans in the vicinity of Lake Chad regularly consumed a blue green algae, *Spirulina platensis*.[6] Another reference, although more general, is made by P. Echlin[7] to a species of *Nostoc* (a blue green algae) that forms balls called "water plums." It has a high protein content and is supposedly eaten in parts of China and South America.

These early accounts of microbial food occurred before the science of microbiology even had its beginning. Once man began to study the microbial world and analyze its composition, he discovered the unity of biochemistry. He found that microorganisms were composed of the same basic building blocks as man. With this in mind, it can be understood how the first thoughts of microbial food developed. Using this same unity of biochemistry, we are today trying to engineer the foods of tomorrow.

2. Basic Principles Underlying the Conversion of Chemicals to Food

2.1. Introduction

The conversion of chemicals to food is a multidisciplinary problem requiring biochemical engineers, chemists, microbiologists, nutritionists, and food scientists as major contributors. The final product of this conversion must not only be safe, nutritious, socially and psychologically acceptable, and palatable, but must also be inexpensive, storable for long periods of time, and functional for use in conjunction with other foods. It has often been said that "a food is not a food until it is eaten"; this simple principle sums up the need for novel food sources to be acceptable. A corollary to this principle is that if a food is to be eaten it must be good enough for someone to want and be willing to buy. Thus, you may have an excellent product according to chemical standards, but if it is unattractive for any reason, e.g., color, taste, texture, etc., or if it is too expensive, it will not find its way into the marketplace.

2.2. Nutritional Value and Safety

One of the primary applications of SCP (as well as other chemically derived food sources) is as a supplemental source of protein. Even though SCP sources contain other nutrients, e.g., vitamins, minerals, carbohydrates, fats, etc., the major need today and in the future is protein.

One of the major reasons for interest in SCP is the naturally high protein content of microorganisms. This is illustrated in Table 2, where several SCP sources are compared with some nonmicrobial and more traditional protein sources. Generally, bacteria have the highest protein content (typically 65%) while yeast are next (typically 50%). Quantity of protein, however, is not the only important factor; protein quality is also of prime concern. As a supplemental protein source, the question of protein quality must be considered in the context of the protein source which is to be supplemented.

There are approximately 20 amino acids in the protein of man and other animals. Of these, eight must be available in the diet of adults and ten in the diet of children since they are not synthesized *de novo* from other food constituents. These essential amino acids are listed in Table 3. Most of the world's protein needs are satisfied by plant sources of protein such as rice, corn, wheat, etc.;

Table 2. Comparison of Protein Content of SCP and Other Protein Sources

Source	Protein (% of dry weight)
Bacteria	
Methylomonas methonalica	64
Bacillus subtilis	63
Yeast	
Hansenula polymorpha	50
Candida utilis (Torula yeast)	50
Saccharomyces cerevisiae	52
Microfungi	
Fusarium graminearum	66
Molds	
Asperigillis niger	35
Penicillium notatum	40
Nonmicrobial sources	
Animal muscle	95
Fish meal	61
Alfalfa meal	43
Soybean meal	44

Table 3. Essential Amino Acids

Amino acid	FAO reference	Source Content (g amino acid/100g protein)				
		Hansenula polymorpha	*Candida utilis*	*Pseudomonas* Sp.	Whole egg	Whole wheat
Valine	4.2	6.21	5.0	5.9	7.3	4.4
Threonine	2.8	4.17	5.5	4.5	5.1	2.9
Lysine	4.2	8.10	7.0	5.3	6.5	2.8
Methionine	2.2	1.15	1.0	1.8	3.2	1.5
Histidine	—	—	1.8	—	—	—
Leucine	4.8	8.34	6.5	7.0	8.9	6.7
Isoleucine	4.2	5.10	4.3	3.9	6.7	3.3
Arginine	—	—	—	—	—	—
Phenylalanine	2.8	4.96	3.8	4.2	5.8	4.5
Tryptophan	1.4	—	—	—	1.6	1.1

these proteins are deficient in one or more of the following amino acids: lysine, methionine, tryptophan, and threonine. The quality of any protein will be limited by the concentration of one of the essential amino acids. If two or more proteins can be mixed in a diet such that the strengths of one coincide with the weaknesses of others, they then act synergistically to upgrade the nutritional value of the diet. SCP can play this role in the formulation of both animal feeds and new human foods.

The essential amino acid content of several SCP sources is also given in Table 3, and is compared with the United Nations Food and Agricultural Organization (FAO) reference level and some traditional protein sources. The possible types of amino acid deficiencies and trade-offs can be seen from this simple comparison. A more detailed analysis is available in the book on the amino acid fortification of foods by Scrimshaw and Altschul.[8]

Lastly, with regard to nutritional value, it is necessary to assess the nutritional availability of the protein.

Frequently, protein is measured simply as the total amount of nitrogen in the sample (e.g., measured by the Kjeldahl method) and multiplied by 6.25; this number comes from the observation that on the average 16% of a protein is nitrogen. In fact, the total nitrogen in a cell is widely distributed, as summarized in Table 4. Thus, it is not at all in the form of protein, and some is structurally unavailable during digestion. As a consequence, one needs to evaluate its actual digestibility and efficiency of utilization in animals and humans. The biological value of protein is most commonly expressed in one of two ways:

The *protein efficiency ration* (PER) is measured by feeding young rats on a diet in which the only source of protein is the particular food whose protein quality is being tested. Conditions are quite carefully controlled and the diet is made up to contain exactly 10% protein. The PER is defined as the average gain in weight of the rats per gram of protein eaten.

Table 4. Distribution of Chemical Constituents in Microbial Cells

Cell fraction	Major components
Capsule	Complex polysaccharide, polypeptides
Cell wall	Lipopolysaccharide, polypeptides, amino sugars
Cytoplasm	
Cytoplasmic membrane	Lipoprotein, RNA
Ribosomal fraction	Protein, RNA
Nuclear bodies	Protein, RNA, DNA
Mitochondria	Lipid, protein
Cell sap	Lipid, protein, polysaccharide, small organic molecules

The *net protein utilization* (NPU) is also measured by feeding experimental animals, usually rats; but instead of assessing the animal's growth, the index of protein quality is measured by weighing the amount of the test food eaten and determining the weight of nitrogen in it by chemical analysis. This weight is taken as a measure of the amount of protein consumed; it is then compared with the amount of nitrogen retained in the rats' bodies, calculated from measurements of the amount of nitrogen which is excreted in the urine and feces of the rats over the test period.

Tests of this sort, although in some ways inexact, do provide a means of comparing the biological values of different proteins. For a more detailed discussion of the nutritional evaluation of SCP, the reader is referred to the guidelines of the United Nations Protein Advisory Group for testing SCP for human use.[2]

2.3. *Organisms and Raw Material Selection*

2.3.1. *Selection of a Microorganism for Human Use*

Bacteria, yeasts, and fungi have all been actively considered for pilot- or commercial-scale processes, and each has its relative advantages and disadvantages. The most popular organisms for food applications appear to be yeasts, which seem more familiar to human experience. They have been used in foods as vitamin additives and flavoring agents for a long time, they have reasonable protein concentrations, and because of their larger size they are easier to recover from fermentation media than bacteria. The yeast genera of particular interest are *Saccharomyces* and *Candida*. In the United States, for instance, *Candida utilis*, *Saccharomyces cerevisiae*, and *Klavaromyces fragilis* are currently accepted as food yeasts.

Bacteria would ultimately possess a number of advantages over other organisms, particularly for animal feed. They have higher growth rates, higher protein contents, and higher content of sulfur amino acids. However, they have

Table 5. Factors Associated with the
Choice of an Organism for SCP

Safety
Nutritive value
Palatability
Available low-cost substrates
Type of process technology available
Yield factors for organism
Energy economy: heating, cooling, etc.
Plant location
Social and psychological acceptance

the disadvantages of being susceptible to phage infection, having high nucleic acid content (a problem as a human food), and being less well known in human nutritional experience. Despite these problems, they have good long-range prospects because of their potential economic advantage.

The higher fungi have received relatively little attention in industrial-scale projects until recently. Ths possibility exists, however, of successful growth in continuous culture on very inexpensive waste carbohydrate sources. Fungi might be of considerable interest for their potent enzymatic capabilities, their ease of harvesting from fermentation media, and their mycelial form which provides natural texture.

The selection of an organism for use as SCP is a complex question and is based on the criteria outlined in Table 5. There is no single, best source of SCP suitable for all parts of the world and all applications; rather, one must consider local conditions in the context of the available technology and the acceptability of the product.

2.3.2. Raw Material Selection: Alternative Carbon-Energy Sources

These same arguments apply to the choice of a substrate or, specifically, a carbon-energy source for SCP production. As will be seen later, about 40–50% of the manufacturing cost of SCP is attributable to the carbon-energy source. As a consequence, the assured availability of a low-cost raw material is essential.

The scientific literature is full of reports on a wide variety of substrates for SCP production. A partial list is given in Table 6. These materials generally fall into the following categories: carbohydrates, alcohols, hydrocarbons, and agricultural wastes. Each has its own merits and problems.

The particular raw material of choice depends, to a very large extent, on locale. Natural gas, or methane, would be of great interest in areas where it is available in abundance, or is even discarded and flared. With regard to *petroleum fractions*, the local economics of the petroleum industry determine to some extent whether crude, semi-purified fractions, e.g., gas oil, or refined fractions, e.g., n-paraffins, are a more desirable substrate for SCP. For

Table 6. Substrates Considered for SCP Production with Different
Types of Organism

Bacteria	Yeast	Molds
Gas Oil	Paraffins	Starch Waste
Methanol	Gas Oil	Molasses
Methane	Ethanol	Cellulose
Ethanol	Sulfite Waste	Ethane, Propane, and Butane
	Liquor	
Manure	(Paper Industry)	
Cheese Whey	Waste Starch	
Cellulose	Coconut Milk	
	Cheese Whey	
	Molasses	
	Methanol	

example, in areas that use large quantities of diesel fuel, it is important to remove the paraffinic fraction from gas oil to lower the pour point of the oil. This dewaxing (removal of n-paraffin) can be done on gas oil simultaneously with the production of SCP, since yeast and bacteria will consume n-paraffins in preference to branched-chain alkanes. The combination of these two processes may greatly aid the economics of each. On the other hand, in areas that use large quantities of gasoline, it is unnecessary to remove the paraffins prior to the cracking operation, and an economic credit would not be allowed for removal of that fraction. Since gas oil contains approximately 10–40% paraffins (the only fraction readily used as a carbon-energy source for microbial growth), a large unmetabolized portion of the initial feed passes through the fermentor, creating problems in the separation of cells and residual unmetabolized hydrocarbons. The removal of this residual material is a significant problem in the use of petroleum fractions as carbon sources and requires that the final biomass product be washed extensively with solvent and/or detergent solutions.

A number of processes which use a relatively pure paraffin distillate for production of either food or feed have been proposed. Since all of the substrate is theoretically metabolizable in this case, it is possible to run the fermentation so that no unmetabolized hydrocarbon remains in the effluent. In practice, however, some hydrocarbon may be adsorbed to the cells and will have to be removed later in the process.

Methane is a substrate which is available at very low cost and in large quantities. Its direct fermentation, however, appears to be limited to bacteria. Since methane is a gaseous substrate, it would have the advantage over paraffinic hydrocarbons of leaving no residue in the final product, and the disadvantage of mass transfer limitations from the gas to the liquid phase (this will be discussed in detail below). As might be expected, this highly reduced hydrocarbon substrate will require much oxygen for microbial mediated oxidation (i.e., resulting in growth) and much heat will also be evolved.

Alcohols are particularly attractive as fermentation substrates, and methanol and ethanol have both received considerable attention. Both of these alcohols are derived from gases by catalytic hydration, which currently appears to be cheaper than biological oxidation. However, in the future this could change. One of the major questions concerning the use of methane and methanol is focused on the relative cost of the biological versus chemical oxidation of methane to methanol.

Both methanol and ethanol can be utilized by a wide variety of micro-organisms. They can be prepared in very high states of purity, they are totally water soluble, and they leave no residues in the cell mass leaving the fermentor. Because of their level of oxidation, they are also intermediate in cell yield (gram of cell/gram of substrate), oxygen demand, and heat load between hydrocarbons and carbohydrates. Since the lower alcohols maintain some of the advantages of both hydrocarbon and carbohydrate fermentations, they have great potential for use in SCP processes, and are particularly attractive for consideration as a source of food protein in both the long and short term.

It is important to note here that methanol may become a common currency of energy and chemical feedstock since it can be synthesized not only from natural gas and petroleum fractions, but also from coal, oil shale, and our primary renewable resource, cellulose. Methanol is easily stored, shipped, and used. If it replaces LNG (liquefied natural gas) as a commodity, thus necessitating large "mega-methanol" plants, its price could become very low. Proposed large-demand uses for methanol include: chemical feedstock, gasoline substitute or additive, gas turbine fuel, methyl fuel to replace or regenerate methane, and food, via SCP. As more large-demand uses are developed, its suitability as a starting material for food will increase since its cost will come down.

The development of coal gasification technology to generate synthesis gas or water gas containing H_2, CO_2, and CO is also important to methanol production, since these are the raw materials for methanol synthesis. In this way, difficult-to-transport coal can be reduced to a simple, easy-to-use commodity—methanol. This is the first step in the conversion of coal to food.

A wide variety of *carbohydrate* (e.g., general formula: $[CH_2O]$) substrates have been considered for SCP. These include well-defined materials such as glucose, sucrose, and starch, and complex mixtures such as molasses, cheese whey, sulfite waste liquor (from sulfite pulping in paper manufacture), cellulose, and hemicellulose, and their hydrolysis products. The ultimate selection of a source again depends on locale. In tropical climates, where molasses from sugar cane is abundant, it might be cost-attractive, whereas in the more temperate regions, where paper manufacture is a major industry, wood waste might fill the need.

Cellulose is one of our most abundant raw materials and is renewable. It is currently available in large quantity in the form of waste paper, wood pulp, and agricultural residues such as bagasse (the residue after sugar is extracted from cane) and cattle manure. From an economic point of view, the most crucial

Table 7. Summary of Cellular Yield Coefficients on Selected Carbon–Energy Sources

Carbon-energy source	Cellular yield (g cell/g substrate)
Glucose	0.5
Methanol	0.5
Ethanol	0.75
Methane	0.62
n-Alkanes	1.0
Cellulose	0.5
Starch	0.5

steps in the utilization of cellulose are its collection, pretreatment for hydrolysis, and conversion to low molecular weight metabolizable sugars. Although proposals for possible processes exist, they have not yet been proven on a suitable scale. In the case of materials such as solid cannery waste and citrus waste, problems exist because of seasonal processing schedules, conversion problems, and the sometimes dilute nature of the material. Biological utilization is frequently rate-limited by the hydrolysis of cellulose, and extensive pretreatment may be necessary.

Starch is much more readily hydrolyzed than cellulose, and the potential exists for direct fermentation of starch by amylolytic fungi in a continuous process. Starch may be of particular interest in tropical areas, which can produce high yields of root crops, such as cassava.

The arguments examined so far for various substrates for SCP production have focused on availability, level of oxidation, solubility, complexity, need for pretreatment (e.g., hydrolysis prior to fermentation), and cost. To place cost in perspective, however, one must examine the conversion yield, which is expressed on the cellular yield coefficient, Y_S (gram of cell dry mass/gram of substrate consumed). A summary of typical yield coefficients on some common substrates is given in Table 7. With these data in hand, as well as some knowledge of the protein content of the organism, one can estimate the carbon-energy source requirement per mass of protein and then estimate the substrate cost per mass of protein. This is a useful way to evaluate alternative substrates.

3. Fermentation Technology for Protein Production

3.1. Introduction

Fermentation processes, like other chemical processes, are a composite of unit operations. In the following discussion we will examine the key aspects of these operations and how they fit into a process for single-cell protein production. The generalized process scheme shown in Fig. 1 is broken down into three

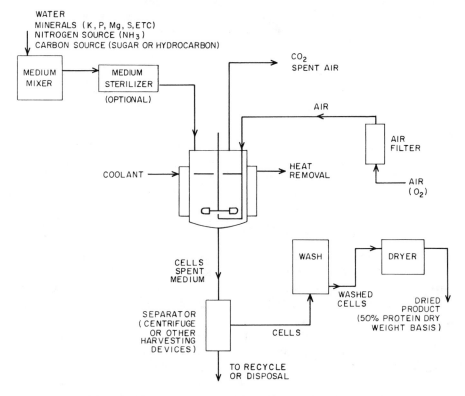

Figure 1. An overall view of a single-cell protein production plant.

phases: raw material preparation, fermentation, and product recovery. This scheme is independent of the raw materials and the type of microorganism employed. Each phase will be considered in detail in the following sections, with particular attention paid to process capabilities and limitations. The ultimate design of such a process requires an integrated effort on the part of both microbiologists and chemical engineers.

3.2. *Raw Material Preparation*

3.2.1. *Raw Material Requirements*

Microorganisms, like all living organisms, have basic nutritional requirements which must be met in order to permit growth. These requirements may be grouped into three categories: elemental requirements, specific biochemical requirements, and energy requirements. All of these categories must be satisfied before growth can proceed, and a limitation in any one of them will act as the limitation for growth.

3.2.1a. *Elemental Requirements for Growth.* The typical elemental composition for a microbial cell is shown in Table 8. Thus any medium for microbial

Table 8. *Elemental Cell Composition for Yeast and Bacteria*

| Element | % of dry cell weight | | Cell yield |
	Yeast	Bacteria	$\dfrac{\text{g cell dry weight}}{\text{g element}}$
Carbon	47	53	—
Nitrogen	7.5	12	8–13
Phosphorus (as PO_4)	1.5	3.0	33–66
Sulfur (as S)	1	1	100
Oxygen	30	20	—
Magnesium	0.5	0.5	200
Hydrogen	6.5	7	—
Total Ash[a]	8	7	—

[a]The ash contains P, Mg, Cu, Co, Fe, Mn, Mo, Zn, Ca, K, Na.

growth must contain, at minimum, these elements in the correct proportions. With the exception of carbon, oxygen, and hydrogen, the formulation of a production medium is based on the data shown in Table 8 expressed as the cell yield. For instance, if it is desired to synthesize 30 g/liter of yeast cell mass using ammonium sulfate as the nitrogen and sulfur source, it is necessary to supply 12 g/liter of $(NH_4)_2SO_4$. This will supply 2.4 g/liter of nitrogen and 3.0 g/liter of sulfur. Using the cell yields in Table 8, one can see that by matching the nitrogen requirement, the sulfur requirement is met in excess. Similar calculations allow one to design the minimal medium shown in Table 9 for the synthesis of 30 g/liter of yeast SCP. A discussion of how to satisfy the requirements for carbon, hydrogen, and oxygen is presented under the topic of energy demand. The trace element requirements are more difficult to estimate precisely since they are required in variable amounts as biochemical cofactors and structural components. However, as a rule, one can begin with the addition of the trace elements shown in Table 9 in 10^{-4}–$10^{-5} M$ concentration for

Table 9. *Growth Medium Designed to Support the Production of 30 g/liter of Yeast on Selected Carbon Sources*

Medium component	Concentration (g/liter)
Carbon-energy source:	
Methanol	60
Ethanol	40
Glucose	60
Hexadecane	30
$(NH_4)_2SO_4$	12
KH_2PO_4	(1.3 minimum)
$MgSO_4$	1.5
Trace Minerals:	
Cu, Co, Fe, Ca, Zn, Mo, Mn	$10^{-4} M$

30–3 g/liter of cells, respectively. It is important to be aware that some organisms may have unusually high requirements for one or more trace elements and that some modification of the initial trace salt solution may be required.

3.2.1b. *Specific Biochemical Requirements.* While many microorganisms are able to grow quite well on simple mineral salts media, many others require one or more specific biochemical compounds: like humans, some microorganisms are unable to synthesize all of their own biochemical components. The most frequent requirements are vitamins and amino acids. Yeast, for instance, frequently have a requirement for biotin, thiamine, or riboflavin. Some organisms are quite fastidious and require 10 or 20 preformed biochemical compounds. Generally, the requirement for vitamins is small and, typically, they are added in milligram-per-liter quantities. Table 10 shows a summary of water-soluble B vitamins found in some bacteria.

3.2.1c. *Energy Requirements for Microbial Growth.* While microorganisms are able to convert basic chemicals into complex molecules and food, they do not violate the second law of thermodynamics in the process. The complex series of synthetic reactions performed in the cell requires the expenditure of energy. This energy is derived, in most cases, from the controlled oxidation of reduced organic compounds. Thus, carbon compounds are used to produce energy for biosynthesis as well as to meet the cell's elemental requirement for carbon. This is illustrated by the following stoichiometry:

$$A(C_aH_bO_c) + B(O_2) + D(NH_3) \rightarrow M(C_xH_yO_zN_w) + P(CO_2) + Q(H_2O)$$

where A, B, D, P, and Q are moles of their respective compounds, M is the moles of a cell unit $(C_xH_yO_zN_w)$ which is derived from the cell composition, and $C_aH_bO_c$ is a generalized carbon-energy source for growth. From this relationship, one can see that the ratio of cell mass formed per unit of substrate consumed (M/A) will depend on the proportion of substrate used for energy

Table 10. The Amounts of B Vitamins in Bacteria[a,b]

Vitamin	Aerobacter aerogenes		Pseudomonas fluorescens		Clostridium butylicum	
	Cells	Medium	Cells	Medium	Cells	Medium
Thiamine	11	8.9	26	48	9.3	30
Riboflavin	44	110	67	310	55	180
Nicotinic acid	240	390	210	350	250	1680
Pantothenic acid	140	640	91	220	93	225
Pyridoxin	6.8	20	5.7	70	6.2	17
Biotin	3.9	44	7.1	61	—	—
Folic acid	14	91	8.8	66	2.8	16
Inositol	1400	—	1700	—	870	—

[a] The data (from R. C. Thompson, *Texas Univ. Publ.* **4237** : 87 (1972)) are based on microbiological assays on cells and medium from cultures in vitamin-free media (except for *C. butylicum*, which requires biotin).
[b] All figures are in parts per million dry weight of cells.

(combusted to CO_2 and H_2O) and for direct incorporation into cell mass. Furthermore, from this balance one can also derive the cell's requirement for oxygen; Mateles[9] derived the following useful expression for this purpose:

$$\frac{g\ oxygen}{g\ cell} = \frac{32C_a + 8H_b - 16O_c}{YM_w} + 0.010O_z' - 0.0267C_x' + 0.017N_w' - 0.08H_y'$$

where C_a, H_b, and O_c are the number of atoms per substrate molecule, C_x', O_z', N_w', and H_y' are the percentages of C, O, N, and H in the cell, Y is the cell yield (g cell/g substrate) for the carbon source. and M_w is the molecular weight of the carbon source. From this expression, it is clear that the oxygen demand is inversely proportional to the cell yield, which is a measure of the efficiency of conversion of the carbon source to cell mass. This relationship is shown graphically for growth on methanol (Fig. 2).

From the above relationship, it is also possible to calculate the heat evolved during growth. The heat evolution is directly correlated with the oxygen demand by the expression

$$Q_F = 0.12 \frac{kcal}{mM\ O_2} Q_{02}$$

where Q_F is the heat of fermentation (kcal/liter-hr), and Q_{02} is the oxygen

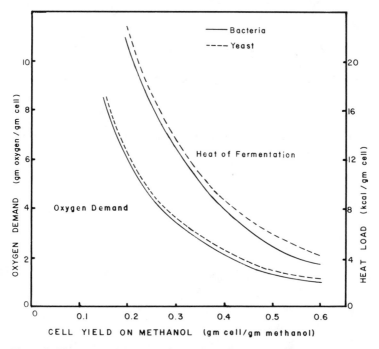

Figure 2. The oxygen demand and heat load for yeast and bacteria as a function of cellular yield on methanol (for additional details see Cooney[10]).

uptake rate (mM O_2/liter-hr). The conversion constant is derived empirically from a wide variety of fermentations.[11] Microorganisms are about 50% efficient in utilizing the available free energy of oxidation in a carbon-energy source. Half of the available energy is incorporated into cell mass, while the rest is evolved as wasted heat. The relationship between the heat load (i.e., heat to be removed) in a fermentation to the cell yield is shown graphically in Fig. 2 for the production of yeast and bacteria on methanol.

From a biochemical point of view, most of the heat is evolved during the stages of terminal oxidation, where energy-carrying cofactors are used to reduce oxygen to water and in the process generate adenosine triphosphate (ATP), an energy-rich compound, for use in energy-requiring biosynthetic reactions. In addition to biosynthesis, the cell requires energy for maintenance of cell structure, active transport of molecules, establishment of concentration gradients between the cell and its environment and, in some cases, cell mobility.

Thus, in order to satisfy the cell's requirement for carbon and energy (as well as become a source of hydrogen for cell mass), one must supply sufficient carbon for biosynthesis and for energy generation. Both of these are included in the cellular yield coefficient, Y. The estimation of carbon source requirements for cell growth are made by dividing desired cell density with the expected cell yield; e.g., to obtain 30 g/liter of cells from methane when the yield is 0.5 g cell/g methanol, one needs 60 g methanol/liter.

The only remaining nutrient is oxygen. Because oxygen is a sparsely soluble gas, its supply to a fermentation is a special problem of mass transfer. For this reason, it will be discussed under the heading of "Agitation and Aeration." Suffice it to say at this point, however, that the demand for oxygen will depend on the nature of the carbon source and the efficiency of its utilization.

3.2.2. *Sterilization*

An important aspect of medium preparation is sterilization. The end product from the fermentation step is a food and, therefore, an acceptable degree of aseptic operation must be maintained during production and subsequent processing. This begins with medium sterilization to destroy potential contaminants of the process, and also includes air sterilization to prevent contamination of the fermentation broth via the airstream.

3.2.2a. *Medium Sterilization.* There are a variety of methods to effect medium sterilization; these methods fall into two categories, as shown in Table 11—removal and destruction. In the large-scale sterilization of media, however, only steam (wet heat) is used as an economic means for sterilization. For this reason, the following discussion will focus on the use of steam for batch and continuous sterilization.

Thermal destruction of microorganisms is a temperature–time phenomenon such that the higher the temperature, the shorter the time required to effect

Table 11. Methods for Media Sterilization

A. Removal of Microorganisms
    1. Filtration[a]
    2. Centrifugation
    3. Flotation
    4. Electrostatic attraction
    5. Ion exchange
    6. Charcoal adsorption of viruses
B. Destruction of Microorganisms
    1. Heat (wet and dry)[a]
    2. Chemical agents[a]
    3. Electromagnetic radiation[a]
    4. Sonic radiation

[a] Industrial or preparative scale application.

cell death. The kinetics of thermal death may be described by first-order kinetics in the following manner:

$$\frac{dN}{dt} = -kN$$

where N is the number of viable cells (cells/ml), t is time (min), and k is the specific death constant (min^{-1}). Separation of variables and integration of the equation for initial and final levels of contamination, N_0 and N, respectively, gives

$$\ln \frac{N_0}{N} = kt$$

At this point several important questions arise. First, how do we determine the initial level of contamination? Second, what final level of contamination is acceptable, and third, what do we really mean by death?

Death is defined in an operational manner, in that an organism may be considered dead if it cannot replicate or perform any deleterious effects in the environment of the fermentor. This means that if the fermentation is being carried out in a very restrictive environment, e.g., high temperature, extreme pH, unusual carbon source, etc., then there is less chance of contamination occurring than if the medium were made from readily utilizable substrates and kept at low-growth temperature (25–32°C) and favorable pH (6–8).

It would be extremely tedious if the initial level of contamination had to be measured for each sterilization cycle. Furthermore, different microorganisms have different resistances to heat. For this reason, one usually designs a sterilization process around a high initial level of the most heat resistant organism that could contaminate the medium. The reference level is typically 10^6 organism/ml. (It is interesting to note that minimum visually detectable levels of cells begin at 10^7 organism/ml.) The most-resistant microbial form is

Table 12. Relative Resistance of Various
Microorganisms to Moist Heat

Microorganism type	Relative resistance
Vegetative bacteria and yeast	1.0
Bacterial spores	3×10^6
Mold spores	2–10
Virus and bacteriophage	1–5

the bacterial spore. Spores are a dormant form of bacteria of the genus *Bacillus* (aerobic) and *Clostridia* (anaerobic). The relative resistance of spores to other microbial forms is shown in Table 12.

The kinetics of thermal death are illustrated in Fig. 3 to show the difference in death rate for a typical bacterial cell and a bacterial spore. From

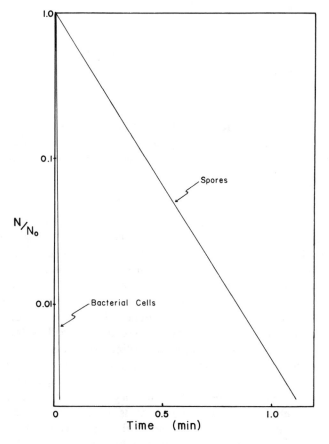

Figure 3. Kinetics of thermal death for typical bacterial vegetative cells and bacterial spores at 121°C.

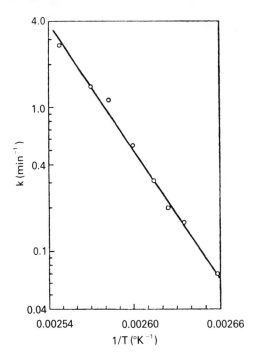

Figure 4. Arrhenius plot for thermal inactivation constant, k, for *Bacillus staerothermophilus*. T is the absolute temperature and the activation energy is 68.7 kcal/mol.

this, it can be seen that if the thermal conditions are sufficient to kill spores, then all of the bacteria, yeast, etc., will also be killed. The specific death rate constant, k, is a function of temperature and obeys the Arrhenius rate law, as shown in Fig. 4, and described by

$$k = A \exp (-E_a/RT)$$

where A is the Arrhenius coefficient, E_a is a pseudo-activation energy (cal/mol), R is the universal gas constant (cal/mol $-^{\circ}$K), and T is the absolute temperature. Typically, E_a for bacterial spores is 60–70 kcal/mol. This is a pseudo-activation energy since it is not clear what the mole refers to; despite this, the relationship is useful for design. From the kinetic behavior of cell death and the dependence of death rate on temperature, it is seen that the degree of sterilization is a temperature–time phenomenon. The higher the temperature, the shorter the time required to achieve sterilization. We shall see later that this is the underlying principle of continuous sterilization of media.

We are assuming in these estimations that the entire medium and all of its constituents are heated to the same temperature for the same length of time. If all the medium components are soluble, this is a good assumption. If, however, there are water-insoluble components such as agricultural waste, cellulose, starch, etc., then the rate of heat transfer to the particles may be important. This may be seen in Table 13, where the time to reach 99% of a final temperature is given for different-size particles.

Table 13. Effect of Particle Size on Heat-Up Time of Solids

Size (diameter, cm)	Time to reach 99% of final temperature[a]
1×10^{-4}	1 μsec
1×10^{-3}	0.1 msec
1×10^{-2}	10 msec
1×10^{-1}	1 sec
1	100 sec

[a] The particle is assumed to have the property $\rho C_P/k = 250$, where ρ is the density, C_P is the heat capacity, and k is the thermal conductivity.

3.2.2b. *Batch Sterilization.* The most common way to sterilize a fermentation medium is to fill the vessel with medium and heat it with steam until the sterilization criteria are met. A typical temperature–time profile for such a sterilization cycle is shown in Fig. 5. For large (e.g., 100,000-liter) vessels the heat-up and cool-down times can be several hours each, while the holding time at 121°C (the temperature for 15 psig steam) is typically 15 min. If a fermentor is to be run continuously, it is necessary to have a separate sterilization vessel as well as a holding tank to maintain a constant flow rate of liquid. The design of the temperature–time profile is based on the integral of the death equation

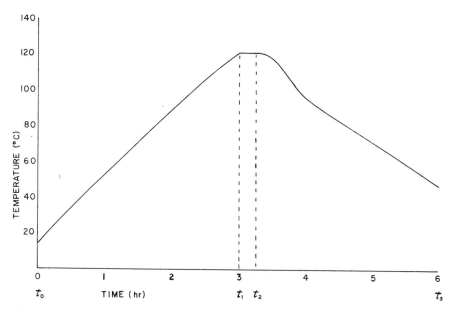

Figure 5. Temperature–time profile for a batch sterilization of an industrial fermentor. t_0, initial time; t_1, time to reach 121°C; t_2, start of cool down; t_3, completion of sterilization cycle.

shown earlier,

$$\int_{N_0}^{N} dN = \int_{0}^{t} k\, dt$$

to meet the desired reduction in viable spores from N_0 to N.

3.2.2c. *Continuous Sterilization.* Most SCP processes are likely to be run as continuous fermentations for reasons to be elucidated below. Thus, it is logical to consider the use of continuous sterilization to interface the medium preparation and fermentation steps. A steam-injection type of sterilizer is shown in Fig. 6, along with the temperature–time profile. This system works by directly injecting high-pressure steam into flowing medium and achieving a very rapid temperature rise. It would be unreasonable to operate this with a 15-min holding time (e.g., as in batch sterilization), so higher temperatures are used; this permits much shorter residence times to be used. Typically, about 140°C for 3–4 min is sufficient to achieve the same degree of sterilization as 15 min at 121°C. Continuous sterilization of solid-medium components is

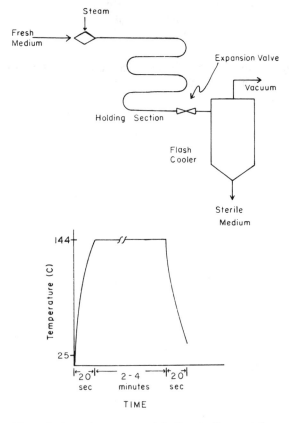

Figure 6. A continuous steam-injection sterilizer and the temperature–time profile in the holding section.

tricky because of the possibility of entrapped spores, which do not get to a high-enough temperature. In this type of situation it is desirable to make the fermentation conditions as selective as possible to minimize problems of contamination. Other advantages of continuous sterilization are the more efficient use of steam and less destruction of heat-sensitive medium components.

Additional reading on sterilization is available in books by Aiba *et al.*[12] and Richards.[13]

3.2.3. Air Sterilization

Because oxygen is a very water insoluble gas, it is necessary to blow large quantities of air through a fermentor to meet the demand for oxygen. For instance, only 20–30% of the oxygen in the airstream may be transferred to growing cells. Thus, the air enters with 20.9% oxygen and leaves with 15–16% oxygen. As a consequence, large volumes of air (typically 1–1.5 volume of air/volume of fermentor/min, or VVM) are required. This air must be sterilized prior to entering the fermentor. Some microbial loadings in various air sources are shown in Table 14. The species present include mold spores, bacterial spores, bacteria, and yeast.

Filtration is the most common method for continuous air sterilization. Two types of filtration are depth filtration and absolute filtration. In depth filtration through columns of packed glass fiber or granular charcoal, the microorganisms are removed by becoming attached to and trapped by the fibers or particles of the filtration medium. The performance of a depth filter is described by the following relationship:

$$\frac{dN}{dL} = -k'N$$

Table 14. Colonies Obtained per Cubic Foot of Air Sampled

Air source	Approx. range colonies/ft
Main city street	
ground level	2–6
75 ft up	1–2
School classroom	40–100
Offices	10–70
Factory workshop	3–350
Small house	
bedroom	2–120
dining room	2–70
Microbiological laboratory "sterile" room	1–2
Hospital ward	6–600
Hospital operating room	2–100

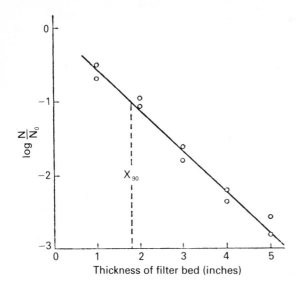

Figure 7. Reduction of microorganisms on passage of air through a packed fibrous filter bed.

where k' is a filtration constant, and N is viable cells at any given distance along the filter L. This equation describes the log-penetration theory for depth filtration; from its solution

$$\ln \frac{N}{N_0} = -k'L$$

we can see that a plot of $\ln N/N_0$ versus L should give a straight line. An example of this is shown in Fig. 7, and its use is described in detail by Humphrey and Gaden.[14] The usual criterion for design is that only one organism in 1000 fermentation runs will pass through the filter.

3.3. Fermentation

The main section of a fermentation is the fermentation step itself. At this point in the process one converts the feed nutrients to biomass (and additional products if they are of interest). The role of the biochemical engineer is to design a system that will support optimum culture growth and product formation. In the following paragraphs, we will examine the primary design problems associated with the fermentation section and establish a set of criteria on which to design a fermentation process.

3.3.1. Design and Maintenance of Culture Environment

Maximum conversion of the carbon-energy source to protein requires that the physical and chemical environment of the microorganism be optimal for growth. The primary requirements that must be met are: complete nutritional requirements (including sufficient oxygen if the process is aerobic), optimal

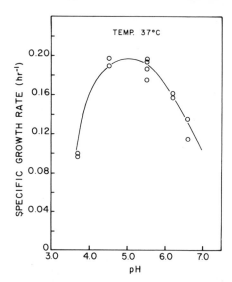

Figure 8. The effect of pH on the specific growth rate of *Hansenula polymorpha* grown in a mineral salts methanol medium.

temperature, optimal pH, and minimum physical shear (in the case of shear-sensitive organisms). Each organism will have its own unique set of chemical–physical conditions for optimal growth, but a few generalizations are possible.

Most microorganisms will grow over a pH range of 4–5 units, with a fairly flat optimum pH of 1–2 units. A typical pH growth rate profile is given in Fig. 8. Generally, bacteria will have their optimum pH in the range of pH 6–7.5, while yeast more often prefer pH 4–5.5, and molds pH 4–7. There are exceptions, such as iron-oxidizing bacteria which grow well at pH 2, but the above criteria are useful as guidelines.

Variation in growth as a function of temperature exhibits some interesting patterns. A series of typical growth–temperature profiles are shown in Fig. 9. The three patterns shown represent psychrophilic (low-temperature), mesophilic, and thermophilic (high-temperature) growth. For a given organism, the temperature range is about 25°C, with both upper and lower limits. In

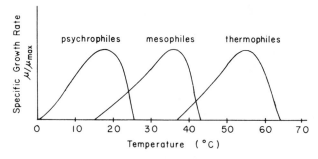

Figure 9. The effect of temperature on the specific growth rate of psychrophiles, mesophiles, and thermophiles.

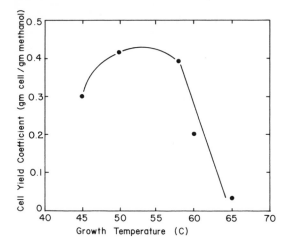

Figure 10. The effect of growth temperature on the cellular yield coefficient for growth of a mixed population of bacteria on methanol.

nature, the ultimate limit is about 0°C (found in Arctic regions) and about 93°C (found in hot springs) for extreme psychrophilic and thermophilic microorganisms, respectively; these are the limits of liquid water. The useful range of an industrial fermentation is more like 20–60°C.

Another interesting feature of the growth–temperature curves is the slow increase in growth rate as temperature is increased to the optimum (e.g., a change in 10°C will approximately double the growth rate), while above this point growth falls off rapidly with increasing temperature; the activation energy in the Arrhenius relationship is about 15 and 65 kcal/mol, for growth and death respectively.

It is also interesting to examine the effect of temperature on the cellular yield coefficient. A typical curve is shown in Fig. 10. The conversion of substrate to cell mass is about constant until the optimal growth temperature is reached, and the yield declines as temperature is increased. This is very likely due to the use of nutrients to repaiɪ and/or maintain cell component damage by the higher temperature. In the production of food from chemicals this places an upper temperature limit on the efficient conversion of carbon source to protein.

3.3.2. Continuous Culture of Microorganisms

Growth in batch culture is a process which always terminates after some finite time interval. The very nature of the batch process is such that it cannot be continued indefinitely. Batch culture may be altered, however, by continually supplying fresh nutrient medium while simultaneously withdrawing the spent broth containing cells and products. In such a continuous microbial culture, growth can be maintained for prolonged and indefinite periods of time. Furthermore, this system can be maintained in a steady-state situation such that the cell concentration, specific growth rate, and culture environment (e.g., nutrient and product concentration) do not change with time. This is in direct contrast to growth in batch culture, where the culture environment changes continually with time. As a consequence, continuous culture provides a unique tool for the continued production of cell mass or other products under optimal environmental conditions. By far the most common type of continuous culture is the chemostat. In such a device, the nutrient medium is designed such that all but a single essential nutrient are available in excess of the amount required to synthesize a desired concentration of cells and products. The single nutrient whose concentration is limited is called the growth-limiting nutrient, and the feed concentration of this nutrient controls the size of the steady-state cell population. Any nutrient required for growth may be used as the growth-limiting nutrient, thus providing considerable flexibility in controlling the physiology of growing cells through design and manipulation of the growth environment. The term "chemostat" stems from the static or constant chemical environment which is characteristic of this type of continuous culture at steady state.

3.3.2a. *Background and Theory of Continuous Culture.* The continuous culture of microorganisms has come a long way since its original introduction, and many reviews and books (Malek and Fencl[15]; Malek and Ricica[16]; Powell *et al.*[17]; Dean *et al.*[18]) have appeared on the theoretical and applied aspects of this technique. In order to analytically describe the steady-state behavior of a chemostat, it is necessary to devise a set of equations relating the cell and limiting nutrient concentration to the primary independent operating variable, medium flow rate; this is done with material balances. A material balance on the cell mass written around the fermentor shown in Fig. 11 is given by

$$\begin{pmatrix} \text{Cells} \\ \text{in} \end{pmatrix} - \begin{pmatrix} \text{Cells} \\ \text{out} \end{pmatrix} + \begin{pmatrix} \text{Cell} \\ \text{growth} \end{pmatrix} - \begin{pmatrix} \text{Cell} \\ \text{death} \end{pmatrix} = \begin{pmatrix} \text{Cell} \\ \text{accumulation} \end{pmatrix}$$

$$\frac{F}{V}X_0 - \frac{F}{V}X + \mu X - \alpha X = \frac{dX}{dt}$$

where X_0 and X are the cell mass (g/liter) in the feed and fermentor, respectively, F is the feed-flow rate (liter/hr), V is the fermentor volume (liter),

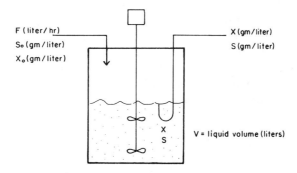

Figure 11. Definition of terms and nomenclature for a single-stage, well-mixed chemostat.

μ and α are the specific growth and death rates (hr^{-1}), respectively, and t is time (hr).

Usually, with a single-stage chemostat, the feed stream is sterile and $X_0 = 0$. Also, in most continuous cultures, the specific growth rate is much greater than the death rate ($\mu \gg \alpha$), so that this equation may be simplified to

$$-\frac{F}{V}X + \mu X = \frac{dX}{dt}$$

As a consequence, at steady state, when $dX/dt = 0$, $\mu = F/V$. Thus, the specific growth is determined by the medium flow rate divided by the culture volume. This ratio is defined as the dilution rate, $D \equiv F/V = \mu$, and at steady state the specific growth rate is equal to the dilution rate.

A second equation may be written by writing a material balance for the limiting nutrient. This is shown as

$$\begin{pmatrix}\text{Nutrient}\\\text{in}\end{pmatrix} - \begin{pmatrix}\text{Nutrient}\\\text{out}\end{pmatrix} - \begin{pmatrix}\text{Nutrient}\\\text{consumed}\end{pmatrix} - \begin{pmatrix}\text{Maintenance}\\\text{requirement}\end{pmatrix} = \begin{pmatrix}\text{Nutrient}\\\text{accumulation}\end{pmatrix}$$

$$\frac{F}{V}S_0 - \frac{F}{V}S \qquad -\frac{\mu X}{Y} \qquad -mX \qquad = \frac{dS}{dt}$$

where Y is the cellular yield coefficient for the limiting nutrient expressed as gram of cells formed per gram of nutrient consumed, S_0 and S are the inlet and outlet concentrations of the limiting nutrient, and m is the maintenance requirement. Frequently the maintenance requirement is low relative to growth (e.g., $mX \ll \mu X/Y$) and it may be neglected such that the steady-state ($dS/dt = 0$) solution is

$$DS_0 - DS = \frac{\mu X}{Y}$$

Recalling that $\mu = D$, we can now obtain

$$X = Y(S_0 - S)$$

Frequently, in this equation, the cell yield is assumed to be independent of the growth or dilution rate. While under many operating conditions this is a good approximation, care is needed in making this assumption because the yield may vary over a wide range of dilution rates.

In order to relate X and S to the dilution rate, a model expressing the growth rate as a function of the limiting substrate is required. The most frequently used model is that suggested by Monod (see reference by Herbert *et al.*[19]):

$$\mu = \mu_{\max}\frac{S}{K_s + S}$$

where K_s is the half-rate saturation constant and equals the substrate concentration, S, when the growth rate, μ, is equal to $0.5\,\mu_{\max}$. When this equation is

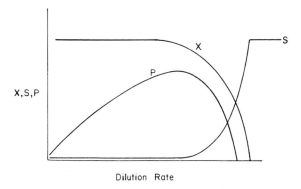

Figure 12. The theoretical behavior of a chemostat. *X*, cell
concentration; *S*, limiting nutrient; *P*, productivity.

applied to continuous culture it becomes

$$D = D_c \frac{S}{K_s + S}$$

where D_c, the critical dilution rate, represents the maximum dilution rate at
which the chemostat may be operated and corresponds to the maximum growth
rate in batch culture. Using these equations, the theoretical behavior of a
chemostat may be depicted as in Fig. 12.

The performance of a fermentation process is evaluated by the final yield
or conversion of substrate to product and the overall productivity. In continuous culture, the productivity P (g/liter-hr) is defined as $P = DX$.

The productivity can be plotted as a function of the dilution rate. Figure 13
is a plot of steady-state cell concentration, residual nutrient concentration, and

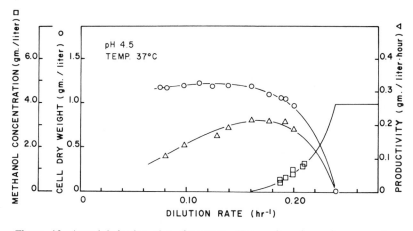

Figure 13. Actual behavior of a chemostat. *Hansenula polymorpha* grown in a
methanol-limited chemostat.

productivity for the production of yeast biomass on methanol. The primary advantage of continuous culture for SCP production is that one can maintain higher productivities than in batch culture. Furthermore, the product is produced in a consistent manner at all times.

3.3.2b. *Cell Composition in Continuous Culture.* The macromolecular and micromolecular composition of the cell is determined not only by the genetic makeup of the cell but also by the growth environment. Cell composition varies as a function of the pH, temperature, ionic strength, and nutrient makeup of the medium. In addition, cell composition varies as a function of the growth rate of the organism. This observation is not surprising, since the metabolic needs of the cell vary with the growth environment and the metabolic activities are reflected in the cell composition.

Herbert[20] examined the composition of *Aerobacter aerogenes* in a nitrogen-limited continuous culture and found (Fig. 14) the RNA content of the cells to increase with increasing growth rate. Most of the RNA (approximately 80%) is ribosomal, and the increase reflects the need to increase the rate of protein synthesis in the cell. Protein and DNA, on the other hand, exhibit a slight decrease with increase in dilution rate. This is at least partly due to the dilution of protein and DNA by the increased RNA. If the DNA were expressed on a per-cell basis, then the DNA level would rise with increasing growth rate because faster-growing cells have more growing points for DNA replication and hence more DNA.

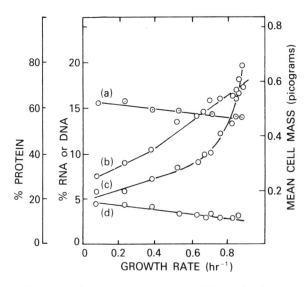

Figure 14. The macromolecular composition of *Aerobacter aerogenes* grown in a nitrogen-limited chemostat: *a* is the % of protein; *b* is the % of RNA; *c* is the mean cell mass; and *d* the % of DNA.

The patterns of compositional changes with growth rate are about the same when the carbon-energy, nitrogen, sulfur, magnesium, or potassium sources are used as the limiting nutrient. However, each of these essential nutrients may be expected to exert different pressures on the cell physiology when used as the limiting nutrient, since they each play a different role in metabolism. It is expected that by choosing different growth-limiting nutrients the investigator can exert considerably more control over the cell than is possible with just the growth rate, pH, and temperature effects. The selection of an appropriate limitation may act to restrict some metabolic activities while allowing others to continue at the same accelerated rate, thus, in effect, uncoupling various aspects of cell metabolism. Relatively little work has been done with the use of other than carbon limitations.

3.3.3. *Aeration and Agitation*

Aeration in biological processes serves two functions. These are to provide adequate oxygen to meet the demand of the microorganisms, and to provide adequate means to insure mixing in order to prevent the settling of solids. While these objectives may be met solely by aeration, this is frequently accompanied by mechanical agitation. For this reason, aeration and agitation are treated as a single-unit process when they are used simultaneously to achieve oxygen transfer.

An examination of some of the physical and biological properties of fermentation processes is shown in Table 15. In the fermentation industry, a variety of devices are used for oxygen transfer. The aeration tower is commonly employed in the food- and feed-yeast industry, and more recently various designs of the airlift fermentor have been used for large-scale production of SCP. The submerged turbine fermentors employed in yeast and pharmaceutical industries are probably the most common design. Figure 15 compares the airlift and mechanically agitated fermentors.

Most processes for the conversion of chemicals to food are aerobic and require large quantities of oxygen. The problem of aeration and agitation is

Table 15. Some Physical and Biological Properties of Fermentation Processes

Temperature range (C°)	20–60
Variation	±0.5
Rheology	Newtonian and non-Newtonian
Variation in viscosity (cp)	1–1000
Initial substrate concentration	5–80 g/liter
Reactor size (gallons)	250–40,000
Power per unit	
Volume (HP/1000 gal)	1–20
Growth rate (hr^{-1})	0.1–1.0

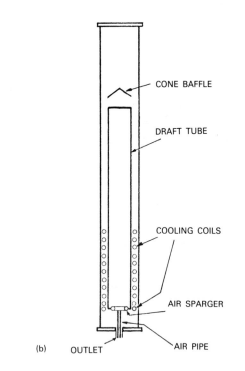

Inoculum from
seed vessel

Exhaust

Air
filter

Gland or
mechanical seal

Manhole

Steam in or
water out

Sample
point

Cooling water in or
condensate out

(a)

Harvest and
drain line

CONE BAFFLE

DRAFT TUBE

COOLING COILS

AIR SPARGER

(b)   OUTLET

AIR PIPE

Figure 15. Schematic drawings of (a) a mechanically agitated, and (b) an airlift type of fermentor.

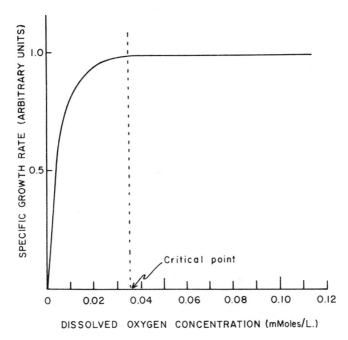

Figure 16. The effect of dissolved oxygen concentration on specific growth rate.

then one of supply and demand; by means of maximizing gas–liquid contact, we need to insure a sufficient supply of oxygen to growing microorganisms. In aerobic growth, oxygen is a required nutrient. This is seen by the curve in Fig. 16, which relates growth rate to dissolved oxygen concentration. As long as the dissolved oxygen is above the critical point, growth is independent of oxygen concentration. Typically, the critical point occurs at 5–10% of air-saturated water for bacteria and yeast.

The transfer of oxygen to a cell may be depicted as shown in Fig. 17. Oxygen passes from the air bubble, across gas and liquid films, and into the bulk

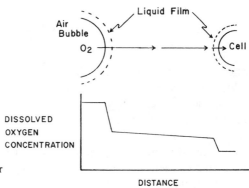

Figure 17. Pathway for oxygen transfer from the air bubble to the cell.

liquid. After diffusion through the broth it crosses a liquid film around the cell, through the cell membrane, and to the site of its utilization. While there are a series of resistances to its transfer, the predominant and rate-controlling resistance is across the liquid film around the bubble (Aiba et al.[12]). As a consequence, we can express the supply and demand problem with the following equation:

$$N_a = k_L a (C^* - C_L)$$

where N_a is the volumetric mass transfer rate (mmoles O_2/liter-hr), k_L is the liquid-phase mass transfer coefficient (cm/hr), a is the specific interfacial area for mass transfer (cm^2/cm^3), C^* is the equilibrium oxygen concentration (mmoles/liter), and C_L is the dissolved oxygen concentration (mmoles/liter).

Knowledge of the exact behavior of k_L and a of a fermentation broth in response to the operating and process variables would be extremely desirable in order to predict and to elucidate some of the necessary criteria leading to successful fermentation operation. In practice, k_L and a are difficult to separate, and we will deal with the overall mass transfer coefficient, $k_L a$. In this light, we will examine the fundamental principles of aeration and agitation. When the supply and demand are met, then N_a is also represented by

$$N_a = \frac{\mu X}{Y_{O_2}}$$

where μ equals specific growth rate (hr^{-1}), X equals cell concentration (g/liter), and Y_{O_2} equals oxygen yield (g cell/g O_2). In continuous culture, the demand becomes DX/Y_{O_2}, where D is the dilution rate (hr^{-1}). Thus, by choosing a desired productivity $P = DX$, the oxygen demand has been established. In order to meet the demand, the term $k_L a \, (C^* - C_L)$ must be sufficient.

The overall mass transfer coefficient, $k_L a$, is a parameter characteristic of the equipment used; it is primarily a function of the power delivered to the broth via mechanical agitation and/or intense aeration. The dependence of the mass transfer coefficient on these parameters is expressed by

$$k_L a \propto \left(\frac{P}{V}\right)^{\alpha} (V_S)^{\beta}$$

where P/V is the power limit liquid volume (kw/m^3), V_S is the superficial gas velocity (m/s), and α, β are exponents. The superficial gas velocity is the ratio of volumetric gas flow to the cross sectional area of the fermentor. For laboratory fermentors, α and β are typically 0.95 and 0.7, respectively, while on conventional scale fermentors, they are reduced to about 0.45 and 0.3. Thus, the efficiency of oxygen transfer decreases with scale-up. A detailed discussion of methods of measurement and calculation of these parameters is given by Aiba et al.[12] and Finn.[21] The important point is the dependence of $k_L a$

on power and gas-flow rate. One can show that power delivered to the fermentor is proportional to the agitation speed cubed. Therefore, the rate of agitation is a much more powerful tool for increasing $k_L a$ than is rate of gas flow.

In addition to increasing the mass transfer coefficient to achieve higher oxygen transfer rates, one can also manipulate and increase the driving force $(C^* - C_L)$. The C_L term cannot go below the critical dissolved oxygen concentration without affecting cell growth. But, C^* can be increased by means of oxygen enrichment of air or by increasing the operating pressure of the fermentor. These approaches provide the engineer with additional tools to facilitate oxygen transfer.

The above discussion has focused primarily on oxygen transfer, but actually will apply to the transfer of any water-insoluble component, such as methane, n-alkanes, and hydrogen. For the purpose of design, it is necessary to be able to meet the demand for the rate-limiting component, which is usually oxygen.

3.3.4. Heat Removal

Microorganisms are relatively efficient in their ability to utilize 40–60% of the available energy from the oxidation of the carbon-energy source for biosynthesis. The rest of the energy is wasted as heat. In order to keep the temperature constant, one must remove the heat from the fermentor. In an earlier section, it was shown that the amount of heat evolved could be directly related to the oxygen consumption and the efficiency of substrate (carbon) utilization. The lower the cell yield relative to its maximum, the greater the heat evolved. This problem is particularly important in the bioconversion of chemicals to food since this is a high-rate process. In industrial antibiotic fermentation, the heat load may be only 2–4 kcal/liter-hr for substrate oxidation, and another 2–4 kcal/liter-hr for mechanical agitation. SCP processes, however, may generate heat loads 10 times these values depending on substrate and productivity. Table 16 is a comparison between the heat load on several substrates. The more reduced the carbon source, the greater the heat load.

Table 16. Heat Production During Growth on Selected Carbon–Energy Sources

Substrate	Cellular yield (g cell/g substrate)	Heat production[a] (kcal/liter-hr)
Glucose	0.5	11
Methanol	0.5	31
Hexadecane	1.0	43
Methane	0.6	115

[a] Calculations are based on a productivity of 6 g/liter-hr.

Heat transfer rates, Q (kcal/liter-hr), may be calculated in the following manner:

$$Q = UA(T_F - T_0)$$

where U is the overall heat transfer coefficient (kcal/m^2-hr-°C), and T_F and T_0 are the temperature of the fermentation and cooling water, respectively. The heat transfer surface is created by having a cooling jacket around the vessel or by coils submerged in the vessel. Typical values for U range from 200–700 kcal/m^2-hr-°C.

From this equation, the problem of heat removal is seen to be one of heat transfer area and/or the temperature driving force for heat removal ($T_F - T_0$). Typical cooling water temperatures are 20–23°C; they are higher in the tropics and lower in colder climates. If a fermentation is run at 25–30°C, the driving force is small. On the other hand, by using thermotolerant or thermophilic microorganisms, one can minimize the problem of heat removal.

3.3.5. Scale-up

One of the attractive features of deriving foods from chemicals is the opportunity to take advantage of the "economy of scale." This means that the larger the plant size, the lower the manufacturing cost. The economy of scale is a consequence of being able to concentrate the capital investment and production activities in a small area; in this way duplication of equipment and facilities is minimized.

A major problem of the engineer is scaling-up the process design that was worked out in the laboratory and/or pilot plant. If the final production plant is to produce 100,000 tons per year of product, then the size of the fermentor and other equipment will be determined by the maximum (economically) achievable volumetric productivity. This in turn is determined by the rate-limiting step of the process. As noted earlier, in the laboratory this may be the rate of mass transfer of oxygen or other water-insoluble components. Likewise, for large-scale fermentors, the rate limitation may also be oxygen transfer. Another major limitation on scale-up, however, is heat transfer. When volume is increased, it goes up by the radius of the vessel cubed. But the heat transfer only goes up by the radius squared. Thus, the greater the scale-up factor, the greater the probability of a heat transfer limitation. In fact, the heat transfer may provide the ultimate limitation to the economy of scale for SCP processes.

Another scale-up and productivity limitation is the total concentration of solids in the fermentor. When agricultural wastes are used for a source of carbon, there may be problems in accumulating a high concentration of fermentable material. Many of these wastes, which are primarily cellulose or starch, contain only 30–50% fermentable substrate, with the remainder as inert solids. These solids must be mixed and pumped along with the cells, and there is a 15–20% total solids limit for biological material, above which one has a semisolid mass. This limit on pumpability may provide an upper limit on biomass concentration achievable in the fermentor.

With the constraints on scale-up, the engineer must be able to insure that the environment surrounding the microorganism is the same at all scales. This is a difficult problem, since scale-up is not linear. Mixing times, for instance, are seconds in the laboratory and minutes in some large fermentors. Temperature control is ±0.5°C or better in the lab and ±1°C or more in some large fermentors. Problems of scale-up require complete knowledge of material and energy balance and of the rate-controlling steps.

3.4. *Product Recovery*

Once the product leaves the fermentor, it must be concentrated, dried, and packaged for storage and distribution. In addition, depending on the nature of the initial raw materials and the end use of the protein, there may be the need for more specialized processing. This section will deal with the basic principles of cell recovery and try to place them in perspective for processes designed to convert chemicals to food.

3.4.1. *Cell Separation*

The stream leaving the fermentor contains microbial cells at a concentration of 20–30 g dry cell mass/liter. Although the process has a high productivity, e.g., 3–6 g/liter-hr, the final concentration is only 2–3%, or quite dilute. The importance of the final concentration is that it will determine the size and operating cost of the recovery plant. A recovery section designed around an effluent concentration of 2% will cost approximately 50% more than one for 3% solids. Therefore, it is essential to not only have a high productivity but also a high final cell concentration as well.

If the SCP organism is a bacterium or a yeast, then the form of the product is small particles. Typical sizes of these organisms are shown in Table 17. In general, yeasts are significantly larger than bacteria; this will be shown to provide a substantial advantage in recovery. Likewise, molds are larger than yeast, and have a long, stringy morphology which greatly facilitates their separation.

By separation, we mean the separation of wet cells from the fermentor broth which contains residual nutrients and soluble by-products of cell growth, i.e., carbon dioxide, organic acids, amino acids, etc. In some cases, when the carbon-energy source contains nonmetabolizable, insoluble solids, the primary

Table 17. Typical Sizes for Bacteria and Yeast

Bacteria:
cocci   spherical with diameter 0.5–1.5 μ
bacilli   rods with length 1–3 μ and diameter 0.5–2 μ
Yeast: spherical or ellipsoidal, with primary dimension 2–5 μ
Molds: branched chains, with a diameter of 5–15 μ and a length of 25–1000 μ

separation also recovers these along with the cells. A secondary separation of noncellular solids for SCP is difficult and will not be dealt with here, except to note it as a major problem in the utilization of substrates such as agricultural waste for SCP production.

The separated product is a mass of cells which contain 75–80% water. Thus, in the initial separation the cells have been concentrated from 2–3% dry solids to 20–25% dry solids. The most common means to effect this separation step are centrifugation and filtration.

3.4.1a. *Centrifugation.* While there are a variety of different types of centrifuges and centrifugal clarifiers, the movement of particles in a gravitational field is the same. It is described by the following expression:

$$V = \frac{gZ \, d_p^2(\rho_c - \rho_M)}{18\mu}$$

where V is the velocity of a particle, g is the gravitational force, Z is the centrifugal effect, d_p is the particle diameter, ρ_c and ρ_M are the specific gravities of the cell and medium, respectively, and μ is the medium viscosity. Generally, μ and ρ_M are similar to water, and ρ_c is typically 1.05–1.1. The predominant terms in the expression are (1) the centrifugal effect, which is proportional to the radius of the centrifuge and the square of the angular velocity, and (2) the diameter of the particle. If we compare a 1-μ bacterium with a 5-μ yeast, we see that the rate of movement of the yeast is 25 times that of the bacterium. This fact gives yeast an advantage over bacteria for use in SCP production. The size effect can be enhanced by the use of aggregation or flocculation to create large particles for facilitation of recovery.

3.4.1b. *Filtration.* The use of filtration for cell separation is very dependent on the size of the cells and the manner in which they pack together. Filtration is commonly used in the pharmaceutical industry for cell separation, and rotary drum filters 12 ft or more in diameter are not unusual.

The rate of filtration is described by

$$\frac{dV}{dt} = \frac{\Delta P g_c}{(r_M + r_c)\mu}$$

where V is filtrate volume, ΔP is the pressure driving force, g_c is the gravitational constant, r_M and r_c are the filtration resistance coefficients for the filter and the filter cake, respectively, and μ is the viscosity of the filtrate. The filter cake is the mass of cells which accumulates on the filter during separation.

If the cells are too small, it is likely that they will pass through the filter and not be recovered. This is a problem with bacteria and yeast. It is possible to use small-pore filters or to add inert filter agents to hold the cells back, but then the resistance of the filter and the cake become so high that a large-pressure driving force is required. This in turn makes the cake less pervious to filtrate. Thus filtration is not generally employed with bacteria or yeast unless aggregation or flocculation is used to greatly increase their size. Molds, on the other hand, are much larger and more easily retained on the filter. The most important term in the filtration equation is the filter cake resistance. Most biological materials are compressible and, as a consequence, the cake resistance can be increased substantially during filtration. The result is the need for large amounts of surface area for filtration.

3.4.2. Washing

The need for slight or extensive washing depends on the nature of the raw material used as well as on the end use of the protein. If all of the initial nutrients are water soluble, then a simple water wash may be sufficient to produce a "clean" product. However, if the substrates are not water soluble and they have low vapor pressure, the problem may be more severe. If n-alkanes or other high-boiling-point hydrocarbons are used, then a solvent wash will be required to remove the residual substrate. One of the advantages of very pure and water-soluble carbon sources, such as methanol and ethanol, is that any residual substrate is easily removed with a water wash. Methane is also easily removed because of its high vapor pressure and water insolubility.

The washing step can frequently be done while the cell paste is still in the centrifuge or on the filter. If it is done separately, then an additional cell-recovery step is required.

It is important to note that both the wash water and spent broth must be treated as waste streams to remove metabolizable material which may contribute to pollution. Thus, some form of particle treatment and/or water-reuse system needs to be incorporated into a food for chemicals plant.

3.4.3. Drying

The cell paste coming from the separation stage is 20–25% solids. In this form it is susceptible to spoilage and deterioration. Also, if it is to be shipped, then one has to pay to ship 4 or 5 lb of water per pound of product. Thus, for reasons of storage and distribution, the product is dried to give a final moisture content of 4–8%. This moisture is not free water but is bound to the constituents of the product.

The most common methods of drying are spray drying, drum drying, and lypholization, or freeze drying. The latter is the most expensive and is only used when essential components of the product are heat labile. The most common large-scale application of freeze drying is in the preparation of instant coffee. Spray and drum drying are less expensive and more likely to be used in the production of SCP.

The nature of the drying operation is important in that it may determine the final physical form of the product, e.g., powder, flakes, etc. Furthermore, it is important that it does not destroy essential components of the food. Because much water must be removed, it is also an energy-intensive operation that requires careful optimization.

4. *Fermentation Process Economics*

In the preceding sections, we have examined the essential components of a process for the bioconversion of simple raw materials into a protein food. The discussion focused on the main design parameters and problems associated with such a conversion. At this point, it is appropriate to look at the economics of the process to identify which parameters and problems deserve the most careful attention in order to minimize the manufacturing cost.

The major economic bottlenecks in any process are characterized by high cost. Therefore, an examination of the economics of single-cell protein production may be used to identify the major problem areas. Several economic analyses have been published and these are compared in Table 18 with our own analysis for the production of SCP from methanol. As may be seen from this comparison, regardless of the nature of the carbon source, it represents 40–50% of the final SCP cost; for this reason, there has been tremendous interest in the use of inexpensive carbon sources.

Another significant cost factor in the process economics is the utilities cost, which may represent as much as 20% of the SCP cost. Most of this cost results from the power required for aeration and heat removal. Thus, the cost of the carbon source, oxygen, and heat removal, may account for as much as 65% of the total SCP cost. For this reason, these problems of SCP represent major engineering problems.

The low labor cost is indicative of the capital intensive nature of the process. The 1975 cost of a 100,000 ton/year SCP plant is approximately $50 million; this translates to $0.25 of capital per pound of product. The major reasons for the high capital cost include the need to use stainless steel, the need for sophisticated instrumentation, and the need for sophisticated sterile piping. Plants for the bioconversion of chemicals for animal food may not need the capital of a plant for human food, but one cannot lose sight of the fact that

Table 18. Comparison of Process Economics for the Production of Single-Cell Protein

Cost item	Percent of SCP manufacturing cost			
	Giacobbie[23] (1973)	Gulf[22] (1975)	ICI[2] (1974)	MIT
Carbon source	42	49	63[a]	53
Type	Alkanes	Alkanes	Methanol	Methanol
Other chemicals	13	6		4
Utilities	13	20	11	10
Labor	2	7	3	6
Depreciation, taxes, insurance, maintenance	30	18	29	27

[a] Includes other chemicals.

regardless of the route, it all eventually ends up as human food, even though poultry, swine, or cattle may act as intermediate converters.

5. *Future Perspectives*

The production of protein from chemicals is not the only process one can employ for converting chemicals to food, but it is representative of one major type of process: fermentation. Microorganisms are able to efficiently produce nutrients, including proteins, fats, carbohydrates, vitamins, etc., with high productivity. With microorganisms, it is possible to intensively convert chemicals to food regardless of climatic variation and environmental pressures. Thus, this route to food production is likely to increase in both developed and developing countries. The needs of the future are to develop more efficient methods of converting chemicals to foods and to develop more applications of the final product. This latter point is especially important when we remember that "a food is not a food until it is eaten," and it is necessary that someone be willing to buy it before it can be sold. In fact, developments in the area of application are likely to be rate-limiting steps in the utilization of these novel foods.

In addition to protein by fermentation, one can make specific products like essential amino acids (e.g., lysine, tryptophan, and threonine) which may be used to supplement plant protein sources as a way to increase their nutritive value. Again, the limitation is frequently in methods of application and/or economics.

There will continue to be a need to tap our widespread but difficult-to-use resources such as coal and oil shale, and to utilize effectively our renewable resources such as cellulose, as initial starting products for food. Microorganisms are quite unique in that they can take a wide variety of raw materials and efficiently convert them to foods. In a sense, they represent miniature farms and factories all in one. The future use of these organisms to overcome food shortages lies in the hands of the creative scientist and engineer.

References

1. R. I. Mateles and S. R. Tannenbaum, eds., *Single-Cell Protein*, MIT Press, 1968.
2. J. S. Gow, J. D. Littlehailes, S. R. L. Smith, and R. B. Walter, in: *Single-Cell Protein II*, S. R. Tannenbaum and D. I. C. Wang eds., MIT Press, 1975, p. 370.
3. A. E. Humphrey, *Chem. Engr.*, July (1966).
4. C. Rainbow and A. H. Rose, *Biochemistry of Industrial Microorganisms*, Academic Press, 1963.
5. W. V. Farrar, *Nature*, **211** : 341 (1966).
6. J. Leonard, *Nature*, **209**: 126 (1966).
7. P. Echlin, *Sci. Amer.*, **214**(6) : 75 (1966).
8. N. S. Scrimshaw and A. Altschul, *Amino Acid Fortification*, MIT Press, 1971.
9. R. I. Mateles, *Biotech. & Bioengr.*, **13** : 581 (1971).

10. C. L. Cooney, *Microbial Utilization of C_1-Compounds*, Soc. Fern. Tech., Japan, 1975, p. 183.
11. C. L. Cooney, D. I. C. Wang, and R. I. Mateles, *Biotech. & Bioeng.*, **11** : 269 (1968).
12. S. Aiba, A. E. Humphrey, and N. F. Millis, *Biochemical Engineering*, University of Tokyo Press, 1973.
13. J. W. Richards, *Introduction to Sterilization*, Academic Press, 1968.
14. A. E. Humphrey and E. L. Gaden, *Ind. Engr. Chem.*, **47** : 924 (1955).
15. I. Malek and Z. Fencl, eds., *Theoretical and Methodological Basis of Continuous Culture of Microorganisms*, Academic Press, 1966.
16. I. Malek and J. Ricica, *Folia Microbiol.*, **9** : 321 (1964).
17. E. O. Powell, C. G. T. Evans, R. E. Strange, and D. W. Tempest, *Microbial Physiology and Continuous Culture*, Her Majesty's Stationery Office, 1967.
18. A. C. R. Dean, S. J. Pirt, and D. W. Tempest eds., *Environmental Control of Cell Synthesis and Function*, Academic Press, 1972.
19. D. Herbert, R. Elsworth, and R. C. Telling, *J. Gen. Microbiol.*, **14** : 601 (1956).
20. D. Herbert, in: *Continuous Culture of Microorganisms*, Society of Chemical Industries, 1961, Monograph No. 12.
21. R. K. Finn, in *Biochemical and Biological Engineering Science*, N. Blakebrough, ed., Academic Press, 1967, p. 69.
22. P. G. Cooper and R. S. Silver, *Chem. Engr. Prog.*, **71**(9) : 85 (1975).
23. F. Giacobbe, P. Puglisi, and G. Langobardi, ACS Symposium on Single-cell Protein, Philadelphia, August, 1975.

Dealing with Sewage

Sewage purification is not unduly stressed in most chemical curricula. However, it provides the basis of one of the vital cycles by which we live. In the future, it will be no longer possible to regard the way we deal with sewage in a casual way. There must be no more rejection into the sea, but methods of treating sewage (and all rejected materials, e.g., rubbish) must be made rational, so that materials are returned into circulation and are not given to the atmosphere, even in the highly purified form of carbon dioxide.

5

The Chemistry of Sewage Purification

R. M. E. Diamant

1. Introduction

Waste effluents from human communities are of two types. *Domestic* sewage flow amounts to about 200 liters per person per day, and is almost totally biodegradable because the impurities in the water are nearly all organic. It is true that domestic sewage can contain other materials as well, such as stones, oils, etc., but these can be separated quite easily. Domestic sewage is treated in municipal sewage works.

Industrial wastes may or may not be biodegradable. Many of them are very poisonous to the organisms which degrade domestic sewage, and in consequence must be treated separately.

Although, in the past, laws regarding river pollution were extremely slack, they are now being tightened up all over the world. In general, industrial waste liquors may not be discharged into public sewers, unless they are free from nonbiodegradable agents and substances which may harm the normal sewage

R. M. E. Diamant • Department of Applied Chemistry, University of Salford, Salford, England

purification processes. The laws regarding discharge into rivers are even tighter. Industrial effluents are nowadays purified to the required degree at the works, using the correct chemical processes to separate harmful constituents.

2. Analytical Examination of Sewage[1-3]

Sewage can be classified according to the following criteria:

1. *Turbidity* expressed in international turbidity units.
2. *Suspended solids* measured in parts per million.
3. *Total dissolved solids* measured in parts per million.
4. *Acidity* in terms of hydrogen ion concentration or pH.
5. *Dissolved oxygen concentration* in parts per million of O_2. This is determined in the same way as the BOD_5 value, by reacting the oxygen dissolved in the sample with manganous hydroxide and potassium iodide, and titrating the iodine liberated with sodium thiosulfate. (See below.)
6. *The potassium permanganate value* (PV). This value is found by adding a certain volume of acidified N/100 potassium permanganate solution to a given volume of the sample, followed by excess ferrous ammonium sulfate and back-titrating with N/100 potassium permanganate solution. The permanganate value is given in parts per million of O_2 and indicates the concentration of unstable organic compounds in the sample of sewage.
7. *The biochemical oxygen demand after 5 days* (BOD_5). This gives the amount of oxygen used up when the sample is allowed to stand in the dark at room temperature for 5 days. The oxygen required is determined by an iodometric determination:

$$2\,Mn(OH)_2 + O_2 \rightarrow 2H_2MnO_3$$

$$H_2MnO_3 + 2KI + 2H_2SO_4 \rightarrow K_2SO_4 + MnSO_4 + 3H_2O + I_2$$

$$I_2 + 2Na_2S_2O_3 \rightarrow Na_2S_4O_6 + 2NaI$$

Both the sample and a blank are shaken up with air and stored for 5 days. At the end of that time, a mixture of manganese sulfate, potassium iodide, and potassium hydroxide are added to both, and the iodine evolved is titrated against standard sodium thiosulfate solution. The BOD_5 value, which is also quoted in parts per million of O_2, gives a measure of the amount of material in the sewage which is readily biodegradable.
8. *The chemical oxygen demand* (COD). In this determination the sewage is reacted with strong-boiling potassium dichromate in the acid state. After the reaction has been completed, the excess potassium dichromate is titrated against ferrous ammonium sulfate solution. This value, which is also quoted in terms of parts per million of O_2, gives the total amount of organic matter present in the sewage. The reaction can be expressed as

$$C_nH_mO_x \text{ (organic matter)} + (n + m/4 - x/2)O_2 \rightarrow nCO_2 + m/2H_2O$$

For any given sample of sewage, the following always applies:

$$COD > BOD_5 > PV$$

9. *The methylene blue value.* This gives the time required for a given sample to start fermenting. It is carried out by adding methylene blue to a sample of sewage and measuring the time in hours required for the blue color of this dye to disappear.

3. *Treatment of Domestic Sewage*

In the past, sewage was simply discharged into rivers, lakes, or the sea with no treatment whatsoever. If the quantities of sewage so discharged are small and do not contain any harmful ingredients, natural processes such as sedimentation and bacterial oxidation are adequate to cleanse natural sewage effluents. Trouble arises when sewage from any but the smallest of human settlements is discharged; natural processes are then completely swamped. Sewage from modern large towns and cities must therefore be adequately treated before discharge.

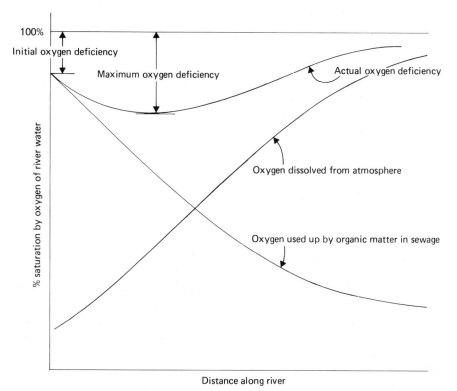

Figure 1. How the continual dissolution of oxygen to river water enables discharge sewage to be treated in a natural way.

Table 1. Constituents of Domestic Sewage

Constituents	Parts per million by weight
Fats	140
Proteins	42
Carbohydrates	34
Detergents	5.9
Amino sugars	1.7
Amides	2.7
Soluble acids	12.6
Total organic constituents	238.9

4. Chemical Composition of Sewage[4]

The actual composition of sewage varies greatly. An analysis of typical standard domestic sewage showed that it contained the constituents shown in Table 1. In addition, the sewage contained suspended mineral matter, bacteria, viruses, and dissolved salts.

Sewage treatment can suitably be classified into three distinct types:

1. *Primary treatment* removes larger suspended particles from sewage, and produces a liquid which contains only colloidal and dissolved matter.

2. *Secondary treatment* removes the bulk of the organic matter present in the fluid, using biochemical oxidation processes, and provides an effluent which can be discharged into the sea and rivers, making use of natural biochemical processes to complete the purification.

3. *Tertiary treatment* processes use the effluent from secondary treatment as feed and produce water which is of drinking water standard. Most sewage plants only go as far as secondary treatment, but in recent years the demand for tertiary treatment has stepped up as the need for complete recycling of water has increased.

5. Primary and Secondary Treatment of Sewage[5–9]

The crude sewage from the sewers is first stored for about one week before treatment. Normal sterilization treatment with chlorine, ozone, or similar substances is extremely effective in killing off bacteria but has little effect on the much smaller viruses. By allowing bacteria to multiply, the viruses are killed off. Bacteria can then be separated, by filtration systems and sterilization, without any undue difficulty.

The crude sewage next passes through a rough screen that separates off large lumps, disintegrates them, and resuspends them in the main sewage flow. Grit and oil removal tanks follow, where insoluble materials which float to the

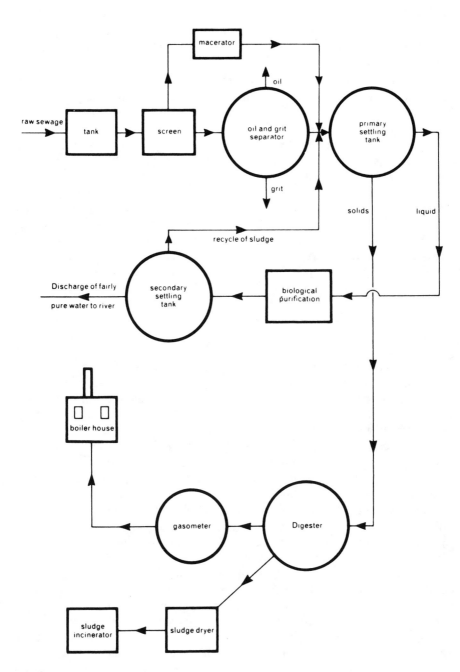

Figure 2. Flow sheet of a conventional sewage plant.

top are taken off, as are heavy stones and grit particles, which sink to the bottom. The sewage now flows into the primary settling tanks. These tanks are invariably circular and are equipped with top and bottom scraping devices to eliminate sludge and scum.

The bottom scraper is connected to a rotating spider's network, which is suspended either at the center of the tank or from a radial bridge above the water surface. In the former case, the drive is from the center of the tank, and in the latter case at the periphery. The radial bridge method permits the transmission of a very considerable force and provides a very flexible assembly of the scraper arms on the framework. The sludge is drawn off at the bottom of the steeply sloping tank by means of telescopic valves which permit easy adjustment of the extraction flow rate and also prevent blockage. Fully automatic diaphragm extraction valves are also widely used. The sludge from these primary settling tanks can then be passed on to a sludge digester if sewage methane is required for combustion purposes. Alternatively, the sludge can be passed on directly to driers and incinerators. The liquid from the primary settling tank passes on to a biological purification plant.

5.1. *The Biological Purification Plant*

Typical traditional bacterial purification beds usually have a diameter of 40 m or more, and head loss in the flow of water usually exceeds 250–300 mm of water. Sprinkler joints which cause little friction and are not sensitive to variations in inlet pressure are used. Traditional bacterial beds need very large quantities of filtering material, yet the purification efficiency of a single bed seldom exceeds 80%. To achieve better purification, either several bacterial beds in series or recirculation flow are necessary.

The tanks are either rectangular or circular in shape. Depths of water varying between approximately 3.5 m and 7 m are employed, and air is supplied by rotary blowers. This air is distributed by means of "nonclog" diffusers fitted with flaps, or through porous media. The amount of air which passes through is automatically adjusted on the basis of the concentration of impurities in the effluent. This is done by means of a Redox potential regulator or a dissolved oxygen probe. In general, the treatment time varies between $1\frac{1}{2}$ and 3 h, which is roughly half the time usually required in the more traditional types of bacterial beds. Excess sludge is removed by means of telescopic or automatic valves and is returned to the primary settling tank. Water which is drawn off the biological purification plant is commonly considered clean enough to be pumped into rivers. If further purification is needed—as, for example, when the water is to be recycled directly into the town's water system—clarification, using such materials as activated carbon, followed by sterilization with chlorine, ozone, or ultraviolet radiation and colloidal silver, is necessary.

5.2. *Digesting of Sludge*

Sludge digesters are used in order to reduce considerably the final volume of sludge produced in the plant and also to obtain valuable sewage methane. During methane fermentation, sludge loses between 40% and 60% of its volatile matter and its volume is reduced by about 60%. In order to increase the capacity of a given size of sludge digester, one of the following techniques is commonly employed:

1. The mixture of sludge and effluent is heated, using steam or hot water, in order to increase the rate of fermentation.

2. The sludge is thoroughly recirculated inside the digester in order to eliminate void spaces and to improve its homogeneity.

In many cases, multistage digestion systems are used to improve efficiency. In general, when planning digestion plants for the sewage outflow of a city, it is reasonable to allow a digester volume of about 30 liters per head of population served. Thus a total digester plant volume of around 30 million liters is desirable for a city of 1 million inhabitants. It is obvious that present practices fall far short of this desirable figure.

The methane produced during the digestion of sewage sludge is burned, primarily for steam production to heat the sludge fed into the digesters. In very large plants, sewage methane can also be used for the operation of gas engines. These operate on the total energy principle to produce electricity for works purposes, and any excess heat is used for sludge heating. However, it is generally uneconomical to install gas engines unless the sewage plant serves at least 300,000 inhabitants; for smaller plants the investment costs tend to be far too high.

5.3. *Treatment of Digested Sludge*

Much of the digested sludge produced is then dried on special drying beds. Modern practice tends to frown on the use of these for several reasons. Drying beds occupy a large ground area, and offensive smells are created which lower the amenity value of the surrounding district. The sludge, even when dry, is difficult to handle and has an unpleasant appearance. New sewage works are therefore preferring a system of thickening followed by filtration.

After emerging from a thickener, the sludge is passed into one or more vacuum filters where most of the water is drawn off. The filtered sludge has a dry matter content of more than 30% and is often pelletized and stored, to be used either directly as an agricultural fertilizer, or for mixing with household refuse prior to composting.

In general, the digested mixture of primary and secondary sludge is filtered directly. However, in the case of a multistage plant, and where biological treatment is carried out later on, filtration may be used at the beginning of the process.

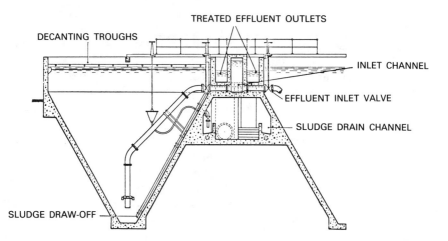

Figure 3. Lay-out of a vertical flow type sludge separator suitable for flocculation treatment of fine precipitates. (Courtesy: Patterson Candy International, Ltd.)

Another system for separating the digested sludge uses a centrifuge, followed by further treatment of the effluent.

5.4. *Filtration and Removal of Colloidal Matter*

The water which emerges from the second filter contains colloidal particles, insect larvae, algae, as well as dangerous amoeboid creatures, all of which are difficult or impossible to kill by standard sterilization procedures.

Filtration is carried out by passing the water through porous material. In most cases filtration media serve not only to remove particles larger than the pores of the medium, but also act as a matrix for entraining microorganisms which secrete enzymes. These enzymes coagulate much of the colloidal material present in the water and facilitate its removal. Often filter media are used which are surface active. These serve to retain colloidal material in the adsorbed state on the filter medium itself.

The main filter media for water are sand and kieselguhr. The advantage of these materials, as compared with more static filter media like cloth or porous porcelain, is that granular filter mediums are much easier to wash. Kieselguhr is mainly employed for smaller filters, like those used in the cleaning of swimming pools. For large-scale industrial practice, sand and gravel beds are preferred, between 600 mm and 1500 mm thick. Once the filter medium is fully loaded, it is agitated by a mixture of compressed air and water and the pure sand is reclaimed by sedimentation.

Both slow and fast filtration are employed in the purification of water obtained from sewage effluents. For slow filtration three stages are usual. The first consists of coarse filters which remove larger particles. These operate at a rate of about 1 m/hr. The second consists of prefilters which remove everything but colloidal particles. The water passes through these at a rate of 500 mm/hr. Finally, there are the filters proper, through which water passes at the rate of only 200 mm/hr. The purpose of these is to coagulate colloidal particles susceptible to bacterial action and to remove them, and to act as surface-active agents for adsorption of other colloids. An essential feature of this form of filtration is the formation of the bacterial membrane, which can take several days after the medium has been washed.

Fast filtration takes place at a rate of up to 20 m/hr. This is incapable of removing colloid particles which have to be coagulated prior to passage of the water into the filter plant.

The filter medium used for fast industrial water filtration consists of 98+% SiO_2, with effective size being between 0.3 mm and 1 mm diameter for standard filters, and 2 mm diameter for prefilters. Many different types of filter plants are on the market, each having its own operating characteristics.

Sewage treated by primary and secondary methods has in the past been discharged directly into rivers without any further treatment, and in the vast majority of cases still is. Table 2 shows that the purification process has certainly not progressed to completion.

It is assumed that the natural cleaning processes which take place in rivers and other waters should be enough to eliminate the residual BOD_5 and kill off the coliform bacteria. It is, however, increasingly felt that sewage purification must go further than the partial state represented by primary and secondary treatment.

Tertiary methods capable of reducing BOD_5 and COD values virtually to zero, combined with adequate sterilization techniques, are being introduced to provide drinking water quality effluents. It is also necessary to remove the bulk of nitrogen and phosphorus compounds from the water; these substances are exceedingly harmful in that they may induce algae formation, clogging up vital water mains. Nitrates also have certain adverse health effects. Phosphates and nitrates are hardly reduced at all in primary and secondary treatment plants.

Table 2. Outflow after Primary and Secondary Treatment[10]

	Feed sewage	Outflow
Turbidity (in units)	100	20
Suspended solids (ppm)	300	20
COD (ppm O_2)	400	50
BOD_5 (ppm O_2)	300	25
Coliform bacteria (per mliter)	150,000	1000

6. Tertiary Methods of Water Treatment[11-14]

These can be summarized as follows:

1. Activated carbon in the form of either granules or powder
2. Microstraining
3. Special selective coagulants
4. High-density aeration processes
5. Removal of phosphates by precipitation
6. Removal of nitrogen compounds
7. Bacterial beds or algae for nutrient stripping
8. Electrochemical methods
9. Sterilization of effluent.

6.1. Activated Carbon[15-17]

One of the most efficient agents for the removal of dissolved and suspended impurities is carbon that has been activated by being heated to 930°C, which drives off any adsorbed materials from its surfaces.

Carbon surfaces are, in general, more active with respect to aromatic than aliphatic compounds, and have a greater adsorption for branched chain compounds than for straight chain types. Amines, carbonyl groups, double bonds, sulfonic groups, and nitro groups all tend to make the compounds concerned adsorb more rapidly.

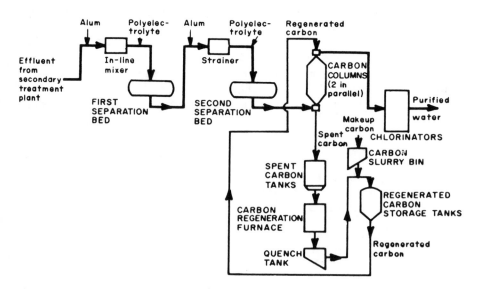

Figure 4. Flow sheet of the tertiary water purification plant at South Tahoe, California. (Courtesy: Pittsburgh Activated Carbon Co.)

Adsorption on carbon surfaces proceeds according to the Freundlich equation, which is normally quoted as

$$\log x/m = \log k + 1/n \log C$$

where x is the weight of impurity adsorbed per m units by weight of carbon, C is the equilibrium concentration of the impurity in question, and k and n are constants whose magnitudes depend on the surface-active nature of the carbon and the nature of the impurity.

A standard tertiary plant employs vertical steel columns filled with granular activated carbon and a water flow not exceeding 5 liters per sec per m^2 cross section. The flow length of carbon must be about 7.5 m.

The carbon particles used should preferably have a surface area of about $1000 \ m^2/g$, which corresponds to a mean particle diameter of between 1.5 mm and 1.7 mm. The bulk density of the carbon charge is around $420 \ kg/m^3$, and it has been found that the capacity of activated carbon is roughly 0.5 kg of COD for each kilogram of carbon used before regeneration is required.

Adsorption on carbon is prevented if the pH of the solution is too high. At pH values above 9, adsorption virtually stops. An activated carbon adsorption plant generally uses continuous withdrawal of the spent carbon charge at the bottom of each column. This is then passed into the regeneration plant where the carbon is heated to 930°C in a steam atmosphere. The steam vaporizes the impurities which have been adsorbed on the carbon surface and drives them off. The furnace gases which carry these impurities are then passed through afterburners and a wet scrubber before being released to the atmosphere. Some 5% of the carbon, representing about 5 $kg/1000 \ m^3$ water treated, has to be discarded. The rest is recycled. The water which emerges from the carbon columns is passed into a chlorinator and is then discharged as drinking quality water.

Starting with a feed concentration of impurities similar to the value given at the end of primary and secondary treatment, the final values obtainable are as shown in Table 3.

Table 3. Outflow after Tertiary Treatment

Feed sewage	Outflow
COD (ppm O_2)	2–10
BOD_5 (ppm O_2)	2–5
Suspended solids (ppm)	less than 0.5
Turbidity (in units)	less than 0.5
Phosphates (ppm)	0.2–1
Coliform bacteria (per m/liter)	less than 0.02

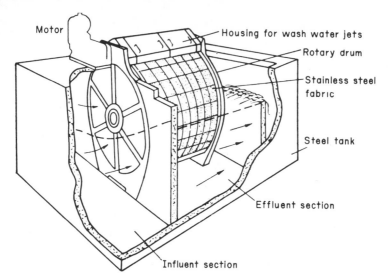

Figure 5. Diagram of an industrial microstrainer. (Courtesy: Crane Ltd.)

6.2. *Microstraining*

This is a form of simple filtration which uses very finely woven stainless steel wire fabrics drawn to diameters that are barely visible. The wire is woven on special high-precision looms to produce these so-called microfabrics. Thin mats of intercepted solids are formed on their surfaces which, together with the fabric support, have relatively high flow ratings at low hydraulic resistance and can intercept solids which are even smaller than the apertures in the woven fabric. The highest grade of such fabrics has 250 apertures per mm^2, and each opening is just 65 μ across. After partial blockage of the apertures by intercepted solids has occurred, such fabrics have been found adequate for removing organisms with a diameter of as little as 7 to 12 μ. Such microstraining techniques can be used for the clarification of potable water and even for filtering finished drinking water because of their efficiency in removing both inorganic particles, such as fine dust, metallic scale, etc., and living organisms, such as algae, diatoms, amoebae, and some of the large bacteria.

The use of two-stage microstraining has been found effective in reducing suspended solids from 53.5 ppm to 7.5 ppm, and the BOD_5 value from 47.1 ppm to 15.0 ppm, at a cost far lower than that of competing tertiary purification systems.

6.3. *Coagulants*

The terminal settling velocity of particles in water is given by Stokes' law, which can be quoted as follows:

$$\text{Terminal velocity} = \frac{gd^2(D_s - D_w)}{18\,\mu}\,\text{m/sec}$$

above. The hydraulic supply is in parallel but the electricity supply is in series. The result of passing an electric current is to produce chlorine gas at the anode and sodium hydroxide and hydrogen gas at the cathode. This forms sludge flocculants which are recirculated and finally separated out by settling. The chlorine concentration is kept at about 100 ppm in order to prevent the sewage from putrifying. The final product has a free chlorine concentration of about 5% which is enough to kill off coliform bacteria. The technique produces a reduction of about 60% in BOD_5.

During the electrolytic action a flocculant of magnesium hydroxide is produced which reacts with any phosphates present and also helps in the coagulation of nitrates. These flocculants tend to be floated to the surface by the hydrogen bubbles. Phosphate removal is about 90–95%, while nitrogen removal is around 70%. The final water has a low turbidity, a negligible coliform bacteria count, and can be discharged without any ill effects into the sea close to recreational areas. Power consumption is of the order of $0.4 \, kWh/m^3$ of sewage treated.

6.9. *Sterilization of Effluent*

The principal bacteria and viruses present in purified sewage effluent that is released to rivers and thence to the sea are the following:

1. Dysenteric bacteria: Flexner's *Bacillus dysenteriae, Shigella dysenteriae, Shigella paradysenteriae,* and *Proteus vulgaris*
2. Koch's bacillus, which causes tuberculosis
3. *Vibrio cholerae,* which causes epidemic cholera; no longer widely distributed in temperate areas
4. Typhoid fever bacilli: *Salmonella typhosa, Salmonella paratyphi,* and Gärtner's, Morgan's, and Schottmüller's bacilli
5. *Leptospira icterohemorrhagiae,* which causes "mud fever"
6. Polio virus
7. Virus of infectious hepatitis
8. Enteric cytopathogenic virus, which causes epidemic diarrhea
9. Adenovirus, which appears to cause eye and other infections

In addition, it is possible for parasitic worms of many kinds to be discharged from purified sewage, but only if filtration is poor. If the filter medium is such that all particles greater than 20 μ in diameter are removed, there should be no danger of worm infestation. Amoebae which can cause a variety of conditions, including liver ulcers and intestinal hemorrhage, have a diameter of 50 μ and should also be removed by efficient filtration.

Bacteria are, on the whole, very sensitive to sterilizing agents and are killed efficiently. Viruses, on the other hand, are much more resistant to oxidizing agents. The best method for destroying viruses is to allow a certain time between the collection and the purification of sewage effluent; virtually all viruses are dead after about a week because they are denatured by living microorganisms. Viruses can also be destroyed effectively by chemical flocculation, because a chemical reaction takes place between the virus protein and the metallic ions of coagulants. It has been estimated that some 95–99% of all viruses contained in a sample of sewage effluent can be destroyed in this way. Viruses can survive for a long time in "purified" water. The main methods for sterilizing water are the following:

1. Chlorine and its derivatives such as hypochlorites
2. Ozone dosing
3. Ultraviolet irradiation
4. Electrolytic methods

6.9.1. *Chlorine and Its Derivatives*

For the sterilization of water, either chlorine gas or one of the two salts, sodium hypochlorite (NaClO) or calcium hypochlorite ($Ca(ClO)_2$), is used. Chlorine, even in small concentrations, effectively kills most bacteria, mainly because it destroys the enzymes needed by the microbes to survive. Enough chlorine is normally added so that some free chlorine remains in the water after 2 hr of contact. This free chlorine—even in the most minute concentrations—is readily detected by the use of potassium iodide and a starch indicator.

For slow sterilization, a slightly acid medium is preferred, together with intimate contact between water and chlorine for at least 2 hr. If the temperature of the water is less than about 10°C, an excess of chlorine compensates for a slower reaction rate.

For rapid sterilization, an excess of chlorine is used in the water and allowed to destroy bacteria and other harmful organic matter for a period of about 10 min. At the end of that time, the excess chlorine is neutralized by the addition of either sulfur dioxide, sodium sulfite, or sodium thiosulfate.

As an alternative, ammonia may be added to the water, which converts excess chlorine into the chloramines: NH_2Cl, $NHCl_2$ and NCl_3. These chloramines are quite odorless and have no specific taste, and they are bactericides in their own right. Their bactericidal power is, however, weaker than that of chlorine on its own.

For large plants, chlorine dosing is effectively carried out with the gas itself. The main drawback is that very strict control is essential; even concentrations as low as 40 ppm of free chlorine in the atmosphere are highly dangerous. For smaller plants hypochlorites are much easier to handle; they can be dissolved or suspended in water prior to dosing.

Viruses such as those of poliomyelitis and infectious hepatitis need quite high concentrations of chlorine for destruction, in excess of 0.4 mg/liter for 30 min or more. Koch's bacillus is even more resistant and needs at least 1 mg/liter for an hour to ensure destruction. The remaining bacteria are, however, easily destroyed at very low concentrations. Chlorine is not too effective in destroying larger organisms; amoebae, for example, need more

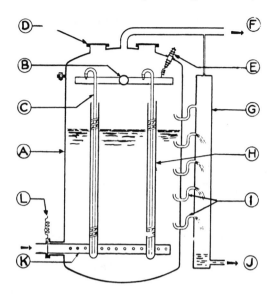

Figure 7. A commercial ozonizer. (A) Plastic body; (B) Branched dry air inlet pipe; (C) Tubular metal electrode; (D) Inspection hole; (E) High-voltage supply; (F) Outlet for ozonized air; (G) Cooling water outlet column; (H) Glass tube; (I) Siphon-shaped overflow; (J) Outlet for cooling water through a siphon or drain cock; (K) Perforated metal inlet pipe for cooling water; (L) Earth connection. (Courtesy: Degremont Laing Ltd.)

than 10 mg/liter for an hour for destruction. But since amoebae have a relatively large diameter, they can be easily removed by good filtration.

6.9.2. Ozone

Ozone has the chemical formula O_3, and is made by passing air through a field of "silent" electric discharge. Ozone cannot be stored because it disintegrates spontaneously; it must therefore be produced on site from air. There are several commercial ozonizers on the market designed to operate at about 12,000 V, obtained by means of a standard stepup transformer. The air entering the ozonizer must be perfectly dry to avoid the formation of nitric acid, which would destroy the electrodes. The ozonized air is then bubbled through water. It is best to use a good power agitator to ensure good distribution.

When ozone is used for water sterilization, an excess of ozone must be present for about 5 min where the temperature is in excess of 10°C, and for about 10 min for temperatures below 10°C. Excess ozone is easily detected by the use of potassium iodide–starch papers. When used correctly, ozone is considered to be more effective than chlorine for sterilization. The main difficulty with ozone is that it is only sparingly soluble in water, and therefore effective dispersion is vital to ensure that there is good contact between the ozone and the bacterial and virus matter to be destroyed.

6.9.3. Ultraviolet Rays

The maximum bactericidal action of ultraviolet rays is when the wavelength of the rays is in the vicinity of 260 nm. Ultraviolet light is commonly produced by a mercury vapor lamp. This method of sterilization is only commercially viable when relatively small quantities of water are to be treated. The water must be very clean, or some of the ultraviolet rays are absorbed and thus rendered inactive. In general, a 25-W ultraviolet lamp can be used for sterilizing about 2000 liters of water per hour. The water is led past the lamp at a depth of about 150 mm.

6.9.4. Electrolysis

Two electrolysis techniques are used. In the first, the anode is of silver and the cathode of either carbon or stainless steel. The electrodes must always be placed where the water is naturally turbulent. On the passage of an electric current silver ions are released in solution, and these have a marked bactericidal effect. The quantity of silver needed for adequate sterilization is minute, normally on the order of one gram of silver to 20 m^3 of water. Faraday's laws apply to the electrolysis of silver, which means that 108 g of silver are liberated by the passage of 96,500 C.

Thus a current of one ampere flowing for one hour liberates

$$108 \times 3600/96{,}500 \text{ g of silver} = 4 \text{ g of silver}$$

or enough to sterilize about 80 m^3 of water per hour. The silver, in the form of its ions, reacts with most bacteria to form a slime which can be filtered off.

7. Methods of Purifying Industrial Effluents[21–23]

Industrial effluents are far more difficult to purify than domestic wastes because they contain such a wide variety of different pollutants, many of which are harmful to the bacteria which carry out biodegradation processes in normal sewage works. It is normally necessary to deal with the purification of industrial effluents separately within the factory complex, and to discharge the effluent only after it has shed any components which would interfere with normal sewage purification operations.

The purification of industrial effluents proceeds along approximately the same paths as normal sewage purification. Sedimentation, coagulation, microfiltration techniques, adsorption, and electrochemical methods are widely used, as are a number of biochemical methods using bacterial organisms immune to specific organic compounds. Certain materials offer, however, special problems.

7.1. Metals[18–20]

The most toxic metals are cadmium, lead, mercury, beryllium, and arsenic, while several others are also exceedingly harmful. It is essential that every trace of such metals be removed. The three main techniques employed are precipitation with reagents, cation exchange techniques, and solvent extraction methods.

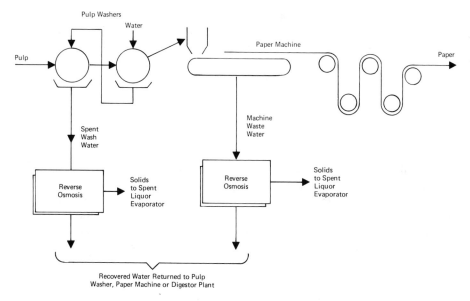

Figure 8. Flow sheet for the purification of wastes from the pulp and paper manufacturing industry. (Courtesy: Aqua-Chem, Inc.)

7.1.1. Precipitation

Precipitation involves the choice of a reagent that can produce an insoluble compound with the noxious metal. The solubility product of the two ions concerned is then of vital importance. The solubility product S is defined as follows:

$$(A^+)(B^-) = S$$

where A^+ is the solubility of the anions in g-ions per liter and B^- is the solubility of the cations in g-ions per liter.

Table 6 gives the solubility products of a certain number of selected insoluble salts, at 18°C.

Table 6. Solubility Products of Selected Salts

Salt	Solubility product S
HgS	2×10^{-49}
Hg_2Cl_2	2×10^{-18}
$Pb(COO)_2$	2.7×10^{-8}
PbS	3.4×10^{-28}
CdS	3.6×10^{-29}
CuS	2×10^{-47}
ZnS	1.2×10^{-23}

In order to precipitate the maximum quantity of the noxious metal ion, one adds excess of the anion concerned. For example, if one wished to precipitate as much as possible of the Hg_2^+ ion from solution, a large excess of chloride ions would be used. Often the precipitates obtained are colloidal and are therefore difficult to filter or to precipitate. Sulfide precipitates are frequently colloidal. Precipitation combined with coagulation methods often helps when particles are too small for efficient filtering and precipitation.

7.1.2. Ion Exchange[24-25]

Many metallic ions can be removed almost totally by the use of cation exchangers. Lead, copper, zinc, mercury, chromium, nickel, and arsenic, all are metals which can be removed from sewage in this way. Ion exchange methods should, if possible, be used in conjunction with precipitation methods, as they work best when the concentration of the metal ions is low. The concentration of most metal ions in effluents can be reduced by a combination of precipitation and ion exchange to less than 0.001 ppm.

7.1.3. Solvent Extraction

These methods are suited to situations where ion exchange is impractical due to liability of exchanger poisoning by other components of the effluent. Nickel ions are removed from solution by the use of di-nonyl naphthalene sulfonic acid (DNSA), zinc and cadmium are removed by di-2-ethylhexyl phosphoric acid (EHPA), and copper can be extracted by quaternary amines.

7.2. Dissolved Salts and Suspended Matter

Industrial effluents often contain dissolved salts which are harmful to the environment. The most frequently employed methods for removal of such salts are ion exchange and electrodialysis.

Many organic solutions and suspensions are best treated by the use of strong oxidation methods, designed to break down the noxious material into carbon dioxide, water, and oxides of sulfur, nitrogen, etc. Even partial oxidation sometimes helps. Complex materials with active groups are separated out by the use of adsorbents like activated carbon, activated silica, etc.

Electrochemical methods are often used for the removal of suspensions such as oil and paint in water. Hydrogen is produced electrolytically, and carries such materials upwards to produce a froth which is easy to separate out. If seawater is added to the effluent, electrolysis produces chlorine, which is capable of oxidizing organic matter.

7.3. Radioactive Wastes

The disposal of radioactive materials involves additional problems not encountered during the disposal of other kinds of industrial wastes. On the other hand, disposal is helped by the fact that radioactivity disappears over time of its own accord.

Radioactivity is a first-order reaction in which $dx/dt = -kx$, where dx/dt is the rate of disappearance of radioactivity, k is the velocity constant, and x is the quantity of radioactive matter present after time t.

Using standard chemical kinetics treatment, one can evaluate that

$$k = 2.303/t \log a/(a - x)$$

where a is the initial quantity of radioactivity present. The most important concept in first-order reactions is the half-life, i.e., the time required for half of the existing radioactivity to be dissipated. This can be evaluated as

$$t_{\frac{1}{2}} = 2.303/k \log 2 = 0.693/k$$

The half-life of radioactive substances can vary from minute fractions of a second to many millions of years. As can be seen, the half-life of radioactive materials is inversely proportional to its activity. From a pollution prevention point of view, radioactive materials are divided into the following:

1. Short-lived materials with half-lives up to 14 days
2. Medium-lived materials with half-lives up to about 100 years
3. Long-lived materials with half-lives over 100 years

7.3.1. *Short-Lived Radioactive Wastes*

These are first concentrated and then stored for a few years under water or inside rock tunnels, during which time the radioactivity drops virtually to zero. At the end of this time the wastes can be disposed of quite normally. Care must obviously be taken if the metals which are present are poisonous (e.g., lead, cadmium, mercury), but the disposal of such substances is the same as of similar waste products in other industries.

7.3.2. *Medium-Lived Radioactive Wastes*

Typical of such materials is strontium 90, which has a half-life of about 23 years. Disposal is by concentrating the solutions, which are then placed into elongated drums to reduce criticality problems. The drums, which are made from tough, corrosion-resistant steel, and are also fitted with gas valves, are then loaded on to ships and dropped into ocean trenches 10 km or more in depth. Although the drums are constructed so that they cannot burst or corrode, no harm would be done even if they did. It can be calculated that it would take about 1000 years for the first trace of radioactive matter to reach the surface of the ocean, even if the drums were to burst on impact. Most of the radioactivity would have disappeared by that time.

7.3.3. *Long-Lived Radioactive Wastes*

Radioactive wastes with half-lives above 100 years are not dangerous, provided they are adequately diluted. In practice such wastes are shipped out to

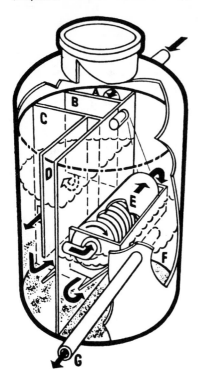

Figure 9. The PARCA mini sewage works. (A) Inlet; (B), (C), and (D) Primary filters; (E) Biological contact rotor; (F) Fiberglass-reinforced plastic casing; (G) Outflow for purified effluent.

sea and poured overboard. If adequate monitoring is carried out, and no concentration of the radioactive isotopes by living organisms occurs, such a practice is totally harmless.

8. Mini Sewage Plants

As one can see, there is very little difficulty in completely purifying sewage if dwellings can be connected to a main sewage line. Problems arise when houses are isolated so that connection to a main sewage system is not economical. In the past the only method was the installation of a "cesspool." A. B. Parca of Linkoping, Sweden, have developed a special mini sewage plant encased in fiberglass-reinforced polyester, in three sizes, suitable for tackling 3, 10, and 30 m^3 of water per 24 hr respectively, and able to achieve a 90% reduction in BOD$_5$.

The main part of the equipment consists of a biological contact rotor of flat spirals within a horizontal cylinder through which the effluent passes. The various microorganisms, whose purpose is the breakdown of the organic impurities within the sewage, are all contained in this unit. A special surface-treatment technique increases the effective contact area between the microorganisms and the effluent water. This method is suitable for serving wastes from farms, small houses, camp sites, and smaller industrial undertakings.

References

1. *Water Treatment Handbook*, Degremont Laing Ltd., 1965.
2. T. H. Y. Tebutt, *Water Quality Control*, Pergamon , Oxford, 1971.
3. G. V. James, *Water Treatment*, Technical Press, Edinburgh, 1971.
4. Metcalf & Eddy, Inc., *Wastewater Engineering*, McGraw-Hill, New York, 1972.
5. L. Klein, *River Pollution*, Butterworths, London, 1962.
6. Technical papers, technical information, and technical brochures supplied by General Engineering Co., Ltd.
7. Technical papers, technical information, and technical brochures supplied by Ames Crosta Mills, Ltd.
8. Technical papers, technical information, and technical brochures supplied by Paterson Candy International, Ltd.
9. E. Nordell, *Water Treatment*, Reinhold, New York, 1972.
10. L. L. Coaccio, *Water and Water Pollution Handbook*, M. Dekker, New York, 1971.
11. P. R. Wathen, M. M. Clemens, and W. Zaaban, "Three-way cooperation develops complete waste treatment system." *Plant Engineering*, June (1967).
12. C. L. Fitzgerald *et al.*, *Coagulants for Waste Water Treatment*, Calgon Corp., Pittsburgh, 1973.
13. Ontario Water Resources Commission, *Microstraining of Sewage Effluents*, 1969.
14. J. R. Herbert, *Technology to Aid Conservation*, Simon Technical Review No. 13, 1970.
15. *Quality of Treated Effluents*, WHO, Geneva, 1973.
16. W. W. Eckenfelder and L. K. Cecil, *Progress in Water Technology*, Pergamon, Oxford, 1972.
17. D. G. Hager and M. E. Flentje, "Removal of organic contaminants by granular carbon filtration." *Journal of the American Waterworks Association*, **57**: 11, Nov. (1965).
18. R. W. Aitken, *Water Purification Processes, Design and Selection Considerations*, Jamaica Seminar on Water Pollution, 1969.
19. D. G. Hager, "Clarification–Adsorption in the Treatment of Municipal and Industrial Wastewater," paper read at Conference on Water Pollution, Dallas, Texas, 1969.
20. Proceedings of the Third International Conference on Water Pollution Research, Munich, 1966.
21. Proceedings of the Fourth International Conference on Water Pollution Research, Prague, 1969.
22. Proceedings of the Fifth International Conference on Water Pollution Research, San Francisco and Hawaii, 1970.
23. Proceedings of the First International Symposium on Water Desalination, U.S. Department of the Interior, Washington, DC, 1965.
24. T. V. Arden, *Water Purification by Ion Exchange*, Butterworths, London, 1968.
25. S. B. Applebaum, *Demineralization by Ion Exchange*, Academic Press, New York, 1968.

Resource Exhaustion

In the famous book Limits to Growth, *Meadows* et al. *showed that many materials, if we continue simply to use them without recycling them, will exhaust themselves in a relatively short time. It is therefore incumbent upon the chemist to attempt to devise recycling methods which will pull back the metals, and indeed many other materials, from the present cycle of oxide–metal–rejected oxide. The efficiency of this process will finally determine the viable (high-living-standard) world population.*

6

Resources

A. W. Mann

... the power of population is indefinitely greater than the power in the earth to produce subsistence for man—Thomas Malthus (1798).

Size of population or the life of an individual will be limited by whatever prerequisite of life is in shortest supply—von Liebig (1870).

1. *Depletion of Resources*

Although it is not yet clear what population the earth is capable of supporting, or what the limiting prerequisite of life will be, these two statements are prophetically important as the twenty-first century approaches. Man is faced, in the space of probably less than 100 years, with the problem of reversing a lifestyle, a sociology, a technology, and an economy that have evolved and have been in constant use for thousands of years. The task is that of progressing from a rate of growth to a "steady-state" situation.

The problems associated with this change will not be trivial; the average annual world population growth rate is currently over 2%, the average growth rate for industrial production during the 1960s was 7% per year, and demand

A. W. Mann • Division of Mineralogy, C.S.I.R.O., Wembley, Western Australia 6014

for major resources currently increases by between 2% and 5.5% per annum.[1,2] Since World War II the United States has consumed more minerals than did the whole world for the remainder of known time before World War II. There are few who would be prepared to argue critically that this situation can continue indefinitely; scientists (and others) do, however, differ in their opinions as to the time scale, the methods, and the consequences involved. In turn, these arguments hinge on the availability of resources and our usage of them.

Three factors are fundamental to understanding an assessment of earth resources:

1. The earth is essentially a closed chemical system: it has fixed mass, and only energy is added or subtracted.
2. The mass of any one element is effectively constant.
3. All resources, apart from those consumed to produce energy, are potentially renewable.

The first factor assumes that the mass of extraterrestrial material added, e.g., meteorites, is negligible on the time scale of our present problems. The second factor assumes that transmutation (natural or artificial) of elements is insignificant compared with the total mass of any one element (either in compound or elemental form). The third factor holds true despite the fact that many metals are often regarded as nonrenewable resources. Iron is not necessarily irretrievably consumed because it is mined and converted to steel for the manufacture of cars or washing machines, to ferrous sulfate for the manufacture of dyes and inks, or to hemoglobin for making blood. These can all be regarded, with varying degrees of suitability, as potential iron reserves, and are part of the total iron resource. The problems involved in utilizing resources of iron or any other resource are those governing its distribution and redistribution, both spatially and chemically. Assessing the "depletion of a resource" does not only involve projecting usage rates against current available reserves, it must also reflect the changes in technology and usage patterns which will inevitably occur as our management of the resource changes with time.

1.1. *Assessment of Resources*

Estimating the usable portion of our available resources is, in itself, a difficult task; this quantity, the *ore reserve*, is dynamic and dependent on many other variables, the cut-off grade, delineation of the extent of the deposit, cost of extraction, and so on. An *ore reserve* is best defined as "a mineral material considered exploitable at present."[3] The resource for iron thus consists of present reserves, plus marginal, submarginal, and latent deposits which may become reserves as the exploration for iron, the technology of its extraction, and the economics for its demand all change with time. Reserves can be further

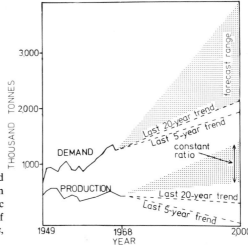

Figure 1. Contingency forecast and trend projections for usage of primary zinc in the United States (1 tonne = 1 metric ton = 10^3 kg). (After U.S. Bureau of Mines, *Minerals Facts and Problems*, 1970.)

categorized as

1. *Measured reserves*: those with a tonnage computed from field and metallurgical data
2. *Indicated reserves*: those with a tonnage extrapolated by geological inference from limited field and metallurgical data
3. *Inferred reserves*: those which are a quantitative estimate based on the geologic character of the deposit

Assessing the depletion rate of world resources, i.e., our usage of these reserves, is an even more difficult task. By far the most comprehensive and sophisticated attempt at this is that due to the U.S. Bureau of Mines.[4] Apparent reserves for major resources, including energy, ferrous, nonferrous, and non-metallic resources, are considered in conjunction with future demand forecasts and likely consumption trends.* The method used for assessing future demand is labeled *contingency*, or *technological forecasting*, and assesses future usage based on technological, economic, social, environmental, and other relevant influences, rather than by extrapolating from present usage trends. Trend projections and the contingency forecast for primary zinc usage in the United States are shown in Fig. 1.

To arrive at the forecast range, the usage of zinc is divided into eight end use categories such as building and construction, transportation, electricity, etc., and low and high forecasts for the year 2000 are calculated using econometric techniques. The significantly higher demand predicted by this technique, as compared with the trend extrapolations from present usage patterns, reflects the expected increase in consumption of zinc due to the

---

*Apparent reserves include, for most metals, the apparent measured, indicated, and inferred reserves.

increase in Gross National Product, and the increasing use of zinc in electric power transmission and communication equipment. The 1970, world zinc reserves were estimated as 113 million tonnes; no attempt was made to assess what might be considered as reserves in the year 2000. In this respect, it is interesting to note that the 1962 estimate for world zinc reserves by the same bureau[5] was 77 million tonnes; the world zinc reserves have increased by 46% in 8 years due to the discovery of new deposits in the United States, Australia, and Canada, and due to a reduction in cutoff grade brought about by increased prices for zinc.

The critically important effect of usage rates on the depletion of resources is shown dramatically (some say overdramatically) by Meadows *et al.*[6] Their assessment of the lifetime of many reserves is examined by extrapolating present and past usage rates in various ways; the diagram for chromium reserves is shown in Fig. 2.

Chromium is one of the less critical examples; the 1970 known reserves are shown by curve *A* in Fig. 2. Curve *B* represents the reserves remaining if the 1970 usage rate is continued (lifetime 420 years), and curves *C* and *D* illustrate the depletion of reserves if the present rate of increase of chromium usage (curve *E*) is maintained. This exponential usage of chromium drops the lifetime to 95 years for present reserves (*C*), or 154 years if reserves are five times those indicated at present (*D*). Herein lies the problem of assessing the validity of their results, however graphic they are, and however important are the conclusions. What are the future reserves for chromium likely to be? Is it possible that they could be one, two, or three orders of magnitude greater than at present as the price structure changes, technology advances, and exploration continues? What effect does recycling and substitution of alternative products

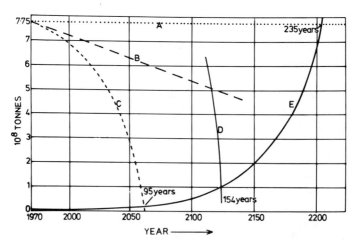

Figure 2. Extrapolation of present and past usage rates for chromium and their effect upon depletion of chromium reserves. (After Meadows *et al.* *Limits to Growth*, Earth Island Limited, London, 1972.)

have on the depletion of a resource? Is it valid to extrapolate present and past usage rates far into the future? We can summarize the contingencies in assessing depletion of resources as

1. Delineation of deposits (cut-off grades, new deposits)
2. Technology of extraction and processing
3. Economics of extraction and processing
4. Usage technology (alternatives, recycling)
5. Usage economics
6. Population growth
7. Consumption patterns (standards of living)
8. Extrapolation of present data

Each of these contingencies in itself is obviously complex and related to other contingencies in the list; each is capable of exerting a marked influence on the overall picture. The difficulty in assessing the depletion of resources either by technological forecasting by by extrapolation is reflected in the variations for high and low estimates in Figs. 1 and 2.

It is pertinent at this stage to review the aims of the two assessment methods which have been considered. The first, by technologically forecasting to a given time—the year 2000—has attempted to assess what resources will be required by the population expected and the standard of living attained at that time. The second, by investigating the effect of various usage schemes on initial reserves in extrapolation "scenarios," has attempted to predict the year in which we will exterminate each resource. Both aims require that the respective method rely heavily on the last three contingencies mentioned: population growth and consumption patterns, both extrapolated from past and present data. There is, however, a way to minimize these problems. What if our aim were either (1) to predict what population could be supported at a given standard of living with a given amount of each resource, or (2) what amount of each resource would be required to maintain a given population at a given standard of living, irrespective of *when* that population was achieved? In this case, calculations could be based on the extent to which our *present* population draws upon *present* resources, i.e., the *per capita mining and consumption figures*, and could eliminate the "rate-of-growth" variable inherent in both population and consumption pattern contingencies.

1.2. *Rates of Depletion of Resources*

Before becoming immersed in such *per capita* calculations, it is first necessary to point out just what information these are likely to yield, how they differ from other methods, and the contingencies involved. *Per capita mining*, calculated by dividing the amount of resource mined by population, will give an indication of the present rate at which we deplete resources. On a regional basis, it may be possible to correlate the level of mining activity to a standard of living (i.e., demand on resource) or to the availability of resource within that region; these are also, naturally, contingencies in the calculation. *Per capita consumption*, calculated by dividing the amount of resource actually processed, by the population, differs in that secondary or recycled material may be included. It represents more accurately the demand per capita on a resource for a given standard of living; it is often difficult, however, to assess the amount of consumption for local use as compared with total consumption, which includes processing of manufactured products for export. Table 1 lists the per capita mining figures for three populations—world, Asian, and North American— and the consumption per capita for the world and the United States.

Despite the fact that many factors contribute to make some of these calculations only approximately quantitative, they exhibit several interesting trends. The world per capita mining figures, for example, show that we mine

Table 1. Per Capita Mining and Consumption Figures*

Resource[a]	Units	Per capita mining			Per capita consumption	
		World	N. America	Asia	World	U.S.
Aluminum	(kgm)	15.1[b]	14.4[b]	1.4[b]	2.8	30.8
Antimony	(gm)	14.8	15.8	1.9	17.3	201
Asbestos	(kgm)	1.0	5.0	0.0	0.9	3.7
Chromium	(kgm)	0.7	0.0	0.3	0.5	2.3
Coal	(kgm)	580	1660	246	624	2274
Cobalt	(gm)	—	—	—	5.6	31.7
Copper	(kgm)	1.6	6.3	0.1	1.5	12.5
Energy (coal equivalent)	(kgm)	1832	7006	721	—	—
Gold	(gm)	0.4	0.1	0.0	0.4	0.4
Iron	(kgm)	110	244	25	109	536
Lead	(kgm)	0.8	2.8	0.1	0.8	6.7
Magnesite	(kgm)	3.1	1.9	1.2	—	—
Manganese	(kgm)	2.0	0.3	0.4	2.2	5.1
Mercury	(gm)	2.6	5.7	0.6	2.5	12.7
Molybdenum	(gm)	20.7	187	0.9	18.5	124
Nickel	(gm)	145	790	3.4	135	834
Petroleum	(kgm)	582	1717	342	471	2872
Phosphate	(kgm)	23.0[c]	108[c]	2.4[c]	3.2	15.6
Potash	(kgm)	4.6	17.1	0.1	3.6	15.6
Silver	(gm)	—	—	—	—	—
Tin	(gm)	50.6[d]	3.5	56.3	70.5	447
Titanium	(gm)	—	—	—	375	1635
Tungsten	(gm)	11.5	22.2	8.6	8.8	36.3
Uranium	(gm)	—	—	—	4.8[e]	12.1[e]
Zinc	(kgm)	1.5	6.3	0.2	1.4	7.9

*Sources: U.S. Bureau of Mines, *Mineral Facts and Problems*, 1970; United Nations, *U.N. Statistical Yearbook*, 1970.

[a]Expressed as metal content unless specified otherwise.
[b]Bauxite equivalent.
[c]As phosphate rock.
[d]Excluding the U.S.S.R.
[e]As U_3O_8.

(per annum) in excess of our own body weight of three resources—coal, crude petroleum, and iron ore—and we mine a sizable proportion of our own weight of at least two others—aluminum and phosphate rock. The fact that the demand on resources is not equally distributed over the entire population is reflected in the contrast between the per capita mining figures for the Asian and North American subcontinents. This difference is apparent, despite the fact that the United States, for example, imports substantial quantities of most of the resources, which if taken into consideration would further accentuate this disparity (e.g., compare per capita mining for North American with the per capita consumption for the United States). The world per capita mining and consumption figures for each resource are comparable in magnitude; the fact that consumption per annum so closely approximates the amount of resource mined implies that secondary usage in consumption is nullified by some stockpiling of mined resources, which of course are not consumed immediately. It is likely that for many resources both of these quantities are small compared with the total consumption figure.

Our primary aim in this exercise, however, was to evaluate depletion of resources, and to do this we must now examine the estimates of world reserves (Table 2), but not without a preliminary note of caution. In examining these figures it should be clearly understood that they represent an estimate of the reserves available under current technology and at current prices. In no way do they, or are they meant to, imply what the future, potential, or ultimate global reserve for any resource might be.

Reserves are quoted on a global basis. However, some of the comments in Table 2 indicate that, as before, there is great variation in the distribution of resources, and problems of resource depletion could be accentuated regionally.

From the data in Tables 1 and 2 it is now possible to calculate a factor indicating by how much we deplete our reserves for each resource per annum, i.e., a *reserves usage index* (RUI), calculated as the percentage of known reserves mined per annum. The indices for various resources for the world 1969 consumption rate are shown in the first column of Table 3.

These calculations contain few contingencies, and show that at our present rate of consumption we use between 12.9% (for gold) and 0.01% (for potash) of our reserves per annum, with usage of most metals falling in the 0.4–6% range. The fact that most common metals have an index in this range should be regarded with some concern, for it implies that without recycling and without the discovery of further reserves, our present reserves of tin (RUI = 5.3%), for example, would be depleted in $100/5.3 \simeq 19$ years. This of course should not and will not occur, but only because more tin will be found and processed, a greater percentage of secondary material will be used in production, and our consumption pattern may alter. Expressing the depletion rate in this way merely indicates what will occur if we do not exercise these options. In this respect, it is apparent that the 1969 world consumption rate is probably not an accurate guide to our future consumption rate because developing nations, which hold the bulk of the population, are rapidly increasing their standard of

Table 2. World Mineral Reserves*

Resource	Reserve[a] (tonnes)[b]	Occurring as	Comments
Aluminum	$1.1×10^9$	$Al_2O_3·nH_2O$	Includes inferred reserves
Antimony	$3.6×10^6$	Sb_2S_3	50% in mainland China
Chromium	$4.4×10^9$	$FeCr_2O_4$	N. America less than 0.5%
Coal	$4.7×10^{12}$	—	58% bituminous and anthracite
Cobalt	$2.2×10^6$	$CuCo_2S_4$, $CoAs_2$	Co-product in Cu, Ni sulfide ores
Copper	$280×10^6$	$CuFeS_2$, Cu_2S	Minor amounts as carbonate ores
Gold	$11.0×10^3$	Au	Some as co-product in Cu, Pb, Zn ores
Iron	$88×10^9$	Fe_2O_3, Fe_3O_4	Vast quantities of potential ore
Lead	$82×10^6$	PbS, $PbCO_3$	Co-product in Cu, Zn ores
Manganese[c]	$635×10^6$	MnO_2, $Mn_2O_3·H_2O$	Excludes undersea deposits
Mercury	$115×10^3$	HgS, Hg	Reserves calculated at \$200/Flask (76 lb)
Molybdenum	$5.2×10^6$	MoS_2	65% in N. America
Nickel	$68.0×10^6$	(Fe, Ni)S	Large reserves of lateritic nickel
Petroleum	$54.1×10^9$	—	Proven reserves, excludes shale oil, natural gas
Phosphate	$19.8×10^9$	$Ca_5(PO_4)_3(F, Cl, OH)$	—
Potash	$98.9×10^9$	KCl, $KMgCl_3·6H_2O$	Vast quantities of potential ore in seawater
Silver	$171×10^3$	Ag, Ag_2S, Ag_3AsS_3	Co-product of Cu, Pb, Zn ores
Tin	$4.76×10^6$	SnO_2	N. America less than 1%
Titanium	$310×10^6$	TiO_2, $FeTiO_3$	
Tungsten	$1.3×10^6$	$CaWO_4$, (Fe, Mn)WO_4	75% in mainland China
Uranium	$749×10^3$	U_3O_8	As U_3O_8; excludes U.S.S.R., China, Japan
Zinc	$112×10^6$	ZnS, ZnO	Often as major co-product with Cu, Pb

*Source: U.S. Bureau of Mines, *Mineral Facts and Problems*, 1970.

[a]Reserves are quoted as metal content in ore unless specified otherwise.

[b]1 tonne = 1 metric ton = 10^3 kg.

[c]Reserves for magnesium from seawater, brines, dolomite, and magnesite are effectively vast and have not been estimated.

Table 3. Reserves Usage Index (RUI) as a Function of Consumption Rate and Population[a]

Resource	RUI (%)			
	Column 1	Column 2	Column 3	Column 4
Aluminum	0.9	10.0	2.5	28.0
Antimony	1.7	19.9	4.8	55.8
Chromium	0.04	0.2	0.1	0.5
Coal	0.05	0.2	0.1	0.5
Cobalt	0.9	5.1	2.5	14.4
Copper	1.9	15.9	5.4	44.6
Gold	12.9	12.9	36.4	36.4
Iron	0.4	2.2	1.2	6.1
Lead	3.5	29.1	9.8	81.7
Manganese	1.2	3.5	2.9	8.0
Mercury	7.7	21.7	39.3	110.4
Molybdenum	1.3	3.6	8.5	23.8
Nickel	0.7	2.0	4.4	12.3
Petroleum	3.1	8.7	18.9	53.1
Phosphate	0.06	0.2	0.3	0.8
Potash	0.01	0.04	0.06	0.16
Silver	5.0	14.0	58.1	163
Tin	5.3	14.8	33.4	93.9
Titanium	0.5	1.2	1.8	5.3
Tungsten	2.4	6.8	9.9	27.9
Uranium	2.3	6.4	5.8	16.2
Zinc	4.5	12.5	25.1	70.5

[a]For columns 1 and 2 the RUIs are calculated for the 1969 world population of 3.56×10^9 people, and for columns 3 and 4 for a projected world population of 10.0×10^9 people. The RUIs in columns 1 and 3 assume the average 1969 world consumption rate for their respective populations, and columns 2 and 4 are indices for the average 1969 U.S. consumption rate for their respective populations. All indices are calculated using the 1969 reserves for each resource (Table 2).

living. For the sake of comparison, a hypothetical RUI, using the 1969 World population of 3.56 billion but assuming consumption at the United States rate, is shown in the second column of Table 3. The increased consumption of antimony, copper, lead, mercury, and zinc arising from a higher standard of living is a distinct threat to the depletion rate of these resources.

Figure 3 is a histogram of the data contained in the first two columns of Table 3, with resources in order of decreasing RUI plotted on the abscissa. For the purpose of simplifying later discussion, and in a rather arbitrary but perhaps informative way, resources can at this stage be divided into two groups according to their present rates of depletion. Using the obvious dividing point between Sb and Mo, which have a difference in RUI of 0.4%, the following two groups can be identified:

1. Those with a rapid depletion rate: RUI $\geqslant$ 1.7% (group 1)
2. Those with a slow depletion rate: RUI $\leqslant$ 1.3% (group 2)

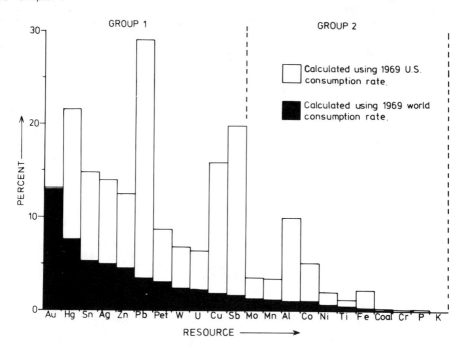

Figure 3.   Reserves Usage Indices (RUI) for various resources plotted in order of decreasing RUI along the abscissa.

The rate of depletion is dependent on two principal factors: our consumption rate and the reserves available. An attempt will be made to distinguish between and determine the relative importance of these two factors in a later discussion of individual resource types.

The other major factor influencing consumption and resource usage is population growth; as pointed out earlier, this is a major contingency in other methods of assessing resource depletion and it is not our intention to allow it to predominate in these calculations. It is apparent, however, that the global population will not stabilize for at least another decade, and resources will have to accommodate this. Demographic projections[7] for the world population by the year 2000 vary between 5.4 and 7.5 billion people, and the ultimate global population is likely to be well in excess of this figure. Were resources faced with the problem of providing for, say, 10 billion people at present world or United States consumption rates with the equivalent of the present reserves, the future would not be healthy, as shown by the hypothetical reserves usage indices in columns 3 and 4 of Table 3. The inference that reserves must be increased, and that secondary usage and consumption patterns must be altered to meet this situation, is all too obvious from the figures in these two columns, where, for many resources, the usage rate per annum would be in excess of 50% of present world reserves.

1.3. Resource Usage Patterns

In discussing present resource usage patterns, it is convenient to consider resources under the following subheadings:

1. Fuel and energy resources
2. Metal resources
3. Nonmetal resources

1.3.1. Fuel and Energy Resources

Coal, petroleum, and uranium have already entered the discussion of resource depletion; the reserve usage indices presented in Table 3 for these three resources, 0.05% for coal, 3.1% for petroleum, and 2.3% for uranium, do not give a clear impression of the complexity of the energy utilization problem. For example, the complete list of energy resource avenues which could be investigated should include:

1. Anthracite
2. Bituminous coal and lignite
3. Natural Gas
4. Peat
5. Petroleum
6. Shale Oil
7. Thorium and Uranium
8. Water (Hydroelectricity)
9. Geothermal
10. Tidal
11. Solar Energy
12. Hydrogen
13. Fusion

As it is not pertinent in this article to present a detailed discussion for each of the above energy types, readers are referred elsewhere[8,9,10] for this. It is relevant, however, to consider the relative importance of each of these types of energy at the moment, and how limitations on reserves and patterns of consumption may alter this picture in the future.

Figure 4 illustrates how the different sources of energy have contributed to the total power production of the United States over the last century or so.

It clearly reflects the transition from energy stored in contemporary plant matter to that stored as fossil fuel, which for coal reached a peak (as a fraction of total power production) in the 1910s, and for crude oil in about 1960. Natural gas is currently providing an increasing percentage of energy for industrial and domestic consumption. Nuclear power generation does not appear as a significant percentage of total power production on Fig. 4. Currently, less than 5% of total electric power generation is from this source, but contingency forecasts[11] predict this to grow rapidly to greater than 20% by 1980 and to 40–60% by 2000.

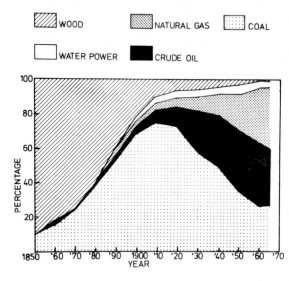

Figure 4. Contribution of the different sources of energy to total power production in the United States since 1850. (After B. J. Skinner, *Earth Resources*, Prentice-Hall Inc., New Jersey, 1969.)

The amount of available reserves has not yet influenced the consumption pattern of energy resources. Reserves of coal and lignite are large (Table 2), even though production and consumption has remained steady for 50 years, while the usage of petroleum-based fuels, with their more limited reserves has increased to the point where depletion of the reserves is a serious problem. Coupled with the fact that the developing nations have a rapidly increasing per capita energy consumption, an alternative to the present high world consumption of petroleum-based fuels will be required before the turn of the century. This alternative in the short term is likely to be nuclear power. Here the problem of assessing the depletion of resources is complicated by two factors:

1. Uranium 235 produces, per gram, heat energy equivalent to 13.7 barrels of crude oil or 2.7 tonnes (= metric tons) of coal.
2. The effectiveness of uranium resources is markedly influenced by the choice of reactor type.

Three reactor types are contemplated as being viable for the period 1970–2000, the light–water reactor (LWR), the high-temperature gas-cooled reactor (HTGR), and the fast breeder reactor (FBR). The LWR consumes only the isotope uranium 235 (0.7% of naturally occurring uranium) and is inefficient in its use of the resource, although it is possible for plutonium, when produced as a fission product, to be recycled. The HTGR and the FBR reactors, by converting nonfissionable uranium 238 and thorium 232 to fissionable plutonium 239 and uranium 233, according to the reactions

$$^{238}_{92}U + n \rightarrow\ ^{239}_{92}U \rightarrow\ ^{239}_{93}Np \rightarrow\ ^{239}_{94}Pu$$

$$^{232}_{90}Th + n \rightarrow\ ^{233}_{90}Th \rightarrow\ ^{233}_{91}Pa \rightarrow\ ^{233}_{92}U$$

are far more efficient in their use of natural uranium, and could, in fact, use

Table 4. Nuclear Capacity by the Year 2000

	Low forecast	High forecast
LWR	400×10^9 W (61%)	360×10^9 W (40%)
HTGR	—	170×10^9 W (19%)
FBR	260×10^9 W (39%)	370×10^9 W (41%)
Totals	660×10^9 W	900×10^9 W

large quantities of uranium, depleted in U-235 by use in LWR reactors. Low- and high-contingency forecasts by the United States Bureau of Mines[11] put the following figures on the nuclear power capacity for each reactor type by the year 2000 (Table 4).

Accurate assessment of the depletion of uranium reserves is also complicated because large amounts of ore, at present subeconomic, would become available if demand increased prices, and by the great amount of secrecy (and speculation?) surrounding information about uranium reserves. Despite this, it appears that sufficient reserves will be available for at least several decades of nuclear power generation. The problems with the use of this energy resource may come from other aspects, namely, waste disposal and pollution.[12]

Among the other energy resources not yet considered, geothermal, hydroelectrical, and tidal, which are at present exploited on a limited scale, do not appear to have the reserves necessary to be considered as major contributors to the world's power requirements. Hydrogen, solar energy, and fusion as energy resources are the long-term "keys" to the energy vault, although at present their development is still only at the experimental stage; their importance and technology is discussed elsewhere in this book.

In the short term it would appear prudent to conserve and consolidate fossil fuel resources for the petrochemical and allied industries of the future, although reserves of coal appear to be in no danger of immediate depletion. Nuclear power will be required to bridge the gap between the present and the time when usage of the continuously available source of energy from the sun becomes a practicality.

1.3.2. Metal Resources

Two groups of resources, decided on the basis of reserves usage index (RUI), were shown in Fig. 3. The metals in each of these two groups are as follows (Table 5).

Group 2, significantly, encompasses all of the so-called ferrous group of metals, plus aluminum, while Group 1 consists of the base and precious metals. That this division, made on the basis of our present rate of depletion of resources, should so neatly distinguish between ferrous and nonferrous metals, is interesting and bears further contemplation. What are the reasons for this distinction? Although we have had a primarily iron- and steel-based technology for several centuries, and although the technology for widespread use of

Table 5. Metal Resources

Group 1 resources (rapid depletion) RUI $\geqslant 1.7\%$	Group 2 resources (slow depletion) RUI $\ll 1.3\%$
Gold	Molybdenum
Mercury	Manganese
Tin	Aluminum
Silver	Cobalt
Zinc	Nickel
Lead	Titanium
Tungsten	Iron
Uranium	Chromium
Copper	Potassium
Antimony	

copper, lead, and zinc is young in comparison, the overriding factor is the large reserves for the ferrous + Al group. Comparison in Table 6 of the average reserves and average per capita consumption figures for these two groups emphasizes this point. The average per capita consumption for the ferrous + Al group exceeds that of the nonferrous metals by a factor of 34, but the average reserves for metals in the two groups are in a ratio 219 : 1. Some geochemical reasons for this discrepancy are suggested later in this chapter.

As was the case for energy resource usage, the consumption pattern for most metals is not yet influenced to a great extent by the availability or size of current reserves. Consumption patterns for copper and iron in the United States over the past 60 years are shown in Fig. 5; the consumption patterns for these two metals are very similar, despite the fact that the present percentages of reserves used per annum are very different (the RUI for copper is 1.9%, for iron 0.4%).

One exception to this principle is the decrease in the amount of silver used in United States coinage since 1965, brought about by excessive demand on world reserves, and in particular on United States stockpiles. In most cases metal price fluctuations, their effect in adding previously subeconomic ores to the reserves list, the increasing proportion of secondary usage, and resource stockpiles, have all combined to efficiently buffer the amount of currently

Table 6. Average Reserve and Per Capita Consumption Figures

	Average reserves	Average per capita consumption
Group 1. Metals (nonferrous)	0.05×10^9 tonnes	0.42 kg
Group 2. Metals (ferrous + Al)	11.8×10^9 tonnes	14.37 kg
Ratio: Group 2/Group 1	219/1	34/1

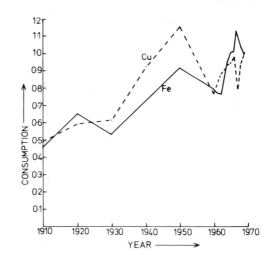

Figure 5. Consumption patterns for copper and iron in the United States between 1910 and 1970. Consumption normalized to 10 for 1969.

available resources. However, as the movements of prices for gold, mercury, silver, and tin (all with high RUIs) might indicate, there must be a limit (in time and extent) beyond which the consumption pattern will inevitably be affected.

1.3.3. *Nonmetal resources*

Major nonmetal resources include; asbestos, carbonates, chlorine, granite, oxygen, phosphate, potash, sand and gravel, sodium compounds and water.

Asbestos, the carbonates—principally those of calcium and magnesium—sand, and gravel, together with granite, constitute the common and most widely used building materials. In the United States, granite comprises 32% of all building stone used, limestone 29%, sandstone 18%, and marble 0.04%; large amounts of carbonates are also consumed in the production of Portland cement. Reserves of these materials are immense (though not infinite) and supply is in most cases limited only on a local scale.

Chlorine, sodium chloride, and magnesium, a metal not considered earlier, are the major representatives of the group of resources with an immense reserve: the oceans. Salt, which is mined as halite from evaporate deposits or obtained by solar evaporation of seawater, has its major uses in the paper and related industries, as a snow and ice control agent in transportation, and in the chemical industry, has a high consumption per capita, surpassed only by coal and iron; reserves are enormous, however, and in no danger of depletion.

Magnesium metal accounts for only 10% of the total use of magnesium, the remainder being in magnesium compounds, chiefly refractory products. For the latter, the magnesium is precipitated as magnesium hydroxide from brine solution; magnesium metal is produced by electrolysis of molten chloride using the Dow process, which yields chlorine as a by-product. The bulk of chlorine production is, however, from electrolysis of brine, which produces sodium hydroxide at the cathode; this electrolysis is performed in either the diaphragm cell or the mercury cell, the latter being a major "consumer" of mercury.

The fertilizers, phosphate and potash, both currently have a very low RUI, 0.06% and 0.01%, respectively. This is principally because large deposits both of phosphate rock, occurring as calcium phosphates, and of slyvite (KCl) are widespread throughout the world. This is fortunate, because the usage pattern for fertilizers renders them virtually nonrenewable; the wide application and distribution of fertilizers and their uptake in plant and ultimately by living matter reduces the possibility of effective secondary usage.

We now have considered all but the two most important resources for life: oxygen and water. The assured supply of these two resources is often taken for granted; this is perhaps justified, since the reserves of these resources are large and remain approximately constant, but it is well to consider why this should be so. The reserves of oxygen and water remain approximately constant, despite our bodily necessity and industrial capacity to inhale, imbibe, combust, excrete, and pollute, because they have a 100% recycle efficiency. For oxygen, photosynthesis is the buffer to depletion; for water, the hydrological cycle, using solar energy, is able to distribute and redistribute the water resource. These two cycles provide model systems for our own efforts to control the distribution and management of resources, and for this reason they will be discussed in detail at a later stage.

2. *Future Reserves and Technology*

Two important conclusions emerge from the foregoing assessment of depletion of resources, if even the minimum of projected populations is to be adequately supported. They are (1) the need to increase reserves and (2) the need to alter consumption patterns for most resources. Discussion of the first of these factors involves not only an investigation of the likely sources for increased reserves, namely those in the lithosphere, the hydrosphere, and the atmosphere, of which those in the first-mentioned category are the most important, but also in an evaluation of the technology associated with better mining, extraction, and processing of the ores obtained in those ways.

2.1. *The Lithosphere*

The earth has a total mass of 5.98×10^{21} tonnes, a volume of 1.08×10^{12} km^3, and an average density of 5.52 g/cm^3. Of this, the outer 7 km (under the oceans) to 40 km (under the continents) is crust, constituting, in all, 0.42% of the total mass, or 0.85% of the total volume. It is this fraction of the earth which up until now has provided all of the earth's mineral resources and will most likely be required to do so far into the future.

Table 7. Geochemical Abundances in the Earth's Crust (in ppm)[a]

O	466,000	F	300	Nb	24	Dy	4.5	Tm	0.2
Si	277,200	Sr	300	Co	23	U	4	Cd	0.15
Al	81,300	Ba	250	La	18	B	3	Ag	0.1
Fe	50,000	Zr	220	Pb	16	Yb	2.7	In	0.1
Ca	36,300	Cr	200	Ga	15	Er	2.5	Se	0.09
Na	28,300	V	150	Mo	15	Ta	2.1	A	0.04
K	25,900	Zn	132	Th	12	Br	1.6	Pd	0.01
Mg	20,900	Ni	80	Cs	7	Ho	1.2	Pt	0.005
Ti	4,400	Cu	70	Ge	7	Eu	1.1	Au	0.005
H	1,400	W	69	Be	6	Sb	1?	He	0.003
P	1,180	Li	65	Sm	6.5	Tb	0.9	Te	0.002?
Mn	1,000	N	46	Gd	6.4	Lu	0.8	Rh	0.001
S	520	Ce	46	Pr	5.5	Tl	0.6	Re	0.001
C	320	Sn	40	Sc	5	Hg	0.5	Ir	0.001
Cl	315	Y	28	As	5	I	0.3	Os	0.001?
Rb	310	Nd	24	Hf	4.5	Bi	0.2	Ru	0.001?

[a]Source: B. Mason, *Principles of Geochemistry,* John Wiley & Sons, N.Y., 1952.

2.1.1. Crustal Abundances and Ore Reserves

Geochemical abundances of elements in the crust are shown in Table 7.
Oxides and silicates of six metals, Al, Fe, Ca, Na, K, and Mg, comprise close to 99% of the earth's crust; several of the metals in group 1—the rapid depletion category—are very low on the abundance list. Lead, for example, is less abundant than several of the "rare earth" elements, and mercury, silver, and gold are extremely rare. It is possible to speculate some sort of correlation between the crustal abundance and reserves of elements; Fig. 6 is a plot of

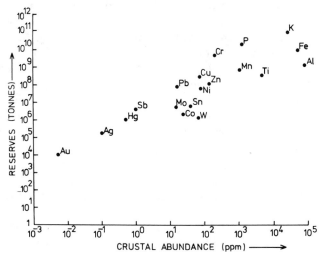

Figure 6. Current ore reserves for some common metals plotted against their crustal abundance.

current reserves versus crustal abundance for 19 of the metals considered earlier. A rough correlation between reserves and crustal abundance is apparent in Fig. 6, although it is hardly sufficient to justify the general application of relationships of the type $R = A \cdot 10^{10} \sim 10^{11}$ (R = reserves, A = abundance) which have been used.[14,15] Intuitively, we would not expect to find such simple relationships applying to a range of resources because (1) the reserve for a particular resource, as pointed out earlier, is a time- and price-dependent term reflecting, to some extent, our demand, and (2) the crustal abundance is only proportionally related to the metal content of the ore which is mined. The latter is a most important point: for each resource we rely on natural enriching processes to produce an ore which has a metal content in most cases significantly above the crustal abundance. Since this enriching process is specific to the element concerned, the *enrichment factors* for different elements (i.e., the constants of proportionality in the relationships between crustal abundance and metal contents of ore) vary significantly. For example, iron ore currently has an *enrichment factor* of 5.0; lead, which is concentrated in the crust by quite a different process, has a ratio of metal content of ore to crustal abundance of ~1250. It would only be fortuitous for the reserves (i.e., total metal amounts) of lead and iron to be related to their crustal abundances by one and the same simple relationship.

2.1.2. Enrichment Factors; Tonnage Versus Grade Relationships

Enrichment factors are obtained by dividing the metal content (grade) or currently mined ore by the crustal abundance. These factors are shown for some important metals in Fig. 7, together with histograms of crustal abundances and their corresponding current minimum mineable ore grades. Enrichment factors vary by three orders of magnitude, from those in the range 2.27–5.0, for titanium, aluminum, and iron, to 1000–4000, for silver, lead, gold, and mercury, a fact which is most relevant in determining their respective reserves. We are now in a position to explain some earlier observations which contrasted the comparatively low reserve depletion rates for the ferrous + Al group of metals with those for the nonferrous group. The ferrous + Al metals, apart from being crustally abundant, have the advantage of being currently mined at close to that abundance, i.e., at low enrichment factor values; metals are arranged along the abscissa in Fig. 7 in order of increasing enrichment values, and 10 of the first 12 metals are from the ferrous + Al group. Fractional decreases in the mineable grades for the resources in this group, and particularly for iron, aluminum, and titanium, add immense tonnages to their reserves; for the base and precious metals for which depletion problems currently exist, this is not so.

Lasky[16] defined a mathematical relationship between ore tonnage and grade which appears to be applicable to the characterization of some deposits of manganese, nickel, gold, and porphyry copper. Referred to as the A : G ratio,[17] the relationship infers that "reserves (ore) increase geometrically as

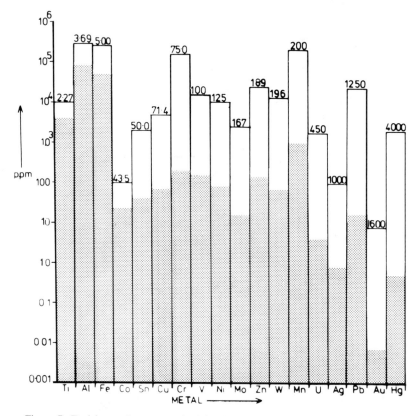

Figure 7. Enrichment factors required for present exploitation of some common metal ores: number above each column is the enrichment factor; unshaded portion of column indicates metal content in ore mineable at present (ppm); shaded area indicates crustal abundance.

grade decreases arithmetically." To consider the implications and limitations of this rule, it is convenient to examine briefly a hypothetical example based on data provided by Lasky. Suppose we are faced with the problem of assessing the tonnage and the profitability of continuing to mine at the low-grade end of a porphyry copper deposit that had been successfully mined over a range in ore grade from 2.0% to 1.0% in copper content. Using the data from these operations, and according to Lasky, we could assess that the tonnage of ore increases at a compound rate of 14.9% for each 0.1% decrement in grade. The pattern for an ore body with an initial tonnage of 100 million tonnes of ore at 2.0% copper, and obeying this law exactly, is shown in Fig. 8. Figure 8a shows that as the grade is decreased the cumulative tonnage of ore increases exponentially; projection below the present cut-off grade is shown as a dotted line. However, if we assess the cumulative *metal content* of this ore, by multiplying tonnage by grade, we obtain the S-shaped curve of Fig. 8b, observed and explained by Lasky, which clearly shows a less rapid increase in tonnage of

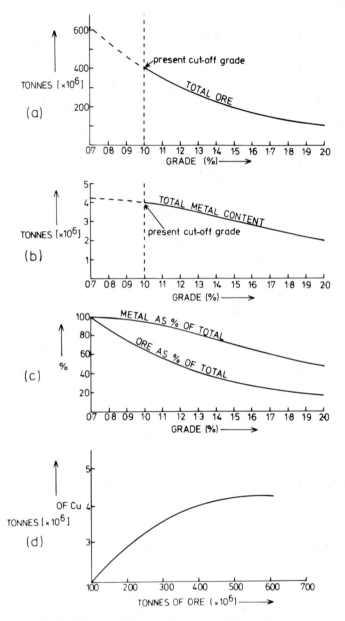

Figure 8. Application of the A : G ratio to a hypothetical porphyry copper deposit. (a) Total ore mined versus grade; (b) Cumulative metal content versus grade; (c) Ore and metal mined at various grades as a percentage of total ore and metal content; (d) Metal content in ore versus ore mined.

metal with decrement in grade. In the instance of contemplating lowering the cut-off grade to 0.7% copper, this curve becomes extremely significant. By nominating the total metal and ore content down to and including 0.7% copper as 100%, we can assess the fraction of metal which could be obtained (and the amount of ore required to be processed) by lowering the cut-off grade a further 0.3% from 1%, as shown in Fig. 8c. At the cut-off grade of 1.0% we have already mined 65.9% of the ore and obtained 94% of the copper available. To extract a further 6% of the total copper we would have to further mine half as much ore as has already been processed. The final curve (Fig. 8d) reflects the fact that the ore body, and the law defining its reserves versus grade relationship, is finite and does not extend continuously down to the crustal abundance for copper.

Although the A : G ratio was proposed, and should only be used, to infer tonnages from grade for specific ore bodies, several generalizations to the wider resources depletion problem arise from this example. First, the frequency, the extent, and the intensity of enrichment processes which produce an ore body obviously play an extremely important role in determining reserves for a particular resource. Second, the folly of contemplating mining down to crustal abundance levels is exposed. For base and precious metals, where enrichment factors are large, the amounts of ore which would require processing would impose a disproportionate strain on our "soft" resources: time, energy, labor, and capital; for the ferrous + Al group, where present enrichment factors are low, the reserves which exist between present cut-off grades and crustal abundance are immense. Third, and in part due to these first two, it is apparent that future exploration will for a long time be directed toward discovering where natural enriching processes have been operating. In almost all ore deposits so far discovered there has been surface expression or indication of this activity; in the future a greater percentage of these will have to be detected "blind."

2.1.3. *Natural Enriching Processes and Ore Deposit Types*

Density differences between the crust (2.8 g/cm^3), mantle (4.9 g/cm^3), and core (10.9 g/cm^3) are but one manifestation of the physical and chemical segregation processes which have occurred in the evolution of the earth from an originally homogeneous melt. Estimates of the average chemical composition of the crust show it to be significantly enriched in silicon, aluminum, and oxygen, and depleted in iron, chromium, and nickel in comparison with the basic, semisolid mantle and the liquid Fe–Ni core. It would be an insult to geology to attempt here a description of the processes responsible for this segregation. Suffice it to say, that many of the complex chemical reactions which could be imagined to accompany the cooling of a multicomponent system under a variety of different physical conditions (as determined by pressure, temperature, and gravity) and chemical influences, including the presence of water and/or the partial pressures of the gases O_2, CO_2, S_2, H_2S,

H_2O, etc., have been observed and described. These reactions and processes have also determined the location, frequency, distribution, composition, and magnitude of mineral deposits, and can be used to classify them.

2.1.4. Classification of Mineral Deposits

A. *Magmatic* or *igneous.*
  1. Orthomagmatic (700–1500°C)
  2. Pegmatitic (±575°C)
  3. Metasomatic (500–800°C)
  4. Hydrothermal (50–500°C)
B. *Metamorphic.*
C. *Sedimentary.*
  1. Residual
  2. Detrital
  3. Chemical precipitation
  4. Evaporite
D. *Solution.*

In the initial cooling of a primary magma—usually a high-temperature mutual solution of various metal silicates and oxides with some additional volatile components—high-temperature chemical processes, such as exsolution, liquid immiscibility, and fractional crystallization, commonly produce *orthomagmatic* deposits. Layered chromite ($FeCr_2O_4$) deposits, where heavy crystallizing minerals have migrated through the melt under the influence of gravity; titaniferous magnetites (Fe_3O_4), thought to be precipitated late in the crystallization sequence of basic igneous rocks; and metallic (Cu, Fe, Ni) sulfides, immiscible in, and originally associated with, an olivine (Mg_2SiO_4)-rich melt, are examples where one or more of these processes has operated. The so-called volatile components of a magma, H_2O, S, Cl, Li, B, F, Zr, Be, Cu, etc., which are not normally crystallographically substituted into the structures of rock-forming minerals, constitute residual hydrous melts which are often squeezed into cracks and fissures of adjacent country rock to form *pegmatite* veins, where enrichments of cryolite ($NaAlF_6$), Zircon ($ZrSiO_4$), monazite (Ce, La, Y)PO_4, cassiterite (SnO_2), fluorite (CaF_2), pitchblende (U_3O_8), etc., occur. *Metasomatism,* or the process by which ore-forming fluids contact, migrate into, and react chemically with country rock, is typified by galena (PbS), sphalerite (ZnS), and fluorite (CaF_2) deposits in limestone. Porphyry copper deposits, in which the copper as chalcopyrite ($CuFeS_2$) is disseminated in a fine-grained (rapidly cooled) groundmass, are the result of shallow, probably rapid injection of melt into surrounding rock. *Hydrothermal* solutions may also transport mineral matter great distances from a magmatic zone; deposits of tin veins and gold in quartz are often of this type. Any rock may be subjected to the subsequent processes of *metamorphism,* which in some cases are instrumental in producing ore minerals, for example, the gem varieties of alumina—ruby and sapphire—in contact metamorphosed limestones.

Sedimentary deposits are, as shown, of four main types. *Residual* deposits are those which contain high concentrations of elements such as aluminum (as the hydrated oxides boehmite, diaspore, and gibbsite), or iron (as hematite (Fe_2O_3), and goethite (FeOOH)), which are not easily leached in secondary or weathering processes. Mechanical processes superimposed on the weathering cycle produce *detrital* deposits such as those for placer gold, cassiterite (SnO_2), diamonds, and ilmenite ($FeTiO_3$). Large and rich deposits of both iron and manganese oxides and hydroxides are of sedimentary origin by *chemical precipitation*. Large sulfide deposits of copper and lead have been formed by the action of sulfate-reducing bacteria in a metal-rich brine, to form stratiform ores characterized by fine-grained mineralization in strata parallel to those of the adjacent sedimentary rock. Phosphorite deposits are also of this sedimentary type. Another fertilizer, potash, is one of the many mineral salts obtained from *evaporite* deposits (obtained as sylvite (KCl) and carnallite ($KCl \cdot MgCl_2 \cdot 6H_2O$)), others being saltpeter ($NaNO_3$), niter (KNO_3), halite (NaCl), gypsum ($CaSO_4 \cdot 2H_2O$), and borax ($Na_2B_4O_7 \cdot 10H_2O$). Finally, we have the *solution* ore deposits, sea water as an ore for magnesium, bromine, and chlorine, and crude petroleum, containing aliphatic, alicyclic and aromatic hydrocarbons.

2.1.5. Techniques for Exploration

Most of the present, economically important ore bodies were discovered because they had some surface expression or indication of their existence. It is probably correct to say that many further discoveries of this type will be made, particularly in the relatively inaccessible and unexplored areas of some continents; it is, however, also correct to say that we have already located and exploited the most obvious, the most amenable, the richest, and the largest of these surface ores, and that to an increasing extent more refined techniques will be required for exploration. The target, particularly in the case of veined and stratiform ores, can be limited in cross section, and when there is no surface expression the problem of location is further compounded. From the frequency of ore bodies which "surface," we can infer that there will also be a large number of mineral deposits within the crust which lie entirely below the surface, i.e., are "blind"; these are likely to play an important, if not critical, role in our resource depletion problems.

The techniques which are likely to be used in defining the location of "blind" ores can be broadly classified as either geophysical or geochemical. The geophysical techniques, namely

1. Seismic reflection and refraction
2. Gravity
3. Magnetics
4. Self potential, induced polarization, resistivity
5. Multiband spectral reflectance
6. Scintillometry

are based on detecting the differences in geological structure, gravity, magnetic susceptibility, electrical properties, reflectance, or radioactivity between the surrounding rock and that containing ore. Most of the methods have already met with minor success, although, since the detection apparatus is usually airborne or ground located, these discoveries have been of surface or near-surface ores. The use of seismics is of course fundamental to petroleum exploration; magnetic anomalies have been instrumental in locating and defining both iron and nickel mineralization and gold and base metal deposits which are associated with magnetically significant intrusive bodies; and electrical methods have had success in locating base metal sulfides. Here *supergene alteration*, or *in situ* corrosion, of the conducting sulfide ore plays a part by producing local galvanic currents and potentials. Visible and infrared reflectance spectra are very obviously surface limited; this is also true of scintillometry, which has, however, been responsible for recent discoveries of some large subsurface uranium deposits.

Geochemical exploration methods are less easily defined. A variety of sampling media can and have been used to detect anomalies consequent to the primary or secondary (e.g., weathering) dispersion of elements associated with ore bodies. Most commonly, soil, natural waters, and vegetation are sampled, but more exotic media such as ice and snow, soil gas, algae, animal organs (e.g., trout livers), and human organs have been suggested as possible trace metal concentrators and indicators. *Pathfinder elements*, those associated with the ore but with dispersion characteristics superior to those of the economic ore minerals, play an important role in exploration geochemistry; some of these pathfinder elements, together with some gases and ions which are currently used or are predicted from research as likely to be useful on pathfinders, are shown in Table 8. In common with some of the geophysical techniques, geochemical exploration methods are not always directly applicable to detection of "blind" ore bodies at great depth. Their greatest value at present, and probably in the future, will be the detection of near-surface mineral deposits obscured by overburden that is often thinly, yet effectively spread, but which can also be hundreds of feet thick.

Recent advances in analytical techniques have played, and will continue to play, a major role in the geochemistry of mineral exploration. One particularly notable example can be cited: the contribution of *atomic absorption spectrophotometry* (AA). Since the application of this technique to chemical analysis was first pointed out in 1955,[18] a revolution in both laboratory and field chemical analyses of mineral specimens has occurred. It is now possible to analyze some 60 or more elements by AA, with detection limits down to the part-per-billion level for some elements. Since the technique is applicable to most metals of economic interest, and cheap and rapid analysis is afforded, comprehensive geochemical programs for sampling soils, streams, sediments, and gases, often on a grid or routine basis, can be planned and carried out. The possibility of using mercury as a pathfinder element (see Table 8) is a case in point. Many sulfide ores of zinc, copper, and lead, and native gold and silver

Table 8. Pathfinder Elements and Compounds[a]

Pathfinder	For	Sampling medium, ore-type, and comments
Sb	Au	galena (PbS); primary haloes around vein-gold deposits
As	Au	rock, soil; primary haloes around vein-gold deposits
	B	rock; endogenic (origin within the earth) boron deposits
	W	rock; skarn (calc-silicate) tungsten deposits
	Cu	rock; skarn copper deposits
Bi	Au	galena (Pbs); primary haloes around vein-gold deposits, also in humus-rich forest soil
B	Au	rock; gold containing xenoliths (inclusions)
Br	evaporites	gas; salt domes
CO_2	Cu, Ni	soil gas; carbonate deposits
	Cu, Mo	
Cl^-	Cu, Pb, Zn	biotites; in intrusive rocks
Co	Ni, Cu	pyrite (FeS_2); in ultramafic rocks, also in sedimentary waters
Fl^-	W	rock; skarn (calc-silicate) tungsten deposits
	Sn	water; CaF_2 containing tin veins
H_2S	Cu, Ni	soil gas; sulfide ores
I^-	Pb, Zn	rock; primary haloes around sulfides
Pb	Cu, Ag	soil; secondary dispersion haloes around sulfides
Hg	Cu, Pb, Zn	soil gas, atmosphere; primary haloes around sulfides
	Au, Ag	stream sediments; secondary dispersion around vein deposits
Mo	Cu	water, sediment, soil; porphry Cu deposits
	U	soil; secondary dispersion from U-Mo deposits
Ni	Cu, Co	pyrite (FeS_2); in ultramafic rocks
K	Au	rock; radiometric det'n. of isotope K^{40}
Ra	U	peat; radiometric det'n. of various isotopes
Rn	U	soil gas, water; radon emanometry of isotope Rn^{222}
Re	Cu, Pb, Zn	rock, soil; oxidation zone of sulfide deposits
Se	Cu	pyrite (FeS_2); Cu deposits in volcanic rocks
	U	pyrite (FeS_2); roll-type uranium deposits
Ag	Au	soil; silver bearing gold ore
SO_4^{2-}	Cu, Ni	water; dispersion from oxidized zones of sulfide ores
Tl	Au	water, sediments; secondary dispersion around veins
W	Au	rocks; primary dispersion haloes around gold-pyrite deposits
U	phosphorites (apatite)	solids; radiochemical det'n. in vein associations
Zn	Cu, Pb	rock, water, sediments; primary and secondary haloes for sulfide deposits

[a]Sources: Association of Exploration Geochemists, Exploration Geochemistry Bibliography 1965–1971. *J. Geochem. Expl.* **2**(1) (1973).

often have native mercury associated with them. For the base metal sulfide ores, the mercury content is commonly up to 0.1% and sometimes higher; native gold and silver can contain over 25% mercury. Mercury vapor has the ability to permeate country rock, soil, and the atmosphere surrounding the deposit, and is thus used as a pathfinder for these ores. In the analysis, gold filters are often used as mercury extractors and preconcentrators; atomic absorption spectrometry at 253.7 nm is capable of detecting nanogram quantities of mercury released from the gold filters subsequent to their being inductively heated.[19] Mercury is therefore a good pathfinder for several reasons: (1) it has a finite vapor pressure at normal temperatures, (2) this vapor forms a widespread *dispersion halo* around ore deposits, and (3) it has the ability to be amalgamated, and therefore preconcentrated, before analysis. *Preconcentration* is widely used in a variety of analytical methods, wherever the detection level required on a sample is initially below the instrument capability. In AA analysis, the preconcentration step is commonly evaporation of a solution, but chelation–extraction, ion exchange, and coprecipitation methods can also be used. Other analysis methods, such as *anodic stripping voltammetry*, use a mercury drop electrode (cathodically polarized) to preconcentrate metals from solution prior to their analysis by polarographic techniques.[20]

Geophysical and geochemical techniques are likely to play an ever-increasing part in minerals exploration, particularly for near-surface deposits. In many cases, a combination of the two techniques—geophysical recognition of a geological anomaly followed by geochemical verification, identification, and delineation—is likely to be adopted. The importance of these techniques is that they provide quantitative data for a scientific approach to exploration and exploitation of resources; the role of the geologist and the prospector in interpreting these data, and data from geological sources, should not be underestimated.

2.2. The Hydrosphere

The sea covers some 71%, or 36.3×10^7 km^2 of the earth's surface, and contains 1.44×10^{18} tonnes of water. Despite this, less than 5% of the total world output of minerals, including oil and gas, comes from the sea, the sea floor, and beneath the sea floor. For certain specific elements the sea is an important source of supply; for most others concentrations are so low that in spite of the huge "reserves" their extraction is not economic. Abundances for the various components in seawater are shown in Table 9. The total salinity for seawater is variable, but close to 3.5%, or 35,000 ppm total dissolved solids. Of these, only two components, chloride and sodium, are present at greater than 1%; two others, sulfate and magnesium, at greater than 0.1%; and only four others, calcium, potassium, bicarbonate, and bromine at greater than 0.001%, or 10 ppm. The remaining elements occur in seawater in very low concentrations, many of them at or below the limits of detectability.

Only three components, magnesium, bromine, and salt, are presently extracted from seawater profitably, and for these the reserves can be afforded the luxury of being termed "temporarily infinite." In the long term it may be possible to profitably extract additional elements such as sulfur, potassium, iodine, fluorine, strontium, and boron.[21] For the base and precious metals, which are at present in shortest supply, seawater does not appear to be the answer. Zinc, for example, occurs at a concentration of 0.01 ppm; extraction of

Table 9. Components of Seawater and Their Concentrations[a]

Component	ppm	Component	ppm	Component	ppm	Component	ppm
Cl	18,980	Rb	0.12	V	0.002	Sn	0.00018
Na	10,556	P	0.07	Al	0.001	Cd	0.00011
SO_4^{2-}	2,649	I	0.05	Ti	0.001	Se	0.0001
Mg	1,272	Ba	0.013	Th	0.0007	W	0.0001
Ca	400	In	<0.02	Co	0.0005	Ge	<0.0001
K	380	Mo	0.013	Ga	0.0005	Cr	0.00005
HCO_3^-	140	Zn	0.01	Ce	0.0004	He	0.00005
Br	65	Fe	0.007	Sb	0.0003	Sc	0.00004
Sr	8.0	Ni	0.005	Y	0.0003	Hg	0.00003
B	4.8	Zr	0.004	Pb	<0.0003	Bi	0.00002
Si	3.0	Mn	0.004	Cs	0.0003	Tl	<0.00001
F	1.3	U	0.003	La	0.0003	Au	0.000004
N	0.5	As	0.003	Ag	0.0003	Ra	3×10^{-11}
Li	0.2	Cu	0.002	Ne	0.0003	Ru	9×10^{-15}

[a]Sources: K. B. Krauskopf, *Introduction to Geochemistry*, McGraw-Hill. N.Y., 1967. K. H. Wedephol, *Geochemistry*, Holt Rinehart & Winston, Inc., N.Y., 1971.

even a modest amount of zinc, e.g., 500 tonnes per annum, would require the processing of 5×10^{10} tonnes or 5×10^{13} liters of water, which again would be an inordinate strain on our "soft" resources.

In 1873, the ship *Challenger* embarked on a voyage around the world to collect data and information on the ocean floor. On this trip, black manganese oxide nodules, since shown to be abundant on the floors of several oceans, were discovered. Now, just over a century later, plans are underway in several countries to mine and exploit these nodules, some of which contain, in addition to·manganese, up to 1.5–2.0% copper and nickel. Mining at depths of 3000–4000 m below sea level obviously requires the development of new concepts in mining techniques. Information on the projects is, for obvious reasons, currently limited, but it does appear that one United States enterprise at least has two submersible vessels, one with mining equipment aboard and the other with a "barge" capable of being submerged to service the first and possible to receive mined material from it. The two most interesting questions to be answered are (1) whether the mining operation is to be performed mechanically (e.g., with a continuous line bucket system) or hydraulically by suction, and (2) whether all vessels are to be manned or whether some parts of the operation are to be remotely controlled. The answers to these questions are likely to be known in the late 1970s when commercial operations begin.

A more recent undersea discovery, that of the Red Sea Brines and their associated sediments, is also attracting a great deal of commercial interest. In 1965, cores of sediments underlying pools of hot brine, discovered some 17 years earlier in the Atlantis II Deep of the Red Sea, were obtained and analyzed. Average assays of metals in the sediments—iron 29%, zinc 3.4%, copper 1.3%, silver 54 ppm, and gold 0.5 ppm—indicated the discovery of a major deposit. It is now generally thought that the hot metal-laden brine pools constitute the outlet of a spring which has its origin at the southern end of the Red Sea and which leaches heavy metals from heated host rock in its 400-km journey to the Atlantis Deep.[22] These metals are deposited as sulfides where the hot brine contacts and mixes with the cold and relatively dilute seawater of the Deep. Because the sediments are fine grained, it should be feasible to pump them to the surface as a slurry. The possibilities are good for further discoveries of this type, either presently active, as in the Red Sea case, or those which have been active in depositing sediment in the past, in the troughs of other ocean basins.

It is hard to evaluate the total future potential of sea floor mining. On the 25% of the total sea floor area regarded as continental margin, various mining operations are already in existence. Petroleum exploration and mining is well advanced in this area, and it is estimated that between one-third and one-fourth of our total recoverable petroleum will be found here.[21] Diamonds, gold, and tin are currently recovered from near-shore submarine placers formed by settlement (due to gravity) of former stream and beach sediments. Phosphate deposits are common on many continental shelves and occur as sediments and nodules, probably chemically precipitated. In the ocean deeps, comprising some 53% of the total earth surface, the problems of exploration and mining are obviously more complicated; as yet mining below the sea bed, for example, has not been seriously contemplated. Manganese nodules, particularly those with high copper and nickel values, are not consistently distributed over the ocean beds, although they are known to occur in specific areas in all major oceans. Vast areas of these oceans are composed of siliceous and carbonate sediments, and the mid-ocean ridges with outcrops of young basaltic rocks may be only poorly mineralized. Despite these reservations, and the technological problems facing exploration and exploitation, the sea bed does offer an attractive proposition for future reserves because of the large total area involved.

2.3. *The Atmosphere*

The atmosphere is the source of supply for the industrial gases argon, carbon dioxide, nitrogen, and oxygen. Helium and hydrogen are at present obtained from natural gas, liquid hydrocarbons, and coal, and therefore have reserves which are more vulnerable to depletion. By far the greatest demand on the reserves of the atmosphere is that on oxygen, due to natural amd artificial combustion processes. Our unconscious and almost contemptuous dependence on this reaction leads to what are probably the most important long-term questions concerning resource depletion. To what extent can the oxygen-supplying reactions (photosynthesis) buffer and counter our demands? Is it possible that we may thermally, chemically, or physically so pollute the environment that photosynthesis is inhibited? What are the likely effects on this reaction of carbon dioxide and solid particle build-up in the atmosphere?

Photosynthesis, or the reaction in which carbon dioxide is reduced on a photosensitizing substrate (e.g., chlorophyll) in the presence of sunlight and a hydrogen acceptor (A), e.g.,

$$A + H_2O \xrightarrow[\text{chlorophyll}]{hv} H_2A + \tfrac{1}{2}O_2$$

$$12H_2A + 6CO_2 \rightarrow C_6H_{12}O_6 + 6H_2O + 12A$$

has, naturally enough, been extensively studied. Because of the complex nature of the many intermediate (charge–transfer) reactions, the complete reaction is not yet fully understood; however, it has been shown that:

1. The reaction is enhanced by an increase in PCO_2
2. The reaction is enhanced by a decrease in PO_2 (the present level of PO_2 in the atmosphere is an inhibitor)
3. At a given PCO_2 and PO_2 the reaction rate is dependent on light intensity*
4. Provided that adequate light is available, the rate of reaction is temperature dependent

These observations are summarized in Fig. 9, which depicts a typical experimental curve on which have been superimposed some approximate figures to relate the curve to natural systems. In darkness, the rate of photosynthesis is zero; as the light intensity is increased, the reaction rate increases at first linearly and finally reaches a plateau (R_{max}), the position of which is set by the prevailing PCO_2 and PO_2. The light intensity on a clear day can amount to 3.3 J/cm^2 min, and on an overcast day may be only one-fifth of this value; the photosynthetic reaction is only reduced to just over one-half for such a decrease in light intensity.

It has been calculated that photosynthesis "fixes" 5×10^{10} tonnes of carbon, or 22.5×10^{10} tonnes of organic matter, per year; close to 50% of this is carried out by *phytoplankton* in the oceans. Within the ocean environment two competing reactions operate, respiration and photosynthesis, with respiration predominant at depth and photosynthesis predominant near the surface, the compensation point being some 100 m in clear water and 1–2 cm in turbid water. What then are the consequences of a reduction in the amount of light incident on the oceans by scattering either from atmospheric dust and smoke or from an oil spill emulsion? Providing that somewhere there exist photosynthesisers which remain unaffected, there is no problem; as the PO_2 falls, the remaining cells compensate by an increased oxygen production. Similarly, any

---

*Photosynthesis is cyclic (day/night) but is effectively the forward reaction of an "equilibrium," the reverse reaction being respiration or combustion; any influence which in the long term increases the rate of photosynthesis shifts the equilibrium to a higher PO_2.

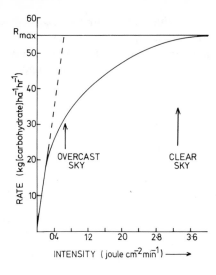

Figure 9. Influence of light intensity on the rate of photosynthesis.

chemical pollutant which affects only selected areas or species of phytoplankton or plants is unlikely to cause a major imbalance in the cycle. However, any effect which decreases light intensity or contaminates and destroys chlorophyll on a scale wider than can be compensated for by meteorological processes, will bring about a decrease in PO_2, although again, fortunately, at the high end of the intensity range (see Fig. 9), a large decrease in light intensity is required to bring about any significant change in photosynthesis rate. Also, fortunately, the effects of CO_2 build-up and a universal temperature increase work in favor of the photosynthesis reaction. Perhaps the greatest long-term threat to the efficiency of photosynthesis is the formation of a gas-impermeable layer between the source of photosynthesis and the atmosphere; hydrocarbons and oils (e.g., from insecticides) are known to have an inhibiting effect on photosynthesis, presumably due to O_2 build-up on the inside of the oil film and lack of supply of CO_2 from without.

These are all hypothetical problems which we may or may not face in the future; there is no immediate indication that we are at present seriously perturbing the photosynthesis cycle. Calculations by the geochemist Broecker[23] show that, compared with the supply of carbon in organic matter and fossil fuel, the reserve of oxygen is large. There are 60,000 mol of oxygen over every square meter of the earth's surface; respiration requires, and photosynthesis replaces, a mere 8 mol per square meter per annum. There are roughly 200 mol of nonfossil carbon per square meter of earth's surface as living tissue and as humus in soils and the sea: complete oxidation of all of this would require only a fraction of one percent of atmospheric oxygen. If we were to burn all known fossil fuel reserves we would use less than 3% of available oxygen. It is comforting also to know that under optimum conditions it takes as little as 1 liter of *chlorella* to provide the 560 liters of oxygen required by one person per day. In terms of food production it has also been shown that the potential photosynthetic productivity of the land surface is not of itself a serious limiting factor to population.[24] The problem is rather one of space: to have a sufficiently large area available for photosynthesis, i.e., food production, apart from that required for dwellings, recreation, and industrial and urban use.

2.4. *Mining and Metallurgical Technology*

Mining and metallurgical techniques contribute to defining ore reserves in two ways. The cost of mining and extracting metal from ore determines, together with the price received for the metal, those mineral deposits which are presently *economically* exploitable and those which are not; this is the basis for our definition of an ore reserve. In a more subtle way the present level of the technology of mining and metallurgy also determines the type of mineral deposit to which we can contemplate applying economic criteria. For example, a deposit containing tens of millions of tonnes of copper ore with a grade as high as 10% copper would be effectively useless (at present) if it occurred at a depth of 7000 m (23,000 ft) below the land surface. In a similar way, metallurgical techniques delimit potential reserves; for example, the recent development of an electrooxidation technique for the organic compounds in "carbonaceous" gold ores has added the gold content of these ores to gold reserves.

Minerals are normally won from the earth by one or more of the following techniques:

1. Open pit mining
2. Underground mining (shafts, drives, declines)
3. Solution and slurry pumping
4. *In situ* leaching

The first two of these methods have until the present attracted the most attention, although for the long-term future they are likely to pose the most problems. Open cutting owes its preeminence to the fact that the economics of surface mining has many advantages, too obvious to enumerate. In the future a greater proportion of mineable ore for the base and precious metals is likely to be found subsurface, although residual materials—iron, aluminum, and titanium oxides—will no doubt continue to be mined from the surface. Problems of competition with other land use functions are only likely to increase. In contrast, the problems for subsurface mining are almost entirely technical in nature. In one sense the problems are difficult to comprehend; technology is sufficiently well developed to send men 400,000 km into space to collect rocks (and retrieve them), and to tunnel horizontally tens of kilometers through the earth, but the current limit to depth of mining stands at a mere 3.475 km (11,400 ft). One problem associated with mining at and below these depths is that conventional mining requires manned machinery; the temperature of the rock at the 11,400-ft level of the South African mine referred to above is 40°C, and the cost of ventilating and airconditioning at this depth is high. In addition there are problems of time and cost with transporting men and materials to and fro between surface and the mining face, and with rock movement due to the pressures at such depths. With remote controlled mining equipment, such as draglines, continuous-line scrapers, tractors, or hydraulic suction systems, depths of mining could be increased perhaps even to the point where the temperature failure of metal components was the limiting factor. The depth corresponding to this temperature is hard to assess, and is dependent on (1) the thermal conductivity of the particular rock type, and (2) the radioactive heat generation from below the crust compared with that in the crust. Two theoretical cases which have been examined are reproduced in Fig. 10.[25] Both cases in Fig. 10 are for a crust comprising granite to a depth of 10 km, and basalt for the remaining 20 km of a 30-km crust, with a normalized surface heat flow of 5.02 μJ cm^{-2} sec^{-1}. Curve *A* is the temperature gradient with depth for the case where all of this surface heat flow originates from a source below the crust, and curve *B* for the case where three-quarters of the surface heat flow originates within the crust; these are likely to be extreme cases. According to this model a crustal temperature of 300°C could occur at depths anywhere between 15 km and 30 km, depending on the characteristics of the rock; at this stage it is therefore very difficult to assess even approximately the likely limits to depths of mining by conventional or semiconventional means. It must also be remembered, of course, that at present there is little direct incentive to mine deeper since there are few cases where exploration has indicated ore at depth; deep mining at present usually occurs more as an extension of the exploitation of a lode which is steeply dipping and has some surface expression.

For the last two mining techniques mentioned, solution and slurry pumping via drill holes and *in situ* leaching—there is tremendous scope for innova-

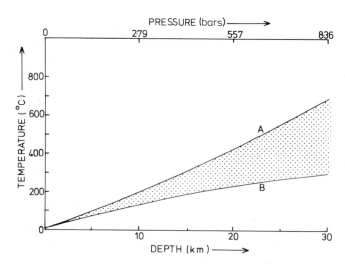

Figure 10. Temperature of the earth's crust as a function of depth. For curve *A* the heat flow originates from below the crust, and for curve *B* three-quarters of the heat flow originates within the crust.

tion and expansion. The first of these techniques is of course used at present for "mining" oil and sulfur, and its extension to pumping out ores (e.g., coal) as slurries is currently the subject of some research. Here the combination of a wide variety of drilling methods and chemical transporting agents suggests that extension of the method to other ores is feasible. Conventional drilling methods (e.g., diamond and percussion) are somewhat restricted by the time and expense involved in replacing cutting tools when the cutting face is at depth. Many alternative techniques for rock drilling, including ultrasonic, electrical, thermal, explosive, flame-boring, and hydraulic (high-pressure water) methods, can produce a hole suitable for transporting hydrothermal solutions of melts from depth. In many cases the choice of the best leachant for extracting minerals from the ore and retaining it in solution as a suspended solid for pumping to the surface, will be known from present hydrometallurgical data, and it is because present solution extraction methods are commonly operated at increased pressures and temperatures that the application of this technique to mining at depth is attractive in principle. The extraction of nickel from sulfide ore, for example, takes place at 70–80°C in autoclaves at 9.00 bars (130 lb/in^2). The reaction,

$$NiS \cdot FeS + 3FeS + 7O_2 + 10NH_3 + 4H_2O$$

$$\rightarrow Ni(NH_3)_6SO_4 + 2Fe_2O_3 \cdot H_2O + 2(NH_4)_2S_2O_3$$

could (in theory) be accomplished at a depth of 4–5 km, using an ammonia solution as leachant and adjusting pressure according to the oxygen or air required for the reaction (there are of course some technical qualifications to this statement). It is also possible that metals could be electrochemically deposited from such leach solutions or melts at depth, and the cathode with nonmetal deposited on it withdrawn to the surface at intervals; in certain cases it may also be possible to regenerate at least some of the leach solution in the electrolysis step.

This technique is in fact one particular example of a more general technique, *in situ* leaching, which is at present in restricted use, for example, in the extraction of copper from chalcopyrite ($CuFeS_2$) ores. There, an oxygenated sulfuric acid leach solution is supplied to the top of the ore body; it percolates under gravity or by aquifer flow to a point where the metal-laden solution collects; from here it is pumped to the surface for processing. The technique has many advantages; it minimizes contamination, there is a short production period, and it is suitable for otherwise inaccessible deposits; however, it is only amenable to those deposits which are hydrometallurgically extractable *and* which have a suitable geology. Many rocks are permeable to solutions, although, without rock fracturing, *in situ* leaching is probably confined to use in sandstones, limestones, and the like. Rocks can and have been "fractured" using acid, water under pressure, gases, explosives, and nuclear detonation.[26] On the scale required for *in situ* leaching, the use of nuclear explosives is the most attractive and has received the most attention.

Many underground detonations have been made and the results partly evaluated; one example is shown in Fig. 11. This test detonation of 3.1 ktons, smaller than that which would be required for a commercial mining operation, was fired at a depth of 361 m in a sandstone–limestone formation. The void created had a total volume of 28,000 m³ and a radius of 19 m of which the

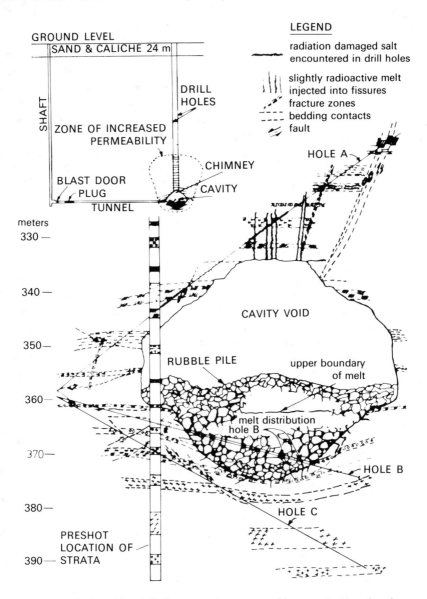

Figure 11. Cavity void and displacement of strata caused by an underground nuclear detonation. (After G. C. Howard and C. R. Fast, *Hydraulic Fracturing*, Soc. of Petroleum Engineers of A.I.M.E., N.Y., 1970.)

bottom half contained approximately 15,000 tonnes of rubble and 2000 tonnes of solidified melt from the explosion. In addition a large volume of surrounding rock was shattered, became as a rule more permeable, and settled, with strata displaced from their original positions (see Fig. 11). The solidified melt contained most of the radioactive material and had a higher temperature than the surrounding rock for a considerable time after the event. Direct radioactive contamination above and around such detonations is relatively small, providing the shot is deep enough; the gas pressure within the cavity at the time of the explosion effectively balances the weight of overlying rock and prevents collapse and fissuring. Samples of solids and fluids from within the cavity are usually radioactive, however, and the extent of the problem is a function of the nature of the reaction (proportion of fission versus fusion) and the local geology; it is a major obstacle to widespread acceptance of nuclear explosives in mining.

Application of this technique to *in situ* leaching is obvious. With the shaft, tunnel, and drill holes used for examination of the test hole (see inset in Fig. 11), one has a perfect circulation system for leachant and gas, and access for removal of the leach solution to the surface. The example shown here was one of many designed for investigating the effect of an increase in permeability on oil-bearing sediments; it was concluded that such fracturing could increase the world reserves of oil considerably. The major problem retarding the implementation of the technique is the radioactive contamination in oil produced; similar contamination of groundwaters and leach solutions, principally due to the production of tritium, will impede the implementation of nuclear fracturing for *in situ* leaching. With the proper technology, and especially if the leachant is continuously recycled, these problems may be overcome. With this proviso, the use of nuclear fracturing for *in situ* leaching appears to be one of the most favorable test cases for the peaceful use of nuclear energy.

Metallurgical techniques for the extraction and purification of metals from ores play an important part in determining use of resources, not only because their efficiency decides the amount of metal which can be extracted from a given amount of ore and the cut-off grade of the ore, but also because they determine those ores which, because of their chemistry, can contribute to reserves irrespective of grade. In the future there will be many major changes in the extractive metallurgical field, and to a great extent this will be inspired by, or be a direct application of, basic chemical research, physical, inorganic, and organic. This is at present evident in the use of organic complexing agents in solvent extraction, new ion exchange resins, and sophisticated electrowinning techniques in the extractive metallurgical industry. Rather than generally examining these and other techniques, the developments which have recently been imposed upon the extraction technology for one metal—copper—will be examined in detail, with the deliberate inference that similar technological advances are likely for other metals in the future.

Changes in extractive technology in the future are most likely to be precipitated by a need to process different ore types, to process existing ores more efficiently, or to handle lower average ore grades. Recent changes in copper extraction technology, however, have been necessitated by a different factor, one which is now generally accepted in principle by the industry as a genuine variable in their operations: pollution. Smelting of sulfide ores produces SO_2, and the emission of this gas has rightly become a target for environmental protection agencies, to the extent that, in the United States, Clean Air Act regulations were foreshadowed in 1970. One result of this has been the development within a short space of time, but with the expenditure of millions of dollars, of several new extraction techniques for copper based on hydrometallurgical methods.

The Arbiter process[27] is a relatively straightforward low temperature–low pressure oxygenated ammonia leach technique; the copper is extracted from the leach solution into an organic solvent, probably a substituted oxime with an exchange mechanism which can be generalized as:

$$2[RH]^0 + Cu^{2+} \rightarrow [R_2Cu]^0 + 2H^+$$

The copper is back extracted from the organic solvent into an acid electrolyte and recovered electrochemically. Sulfate by-product from the operation is removed as gypsum after treatment with lime, or as ammonium sulfate.

The Cymet process[27,28] is a hydrometallurgical process with potential for general use on base metal sulfides; it illustrates beautifully the principles of "chemical conservation." In the case of copper, chalcopyrite ($CuFeS_2$), previously concentrated by froth flotation, is leached with ferric chloride (see Table 10) to produce a leach solution of cuprous and ferrous chlorides (and sulfur) which is thickened with more chalcopyrite feed for electrodissolution. At the anode, further leaching of chalcopyrite occurs, and at the cathode, copper is reduced in the leach solution and is precipitated. Ferric chloride for the initial leach step is regenerated in the Fe cells at the anode, and high-purity iron is produced at the cathode; this is an added economic advantage of the process. High-purity sulfur from the chalcopyrite leach is obtained by filtering and washing the leach slurry. The process, which needs additional development before becoming commercial, has, in summary, the capability to produce

Table 10. Chemistry of the Cymet Process for Copper Extraction

Chalcopyrite electrodissolution
Leach Step:	$6FeCl_3 + 2CuFeS_2 \rightarrow 2CuCl + 8FeCl_2 + 4S°$
Anode:	$CuFeS_2 + 3HCl \rightarrow CuCl + FeCl_2 + 2S° + 3H^+ + 3e^-$
Cathode:	$3CuCl + 3e^- \rightarrow 3Cu° + 3Cl^-$

Fe Cells
Anode:	$6FeCl_2 + 6Cl^- \rightarrow 6FeCl_3 + 6e^-$
Cathode:	$3FeCl_2 + 6e^- \rightarrow 3Fe° + 6Cl^-$

Summary of process
$$3CuFeS_2 \rightarrow 3Cu° + 3Fe° + 6S°$$

high-purity copper, iron, and sulfur from chalcopyrite with minimum pollution, and involves an electrical energy expenditure which will make it competitive.

Apart from these two major innovations, many other techniques are being applied to copper extractive metallurgy. Treatment of sulfides with molten elemental sulfur or substituted quinolines; electrowinning copper from melts and organic solvents; and cleaning up conventional smelter methods by redesigning furnaces and introducing catalytic reagents into exhaust stacks, are but a few examples. These techniques are likely to enter the metallurgy of other metals as both pollution and supply problems arise. As in the case for copper, the long-term emphasis for cleaner, more efficient extraction is likely to be on solution metallurgy—solvent extraction, ion exchange, and electrochemistry—where greatest chemical control over the concentration and distribution of metal components can be achieved.

Chemical control over metallurgical processes is also vital to another important aspect of resource usage, that of maintaining high efficiency in the extraction of metal from ores. In the past there have been many examples, particularly in the metallurgical extraction of tin, tungsten, and gold, of poor extraction efficiency. The unfortunate result in many cases has been the extraction of only part of the usable material from a high- or moderate-grade ore, leaving a significant metal content in the residue which is not economical to extract and which is, at least temporarily, waste. With greater process control, and with modifications on extraction procedures, better use could have been made of these resources. The tin mines of Cornwall are one example where improvements in metallurgical methods and procedures now allow us to make better use of resources. For many years the only real measure of extraction efficiency (or rather, of extraction inefficiency) for these mines was the number of streaming plants which could operate profitably downstream on the tailings from the initial extraction plant. The minute particle size of the cassiterite (SnO_2) and the high concentration of sulfide minerals in the ore combined to make efficient extraction by conventional gravity separation methods extremely difficult, to the extent that many of the mines became uneconomical around the turn of the century. Today, in one reopened mine, the traditional gravity treatment, which collects approximately 50% of the tin concentrate, is supplemented by a flotation procedure to recover the finer-grained material; the iron, zinc, copper, and arsenic sulfides which would normally interfere with this separation are removed (by froth flotation) at an earlier stage. The entire process is carefully monitored by an on-line X-ray spectrometer system that analyzes for tin, copper, zinc, sulfur, arsenic, and iron to ensure optimum control. Not only is the percentage recovery of tin greater than that previously obtained, but, in addition, the production of copper–zinc concentrate makes a contribution to the overall economy and efficiency of the process. In the future, as the availability of many resources becomes more and more critical, even greater care will have to be taken to ensure that metallurgical processes, apart from being economical, are also efficient in resource usage terms, in their recovery of metal from ore.

3. Recycling and the Use of Alternative Materials

Future reserves and the availability of materials represent one aspect of the resources depletion problem. We are now in a position to examine the second important variable: our consumption and usage of the material derived from these reserves.

3.1. The Hydrological Cycle: A Model

Water is an important resource which has not yet been discussed, and which is likely to play an important role in limiting population, at least on a local scale. It also provides an excellent natural example for management principles of resources in general. The hydrological cycle is shown schematically in Fig. 12. The three major reservoirs for water are the sea, the atmosphere and, the lithosphere, the sea being, by two orders of magnitude, the most important. Transfer of water between these reservoirs via gravity or solar energy is by evaporation, condensation, wind, precipitation, and runoff, as shown in Fig. 12; the importance of each transfer process is indicated by the mass of water transported annually.

Three reservoir types can also be isolated for most nonenergy mineral resource cycles, as shown in Fig. 13, and in many cases these are more or less analogous to the reservoirs of the hydrological cycle. In the particular case of water, the oceans are analogous to the *ore reserve*, atmospherically held water is analogous to the processed ore or the working *supply reservoir*, and ground, surface, and biologically held water is analogous to the retention of *processed material* (particularly of metals) in use in a mineral resource cycle. As in the hydrological cycle, where precipitation on land is balanced by evaporation plus recharge by transport of water vapor from the sea (equal to runoff), usage of a mineral resource must be balanced by the amount of secondary or recycled

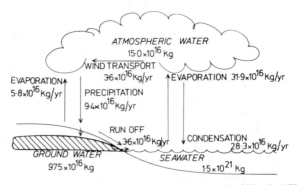

Figure 12. The hydrological cycle. (After G. Borgstrom, *Too Many*, Macmillan Co., N.Y., 1969.)

Figure 13. Reservoirs and material movements in an idealized mineral resource cycle.

metal *plus* a replenishment from ore, which in a steady state case is equal to the "runoff" as unrecoverable waste, i.e.,

Usage = Replenishment (Primary) + Recycle (Secondary)

The importance of the hydrological cycle as an example is that is has the in-built capability to return all waste (runoff) to the ore reserve and thence back to the working supply; this runoff from the land is 38.3% of the precipitation which falls onto land, the remaining 61.7% being recycled directly to the working supply via evaporation.

Despite the perfect recycle efficiency of the hydrological cycle, water is, and is likely to continue to be, a major factor in limiting population. This is because at present only the runoff portion of the cycle is used, albeit ineffectively due to wide geographical and seasonal variations in rainfall distribution, although in time it appears likely that we will be able to increase the amount of water in circulation by using saline water distillation plants. Greater use of groundwater in the long term does not necessarily solve the problem, as it constitutes merely a temporary storage in the runoff part of the cycle; it may in local situations improve the efficiency of runoff water use. The water usage problem is further compounded by increases in the standard of living, as water consumption figures in Table 11 for the production of a variety of goods indicate.[7] Thus the actual water requirement for an adult male is approximately 2.5 liters per day in fluids and food, yet it takes a further 14,000 liters of water to grow that food for the average North American person; for an Indian, less than 2000 liters of water are required to grow the average daily food intake.

Table 11. Water Consumption for Commodity Production

Commodity	Water required (liters)
1 kg wheat	600
1 kg rice	2,000
1 kg meat	25,000
1 liter milk	4,000
1 automobile	450,000

3.2. *Recycling*

It has already been mentioned that supply of a material is from two major sources: *secondary* or *recycled material*, and *primary material* from the processing of ore. In a steady-state mineral resource cycle analogous to the hydrological cycle, the waste deficit which is not recycled must be replenished by primary material; in the present growth-rate situation, additional material for growth must also be provided from the primary source. For either case, the percentage of recycled material obviously plays a vital role in determining rates of depletion of resources. The importance of the *recycle percentage*, particularly for metals with rapid depletion rates, can be seen from the graphs in Fig. 14. In Fig. 14a, the rate of depletion of zinc, expressed as the number of years to (hypothetical) extinction, has been calculated as a function of the percentage of recycled or secondary zinc used, assuming that present reserves and usage rates (RUI = 4.5%) will hold. In Fig. 14b, a similar curve for iron (RUI = 0.4%) is shown. The curves are in fact identical in shape, but the magnitude of the depletion problem and the importance of recycling is clearly different in the two cases. For zinc, at recycle percentages below 60–70%, we add very few years to the "lifetime" of zinc reserves by increasing the recycle efficiency; it is only at very high recycle percentages that we use considerably less of the zinc reserves per unit of time. Small changes in the efficiency of zinc recycling play a significant part only at these higher recycle percentages. For iron, where the usage rate is a smaller percentage of reserves, the need to achieve high recycle percentages is, in the short term, less acute, and an increment in recycle efficiency of 40–50% adds 100 years to the iron reserves. Time, ultimately, is the all-important factor, since it decides how much new ore can be found, how much low-grade material can be processed, and what new technology can be evolved to cope with resource consumption.

As we have seen, the curve for the depletion of iron with recycle percentage is identical to the curve for zinc: only the y-axis scale differs. In Fig. 15, the

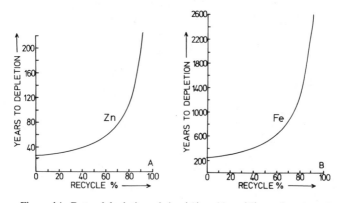

Figure 14. Rate of depletion of zinc (A) and iron (B) as a function of recycle percentage.

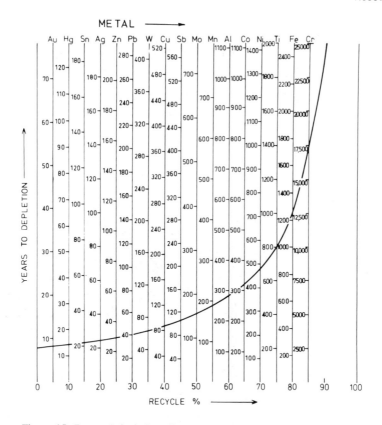

Figure 15. Rates of depletion of reserves of some common metals as a function of recycle percentage.

depletion versus recycle percentage curve is shown again, with ordinate scales appropriate to 17 of the metals discussed earlier superimposed. From this curve it is possible to observe, for any of the metals, the projected hypothetical "lifetime" of present reserves at various recycle or secondary metal usage percentages, assuming reserves usage levels as shown in column 1 of Table 3. It should again be stressed that these figures are not presented as a prediction of the actual lifetimes of these resources, but merely as indicators to the rates of depletion, and as a guide to the areas to which our attention in solving resource depletion problems should initially be directed.

Recycle percentages and secondary usage classifications of consumption patterns, indicating the present recycling situation for some common metals, are presented in Table 12. Present recycling percentages are in general lower than the levels indicated in Fig. 15 as being required to significantly halt the rate of depletion of resources. It is also apparent that, in general, the present recycle percentage does not correlate with the rate of depletion, because silver, copper, and lead have significantly higher recycle percentages than gold, mercury, and

Table 12. Present Recycle Percentages and Secondary Usage Patterns

Metal	Percentage recycled (1969)	Uses with good recovery	Uses with poor recovery	Accompanying contaminants
Aluminum	11.7%	New scrap (81%) Electrical Construction Packaging	Chemical Catalytic reagent Refractory	Mg, Cu, Mn, B
Copper	40.9%	Brass, bronze, alloys Coinage Electrical	Chemical Fungicide Fertilizer (trace element)	Zn, Sn, Pb
Gold	15.9%	Alloys	Jewelry Electronics Bullion	Ag, Pt group
Iron	27.9%	New scrap (75%) Construction Transport Packaging	Chemicals Fertilizer	Cr, Mn, Ti, C

Lead	40.0%	Storage batteries Pb- and Cu-based alloys	Gasoline additives Solder Pigments	Sb, Pb, Cu
Mercury	20.6%	Hg cells in Cl_2 plants Electrical Amalgams	Fungicides Germicides Paints	organic matter Ag
Nickel	19.1%	New scrap	Stainless steel alloys	Fe, Co, Cr
Silver	47.2%	Metal stampings Bimetal scrap Batteries	Photographic solutions	Cu, Ni
Tin	20.4%	Tin alloys, bronze	Solder Tin cans	Cu, Pb
Tungsten	4.1%	Superalloy scrap Cutting tools	WC cutting dust Welding rods Lamp filaments	Th, C
Zinc	27.0%	New scrap (75%) Brass, bronze alloys Batteries	Galvanized products Pigments	Cu

tin, for which we currently consume a higher proportion of known reserves. Rather, it appears that consumption patterns determine recycle percentage: the large amounts of copper and silver used in coins and electrical equipment ensure a rapid and efficient recovery, while the use of gold in jewelry and as a monetary standard, of mercury in thermometers, electrodes, fungicides, and germicides, and of tin in tin plating and as solder, ensures a poorer recovery for these metals.

The problems of scrap recycling, which must be at least partially overcome if recycling percentages are to be increased and depletion of reserves reduced, are threefold:

1. Distribution of potential scrap material
2. Chemical composition of scrap material
3. Lifetime of products containing potential scrap

Distribution of scrap, both geographically and as different size fractions, is a problem which has examples in the secondary recovery of most metals, and is a major problem for gold, lead, tin, and tungsten. The larger and more locally confined a manufactured article, the easier is scrap recovery. Gold contacts and gold foil as used in electronics, the presence of lead in solder at joints in other metals, tin coatings on steel cans, and tungsten in welding rods, are all examples of this ideal not being satisfied, and scrap recovery in these cases is difficult. This is also one of the reasons why new scrap (industrial scrap offcuts, trimmings, and casting waste) plays a more significant recycling role than old (or obsolescent article) scrap for many metals (see Table 10). The chemical composition of scrap, principally in terms of alloy components, is a major basis for classification of scrap. For example, copper scrap is divided into at least ten categories on the basis of content of other metals (usually alloy components), and aluminum into at least five categories according to its iron, magnesium, tin, and zinc content. The formation of chemical compounds usually signifies the end to secondary recovery of metal, although, in the case of silver, the recovery of metal from spent photographic solution is now profitable. The lifetime of manufactured articles is an important and complex variable; in terms of secondary recovery, the shorter the lifetime, the sooner the material can be pressed back into service. However, for articles which must then be replaced, the short lifetime is of no consequence to the overall resource usage and depletion problem. While the use of materials continues to grow, as in our present economy, the time for a manufactured article to become obsolescent is critical; material which stays in use, for example, for 10 years constitutes a far smaller fraction of the total metal requirement when the time for recycling arrives. Some idea of the approximate average lifetime of iron in use in various manufactured iron and steel goods, and the expected percentage recovery of iron from them, can be obtained from Table 13.

Chemistry presently plays only a minor role in secondary metal recovery and recycling processes. In the main, scrap is simply classified and sorted prior to reprocessing, and in many cases secondary metal (with impurities) is used

Table 13. Percentage Iron Recoverable and Approximate Lifetimes for Various Iron-Containing Manufactures[a]

Market classification	% of total Fe recoverable	Lifetime of product (years)
Shipping, marine	100	29
Rail transport equipment	86	27
Fabrication	87	27
Foundry products	100	27
Military equipment	36	21
Electrical	75	17
Mining, quarrying equipment	91	16
Machinery, industrial tools	94	16
Agricultural equipment	99	15
Containers (all types)	13	14
Automotive	100	13
Domestic, commercial apparatus	57	12
Drilling equipment	100	11
Appliances, utensils, cutlery	76	11
Aircraft	100	10

[a]Source: J. McHale, *The Ecological Context*, Braziller, New York, 1970.

again only for its original purpose. Chemical separation of elements is not yet widely practiced in the industry. For the future, the application of molten salt technology for selective dissolution of melting appears to be most encouraging. Copper, for example, can be recovered from automotive starters, generators, and armatures in a molten $CaCl_2$ bath. An electrochemical method for the recovery of aluminum, copper, lead, tin, silver, and gold from electronic scrap has recently been published.[29] Electronic scrap is now a significant component of all scrap for many metals; copper pins, contacts, and connectors are often plated with silver, gold, and other precious metals for better contact and corrosion resistance; in many cases the electronic components are in an aluminum case. Conventional sweating processes, used to recover aluminum, cannot be used to recover copper and precious metals since they are soluble in molten aluminum. The basis of the molten salt electrorefining process is a three-layer electrolytic refining cell shown in Fig. 16.

The three layers of the cell, formed at the operating temperature of 750–850°C, are:

1. The electronic scrap anode, S.G. 3.3
2. The electrolyte ($BaCl_2$ 60 wt %, NaF 17 wt%, AlF_3 23 wt %), S.G. 2.7
3. Refined aluminum, S.G. 2.3

Typically, at the conclusion of a run, the cathode is refined aluminum with traces of copper and zinc, and the anode is an alloy of copper (62%), zinc (11.2%), tin (6.8%), aluminum (6.2%), silicon (6.0%), and lead (1.5%) containing significant amounts of gold, silver, nickel, and iron. This anode

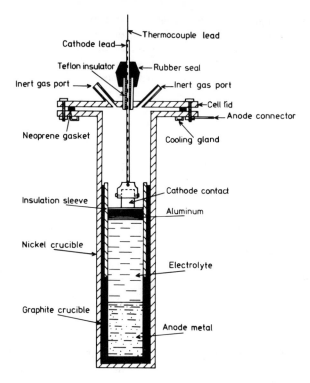

Figure 16. Three-layer electrolytic refining cell used for recovery of Al, Cu, Pb, Sn, Ag, and Au from electronic scrap. (After E. L. Singleton and T. A. Sullivan, *J. Metals*, **25**(6) : 31, 1973.)

material must be subjected to further conventional metallurgical processing in order to extract the metals of value; 99% of the total copper, 97% of the gold, and 94% of the silver is recovered in this way.

Electrochemical methods constitute a major metallurgical process technique, and their use in the refinement of primary metal, as well as their potential use in secondary metal recovery (from melts and solutions), is dependent on the economics of the supply of energy in the form of heat and electrons. The energy requirements for producing equal weights of the competing materials steel and aluminum are often compared; these comparisons usually suggest that it requires between 6 and 20 times as much energy to produce a given weight of aluminum as it does to produce the same weight of steel. It has been pointed out, however, that these calculations are erroneous if no account is taken of recycling.[30] For aluminum, the presence of recycled material makes a very large and favorable contribution to the energy balance, because the major consumption of energy in aluminum production is in the original electrolytic smelting of alumina with molten cryolite to produce aluminum metal. A summary of the energy balance for two aluminum can cycles is shown in Table 14. Cycle 1 is a scheme for the production of

Table 14. Summary of Energy and Material Balance for Aluminum Recycle[a]

Process	Cycle 1 (no recycle)			Cycle 2 (25% recycle)		
	Amount of metal (tonnes)	Electrical energy (MWH)	Thermal energy (G Joules)	Amount of metal (tonnes)	Electrical energy (MWH)	Thermal energy (G Joules)
Electrolysis of new metal	22.81	425.0	1314.3	18.37	342.2	1058.2
Remelting scrap—new	6.53	0.6	32.6	6.43	0.9	54.8
—old	0			4.44		
Remelting ingot waste	20.80	2.3	240.0	20.80	2.1	215.4
	50.14			50.14		
Delacquering scrap	3.23	0.07	9.4	7.67	0.2	22.3
Sheet processing	28.09	27.0	293.7	28.09	27.0	293.7
		455.0	1890.0		372.4	1644.4
Finished cans	20.86			20.86		
(Recycle) before losses = cans	(0)			(5.22)		
after losses = old scrap	(0)			(4.44)		

[a]Source: P. R. Atkins, *Engng. Min. J.*, **174**(5) : 69 (1973).

20.86 tonnes of aluminum, without recycling the used cans but including recycle of new scrap from the ingot and sheet processing. The energy and material balance for cycle 2, in which 25% of the cans output is recycled, shows an 18% reduction in electrical energy and a 13% reduction in thermal energy because of the smaller amount of primary metal required. Whether or not this makes aluminum a better choice than steel for packaging is another complex question. The important point is, as was stressed earlier, that our usage of metal resources will ultimately be limited by the "soft" resources, i.e., energy, time, and money, the former of course being a major factor in the latter two. Any process which minimizes our call on these "soft" resources is advantageous. The production of an aluminum can is an investment of energy within the aluminum metal cycle. This is a very sound reason for aiming, in the case of aluminum, at a high recycle efficiency, higher than the level used in the example above. While it is true to say that no other material benefits so clearly in terms of energy from secondary metal usage as does aluminum, this principle also holds for most other metals.

3.3. *Use of Alternative Materials*

We have already investigated, in a general way, the influence of recycling on the depletion of reserves, and have seen that, for many metals with high present usage rates, efficient recycling will greatly affect their future reserves and usage patterns. It has also been pointed out that our usage of resources varies markedly from mineral to mineral, as shown by a comparison of the highest RUI, 12.9%, for gold, with that for chromium, 0.04%. The potential thus arises for substituting a material which has a high reserves usage index with an alternative that has a low index. This can be construed in the long term as the principle of making best use of those materials with greatest reserves; in the short term, more importantly, the use of alternative materials has the potential to divert demand from those materials under greatest usage pressure, and to provide time for fuller exploration, evaluation, and exploitation of their as yet undiscovered reserves.

Obviously enough, there are usage patterns for many resources for which it appears no suitable alternatives can be used. The combined hardness and low S.G. of aluminum, the high-temperature properties of molybdenum, the low melting point of tin for use in solder, the fluidity of mercury, and the radioactivity of uranium, are all examples for the case in point. The properties mentioned for these metals are difficult to duplicate, but in no case can it be categorically stated that a technically suitable alternative cannot or will not be found. Technical considerations such as these do not as a rule completely specify the material used for a particular purpose; the availability and the processing required—both reflected in the cost of the final article—also play a part in the choice of material, and are, in many cases, the reasons why the technology for their use has been so well developed and established.

The crucial question is, however, what is likely to happen to material consumption patterns in the coming years? Are our reserves of the base and precious metals likely to be eroded to the point where our use of these is drastically curtailed? Will we have found suitable alternatives if and when this time arrives? Should we commence planning for high recycling percentages and for use of alternative materials in anticipation of such events, or wait and see, hoping that material prices will naturally dictate the introduction of recycling and alternatives? There are preliminary indications, and direct evidence in the case of crude petroleum, that events may not be as orderly and as natural as we might expect. Table 15 lists some uses for the common metals for which alternative materials can be substituted. It is not always possible in this very general way to assess immediately whether an alternative imposes a smaller or larger strain on resources. In many cases a smaller or larger quantity of the alternative material must be used, processing is different, and energy consumption and use of subsidiary resources is required. As a guide to present processing requirements, the market prices of metals are included in Table 15, and, as a qualitative indication as to whether use of an alternative is advantageous as far as overall use of reserves is concerned, the *reserves usage status* of the alternative and the original metal are quoted (RUI status = $RUI_{metal} - RUI_{alternative}$). Thus, in the case of aluminum replacing copper in electrical equipment, we find that the original metal, copper, has a reserves usage index 1.0% higher than that of aluminum, and that as far as the overall resources picture is concerned—subject to the simplicity of this analysis—the replacement of copper by aluminum is desirable. On the other hand, the replacement of zinc as an anticorrosion agent with tin is not favorable; the RUI for zinc is 0.8% less than that for tin at the present time.

Two points of more than usual significance derive from Table 15. First, the question as to whether aluminum should be used to replace iron, or vice versa, in the packaging, construction, and automotive spheres, again presents itself. In terms of the depletion of reserves, we currently consume 0.5% less of iron reserves than of aluminum reserves per annum. However, to give iron corrosion-resisting properties similar to aluminum, alloys containing Cr, Ti, Mn, V, or Ni (none of which is in very short supply) must be used; the total resource depletion comparison is also compounded by the energy requirements and the effective lifetimes and recycling efficiencies of the two cycles. Fortunately, in the short term, neither of these two elements is likely to be in more than momentary short supply. In the long term, alloys containing high percentages of magnesium, which has an almost infinite reserve, are likely to play an increasingly important part in the consumption of materials for structural, automotive, aircraft, and packaging purposes.

The second important conclusion evident from Table 15 concerns the use of plastic as an alternative material to many metals in packaging, plumbing, and tubing, and of organic chemicals as substitutes for compounds of the common metals in insecticides, fungicides, and pharmaceuticals. The future importance of plastics and other organic chemicals is beyond question; however, it is

Table 15. Alternative Materials

Metal	Price[a] ($U.S./kg)	Uses	Alternatives	RUI status ($RUI_{metal} - RUI_{alt.}$)
Aluminum	0.55	automotive aircraft packaging	steel, stainless steel Mg, Mg alloys, Ti plastic	0.5 (Fe) ~0.9 (Mg) -2.2 (Petroleum)
Copper	1.32	electrical construction plumbing	Al, Na alloys stainless steel, Al plastic	1.0 (Al) 1.5 (Fe) -1.2 (Petroleum)
Gold	3311.00	jewelry electronics dental	Pt, Pd Ag, Al stainless steel, plastics	— 7.9 (Ag) 9.8 (Petroleum)
Iron	0.52 (stainless steel)	construction automotive packaging	Al, cement, wood Al, plastic Al, glass	-0.5 (Al) -2.7 (Petroleum) -0.5 (Al)
Lead	0.36	storage cells gasoline additive pigment cable coverings plumbing, piping	Cd, Hg, Ni, Ag, Zn Ni, catalytic reforming Ti, Zn plastics plastics, cement, asbestos	-1.0 (Zn) 2.8 (Ni) 3.0 (Ti) 0.4 (Petroleum) 0.4 (Petroleum)
Magnesium	0.84	structural refractories	Al Al$_2$O$_3$, ZrO$_2$, SiC	-ve -ve
Manganese	0.73	desulfurizing steel alloying	Ti, Zr Ni, Cr, Mo	0.7 (Ti) 1.16 (Cr)

Metal	Price	Uses	Substitutes	
Mercury	8.00	cathode in Cl_2, NaOH prod'n.	S, organics	4.6 (Petroleum)
		medicinal	organics	4.6 (Petroleum)
		germicide		
		protective paint	plastics, Cu paint	5.8 (Cu)
Molybdenum	8.81	stainless trim	Cr	1.26
		refractory	graphite, W, Ta	—
		hardened steels	B	—
Platinum	4952.00	alloys	Pd, Ir, Rh, Ru	—
		electrical contacts	alloys, Au, Ag alloys	—
		catalyst	V	—
Silver	85.85	photocopying, photographic	Se	
		reflectors	Al, Rh	4.1 (Au)
		coinage	Cu–Ni, Mn alloys	3.1 (Cu)
Tin	5.29	tinplate	Al, paper, plastics, steel	4.4 (Al)
		roofing	Cu, Al, Zn coated products	0.8 (Zn)
		fungicides, insecticides	organics	2.2 (Petroleum)
Titanium	3.13	pigment	Zn, Si, Al	−4.0 (Zn)
		structural	low alloy steel, Al	−0.4 (Al)
		welding electrodes	Na, K silicates, $CaCo_3$, CaF_2	
Tungsten	9.91	cutting material	diamond, Al_2O_3	1.5 (Al)
		abrasive	Ta, Ti, Cr carbides	1.9 (Ti)
		tool steel	Mo	1.1 (Mo)
Vanadium	3.30 (V_2O_5)	Cr-V steels	Ni steels	—
		catalyst	Pt	—
Zinc	0.45	diecasting	Al, Mg, plastics	3.8 (Al)
		anti-corrosion agent	Al sheet, ceramic, Sn	−0.8 (Sn)
		reducing agent	Al, Mg	~4.5 (Mg)

[a]Prices for September, 1973.

Table 16. Principal Petrochemicals and Their Uses[a]

Product	Uses
Ethylene derivatives	
Polyethylene	Films, mouldings, cable coverings, etc.
Ethylene oxide	Intermediate for ethylene glycol
Ethylene glycol	Anti-freeze, terylene
Styrene	Polystyrene plastics, synthetic rubber
Ethyl alcohol	Industrial solvent, chemical intermediate
Acetaldehyde	Intermediate for acetic acid
Acetic acid	Cellulose acetate fibers, plastics, adhesives
Ethylene dichloride	Intermediate for vinyl chloride
Vinyl chloride	PVC plastics, piping, guttering
Propylene derivatives	
Isopropyl alcohol	Solvent, chemical intermediate for acetone
Butanol	Solvent, chemical intermediate for oxo-alcohols
Ethyl hexanol	Plasticizers
Heptenes	Plasticizer alcohols
Propylene tetramer	Reacted with benzene for dodecyl benzene (detergents)
Cumene	Raw material for acetone, phenol production
	Intermediate for plastics, nylon
Propylene oxide	Intermediate for manufacture of plastic foams
Propylene	Films, fibers, plastic mouldings
Glycerol	Pharmaceutical, resin manufacture
Acrylonitrile	Acrylic fibers, adiponitrile, and nylon
C_4 olefins	
n-Butenes	
Sec. butanol	Solvent
Methyl ethyl ketone	Solvent, lube oil, dewaxing agent
Heptenes	Plasticizer alcohol intermediate
iso-Butene	
Butyl rubber	Inner tubes
Polyisobutene	Lube oil additives, sealing compounds
Di-isobutene	Plasticizer alcohol intermediate
Butadiene	
Copolymers (styrene)	Synthetic rubber, shoe-soling compounds
Polybutadiene	Synthetic rubber
Copolymers (acrylonitrile)	Speciality rubbers, plastics
Chloroprene	Oil resistant rubbers
Adiponitrile	Raw material for nylon
Aromatic derivatives	
Benzene	
Styrene	Polystyrene plastics, synthetic rubber
Phenol	Intermediate for resins
Cyclohexane	Intermediate for nylon production
Dodecyl benzene	Detergents
Aniline	Dyestuffs, rubber additives
Maleic anhydride	Polyester fiberglass plastics

Table 16—continued

Product	Uses
Toluene	
Toluene diisocyanate	Plastic foams
Phenol	Intermediate for resins
TNT	Explosives
Solvents	Paints and lacquers
Xylenes	
Solvents	Paints, lacquers, insecticides
Terephthalic acid	Polyester (terylene) fibers
Phthalic anhydride	Phthalate plasticizers and resins
Isophthalic	Resins
Chlorinated Methanes	
Monochloromethane	Silicones
Dichloromethane	Paint stripper, solvents
Trichloromethane	Pharmaceuticals
Tetrachloromethane	Aerosol propellants

[a]Source: *Our Industry Petroleum*, The British Petroleum Company Ltd., 1970.

apparent, on the basis of the present reserves usage index for petroleum (3.1%), that the increasing use of these materials presents some problems. At present in Europe, some 87% of all organic chemical production is based on petroleum, 10% on coal, and the remaining 3% on miscellaneous sources, yet the use of petroleum products for energy production is the principal reason for the present high reserves usage index for petroleum crude. Should this crude petroleum be reserved for the petrochemical industries of the future, or should alternative source materials for the petrochemical industry be contemplated?

Coal, suitably processed, is a rich source of an extremely wide variety of organic compounds and, in direct contrast to petroleum, its reserves are in no immediate danger of depletion (RUI = 0.05%). Since 1950, however, coal has been steadily decreasing as a source of raw material for the chemical industry due to the directness and cheapness of the products derived from crude petroleum. Some of the major petroleum products and their principal uses are shown in Table 16. Manipulation of the natural aliphatic, alicyclic, and aromatic hydrocarbons of crude petroleum to produce the desired quantities of basic chemicals (as indicated in Table 16) is achieved by one or more of the following reactions:

1. Dehydrogenation
2. Oxidation
3. Chlorination
4. Sulfuration
5. Cracking

It is because such reactions are facilitated with natural crude and can be controlled and varied according to demand, that petroleum provides a cheaper avenue to most of these basic chemicals.

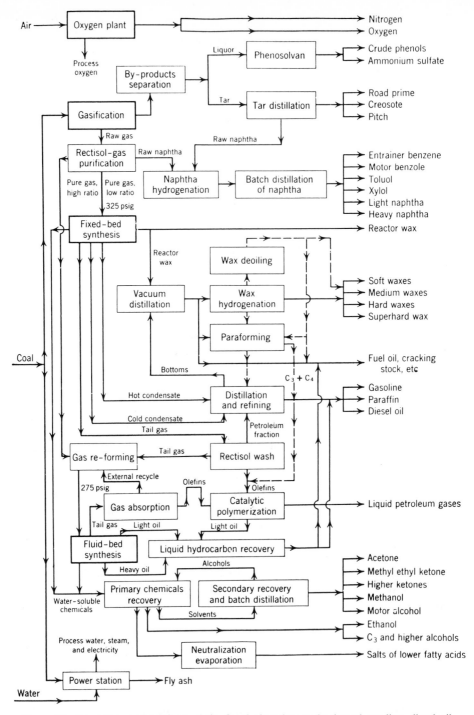

Figure 17. Operations and products of the Sasol plant for production of gasoline, diesel oil, solvents, and chemicals from coal. (After H. F. Mark, J. J. McKetta, and D. F. Othmer, eds., *Kirk–Othmer Encyclopedia of Chemical Technology*, Vol. 4, John Wiley & Sons Inc., N.Y., 1963, p. 446.)

Carbonization of coal, which accounts for approximately 20% of all coal consumption, produces coke, coke-oven gas (principally methane, hydrogen, and carbon monoxide), light oil, coal tar, and ammonia, usually recovered as ammonium sulfate. Not only is the quantity of light oil produced small in quantity (11 liters/tonne of coal) and in value, it also consists predominantly of aromatic hydrocarbons. Although the exact nature of the product is dependent on the coal and on the processing plant, the composition of light oil from carbonization is usually close to 70% benzene, 15% toluene, 4% *o*-, *m*-, and *p*-xylenes, 1–5% naphthalene, 2.5–3% styrene-indene, and small amounts of paraffinic naphthenic hydrocarbons, thiophene, and carbon disulfide. To produce a range of aliphatic and alicyclic hydrocarbons similar to those in crude petroleum, coal must be subjected to gasification, to produce carbon monoxide, hydrogen, and methane, and then to hydrogenation by the Fischer–Tropsch process.

The Sasol plant in South Africa is such a unit, capable of producing, economically, in excess of 250,000 tonnes of gasoline, diesel oil, solvents, and chemicals from just over one million tonnes of coal per annum.[31] Figure 17 shows the operations and the products of the Sasol plant. The process has three principal stages: (1) gasification of the coal (Lurgi process), (2) purification of the gases, and (3) synthesis using iron as a catalyst. The distribution of products is dependent on temperature, pressure, and chemical composition and size of catalyst; two distinct sets of operating conditions are used, depending on whether the catalyst is circulated, as in the fluid-bed process, or operates on a fixed bed. Typical product distributions for these two conditions are shown in Table 17. The fixed-bed process produces mainly straight-chain, high-boiling oils, but relatively low amounts of gasoline, liquid petroleum gas, and oxygenated compounds. The fluid-bed process, on the other hand, produces mainly hydrocarbons and gasoline with little high boiling point material.

Coal can thus be made to produce a range of hydrocarbons similar to crude petroleum. The principal reason for the predominance of crude petroleum in producing gasoline and chemicals over the period 1950–1970 has been price;

Table 17. Conditions and Distribution of Products in Sasol Process

	Fixed-Bed Process	Fluid-Bed Process
Catalyst	precipitated iron	reduced magnetite
Temperature	220–255°C	320–330°C
Pressure (kPa)	2482	2275
Products	Percentages	
liquefied petroleum gas (C_3–C_4)	5.6	7.7
gasoline (C_5–C_{11})	33.4	72.3
middle oils (diesel, etc.)	16.6	3.4
waxy oil or Gatsch	10.3	3.0
wax	29.8	
alcohols and ketones	4.3	12.6
organic acids	traces	1.0

the cost of a ton of coal was for a long time similar to that of a ton of petroleum crude. There is little doubt as to which of these commodities is likely to show the greatest price increases in coming years. Do we substitute coal as an alternative fuel for petroleum and use petroleum crude for petrochemicals? Or, alternatively, should the conversion of coal gas to gasoline-type products be encouraged? In the long term the answer is, of course, that neither crude petroleum nor coal should be burned as fuel; their value as basic materials for the chemical industry is inherently far greater and more permanent than their calorific value. In the meantime, while the pollution aspect of burning coal directly mitigates against its use as an energy source, the production of gasoline-type products from coal seems destined to play an important role in the depletion of energy resources.

4. Resource Planning for the Future

The so-called "energy crisis," or crude petroleum shortage of 1973, represented a milestone in the resources depletion area. For the first time "disbelievers" saw concrete evidence of the predictions of resource shortages made 5–10 years previously; those who had foreseen the problems but maintained that economic policies and technological advantages would cope with them in an orderly fashion must also have been shaken by the scramble on the part of nations and corporations to regain energy security. It is more than likely that unless deliberate and premeditated planning—both economic and technological—is taken, the crisis will be repeated in the future for other resources. One economic measure of obvious benefit would be the imposition of a *resources tax* on the use of primary materials, in proportion to the reserves usage index. The revenue of such a tax, which would encourage the use of secondary material, and of materials with a low reserves usage rate, could be directed to solving technological problems associated with better resource utilization.

It is very easy to state the general principles which should be followed to minimize the consequences of resource depletion, initially as the "steady-state" usage situation is being approached, and subsequently as the growth of population and usage of resources is rationalized. To say that all we have to do is to expand reserves, consume less, and recycle a higher percentage of material is about as explicit and meaningful as many a political preelection promise. Of course, these are the broad aims which should be alluded to, but there are many specific recommendations which can be made within this broad framework. Some which are suggested from the foregoing assessment of resource depletion, future reserves, and technology and resource utilization are shown below, in order of chronological urgency rather than in order of overall importance:

1. That the conversion of coal-gas to gasoline-type products be given positive incentive.
2. That improvements in nuclear power technology (particularly in the area of waste disposal) be made, to ensure the use of nuclear fission as a guaranteed safe (though perhaps temporary) energy source.

3. That research and development be directed to the development of solar and fusion energy as the ultimate energy resources.

4. That economic penalties be imposed on usage of the base metals Cu, Pb, Zn, and Sn in areas with low recycle percentage, e.g.,

    a. Pb as an additive in gasoline
    b. Cu as a fungicide
    c. Zn as a pigment

5. That the substitution of alternative materials for the base metals Cu, Pb, Zn, and Sn, be encouraged, e.g.,

    a. The replacement of Cu by Al in electrical equipment
    b. The replacement of Zn and Sn protected iron by stainless steel and aluminum

6. That the "immobilization" of precious metals in jewelry and as monetary standards be discouraged.

7. That the use of magnesium and high magnesium content alloys be promoted wherever possible.

These recommendations are given purely from the standpoint of rationalizing usage of those materials in shortest supply and optimizing our overall use of resources. The list is, of necessity, incomplete. For example, higher recycle efficiencies in the iron and aluminum cycles are desirable for many reasons from a resource depletion standpoint alone. However, the case of iron and aluminum is, in the short term, not as urgent as many others.

Our management of resources for the future, and the course which is to be taken to overcome and alleviate resource depletion problems, will be decided by a combination of two main factors: policy and technology. Policy and politics will be required to determine and direct technological research and development into the correct areas of enquiry, both for expanding our known reserves and for using the acquired raw materials. Technological advances will be required to detect, delineate, and exploit unfound reserves; to process low and lower grade ore; to incorporate greater quantities of less-pure secondary metal in processing cycles; and to generate alternative materials with properties akin to the original. Taken literally, the very definition of chemistry as, "a science that deals with the composition, structure, and properties of substances and the transformations that they undergo,"[32] implies that the chemist and chemical technology are bound to play a vital role in the future management of earth resources.

References

1. United Nations Statistical Office, *Statistical Yearbook 1970*, United Nations, 1973.
2. U.S. Bureau of Mines, *Mineral Facts and Problems*, P3, U.S. Dept. of the Interior, Bureau of Mines, 1970.
3. F. Blondel, S. G. Lasky, *Econ. Geol.*, **51** : 686 (1953).

4. U.S. Bureau of Mines, *Mineral Facts and Problems*, U.S. Dept. of the Interior, Bureau of Mines, 1970.
5. U.S. Bureau of Mines, *Mineral Facts and Problems*, P1094, U.S. Dept. of the Interior, Bureau of Mines, 1970.
6. D. H. Meadows, D. L. Meadows, J. Randers, and W. W. Behrens III, *Limits to Growth*, Earth Island Ltd., London, 1972.
7. P. R. Ehrlich and A. H. Ehrlich, *Population, Resources Environment*, W. H. Freeman & Co., San Francisco, 1972, Ch. 3.
8. Committee on Resources and Man, *Resources and Man*, W. H. Freeman & Co., San Francisco, 1969, Ch. 8.
9. B. J. Skinner, *Earth Resources*, Prentice Hall, Inc., New Jersey, 1969.
10. M. King Hubbert, *Can. Min. Metall. Bull.*, **66** (735) : 37 (1973).
11. U.S. Bureau of Mines, *Mineral Facts and Problems*, P219, U.S. Dept. of the Interior, Bureau of Mines, 1970.
12. E. R. Mitchell, *Can. Min. Metall. Bull.*, **66** (734) : 65 (1973).
13. *Metal Statistics*, 63rd edition, American Metal Market Co., N.Y., 1970.
14. V. E. McKelvey, *Am. J. Science*, **258A** : 234 (1960).
15. G. J. S. Govett and M. H. Govett, *Earth Sci. Reviews*, **8** : 275 (1972).
16. S. G. Lasky, *Engng. Min. J.*, **151** (4) : 81 (1950).
17. Committee on Resources and Man, *Resources and Man*, W. H. Freeman & Co., San Francisco, 1969, Ch. 6.
18. A. Walsh, *Spectrochim. Acta*, **7** : 108 (1955).
19. B. G. Weissberg, *Econ. Geol.*, **66** : 1042 (1971).
20. J. B. Flato, *Anal Chem.*, **44** : 75A (1972).
21. Committee on Resources and Man, *Resources and Man*, W. H. Freeman & Co., San Francisco, 1969, Ch. 7.
22. J. L. Mero, *Can. Min. Metall. Bull.*, 65 (720) : 21 (1972).
23. W. S. Broecker, *Science*, **168** : 1537 (1970).
24. A. San Pietro, F. A. Greer, and T. J. Army, ed., *Harvesting the Sun*, P315, Academic Press, N.Y., 1967.
25. A. Poldervaart, ed., *Crust of the Earth*, P101, Geol. Soc. of America, Special Paper 62, 1963.
26. G. C. Howard and C. R. Fast, *Hydraulic Fracturing*, Soc. of Petroleum Engineers of AIME, N.Y., 1970.
27. F. C. Price, *Engng. Min. J.*, **174** (4) : RR (1973).
28. P. R. Kruesi, E. S. Allen, and J. L. Lake, *Can. Min. Metall. Bull.*, **66** (734) : 81 (1973).
29. E. L. Singleton and T. A. Sullivan, *J. Metals*, **25** (6) : 31 (1973).
30. P. R. Atkins, *Engng. Min. J.*, **174** (5) : 69 (1973).
31. H. F. Mark, J. J. McKetta, D. F. Othmer, eds., *Kirk–Othmer Encyclopedia of Chemical Technology*, Vol. 4, John Wiley & Sons, Inc., N.Y., 1963, p. 446.
32. P. B. Gove, ed., *Webster's Third New International Dictionary*, G. & C. Merriam Co., Mass., 1966.

Air Pollution

This book began with considerations of air pollution and smog. There is hardly any need therefore to connect the following two chapters on air pollution with the former themes of population reduction and resource recycling. Air pollution chemistry is vital for the future, both for removing those pollutants which invade our atmosphere and for reducing the continuation of that invasion.

7

Formation and Control of Air Pollutants

W. Strauss

1. Introduction

Air pollutants are those substances—gases, mists, and particulate aerosols—which are present in the atmosphere in such concentrations that they can adversely affect man and his environment. Most of these substances—sulfur dioxide, the oxides of nitrogen, and many others—are naturally present in the atmosphere in low (background) concentrations. They are produced in nature, as well as from anthropogenic sources, and generally the natural background concentrations are such as to be harmless. Indeed, they may play a vital part in the natural cycles of growth and decay. Unusually high natural concentrations do occur; examples being methane (as "marsh gas") and hydrogen sulfide from geothermal sources, but these are exceptional circumstances. So when we refer to air pollutants we consider these to be those in relatively high concentrations (compared with background values) which result from the chemical and

W. Strauss • Department of Industrial Science, University of Melbourne, Parkville, Victoria, Australia 3052

Table 1. Composition of Dry Air in the Lower Troposphere (Free of Water Vapor)

Gas	Chemical symbol	Concentration[a] (%)	Calculated residence time
Principal Gases			
Nitrogen	N_2	78.0	continuous
Oxygen	O_2	20.9	continuous
Argon	A	0.93	continuous
Carbon dioxide	CO_2	0.032[b]	2–4 years
Trace Gases			
Permanent Gases		(ppm)	
(Nonreactive)			
Helium	He	5.2	about 2 million years
Neon	Ne	18	continuous
Krypton	Kr	1.1	continuous
Xenon	Xe	0.086	continuous
Reactive Gases			
Carbon monoxide	CO	0.1	0.5 years
Methane	CH_4	1.4	4–7 years
Nonmethane hydrocarbons	"HC"	$<1 \times 10^{-3}$	?[c]
Nitrous oxide	N_2O	0.25	4 years
Nitric oxide	NO	$0.2–2.0 \times 10^{-3}$	5 days
Nitrogen dioxide	NO_2	$0.5–4 \times 10^{-3}$	several days
Ammonia	NH_3	$6–20 \times 10^{-3}$	7 days
Hydrogen sulfide	H_2S	0.2×10^{-3}	2 days
Sulfur dioxide	SO_2	0.2×10^{-3}	4 days
Chlorine	Cl_2	$3–15 \times 10^{-4}$	several days
Iodine	I_2	$0.4–4 \times 10^{-5}$	?
Hydrogen fluoride	HF	$0.08–18 \times 10^{-3}$	?
Hydrogen	H_2	0.58	?
Ozone	O_3	0–0.05	about 60 days

[a]This is the atmospheric background concentration, and not the concentrations found in polluted areas. When a range of concentrations is given, it indicates that these have been measured by different workers at different places.
[b]Minimum concentration of CO_2 measured away from centers of population. In population centers CO_2 concentrations vary from about 0.034% to 0.035%.
[c]Little is known about the residence time of the gas.

biological processes used by man. Most notable among these are fuel combustion—for heating, cooking, and industrial processing—and the generation of electricity.

The natural (background) levels of the naturally occurring gases and their estimated residence times* are given in Table 1.[1] The levels above which these could have an effect on man and the environment have been called "Air Quality Standards." The "Primary Standard" is based on information about

*In this context, the residence time is the average time that the substance remains in the atmosphere before being adsorbed on the earth's surface, or interacts in the atmosphere, rendering it harmless.

human health, and generally specifies a higher concentration (usually 25% or more) than the "Secondary Standard," which protects the environment: property, materials, climate, economic values, etc. The present Air Quality Standards are two or more orders of magnitude higher than the background levels.

2. Air Pollutants from Combustion

The most common fuels are fossil fuels: oil, coal, and natural gas. In some less-developed countries wood is also widely used. The major constituents of coal, oil, and wood are carbon, hydrogen, and oxygen, with lesser amounts of sulfur and nitrogen, as well as traces of other materials such as metal compounds (sulfides and oxides). Natural gas, as pipeline gas, has had the sulfur compounds removed, as they corrode pipelines. The actual quantities of sulfur compounds depend both on the fuel type and on the source. Thus the sulfur content in oil and coal can vary between less than 1% to 5%. In the case of oil, virtually all sulfur is tied into the organic chemical structure (organic sulfur), while for coal about half the sulfur is "organic" and the rest is "inorganic," distributed as fine crystals of pyrites (FeS_2), ferric sulfate ($Fe_2(SO_4)_3$), magnesium sulfate ($MgSO_4$), calcium sulfate ($CaSO_4$), and other compounds.

During combustion, the hydrogen and carbon compounds are oxidized, with the evolution of energy:[2]

$$C + O_2 \rightarrow CO_2 + 94.0 \text{ kcal/mol (kg mol) (395 MJ/mol)}$$

$$H_2 + \tfrac{1}{2}O_2 \rightarrow H_2O + 68.3 \text{ kcal/mol (287 MJ/mol)}$$

If insufficient oxygen is present for complete oxidation of the carbon, some of the carbon dioxide formed may react, to give the monoxide

$$C + CO_2 \rightarrow 2CO - 410 \text{ kcal/mol (172 MJ/mol)}$$

or some of the carbon may react with less oxygen:

$$C + \tfrac{1}{2}O_2 \rightarrow CO + 26.4 \text{ kcal/mol (111 MJ/mol)}$$

This shows that when combustion occurs with insufficient oxygen, much carbon monoxide is produced, and the overall heat release is far less than when complete combustion is achieved.

During incomplete combustion of oil or coal, volatile organic compounds that are complex hydrocarbons can be driven off, and these form one component of smoke, particularly from small domestic furnaces. In larger furnaces the volatiles, which are highly combustible, are ignited by radiation and convection from the hot walls of the furnace, and are burned to carbon dioxide and water.

The minor constituents of the coal and oil, sulfur and nitrogen, also form oxides during combustion. For sulfur

$$S + O_2 \rightarrow SO_2$$

and to a lesser extent in the flame

$$SO_2 + O^* \rightarrow SO_3$$

and also

$$SO_2 + \tfrac{1}{2}O_2 \rightarrow SO_3 + 85 \text{ MJ}$$

In fact, in the normal flame SO_3 only represents about 1% of the total sulfur oxides. Although SO_3 is the stable form at lower temperatures, the rate of formation, without a catalyst present, is so slow as to be insignificant, while at flame temperatures, the dioxide is more stable. Nitric oxide (NO) is also formed during combustion. In part this is from nitrogen in the fuel, of which, according to different experimenters, between 18% and 80% is oxidized; the rest is due to reactions with atmospheric nitrogen in the flame and post-flame regions. The following reactions, referred to as the *Zeldovich reactions*, are considered primarily responsible

$$N_2 + O^* \rightarrow NO + N^* - 75 \text{ kcal/mol (315 MJ/mol)}$$

$$N^* + O_2 \rightarrow NO + O + 31.8 \text{ kcal/mol (133 MJ/mol)}$$

After being emitted to the atmosphere, the nitric oxide is turned into the dioxide comparatively slowly—because of the very dilute concentrations—by the complex series of reactions in photochemical smog. These, simplified, are

$$NO + \tfrac{1}{2}O_2 \rightarrow NO_2$$

High flame temperatures also favor nitric oxide formation because of the possibility of forming active nitrogen and oxygen atoms in the flame, as well as hydroxyl radicals, OH^*, which can react very quickly

$$N^* + OH^* \rightarrow NO + H^* + 39.4 \text{ kcal/mol (165 MJ/mol)}$$

In the case of gas firing, or firing with sulfur-free (and nitrogen-free) oils, no sulfur oxides occur in the combustion gases, and the nitrogen oxides are solely from the fixation of atmospheric nitrogen. The amounts produced vary over a wide spectrum (see Table 2). But while the sulfur oxides are wholly a function of the sulfur in the fuel, the nitrogen oxides depend largely on the method of firing and the resultant flame temperature.

Particulate emission from combustion, i.e., smoke, is a function of the solid incombustible materials in the fuel and of incomplete combustion of the carbon. Particularly with small industrial, commercial, or domestic boilers and furnaces, when these are overloaded, or burning at very low levels, both unburned carbon particles and inorganic matter are emitted. There is generally very little smoke from oil- or gas-fired sources, at optimum rates, and with well-adjusted burners. However, coal-fired units invariably produce some smoke, especially when fired with pulverized fuel, as in modern, large-scale operating practice, and some system of particulate control is essential. In older coal-fired systems, where the coal was burned on grates, it was possible to minimize emissions since much of the ash would fall down into the ash pit and was not entrained in the flue gases passing to the chimney.

Table 2. *Gaseous Emissions from External Combustion Plants (Parts per Million in the Waste Gases)*

Type of fuel	Oxides of nitrogen	$SO_2{}^c$	SO_3
Residual oil	70–500 (525^a, 225^b)	200–700	3–12
Coal	200–1200 (225^a, 225^b)	1100–2100	10–80
Natural gas	60–1600 (150^a, 125^b)	—	—

(Note: A fuel oil fired boiler emits 12 to 15% carbon dioxide. In good combustion practice carbon monoxide is less than 2% of the carbon dioxide, i.e., less than 0.024–0.030%.)
[a]Maximum permitted under U.S. Federal Regulations (December 23, 1971) for large boilers.
[b]Maximum permitted in California since December 31, 1974.
[c]The sulfur oxides in the waste gases depend directly on the sulfur content in the fuel, and may be outside the range indicated here.

For stationary combustion systems, where firing takes place in a chamber, and in which there is a stable flame and an appreciable residence time for the oxidation reactions to take place, the air pollutants are the oxides of sulfur and nitrogen, and particulate matter. The use of controls is usual practice for particulate matter and to a limited extent for the oxides of sulfur and nitrogen.

This is in contrast with "internal combustion" systems used almost invariably for automobiles.[3] In these, a charge of fuel is burned in a chamber with cooled walls, and this drives a piston, which in turn moves a connecting rod, crank, and drive shaft. The time for combustion is limited to a fraction of a second, and the cooled walls prevent complete combustion of the charge. Unburned and incompletely burned combustion products are then emitted. There are two principal types of internal combustion systems: the spark-ignition (or Otto) engine, which is the usual automobile power unit, and the compression ignition (or Diesel) engine. The cycles in these engines, and the methods of controlling the emissions, will be dealt with fully in Chapter 8. It should be stated here however that in addition to the oxides of nitrogen, internal combustion engines produce considerable quantities of unburned "active" hydrocarbons and carbon monoxide (Table 3).

All combustion processes using carbonaceous fuels produce carbon dioxide, which until recently was not considered an air pollutant *per se*. It was considered that the carbon dioxide from combustion was used and recycled by plants during photosynthesis. Measurements of carbon dioxide levels over the past century have shown a steady rise, amounting to about 10% in carbon dioxide levels in the atmosphere at remote locations; in fact, this rise corresponds to about half the carbon dioxide produced by combustion of 2 fossil fuels. This problem and its possible consequences will be discussed fully in Chapter 10.

Table 3. Emissions from Combustion Sources (kg/tonne fuel)

| Emission | Internal combustion | | External combustion | | | |
| | Otto engines | Diesel engines | Fuel oil | | Coal | |
			Power Gener.	Commer. Domest.	Power Gener.	Commer. Domest.
Carbon monoxide	395	9	0.005	0.025	0.25	25
Nitrogen oxides	20	33	14	10	10	4
Sulfur oxides	· 1.55	6.0	20.8S	20.8S	19S	19S
Hydrocarbons	34	20	0.42	0.26	0.1	5
Aldehydes, organic acids	1.4	6.1	0.08	0.25	0.0025	0.0025
Particulates	2	16	1.3	1–12	8A	2–8A

Note: Most coal-fired units are fitted with control devices and emissions are of the order of 1–10% of these.
S: To obtain sulfur oxides produced, multiply number by %S in fuel.
A: To obtain particulates produced, multiply number by %A in fuel.

3. Air Pollutants·from Industry

Apart from combustion systems for generating heat or electricity, or providing motive power, the major sources of air pollutants are processes in the chemical, metallurgical, and petroleum industries. Thus a major, and often quite concentrated source of sulfur dioxide is the roasting of nonferrous sulfide minerals, principally lead, zinc, and copper ores. The petroleum industry, directly and indirectly, is responsible for hydrocarbon and sulfur oxide emissions, and the chemical industry for a wide range of air pollutants, usually specific to the processes in which they are formed.

3.1. The Nonferrous Metallurgical Industry

A most important source of sulfur dioxide emissions, second only in total quantity to the sulfur dioxide from combustion of fuels, and representing over 10% of this, is the roasting and subsequent treatment of the sulfide ores of copper, lead, and zinc. This industry has sources which produce concentrations often of the order of 4–10% which are sufficiently high to be converted to the trioxide and to sulfuric acid with conventional catalysts. Unfortunately, many of the world's primary treatment plants for nonferrous ores are far from places where there is a demand for the sulfuric acid for fertilizers or chemical processing.

It has been pointed out, for example, that Australia, which imports large quantities of sulfur for fertilizer for its agricultural industry, would be self-sufficient if it used all the sulfur produced by metallurgical processing.[4]

Another factor resulting from these high concentrations is that considerable damage to vegetation has been caused by the waste gases from smelters, particularly copper smelters. The initial destruction of plants, followed by soil erosion, has given rise to the classical barren landscapes of Ducktown, Tennessee; Queenstown, Tasmania; and elsewhere, where copper and other nonferrous smelters have been allowed to operate without controls.

In conventional copper smelting (Fig. 1), the ore is first roasted (oxidized) to remove some of the sulfur, which is emitted in concentrations of about 8%. The resultant calcine, a mixture of the sulfides and oxides of copper and iron, is fed to a reverberatory furnace, together with limestone and/or silica as fluxes. This combines with metal oxides and other impurities to float on top of a layer of copper and iron sulfides, which is tapped off to form the "matte." The waste

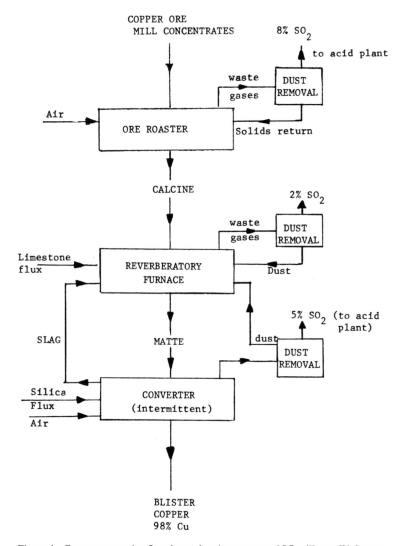

Figure 1. Copper processing flowsheet, showing sources of SO_2. (From: W. Strauss, *Air Pollution*, E. Arnold, 1976.)

gases from this process contain only 1–2% sulfur dioxide, and so require special treatment if SO_2 is to be removed. The matte is subsequently heated in a converter, where the iron is oxidized, along with the remaining sulfur, to sulfur dioxide, which comes off in concentrations of about 3–5%. The problem of SO_2 recovery in this case is the batch nature of the conventional process, although a continuous process, such as the WORCRA process, would help to give a steady supply suitable for acid manufacture.

Lead and zinc are generally obtained from the sulfide minerals galena (PbS) and zinc blende (ZnS). Usually, as in the case of ores from Broken Hill, Australia, they are found together. The ores are roasted, either on a moving grate or in a fluidized bed roaster, and the waste gases, containing sulfur dioxide in concentrations of about 6%, are used, when possible, for the manufacture of acid. The major problems in this case are that the roaster off-gases contain impurities such as arsenic trioxide (As_2O_3), hydrochloric acid gas, hydrogen fluoride, and similar compounds, which are volatilized from the ore and are catalyst poisons. They would prevent the catalytic oxidation stage from operating satisfactorily, and so must be removed, together with other particulate impurities in the gas stream.

3.2. Ferrous Metallurgy

Iron ore, which is generally quite pure ferric oxide (hematite) is essentially reduced with carbon, to give iron. In practice, the reduction is carried out on a very large scale in blast furnaces. In modern plants, the ore and carbon are introduced into the blast furnace sintered together, whereas older processes introduced the ore and carbon (as coke) separately. The product is still relatively impure iron, containing some 6% carbon. This can now be used, together with some scrap iron, for cast iron products, or it can be further treated to give steel in one of a number of processes such as the "basic oxygen converter," the open hearth furnace, or the electric furnace. In each case, scrap metal forms an appreciable part of the charge. The final product, which now contains much less carbon, also has small traces of substances—alloying elements—which impart very specific properties to the steel, such as hardness, tensile strength, corrosion resistance, etc. There are many sources of air pollution in the iron and steel industry, originating from the large-scale handling and high-temperature processing which is involved. The mining and handling of the ore is a serious control problem, as is the dust produced during the sintering of coal and ore before it passes into the blast furnace. The blast furnace itself was a traditional source of pollution, but the high carbon monoxide content of the "waste" gases makes them a valuable fuel for other steel works operations, and so it is carefully collected, cleaned, and used. The steel making is carried out in converters, in reverberatory furnaces, or in electric furnaces. Frequently, and particularly in modern converter practice, pure oxygen is used to speed up the reactions, and this is accompanied by a tremendous evolution of very fine orange brown fume which must be controlled. These converter gases also have a high, although intermittent, carbon monoxide content which in the most modern plants is collected for subsequent use.

Coke ovens are a further serious pollution source because of the dust caused in filling the ovens, the evolution of steam and fine particles in coke quenching, and the gases emitted through leaks in the ovens, which can contain sulfides and mercaptans and produce an odor problem associated with coke oven batteries.

3.3. *Nonmetallic Minerals*

Materials such as cement, glass, ceramics, asbestos, sand, crushed rock (aggregate) for concrete and road surfacing, and coal are mined and quarried in very large quantities. Many of the materials have a low unit value, and so must be produced relatively close to the point of consumption, as with, for example, sand and aggregate. With the exception of some underground coal mining, most of the materials are quarried, which in the case of coal means "open-cut" mining. The material is usually blasted, crushed, and transported to storage bins, from which it is carried in trucks to the point of use. Coal may receive additional treatment that separates out noncoal constituents by washing.

Cement and glass are obtained by chemical treatment mainly of limestone and sand, respectively (as well as other materials). In cement manufacturing, the crushed limestone, after blending, is calcined in rotary kilns, with emission of considerable amounts of fine dust, chiefly lime dust. In glass manufacturing, the sand and crushed broken (used) glass is melted down in a furnace with the evolution of fumes and gases containing (besides silica), carbonates, nitrates, chlorides, and fluorides, all of which present a specific air pollution source.

In all of these industries the mining, crushing, transport, and storage of the materials is associated with the generation of large amounts of dust, both coarse and relatively fine. As the dusts contain silica, and in some instances, asbestos, they present a danger to workers in these industries, and considerable "in-plant" efforts are made to control them for purposes of industrial hygiene. However, careless plant operation and transport can cause dust problems elsewhere.

3.4. *Petroleum Refining*

The refining and subsequent treatment of crude petroleum gives a vast variety of product hydrocarbons, from jet fuel and motor oil to furnace oils and bitumen, including the raw materials for the petrochemical industry. Essentially, crude petroleum is distilled, and the different fractions are either used directly or are treated chemically, breaking down and reforming the molecules to give the products in the quantities and proportions required by consumers. The great changes in refining practice since 1940 have been a decreasing dependence on simple distillation and an increase in the degree of chemical treatments, both of which reduce, to a large extent, the dependence of the refiner on the source and constitution of his crude.

Crude petroleum contains hydrocarbons, together with organic sulfur compounds, inorganic compounds (iron, vanadium, etc.), and some alkali and alkali earths, etc., which form ash. However, ash in oils is low—of the order of 0.1% or less—compared with 1% or more as is common in coal. Sulfur content in oils varies from 0.5% (or less) in Indonesian or Bass Strait crudes to 4% (or more) in some Middle Eastern and Mexican crudes.

The following are the main sources of air pollution in a refinery. First, the pollutants from combustion gases: a refinery tends to use the lowest grade, high-sulfur fuels in its own furnaces and boilers. Second, there are fine particles from badly operated "flares," used to cope with sudden surges of hydrocarbon gases. Third, there are particulate emissions from the catalytic cracking, and particularly the catalyst recovery unit. These are usually fluidized bed reactors, and as the catalyst particles are reduced in size by constant cycling, they are blown out of the catalyst fluidized beds. Fourth, there are many sources of hydrogen sulfides and mercaptans, which are stripped from the lighter grades of fuel (motor oil, kerosene, light diesel oil, etc.) since they would corrode the engines in which they are used. In good refinery practice these are collected and sulfur is recovered. Finally, in a refinery there are many sources of hydrocarbon emissions from leaks in valves and pipelines, in reactors, and in storage tanks. In correctly operated refineries these are reduced to a minimum, and the unavoidable emissions are recovered and used, since they are valuable.

3.5. Chemical Process Industries

The chemical process industries (of which petroleum refining is of course a major part) produce such a wide variety of chemicals from such a wide range of raw materials, that it is not possible to discuss here any individual processes in detail. However, some of the major air pollutants can be considered and their sources indicated.

Sulfur dioxide is emitted by plants manufacturing sulfuric acid, as the catalytic conversion of the sulfur dioxide to the trioxide is incomplete at the temperatures commonly used in the reactors (450–600°C). Conventional practice (which uses three catalyst passes, followed by trioxide absorption) uses about 98% of the sulfur dioxide, emitting the remainder. With small plants, producing on the order of 200 tons of acid each day, the emissions could be tolerated, but some modern acid plants are ten times larger. Here, an "interpass absorption" stage has been introduced, reducing the emissions to about 0.2% and increasing the acid yield.

Some hydrogen fluoride and other fluorides are emitted from "superphosphate" fertilizer manufacture, when sulfuric acid is reacted with crushed phosphate rock. Other sources of fluoride are from electrolytic smelting of aluminum, where the "flux" used is cryolite, an aluminum sodium fluoride.

Fluoride emissions have also occurred from brick kilns when there have been traces of fluoride in the clay used for brick making.

Ammonia and oxides of nitrogen are emitted in some chemical processes using ammonia or nitric acid as reagents or catalysts.

Hydrogen sulfide and mercaptans, besides being emitted by hydrocarbon processors, are also emitted in the paper or cellulose pulp industry. Wood, after debarking and chipping, is treated in reactors with either neutral (neutral sulfite pulping) or alkaline ("Kraft" pulping) reagents. In the former case, which is the most common, the reagent is a bicarbonate-buffered solution of sodium sulfite, while in Kraft pulping it is a mixture of sodium hydroxide and sodium sulfide. After reacting with the lignin in the wood, the sulfur forms complex mercaptans, and hydrogen sulfide with low odor threshold forms part of the "black liquor," which is filtered off the pulp. Most of the odors are emitted when the "black liquor" is concentrated and then burnt in special recovery furnaces. Another air pollution source in the wood and paper industry is associated with the difficulty of burning bark and sawdust without emitting smoke.

Hydrocarbon vapors are produced in the paint industry, first during paint manufacture in cooking the resins, and to a lesser extent in the mixing of the paint. Large total quantities of hydrocarbons, although low in individual concentrations, are given off from paints when applied to surfaces. When used in enclosed spaces these can occasionally cause headache and nausea. More serious is the problem of hydrocarbon vapors, which may be partly "cracked" from the baking of enameled surfaces, such as car bodies, tinplate, etc.

3.6. *Food and Foodstuffs Processing*

Odors are the main air pollution problem in the food and related industries. In many cases the odors from some types of food processing, e.g., bread baking and coffee roasting, are very pleasant, as long as the concentrations are low and last only for limited periods. But large-scale processing of these and other foodstuffs produces odors in such concentrations and for such extended periods as to be extremely nauseous. The odors are complex chemical mixtures of aldehydes and related compounds, with traces of amines.

The processing of different natural products, not for edible foods, also causes serious odor problems. Essential oils have their characteristic odors which are nauseous in large concentrations.

The processing of animal residues, an important source of protein feed for poultry, and of edible or inedible fats (depending on the source of the residues), using medium-pressure steam distillation (rendering), produces particularly nauseous odors which are mainly ethylamines.

More complaints are received by air pollution control authorities about odors than about any other form of air pollution. While odors are rarely dangerous or toxic, in the usually accepted sense, they can cause headaches, nausea, and disturbing psychological effects.

3.7. Incineration of Wastes

Small-scale incineration of garbage, garden cuttings, and leaves results in serious smoke pollution and acrid odors when carried out by a multiplicity of sources. Some industrial incineration, such as the burning of old tires or of the covering from electrical cables to recover the metal values, has similar results. In many places this is now prohibited.

Large-scale incineration by public authorities, when carried out in modern units, has relatively few problems of this sort, since complete combustion is achieved and effective particulate collection systems are installed. However, a certain proportion of current domestic wastes consists of polyvinylchloride (PVC) which on combustion releases hydrochloric acid gas, and this can be a serious problem.

4. Control of Air Pollutants (General)

Air pollutants are generally produced in dilute concentrations, together with large volumes of a carrier gas. One way of controlling air pollutants is to remove the source, either by changing the process or by stopping the process altogether. Examples of these measures are numerous, and can involve electrical generation by nuclear-fueled reactors instead of coal-fired boilers (see also Chapter 13), electric cars instead of internal combustion engine vehicles, or the alteration of a mineral extraction process from ore roasting to alkaline leaching. Complete elimination of a process may in some cases be necessary, although if the product is essential and there is no substitute (either process or product), then the process may have to be carried out at a new location, with an adequate control system and with supplementary precautions. It may be noted that the decision to use a cleaner, more complex, more sophisticated, and more costly basic process can be encouraged by the enforcement of stricter pollution controls. The application of "add-on" controls may be more costly and difficult for conventional processing technology than changing to the new technology.

The control of air pollutants from industry and electric power generation will be discussed here, while the control of emissions from automotive internal combustion engines will be considered in Chapter 8. Nonetheless, the different processes that are combined for internal combustion engine controls are merely special applications of industrial control technology.

Gaseous and particulate air pollutants can be considered as two stages in the size spectrum of airborne material: gaseous pollutants are molecular in scale, and particulates range from about $0.05\ \mu$ to $100\ \mu$.* For control of gases, only the diffusion of molecules into a "control region" can be used, while particles can be removed from a gas stream by using one or more of the forces of gravity, inertia, and electrostatic or thermal attraction, as well as Brownian diffusion of the particles. For particles larger than about $1\ \mu$ Brownian diffusion becomes a mechanism of negligible importance.

*The sizes refer to the diameters of particles, treated, if nonspherical, as equivalent spheres.

5. *Control of Gaseous Pollutants*

For gaseous air pollutants to be removed from a carrier gas, they must diffuse to a region or surface. At a solid surface they can be retained or, if the surface is catalytic, changed to a nonpolluting species. At a liquid interface pollutant molecules can be absorbed and carried away, while in a "control region," such as a flame, pollutant molecules can react homogeneously and be transformed into other molecules which are not air pollutants in the accepted sense.

In more specific terms, control of gaseous pollutants consists of either liquid absorption, an adsorption on an inert catalyst surface, or flame combustion. All these are used in industrial air pollution control systems in different applications.

Gas absorption is a standard chemical engineering unit operation, which can be carried out either countercurrently, with gas and liquid flowing in opposite directions, or co-currently, with both fluids flowing in the same direction. When dealing with relatively high concentrations of pollutant gases (of the order of one percent or more), then countercurrent flow is the common practice, since a reasonable concentration difference, resulting in an optimum "driving" force, occurs throughout the whole system, the lowest gas stream concentration contacting the "weakest" liquid concentration (referring to the dilute concentration of gas in the liquid (Fig. 2)). However, in the removal of pollutant gases already in relatively low concentration in the carrier gas, co-current scrubbers, with special design features giving large contact surfaces for short contact times, are often used. These contactors may be spray scrubbers in which the liquid is dispersed in the gas, or vice versa, intimate contact being achieved by having small bubbles of gas in the liquid or small droplets of absorbing liquid in the gas.

The rate of transfer (mass transfer) of the pollutant gas (N_A) to the absorbing liquid is a complex function of the available surface of absorbent (A), the driving force (Δc), and the "mass transfer" coefficient (k_G)

$$N_A = k_G A \, \Delta c$$

The surface area depends on the amount of scrubbing liquid per unit volume of gas, droplet size (for spray units), surface area of packings (for packed towers), and other factors concerned with the physical characteristics of the absorber; the driving force is the concentration difference (Δc) between the pollutant gas in the gas stream and at the surface of the liquid, while the mass transfer coefficient is a function of the diffusion of the gas molecules, the thickness of the "boundary" layer, the concentration difference, and the temperature and pressure of the system.

These parameters can be determined experimentally or calculated, and the methods of sizing and designing suitable gas absorption units can be found in chemical engineering textbooks concerned with industrial gas cleaning, gas absorption, or more generally, with mass transfer processes (e.g., see Kohl and

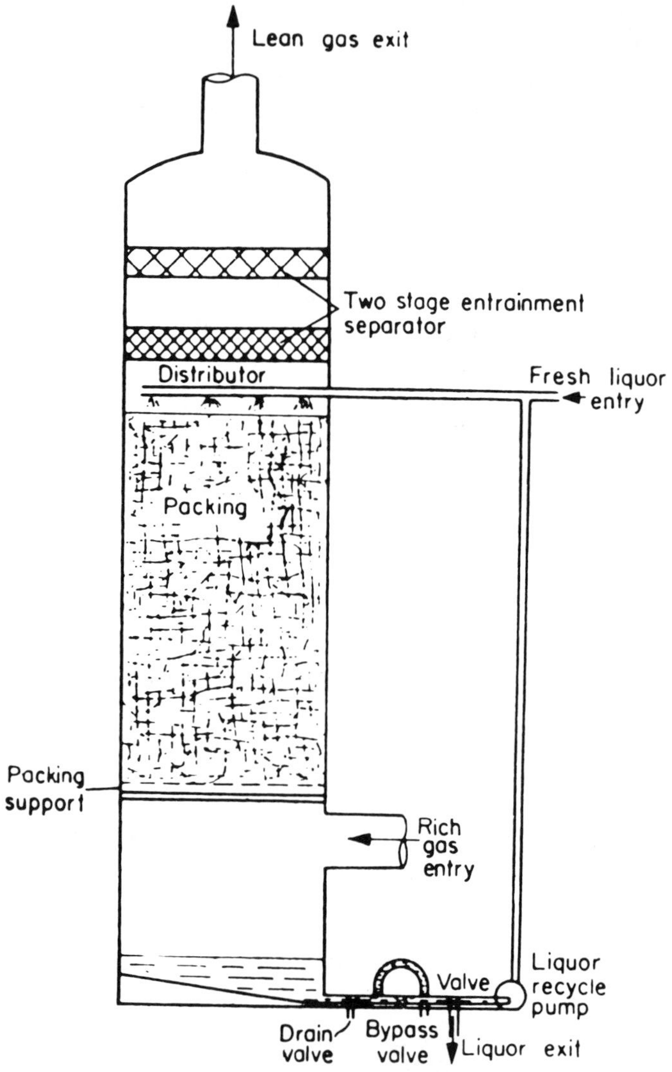

Figure 2. Countercurrent absorption column (packed tower), used in liquid absorption of air pollutants. (From: W. Strauss, *Industrial Gas Cleaning*, Pergamon Press, 1976.)

Riesenfeld[8]). It should be noted, however, that the empirical determination of mass transfer coefficients and an accurate knowledge of system behavior are essential if precise and realistic designs are to be developed.

In general, gaseous pollutants present in concentrations down to 0.1% (by volume), i.e., 1000 ppm, are effectively handled by liquid scrubbing systems. These include hydrogen sulfide, sulfur dioxide, ammonia, nitrogen oxides, and others, particularly hydrocarbons. The gases can often be recovered, possibly by steam distillation ("stripping") and the absorbing liquid recycled. The processes prove economically viable under some conditions, and will be discussed in some detail for specific pollutant gases.

The processes involved in adsorption of a gas on a solid are fundamentally similar to their absorption in a liquid, as the molecules diffuse across a boundary layer to the solid–gas interface. At the solid surface, however, the molecules are retained, and in time build up a concentration such that the concentration difference diffusion driving force is lost, and only an interchange of molecules, rather than net removal from the gas phase, takes place.

The surface available for adsorption may, of course, be very large; in the case of some materials such as the diatomaceous earths or activated carbon, they amount to several square meters per gram for the former, to many hundreds of square meters per gram for the latter. If the air pollutant is not present in large concentrations, as is the case with odor molecules, it may take a long time—many months—before the adsorbent surface is occupied, and the surface must be replaced because it becomes no longer effective.

If the surface is that of a catalyst, the pollutant molecules will settle on its "active sites" for a period, undergo a reaction, and then leave, the site being free for further reactions. After a time, of course, the catalyst loses activity (becomes "poisoned") and has to be replaced and possibly reactivated. Catalyst poisons are substances which rapidly deactivate catalysts.

Some solid adsorbents are replaced continuously, in a countercurrent (or cross-flow) system, and this may lead to systematic recovery of the "pollutant" gas. Such processes are used for sulfur dioxide and hydrogen sulfide recovery. When molecules are small and polar, as in water vapor, ammonia, or sulfur dioxide, a polar surface adsorbent such as diatomaceous earth or silica gel is suitable, while larger, nonpolar molecules, such as organic vapors (including odor molecules) are best adsorbed on a nonpolar surface such as activated carbon. However, activated carbons can be specially made and treated so that they attract some polar molecules (such as sulfur dioxide), and they also have the advantage over polar surface adsorbents of a much greater surface area.

Homogeneous gas reactions can also be used for control of some gaseous pollutants, and in particular for hydrocarbons and organic substances containing oxygen and nitrogen. The common reaction used is of course combustion, where the molecules are oxidized to carbon dioxide and water, with possibly some small quantities of nitrogen oxides. Sulfur and chlorine containing organic materials are not usually treated by combustion because sulfur dioxide, hydrogen sulfide, and hydrogen chloride result.

When using flame combustion it is important that the gases be completely oxidized to CO_2 and H_2O and not to some intermediate organic molecules which could remain a pollution (and possibly an odor) problem. This requires the gases to be retained in the combustion chamber for a reasonable time—0.3–0.5 sec—at the flame temperature. As the whole gas stream has to be heated, the intrinsic calorific value of the pollutant gas is of great importance, since supplementary heat would otherwise have to be provided. Provision of heat exchangers can help reduce the additional heat input, but adds to the cost and size of the equipment. If the intrinsic calorific value of the pollutant gas is such that the reaction is self-supporting, then the heat output can be used for process or space heating.

Catalytic combustion is the use of a heterogeneous catalyst which can reduce the reaction temperature for oxidation from about 800°C to 300–400°C. The catalysts usually used are active materials (frequently metals of the platinum group) supported on a base metal (nickel alloy ribbon) support, or a metal oxide (alumina) support, or active metal oxides, usually copper or manganese oxides.

6. Controls for Specific Gaseous Air Pollutants

6.1. Sulfur Dioxide

Sulfur dioxide, from smelters or from combustion of fossil fuels, is not easily soluble in water, and it is necessary to use alkaline solutions or alkaline solids. Many processes have been suggested, and they can be classified into those that use a cheap reagent which, along with the adsorbed and reacted sulfur dioxide, is subsequently thrown away, and those that use an expensive reagent which can be recovered, in the latter; the sulfur is also recovered and then used for making either pure sulfur dioxide, sulfuric acid, sulfur, or, if ammonia is used, ammonium sulfate.

Limestone ($CaCO_3$), lime (CaO), or dolomite ($CaCO_3 \cdot MgCO_3$) has been added to the gases in the combustion chamber, and the sulfate formed (e.g., Still, Wickert, Jüngten and Peters, Combustion Engineering Processes)[5]

$$CaO + SO_2 + \tfrac{1}{2}O_2 \rightarrow CaSO_4$$

The dry sulfate powder can then be taken out by particle recovery, or can be washed out in a scrubber. Removal efficiencies of 40–70% have been achieved with dry recovery, and over 90% with wet collection. The alternative is using lime or magnesium oxide slurries, the latter producing sulfite (e.g., Mitsubishi, Bahco, Chemico, Combustion Engineering, Nippon Kokan, Research Cottrell)

$$MgO + SO_2 \rightarrow MgSO_3$$

which led to a considerable number of plants used on a large scale in the United States and Japan. Most, however, suffer from mechanical difficulties.

A modification which has led to several successful commercial applications is the "double alkali" process, where a more active alkali, such as ammonia solution, is used to remove the sulfur dioxide, and this is then recovered by using a second alkali, lime, to produce the sulfate $CaSO_4$ either for sale or disposal, and the active alkali, for reuse (e.g., Monsanto Calsox, Envirotech, EDF-Kuhlman, General Motors, etc.).

Ammonia in the vapor state has also been advocated as reacting effectively with SO_2 at 150°C (USBM Process)[6]:

$$2NH_3 + SO_2 + H_2O \rightarrow (NH_4)_2SO_3$$

$$NH_3 + SO_2 + H_2O \rightarrow NH_4HSO_3$$

The gases are then scrubbed with water, and the resultant solution is reacted with NaOH, lime (CaO), or zinc oxide (ZnO) slurries, to recover the ammonia, e.g.,

$$(NH_4)_2SO_4 + 2NaOH \rightarrow Na_2SO_3 + 2NH_3 + 2H_2O$$

$$NH_4HSO_4 + NaOH \rightarrow NaHSO_3 + NH_3 + H_2O$$

Such a process may be well over 95% effective in removing sulfur dioxide.

Water itself, while a poor adsorbent for sulfur dioxide, can be used in combination with a catalyst of active carbon. Here oxidation takes place on the moist catalyst surface, and dilute (10–15%) sulfuric acid is produced (e.g., Lurgi Sulfacid Process) (Fig. 3).

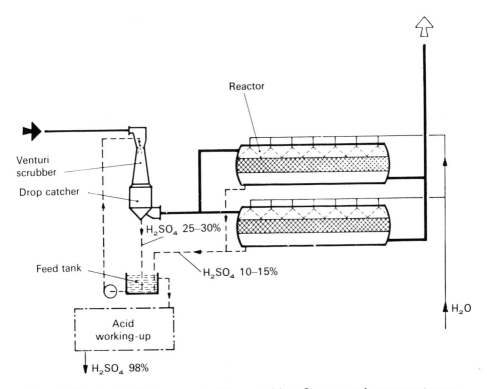

Figure 3. The Lurgi Sulfacid process for SO_2 removal from flue gases and process waste gases. (From: W. Strauss, *Industrial Gas Cleaning*, Pergamon Press, 1976.)

Manganese oxide is also a good adsorbent, and this is being used in several pilot plants (e.g., Mitsubishi, Grillo-AGS), as are sodium sulfite (e.g., Wellman-Lord), sodium citrate (e.g., USBM), and several other adsorbents.

The simplest of all processes is of course the direct oxidation of SO_2 to SO_3

$$SO_2 + \tfrac{1}{2}O_2 \rightarrow SO_3$$

This, the standard catalytic oxidation in sulfuric acid manufacturing, is generally carried out at 450–600°C where the conversion is satisfactory and the rate fast, and is self-supporting if the gases are at 450°C, the reaction temperature. Low concentrations—on the order of 0.1–0.4%—common in flue gases require special catalysts, which have recently been developed (e.g., Monsanto Catox). At lower temperatures, certain activated carbons are also suitable for the oxidation (e.g., Hitashi, Westvaco, Reinluft, Strauss) and sulfur dioxide can be recovered for subsequent use in an acid plant.

Chemically interesting is the process where, after absorption by ammonia, sulfur as well as ammonium sulfate is recovered from an ammonium hydrogen sulfate and thiosulfate solution by a disproportionation reaction carried out in a high-pressure autoclave (e.g., Fulham–Simon–Carves)

$$2NH_4HSO_3 + (NH_4)S_2O_3 \rightarrow 2(NH_4)_2SO_4 + 2S + H_2O \;(170°C, 1.4\,MPa, 3\,hr)$$

Both the sulfur and ammonium sulfate can be used commercially.

A mixture of molten alkali carbonates is also being used for SO_2 recovery (e.g., Atomics International),[7] the mixture being 32% Li_2CO_3, 33% Na_2CO_3, and 35% K_2CO_3 (by weight). These give a eutectic mixture, melting at 397°C, which appears at 450°C as a clear liquid, with a viscosity about ten times that of water at ambient temperatures (9.82 cp). The density at 450°C is 2120 kg/m^3 and surface tension is 236.9 dyn/cm. If the alkali atoms are written as M, the absorption stage is

$$SO_2 + M_2CO_3 \rightarrow M_2SO_3 + CO_2$$

The sulfite can oxidize

$$M_2SO_3 + \tfrac{1}{2}O_2 \rightarrow M_2SO_4$$

After absorption, the sulfate and sulfite are reduced, using carbon (in the form of petroleum coke) at 800°C

$$2M_2SO_3 + 3C \rightarrow 2M_2S + 3CO_2$$
$$M_2SO_4 + 2C \rightarrow M_2S + 2CO_2$$

Some sulfate and sulfide can also arise from the reaction, and is known to be important at 800°C in the reducer

$$4M_2SO_3 \rightarrow M_2S + 3M_2SO_4$$

The carbon dioxide is used to regenerate the carbonate, and, with water vapor at 450°C

$$M_2S + CO_2 + H_2O \rightarrow M_2CO_3 + H_2S$$

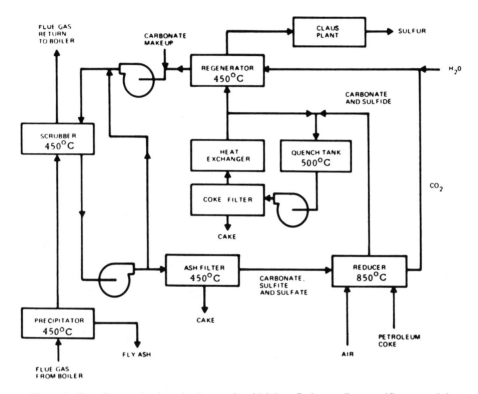

Figure 4. Flow diagram for Atomics-International Molten Carbonate Process. (Courtesy of the American Chemical Society.[7])

The hydrogen sulfide produced is in turn oxidized to sulfur, over a catalyst, in a Claus plant

$$H_2S + \tfrac{1}{2}O \rightarrow S + H_2O$$

Difficulties with this process are the highly corrosive molten carbonates which have to be handled at high temperatures. Special stainless steels are used at the lower temperatures, where a protective $LiCrO_2$ film forms, and at the very high temperatures (to 900°C) the materials of construction are high-density alumina. The lithium carbonate is a relatively expensive material, and considerable care must be taken to reduce losses and recover the lithium from process wastes (Fig. 4).

6.2. Hydrogen Sulfide

Hydrogen sulfide is produced during coal gasification, in biological reactions involving sulfur components, and in some chemical reactions. The removal of H_2S from coal gas (and from natural gas) is important because first, it is highly corrosive and has to be reduced to extremely low concentrations if it is to be transmitted by pipeline, and second, when these gases are burnt, the oxidation product is sulfur dioxide. However, it is chemically easier to take H_2S out of the gas stream before combustion (and dilution with air) than it is afterward, as SO_2. The efficiency of removal is also much higher; while it is possible to reduce SO_2 in waste gases to 500 ppm, with effective (90%) treatment processes, hydrogen sulfide can be easily reduced to 1–5 ppm, which results in less than 1 ppm SO_2 after combustion.

Coal gas also contains some ammonia (about 1%), but treatment processes involving the coal gas ammonia alone remove less than half the H_2S present. However, with additional ammonia, hydrogen sulfide can be effectively removed (e.g., Collin process). The gases need, however, to be further scrubbed to remove entrained ammonia. The product is usually ammonium sulfate and H_2S, which is by this time in sufficiently high concentration to be used for sulfur recovery by the Claus process.

As an alternative to ammonia, the ethyl- and methylamines are used in a range of processes. As the reagent is relatively expensive, it is recovered after use as an absorbent by steam stripping, and is then recycled (Girbotol processes). The first compound to be used was triethanolamine (($(C_2H_4OH)_3N$)(TEA), but this proved to be difficult to regenerate, and so a 15–20% aqueous solution of monoethanolamine (MEA) was brought into use. Its disadvantage is largely that some reagent is lost because a heat-stable compound is formed with carbonyl sulfide (COS), diethanol urea ($CO(NHCH_2CH_2OH)_2$). MEA is therefore used for natural gas purification, when virtually no COS impurity is present, while diethanolamine (DEA) is used for coal and refinery gases.[8]

In a number of processes it is possible to obtain sulfur directly in the regeneration stage, using atmospheric oxygen as the oxidizing agent. In the first series (Thylox, Giammarco-Vetrocoke), sodium thioarsenates and complex arsenic compounds are used. The absorption stage is

$$H_2S + Na_4As_2S_5O_2 \rightleftharpoons Na_4As_2S_6O + H_2O$$

and regeneration:

$$2Na_4As_2S_6O + O_2 \rightleftharpoons 2NA_4As_2S_5O + 2S$$

In the Vetrocoke process, the arsenite-to-thioarsenite reaction is used, and the thioarsenite is subsequently decomposed using carbon dioxide under pressure.

A very widely applied process in recent years uses organic dyestuffs in the critical oxidation–reduction stages (Stretford process). The hydrogen sulfide (in concentrations of 10–10,000 ppm (1%)) is first absorbed in an alkaline solution maintained at pH 8.5–9.5, and containing sodium carbonate, sodium vanadate, anthraquinone 2:6, and 2:7 disulfonate (ADA). Rochelle salt (potassium sodium tartrate) is added as a sequestering agent to prevent precipitation of the vanadate. The reactions are as follows:

absorption: $H_2S + Na_2CO_3 \rightarrow NaHS + NaHCO_3$
sulfur production: $2NaHS + H_2S + 4NaVO_3 \rightarrow Na_2V_4O_9 + 4NaOH + 2S$

recovery of vanadate:

$$Na_2V_4O_9 + 2NaOH + H_2O + 2ADA \rightarrow 4NaVO_3 + 2ADA \text{ (reduced)}$$

oxidation of the ADA: $2ADA \text{ (reduced)} + O_2 \rightarrow 2ADA + 2H_2O$

The process plant, in spite of these complex reactions, is quite simple, a packed column being used for the absorption stage, followed by a delay tank and a sulfur extraction plant. The exit concentrations of hydrogen sulfide are of the order of 1 ppm.

Ferric oxide is the other reagent material that has been widely used for hydrogen sulfide removal. In aqueous suspension, the ferric oxide acts largely as a catalyst (e.g., Ferrox, Gluud, and Manchester processes)[9]. Again, the suspension passing down a packed column is the absorbent, and the sulfur is recovered by blowing atmospheric oxygen through the sulfide-containing solution. Unfortunately, the sulfur from this process tends to be contaminated with iron oxide, reducing its subsequent usefulness. Ferric oxide, deposited in a finely divided state on peat or fibrous material, or in pellets, has long been used in "oxide boxes" for the removal of small amounts of hydrogen sulfide. The oxide may be subsequently recovered by combustion of the carrier material or by solvent extraction, using, for example, perchlor-ethylene. Here, in general terms, the reactions are

$$\text{absorption: } 6H_2S + 2Fe_2O_3 \rightarrow 2Fe_2S_3 + 6H_2O$$
$$\text{regeneration: } 2Fe_2S_3 + 6O_2 \rightarrow 2Fe_2O_3 + 3SO_2$$

The reactions can also be carried out at high temperatures (e.g., Appleby–Frodingham process) (Fig. 5). Hydrogen sulfide, at concentrations of 14 g/m^3 in gas, has been reduced to 0.01–0.03% by passing the gas through a four-stage fluidized bed maintained at 340–360°C. The recovery stage is achieved by heating the ferric sulfide to 800°C in air. The sulfur dioxide driven off can be

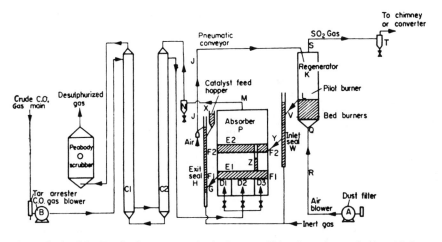

Figure 5. Appleby–Frodingham process for hydrogen sulfide adsorption on ferric oxide in a fluidized bed at 350°C. (From: W. Strauss, *Industrial Gas Cleaning*, Pergamon Press, 1976.)

used in making sulfuric acid. Unfortunately, in coal gas, hydrogen sulfide is usually accompanied by organic sulfides and thiophene, and the degree of removal is 70–80% for the sulfides and only 30–45% for thiophene. By placing a pretreatment stage into the system, which turns the sulfides and thiophenes to hydrogen sulfide, using special catalysts (chromic alumina, cobalt–molybdenum, zinc oxide, etc.), the overall process can be improved. There are many other processes that are used for hydrogen sulfide removal, and they are discussed at length in Yosmin *et al.*,[7] and Kohl and Riesenfeld.[8]

6.3. *Ammonia, Amines, and Pyridine Bases*

These are all formed in coal carbonization, and the former two are also found in the waste gases from rendering of meat, offal, fish, etc., and are characterized by their pungent and nauseous odor. Ammonia and the amines are very soluble in water and strong oxides, because of the reaction giving ammonium or alkyl ammonium ions. However, the solubility of these compounds in water falls off rapidly with rising temperature. In gas works practice one method is therefore a two-stage system, the first being a cooler of some sort (shell and tube exchanger or spray cooler) where the gases are cooled to 30–50°C, followed by a water scrubber. The other system is a single-stage absorber and crystallizer, into which sulfuric acid is sprayed, ammonium sulfate crystallizing out in the bottom. The pyridine bases (pyridine, picoline, aniline, and quinolene) are removed in the same process.

When ammonia and the ethylamines, and other amines in smaller concentrations, are produced in rendering (steam heating) of animal and fish wastes, they can be largely removed by dissolving them in the water given off by the rendering, when this is condensed in shell and tube condensers of special design.

In all these cases the final traces of the amines (although not ammonia itself), can be efficiently removed by passing the waste gases through deep beds of activated carbon.

6.4. *Nitrogen Oxides*

There are six stable oxides of nitrogen (N_2O, NO, NO_2, N_2O_4, N_2O_3, and N_2O_5), and of these, NO is mainly produced by combustion, while NO_2, N_2O_3, and N_2O_5 are given off in some industrial operations such as pickling of metals prior to other treatment such as anodizing. The volumes from the latter processes are comparatively small, and scrubbing the gases with alkaline solution (calcium hydroxide, sodium hydroxide, sodium carbonate, ammonium carbonate, etc.) has been tried, but only with limited success, as few processes remove more than 70%. What does appear promising are those systems where the gases are passed rapidly over the liquid surface on special wave-formed plates, and such systems have reported removal efficiencies around 95%. Systems where the gases are passed through moist activated carbon also appear to be effective, although removal efficiencies are not quoted.

When nitrogen oxides are produced by combustion, the very large volumes and low concentrations preclude their removal by scrubbers. However, it has been found possible by careful attention to burner design and operation (using tangential instead of opposed firing), two-stage combustion, partial recirculation of combustion gases (up to 25%), and in some cases, derating the boiler, to reduce oxide of nitrogen production from well over 1000 ppm to 42 ppm (required under Rule 67, Los Angeles Air Pollution Control District).

6.5. *Fluorine and Fluorides*

Very low concentrations of fluorides (of the order of parts per thousand million) can damage vegetation, and so the removal of fluorides from waste gases from aluminum electrolysis plants, fertilizer plants, etc., is essential. Hydrogen fluoride is very soluble in water, and simple counterflow scrubbers, wet cell washers consisting of a succession of packed boxes filled with straw and other materials and sprayed with water, crossflow packed scrubbers, and floating bed scrubbers where plastic balls are fluidized in air and water, have all been found effective.

When fluorine is present, water should be avoided, and a 5–10% sodium hydroxide solution is substituted. In aqueous and very low caustic (less than 2%) solutions, the very poisonous fluorine oxide, OF_2, is formed. In packed towers, with liquid gas contact times of about 1 minute, the following reactions occur:

$$F_2 + 2NaOH \rightarrow \tfrac{1}{2}O_2 + H_2O + 2NaF$$

$$HF + NaOH \rightarrow NaF + H_2O$$

Sodium fluoride is then treated with lime to regenerate the hydroxide

$$2NaF + CaO + H_2O \rightarrow CaF_2 \text{ (ppct)} + 2NaOH$$

A very elegant method of recovering gaseous and particulate fluorides from aluminum smelters has been developed, where the waste gases are passed through a fluidized bed of alumina—which will subsequently be used in the smelter—and so return the emitted fluoride (Alcoa process A398). The temperature of absorption is 65–160°C, and the bed thickness 50–300 mm. The efficiency of the process is 99.2% or better for gaseous fluorides, and over 90% for particulate fluorides, with outlet concentrations being reduced from 100–200 mg/m^3 to less than 3 mg/m^3.

6.6. *Odors*

Unlike the gaseous air pollutants discussed above, odors generally occur in very low concentrations—parts per million or less—and must be reduced to much lower concentrations. Two methods are commonly employed, besides scrubbing, which was considered with reference to the amines; these are, either adsorption on activated carbon, or combustion in a flame or on a catalyst surface.

Adsorption of odors on activated carbon is widely used, for food and cooking smells, for the remaining traces of amine, and for organic materials such as mercaptans, cresols, butyric acid, etc. The sources of active carbon can be coconut shells, wood, brown coal, coal, and other carbonaceous materials, which undergo destructive distillation and subsequent treatment by steam at elevated temperatures. The adsorptive capacity depends on the surface area of the carbon (200–1200 m^2/g), the temperature of adsorption, and the material. Usually, active carbon beds last from a few weeks to several months before they require renewal.

The activated carbon can be reactivated by heating in an inert gas to 600–650°C and then treating with steam. However, the reactivated material is unlikely to be as satisfactory as the original carbon.

Flame combustion is widely used for disposal of waste hydrocarbons, odors, etc. In oil refineries and chemical works, where the (fume energy) concentration of the material is such as to produce self-supporting combustion, "flares" are used, but if these are fed with an aromatic material and air they tend to burn with a smoky flame. This is overcome by injecting steam into the flame which results in the water–gas reaction, producing hydrogen and carbon monoxide, both of which burn smokelessly (Fig. 6). Steam-injected flares tend to be noisy, and considerable attention is now being paid to developing efficient flares with very limited noise.

An alternative to the open flare is the closed combustion chamber, which can also be used if the fume energy in the pollutant gas, such as odor, is not sufficient to produce a self-supporting flame. In some cases, such as the black liquor drying odors in Kraft mill black liquor recovery, these can be passed through the combustion chamber of a boiler. Great care has to be taken, however, so that corrosion of boiler tubes is not accelerated, or the combustion temperature reduced so to make the unit ineffective. Special fume incinerators using diffusion or premixed flames have been developed, in which adequate retention time for complete oxidation is achieved.

Catalytic combustion[10] has the advantage that it occurs at 300–400°C instead of at more than twice this temperature, as in direct-flame incineration, and so the problems of high-temperature heat exchange, and additional energy input in the case of low-fume energy, are reduced. Unfortunately, the catalysts used are expensive, and need occasional regeneration due to "poisoning" (deactivation). This can be due to iron oxide, silica, or alumina dusts which permanently block the catalyst, or carbon, which can be burned off at intervals by heating the catalyst to a higher temperature. Some compounds can deactivate specific catalysts; for example, chlorides form cuprous chloride with copper oxide catalysts, sulfur compounds affect alumina catalysts, and organic phosphorus compounds from lubricating oils form phosphoric acid. Heavy metals also deactivate by covering the active sites. Thus, the use of catalytic oxidation should be restricted to those applications where there is little possibility of catalyst poisoning, as in treating the vapors from coffee roasting, bread baking, varnish cooking, etc.

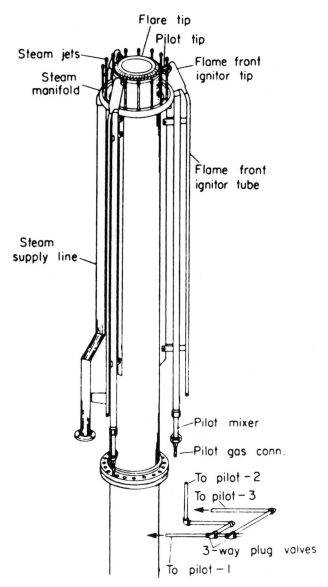

Figure 6. Steam injection flare for smokeless hydrocarbon combustion. (From: W. Strauss, *Industrial Gas Cleaning*, Pergamon Press, 1976.)

7. *Control of Particulate Air Pollutants*

The first widely recognized form of air pollution was smoke, the fine particulate matter consisting partly of entrained ash in fuels and partly of incomplete combustion. Even quite heavy smoke—with concentrations of the order of 2–3 g/m^3—represents only a volume concentration of about 1 ppm. Clean air regulations in most countries require abatement to concentrations of 0.1 g/m^3 or less.

In the most general terms, the control of particulate emissions involves passing the gases through a chamber in which a force acts on the particles and removes them from the gas stream. The force may be gravitational, electrostatic, thermal, centrifugal, or inertial, while for very small particles (less than 0.5 μ) the particle can also diffuse out of the gas stream. While the chamber can have one of a number of configurations, in nearly all cases the model for the collection process is the speed of the cross-stream movement of the particles, under the action of one of these forces, against the resistance of the gas during the residence time of the gas in the chamber. Thus, particles of the size (as measured by their falling speed) able to traverse the whole depth of the chamber (or its equivalent) are likely to be fully collected, while smaller particles are collected in proportion to the fraction of the depth of the chamber they are able to cross during the gas stream residence time.

Except for particles larger than about 50 μ, the gas resistance is usually simplified to a Stoke's law resistance, and turbulent deposition, which may enter very sophisticated models, makes only a minor contribution to most collecting systems. The "depth" of the chamber can in some cases imply an actual chamber, but in most calculations this is an equivalent distance specific to the apparatus, such as the distance from wire to plate in an electrostatic precipitator, or the width of the entry into a cyclone.

7.1. Settling Chambers

In simple gravitational settling chambers (Fig. 7), which may simply be an enlargement in a duct carrying the gases, the efficiency of the chamber depends on the residence time of the gases in the chamber and the distance the particles travel under a gravitational force. The residence time in turn depends on the

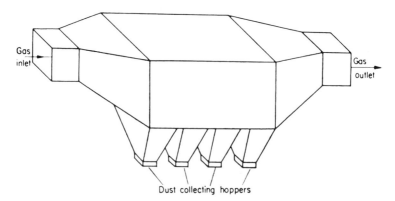

Figure 7. Simple gravity settling chamber. (From: W. Strauss, *Industrial Gas Cleaning*, Pergamon Press, 1976.)

chamber volume and on the volume flow. It is furthermore essential that the gas flow velocity is below that for reentrainment, which for most materials, except some light flaky substances, is greater than 3 m/sec.

The simple settling chamber can be modified by inserting trays, which reduces the distance particles travel before collection. However, the trays are difficult to clean, so that the "Howard" multitray settling chamber has not proved popular.

7.2. Electrostatic Precipitators[11,12]

A similar theoretical concept can be applied to a plate-type electrostatic precipitator, but here the problems of the multiplate settling chamber have been overcome. First, the relatively weak gravitational force is replaced by the much stronger electrostatic force, acting on particles which have been charged, and these act horizontal to the gas stream into which vertical collection plates have been inserted. Second, after deposition of a layer of dust on these vertical plates, they are easily removed by shaking (or hitting) the plates, allowing the dust layer "cake" to fall down.

In practice (Fig. 8), in an electrostatic precipitator, wires are equispaced between plates, arranged at about 200–300 mm apart and in line with the direction of the gas stream. The wires are charged negatively, in industrial practice, to 40–60 kV, and the plates are earthed. If electrostatic precipitation is used for cleaning the air in air-conditioning systems, positive-charged wires (16 kV) are used, as these produce far less ozone and nitrogen oxides, which would be unpleasant in conditioned air.

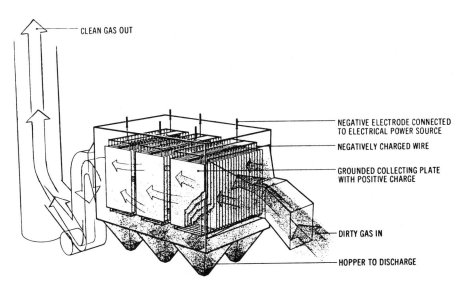

CLEAN GAS OUT

NEGATIVE ELECTRODE CONNECTED TO ELECTRICAL POWER SOURCE

NEGATIVELY CHARGED WIRE

GROUNDED COLLECTING PLATE WITH POSITIVE CHARGE

DIRTY GAS IN

HOPPER TO DISCHARGE

Figure 8. Negative corona, cross-flow plate type electrostatic precipitator. (From: *Controlling Air Pollution*, courtesy of the American Lung Association.)

In an electrostatic precipitator the gases immediately around the high-voltage discharge wire are broken down electrically, and gas ions are formed (corona). The gas ions move toward the earthed plates, colliding with dust and other particles and charging them. The charge acquired is a function of the field strength, the surface area of a particle, and its dielectric constant. When the particle has been charged either by bombardment or by diffusion of the gas ions, it then migrates to the earthed plates. The velocity of this migration is a function of the charge on the particle, the field strength (the driving force), and the gas resistance. For a particle in the micron sizes it is of the order of 10–100 mm/sec, which must be compared with the much lower velocity—on the order of 0.1 mm/sec—of such a particle in a gravitational field. The particles, on settling on the plates, are discharged. When a cake has been formed, it is removed by hammering the plates, a technique called "rapping."

This model of electrostatic precipitation is much simplified, and ignores effects such as the electric wind (the resultant effect of the moving gas ions); the problems involved in a charge build-up in the layer of deposited dust, which occurs when this is a poor conductor such as cement dust, or fly-ash from some low-sulfur coals; and the reentrainment of dust dislodged during "rapping."

Electrostatic precipitation is widely used for gas cleaning on a large scale in industry, outstanding examples being cement kilns and electricity generation boilers.

7.3. Cyclones[13]

In cyclones, particles are spun in a rotating gas stream, and because of the centrifugal force on the particles they move to the walls, where they are removed by a secondary movement of the gas stream. There are two basic types of cyclones. The first is the straight-through type, where gases are spun by a bladed stator, and the dust is then concentrated in the outer layers of the gas as this spins through a chamber. The outer layer, with concentrated and partially agglomerated particles, is stripped off and led to a secondary collector, such as a settling chamber or a "reverse-flow" type cyclone.

In the second type of cyclone, which is the conventional "reverse-flow" type (Fig. 9), the spinning gases are first moved toward a hopper (or other collector), and then the gases, still spinning in the same direction, are reversed and led through a centrally placed exit pipe. The particles from the outer layer settle in the hopper. The gas entry can be through either a bladed, in-line stator, as in the first type, or, more frequently, through a tangential entry.

There are three ways of calculating the effectiveness of a cyclone. The first is the model of a settling chamber, with a cross section equivalent to the tangential entry and a length equivalent to the "screw" path of the gas stream along the wall. In this "settling" chamber a force equal to the centrifugal force on the particle acts on the mean position of the particle at the entry. This model, first used in 1932, enables a range of efficiencies for particles of different sizes to be predicted.

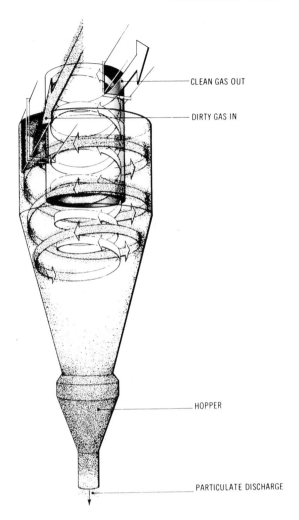

CLEAN GAS OUT

DIRTY GAS IN

HOPPER

PARTICULATE DISCHARGE

Figure 9. Reverse-flow cyclone collector. (From: *Controlling Air Pollution*, courtesy of the American Lung Association.)

The two other theories for predicting cyclone efficiencies consider, first, the movement of the gas stream as a vortex, and second, as a particle held in balance between its intrinsic capacity to move towards the wall and the inward movement of the gases toward the exit. Both theories require empirical correcting parameters, which enable the theoretical approaches to agree with experimental determinations of efficiency.

Cyclones are widely used to remove particles above 10 μ; they have the advantage of being simple to construct, relatively small in comparison with the volumes of gas handled, and they have no moving parts. As a result, they can be built from a wide range of materials that are able to withstand high temperatures and erosive conditions. A problem with cyclones is that about 40% of the energy input is converted to rotational energy, which, at present, is not recovered. However, special shapes have been developed which can reduce this by an appreciable amount.

7.4. *Filtration and Scrubbing*

In both gas filtration and scrubbing for removing small particles, the inertial forces possessed by the particles play a vital part. The gas stream passes by the fiber (in a filter) or droplet (in a scrubber), and in passing the streamlines are diverted. The inertia possessed by the particles enables them to cross the gas streamlines and impinge on the collecting droplet or fiber. As a secondary mechanism, the collector also intercepts the particle, and very small particles (of the order of 0.5 μ and less) diffuse, to a significant extent, to the surface of the collector. All of these mechanisms act simultaneously, and they can be assisted (or hindered) by electrostatic and thermal forces acting between the particle and the collector.

In general, the smaller the collecting body, the more effective it is likely to be as an interceptor. Thus the most efficient fabric and fiber filters consist of very fine fibers, of sizes 5–10 μ, of spun glass, of asbestos, or of synthetic fibers. Once particles have reached the fiber surface they may bounce back into the gas stream or they may be retained on the fiber surface, mainly by van der Waals' forces, although electrostatic forces and surface tension forces can also play a part. Once deposited, the particles as well as the fibers can act as interceptors for other particles in the gas stream.

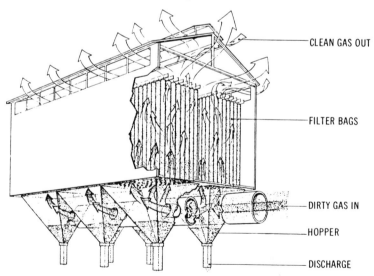

CLEAN GAS OUT

FILTER BAGS

DIRTY GAS IN

HOPPER

DISCHARGE

Figure 10. Industrial fabric filter with tube sleeves. (From: *Controlling Air Pollution*, courtesy of the American Lung Association.)

In time a cake can build up on the fiber surface which is, at intervals, shaken or blown off. For filtering particles in low concentrations, as, for example, in atmospheric air, a deep bed of fine fibers is very effective. This is kept in position, without particle removal, for periods on the order of several months. On the other hand, high concentrations of particles in industrial waste gases are deposited on filter cloths formed into tubes or into sleeves supported on wire (Fig. 10). The dusts are deposited on these in a cake, which is, eventually, the effective method of filtration. At intervals, the cake is blown or shaken off. Filtration efficiency immediately after this is somewhat lower than the maximum because of the time required for the cake to reform. On the other hand, the pressure loss at this stage is low and the filter is able to handle a large throughput. In practice, an optimum has to be found between the permissible pressure loss, the volume of gas handled, and the efficiency, as well as the size, of the unit.

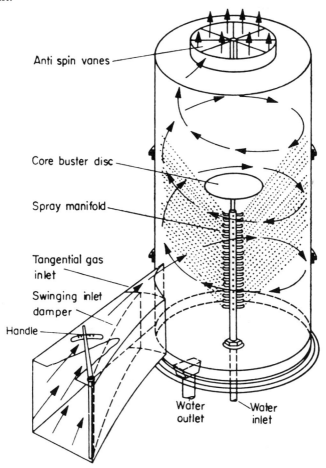

Figure 11. Cyclonic spray scrubber. (From: W. Strauss, *Industrial Gas Cleaning*, Pergamon Press, 1976.)

In liquid scrubbing[14] for particles, these units can range from simple countercurrent spray towers to those where the gas stream breaks through a wall of water. In all cases, the finer the droplets used (or produced in the scrubber), the more effective the scrubber. However, fine particles are entrained in the gas stream and must be removed by some method also used for particles. In simple spray towers the optimum droplet size is 800 μ, which is not effective in removing particles smaller than about 50 μ, or on the order of one magnitude smaller. In a cyclonic spray scrubber (Fig. 11), where the droplets are sprayed into a cyclone, the optimum droplet size is about 100 μ, and particle collection is again effective down to one order of magnitude smaller (i.e., 10 μ). Extensive experiments have shown that secondary effects, such as evaporation or condensation of the liquid, play virtually no part in the collection because of the brief period of contact.

The most effective scrubbers for fine particles are the venturi scrubbers, where the gas stream breaks a curtain of water at high velocity. The small droplets formed, the high degree of turbulence, and possibly other effects such as sweep diffusion, enable even submicron particles to be collected. However, the energy consumption of these units is very high.

All of these collectors for particles are widely used in industry: in the chemical industry, in minerals extraction, etc., whenever fine particles are produced. A general conclusion is that the finer particles generally require larger, technologically more advanced equipment, with higher energy consumption. Much can be achieved by process modifications that produce less air pollution. However, this may require further technical developments, or new processes, rather than an added-on control system.

References

1. C. E. Junge, *Air Chemistry and Radioactivity*, Academic Press, New York, 1963, pp. 1–4. H. Israel, *Trace Elements in the Atmosphere*, Ann Arbor Science Publishers, 1973, pp. 5–8.
2. J. B. Edwards, *Combustion: Formation and Emission of Trace Species*, Ann Arbor Science Publishers, 1974.
3. D. J. Patterson and N. A. Heinein, *Emissions from Combustion Engines and their Control*, Ann Arbor Science Publishers, 1972.
4. T. R. A. Davey, "The Sulphur Problem and the Non-Ferrous Metal Industries in Australia," *Process and Chem. Eng.* (Aust.) **26**(8) : 20–27 (1973); **26**(9) : 9–11 (1973).
5. R. J. Gleason and F. Heacock, "Limestone wet scrubbing of sulphur dioxide from power generation flue gases for high and low sulphur fuels," in *Pollution Control and Energy Needs*, R. M. Jimeson and R. S. Spindt, eds., A.C.S. Advances in Chemistry Series 127, Washington, D.C., 1973, pp. 152–160.
6. C. C. Shale, "Ammonia Injection: A Route to Clean Stacks," in *Pollution Control and Energy Needs*, R. M. Jimeson and R. S. Spindt, eds., A.C.S. Advances in Chemistry Series 127, Washington, D.C., 1973, pp. 195–205.
7. S. J. Yosmin, L. F. Grantham, D. E. McKenzie, and G. C. Stegmann, "The Chemistry of the Molten Carbonate Process for Sulphur Oxides Removal from Stack Gases," in *Pollution Control and Energy Needs*, R. W. Jimeson and R. S. Spindt, eds., A.C.S. Advances in Chemistry Series 127, Washington, D.C., 1973, pp. 174–182.

8. A. L. Kohl and F. C. Riesenfeld, *Gas Purification*, McGraw-Hill, New York, 1960, pp. 18–86.
9. Ibid., pp. 286–303.
10. G. Vollheim and G. Domin, *Catalytic Afterburning: Theory and Application*, Schilde AG, Bad Hersfeld, 1966.
11. H. E. Rose and A. J. Wood, *An Introduction to Electrostatic Precipitation*, Constable & Co., London, 1956.
12. H. J. White, *Industrial Electrostatic Precipitation*, Addison Wesley, Reading, Mass., 1963.
13. R. Jackson, *Mechanical Equipment for Removing Grit and Dust from Gases*, B.C.U.R.A., Leatherhead, 1963.
14. E. Weber and W. Brocke, *Nassabscheider* in *Apparate und Verfahren der Industriellen Gasreinigung*. R. Oldenburg, München, 1973, pp. 174–308.

Further Reading

General

Books

W. Strauss, *Air Pollution*, E. Arnold, London, 1976.
A. C. Stern, ed., *Air Pollution*, 3rd ed., 4 volumes, Academic Press, N.Y., 1976.
W. Strauss, ed., *Air Pollution Control* (Series), Wiley–Interscience, New York, Pt. I : 1971, Pt. II : 1972, Pt. III : 1976, Pts. IV & V in preparation for publication in 1977 and 1978.

Many other books on air pollution have been published and cover the ground in slightly different ways, depending on the author's specialization.

Journals

Journal of the Air Pollution Control Association (4400 Fifth Ave., Pittsburgh, Pa. 15213).
Environmental Science and Technology (American Chemical Society, 1155 Sixteenth St., N. W., Washington, D.C. 20036).
Clean Air (journal of the Clean Air Society of Australia and New Zealand, 1 Fernhill Avenue, Epping, N.S.W. 2121, Australia).

Sources of Air Pollution

R. N. Schreve, *Chemical Process Industries*, 3rd ed., McGraw-Hill, New York, 1967.
A. C. Stern, ed., *Air Pollution*, Pt. III (2nd ed.) or Pt. IV (3rd ed.), Academic Press, New York, 2nd ed: 1968, 3rd ed: 1975.
Publications covering different industries, published by the U.S. Environmental Protection Agency, NERC, Research Triangle Park, N.C. 27711.

Control of Air Pollution

W. Strauss, *Industrial Gas Cleaning*, 2nd ed., Pergamon Press, Oxford (1976).
W. Strauss, ed., *Air Pollution Control*, Pts. I, II, IV, and V.
A. C. Stern, ed., *Air Pollution*, 2nd ed., Vol. III, or 3rd ed., Vol. IV.

Publications covering control processes, published by the U.S. Environmental Protection Agency. In particular,

John A. Danielson, ed., "Air Pollution Engineering Manual," 2nd ed., 1975. (Initially compiled for the Air Pollution Control District, County of Los Angeles.)

E. Weber and W. Brocke, *Apparate und Verfahren der Industriellen Gas Reinigung,* Pt. I "Feststoffabscheidung," Pt. II. "Niederschlagung von gasförmigen Komponenten," R. Oldenburg Verlag, Munich 1973–75.

Robert M. Jimeson and R. S. Spindt, "Pollution Control and Energy Needs," *Advances in Chemistry Series No. 127,* A.C.S. Publications, American Chemical Society, 1155 Sixteenth St., N.W., Washington, D.C. 20036.

The formation and control of air pollutants, the subject of Chapter 7, tells half of the pollution story in the sense that we have to find out what pollutants do to the earth. For example, are they noxious like sulfur dioxide, or are they insidious like carbon dioxide which may not have an effect for decades and could eventually alter the climate in a decisive way?

Dr. Spedding has written an article in which he discusses these actions of gaseous pollutants with materials on the surface of the earth, and his chapter must be studied particularly in conjunction with that of Dr. Strauss.

8

The Interaction of Gaseous Pollutants with Materials at the Surface of the Earth

D. J. Spedding

1. Introduction

Gaseous materials released into the atmosphere as the result of the activity of man suffer a variety of fates, depending upon their chemistry in the atmosphere, and this in turn influences their lifetime in the atmosphere. Sulfur dioxide, for example, has a mean lifetime in the atmosphere of only a few days, while methane has a mean lifetime of about 20 years.

One of the ways of removing a gaseous pollutant from the atmosphere is through interaction with materials at the surface of the earth. Such interactions may result only in the destruction of the pollutant with no obvious damaging effects on the environment, or, on the other hand, the interactions may lead to quite noticeable effects on the environment. In the following sections of this

D. J. Spedding • Department of Chemistry, University of Auckland, Auckland, New Zealand

chapter some rather arbitrarily selected interactions of gaseous pollutants at the earth's surface will be discussed to illustrate both types of interaction.

2. The Oceans: Sinks and Sources of Atmospheric Gases

2.1. Introduction

Since the oceans occupy more than 70% of the total surface area of the earth, interactions at the surface of the ocean are of potential importance for all pollutants that are widely dispersed in the troposphere. If a gas is to be transferred from the troposphere, it must first diffuse in the gas phase to the surface of the ocean, cross the gas–liquid boundary, and finally diffuse into the bulk of the ocean. There are two basic theoretical models that can be used to predict these diffusion processes: the stagnant film model, and the more recent film replacement model. It has recently been shown that these two models produce essentially similar results[1]; the commonly used stagnant film model will be discussed here.

2.2. Theory

In the stagnant film model the bulk of both phases is taken as being well mixed by the turbulent diffusion, but at either side of the boundary between the phases there is a thin film of fluid through which transfer can take place only by molecular diffusion (see Fig. 1). It is the rate of transfer through these layers that determines the overall rate of transfer from one phase to the other.

The flux J of gas passing through either of the boundary layers is described in the general case by Fick's first law of diffusion:

$$J = -D\frac{\delta c}{\delta x} \tag{1}$$

where J is the number of moles of gas crossing the unit area of a plane perpendicular to the direction of flow in unit time; the concentration gradient, $\delta c/\delta x$, is the rate of increase of concentration c, with distance x, measured in the direction of the flow; and D is the diffusion coefficient. In gas exchange studies, where it is assumed that the transfer across the interface is a steady-state process, equation (2) is the form of Fick's law that is used:

$$J = D\frac{\Delta c}{x} \tag{2}$$

where Δc is the concentration difference measured from the interface to the outer edge of the boundary layer, at distance x from the interface. A parameter often found in gas exchange studies is the gas exchange constant, k, which is defined as

$$k = D/x \tag{3}$$

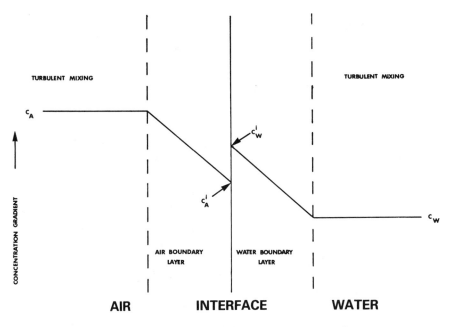

Figure 1. Diagram illustrating the stagnant film model.[2]

Equation (2) may thus be written in simplified form as

$$J = k \Delta c \qquad (4)$$

The concentration of the gas being transferred across the boundary layer is c_A in the bulk air and c_W in the bulk water. It is assumed that these concentrations are constant up to the edges of the respective boundary layers. The concentration of the transferring gas in the gas phase at the interface is c_A^i, which may be greater or less than c_A, depending upon whether the flux of gas is out of or into the water. The corresponding liquid phase concentration of the gas in the liquid phase at the interface is c_W^i. Since, in a steady-state process, the flux of gas across both boundary layers is identical, equation (5) arises from the equality of equation (4) for each layer

$$J = k_g(c_A - c_A^i) = k_l(c_W - c_W^i) \qquad (5)$$

where k_g and k_l are the exchange constants for the gas and liquid phases, respectively.

There also exists a relationship between c_A^i and c_W^i, assuming that the gas involved in the transfer obeys Henry's law (H is the Henry's law constant):

$$c_A^i = H c_W^i \qquad (6)$$

Using equation (6), c_A^i can be eliminated from equation (5) and following an algebraic manipulation to eliminate c^i_W,

$$J = \frac{k_l k_g (c_A - Hc_W)}{k_l + k_g H} \tag{7}$$

Rearrangement of equation (7) gives the equalities expressed as

$$J = \frac{c_A - Hc_W}{1/k_g + H/k_l} = \frac{c_A/H - c_W}{1/k_l + 1/Hk_g} \tag{8}$$

Finally, if the denominators in equation (8) are replaced by $1/K_g$ and $1/K_l$, it can be written as

$$J = K_g(c_A - Hc_W) = K_l\left(\frac{c_A}{H} - c_W\right) \tag{9}$$

It is helpful to visualize the transfer of a gas from one phase to another as being subjected to a resistance R to its transfer. This resistance is analogous to electrical resistance and, continuing the analogy, is equivalent to the reciprocal of the conductance which in this case is the transfer coefficient. As we have seen, the transfer coefficient may be expressed either on a gas phase basis, K_g, or on a liquid phase basis, K_l. The total resistance R to gaseous transfer is made up of two components: resistance in the gase phase, r_g, and resistance in the liquid phase, r_l. The relationship between the parameters in equation (8) and these resistances is expressed as

$$R = 1/K_g = 1/K_l = (1/k_l + 1/Hk_g) = (r_l + r_g) \tag{10}$$

The relative contributions of r_l and r_g to R depend to a considerable extent on the solubility and solution chemistry of the transferring gas. In almost all cases one or other of r_g or r_l predominates and this has a profound effect on the behavior of the gas at the surface of the ocean. We will now consider two pollutant gases whose properties are such that r_g dominates R (sulfur dioxide), and r_l dominates R (carbon monoxide).

2.3. *Sulfur Dioxide Transfer*

In order to understand the transfer of sulfur dioxide from the troposphere to the ocean, the solution chemistry of SO_2 must be considered:

$$(SO_2)_g + H_2O \rightleftharpoons (SO_2)_{aq} \tag{11}$$

$$(SO_2)_{aq} + H_2O \rightleftharpoons H_2SO_3 \tag{12}$$

$$H_2SO_3 + H_2O \overset{K_1}{\rightleftharpoons} H_3O^+ + HSO_3^-; K_1 = 1.6 \times 10^{-2} \tag{13}$$

$$HSO_3^- + H_2O \overset{K_2}{\rightleftharpoons} H_3O^+ + SO_3^{2-}; K_2 = 1.0 \times 10^{-7} \tag{14}$$

$$2HSO_3^- \overset{K_3}{\rightleftharpoons} S_2O_5^{2-} + H_2O; K_3 = 7 \times 10^{-2}\,\text{mol}^{-1} \tag{15}$$

pH and total S(IV) concentration define the amounts of the above S(IV) species that exist in a solution of SO_2. Inspection of the values of K_1 and K_2 indicates that only in solutions of pH$\leq$2 does molecular SO_2 exist in high concentration. At higher pH values, transport of SO_2 across the solution boundary layer involves transfer of ionic species across the layer. These ionic species have a higher solution mobility than molecular SO_2, so as pH increases, the value of r_l decreases. This is very well illustrated in Fig. 2.

The constant value for R at pH 4 can be shown to be completely dominated by r_g. As most natural waters are of pH 4–9, and as the oceans are at about pH 8, it can be seen that with SO_2 the ability to ionize in solution has a dominating influence on its transfer properties from the troposphere to the oceans.[2] The thickness of the air boundary layer which determines r_g varies with wind velocity, decreasing as the wind velocity increases. This has the effect of increasing K_g (or decreasing r_g), as may be seen in Fig. 3.

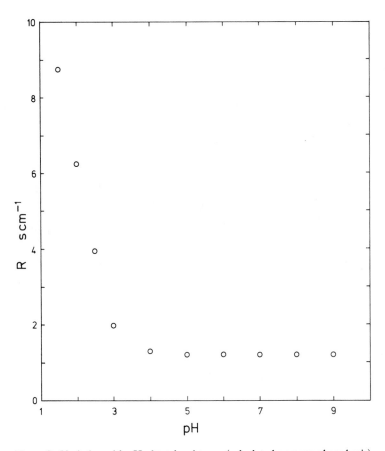

Figure 2. Variation with pH of total resistance (calculated on a gas-phase basis) to SO_2 exchange into water.[11]

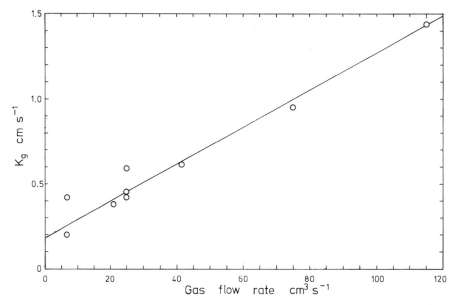

Figure 3. Variation of exchange constant with gas flow rate for SO_2 exchange into seawater.[12]

If the value for K_g at the highest wind velocity in Fig. 3 is taken as representative of the exchange constant under average wind conditions over the ocean, it is possible to calculate the potential transfer of SO_2 from the atmosphere to the ocean. Taking K_g as 1.4 cm/sec, the total area of the oceans as 3.7×10^{18} cm^2, and the average SO_2 level in the ocean air as 0.5 μg/m^3, the transfer of SO_2 from the atmosphere to the oceans is 8.2×10^7 tonnes/year. The importance of the ocean as a sink of atmospheric SO_2 can be seen when the oceanic absorption of SO_2 is compared with the estimated 14×10^7 tonnes/year of SO_2 emitted as the result of the combustion of fossil fuels.[3,4]

During 1974–75 several field measurements of the gas transfer coefficient K_g have been made. Whelpdale and Shaw[107] foung K_g values of the order of 1 cm/sec for SO_2 transfer to fresh water and snow surfaces. Experiments[108] using radioactive SO_2 to measure the deposition of SO_2 into artificial ponds of seawater indicated that K_g was 0.5 ± 0.1 cm/sec. A large-scale aircraft sampling program[109] over and about the United Kingdom produced data that predicted that K_g was in the range of 0.6–0.8 cm/sec. These recent data suggest that the estimate of the annual transfer of SO_2 to the oceans is probably overestimated by about a factor of two, and is more likely to be 4×10^7 tonnes/year.

2.4. *Carbon Monoxide Transfer*

The study of the exchange of carbon monoxide from the atmosphere to the oceans has an interesting history. Early studies[5] were able to identify only a man-made combustion source of CO. Concern for the possibility that CO may accumulate in the atmosphere led to a study of the ocean as a potential sink for the gas. However, in recent years, a number of large natural sources of CO have been found, including the surface of the ocean.[6] It now appears that much of the surface of the ocean is supersaturated with respect to CO, and that the concentration gradient of the gas is from the ocean to the atmosphere.[7] Sinks for CO exist in the form of gas-phase atmospheric reactions[8] and utilization by microorganisms at the surface of the earth.[9] The magnitude of the flux of CO from the ocean to the atmosphere is of importance in assessing the relative importance of the contribution of pollutant CO to the total natural CO cycle.

Unlike SO_2, carbon monoxide does not react chemically with water, remaining in a molecular form. With this gas there is no enhancement of transfer across the liquid boundary layer and it is found that r_l dominates the transfer process.[2] The numerical value of K_l can be obtained using the estimated[1] oceanic boundary layer thickness of 30 μ in equation (3), together with the diffusion coefficient for CO of 2.2×10^{-9} m^2/sec

$$K_l = \frac{1}{r_l} = \frac{D}{x_l} = \frac{2.2 \times 10^{-9}}{3 \times 10^{-5}}$$

Use of this value in equation (4) to obtain the flux of CO from the ocean to the atmosphere requires a knowledge of the CO concentration gradient from the bulk ocean to the ocean surface. It is assumed that the CO concentration in solution at the ocean–air interface is essentially the same as that of the air at the surface of the ocean.[2] Measurements have been made of both air and ocean CO concentrations in cruises in research ships in both hemispheres in the Atlantic and Pacific oceans.[7] A considerable difference has been found between the atmospheric concentration of CO in the Northern Hemisphere (0.13 ppm) and that in the Southern Hemisphere (0.04 ppm). The concentration in the ocean surface water is relatively constant over all the oceans in both hemispheres, averaging 6×10^{-8} cm^3 CO per cm^3 water. The concentration gradient in the ocean is thus $(6 \times 10^{-8} - 2.6 \times 10^{-9})$ cm^3 CO/cm^3 H_2O. The latter figure is obtained by expressing the aerial CO concentration, 1.3×10^{-7} cm^3 CO/cm^3 air, on a liquid phase basis by dividing by the Henry's law constant for CO. Use of equation (4) leads to a value of 5×10^{-9} $g\,m^{-2}\,sec^{-1}$ for the flux of CO from the Northern oceans to the atmosphere. Calculations using the southern hemisphere aerial CO concentration show only a small change in the flux, because the concentration difference in equation (4) is in this case dominated by the ocean surface CO concentration. A value for the total flux of CO from the oceans may thus be obtained assuming the flux to be constant over all the oceans. This flux is 5.85×10^7 tonnes/year. This may be compared to the estimated global CO emissions from combustion sources[6] of 46×10^7 tonnes/year, and from the natural oxidation of methane by hydroxyl radicals in the troposphere of 150×10^7 tonnes/year. The ocean has thus passed from being a potential sink for combustion CO to a minor natural source of the gas, although it has recently been suggested[110] that it may act as either a source or a sink in different areas.

2.5. Dimethyl Sulfide Transfer

The cycle of sulfur in the environment is usually balanced by an emission of hydrogen sulfide from the ocean to the atmosphere. The total mass of sulfur transferred as H_2S is "guessed" as being that necessary to balance the sulfur cycle.[10] Early forms of the sulfur cycle[3] suggested that H_2S was emitted generally from the ocean surface. More recently[4] this has been modified to include only areas where ocean deep water upwells from anoxic depths and to tidal mudflats. Several workers[4] have warned against overestimating the H_2S contribution to the cycle noting that even one meter coverage of ascending H_2S with aerated seawater will oxidize it quantitatively and prevent its emission.

In 1972,[10] dimethyl sulfide was found in the ocean surface water at an average concentration of 1.2×10^{-11} g/cm^3. This discovery led to the postulation that $(CH_3)_2S$ may be the gas necessary to carry sulfur from the ocean to the atmosphere in order to balance the sulfur cycle. Unfortunately, no quantitative measurements of the gas have been made in the air over the ocean, but it is well known over the land largely as an emission from paper manufacturing processes. Rasmussen[111] has used gas chromatography to show that dimethyl sulfide and dimethyl disulfide dominated over H_2S emission from aerobic environments in soils, intact foliage, and waters. Only in anaerobic conditions was H_2S found to be dominant.

A calculation has been made of the potential of the ocean as a source of atmospheric $(CH_3)_2S$, assuming the maximum possible concentration gradient over the water boundary layer, i.e., a zero concentration of the gas at the ocean–atmosphere interface.[2] This calculation suggests that $(CH_3)_2S$ would constitute a maximum emission of 4% of the total sulfur necessary to balance the sulfur cycle. $(CH_3)_2S$ must thus be produced in large quantities outside the oceans, or else some other sulfur compound, perhaps even H_2S, acts to complete the cycle.

3. Vegetation as a Sink for Gaseous Pollutants

3.1. Introduction

Although land forms only about 30% of the total area of the globe, it offers a large surface area of vegetation to the gases in the atmosphere. Plants may take up atmospheric gases without active metabolism in the same manner as inorganic materials or, more importantly, they may actively metabolize the gas, thus creating a favorable concentration gradient for further absorption. Carbon dioxide is a very well known example of the latter case. It is also a pollutant gas, being the major end product of the complete combustion of carbon in fossil fuels.

3.2. *Theory*

The rate of transport of a gas from the atmosphere to plant leaves may be expressed in a manner similar to that discussed earlier for transport to water surfaces. The total resistance R to the transport can be written as the sum of individual resistances to transport to the final deposition sites for the gas in the leaf.[13] Figure 4 outlines these resistances, which are summarized as

$$\frac{1}{R} = \frac{1}{r_a + r_{\text{cut}}} + \frac{1}{r_a + r_s + r_{\text{mes}}} \tag{16}$$

r_a is the resistance to transfer across the air boundary layer over the surface of the leaf. This is calculated in the same way as the comparable resistance over the surface of water. For the purposes of this discussion, the leaf will be

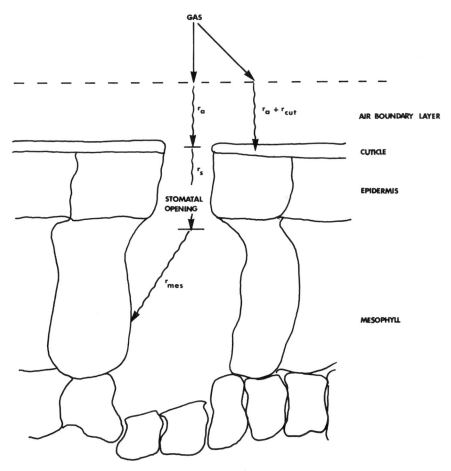

Figure 4. Simplified diagrammatic representation of resistances to gas transport into plant leaves.[13]

presumed to have identical upper and lower surfaces and the resistances will be assumed to be the same for both surfaces, and to be the sum of both surfaces.

The surface of leaves is covered by a waxy layer known as the cuticle. A number of gases are sorbed by the cuticle or by the thin film of water on the cuticle at high air humidities. The resistance to transfer to this surface is expressed as r_{cut}.

The leaf surface is penetrated by a large number of openings, the stomata. These are under metabolic control and may be opened or closed in response to physiological conditions. In light and in high air humidity the stomata open, and in dark and in low air humidity they close. The resistance to transfer through these openings, r_s, thus varies from a low value for open stomata to a very high value for closed stomata.

After it has traversed the stomata, the gas can diffuse to the surface of cells within the leaves. The cell surfaces are covered by a film of water, the air spaces within the leaf being at 100% relative humidity. Transfer to these mesophyll cell surfaces is subject to a resistance, r_{mes}.

The resistance to transport across the air boundary layer depends on wind speed (which affects the thickness of the layer) and on the diffusivity of the transferring gas.[14] The value of r_a may thus be calculated from simple physical parameters. Under conditions when the stomata are closed, the value of r_{cut} can be calculated if R and r_a are known. The sum of r_s and r_{mes} may be obtained (following the determination of R, r_a, and r_{cut}) when the stomata are fully opened. These two resistances may be separated by measuring all the resistances to transfer for water vapor. This gas is subject to resistances $r_a + r_s$. There is zero resistance to the transport of water vapor to the thin film of water on the mesophyll surfaces and to the cuticle.[14] A measurement of R and r_a hence enables r_s to be calculated for a particular leaf. It is assumed generally that r_s for water vapor is about the same as r_s for most other gases.

The basic measurements necessary for obtaining quantitative data for the resistances to gaseous transport from the atmosphere to plant leaves are (1) the total resistance R, and (2) the air boundary layer resistance, r_a. The quantity normally measured is not the total resistance R but its reciprocal, v_g, which is known as the velocity of deposition and is defined as[15]

$$v_g = \frac{1}{R} = \frac{J}{C} = \frac{\text{Total deposition per unit area}}{\text{Dosage (concentration} \times \text{time)}} \qquad (17)$$

Here, J, the flux of the gas to the surface, is usually expressed in $\mu g/cm^2$ sec. C is the volumetric concentration of the gas in the air at some appropriate reference point clear of the air boundary layer, and is expressed in $\mu g/cm^3$. The units of v_g are thus cm/sec. The evaluation of v_g under field conditions is normally carried out using the first form of v_g, i.e., J/C.

The velocity of deposition may also be measured using the second form in equation (17), where the total deposition per unit area is expressed in $\mu g/cm^2$ and the dosage is expressed in μg sec/cm^3. The use of dosage, which is the time integral of the gas concentration, allows direct measurement of v_g even though the gas concentration varies during the experiment. This form of v_g is often used in laboratory determinations of gaseous deposition and in field measurements using radiolabeled gases which give a direct measure of deposition per unit area. It can be seen that $v_g(=J/C)$ is analogous to the exchange constant $K(=J/\Delta C)$ when it is assumed that the deposition is an irreversible process and the concentration of the gas at the solid surface is zero.

3.3. Experimental Observations

In recent years an increasing number of experiments have been carried out to determine v_g for plant leaves. Initial efforts involved laboratory studies on gas deposition to individual leaves (Table 1); more recently, techniques have been developed to obtain field data on natural vegetation (Table 2).

It is quite clear from these data that many pollutant gases are actively taken up by plants. This is particularly obvious when large differences in resistance are found between conditions promoting open and closed stomata. In Tables 1 and 2, stomatal resistance (together with other internal resistances) is represented by $(R - r_a)$. All of the major pollutant gases illustrated, except carbon monoxide, appear to be involved in plant metabolism to a considerable extent.

Comparison of data obtained in the laboratory on individual leaves with field data for a complete plant canopy offers some difficulty. The laboratory gas uptake data are usually expressed as mass of gas sorbed per unit plan area of plant leaf surface, while field data are expressed as uptake per unit plan area of plant canopy. With the layering of leaves through the depth of the canopy, the v_g value obtained in the field appears larger than the laboratory v_g value. Measurements have been made of the average area of leaf per unit plan area of vegetation canopy for a number of types of vegetation. This is usually expressed as a ratio known as the leaf area index, which varies from one type of vegetation to another as the density of leaves through the canopy changes. The use of the leaf area index is illustrated in Table 3.

Experimental data of the type given in Tables 1 and 2 have been used to explore the role of vegetation as a sink for pollutant gases on a global scale. The calculations involved in such studies include, in the best case, an extrapolation from gas uptake by the plant canopy in a field to total global uptake. In the worst case, the extrapolation is from gas uptake by a single leaf to total global uptake. The reader is thus warned to exercise caution when studying these types of data.

In Table 4, vegetation uptake of gases at concentrations typical of a polluted area is calculated. The final column of the table indicates the height of the column of polluted air that would be depleted by one square meter of the named vegetation, if it is assumed that the v_g value is correct. It is obvious that the grass-like vegetation for which deposition data are available have a considerable capacity to reduce O_3, SO_2, and NO_2 concentrations in polluted areas.

Table 1. Deposition Velocities for Some Gases onto Vegetation under Laboratory Conditions

Gas	Plant	Conditions	v_g(cm/sec)	R(sec/cm)	r_a(sec/cm)	$(R - r_a)$	Reference
CO_2	bean	light	0.025	40	[a]	[a]	17
CO_2	bean	dark	0.0004	2500	[a]	[a]	17
CO	bean	light	0.0025	400	[a]	[a]	17
CO	bean	dark	0.0020	500	[a]	[a]	17
O_3	maize	light	0.33	3	0.5	2.5	18
O_3	maize	dark	0.004	271	0.5	270.5	18
SO_2	barley	stomata open	0.066	7.5	1.2	6.3	19
SO_2	barley	stomata closed	0.011	45	1.2	44	19
I_2	bean	light, high humidity	0.42	2.4	1.0	1.4	14
I_2	bean	dark, low humidity	0.024	42	1.0	41	14
NH_3	soybean	optimum	0.46	2.2	[a]	[a]	20

[a] Data not available in original papers.

Table 2. Deposition Velocities for Some Gases to Plant Canopies

Gas	Vegetation	v_g(cm/sec)	R(sec/cm)	r_a(sec/cm)	$(R-r_a)$	Reference
SO_2	grassland	0.8	1.3	0.4	0.9	21,108
SO_2	grassland (summer)	0.8	1.3	0.8	0.5	112
SO_2	grassland (autumn)	0.3	3.3	3.0	0.3	112
HF	alfalfa	0.3	3.3	[a]	[a]	113
I_2	grassland	0.66	1.5	0.3	1.2	23
O_3	alfalfa	1.66	0.6	[a]	[a]	24
NO_2	alfalfa	2.0	0.5	[a]	[a]	24

[a] Data not available in original papers.

Table 3. Comparison of V_g Determined in Field or Laboratory Using the Leaf Area Index

Gas	Plant	Conditions	Leaf area index	v_g (laboratory)[a] (cm/sec)	v_g (field)[a] (cm/sec)
O_3	maize	light	3.8	0.33	1.25
O_3	alfalfa	optimum	4.7	0.35	1.66
SO_2	barley	high humidity	8	0.066	0.53
SO_2	wheat	crop wet	8	0.05	0.4

[a] The underlined v_g values were determined experimentally; the adjacent v_g value was calculated using the leaf area index.

Table 4. Calculated Uptake of Pollutant Gases by Vegetation

Gas	Vegetation	v_g(cm/sec)[a]	Gas concentration ($\mu g/m^3$)	Uptake[b] (g/m^2)	Volume depleted (m^3)
O_3	maize	1.25	400	0.2	500
SO_2	grassland	0.8	500	0.17	340
CO	bean	0.012	10^5	0.52	5
NO_2	alfalfa	2.0	200	0.17	850

[a] Expressed on a canopy basis.
[b] Calculated for a 12-hr exposure to the gas.

In a report to Charles II in the seventeenth century, John Evelyn suggested that the planting of trees would reduce the pollutant concentration in the air of London.[16] Although trees have a leaf area index 60–70%—that of the grasses—it can be seen that the advice of John Evelyn was indeed valid.

4. *Plant Damage by Gaseous Pollutants*

4.1. *Introduction*

In the previous section it was noted that most gaseous pollutants are actively metabolized by plants. However, many plant species are susceptible to damage by very low concentrations of pollutant gases. It is an interesting observation that in general plants are damaged by pollutant gases at lower concentrations than are animals. This is especially true with ozone, PAN, and sulfur dioxide. As is usual with a general statement, there are exceptions, the most notable being carbon monoxide, which affects plants at concentrations higher than those affecting animals.

There is a large body of information on the nature of the damage to plant leaves by pollutant gases. Formulas have been derived to relate pollutant gas dosage to the degree of plant damage.[25] Knowledge of the nature of the damage makes it possible for experienced plant pathologists to detect the presence of a pollutant gas and to obtain a rough estimate of its concentration by observing the damage to a variety of selected plant species.[26,27]

It is unfortunate that studies on the biochemical basis of plant damage have lagged behind the plant pathological studies. It is only in recent years that information has emerged on some of the points of attack of gaseous pollutants in the metabolism of plants. In this section this information will be summarized and discussed.

4.2. *The Problem of Selecting Suitable Dosages*

It has been well established that the uptake of pollutant gases by plants is related to the solubility of those gases in the water film on the mesophyll cells.[28] Since solution of the gases usually results in ionization, the resulting ions may also be involved in metabolic disruption. Many investigations of the metabolic behavior of gaseous pollutants in plants have been carried out in aqueous solutions containing these ions. A major difficulty in this type of investigation in the choice of the concentration of the ion in solution is that it is directly related to the atmospheric concentration of its parent gas. There is no method for adequately calculating the ion concentrations, especially as most are continuously depleted by their normal metabolism within the cells. The oxyanions of nitrogen and sulfur are specific examples of normal cell metabolites that also arise from the solution of pollutant gases in cell water.

Individual workers have made their own choices of suitable concentrations for their investigations, and as a consequence it is difficult to compare results because of frequent wide divergences in the concentrations studied. In Tables 5, 6, and 7 an indication is given of pollutant gas dosages that lead to visible effects on sensitive plants. This dosage, together with the average concentration of the given gas in a polluted atmosphere, should be kept in mind when studying the data in the tables.

Table 5. Effect of Ozone on Plant Biochemistry[a]

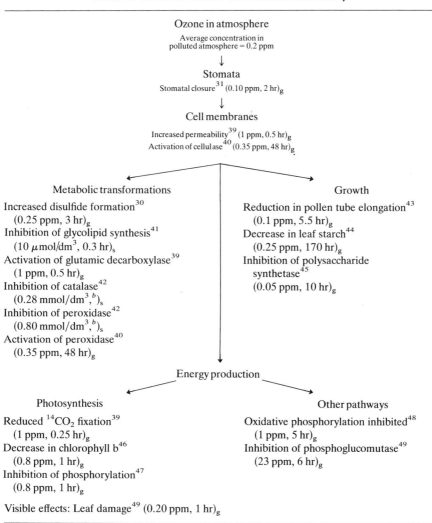

Ozone in atmosphere

Average concentration in
polluted atmosphere = 0.2 ppm

↓

Stomata

Stomatal closure[31] (0.10 ppm, 2 hr)$_g$

↓

Cell membranes

Increased permeability[39] (1 ppm, 0.5 hr)$_g$
Activation of cellulase[40] (0.35 ppm, 48 hr)$_g$

Metabolic transformations

Increased disulfide formation[30]
(0.25 ppm, 3 hr)$_g$
Inhibition of glycolipid synthesis[41]
(10 μmol/dm^3, 0.3 hr)$_s$
Activation of glutamic decarboxylase[39]
(1 ppm, 0.5 hr)$_g$
Inhibition of catalase[42]
(0.28 mmol/dm^3,[b])$_s$
Inhibition of peroxidase[42]
(0.80 mmol/dm^3,[b])$_s$
Activation of peroxidase[40]
(0.35 ppm, 48 hr)$_g$

Growth

Reduction in pollen tube elongation[43]
(0.1 ppm, 5.5 hr)$_g$
Decrease in leaf starch[44]
(0.25 ppm, 170 hr)$_g$
Inhibition of polysaccharide
synthetase[45]
(0.05 ppm, 10 hr)$_g$

Energy production

Photosynthesis

Reduced $^{14}CO_2$ fixation[39]
(1 ppm, 0.25 hr)$_g$
Decrease in chlorophyll b[46]
(0.8 ppm, 1 hr)$_g$
Inhibition of phosphorylation[47]
(0.8 ppm, 1 hr)$_g$

Other pathways

Oxidative phosphorylation inhibited[48]
(1 ppm, 5 hr)$_g$
Inhibition of phosphoglucomutase[49]
(23 ppm, 6 hr)$_g$

Visible effects: Leaf damage[49] (0.20 ppm, 1 hr)$_g$

---

[a] Data in parentheses represent dosage used.
[b] Undetermined.

Table 6. Effect of PAN on Plant Biochemistry[a]

PAN in atmosphere

Average concentration in polluted
atmosphere = 0.03 ppm

↓

Stomata

↓

Cell membranes

$$CH_3COOONO_2 + 2OH^- \longrightarrow CH_3COO^- + O_2 + NO_2^- + H_2O$$

Metabolic transformations

Decrease in total—SH groups[50]
(1 ppm, 0.5 hr)$_g$
Epoxidation of olefins[51]
(0.1 mol/dm^3, 1 hr)$_s$
Oxidation of glutathione[52]
(1 μmol/dm^3, 0.3 hr)$_s$

Growth

Inhibition of cellwall polysaccharide
synthesizing system[53]
(0.3 μmol/dm^3, 4 hr)$_s$
Inhibition of metabolism of cellwall
glucans and cellulose synthesis[54]
(1 μmol/dm^3, 6 hr)$_s$

Energy production

Photosynthesis

$^{14}CO_2$ fixation inhibited[55]
(1 μmol/dm^3, 0.1 hr)$_s$
Decrease in chlorophyll[56]
(1 μmol/dm^3, 0.2 hr)$_s$
Inhibition of photophosphorylation[57]
(10 μmol/dm^3, 0.2 hr)$_s$
Inhibition of Hill reaction[58]
(1 μmol/dm^3, ^{b})$_s$

Other pathways

Inhibition of phosphoglucomutase[59]
(0.1 μmol/dm^3, 6 hr)$_s$
Inhibition of isocitrate dehydrogenase[60]
(1 μmol/dm^3, 0.02 hr)$_s$
Inhibition of glucose-6-phosphate
dehydrogenase[60]
(1 μmol/dm^3, 0.02 hr)$_s$

Visible effects: Leaf lesions[61] (0.014 ppm, 4 hr)$_g$
Chloroplast damage[62] (1 ppm, 0.5 hr)$_g$

^{a}Data in parentheses represent dosage used.
^{b}Undetermined.

Table 7. Effect of Sulfur Dioxide on Plant Biochemistry[a]

Sulfur dioxide in atmosphere

Average concentration in
polluted atmosphere = 0.3 ppm

↓

Stomata

Increased stomatal opening[32]
(0.1 ppm, 24 hr)$_g$

↓

Cell membranes

$$SO_2 + H_2O \nrightarrow H_2O \cdot SO_2$$

$$H_2O \cdot SO_2 \nrightarrow HSO_3^- + H^+$$

$$HSO_3^- \nrightarrow SO_3^{2-} + H^+$$

Energy production

←——————   ——————→

Photosynthesis	Other pathways
Decreased $^{14}CO_2$ fixation[63] (0.3 mmol/dm³, 0.3 hr)$_s$	Glycollate oxidase inhibited[37,38,116] (5 ppm, 0.5 hr)$_g$
Increased $^{14}CO_2$ fixation[34] (0.5 mmol/dm³, [b])$_s$	Peroxidase activated[114] (1 ppm, 6 days)$_g$
ATP concentration decreased[34] (2 mmol/dm³, [b])$_s$	Glutamate dehydrogenase activated[115] (0.3 ppm, 7 days)$_g$
Hill reaction stimulated[34] (3 mmol/dm³, [b])$_s$	Glutamate-pyruvate transaminase activated[114] (0.2 ppm, 6 days)$_g$
Ribulose diphosphate carboxylase inhibited[35,114] (0.2 ppm, 6 days)$_g$	Glutamate-oxaloacetate transaminase activated[114] (0.2 ppm, 6 days)$_g$
Phosphoenolpyruvate carboxylase inhibited[36,117] (10 mmol/dm³, 0.1 hr)$_s$	Glutamate-oxaloacetate transaminase inhibited[116] (4 mmol/dm³, [b])$_s$
Malate dehydrogenase inhibited[36,118,121] (0.5 mmol/dm³, [b])$_s$	Increase in total free amino acids[120] (0.7 ppm, 72 hr)$_g$

Visible effects: Reduction in yield[64] (0.12 ppm, 1500 hr)$_g$
Leaf damage[65] (0.10 ppm, 1350 hr)$_g$
Pollen tube elongation depressed (0.30 ppm, 4 hr)$_g$

[a] Data in parentheses represent dosage used.
[b] Undetermined.

4.3. *Effects of Ozone and PAN*

Both ozone and PAN act as oxidizing agents in solution, and their main effect in plants appears to be the oxidation of sulfydryl groups.[29] These groups are important in the structure of proteins, and enzyme activity is often related to the presence of sulfydryl groups in the protein chain. Oxidation by ozone and PAN largely results in the formation of disulfides, and it is likely that the inhibition of individual enzyme activity by these gases is related to this oxidation.[30]

The solubility of ozone in water[28] at 20°C, and with an ozone partial pressure of one atmosphere, is 12 mmol/dm^3. If it is assumed that Henry's law applies, then at an ozone partial pressure of 1 ppm, the ozone concentration in solution is 0.012 μmol/dm^3. It is obvious that inhibition studies using ozone gas concentrations of about 1 ppm are much closer to the real situation than are studies using ozone solution concentrations in the mmol/dm^3 range. It is therefore most likely that in the real situation peroxidase is activated rather than inhibited by ozone (see metabolic transformations, Table 5). Similar arguments can undoubtedly be applied to PAN, although ionization of this gas in solution yields normal cell metabolites.

Photosynthesis (and possibly other energy-producing pathways) is affected by both PAN and ozone. The nature of the influence on photosynthesis is not yet obvious but it is to be hoped that the increasing activity in this field will lead to an understanding of the effect. Ozone is known to cause stomatal closure at realistic dosages[31]; this undoubtedly has an overall effect on photosynthesis in limiting gas exchange, but is unlikely to be the real effect on the photosynthetic pathway.

4.4. *Effects of Sulfur Dioxide*

With sulfur dioxide a similar physiological effect has been observed with the stomata; however, stomatal opening is promoted by low dosages and closure is promoted by high dosages.[32] Increased stomatal opening would lead to water stress under appropriate conditions, but the relationship of this physiological effect to leaf metabolism is not at all clear. Most of the studies of sulfur dioxide effects on leaf metabolism have been carried out with sulfite solutions in the millimole concentration range. An attempt has been made to relate gas-phase sulfur dioxide concentration with its equilibrium concentration, taking into account reactions (excluding oxidation) in solution. This is shown in graphic form in Fig. 5. At SO_2 concentration of 100 μg/m^3 (about 0.3 ppm) in the atmosphere the equilibrium solution concentration of all S(IV) species is 36 μmol/dm^3. This figure would be higher if the buffer capacity of mesophyll water could be estimated and taken into account. On the other hand, oxidation and metabolism have not been considered, and these would act to reduce the concentration. A sulfite concentration in the range 0.1–0.5 mmol/dm^3 would thus seem reasonable for experimental studies of sulfur

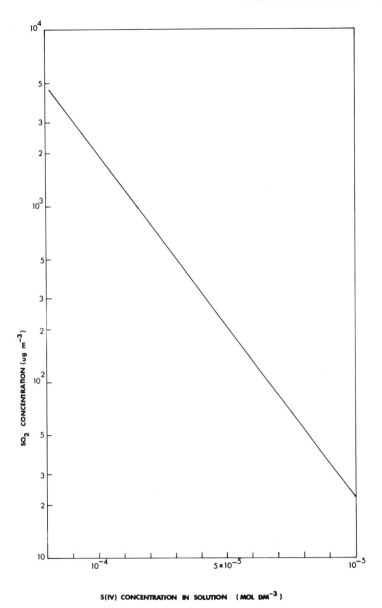

Figure 5. Relationship between $(SO_2)_g$ and S(IV) in solution.[33]

dioxide effects on leaf metabolism. The sulfite concentrations investigated have almost invariably been in excess of $1\,mmol/dm^3$, and because the effects investigated have been on the photosynthetic pathway, this choice of concentration is of considerable importance. It has been observed[34,116] that $^{14}CO_2$ fixation by isolated chloroplasts is increased by sulfite concentrations of $<1\,mmol/dm^3$, but decreased by concentrations $>1\,mmol/dm^3$. It could thus be said that many of the investigations of sulfur dioxide effects on photosynthesis are invalid, in that the plants may not experience the sulfite concentrations used. Experimentation to obtain direct measurements of cell water sulfite concentrations in plants exposed to realistic atmosphere sulfur dioxide concentrations is thus of great importance.

If it is assumed that the dosages of sulfite used in the investigations summarized in Table 7 are realistic, then it can be seen that photosynthesis is indeed affected by sulfur dioxide fumigation. The enzyme inhibition is either by competition between SO_3^{2-} and HCO_3^- as a substrate,[35,36] or by competition between glycollate and sulfoglycollate.[37] The latter compound has been isolated from rice plants exposed to sulfur dioxide.[38]

Recent reviews have emphasized the need for more work on the biochemical nature of pollutant gas effect on plant metabolism.[66,67] The data outlined in Figs. 5 and 6 illustrate this need, which must be coupled with a soundly based selection of pollutant gas concentration for use in the investigations. Finally, Pahlich[122] has attempted to unify most of the data pertaining to SO_2 effects on plants in a general theory of the mode of action of gaseous pollutants. Basically, he suggested that at low dosages SO_2 is converted to organic compounds after absorption, and that enzymes are unaffected. If the dosage is such that conversion to organic sulfur is not complete, the inorganic sulfur concentrations may cause enzyme inhibition.

5. *Sulfur Dioxide and Metal Corrosion*

5.1. *Introduction*

One of the most costly effects of atmospheric pollution by sulfur dioxide is its accelerating influence on the corrosion of metals. The cost must be measured not only in terms of replacement or repair costs of corroded structures, but also in terms of the cost of anticorrosion treatment. Most of the costs are associated with the corrosion of iron and its alloys, but other important metals used in outdoor construction—zinc, copper, and aluminum—also suffer accelerated corrosion in the presence of sulfur dioxide.

In the past, most scientific and technological investigations have been concerned with methods of reducing corrosion, largely by the application of surface coatings. In recent years, however, some understanding has been obtained of the nature of the involvement of sulfur dioxide in the corrosion process. This section will deal for the most part with a discussion of the role of sulfur dioxide in the atmospheric corrosion of iron.

5.2. *Atmospheric Corrosion of Iron*

When test panels of iron (or steel) are exposed to the atmosphere at sites with different environments, it is found that the greatest corrosion occurs in industrial environments and the least in inland rural environments.[68] The difference in corrosion rate has been attributed to the higher sulfur dioxide concentrations in industrial atmospheres.[69] In atmospheres with very low sulfur dioxide concentrations, the main corrosion products on the surface of iron are α-FeOOH, γ-FeOOH, Fe_3O_4, and some amorphous material.[70]

It is now generally agreed[69,71,123] that the mechanism of the formation of these compounds is electrochemical, and takes place in a thin water film on the metal surface. In simplified form, the initial corrosion reactions are

$$Fe \rightarrow Fe^{2+} + 2e \text{ (anodic reaction)} \qquad (18)$$

$$\tfrac{1}{2}O_2 + H_2O + 2e \rightarrow 2OH^- \text{ (cathodic reaction)} \qquad (19)$$

The reactions following reactions (18) and (19) depend upon the pH of the water film. If the pH is very alkaline, reaction (20) follows, while if it is slightly alkaline, reaction (21) is found:

$$4Fe(OH)_2 + O_2 \rightarrow 4\alpha FeOOH + 2H_2O \qquad (20)$$

$$6Fe(OH)_2 + O_2 \rightarrow 2Fe_3O_4 + 6H_2O \qquad (21)$$

When the pH is neutral or slightly acidic, $Fe(OH)_2$ cannot be formed, and its place is taken by various Fe(II) hydroxo complexes involving other anions in solution. These anions reflect the atmospheric environment to which the iron was exposed, e.g., chloride is found in marine environments, and sulfate is found in industrial environments. The Fe(II) hydroxo complexes are also oxidized by dissolved oxygen, resulting in the γ-modification of FeOOH.

On the microscale, the surface of iron exposed to the atmosphere can be seen to be covered by a two-layer film of oxidation products (Fig. 6).[69] At the surface of the pure metal, the anodic reaction (18) occurs, while at the interface of the Fe_3O_4 and FeOOH layers, the cathodic reaction (19) takes place. This latter reaction is followed by oxidation with molecular oxygen, yielding FeOOH or Fe_3O_4, depending upon the hydrogen ion activity. The overall effect of the anodic plus cathodic reactions is that one Fe atom is transferred from the pure metal and is deposited as FeOOH at the bottom of the surface layer. As the anodic and cathodic sites are separate in space, conductors are necessary to complete the electrochemical circuit. The Fe_3O_4 layer serves as an electron conductor, while soluble ferrous salts in solution in the Fe_3O_4 layer act as an ionic conductor.

Of the soluble ferrous salts, $FeSO_4$ is the most important. The source of the sulfur in this salt is atmospheric sulfur dioxide. The initial step in the corrosion process may thus be considered to be the absorption of sulfur dioxide by the metal surface. The amount of sulfur dioxide absorbed is increased by the presence of rust,[72,74] surface particles,[73] and relative humidity.[74,75] In the absence of rust and surface particles, sulfur dioxide is absorbed at specific sites on the iron surface[76] which are not related to any structural dislocation such as

ATMOSPHERE

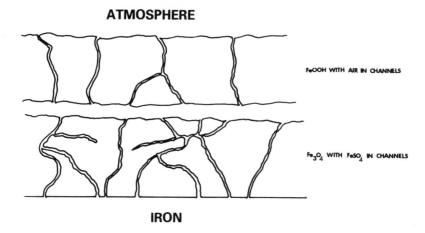

FeOOH WITH AIR IN CHANNELS

Fe₃O₄ WITH FeSO₄ IN CHANNELS

IRON

Figure 6. Diagrammatic representation of two-layer film involved in the atmospheric corrosion of iron.[69]

grain boundaries.[77] It has been proposed that the absorption sites are activated by water vapor, and that there sulfur dioxide is adsorbed by a surface oxide and is oxidized to sulfate in the adsorbed state:[78,79]

$$Fe(II)oxide + SO_2 + O_2 \rightarrow FeSO_4(ads)oxide \qquad (22)$$

The electrochemical mechanism of rusting requires $FeSO_4$ only as an ionic conductor. Once sufficient is available for this purpose, rusting proceeds apparently without the need for further addition of sulfur dioxide to form further $FeSO_4$.[80] It would thus be expected that an infinite amount of FeOOH could be formed in the presence of a finite amount of $FeSO_4$. This is not found to be the case because a slow reaction resulting in the formation of insoluble basic iron sulfates is continually removing ferrous and sulfate ions from solution.[69] The accelerated corrosion of iron in atmospheres of high sulfur dioxide concentration is hence closely related to the maintenance of an $FeSO_4$ solution in the lower oxide layer. Similarly, increased rusting in high humidities is related to the maintenance of an aqueous solution in the lower oxide layer. At low atmospheric humidities this solution dries out, and the electrochemical process is disrupted, lacking an ionic conductor.

It should be noted here that Misawa *et al.*[71] have cast doubt on the suggestion that Fe_3O_4 is the basic constituent of the lower oxide layer on iron. They suggest that it is an amorphous ferric oxyhydroxide, $FeO_x(OH)_{3-2x}$, which in aqueous solution is a good electronic conductor. It can be seen that although there is some conjecture over the nature of the oxide layers, the electrochemical mechanism would operate under both conditions and the involvement of sulfur dioxide would be the same.

5.3. *Atmospheric Corrosion of Other Metals*

5.3.1. *Zinc*

The main use of zinc in outdoor structures is that of a sacrificial surface coating on iron and steel. The surface film of insoluble basic zinc carbonate formed on zinc in the atmosphere[81] is its normal means of protection. Atmospheric sulfur dioxide concentration is a major factor in controlling the rate of corrosion of zinc.[82] Sulfur dioxide is absorbed uniformly over a zinc surface,[83] the mass absorbed increasing with relative humidity.[74,75] The accelerated corrosion of zinc by sulfur dioxide may be due to the formation of soluble $ZnSO_4$ at the outer surface of the protective film,[84] or to the action on the surface film of the acidic solution produced when sulfur dioxide dissolves in the surface water layer. The uniform uptake of sulfur dioxide at all humidities on zinc suggests that the former mode of attack is more likely. It is of interest to note that the corrosion of zinc occurs at the outer surface of the protective film, while that of iron occurs at the interface between two oxide layers on its surface.[83]

5.3.2. *Copper*

The corrosion of copper in the atmosphere is very slow, resulting in the formation of a blue–green patina. Observations of copper roofs in Scandinavia have shown that the rate of production of the patina has increased as the atmospheric sulfur dioxide concentration has increased.[85] Newly formed patina has a composition of $CuSO_4 \cdot 6Cu(OH)_2$, while older patina becomes $CuSO_4 \cdot 3Cu(OH)_2$. Obviously, sulfur dioxide has become incorporated into the corrosion product. Little is known about the mechanism of the formation of the patina or about the involvement of sulfur dioxide in the mechanism. Although the mass of sulfur dioxide sorbed by copper increases with relative humidity, it is much lower than that for iron and zinc[74] and is confined to specific sites on the surface.[83]

5.3.3. *Aluminum*

Observations of aluminum panels exposed for long periods of time to a variety of atmospheric environments have shown a strong correlation between the corrosion rate of aluminum and the sulfur dioxide content of the atmosphere.[86] The sulfur-containing corrosion product found on aluminum in industrial atmospheres is $Al_2(SO_4)_3 \cdot 18H_2O$.[87] At low relative humidities, very little sulfur dioxide is taken up by aluminum and most of this is physically adsorbed.[87] At higher humidities, an increasing mass of sulfur dioxide is taken up, although the total mass adsorbed is much less than that for iron and zinc.[74,87] The suggested mechanism for the accelerated corrosion of aluminum in the presence of sulfur dioxide is an acid attack on the protective oxide film, the acid being derived from the solution and oxidation of sulfur dioxide in the water film on the metal surface.[87]

6. *Indoor Reactions of Pollutant Gases*

6.1. *Introduction*

The concentration of pollutant gases is highest in urban areas, and it is in these areas that the population spend most of their lives indoors. Studies of the fate of pollutant gases in the enclosed volumes found in indoor environments are thuse of considerable importance. The nature of indoor surfaces is quite different than those found outdoors, which suggests that the rate of deposition of pollutants from the indoor atmosphere may differ from that measured outdoors. Also, gases indoors have surfaces in three dimensions with which to interact, whereas outdoors the interaction is possible only in one dimension, if one ignores the reactions within the gas phase which occur in both environments. In this section, the effect of these interactions on the indoor concentration of some gaseous pollutants will be discussed, together with a survey of the possible damaging effects of the interactions.

6.2. *Indoor Pollutant Gas Concentrations*

Health physicists, who have a responsibility for assessing the damaging effects of airborne radionuclides, have recognized for some time that man is exposed to a lower radiation dose from nuclear debris indoors than he is outdoors. Iodine-131 is the major gaseous product arising from a fission process, and studies on this gas have shown it to be drastically reduced in concentration indoors.[88] It is thus likely that air pollutant gases of a chemical reactivity similar to iodine also show a reduction in concentration indoors. Table 8 shows that both sulfur dioxide and ozone are reduced in concentration indoors, while carbon monoxide, nitrogen dioxide, and PAN show no reduction. It should be emphasized that data for all gases except sulfur dioxide are very limited.

Studies have been made of the rate of removal of ozone from a variety of containers and rooms. It has been found that the rate of removal increases as relative humidity increases,[92] and that the nature of the container walls has an

Table 8. Relationship between Indoor and Outdoor Concentrations of Pollutant Gases

Gas	Indoor concentration / Outdoor concentration	First-order half-life for disappearance from an enclosed volume	Reference
Carbon monoxide	1.0	—	89
Nitrogen dioxide	1.0	—	90
Ozone	0.6	11 min for office, 6 min bedroom	90
PAN	1.0	—	90
Sulfur dioxide	0.2	40–60 min for empty laboratory	91

influence on the rate[90]: the first-order half-life for the removal of ozone was 28 min in a stainless steel container and only 13 min in an aluminum container. Further, studies indicate that not all collisions of ozone with an aluminum container at high relative humidity result in the removal of ozone from the gas phase.[92]

Ozone is well known as a strong oxidizing agent and this, together with its reactivity toward many organic molecules, probably accounts for its destruction at solid surfaces. The increased removal of ozone from the gas phase with increased relative humidity suggests that oxidation in the thin water film on solid surfaces is the major removal mechanism. Differences in removal rate by different surfaces could be explained by the presence of different materials in the water films.[92] Use of the activated carbon filters associated with air conditioning systems has been shown to be very effective in reducing the ozone concentration in air used to ventilate buildings.[90]

A number of studies have been made of the rate of removal of atmospheric sulfur dioxide by indoor surfaces. The results obtained are rather variable, which is not surprising since the rate of deposition of sulfur dioxide on solid surfaces is very dependent upon humidity, e.g., the first-order half-life for sulfur dioxide removal from a gloss-painted box at 30% relative humidity was 163 min, while at 86% relative humidity it was 4.8 min.[92] All measurements made on rooms and houses have omitted a consideration of this problem, although it is partially expressed in the finding that in two separate buildings the outdoor concentration was reduced indoors to 15–20% in winter and to 40–50% in summer.[93] At very high relative humidities, laboratory experiments suggest that each collision of a sulfur dioxide molecule with a wall results in its removal from the air.[92] Very high humidities are unusual indoors, so this condition is seldom attained.

Removal of sulfur dioxide by indoor surfaces is dominated by deposition to the ceiling,[91,94] and as the surfaces within the building age, the proportion of sulfur dioxide removed from admitted outdoor air decreases.[93,95] Aesthetic considerations usually dictate the renewal of indoor surfaces at frequent intervals by painting, wallpapering, etc., so that a more reactive surface is provided and deposition again increases. Radioactive sulfur dioxide has been used in laboratory studies to assess the relative efficiencies of typical indoor surfaces in absorbing sulfur dioxide.[96–99] Table 9 outlines the results of these experiments, which indicate that natural fibers and emulsion paint are particularly effective in removing sulfur dioxide from the air.[100] An assessment of the relative area of each of the surfaces in a house suggests that emulsion-painted surfaces would remove about 70% of the sulfur dioxide admitted to indoor air.[100] Emulsion paint is used largely for ceiling decoration, and thus the field observations of high uptake by ceilings is confirmed. The other major surface for the removal of sulfur dioxide indoors is furniture fabric because of its relatively high deposition velocity and overall surface area in a house.

Carbon monoxide, nitrogen dioxide, and PAN concentrations indoors follow the outdoor air concentrations.[89,90] Carbon monoxide is certainly a

Table 9. Average Deposition Velocities for Sulfur Dioxide
onto Indoor Surfaces[a]

Surface	Velocity of deposition (cm/sec)
Carpet, wool	0.02
Carpet, nylon	0.007
Furnishing fabric, cotton	0.22
Furnishing fabric, wool	0.29
Furnishing fabric, artificial	0.03
Leather, upholstery	0.16
Linoleum, waxed	0.0006
Paint, emulsion	0.24
Paint, gloss	0.02
Wallpaper, cellulose	0.02
Wallpaper, PVC	0.003

[a]Exposures carried out at 60–75% RH, 16–22°C.[100]

relatively unreactive gas, and the reactivity of nitrogen dioxide and PAN must be sufficiently low for the inward diffusion of outside air to balance any removal by indoor surfaces. All three gases are efficiently removed from ventilating air by activated carbon filters.[90]

Although it has been presumed that the concentration gradient of pollutant gases is from the outside to the inside air, it is possible for the reverse to occur. This is particularly the case with carbon monoxide and sulfur dioxide, which may occur at quite high concentrations in indoor air through the faulty operation of gas cookers, gas heaters, and coal-fired furnaces.[89]

6.3. Effects of Pollutant Gases on Indoor Surfaces

At the concentrations found in polluted atmospheres, carbon monoxide does not have any effect on nonliving materials, and no effect is known for PAN, while it is possible that a long exposure to nitrogen dioxide could cause bleaching of some dyestuffs. Ozone is known to degrade some organic polymers and to influence dyestuffs that are susceptible to oxidation, but sulfur dioxide would appear to have the greatest damaging effects on indoor materials. Unpainted stone and concrete masonry, mortar, and bricks absorb sulfur dioxide, resulting in the formation of sulfates.[94,101] Materials containing calcium carbonate are especially affected by sulfate formation, since calcium sulfate has a greater molecular volume than calcium carbonate, leading to mechanical stress.[102] This effect is particularly important in ancient buildings, where accelerated damage to frescos is becoming a serious problem.[103]

Fibers of nylon,[104] cotton,[105] and rayon[105] when exposed to 0.1–0.2 ppm sulfur dioxide show a loss in mechanical strength thought to be due to acid hydrolysis, the acid arising from the absorption of sulfur dioxide in a water film on the surface of the fiber. In the form of papers, both cellulose and artificial fibers are degraded by very low concentrations of sulfur dioxide.[96,106] The degradation is thought to be due both to acid hydrolysis and, in the case of cellulose papers, to the formation of lignosulfonic acids. Documents and books are particularly susceptible to attack by sulfur dioxide because of their long useful life. Even a very low concentration of sulfur dioxide becomes a high dosage when the time factor in the dosage parameter becomes large. Similarly, the leather bindings of books and documents are degraded by long exposures to low levels of sulfur dioxide. Once again, it is believed that the degradation is due to the formation of sulfuric acid in the surface water film, followed by acid hydrolysis.[98]

A number of techniques are available to reduce the damage of indoor materials by sulfur dioxide. The most obvious is the provision of air conditioning for all the air within the building by passing the ventilating air through an appropriate filter or scrubber. In real terms this technique is practicable only for large buildings. Books and documents may be protected by the choice of bindings and papers of low sulfur dioxide sorption capacity,[106] although the current trend is to store old and valuable books in air-conditioned buildings. In the private home, aesthetic considerations usually dictate the removal of carpets, furnishing fabrics, wallpapers, etc., before any sulfur dioxide damage makes itself obvious.

References

1. W. S. Broecker and T-H Peng, *Tellus*, **25** : 21 (1974).
2. P. S. Liss and P. G. Slater, *Nature*, **247** : 181 (1974).
3. E. Robinson and R. C. Robbins, *J. Air. Pollut. Contr. Assn.*, **20** : 233 (1970).
4. W. W. Kellogg, R. D. Cadle, E. R. Allen, A. L. Lazrus, and E. A. Martell, *Science*, **175** : 587 (1972).
5. L. S. Jaffe, *J. Air Pollut. Contr. Assn.*, **18** : 534 (1968).
6. W. Seiler, *Tellus*, **26** : 116 (1974).
7. J. W. Swinnerton and R. A. Lamontagne, *Tellus*, **26** : 136 (1974).
8. H. Levy, *Science*, **173** : 141 (1971).
9. R. B. Ingersoll, R. E. Inman, and W. R. Fisher, *Tellus*, **26** : 151 (1974).
10. J. E. Lovelock, R. J. Maggs, and R. A. Rasmussen, *Nature*, **237** : 452 (1972).
11. P. Brimblecombe and D. J. Spedding, *Nature*, **236** : 225 (1972).
12. D. J. Spedding, *Atmospheric Environment*, **6** : 583 (1972).
13. J. H. Bennett, A. C. Hill, and D. M. Gates, *J. Air Pollut. Contr. Assn.*, **23** : 957 (1973).
14. P. J. Barry and A. C. Chamberlain, *Health Physics*, **9** : 1149 (1963).
15. A. C. Chamberlain, *Int. J. Air Pollution*, **3** : 63 (1960).
16. J. Evelyn, *Fumifugium*, London, 1661.
17. R. G. S. Bidwell and D. E. Fraser, *Can. J. Bot.*, **50** : 1435 (1972).
18. N. C. Turner, P. E. Waggoner, and S. Rich, *Nature*, **250** : 486 (1974).
19. D. J. Spedding, *Nature*, **224** : 1229 (1969).
20. G. L. Hutchinson, R. J. Millington, and D. B. Peters, *Science*, **175** : 771 (1972).
21. J. A. Garland, W. S. Clough, and D. Fowler, *Nature*, **242** : 256 (1973).
22. D. Fowler and M. H. Unsworth, *Nature*, **249** : 389 (1974).
23. A. C. Chamberlain and R. C. Chadwick, *United Kingdom Atomic Energy Authority Research Report*, AERE-R4870, 1965.

24. J. H. Bennett and A. C. Hill, *J. Air Pollut. Contr. Assn.*, **23** : 203 (1973).
25. H. C. Perkins, *Air Pollution*, McGraw-Hill, New York, 1974, p. 320.
26. W. W. Heck, *Air and Water Pollut. Int. J.*, **10** : 99 (1966).
27. R. Guderian and H. van Haut, *Staub-Reinhalt Luft*, **30** : 22 (1970).
28. A. C. Hill, *J. Air Pollut. Contr. Assn.*, **21** : 341 (1971).
29. J. B. Mudd, *Proc. 1st European Congress on Influence of Air Pollution on Plants and Animals, Wageningen 1968*, 1969, p. 161.
30. H. Tomlinson and S. Rich, *Phytopathology*, **60** : 1842 (1970).
31. A. C. Hill and N. Littlefield, *Environ. Sci. and Technol.*, **3** : 52 (1969).
32. P. V. Biscoe, M. H. Unsworth, and H. R. Pinckney, *New Phytologist*, **72** : 1299 (1973).
33. P. Brimblecombe, Ph.D. Thesis, University of Auckland, 1973.
34. W. Libera, H. Ziegler, and I. Ziegler, *Planta*, **109** : 269 (1973).
35. I. Ziegler, *Planta*, **103** : 155 (1972).
36. I. Ziegler, *Phytochemistry*, **12** : 1027 (1973).
37. D. J. Spedding and W. J. Thomas, *Aust. J. Biol. Sci.*, **26** : 281 (1973).
38. H. Tanaka, T. Takanashi, M. Kadota, and M. Yatazawa, *Water, Air and Soil Pollution*, **1** : 343 (1972).
39. H. Tomlinson and S. Rich, *Phytopathology*, **57** : 972 (1967).
40. H. C. Dass and H. M. Weaver, *Atmospheric Environment*, **6** : 759 (1972).
41. J. B. Mudd, T. T. McManus, A. Ongun, and T. E. McCullogh, *Plant Physiol.*, **48** : 335 (1971).
42. G. W. Todd, *Physiol. Plantarum*, **11** : 457 (1958).
43. W. A. Feder, *Science*, **160** : 1122 (1968).
44. W. M. Dugger, J. Koukol, and R. L. Palmer, *J. Air Pollut. Contr. Assn.*, **16** : 467 (1966).
45. R. L. Barnes, *Can. J. Bot.*, **50** : 215 (1972).
46. H. W. de Koning and Z. Jegier, *Atmospheric Environment*, **2** : 615 (1968).
47. H. de Koning and Z. Jegier, *Arch. Environ. Health*, **18** : 913 (1969).
48. T. T. Lee, *Plant Physiol.*, **42** : 691 (1967).
49. F. D. H. MacDowall and A. F. W. Cole, *Atmospheric Environment*, **5** : 553 (1971).
50. W. M. Dugger and I. P. Ting, *Phytopathology*, **58** : 8 (1968).
51. K. R. Darnall and J. N. Pitts, *Chem. Commun.*, 1305 (1970).
52. J. B. Mudd, *J. Biol. Chem.*, **241** : 4077 (1966).
53. L. Ordin and M. A. Hall, *Plant Physiol.*, **42** : 205 (1967).
54. L. Ordin and B. P. Skoe, *Plant Physiol.*, **39** : 751 (1964).
55. W. M. Dugger, J. Koukol, W. D. Reed, and R. L. Palmer, *Plant Physiol.*, **38** : 468 (1963).
56. R. E. Gross and W. M. Dugger, *Environ. Res.*, **2** : 256 (1969).
57. J. Koukol, W. M. Dugger, and R. L. Palmer, *Plant Physiol.*, **42** : 1419 (1967).
58. W. M. Dugger, J. B. Mudd, and J. Koukol, *Arch. Environ. Health*, **10** : 195 (1965).
59. L. Ordin and A. Altman, *Physiol. Plantarum*, **18** : 790 (1965).
60. J. B. Mudd, *Arch. Biochem. Biophys.*, **102** : 59 (1963).
61. O. C. Taylor, *J. Air Pollut. Contr. Assn.*, **19** : 347 (1969).
62. W. W. Thomson, W. M. Dugger, and R. L. Palmer, *Bot. Gaz.*, **126** : 66 (1965).
63. H. Tanaka, T. Takanashi, and M. Yatazawa, *Water, Air and Soil Pollution*, **3** : 11 (1974).
64. J. N. B. Bell and W. S. Clough, *Nature*, **241** : 47 (1973).
65. L. S. Dochinger, F. W. Bender, F. L. Fox, and W. W. Heck, *Nature*, **225** : 476 (1970).
66. J. B. Mudd, *Advances in Chemistry Series*, **122** : 31 (1973).
67. I. Ziegler, *Environmental Quality and Safety*, F. Coulston and F. Korte, eds., **II** : 182 (1973).
68. C. H. Giles, *Chem. & Ind.*, 770 (1964).
69. U. R. Evans and C. A. J. Taylor, *Corrosion Science*, **12** : 227 (1972).
70. T. Misawa, T. Kyuno, W. Seutaka, and S. Shimodaira, *Corrosion Science*, **11** : 35 (1971).
71. T. Misawa, K. Asami, K. Hashimoto, and S. Shimodaira, *Corrosion Science*, **14** : 279 (1974).
72. I. Matsushima and T. Veno, *Corrosion Science*, **11** : 129 (1971).
73. R. Ericsson, B. Heimler, and N-G Vannerberg, *Werkstoffe und Korrosion*, **24** : 207 (1973).
74. T. Sydberger and N-G Vannerberg, *Corrosion Science*, **12** : 775 (1972).

75. J. R. Duncan and D. J. Spedding, *Corrosion Science*, **13** : 993 (1973).
76. J. R. Duncan and D. J. Spedding, *Corrosion Science*, **13** : 69 (1973).
77. B. Heimler and N-G Vannerberg, *Corrosion Science*, **12** : 579 (1972).
78. J. R. Duncan and D. J. Spedding, *Corrosion Science*, **14** : 241 (1974).
79. J. R. Duncan, *Werkstoffe und Korrosion*, **25** : 420 (1974).
80. U. R. Evans, *Br. Corros. J.*, **7** : 10 (1972).
81. H. Guttman, *Metal Corrosion in the Atmosphere*, ASTM STP, **435** : 223 (1968).
82. F. H. Haynie and J. B. Upham, *Mater. Protect. & Perform.*, **9** : 35 (1970).
83. J. R. Duncan and D. J. Spedding, *Corrosion Science*, **13** : 881 (1973).
84. K. Barton, *9. Internationale Verzinkertagung, Düsseldorf 1970*, 1 (1970).
85. M. Schmidt, *J. Inst. Metals*, **98** : 238 (1970).
86. V. E. Carter, *Metal Corrosion in the Atmosphere*, ASTM 435, 257 (1968).
87. P. M. Aziz and H. P. Goddard, *Corrosion*, **15** : 39 (1959).
88. A. C. Chamberlain, *Contemp. Phys.*, **8** : 561 (1967).
89. J. E. Yocom, W. L. Clink, and W. A. Cote, *J. Air Pollut. Contr. Assn.*, **21** : 251 (1971).
90. C. R. Thompson, E. G. Hensel, and G. Kats, *J. Air Pollut. Contr. Assn.*, **23** : 881 (1973).
91. M. J. G. Wilson, *Proc. Roy. Soc. A*, **300** : 215 (1968).
92. R. A. Cox and S. A. Penkett, *Atmospheric Environment*, **6** : 365 (1972).
93. A. Derouane and G. Verduyn, *Atmospheric Environment*, **7** : 891 (1973).
94. D. J. Spedding, *Atmospheric Environment*, **3** : 341 (1969).
95. K. Biersteker, H. de Graff, and A. G. Nass, *Air and Water Pollut. Int. J.*, **9** : 343 (1965).
96. D. J. Spedding and R. P. Rowlands, *J. Appl. Chem.*, **20** : 143 (1970).
97. D. J. Spedding, *J. Appl. Chem.*, **20** : 266 (1970).
98. D. J. Spedding, R. P. Rowlands, and J. E. Taylor, *J. Appl. Chem. Biotechnol.*, **21** : 68 (1971).
99. D. J. Spedding, *J. Appl. Chem. Biotechnol.*, **22** : 1 (1972).
100. D. J. Spedding, *Proc. Int. Clean Air Conf., Melbourne*, 187 (1972).
101. R. C. Braun and M. J. G. Wilson, *Atmospheric Environment*, **4** : 371 (1970).
102. D. J. Spedding, *Atmospheric Environment*, **3** : 683 (1969).
103. G. Thomson, *Atmospheric Environment*, **3** : 687 (1969).
104. S. H. Zeronian, K. W. Alger, and S. T. Omaye, *Textile Res. J.*, **43** : 228 (1973).
105. S. H. Zeronian, *Textile Res. J.*, **40** : 695 (1970).
106. J. B. Atherton, F. L. Hudson, and J. A. Hockey, *J. Appl. Chem. Biotechnol.*, **23** : 407 (1973).
107. D. M. Whelpdale and R. W. Shaw, *Tellus*, **26** : 196 (1974).
108. M. J. Owers and A. W. Powell, *Atmospheric Environment*, **8** : 63 (1974).
109. F. B. Smith and G. H. Jeffrey, *Atmospheric Environment*, **9** : 643 (1975).
110. M. H. Bortner, R. H. Kummler, and L. S. Jaffe, *Water, Air and Soil Pollution*, **3** : 48 (1974).
111. R. A. Rasmussen, *Tellus*, **26** : 259 (1974).
112. J. G. Shepherd, *Atmospheric Environment*, **8** : 69 (1974).
113. G. W. Israel, *Atmospheric Environment*, **8** : 1329 (1974).
114. D. C. Horsman and A. R. Wellburn, *Environ. Pollut.*, **8** : 127 (1975).
115. H-J Jäger and E. Pahlich, *Oecologia*, **9** : 135 (1972).
116. W. Libera, I. Ziegler, and H. Ziegler, *Z. Pflanzenphysiol*, **74** : 425 (1975).
117. S. K. Mukerji and S. F. Yang, *Plant Physiol.*, **53** : 829 (1974).
118. I. Ziegler, *Biochim. Biophys. Acta*, **364** : 28 (1974).
119. D. F. Karnosky and G. R. Stairs, *J. Environ. Quality*, **3** : 406 (1974).
120. S. Godzik and H. F. Linskens, *Environ. Pollut.*, **7** : 25 (1974).
121. I. Ziegler, *Phytochemistry*, **13** : 2411 (1974).
122. E. Pahlich, *Atmospheric Environment*, **9** : 261 (1975).
123. K. Barton, S. Bartonova, and E. Beranek, *Werkstoffe und Korrosion*, **25** : 659 (1974).

Smog from Cars

Many of us regard automobile exhaust as the chief "devil" in the air pollution problem. The reaction of most people to this problem is to talk about electric cars. However, there is much that can be done in the direction of reducing the undesirable materials emitted from automobile exhausts, and, for the next two decades, it may be necessary to continue to apply these methods to reduce the effects of burning residual fossil fuels.

9

Control of Noxious Emissions from Internal Combustion Engines

B. G. Baker

1. Introduction

1.1. A Well-Defined Problem?

Automotive air pollution has been recognized as an environmental hazard for over 30 years. The nature of the problem has been investigated from almost every angle. The relationship between photochemical smog and the components of engine exhausts has been demonstrated. Methods of measuring exhausts and atmospheric pollutants have been developed. Consideration has been given to the effect of traffic density, driving mode, and vehicle maintenance. Solutions have been sought by changes in vehicle design, in the road system, and in the fuels used. Potential new methods of powering automobiles and new methods of public transport have been evaluated.

B. G. Baker • School of Physical Sciences, Flinders University of South Australia, Bedford Park, South Australia 5042

The economic, social, and legal aspects of the problem have received greater public recognition than any other environmental issue. With this concentration of technology and public interest it should be hoped that a total and lasting solution can be found. And yet few people would be sure that at this stage we have found the correct solution, even on a short-term basis.

What appears to be a well-defined environmental problem has proved to be complex and difficult to solve. Is this to be the pattern for our handling of other environmental issues? Or can we learn from this experience and develop an approach which reconciles immediate action with long-term goals?

1.2. The Major Contributor to Air Pollution

Motor vehicles are only one of the many sources of air pollution from combustion, but in urban industrialized countries they are the major source.

The relative amounts of atmospheric pollutants can be seen in Table 1.[1] These data for the United States in 1970 precede the stringent control of motor vehicle exhausts. Motor vehicles produce about 92% of the carbon monoxide, 63% of the hydrocarbons, and 46% of the oxides of nitrogen. Industry is generally the major source of sulfur and particulates but, whereas the particulates from industry are generally flyash and similar materials, the particulates from motor vehicles contain lead.

Estimates of emissions totaled over an entire country, while illustrating the magnitude of the problem, are not the correct method of assessing this type of air pollution. It is the concentrations of pollutants that concern us: motor vehicle density rather than total numbers. Automotive air pollution is localized in urban areas.

The need for action and the type of action to be taken could therefore be determined by each city. In practice, only decisions relating to the road system and traffic density can be made at this level. Regulations requiring the manufacture of vehicles with controlled emission require legislation at a national level.

Table 1. Atmospheric Pollutants in the United States, 1970

	Annual emission in millions of tonnes	
Pollutant	Motor vehicles	Power plants, factories, etc.
Carbon monoxide	59.7	5.2
Hydrocarbons and other organic materials	10.9	6.4
Oxides of nitrogen	5.5	6.5
Sulfur compounds	1.0	22.4
Particulates	1.0	9.8

1.3. *The Role of Traffic Engineering*

Traffic congestion was early recognized as a major factor in determining engine emissions. Each vehicle generates the minimum emission when driven at uniform speed. When traffic density reaches the point where vehicles repeatedly stop and start, emissions increase sharply. If the emission is measured per unit distance traveled, then time spent with the engine idling is obviously detrimental.

The emission of carbon monoxide is shown in Fig. 1 as being dependent on the time taken to travel unit distance.[2] Such low average speeds result from traffic congestion. The increased frequency of stops and starts causes the increased emission. Similar data have been found for emission of hydrocarbons.

It is apparent that estimates of future pollution levels cannot be simply arrived at by multiplying the number of vehicles by an estimate of average emission. The number of vehicles has been growing exponentially in most urban and industrialized countries. The road systems have generally not kept pace with this growth and hence the emissions per vehicle have increased.

One approach to this aspect of the problem is to provide increased road capacity. Freeways certainly allow traffic to move freely and with minimum exhaust emission per vehicle mile. But, as has been found in Los Angeles and in many other cities, a freeway system encourages more cars to travel greater

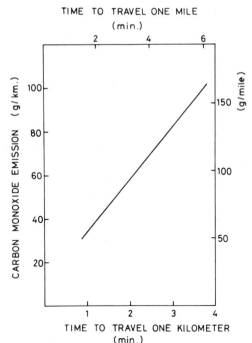

Figure 1. Carbon monoxide emission per vehicle as a function of traffic congestion.

distances. As the freeway system becomes more heavily loaded the problem reappears with greater severity.

The provision of an adequate road system is an important part of the treatment of exhaust emissions as long as the internal combustion engine is in use. Without other provisions, however, traffic engineering cannot provide a lasting solution.

1.4. *Legislation before Technology*

The development of devices to control automobile exhausts has followed the establishment of emission regulations. It is important to recognize that the legislation preceded the development of the technology and so was passed without regard to the difficulty and cost of meeting the limits set.

California began enacting emission control legislation in the 1960s. It required all new 1963 model automobiles to be equipped with closed crankcases, and for the 1966 model year required that unburned hydrocarbons and carbon monoxide be controlled in the exhaust emission of all new automobiles. The United States Air Quality Act of 1967 called upon the National Air Pollution Control Administration to report on air quality criteria. These criteria were to serve as a basis by which States could establish air quality standards for various air quality control regions.

The Clean Air Amendments of 1970 called for the establishment of national air quality standards. These were to be met by 1975 unless the Administrator of the newly created EPA set back the date to 1977. The amendments also set emission limits on hydrocarbons and carbon monoxide emissions for the 1975 model year. The limits were aimed at reducing the air pollution levels prevailing in 1970 by 90%.

Other countries have enacted legislation to limit exhaust emissions, but the limits set are less stringent than those for the United States. The European and United States limits are compared in Fig. 2. The control of nitric oxide is not demanded in Europe, and the emissions of carbon monoxide and hydrocarbons proposed for 1976 represent a decrease of only 50% and 30%, respectively. These measures do not amount to an improved air quality when the growth of vehicle numbers is taken into account. On the other hand, the United States proposal anticipates a technology capable of 90% improvement.

Criticism of the United States proposal was immediate. How could air quality standards be achieved in 1975 or 1977 when the vehicles specifically designed to meet these standards would go on the market in late 1974? The intention was to encourage the development of new technology, but manufacturers were forced to depend on modifications to engines already available. The cost of these controls in terms of the consumption of natural resources has become more evident, and the economic burden is now questioned. Will we achieve an improved air quality standard at reasonable cost? Has any new technology of enduring worth been developed?

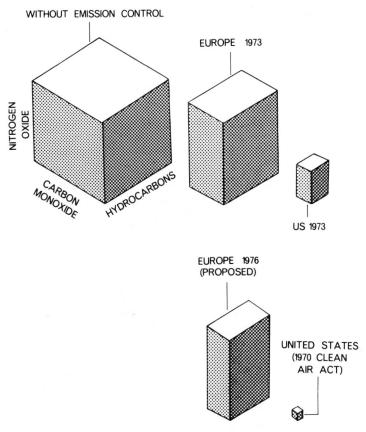

Figure 2. Comparison of United States and European emission control programs.

At this time it is still not certain that the limits set by the United States Clean Air Act of 1970 can be best met by the adaptation of existing engines rather than by a new propulsion system. The requirement for effectively emission-free engines by 1975/76—now delayed to 1976/77—hardly allows time for radical developments in technology.

2. Noxious Emissions

2.1. Safe Limits

The constituents of motor vehicle exhausts which may be regarded as toxic are carbon monoxide, hydrocarbons, oxides of nitrogen, lead compounds, and sulfur dioxide.

The amounts of these depend markedly on the mode of operation of the engine and on the fuel used. While these compounds are known to be toxic at certain concentrations, it is important to establish the conditions under which actual emissions are a hazard to health.

The exhaust from a vehicle is rapidly dispersed and presents no hazard to its occupants. Consideration of the exhaust composition is therefore secondary to the assessment of accumulated concentrations in the atmosphere. The traffic density and climatic factors determine the overall buildup of pollutants in an urban area.

Which of these pollutants represent a serious health hazard and in what quantity? Can we afford to control some and ignore others? The cost of control is high in both money and resources. The cost may rise more steeply than the degree of control and overspecification is undesirable.

Unfortunately, it is not easy to draw up safe limits of concentration for any of the above compounds. The effects of atmospheric pollution on health have not been fully determined. The potential hazards are well known from experiments with the individual constituents, but their combined effects at low concentration and prolonged exposure are difficult to measure.

2.2. *Effects of Exhaust Pollutants*

The human respiratory system has the function of removing oxygen from the air and excreting carbon dioxide from the body. Oxygen is only sparingly soluble in water but reacts with hemoglobin in the blood. The air passages and lungs are designed to maximize the surface area of tissue in contact with the air. Pollutants affect the respiratory system if they are absorbed into the blood stream more efficiently than oxygen; irritate and cause scar tissue, which lessens the effective lung area; and close air passages by promoting secretions or reflex actions.[3]

Carbon monoxide is the most abundant exhaust emission and has a relatively long residence time in the atmosphere (up to 3 years). Local concentrations readily accumulate. Carbon monoxide combines with hemoglobin more than 200 times as readily as oxygen, so that even low concentration levels have adverse health effects. Exposure to 100 ppm results in headache and reduced mental acuity. After about 2 hr at this level of exposure the oxygen-carrying capacity of the blood is below 90% of normal. While this concentration is higher than that usually encountered in practice, the threshold of effect must be much lower. Impairment of physiological functions is observed at a concentration of 17 ppm, which is less than the level measured in many urban areas. The effects include impaired time interval discrimination, decreased performance on psychometer tests and, at higher levels, decreased tissue oxygenation leading to cardiovascular changes. The effects are most pronounced at high altitudes and in people with heart, lung, or central nervous system diseases.[4]

Sulfur dioxide is readily soluble and dissolves in the larger airways above the larynx. This stimulates a contraction of the airway at concentrations. of 2–5 ppm. At higher concentrations severe contraction restricts ventilation.

Nitrogen dioxide, usually formed by atmospheric oxidation of nitric oxide, is less soluble than sulfur dioxide, and so penetrates to the smaller airways and the alveoli, the air sacs where gas exchange occurs. The alveoli have linings with a surfactant which lowers surface tension. The nitrogen oxides destroy this material and the alveoli then either collapse or fill with fluid.

Photochemical smog is a reaction product from nitrogen oxides and hydrocarbons.[5] Ultraviolet light from the sun interacts with nitrogen dioxide to give nitric oxide and free oxygen atoms. These highly reactive entities react with hydrocarbons to initiate chain reactions leading to peroxyacyl nitrates and other organic compounds with irritant properties. Because it is visible and irritating, much emphasis has been placed on the control of smog by the limiting of automotive emissions. Since both nitric oxide and hydrocarbons are involved, it is relevant to ask which is the more important to control.

The rates of the reactions leading to photochemical smog are not simply proportional to the concentrations. Nitric oxide formed in the engine is relatively nontoxic, does not absorb ultraviolet light, and does not initiate photochemical reaction. It is, however, readily oxidized by air to nitrogen dioxide:

$$2NO + O_2 \rightarrow 2NO_2$$

The rate of this reaction is controlled by $[NO]^2 [O_2]$ (i.e., a third order reaction). Since air is present in excess after exhaust, the important term is the square of the nitric oxide concentration. This reaction is most important in the early stages of dilution of the exhaust gases. Nitrogen dioxide absorbs light at wavelengths <430 nm:

$$NO_2 + h\nu \rightarrow NO + O$$

The highly reactive oxygen atoms so produced react further:

$$O + O_2 + M \rightarrow O_3 + M$$

where M is a third body needed to remove energy. Ozone may further react with nitric oxide:

$$O_3 + NO \rightarrow O_2 + NO_2$$

The net effect of this series of reactions is

$$NO_2 + O_2 \overset{h\nu}{\rightleftharpoons} NO + O_3$$

Both ozone and free oxygen atoms can react with hydrocarbons, particularly those containing a double bond. For example, 2-butene reacts with ozone as follows:

$$CH_3CH{=}CHCH_3 \overset{O_3}{\rightarrow} CH_3CH(\dot{O})CH(\dot{O}_2)CH_3 \rightarrow CH_3\dot{C}H\dot{O}_2 + CH_3CHO$$

The free radical centers (marked with dots) are highly reactive species capable of further reaction with oxygen or nitric oxide. The stable molecules, aldehydes, acids, and peroxyacyl nitrates (e.g., $CH_3COO_2NO_2$), are the ultimate products of these reactions, but many intermediates are free radicals. These include $H\dot{O}_2$ and $H\dot{O}$, which are capable of initial reaction with hydrocarbons. A chain reaction is set up which converts the hydrocarbons into the irritating oxygenated products. Hydrocarbons are the fuel of this reaction, and so at first sight it would appear more important to control hydrocarbon emission. However, the high dependence of the rate of smog formation on nitric oxide concentration and the toxic nature of NO_2 demand that the oxides of nitrogen be controlled also.

The smog reaction sequence is reflected in the concentrations measured during one day in a city with photochemically active conditions. In Fig. 3 it can be seen that NO peaks before NO_2, and that total oxidant, $O + O_3$, peaks only after several hours of sunlight. The hydrocarbon concentration is at a maximum for the morning peak traffic but is consumed by reaction with oxidant as the day progresses.

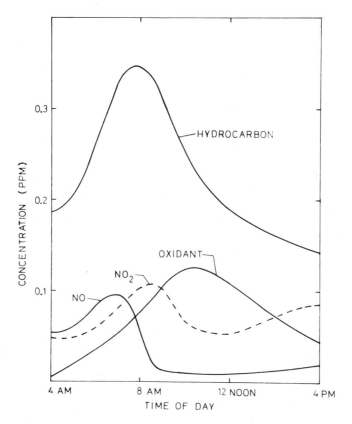

Figure 3. Diurnal concentrations of hydrocarbons and nitrogen oxides under photochemical smog conditions.

Hydrocarbons are not generally toxic in low concentrations, but their participation in smog formation requires control. The unsaturated hydrocarbons (those containing a double bond) are most reactive. The order of reactivity is olefins > aromatics with side chains > saturated hydrocarbons.[6] These are present in varying amounts in gasolines and reach the atmosphere by evaporation and as unburnt fuel at the exhaust. Regulations in the United States currently limit the olefin content of gasoline.

Polynuclear aromatic hydrocarbons such as benzopyrene are known carcinogens. Motor vehicles account for about 20% of the production of this material and, considering the insidious nature of malignant cell growth, there is concern over control of this emission. The production of polynuclear aromatic hydrocarbons in the engine is not directly related to the aromatic content of the fuel. Engine characteristics, oil consumption, and mode of operation have a greater effect on the amount of these hydrocarbons in the exhaust than does the fuel type.[7]

Lead-containing particulates from motor exhausts are the chief source of atmospheric lead. Lead added to the fuel as a volatile liquid (lead tetraethyl or lead tetramethyl) improves octane rating and acts as an antiknock agent. After combustion it is exhausted as lead chlorobromide in particulate form.[8] The particles are of various sizes, but about 80% are less than $0.9\ \mu$ in mean diameter.[9] These become airborne as an aerosol. Heavy particles ($5\ \mu$ and larger), which might be expected to settle rapidly, represent about 30% by weight of the exhausted lead. The average composition of the airborne particles has been determined as in Table 2.

While there are other sources of lead from industry which contribute to the lead found in dust and water, the lead pollution of our urban atmosphere originates primarily from motor vehicle exhausts. Data from Britain and the United States estimate that less than 10% of the airborne lead can be attributed to these other sources.[10] The correlation has been supported by studies of the dependence of atmospheric lead concentrations on proximity to roads and on

Table 2. The Composition of Airborne Material Arising from Vehicle Particulate Emissions

Element or substance identified	%wt
Lead	24.5
Iron	0.9
Chlorine	8.6
Bromine	4.0
Carbon	28.0
Hydrogen	5.8
Nitrate ion	7.3
Ammonia	5.4
Alkaline earths	2.6
Total	87.1

traffic density.[11] The source of lead in the atmosphere is also indicated by the correlation of lead concentration with particulate and gaseous bromine concentrations. Ethylene dibromide is contained in the antiknock fluid, so that lead and bromine in the atmosphere have this common source.[12] Some lead is also detected as gaseous lead alkyl, presumably from evaporation and from incomplete combustion of gasoline.

The atmospheric concentration of lead varies with meteorological conditions. Thermal inversion layers which result in high photochemical smog concentration also cause high lead aerosol concentrations.

The concentration levels of lead in the atmosphere and the resulting levels in the plant and animal tissue are discussed in Chapter 12. There is widespread concern over the high lead levels found in human blood and tissue. Because of the public health hazard, control of the use of lead additives is proposed or in force in several countries. The United States regulations limiting lead additives are due in part to recognition of the health hazard, but also to the requirements of the motor industry to protect catalytic exhaust gas converter systems.

2.3. Air Quality and Emission Standards

The ultimate objective of clean air legislation is to achieve concentrations at which there is no measurable effect on public health. This requires the determination of thresholds for each pollutant, a safe level at which there is no effect on the exposed population.

As methods of measurement and diagnosis improve, such thresholds tend to lower levels. The onset of the clinical symptoms of poisoning by lead or carbon monoxide cannot be regarded as defining appropriate levels when metabolic disturbance is detected at much lower concentrations. The effect of long-term exposure to pollutants cannot be simply assessed by examining adults since they were not exposed in their childhood to present pollution levels. Rather than seek to define a threshold concentration we should consider that a fraction of the population will suffer some disability at all concentrations.

The decision-making process regarding automotive emission standards has, for administrative convenience, depended on the setting of levels of air quality to be achieved. Then, having regard to other sources of air pollution and possible improvements in transportation patterns, the emission standards have been determined from the existing air quality, background levels, meteorology, and automobile population.

At this point, their feasibility in technological, administrative, economic, and political terms may demand reconsideration of the air quality standards set. The United States is at present caught in this loop. The desired standards for emission are not yet achieved. This is not necessarily a bad situation, provided that continued effort is encouraged to meet these standards.

Considering the lack of information on the long-term effects of low levels of pollutants, it is likely that further lowering of air pollution levels will be progressively required.

If, as appears inevitable, the cost of emission control in money and natural resources becomes limiting, a possible saving can be achieved by prescribing emission limits on a regional basis. The United States has set standards to deal with the worst situations in the country. For carbon monoxide, the worst case was Chicago, and for oxidant, Los Angeles. The administrative convenience of nationwide legislation has to be weighed against the cost of overcontrol in less-polluted areas.

3. *Engine Exhaust Gases*

3.1. *Internal Combustion Engines*

Fuel for the internal combustion engines, whether gasoline or diesel distillate, is composed largely of hydrocarbons, compounds of carbon and hydrogen. These burn in air to produce carbon dioxide and water with the release of energy. The internal combustion engine is a pulsed chemical reactor which aims to convert the heat energy of this reaction into mechanical work. The maximum heat energy is released by complete combustion to carbon dioxide and water. For example, combustion of octane would be represented by:

$$C_8H_{18} + 12.5O_2 + 47N_2 \rightarrow 8CO_2 + 9H_2O(g) + 47N_2 \quad \Delta H = -5062 \text{ kJ mol}^{-1}$$

The nitrogen has not entered into the chemistry of the combustion and has not influenced the total energy, but it is present along with oxygen in the air and has to be heated in the combustion process. It therefore affects the maximum temperature of the reaction flame.[13]

The heat content per mole of hydrocarbon increases with molecular weight, but, expressed as heat of combustion per unit weight, the values for liquid hydrocarbons are approximately constant at 44.4 kJ g^{-1}. This does not mean that all hydrocarbons are equally good fuels for the internal combustion engine. The rate of propagation of the flame determines the efficiency with which the heat can be converted to work in the engine. Hydrocarbons of differing molecular structure may have very different combustion properties in an engine.

The operating cycle for the spark ignition engine (the Otto cycle) is shown in Fig. 4a. Gasoline consisting of hydrocarbons in the range C_4–C_{14} is mixed with air in the carburetor. The fuel–air ratio is close to the stoichiometric ratio for complete combustion to CO_2 and water. A range of fuel from about 10% deficient (lean) to about 40% (rich) can be tolerated.[14]

The piston compresses the gases prior to ignition. Compression ratios of about 7–11 are currently in use. Ideally, the ignition of the fuel–air mixture should occur at the top of the piston stroke and result in an instantaneous rise in pressure as depicted by the full line in Fig. 4a. The rate of the chemical reaction, however, is limiting. Ignition should occur before the upstroke is complete in order to allow time for the flame front to move from the spark through the gas

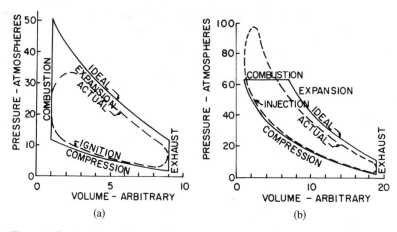

Figure 4. Combustion cycles from (a) the spark ignition engine, and (b) the diesel engine.

mixture. As a result of this reaction rate limitation, the ultimate pressure falls short of the ideal.

The compression ignition engine (the diesel cycle) is shown in Fig. 4b. The compression stroke involves air only, the fuel being injected near the top of the stroke. The fuel is a heavier hydrocarbon fraction and the amount is always less than the stoichiometric equivalent of the air. That is, the diesel engine runs with an excess of air. The fuel–air ratio is in the range 10–70% of stoichiometric.[14]

Ignition occurs without sparking because a much higher compression has occurred. Compression ratios are typically 15–20. The pressure rises beyond the ideal because in this case the reaction in the precompressed air is too rapid for the downward movement of the piston.

Temperatures in both types of engine are comparable, but the pressures are higher in the diesel. Exhaust problems are inherently different, and since the spark ignition engine has come under greater attention in emission control legislation it will be discussed first.

3.2. Operating Conditions in the Spark Ignition Engine

The stoichiometric ratio for complete combustion of gasoline in air is about 15 parts by weight of air to 1 part by weight of fuel.[15] This ratio should give best fuel economy, but because of the nonideal reaction conditions in the engine cylinder, maximum power is only obtained by using excess fuel.

It is inevitable that with a deficiency of air some hydrocarbon will remain unburned and that incomplete combustion will result in some carbon monoxide formation. The dependence of the concentrations of these gases on the fuel mixture is shown in Fig. 5.

The formation of nitric oxide occurs by high-temperature reaction of the nitrogen in the air with oxygen. The burning of the fuel in the engine is not a well-controlled process, and the flame front advancing from the spark achieves

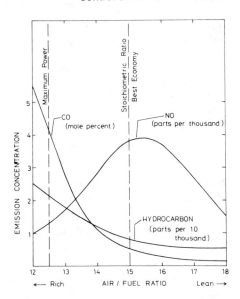

Figure 5. Dependence of gaseous emissions on engine air–fuel ratio.

temperatures of about 2500°K. This is an unnecessarily high temperature for complete combustion of the hydrocarbon. It is determined by the heat capacity of the gases present. If more nitrogen were present, a lower flame temperature would result because this additional gas would have to be heated.

The equilibrium concentration of nitric oxide in air is shown in Fig. 6. At the flame front temperature, concentrations of more than 2% NO could form if

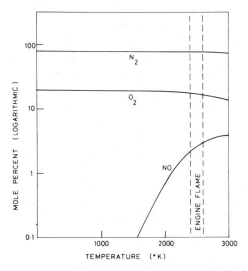

Figure 6. Equilibrium concentrations of nitric oxide in air as a function of temperature.

Table 3. Typical Exhaust Composition from an Unmodified Automobile

Mode of engine operation	Unburnt hydrocarbon (ppm)	Carbon monoxide (%)	Nitrogen oxides (ppm)
Cruise	200–800	1–7	1000–3000
Idle	500–1000	4–9	10–15
Accelerate	50–800	0–8	1000–4000
Decelerate	3000–12,000	2–9	5–50

there were time for reaction.[16] The reaction of oxygen with nitrogen is very slow at low temperatures, but is fast enough at 2500°K to form significant amounts of NO as the flame propagates. At lower temperatures the nitric oxide is thermodynamically unstable; that is, it should revert to oxygen and nitrogen but the rate of decomposition is now very slow. Thus nitric oxide is kinetically stable and is emitted at the exhaust. The amount of nitric oxide is a maximum with slight excess of air (Fig. 5); with greater excess, flame temperatures are lower, so that less is formed.

Mixture ratio is not the only factor in gaseous emission; the timing of the spark and the compression ratio are also important. Advancing the spark markedly increases NO emission, and both NO and hydrocarbon emissions increase as compression ratio increases.[14]

3.3. Effect of Driving Mode on Exhaust Composition

Since the exhaust gas composition depends on the fuel–air mixture and spark timing, it will also depend on the driving mode.

Maximum power demands a mixture about 10–15% rich, while cruising is most economical at slightly lean mixtures. At idle, most engines require rich mixtures because combustion products tend to remain in the cylinder. Under acceleration, fuel tends to condense on the manifold walls because the pressure decreases. To prevent too much weakening of the mixture, the carburetor is set to provide more fuel under accelerating conditions. When decelerating, the closed throttle causes an increase in the manifold vacuum. Reduced air intake and exceedingly rich mixtures are the result.

With these fluctuations it is not surprising to find that emissions are highly dependent on the demands being made on the engine. The range of exhaust gas compositions is shown in Table 3. Acceleration and cruising are the worst conditions for nitric oxide emission, while idling and deceleration cause the highest hydrocarbon and CO emission.[6] It is apparent, therefore, that exhaust emission limits cannot be specified without prescribing the mode of engine operation.

3.4. Driving Cycles and Emission Testing

Exhaust limits are expressed as a total emission over unit distance, driving in accordance with a prescribed driving cycle. This cycle is intended to simulate the succession of acceleration, cruising, deceleration, and idling, as encountered when driving in traffic.

Three driving cycles are shown in Fig. 7. The seven-mode California cycle (1971) is over a short time-span and is less representative of normal traffic than

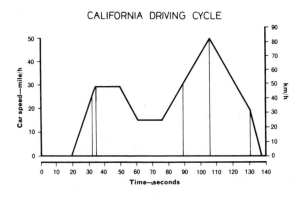

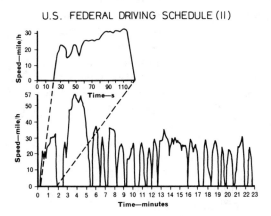

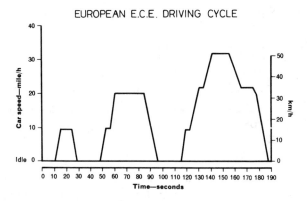

Figure 7. Test driving cycles for the measurement of vehicle emissions.

Table 4. Legislated Emission Levels, United States (Federal)

Year	Hydrocarbon g/mi	Hydrocarbon g/km	Carbon monoxide g/mi	Carbon monoxide g/km	Nitrogen oxides g/mi	Nitrogen oxides g/km
1973	3.4	2.1	39	24.2	3	1.9
1974	3.4	2.1	39	24.2	3	1.9
1975 (interim)[a]	1.5	0.93	15	9.3	3.1	1.9
1975 (original)	0.41	0.25	3.4	2.1	3.1	1.9
1976 (deferred)	0.41	0.25	3.4	2.1	0.4	0.25

[a]Change in test procedure.

the nonrepetitive federal cycle (1972). Different short-time cycles are used in Europe, Japan, and Australia.

The vehicle to be tested is driven in a laboratory on a dynamometer equipped with a power absorption unit and inertia flywheels to represent the load of the vehicle under road conditions. The exhaust gases are either analysed continuously (the United States and Japan) or collected in large plastic bags for subsequent analysis (Europe and Australia).

Nondispersive infrared analysis is used to determine carbon monoxide content. This method is also used to determine hydrocarbon content, except in the United States where it has been replaced since 1972 with flame ionization detection. The change from the infrared technique was due to its poor sensitivity to the unsaturated hydrocarbons active in photochemical smog formation. Oxides of nitrogen are determined by a chemiluminescence technique involving reaction with ozone.[17]

The methods of sampling, the preconditioning of the vehicle, and the analytical technique must all be carefully standardized to achieve reproducible test results. The slow test procedures and extensive laboratory facilities needed allow only a small number of vehicles to be tested. These are taken as representative of a particular model.

The results of the test analyses can be computed as an emission in grams per mile or grams per km. The United States legislated emission levels are shown in Table 4. Japan has a schedule of comparable severity. Australia has adopted the United States 1973 standards to apply from 1976. Europe has controls, but excludes the nitrogen oxides.

The stringency of the 1976 standards proposed in the United States can be judged by comparing the data in Table 3 with an estimate of the actual average exhaust composition of a 4000-lb car that meets the standards: hydrocarbons, 33 ppm; carbon monoxide, 0.2%; nitrogen oxides, 100 ppm.

3.5. The Diesel Engine

The most important difference between the spark ignition and compression ignition engines is that the latter always operate with an excess of air. Combustion is generally more complete and hence emissions of carbon monoxide and unburned hydrocarbons are lower.

Diesel fuel consists of higher molecular weight hydrocarbons, and although less are emitted there is evidence that a higher proportion are carcinogenic.[14]

Combustion temperatures are at least as high as in the spark ignition engine, and, since there is excess air, nitric oxide formation is generally greater.

The methods of testing diesel emissions are not well developed and are necessarily different than those used for spark ignition engines. The presence of excess air allows postcombustion to occur in the exhaust and after sample collection. The excess air also acts as a dilutant, so that more sensitive analysis techniques are required.

The most noticeable emissions from a diesel engine are smoke and odor.[6] Regulations have been mainly concerned with these because they are obvious and irritating.

Black smoke consists primarily of unburned carbon particles, 0.1–0.3 μ. They arise from incomplete combustion and so are an indication of inefficient fuel conversion. Good maintenance of the fuel injection system minimizes smoke, but maximum engine power is only obtained under visible smoke conditions. The elimination of visible smoke requires operation at about 90% maximum power.

White smoke is a fine mist of partly vaporized fuel and water droplets. It is produced at cold start and when the engine misfires. It contains irritants, mainly aldehydes, but since emission is not continuous it presents no serious problem.

Blue smoke consists of droplets of liquid hydrocarbon, smaller in size than those in white smoke or than the particles in black smoke. The smoke forms only after the exhaust gases cool in the air. The formation of blue smoke is dependent on the fuel composition and on the engine load.

The odor of diesel exhaust is unpleasant. Its active constituents are aldehydes and other oxygenated organic compounds. Nitric oxide and nitrogen dioxide contribute to the odor and are probably active oxidants in the formation of the other odoriferous constituents. The measurement of odor is subjective and depends on the human nose used under, it is hoped, standardized conditions.

There is generally less concern about diesel exhaust emissions because they are not major contributors to the CO and gaseous hydrocarbon emission, and also because in most places they constitute a minority of the vehicle population. In cities with a large proportion of diesel automobiles, e.g., Hong Kong, the serious problems of smoke, odor, and noise are evident; there are no grounds for regarding a change to the compression ignition engine as a cure for air pollution.

4. Emission Control by Engine Modification

4.1. Engine Design Characteristics for Low Emission

The main cause of gaseous emissions from spark ignition engines is the lack of control of the actual combustion process. The most direct and reliable approach to reduced emission is to modify the conventional engine design. The air–fuel ratio, ignition timing, compression ratio, and combustion chamber design are obvious parameters for consideration.

Carbon monoxide emission is minimized by an air–fuel ratio close to stoichiometric (15 : 1), lowered compression ratio, and high exhaust temperature.

Hydrocarbon emission is decreased by an air–fuel ratio 15 : 1, retarded ignition, high air inlet temperature, high coolant temperature, low compression ratio, low surface area in the combustion chamber, high exhaust temperature, and minimum valve overlap.

Nitric oxide control requires an air–fuel ratio either rich (<13 : 1) or lean (>17 : 1) (see Fig. 5), retarded ignition, low air inlet temperature, low coolant temperature, low compression ratio, and possibly high surface area in the combustion chamber.[18]

It can be seen that there are counter-requirements, particularly between nitric oxide and the other two gases. There is therefore a limitation on the improvements that can be made by this course of action.

Because the emission control regulations have been introduced in a stepwise fashion with the initial emphasis on CO and hydrocarbon, it has been possible to meet immediate limits on them by the engine modification approach. In fact, given the short time available for development this approach was mandatory.

4.2. *Positive Crankcase Ventilation*

The first move against hydrocarbon emission was to eliminate the escape of gases from the crankcase of the engine. This is gas that has leaked past the pistons and is essentially unburned fuel–air mixture. It has been shown to contain more of the unsaturated hydrocarbons, which are active in smog formation, than do exhaust gases.

Californian legislation required 1963 cars to eliminate the discharge into the atmosphere of cylinder blow-by gases by recycling them back through the induction system into the combustion zone of the engine.

Positive crankcase ventilation is achieved either by a tube connecting the crankcase to the air cleaner or by a tube leading directly to the induction manifold. In the latter case, a ventilation valve is necessary to automatically meter the gas flow when the manifold vacuum varies. This valve must also act to isolate the crankcase in the event of backfire. A combination of the two ventilator systems is also in use; this allows gas into the manifold at low speeds and into the air cleaner at higher speeds. Positive crankcase ventilation has reduced hydrocarbon emission by about 20% of the initial level.[6]

4.3. *Evaporative Loss Control*

Evaporative loss from the fuel tank and carburetor amount to about 15% of the hydrocarbon emission. This loss is dependent on temperature, fuel volatility, and venting arrangements. The evaporative loss is predominantly of the lighter hydrocarbons: butanes and pentanes.

The requirement to eliminate this source of hydrocarbon (United States, 1971) has been met by an evaporative loss control device depending on vapor adsorption. A canister containing charcoal adsorbent collects vapor from both the fuel tank and the carburetor when the engine is stopped. A pressure-balancing valve closes external vents when the engine stops, but maintains correct metering pressures at the carburetor when the engine is running. The valve is actuated by induction manifold vacuum. While the engine is running the air intake passes through the adsorbent canister and sweeps the collected hydrocarbon vapor into the engine.[6]

An alternative system, which uses the crankcase air space to collect evaporated fuel, has also been employed.

These evaporative loss systems are claimed to have no significant adverse effect on engine operation or exhaust emission.

4.4. *Combustion Chamber Improvements*

The wall of the combustion chamber is cooler than the burning gas and a thin layer of mixture in contact with it does not burn. This incomplete combustion is a source of unburned hydrocarbon in the exhaust.

The effect is minimized by reducing the surface-area-to-volume ratio and thereby reducing the extent of the cool surface. Compact combustion chambers are achieved in engines that have a hemispherical head or a bowl-in-piston design. These engines also have an advantage in minimizing fuel octane requirements. Lower compression ratios decrease hydrocarbon and nitrogen oxide emissions.[14]

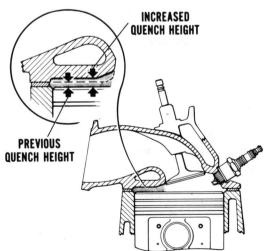

Figure 8. Increased clearance in combustion chamber to reduce the tendency to quench the flame. (Courtesy of Chrysler Corp.)

Detailed modifications to the combustion chamber to eliminate pockets and close clearance spaces reduce the tendency to quench the flame before combustion is complete (Fig. 8).

A more radical design change is the Honda system, which incorporates an auxiliary combustion chamber around the spark plug and an additional intake valve.

4.5. *Fuel Mixture Control*

The air–fuel ratio has the most significant effect on gaseous emission. A setting close to the stoichiometric ratio for complete combustion is essential to limit the emission of hydrocarbons and carbon monoxide (Fig. 5). This mixture is leaner than is required for maximum power and so performance must to some extent suffer. This change also increases nitric oxide formation and so this emission must be dealt with by other means. It is impractical to operate the engine at the very high air–fuel ratio needed to decrease nitric oxide formation.

The regulation of the air–fuel ratio is generally achieved by a carburetor which adjusts the mixture to account for variations in engine speed and load. Smaller fuel jets are now fitted and much greater attention is being paid to reducing manufacturing tolerances. The greatest difficulty encountered is under idling conditions. Engines tend to run roughly and stall when set to a leaner idle.

The carburetor system depends on an induction manifold to carry the fuel mixture to the cylinders. One serious defect is that closure of the throttle cuts off the fuel supply at the carburetor but cannot prevent liquid fuel from the manifold walls entering the cylinders. This ineffective cut-off accounts for much of the excessive hydrocarbon emission during deceleration. Attempts to minimize this effect involve delaying throttle closure or limiting manifold vacuum.

Fuel injection systems have inherent advantages over carburetor systems. The fuel can be measured more precisely, each cylinder can receive the correct amount of fuel, and the wall effects of the manifold are avoided. Problems are encountered in practice in achieving reliable operation. The amounts of fuel at idle are very small and are difficult to control with sufficient accuracy, and precise cut-off when decelerating is not easily achieved.

The cold start condition with the choke operating results in high hydrocarbon emission; fully automatic choke operation is necessary to avoid unnecessary use. The warmup period presents a major difficulty in meeting emission standards. Closer control of fuel volatility specifications can help.

The design of the induction manifold has an important bearing on emissions. An unsatisfactory manifold does not distribute the air–fuel mixture uniformly to the cylinders. The power output and emission characteristics of individual cylinders will therefore differ, and no adjustment of the carburetor can compensate for this imbalance. The improved design and manufacture of

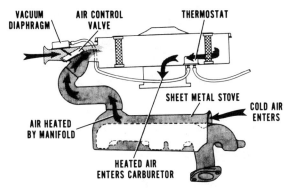

Figure 9. Inlet air heater to improve fuel vaporization.
(Courtesy of Chrysler Corp.)

manifolds to achieve equal gas flow to all cylinders improves engine power performance as well as reducing emissions. There is a partly conflicting requirement to decrease the total manifold volume between carburetor and cylinders in order to reduce the manifold wall effects referred to above.

Improved vaporization of the fuel and a more homogeneous air–fuel mixture are achieved by heating the inlet air (Fig. 9). This reduces emissions and gives better engine operation with lean fuel mixtures.[19] A thermostatically controlled valve on the air cleaner regulates the intake of air which has been heated by passing over the exhaust manifold. The direct heating of the induction manifold has a similar effect, but excessive heat reduces the power output and may increase nitric oxide formation.[20]

4.6. *Ignition Timing Control*

The retarding of the spark timing reduces hydrocarbon emission, but excessive retardation results in loss of power and excessive fuel consumption. As with the leaner carburetor settings, the requirement is for closer control and reduced tolerances in manufacture.

The maintenance of timing accuracy and the design of systems that give optimum timing over the whole range of engine operation has received much attention. Control of manufacturing operations and testing of components does not ensure the long-term accuracy of timing in a system using conventional contact breaker points: these wear and consequently the spark gap changes. Electronic ignition systems have been developed which avoid the mechanical action of points and are hence less susceptible to the effects of wear.

4.7. *Exhaust Gas Recirculation*

The above engine modifications have generally reduced carbon monoxide and hydrocarbon emissions. These modifications, with the exception of ignition

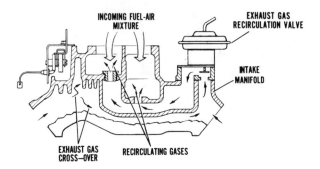

Figure 10. Proportional exhaust gas recirculation to decrease
nitric oxide formation. (Courtesy of Chrysler Corp.)

retardation, do not decrease emissions of the nitrogen oxides. The leaner fuel
mixtures make them substantially worse. A further approach has been
developed to control the nitric oxide formation.

The formation of nitric oxide in the engine was discussed above. The high
flame temperature arises from the amount of fuel and the composition of the air
used to burn it. A lower flame temperature—and hence less nitric oxide—
would result from combustion with air that is less rich in oxygen than normal
air. The obvious source of a dilutant for the air is exhaust gas from which most
of the oxygen has been removed by combustion. The technique is known as
exhaust gas recirculation or recycling. A metered amount of the exhaust gas is
returned to the intake of the carburetor and mixes with the incoming air and
fuel (Fig. 10). In this way, up to about 15% of the exhaust can be recycled and
nitric oxide emissions are reduced up to 60%.[21]

The success of these systems depends on the control valve automatically
adjusting to the driving mode of the vehicle. Under some conditions there is a
falling off in engine performance.

4.8. *Effects of Engine Modifications*

A typical emission control system to meet the United States standards for
1973 (Table 4) is shown in Fig. 11. Systems of this kind, based only on engine,
fuel, and air modifications, are not capable of improving emission levels much
beyond the 1973 standard. While the initial cost of the equipment is moderate,
the detrimental effects on fuel consumption and on the general drivability of
the vehicle have been unpopular with motorists.

Motorists have been accustomed to vehicles tuned to optimize performance, ease of starting, idling, and economical operation. Tuning to limit
exhaust emission is not consistent with these goals. In the belief that motorists
would more readily accept the cost rather than a loss of performance, the
automobile industry in the United States chose to increase engine size to
compensate for the decline in performance. The resultant steep increase in fuel

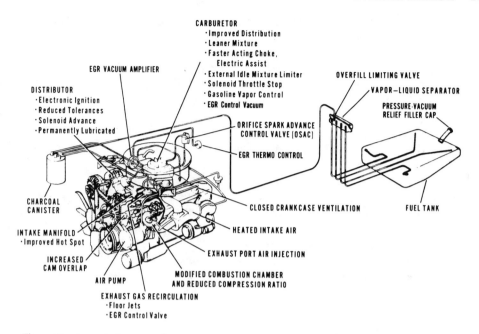

CARBURETOR
·Improved Distribution
·Leaner Mixture
·Faster Acting Choke,
 Electric Assist
·External Idle Mixture Limiter
·Solenoid Throttle Stop
·Gasoline Vapor Control
·EGR Control Vacuum

EGR VACUUM AMPLIFIER

DISTRIBUTOR
·Electronic Ignition
·Reduced Tolerances
·Solenoid Advance
·Permanently Lubricated

OVERFILL LIMITING VALVE

VAPOR–LIQUID SEPARATOR

PRESSURE-VACUUM
RELIEF FILLER CAP

ORIFICE SPARK ADVANCE
CONTROL VALVE (OSAC)

EGR THERMO CONTROL

CHARCOAL
CANISTER

CLOSED CRANKCASE VENTILATION

FUEL TANK

INTAKE MANIFOLD
·Improved Hot Spot

HEATED INTAKE AIR

INCREASED
CAM OVERLAP

EXHAUST PORT AIR INJECTION

AIR PUMP

MODIFIED COMBUSTION CHAMBER
AND REDUCED COMPRESSION RATIO

EXHAUST GAS RECIRCULATION
·Floor Jets
·EGR Control Valve

Figure 11. An emission control system to meet the United States 1973 standards. (Courtesy of Chrysler Corp.)

consumption in 1971–74 cars proved a serious embarrassment when petroleum shortage and price increases occurred.

For two reasons—the need to make further decreases in emissions and the need to achieve these without fuel wastage—an alternative approach to control has developed. This is the direct treatment of exhaust gases to remove active pollutants by chemical reaction.

5. *Treatment of Exhaust Gases*

5.1. *Basic Problems in Exhaust Treatment*

Methods of reacting exhaust gases have been extensively investigated, but earlier efforts were overshadowed by the work on mechanical modifications to reduce engine emission. Now that those methods seem to have reached their limit, interest in practical reactor systems is running high.

The objective is to convert carbon monoxide and hydrocarbons by oxidation in excess air and to react nitric oxide with carbon monoxide or decompose it to nitrogen. The oxidation of NO to NO_2 is not desirable, so the treatment of NO must be considered separately from the oxidation reaction. The reactions can in principle be achieved by either thermal reaction or by the use of a catalyst.

The most immediate difficulty is that the active species are present in low concentration in a fast-flowing gas. The space velocity in an exhaust system is high (5–15 volume per volume per second), and therefore very fast chemical reaction is required in the controlled reaction zone.

The composition of the exhaust gas varies markedly with engine operating conditions (Table 3). These variations in concentration must be coped with by the reactor.

The reactions require elevated temperatures, but when starting from cold the reactor will be ineffective. Water will condense in the cold reactor during the warmup period.

To these technical difficulties is added the requirement that the reactor system must operate without attention for a long period. Unlike other functions and accessories of the car, the motorist is not motivated to pay attention to an inoperative reactor system and would probably have no indication that the system had become ineffective.

Exhaust emissions are difficult to measure, and it is less feasible to regularly test and police effectiveness than to insist on a certain average level of reliability.

5.2. *Thermal Oxidation Reactors*

The essential feature of these systems is to add excess air after the combustion products leave the cylinder, but before they have a chance to cool down. Complete combustion is then achieved without interfering with the air–fuel ratio in the engine. Because the supplementary air must be added close to the engine, where the exhaust pressure is high, a pump is required to force the air in.

An early version of an exhaust manifold reactor was the Man-Air-Ox System (General Motors) used in the late 1960s in the United States. The pump, driven by the engine, injects air into the exhaust manifold close to the exhaust valves. Hydrocarbons and CO are oxidized and, by running the engine on a rich mixture, nitric oxide formation is minimized. While this method was successful, it was superseded by engine modifications, mainly because of the extra cost of the pump and because of the excess fuel consumption resulting from the rich mixture and the power required to drive the pump.

More recent exhaust manifold reactors are larger than conventional manifolds and are fitted with internal baffles. Radiation shields and insulation ensure that high-reaction temperatures, up to 1000°C, are reached. By holding the exhaust gases mixed with injected air at high temperature, near complete oxidation is achieved. The device depends on having a mixture rich enough to ensure that sufficient fuel is being burned in the reactor to maintain high temperature.[22] This results in a 10–15% loss in fuel economy.

The high temperatures require that the reactor be constructed of expensive materials. Failure of a spark plug would allow excessive fuel to enter the reactor and could allow even higher temperatures to damage the reactor. One method of preventing this is to have a temperature-actuated cut-out on the reactor air supply.

The presence of lead in gasoline is compatible with this reactor, but high-temperature corrosion by lead salts decreases reactor life.

The control of carbon monoxide and hydrocarbons is very good except at a cold start. Nitrogen oxides are not reacted but some degree of control results from the use of a rich mixture. The system is compatible with exhaust gas recirculation which can further control nitric oxide formation.

An alternative type of thermal reactor has the engine use a lean air–fuel ratio (17 : 1).[23] Because there is already excess air in the exhaust, no auxiliary air pump is necessary. The reactor temperature is lower and the construction of the reactor is therefore cheaper. The running of an engine on such a lean mixture requires a special high-velocity carburetor. Fuel economy is improved by this reactor, but it is less efficient than the rich reactor in limiting emissions.

5.3. *Catalyst Properties for Exhaust Gas Reaction*

Since the first proposals for exhaust emission controls were made, much effort has been expended in the development of catalytic exhaust reactors. While engine modifications and thermal reactors have met the intermediate stages of control, it is considered that the catalytic systems will be needed to meet the proposed 1976 controls.[7]

A catalyst increases the rate of a chemical reaction, allows it to proceed at a lower temperature, and may selectively influence the formation of preferred products. The chemical industry is heavily dependent on catalytic processes, and there is much practical knowledge about catalysts, although a limited theoretical understanding of them.

The action of a solid catalyst in a gas reaction occurs only at the surface. The catalyst may be a porous material with high surface area but nevertheless it is easily contaminated by relatively small quantities of material. In an industrial process, the catalyst is prepared by carefully controlled procedures and is then maintained in the reaction environment with constant temperature and gas composition.

In the automobile exhaust the catalyst must withstand wide variations in temperature and exhaust gas composition. Clearly this catalyst cannot function effectively under all conditions and very likely will cease to function at all due to irreversible contamination.

The desired characteristics of a catalyst for the severe conditions in an exhaust system are a high reactivity for the specific reaction under all operating modes, and a long service life. The properties of a catalyst likely to meet these conditions are as follows:

1. *High surface area* is achieved by having a pore size suitable to admit gas rapidly to internal surfaces. This can be achieved by the choice of a catalyst which inherently has this property (an "unsupported catalyst"), or by deposit-

ing the active material in a finely dispersed form on a porous support material (a "supported catalyst").

2. *High melting point* is needed to withstand the hot exhaust gases. The oxidation reactions are strongly exothermic and catalyst temperatures may rise above the optimum. In general, an operating temperature up to 600°C is to be expected.

3. *Resistance to poisoning* is effectively a chemical inertness to chemicals other than the reactant gases. The most troublesome catalyst poisons are lead salts, phosphorus, and sulfur. Particulate matter may also reduce catalyst activity by mechanically blocking the catalyst pores.

4. *Resistance to attrition* is most important if a stable catalyst bed with good gas contact is to be maintained. Materials with the high-area properties required in condition (1) above are generally ceramic oxides. These tend to break down with repeated temperature cycling, vibration, and the abrasive effect of the hot gas flow. The effect for a material of given hardness is minimized by design of the gas inlet and flow to achieve relatively low gas velocity in the catalyst bed.

With these criteria in mind, the search for catalysts of suitable activity has been conducted by empirical methods. Laboratory experiments have been conducted on large numbers of samples of supposed catalytic activity. Synthetic gas mixtures or actual exhaust from an engine may be used on a test rig. The latter type of experiment is the better simulation of the service condition, but is not sufficiently controlled to determine the ultimate potential of a newly investigated catalyst. Basic catalyst research has not kept pace with the intensive testing program of the automobile industry.

5.4. *Catalytic Oxidation of CO and Hydrocarbons*

The objective of complete oxidation of hydrocarbons to carbon dioxide and water is quite different than the aim of industry to produce partial oxidation products of commercial value. New catalysts have had to be investigated. Some of these catalysts are found to have activity for the reaction of nitric oxide with carbon monoxide. However, this reaction requires reducing conditions and will not occur when excess air is introduced to oxidize the carbon monoxide.

For the oxidation reactions, laboratory tests have shown catalytic activity for the oxides of copper, chromium, manganese, and cobalt. These catalysts are effective for both hydrocarbons and carbon monoxide. Vanadium pentoxide is a poor catalyst for CO but effective for hydrocarbons.[24] Mixed catalysts, consisting of copper, cobalt, and vanadium oxides supported on alumina, have shown good performance in stationary engine tests.[24]

The noble metals, platinum and palladium, supported on either silica or alumina, have very good conversions (>80%) for the oxidation of both

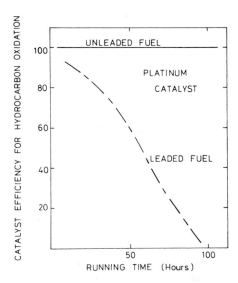

Figure 12. Poisoning of a platinum oxidation catalyst by lead.

hydrocarbons and carbon monoxide. These catalysts are susceptible to poisoning by the lead additive in gasoline. In Fig. 12, a platinum catalyst of almost 100% efficiency is shown to become ineffective in less than 100 hr when run on leaded gasoline.[24] It is generally agreed in the catalyst and automobile industries that cars with catalytic converters will need lead-free fuel.

The use of noble metals as catalysts raises questions of cost and supply. Platinum is certainly expensive and has become scarce, but the amount needed to provide a car with an effective catalyst is small. It should be remembered that only the surface is needed as a catalyst and by very fine dispersal a high area can be covered with very little metal. It has been shown that a car (V8 engine, 5.7 liter) equipped with an oxidation catalyst containing only 0.56 g of total metal can meet the 1975 emission standards for both CO and hydrocarbons for over 50,000 mi of operation.[25] This test involved running under regulated conditions rather than under normal road usage.

The catalytic systems have been added to cars already modified for leaner fuel operation. There is evidence that better results with catalysts can be achieved by reverting to stoichiometric air–fuel rations and normal ignition timing. Thus normal drivability and good fuel economy may be regained by the change to catalytic emission control systems.

5.5. *Comparison of Oxidation Systems*

The competitive systems, thermal and catalytic oxidation converters, are almost equally efficient in dealing with CO and hydrocarbons. The initial cost of the installed systems is about equal. The thermal reactor runs on leaded gasoline, which is slightly cheaper than unleaded. Fuel consumption is higher for the thermal reactor system because of the power needed for the pump and the rich fuel mixture.

The catalytic system gains in fuel economy by more than the added cost of unleaded gasoline and gains by the lower engine maintenance costs claimed for unleaded fuel.[7] The conclusion of this study is that the catalytic system will prove more efficient and more economical than the thermal reactor.

5.6. *Catalytic Reaction of CO and NO*

The most promising approach to the nitric oxide problem is to react this gas with the carbon monoxide already present in the exhaust gas:

$$2NO + 2CO \rightarrow N_2 + 2CO_2 \qquad \Delta G^\circ_{298} = -690 \text{ kJ mol}^{-1}$$

This reaction is thermodynamically very favorable for the composition and temperature of the exhaust gases.

The presence of added air usually results in preferential oxidation of the CO by oxygen. This may, however, be a matter of catalyst selectivity as both reactions can, in principle, occur.[24,26]

Other possible reactions are the formation of nitrous oxide:

$$2NO + CO \rightarrow N_2O + O_2$$

and the reaction of CO with the water in the exhaust gas:

$$CO + H_2O \rightarrow CO + H_2$$

This hydrogen so formed can react with nitric oxide to produce ammonia:

$$2NO + 5H_2 \rightarrow 2NH_3 + 2H_2O$$

This is not a desirable reaction path, since the ammonia is reoxidized to NO when the gases pass through the oxidation catalyst.[27]

A further reaction possibility, observed on some catalysts, is the reduction of NO by hydrocarbons to yield nitrogen, water, and carbon monoxide.

The most effective catalysts for the reduction of nitric oxide are unsupported metal alloys (Monel and stainless steel) or mixtures of base metal oxides (chromia, copper oxide, ferric oxide). Various physical forms of the unsupported metal have been tried. Metal felts made from fine metal fibers, and metal foams with ~95% porosity, have the advantage of high area, but practical systems have generally used metal chips, which are less sensitive to sintering.

5.7. *Catalytic Decomposition of NO*

The catalytic reduction of nitric oxide depended on the presence of a reducing atmosphere containing an excess of CO. For the spark ignition engine conditions are favorable, provided that the catalyst precedes the introduction of secondary air. For the diesel engine the condition is never met because excess air is already present in the exhaust from the engine. For this reason there has been considerable interest in the direct decomposition of NO:

$$2NO \rightarrow N_2 + O_2$$

The equilibrium constant for this reaction is

$$K = \frac{([N_2][O_2])^{\frac{1}{2}}}{[NO]} = 3 \times 10^5 \text{ (at 500°C)}$$

Thus the decomposition reaction is highly favorable at exhaust gas temperatures. However, as explained above, the rate of decomposition is slow at this temperature: nitric oxide is kinetically stable.

Trials of catalysts to achieve high-enough rates of decomposition are generally discouraging.[28] There is evidence that the reaction is inhibited by the presence of water vapor in the exhaust gas but that at low temperatures (<100°C) the reaction is more favorable.[29] Further work is needed on this reaction to develop control of nitric oxide emission from diesel engines.

5.8. *Compatible Control Systems*

The methods of treating exhaust gases are not all compatible with each other or with the engine modification approach to emission control. Three overall systems which have shown promising performance are illustrated in Fig. 13.

System A is based on a manifold thermal oxidation reactor to deal with CO and hydrocarbons. This requires a large-capacity secondary air pump and enriched carburetion. Nitric oxide control is achieved by exhaust gas recirculation to lower the combustion temperature. A possible over-temperature condition is protected by a thermal cutout on the secondary air. Evaporation controls on the fuel tank and the carburetor can be included, and ideally the engine would run on low-lead or unleaded fuel to prolong reactor life.

System B is based on an oxidation catalyst to remove CO and hydrocarbons. The secondary air pump in this case is smaller because the device operates at lower pressure. A programmed protection system operates a valve to allow the exhaust to bypass the catalytic converter whenever conditions are likely to damage the catalyst. Enriched carburetion and exhaust gas recirculation are still required to control nitric oxide formation. In this case, the exhaust gas must be drawn off close to the engine before the secondary air is introduced. A cooler is therefore needed at the recirculating valve. Evaporative controls are included and unleaded or very low lead fuel is required to preserve catalyst life.

System C is fully catalytic. The first catalyst in the dual-bed system reacts NO and CO. At this point secondary air is introduced and the second catalyst oxidizes the remaining CO and hydrocarbons. Programmed protection and evaporative controls are included as before, and unleaded fuel is essential.

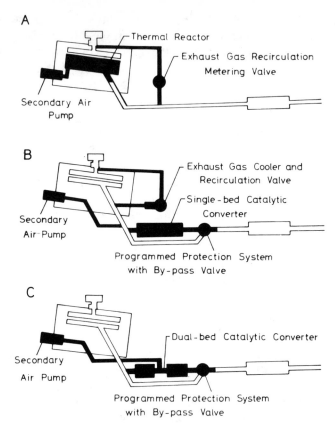

Figure 13. Emission control systems involving thermal and catalytic reactors.

Recent developments in control systems have also involved the monitoring of exhaust gas composition to provide an immediate feedback control of engine operation.[30] Oxygen sensors in the exhaust provide a means of controlling the air intake to the engine and so maintaining constant reaction conditions in the catalyst. One type of sensor involves a solid electrolyte, such as zirconia (ZrO_2) with yttria (Y_2O_3), which forms part of an electrochemical cell. The cell potential is dependent on the oxygen concentration in the electrolyte which, at elevated temperatures, is dependent on the oxygen content of the exhaust gas. Another type of sensor depends on the electrical conductivity of semiconductor solids.

6. Fuel Specifications and Lead Emission

6.1. Demands on the Petroleum Industry

The development of devices to control vehicular emissions has absorbed a significant proportion of the development resources of the automotive industry, but there are aspects of emission control which cannot be solved by this means because they involve the composition of fuel. An environmental responsibility then falls on the petroleum industry to produce fuels which enable emission standards to be met.

The quality of motor fuels has been determined in the past largely by competitive economic pressures on the automotive and petroleum industries.

On a worldwide scale, the petroleum industry now anticipates that motor fuel properties will be prescribed by decisions made by governmental authorities and by the automotive industry.

The forward planning of the petroleum industry has, for several years, been concerned with meeting the new requirements.

Some of the changes anticipated are:[31]

1. *Volatility.* Closer controls of the vapor pressure range are being considered in the automotive industry in order to allow better control of hydrocarbon emissions in the engine warmup period.
2. *Octane rating.* Based on new car requirements in a number of countries, octane number levels have changed.
3. *Composition.* Limitations have been considered on olefins and aromatics because of their contribution to photochemical smog and, in the case of the latter, possible carcinogenic activity.
4. *Antiknock additives.* Lead compounds are now regulated by governments in several countries. Vehicle emission control devices which utilize catalytic systems preclude the use of lead.
5. *Sulfur content.* Possible environmental regulation.
6. *Odor.* Possible environmental regulation.
7. *Additive use.* Based on new car requirements and effects on catalytic converters, new or additional additives are likely.

The petroleum industry has made some progress in meeting these demands, but the recent steep price increases and shortage of supplies have introduced many uncertainties. The refining developments needed to meet environmental regulations will add to already high costs. In the following discussion of the developments contemplated by the petroleum industry it should be remembered that their economic feasibility will have to be reconsidered as further changes occur in the pricing and supply of petroleum.

6.2. Volatility and Composition Specification

The boiling range of gasoline is controlled by the composition of the mixture. Very volatile components contribute to evaporative losses and hence to hydrocarbon pollution. However, the least volatile components are also cause for concern. In 1972, General Motors Corporation proposed that the petroleum industry increase fuel volatility so that 90% boiled before 121°C. This was later amended to 160°C.

The special emphasis on volatility characteristics is in connection with the high level of emissions which occurs during warmup. The carbon monoxide and hydrocarbon emissions are high because choking is required for starting and satisfactory drivability, and catalytic devices are not efficient until they reach operating temperatures.

The significance of the early part of the test cycle on total hydrocarbon and carbon monoxide emissions is that even with the 1975–1976 systems designed to minimize this effect, 33% of the total hydrocarbons and 53% of the total carbon monoxide are produced in the first 30 sec of engine operation. A motor fuel of high volatility would permit the use of a shortened choking period which would minimize emissions during warmup.

While the industry has not met this new requirement, studies have shown that, at some cost, it could be met with the range of process equipment needed to produce the higher-octane fuels.

Other controls of composition have been under consideration. The limiting of olefin and aromatic concentrations would reduce photochemical smog, but it would be impractical, with the present refining techniques and octane requirements, to impose specific limits.

A level of allowable sulfur in gasoline lower than present average levels is probable. Most of the new refining techniques automatically result in lower sulfur levels and, therefore, little difficulty is expected in meeting future requirements.

6.3. *Gasoline Octane Rating*

The important characteristic of the fuel in the combustion chamber is its ability to withstand preignition. The sound of knocking from the engine is an indication that an explosive burning has occurred ahead of the flame front and that this energy has been wasted.[32]

The knock rating of a fuel is determined in a standard single-cylinder test engine of variable compression ratio. The rating is specified by an octane number. This is an empirical scale, introduced in 1927, that assigns a rating of zero to normal heptane, which is very prone to knock, and of 100 to isooctane (2,2,4-trimethylpentane), which is a branched paraffin resistant to knocking.

The octane number of a given fuel is the percent of isooctane in a blend with *n*-heptane that has the same knocking characteristics as the fuel under examination in the test engine and operated under specific conditions. Since the scale was devised, fuels superior to isooctane have been found and the octane scale is now extrapolated to about 120.

Determination of the octane ratings of a large number of synthetic hydrocarbons has shown that in the paraffin series octane number decreases as the chain length increases, but increases with branching of the chain.[33] Olefins have higher ratings than the corresponding paraffins, and the octane number improves as the double bond shifts to the center of the molecule. Naphthenes have higher octane numbers than normal paraffins, and aromatics have exceptionally high octane numbers. Some values are shown in Table 5.

Table 5. Research Octane Ratings of Hydrocarbons

Name	Formula	Octane number
	ALKANES	
C_4 n-Butane	$CH_3CH_2CH_2CH_3$	94
Isobutane	$CH_3CH(CH_3)_2$	101
C_5 n-Pentane	$CH_3(CH_2)_3CH_3$	62
2-Methylbutane	$CH_3CH_2CH(CH_3)_2$	92
C_6 n-Hexane	$CH_3(CH_2)_4CH_3$	25
2-Methylpentane	$CH_3(CH_2)_2CH(CH_3)_2$	73
3-Methylpentane	$CH_3CH_2CH(CH_3)CH_2CH_3$	75
2,2-Dimethylbutane (neohexane)	$CH_3CH_2C(CH_3)_3$	92
2,3-Dimethylbutane	$CH_3CH(CH_3)CH(CH_3)CH_3$	102
C_7 n-Heptane	$CH_3(CH_2)_5CH_3$	0
C_8 3-Methylheptane	$CH_3(CH_2)_3CH(CH_3)CH_2CH_3$	27
3-Ethylhexane	$CH_3(CH_2)_2CH(C_2H_5)CH_2CH_3$	34
2,2-Dimethylhexane	$CH_3(CH_2)_3C(CH_3)_3$	73
2-Methyl-3-ethylpentane	$CH_3CH_2CH(C_2H_5)CH(CH_3)_2$	87
2,2,4-Trimethylpentane (isooctane)	$(CH_3)_3CCH_2CH(CH_3)_2$	100
	ALKENES	
Butene-1	$CH_3CH_2CH=CH_2$	97
Butene-2	$CH_3CH=CHCH_3$	100
Pentene-1	$CH_3CH_2CH_2CH=CH_2$	91
2-Methylbutene-2	$CH_3C(CH_3)=CHCH_3$	97
Hexene-2	$CH_3(CH_2)_2CH=CHCH_3$	93
	NAPHTHENES	
Methylcyclopentane	$C_5H_9CH_3$	91
Ethylcyclopentane	$C_5H_9CH_2CH_3$	67
Cyclohexane	C_6H_{12}	83
Methylcyclohexane	$C_6H_{11}CH_3$	75
	AROMATICS	
Toluene	$C_6H_5CH_3$	107
Ethylbenzene	$C_6H_5CH_2CH_3$	104
n-Propylbenzene	$C_6H_5CH_2CH_3$	105

The straight-run gasoline obtained by the direct distillation of petroleum consists mainly of paraffins and its octane rating is generally in the range 50–70. This low-octane fuel would limit the compression ratio and thermodynamic efficiency of engines. In modern gasoline this limitation is overcome by refinery processing, which makes hydrocarbons of more favorable structure, and by the use of additives to improve octane number.

6.4. Antiknock Additives

The first commercial antiknock agent in gasoline was lead tetraethyl. It was first marketed in 1923 and is still the most important. Other lead alkyl compounds are also effective. Since 1960, lead tetramethyl and the mixed methyl-ethyl compounds of lead have also been used.

The only other successful antiknock agents found are carbonyls of transition metals. Iron carbonyl was used for a time in Europe, but caused problems of excessive wear due to the formation of abrasive iron oxide in the engine. Another compound of this class, an organomanganese carbonyl (methylcyclopentadienylmanganese tricarbonyl), was marketed in the United States in 1959 under the trade name AK-33X. Cost has limited its use to supplementing lead tetraethyl rather than replacing it.[34]

Carbonyls are highly poisonous, and the manganese compounds may also introduce a problem of metal poisoning comparable to lead. There are no known effective additives to improve octane number which do not introduce problems of toxicity.[35]

The mechanism of the antiknock action of the lead alkyls is not completely understood. The antiknock compound is a volatile liquid and enters the engine as vapor mixed with the fuel–air intake. Decomposition then occurs due to the engine heat and a fog of solid lead monoxide particles is formed. It is either these solid particles or gaseous lead oxide that is responsible for trapping active oxygen atoms which would otherwise enhance the chain reactions which result in knocking.[36]

The antiknock fluid also contains ethylene dibromide and ethylene dichloride. These act as scavengers by reacting with lead oxide to form the more volatile lead chlorobromide which is swept from the cylinders in the exhaust. This prevents buildup of lead deposits in the engine.

The effect of a given quantity of antiknock agent is strongly dependent on the fuel composition. Apart from the effects of impurities, notably sulfur, which reduce antiknock effectiveness, the response of the different hydrocarbon types is inherently different.[37]

Paraffins, both straight chain and branched, have high susceptibility to lead. Naphthenes also show a large gain in octane number for added lead. The olefins and aromatics, which generally have higher initial octane numbers, have low lead susceptibility.

Modern refining methods produce olefins and aromatics so that the octane rating of the pool gasoline clear (i.e., without lead) is relatively high.

The effects of tetraethyl lead on the octane ratings of some refinery products are shown in Fig. 14.

Catalytically cracked gasoline comprises up to 30% of some premium grades. High-severity (H/S) reformate is composed largely of aromatic hydrocarbons. It has a high octane number but low susceptibility to lead. This material generally comprises in excess of 50% of premium grades of gasoline. High severity refers to the high temperature of the reforming process. Increased temperature produces a higher octane product but lower yield. Low-severity (L/S) reformate has a lower octane rating and is generally used in the regular grades. Naphtha, consisting largely of C_5 and C_6 paraffins and naphthenes, is also blended in regular grade. Note that these grades have high susceptibility to lead.

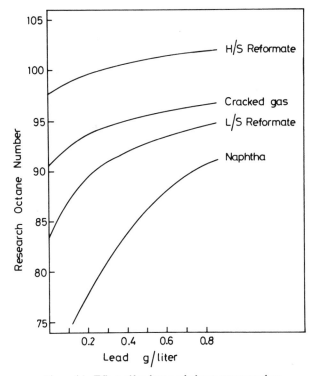

Figure 14. Effect of lead tetraethyl on octane number.

6.5. *Petroleum Refining Developments*

Recent developments in refining technology have been directed toward the production of high-octane gasoline. Part of the incentive for this development is the demand for lead-free fuels.

The chemical processes in refining are of several types. Large molecules are broken down to small ones (cracking). Gases—molecules too small for gasoline—are reacted to form large molecules (polymerization and alkylation). Straight-chain structures are changed to more favorable branched chains (isomerization), and naphthenes are changed to aromatics (reforming).

The general aim is to optimize the yield of gasoline and achieve a maximum octane rating over the required range of boiling points.

The processes in current practice are as follows:

Thermal cracking involves heating under pressure to convert a heavy fraction into light compounds. Olefins are formed, but a considerable amount of the product is gaseous. Some of this gas can be recovered by polymerization. Thermal cracking is now mainly associated with the petrochemical industry, which utilizes the light olefinic products.

Catalytic cracking takes a feed stock boiling in the range 250–500°C and passes it over an acidic solid catalyst at 450–550°C. The catalysts are synthetic silica–alumina or, most recently, zeolites. This is a most important process, particularly for crudes that are rich in the heavier fraction needing cracking. The product of catalytic cracking contains large amounts of olefin. It is superior to thermal cracking because the isomers with central double bonds and branched chains are predominant. The yield also has a boiling point range most favorable for gasoline.

Polymer gasoline is made from the gaseous olefins derived from the cracking process, e.g., isobutene (C_4) is converted to isooctane (C_8) by acid catalysts (phosphoric or sulfuric acid), followed by hydrogenation

$$2CH_3-\underset{\underset{CH_3}{|}}{C}=CH_2 \xrightarrow{Acid} CH_3-\underset{\underset{CH_3}{|}}{\overset{\overset{CH_3}{|}}{C}}-CH_2-\underset{\underset{CH_3}{|}}{C}=CH_2 \xrightarrow{H_2} CH_3-\underset{\underset{CH_3}{|}}{\overset{\overset{CH_3}{|}}{C}}-CH_2-\underset{\underset{CH_3}{|}}{CH}-CH_3$$

isobutene   isooctene   isooctane

Alkylation is a low-temperature process which reacts an olefin with a paraffin. The catalyst is sulfuric acid or anhydrous hydrofluoric acid

$$CH_3-\underset{\underset{CH_3}{|}}{\overset{\overset{CH_3}{|}}{C}}-H+CH_2=\underset{\underset{CH_3}{|}}{\overset{\overset{CH_3}{|}}{C}}-CH_3 \rightarrow CH_3-\underset{\underset{CH_3}{|}}{\overset{\overset{CH_3}{|}}{C}}-CH_2-\underset{\underset{CH_3}{|}}{CH}-CH_3$$

isobutane   isobutene   isooctane

Isomerization converts straight-chain paraffins to the branched structures of higher octane number, e.g.,

$$n\text{-hexane} \rightarrow 2 \text{ methylpentane}$$

$$n\text{-butane} \rightarrow \text{isobutane}$$

The catalysts for gas phase isomerization are metals on zeolites or alumina, and for the liquid phase process, aluminum chloride in antimony trichloride.

Isomer separation processes are often run in conjunction with isomerization to allow recycling of the low-octane stream.

Catalytic reforming converts naphthenes and some paraffins to aromatics. Reforming catalysts consist of platinum and a second metal on a zeolite or alumina base. Rhenium is used as the second metal in many catalysts, but has been replaced recently because of its high cost and scarcity. Reforming is a most important process for the production of high-octane fuel.

The processes which have undergone the most development are catalytic reforming, isomerization, and alkylation. There have been consequential developments in catalytic cracking and isomer separation.

Two types of hydrocarbons are of notably high octane number: aromatic hydrocarbons and branched-chain paraffins (see Table 5). Either of these can be chosen as the goal of the refining process to produce a high-octane gasoline.

The most straightforward method for increasing gasoline pool octane level in most refinery situations is to increase reforming severity, which in turn increases the aromatic level of the gasoline. As the severity of reforming is increased to obtain octane quality improvement, conversion of naphthenes to aromatics occurs first, with the reaction generally being complete at clear Research octane numbers (i.e., without lead) in the high 80s or low 90s for most naphthas. Further increases in octane level are obtained by converting paraffins to aromatics and by hydrocracking paraffins to lighter paraffins and gas.

This is one approach to the production of lead-free high-octane gasoline. It suffers from the fact that there is a loss of liquid yield and the product is rich in aromatic hydrocarbons which have a high propensity for smog formation.

The alternative approach is to isomerize straight-chain paraffins of low octane number to branched-chain paraffins such as isooctane. This method gives a superior yield of high-quality gasoline but involves larger capital outlay for a refinery plant.[38]

The present gasoline pool without lead has a Research octane number of 90–91. Improved refining can increase this to 95 or more. With the obvious consideration that increased refining cost would have to be met by the consumer, it is relevant to ask if this is worthwhile.

It has been shown that the inherent efficiencies of the higher compression engines permitted by pool octane ratings up to 97 Research clear can result in fuel economies to the consumer in excess of 15% when two grades of gasoline are made available—somewhat less for one grade—when compared with a 91 octane pool.[39,40] These studies have also indicated that crude requirements for satisfying the demand for gasoline decrease with increasing octane up to a 97 octane pool, an important energy consideration.

6.6. *Removing the Lead*

6.6.1. *Effects on Engine Operation and Emissions*

There has been much study of the effects of lead removal on engine operation. It is important to distinguish the effects of low-lead gasoline from those of lead-free gasoline. Many of the United States data relate to the latter.

The complete removal of lead can cause a problem of exhaust valve recession. In the absence of lead deposits, which act as a solid lubricant, the exhaust valve seat may wear rapidly. The problem is most acute in engines which deliberately provide for valve rotation. The effect is eliminated by very low lead concentrations.[41]

Lead-free gasoline is claimed to double the life of spark plugs, silencers, and exhaust pipes, and to reduce corrosion.[42] Proportionate effects should not be assumed for low-lead fuel, but deposit formation and the resulting octane requirement increase should generally be less than for the high-lead fuel.

If lower-lead gasoline has a reduced octane rating, then the engine emission of gaseous nitrogen oxides may be affected. Marginal operation on lower-octane fuel may increase NO_x emissions. If compression ratios are lowered to achieve satisfactory performance on this fuel, then NO_x emissions will decrease.[14]

Leaded gasoline causes hydrocarbon emissions up to 30% greater than for unleaded gasoline.[42] The removal of lead then aids the control of hydrocarbons. It should be emphasized that this result was obtained for lead-free gasoline. The improvement is apparently due to the oxidation of unburned hydrocarbons in the exhaust gas, a process inhibited by lead. Low-lead gasoline offers no advantage in this regard; hydrocarbon emissions increase to their maximum level when minimal amounts of lead are added.

6.6.2. Satisfying Vehicle Octane Requirements

There are then several grounds for ceasing to use lead additives in gasoline. As a health hazard it is insidious but potentially more serious than gaseous emissions. Engines will run cleaner and last longer if its use is discontinued. Catalytic systems demand lead-free fuel. Increased refining costs will have to be met and the forward planning of automobile octane requirements has to be geared to changes in the petroleum industry.

The effect of the stepwise removal of lead can be gauged from Fig. 15, which shows the distribution of octane requirement in the United States in 1970 just prior to the manufacture of cars with low octane specifications. This graph shows that the lead-free 91 octane gasoline marketed from 1971 could satisfy about 30% of existing cars, but over 60% if spark timing were retarded within allowed limits.[42] From Fig. 14 it can be seen that removal of part of the lead has a relatively small effect on octane rating. The vehicle populations of most countries could readily adapt to partial removal of lead. Ideally, the automotive industry should design for an octane requirement that can ultimately be met without lead. Only by this action can the catalyst technology for emission control be generally applied.

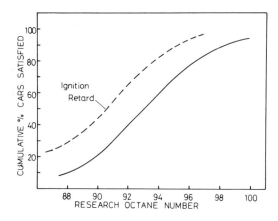

Figure 15. Octane requirement of United States cars, 1970, and the effect of retarded ignition timing.

6.6.3. *The Petroleum Industry Viewpoint*

The petroleum industry has been most defensive in its attitude to the lead-limiting proposals.

The addition of lead to gasoline is the cheapest method of obtaining the top 6–8 octane numbers. The present market is geared to 98 octane, which is close to the optimum cost benefit when fuel consumption, performance, and refinery and crude costs are taken into account. The industry has therefore promoted lead as a "energy extender," and has warned that higher costs and waste of crude oils will result from any decrease in its use.[43]

The sources of cost increase that have been identified are increased capital cost of refinery equipment and lower yield of gasoline from a given crude. And there is also the fuel economy loss in the engine with a lowered compression ratio. If the octane rating is maintained by improved refining, then clearly the economy loss in the car is avoided.

If octane numbers fall and lower-compression engines result, then developed engine power will fall. Only if engine size is increased to recover this power loss will fuel economy be less. This latter course of action took place in the United States during 1971–73 and is now regretted.

The United States oil industry is now adapting to the new market, and detailed surveys have appeared to establish new optimum industry standards, based on lead-free gasoline.[40]

6.6.4. *Alternative Proposals*

Lead is not a fuel but an additive in gasoline. While there are no acceptable alternative additives, it has been suggested that octane ratings could be improved by additions of either ethanol or methanol to gasoline. These are fuels of high octane rating but would need to be added in quantity to produce a significant effect. About 10% would be needed to achieve a useful octane increase.

The production of ethanol from grain has been investigated and found to be uneconomic.[44] Another source of ethanol is sugar molasses. These fermentation processes are not compatible with the petroleum industry, and the production of ethanol would be a large undertaking difficult to coordinate with petroleum refining. The use of food materials as an energy source is also open to question.

Methanol can be produced by the distillation of wood. A better proposal is to convert natural gas to methanol by a catalytic process. The advantage is gained if the process is carried out in a remote location. Gas pipeline costs are avoided by transporting the liquid methanol. With rising petroleum prices this process may be cost competitive and deserves further study. Tests of methanol fuels have shown the feasibility of the product.[45]

A different method of controlling lead emissions that has been suggested is to attach a filter or trap to the exhaust system to catch lead-containing particles. The nature of the particulate emission was discussed briefly above. The size range most important in the formation of lead aerosols is $<1\ \mu$. Particles of this size are the most numerous but have been claimed to account for only 5% of the weight of exhausted lead. Since the average automobile running on leaded gasoline emits about 3 kg of lead particulates annually, it is apparent that the life of a filter which traps these small particles effectively would be short.

7. A Review of Progress

The objective of the attempts to control vehicle emissions is to ensure cleaner air and better health. The role of the automobile in affecting these qualities could not be measured precisely at the outset, so that it would be difficult to establish the extent of improvement in air and health. In localized situations some air analyses have improved, but the growth of vehicle numbers, changing traffic patterns, and other pollution sources make overall assessment difficult.

The most practical criterion is to ask whether emission standards are being met. The answer is yes, if one accepts the very small sample of vehicles tested as representative of the total car population in normal service, and if one accepts the deferment of the standards which the industry cannot yet meet.

It is apparent that each level of control has been met by a technique already at its limit, and that the succeeding stage of control demands further technology. The present developments in catalytic systems to take over from the engine modification packages is an example of this process.

The cost of improvement increases as greater control is sought. The increase is not monotonic, but is represented in Fig. 16. A substantial initial

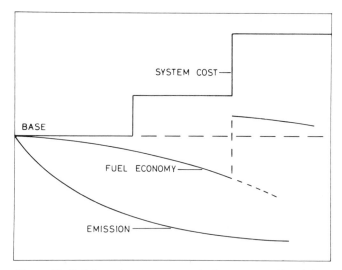

Figure 16. Relation of system cost to fuel economy and emission
control (schematic).

decrease in emission was obtained at little cost by minor engine modifications. A further improvement was made by installing devices which decrease fuel economy. At much higher system cost, it is now possible to achieve further improvement in emissions but at regained fuel economy.

Have we found the correct answers to the control of automotive emissions? It is more pertinent to ask whether we have sought the right actions. The petroleum industry is the largest purveyor of energy and the automotive industry the largest provider of transport.

It was inevitable that established industries would seek solutions to the problem in terms of their own technology. The short-term planning in these industries involves marketing essentially the same product, medium-term planning involves only modest changes, and long-term planning, involving radical changes in technology, hardly exists.

The stepwise imposition of controls has ensured only a stepwise change in existing technology. It can be shown that the present internal combustion engines, even with efficiency reduced by emission control devices, still have the advantage over other types of power units in their present stages of development.[18] But we might ask whether the expense and effort in the automotive industry to produce emission control devices has dissipated a chance to develop new power sources.

The community pays for the development of pollution control measures. In this case we are paying the automotive industry to add some accessories to the car and paying the petroleum industry to modify the composition of gasoline. It is likely that we will have to pay again to achieve the objective of clean air.

References

1. E. S. Forest, "Pollution," *Commercial Car Journal,* 87 (1970).
2. A. H. Rose *et al.,* Paper presented to Annual Meeting, Air Pollution Control Association, Houston, Texas, June, 1974. R. A. Dunstan, Automotive Air Pollution Analysis Discussions, Australian Road Research Board, *Proceedings,* 6 (2) : 434 (1972).
3. J. A. Nadel, in *Combustion-Generated Air Pollution,* E. S. Starkman, ed., Plenum Press, 1971, p. 239.
4. "Air Quality Criteria for Carbon Monoxide," U.S. Dept. of HEW, Washington, D.C., March 1970, Publication AP-62.
5. J. A. Kerr, J. G. Calvert, and K. L. Demerjian, *Chemistry in Britain,* 8 : 252 (1972).
6. R. G. Temple, in *Air Pollution Control,* Vol. 1, W. Strauss, ed., Wiley-Interscience, 1971, p. 177.
7. R. P. Schuller, D. E. Benson, and R. M. De Veirman, "Impact of Automotive Emissions on the Petroleum Industry," Universal Oil Products, Technical Seminar, 1971.
8. D. A. Hirschler and L. F. Gilbert, *Arch. Envir. Health,* 8 : 297 (1964).
9. T. J. Chow, *Chemistry in Britain,* 9 : 258 (1973).
10. R. M. Hicks, *Chem-Biol. Interactions,* 5 : 361 (1972).
11. R. H. Daines, H. Motto, and D. M. Chilko, *Envir. Science and Technol.,* 4 : 318 (1970).
12. J. L. Moyers, W. M. Zoller, R. A. Duce, and G. L. Hoffman, *Envir. Science and Technol.,* 6 : 68 (1972).

13. M. L. Smith and K. W. Stinson, *Fuels and Combustion*, McGraw-Hill, New York, 1952.
14. E. S. Starkman, in *Combustion-Generated Air Pollution*, E. S. Starkman, ed., Plenum Press, 1971.
15. T. R. Wildeman, *J. Chem. Educ.*, **51** : 290 (1974).
16. J. Bagg, in *Air Pollution Control*, Vol. 1, W. Strauss, ed., Wiley-Interscience, 1971, 35.
17. J. P. Soltau and R. J. Larbey, Symposium on Air Pollution Control, Inst. of Mech. Engineers, Nov., 1971.
18. H. C. Hottel and J. B. Howard, *New Energy Technology: Some Facts and Assessment*, M.I.T. Press, 1972.
19. W. G. Agnew, 34th Mid-year meeting of the A.P.I., May 1969, Paper 15-69.
20. G. F. Harrison, Soc. of Automotive Engineers, Australasia, National Convention, Oct., 1971, Paper 19.
21. C. D. Haynes and J. W. Weaving, Inst. Mech. Engineers, Solihull Conference, Nov., 1971, Paper C-144/71.
22. A. Jaimee, D. E. Schneider, A. I. Rozmanith, and J. W. Sjoberg, S.A.E. Congress, Detroit, Jan., 1971. SAE No. 710293.
23. D. A. Hirschler and F. J. Marsee, National Petrol. Refiners Assoc. Meeting, April, 1970.
24. R. S. Yolles and H. Wise, *C.R.C. Critical Reviews in Environmental Control*, **2** : 125 (1971).
25. V. Haensel, U.O.P. paper, "Refineria de Petroleos de Escombreras S.A.," Madrid, 1972.
26. L. Riekert, Euchem. Conference, "The Role of Catalysis in Problems of Pollution," Santander, Spain, 1973.
27. J. D. Butler, *Chemistry in Britain*, **8** : 258 (1972).
28. M. Sheleff, K. Otto, and H. Gandhi, *Atmos. Environ.*, **3** : 107 (1969).
29. A. Lawson, *J. Catal.*, **24** : 297 (1972).
30. E. M. Logothetis, *J. Solid State Chem.*, **12** : 331 (1975).
31. F. H. Adams and A. P. Krueding, Universal Oil Products, Technology Conference, 1974.
32. B. Lewis and G. von Elbe, *Combustion Flames and Explosions of Gases*, 2nd Edit., Academic Press, 1961.
33. H. K. Livingston, *Ind. Eng. Chem.*, **43** : 2834 (1951).
34. M. R. Barusch and J. H. Macpherson, *Advances in Petroleum Chemistry and Refining*, Vol. 10, J. J. McKetta, ed., Interscience, 1965.
35. D. Bryce-Smith, *Chemistry in Britain*, **7** : 54 (1951).
36. H. Shapiro and F. W. Frey, *The Organic Compounds of Lead*, Interscience, 1968.
37. W. I. Zang and W. G. Lovell, *Ind. Eng. Chem.*, **43** : 2826 (1951).
38. F. H. Adams, Universal Oil Products, Technical Seminar, 1971.
39. W. R. Epperley, Paper presented to California Assembly, Transportation Committee, Los Angeles, Nov., 1970.
40. L. O. Wagner and L. W. Russum, 38th Mid-year meeting of American Petroleum Institute, May 1973.
41. R. P. Doelling, *S.A.E. Transactions* 710841 (1971).
42. D. O'Neill, *Automotive Engineering*, **78** : 17 (1970).
43. L. E. Blanchard, *Automotive Engineering*, **78** : 24 (1970).
44. Y. Ip, *Australian Chemical Processing and Engineering*, **25** : 11 (1972).
45. R. E. Fitch and J. D. Kilgroe, Consolidated Engineering Technology, Corp., 01800-FR, Feb., 1970.

Whatever may be done to control noxious exhausts and to remove the unsaturates from them, internal combustion engines will produce carbon dioxide. It has long been stated that such carbon dioxide will eventually cause a "greenhouse effect" because light is absorbed and converted to heat by CO_2. It is important not to regard this possibility as some kind of legend but to investigate it scientifically and see what it may mean for our choices of energy supplies in the near future.

10

Possible Climatic Changes from Carbon Dioxide Increase in the Atmosphere

K. Sekihara

1. Introduction

It is now well understood that the main heat sources of the troposphere are given from the earth's surface in the low-latitude zones and the thermal energy thus given to the atmosphere goes up by convection, making clouds and rain, and in the middle and high latitudes the atmosphere loses energy through radiative cooling to space via the infrared emission bands of water vapor. However, in the stratosphere, the density of water vapor becomes extremely low, and heating due to ultraviolet radiation absorption by ozone and cooling due to infrared emission by ozone and carbon dioxide become predominant. In the mesosphere, above 50 km, the cooling effect of carbon dioxide is the most important.

K. Sekihara • On leave from the Meteorological Research Institute, Tokyo, Aerological Observatory, Tateno, Ibaraki-ken, 300-21, Japan

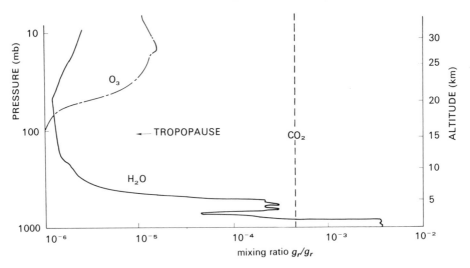

Figure 1. Mixing ratios of the main atmospheric gaseous components causing radiative activity, adopted from an example of real observation (H_2O and O_3: observed at TATENO (36°03'N, 140°08'E) January 18, 1972; CO_2: assumed).

These features are shown in Fig. 1. It should be noted that water vapor exists mainly in the lower part of the troposphere and that ozone occupies the stratosphere, but that carbon dioxide is distributed uniformly in every layer.

In Fig. 2 are shown the energy distribution of solar and terrestrial radiation outside the earth's atmosphere, together with the percentage absorption of the earth's atmosphere in different wavelengths. We can see here that each atmospheric constituent *absorbs* solar radiation and reemits *infrared* radiation.

In Table 1 are shown the percentage contributions of each wavelength range and consequently each component of the atmosphere to the total downward flux of atmospheric radiation at different heights, calculated

Table 1. Percentage Distribution of Various Emitting Components in the Atmospheric Downward Radiation Flux at Varying Levels[a]

Band level (mb)	H_2O (6.3 μ)	O_3 (9.6 μ)	Window (11 μ)	CO_2 (15 μ)	H_2O (rot.)
Surface	12.4	7.4	2.7	24.3	51.5
511	5.9	8.0	0.1	39.1	47.3
204	0.2	16.1	0	68.1	15.4
153	0.3	18.0	0	68.2	13.8
102	0	21.4	0	66.7	11.6
51	0	29.5	0	61.0	8.6
10	0	55.8	0	40.0	4.2

[a] Calculated by Sekihara, Kano, and Miyauchi (unpublished) using the method of Rodgers and Walshow.[16]

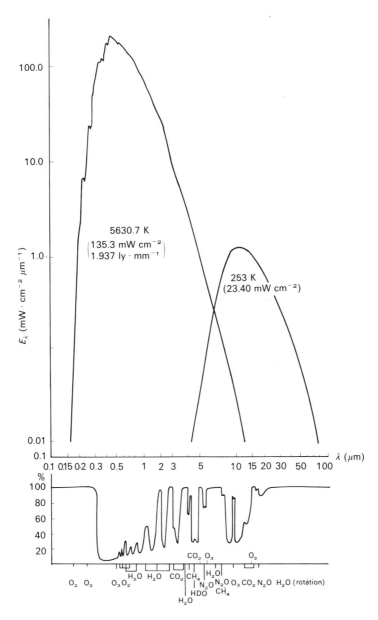

Figure 2. Solar and terrestrial radiation outside the earth's atmosphere, together with the percentage absorptivity of atmospheric constituents. Solar radiation curve was adopted from Thekaekara.[1] Terrestrial radiation has been computed by the author on the assumption of blackbody having the assigned effective temperature and integrated energy in accordance with London and Sasamori.[2] Absorptivity curve from Goody.[3]

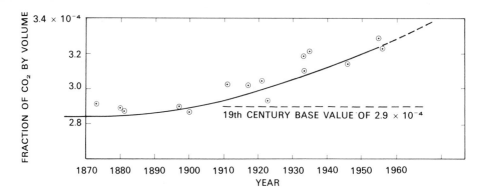

Figure 3. The amount of CO_2 in the free air of the North Atlantic region. The full curve is obtained from measurements on the fossil fuels. (After Callender.[4])

recently with the usual conditions for the temperate zone, which are not very different from those assumed in Fig. 1. We can see that the largest contribution of water vapor occurs near the surface, whereas the contributions of CO_2 and O_3 become significant in the stratosphere, as mentioned above.

These facts give the background of the radiative situation in our atmosphere.

2. The Increase of Carbon Dioxide

The idea of a climatic effect due to an increase in carbon dioxide is not new. To find the origin of such a suggestion we must go back to the last century, to Tyndall (1861), and then to the more definite theory which was proposed independently by Chamberlain (1899) and then by Arrhenius (1903). But actual and accurate measurements of CO_2 content in the real atmosphere were comparatively few until the period of the IGY (1957), when systematic measurements by many authors began. The data of earlier days were summarized by Callender,[4] as shown in Fig. 3, where there was already shown a gradual increasing tendency with time. After that, several authors reported on CO_2 measurements in various regions and in varying time frequencies. Results are summarized in SCEP[5] and SMIC[6] Reports, from which we can assume that CO_2 has been increasing at about 0.2% per year, or 0.64 ppm out of 320 ppm by volume during the earlier period of the 1960s.

The most continuous observations have been carried out at Mauna Loa and also at the South Pole by Keeling.* They are shown in Figs. 4 and 5, respectively. Very recently, Keeling corrected the data taking into consideration the pressure-broadening effect of gas mixtures in CO_2 measurement. The newest data thus corrected are given in Figs. 4 and 5. Also, according to recent

*The author is indebted to Dr. Keeling for this newest data, available by courtesy of Dr. Pack, Air Resources Laboratories, NOAA.

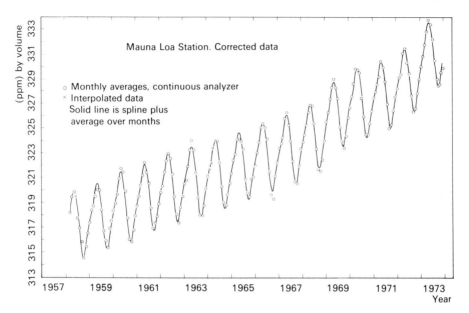

Figure 4. The amount of CO_2 in the free air at Mauna Loa Station during the period from IGY through 1973. (Courtesy of Dr. Keeling via Dr. D. H. Pack of Air Resources Laboratory, NOAA, U.S.A.)

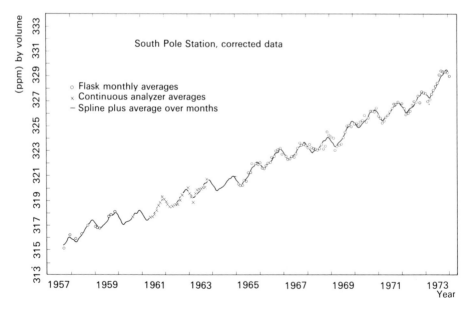

Figure 5. The amount of CO_2 in the free air at South Pole Station during the period from IGY through 1973. (Courtesy of Dr. Keeling via Dr. D. H. Pack of Air Resources Laboratory, NOAA, U.S.A.)

information from NOAA, a first full year of continuous CO_2 measurement was completed in August, 1974, at Point Barrow, Alaska, and showed an increase of 2 ppm from 1973 to 1974, which is essentially the same as the 1972 to 1973 Mauna Loa increase. The seasonal change apparent in Figs. 4 and 5 can be attributed to the photosynthetic activity of plants, but at present no *quantitative* interpretations have been given to this phenomenon.

With respect to the annual increase, there are some suggestions of possible mechanisms and predictions for the future. Among them, those of Machta[7] and Revelle[8] may be described briefly, and some recent observations regarding the transfer processes will be mentioned here.

The transfer mechanisms proposed by Machta[7] are shown schematically in Fig. 6, where stratosphere, troposphere, and mixed layers of the ocean and deep ocean are considered as the main reservoirs of CO_2, and various types of biospheres are also considered as other temporary reservoirs. The exchange rates are given by numbers adjacent to the arrows between each box representing reservoirs. Also, there is a time lag and a net production to be considered for the boxes representing the biosphere. As to the rate between troposphere

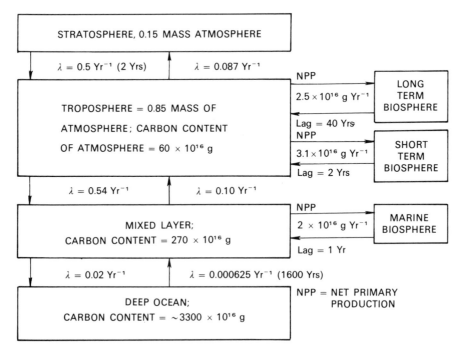

Figure 6. The model of the three reservoirs of CO_2. The quantity λ denotes the fraction of the CO_2 in one reservoir being transferred to an adjacent reservoir according to the sense of the arrow. The transfer between troposphere and mixed layer is calculated from the history of the C^{14} data; all others are preassigned. The carbon content of the mixed layer and the exchange constant from mixed layer to deep ocean are derived from other assumed and computed parameters. (After Machta.[7])

and ocean mixed layer, it has been determined by means of bomb-produced radioactive ^{14}C, in the form of CO_2 as a tracer. Other figures are inferred on the basis of various assumptions. First-order kinetics have been assumed in all processes. In this way, Machta[7] made a prediction of future CO_2 concentration in the troposphere. The shape of the increasing curve is exponential and the estimate for the year 2000 is about 380 ppm.

Revelle[8] also discussed the future increase of global atmospheric CO_2. Using present fossil fuel consumption and estimates of future consumption, he estimated the amount of CO_2 increase in the year 2000 to be 21% of the present atmospheric CO_2 content. This would be approximately 387 ppm, as compared with the present value of 320 ppm, but Revelle considers this to be the minimum amount for the following reasons:

1. The denudation of forests, which diminishes one of the major reservoirs for CO_2, will continue in the future.
2. The activity of the biosphere consuming CO_2 in photosynthesis, will tend to saturate when compared with the speed of oxidation of plant materials.
3. If an increase of CO_2 in the atmosphere causes an increase in the sea surface temperature, it will cause a decrease of CO_2 solubility in seawater, i.e., a positive feedback mechanism will be set up.
4. The partial pressure of CO_2 in the deep-ocean water is higher than that of the water near the surface.

With respect to point (4), Revelle[8] points out that the northern North Atlantic, the far northwestern Pacific, and the Weddell Sea are the main areas where deep ocean comes to the surface and releases CO_2 to the atmosphere. He infers qualitatively the climatic effect of a CO_2 increase. Because this effect would be most pronounced in cold areas where water vapor is less effective, the resultant influence will diminish the temperature gradient between high and low latitudes. This point is discussed in more detail in a paper by Manabe and Weatherald.[27]

It is of interest to discuss recent measurements which may be helpful in understanding the process of transfer between ocean and atmosphere. The first one is that of Miyaka *et al.*,[9] who made measurements of both atmospheric CO_2 and the CO_2 partial pressure of surface ocean water in a wide area covering almost the whole of the Pacific Ocean. This was done on a ship during 1968–72. The result is shown in Fig. 7. In this figure, whereas the CO_2 content of the atmosphere varies in the range 319–335 ppm, the partial pressure of surface ocean water varies in wider ranges of 295–410 ppm. This result supports Revelle's assertion that, in some areas, deep ocean water which contains more CO_2 comes out to the surface.

Ben-Yaakov *et al.*[10] made a measurement of the degree of saturation of seawater at varying depths, with respect to carbonate materials, together with the vertical distribution of carbonate content in adjacent sediment. The measurement was done near Hawaii where the depth of the sea is about

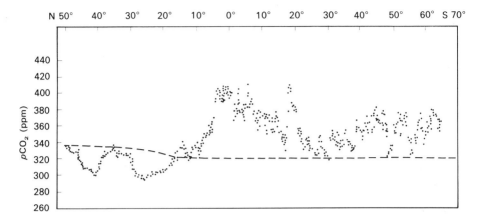

Figure 7. Carbon dioxide content in surface water and air along 146°W. Dashed line indicates air; dotted line indicates water. (After Miyake, Sugiura, and Saruhashi.[9])

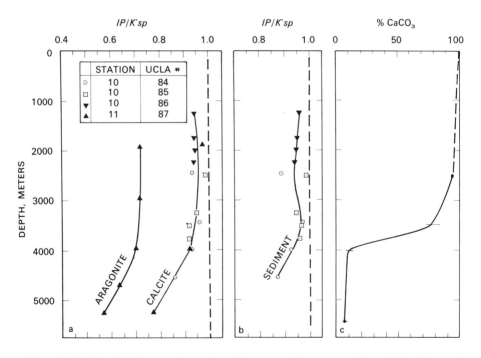

Figure 8. Profiles of the degree of saturation of seawater with respect to (a) standard calcite and aragonite and (b) local sediment (all separated by sieving to the size range 0.35–0.5 μm) compared with the carbonate content in adjacent sediment. The degree of saturation is given by $IP/K'sp$ where IP is the product of the concentrations of total dissolved calcium and carbonate ions, and $K'sp$ is the apparent solubility product. The observations were done near the Hawaii Islands by Ben-Yaakov, Ruth, and Kaplan.[10]

5000 m. The results showed that above 3750 m seawater was supersaturated with respect to carbonate, but that below that level it was undersaturated, as seen in Fig. 8. This kind of measurement is still scarce and many more observations are needed, covering a wider area.

Finally, Robinson[11] pointed out that there is a small but distinct distribution in CO_2 content of the atmosphere in passing through the tropopause, the amount being about 0.2%. This is an important topic which needs verification as soon as possible.

3. *The Beginning of Modern CO_2 Climatic Theory*

Although it has been suggested from earliest times that a CO_2 increase may change the climate because of the "greenhouse" effect, there has been doubt in the concept because the absorbing wavelength region overlaps with that of water vapor. It was not until Plass[12] formulated his theory that the problem attracted scientists' attention to try to formulate the processes of climatic change in terms of a CO_2 increase.

Plass' theory is based on an accurate calculation of upward and downward radiation fluxes from the earth–atmosphere system, relying on the measured absorption bands of CO_2—15 μ—reported by Cloud.[13] The calculation was done line by line (in respect to the spectrum) and the height interval was from 1 km above the surface to a height of 75 km. Plass deduced that the surface temperature will rise 3.6°C if atmospheric CO_2 concentration is doubled, and will fall 3.8°C if it is halved, thus maintaining radiative equilibrium at the earth's surface.

Plass showed that water vapor is concentrated in the lower troposphere, in contrast to CO_2, which is uniformly mixed as high as 70 km, and the absorption lines of both species are located at random wavelengths, so that the masking effect of H_2O is not so effective, as discussed by previous authors.

The figures of 3.6°C and 3.8°C given above will be reduced to 2.5°C and 2.7°C if the effect of average cloudiness is taken into account.

Starting from these radiation calculations, Plass' discussion is mainly focused on the processes of geological climatic changes in glacial and interglacial epochs, at intervals of several tens of thousands of years. The assumptions on which Plass' discussion stands are that the partial pressure of CO_2 between atmosphere and ocean comes to equilibrium within some 50,000 years, since the turnover time of the ocean is about 10,000 years. He considers that the equilibrium of solid $CaCO_3$ and CO_2 partial pressure in the ocean may take a much longer time than that between atmosphere and ocean only. The effect of biospheric activity is estimated as coming to equilibrium within 10 years, so that it may be rapid enough to be neglected in CO_2 balance considerations.

In addition, ice can contain only negligible amounts of CO_2 as compared with seawater. Plass discusses the effect of change of volume of seawater, caused by the development of glaciers, which will finally limit the amount of CO_2 that can be dissolved in the ocean.

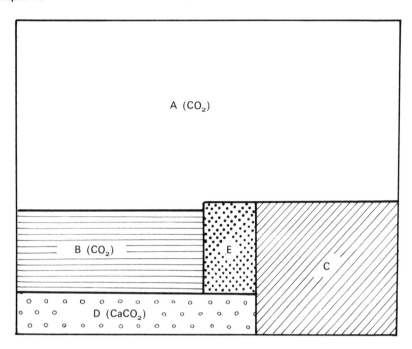

Figure 9. Schematic diagram of CO_2 distribution during glacial–interglacial periods.[12]
A = atmosphere; B = ocean; C = land; D = sediments; and E = ice.

Thus, the problem of long-term climatic change—which in the Plass theory is supposed to be controlled by the CO_2 radiation effect—will be in effect controlled by the CO_2 concentration in the atmosphere, which is in equilibrium with the ocean water and with the solid $CaCO_3$ at the bottom of the ocean. These systems are shown schematically in Fig. 9. Plass states that, at present, the equilibrium of CO_2 distributed between A, B, and D in the figure is nearly established. But, if for some reason, such as the build-up of coal deposits in the Carboniferous period, a significant amount of CO_2 in the atmosphere is removed from this system, the earth's surface will cool down and a glacial period will begin. This will result in the shrinkage of the volume of the liquid-phase ocean. The shrinkage of ocean volume will give an increase in the partial pressure of CO_2 in the ocean. The increased partial pressure will result in the transfer of CO_2 from ocean to atmosphere, which will restore the temperature depression, thus explaining the periodicity of glacial–interglacial epochs.

Although disturbances by man-made CO_2 are mentioned by Callender,[4] Plass' paper seems to have triggered the later intense activity in the study of climatic change for the purpose of investigating more realistic climatic modeling.

One direction in which Plass' theory can be improved is that of the radiation calculation. Thus Kaplan[14] tried to take the effect of cloudiness into account, as well as to revise the calculation scheme. Kondratyev and Niilisk[15] considered the effect of overlapping between the CO_2 15-μ band and the H_2O-rotation band. Both author's estimates of the CO_2 effect turned out to be significantly less than that of Plass (see also Rodgers and Walshow[16]).

4. The Feedback Mechanism and Competing Causes

4.1. The Humidity Feedback

A realistic modeling of climatic change must incorporate factors other than radiation. Möller's paper[17] was a pioneering work which stimulated the activities of later authors; he introduced feedback mechanisms which have to be taken into account before any decisive conclusions can be deduced concerning an increase in CO_2.

Thus, if the surface temperature rises, it will cause an increase in air humidity, and the increased water vapor will absorb solar radiation more intensely, causing more heating of the air. This is a self-amplifying effect which depends on the surface temperature because of the temperature dependence of the vapor pressure of H_2O. Möller pointed out the importance of a change in cloudiness which may be caused by the surface temperature change. Thus, a changed CO_2 concentration can be compensated for completely, without variation in surface temperature, when cloudiness is increased by the amount +0.006 l.e., or when the water-vapor content is decreased by −0.07 cm (l.e. = liquid equivalent). Möller was not able to estimate the influence of the expected humidity distribution of the atmosphere; for this, we have had to await the appearance of the paper by Manabe and Weatherald.[18]

The starting point, and the main point of Manabe and Weatherald's paper, is the consideration of the latent and sensible heat at the earth's surface. This is the main way in which the paper differs from those of previous authors, for they only considered radiative processes. Here, radiative equilibrium was required at the top of the atmosphere instead of the earth's surface.

The paper by Manabe and Weatherald is an extension of that of Manabe and Stricker,[19] who developed a method of calculating the temperature profile of the atmosphere with a radiative–convective equilibrium by successive time integrations. Guided by Möller's suggestion, and using the investigation of the global humidity distribution made by Telegadas and London,[20] they introduced an important assumption, i.e., that the vertical distribution of the relative humidity is constant. Taking into account the effect of condensation, the flow-chart cycle of calculation for determining the temperature profile is shown in Fig. 10. In this figure, $e_s(T)$ denotes the saturation vapor pressure of water vapor as a function of temperature T, and h denotes the relative humidity; τ denotes the number of time steps in the numerical integration, and I the indexing of the finite differences in the vertical direction. We can see schematically how the radiation, convection, and condensation processes are incorporated during the calculation of a vertical temperature profile which eventually goes to the asymptotic equilibrium state. The assumptions behind the calculation can be detailed as follows:

1. At the top of the atmosphere, the net incoming solar radiation should be equal to the net outgoing long-wave radiation.
2. No temperature discontinuity should exist.
3. Free and forced convection, and mixing by large-scale eddies, prevents the lapse rate from exceeding a critical lapse rate equal to 6.5°C km^{-1}.
4. Whenever the lapse rate is subcritical, the condition of local radiative equilibrium is satisfied.
5. The heat capacity of the earth's surface is zero.
6. The atmosphere maintains the given vertical distribution of relative humidity.

The time scale of convergence to the equilibrium values is estimated to be 500 days at the longest, and according to the calculations, depends upon the initial value of the temperature profiles given. The way the temperature profiles converge to the same final values from the initial isothermal conditions can be seen in Fig. 11. That such calculations give reasonable results is shown by the example in Fig. 12. Here we can see three calculated temperature profiles, with radiative equilibrium at a given relative humidity and at a given absolute humidity, respectively, and the radiative–convective equilibrium at a given relative humidity. The third calculation gives the most reasonable result. By means of their scheme of calculation, the authors discussed many meteorological problems other than the CO_2 concentration, e.g., the effect of variations in the solar constant, change of cloudiness, change of surface albedo,

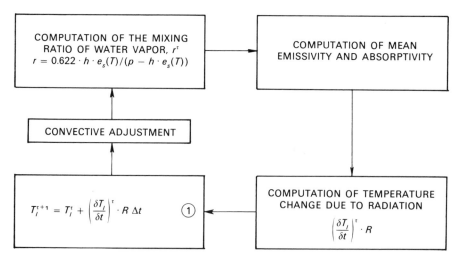

Figure 10. Flow chart for the numerical time integration in the radiative–convective adjustment theory. (After Manabe and Weatherald.[18])

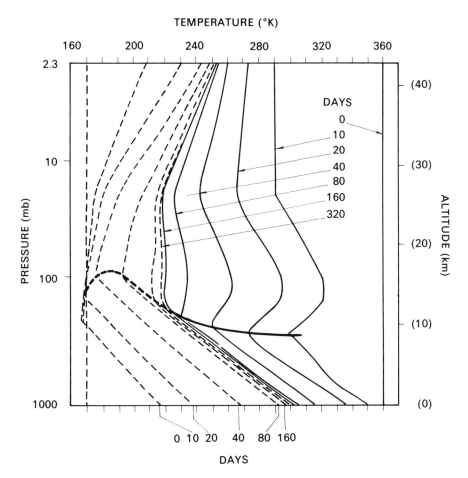

TEMPERATURE (°K)

Figure 11. Approach toward the state of radiative–convective equilibrium. The solid and dashed lines show the approach from a warm and a cold isothermal atmosphere. (After Manabe and Strickler.[19])

etc. Among the results obtained, that of changing the amount of CO_2 is the most remarkable, and this is shown in Fig. 13 and Table 2. The results can be summarized as follows:

1. The larger the mixing ratio of carbon dioxide, the warmer is the equilibrium temperature of the earth's surface and troposphere.
2. The larger the mixing ratio of carbon dioxide, the colder is the equilibrium temperature of the stratosphere.
3. The dependence of the equilibrium temperature of the stratosphere on CO_2 is larger than that of the tropospheric temperature on CO_2.

The table shows the results for the assumption of constant absolute humidity, in addition to constant relative humidity. We can see the slightly

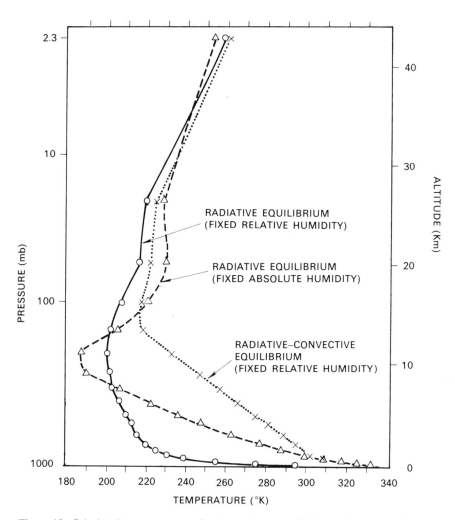

Figure 12. Calculated temperature profiles by radiative equilibrium with a given absolute humidity distribution (solid line), by radiative equilibrium with a given relative humidity distribution (dashed line), and by radiative–convective equilibrium with a given relative humidity distribution (dotted line). (After Manabe and Weatherald.[18])

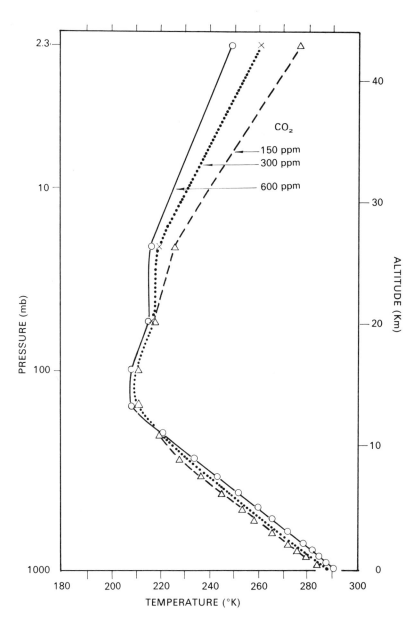

Figure 13. Vertical distributions of temperature in radiative–convective equilibrium for various values of CO_2 content. (After Manabe and Weatherald.[18])

Table 2. Change of Equilibrium Temperature of the Earth's Surface Corresponding to Various Changes of CO$_2$ Content of the Atmosphere Computed by Means of Radiative–Convective Equilibrium[a]

Change of CO$_2$ content (ppm)	Fixed absolute humidity		Fixed relative humidity	
	Average cloudiness	Clear	Average cloudiness	Clear
300 → 150	−1.25	−1.30	−2.28	−2.80
300 → 600	+1.33	+1.36	+2.36	+2.92

[a] After Manabe and Weatherald.[18]

larger amplification effect on the latter. Thus, Manabe and Weatherald have quantitatively solved the problem proposed by Möller.[17]

4.2. The Polar Ice Cap and Aerosols

Probably the main reason for the appearance of so many theories of climate during the past decade is because the actual global mean temperature is falling in spite of the increase in CO$_2$ concentration, as observed accurately since 1940. This tendency is shown in Fig. 14, where we can see at the same

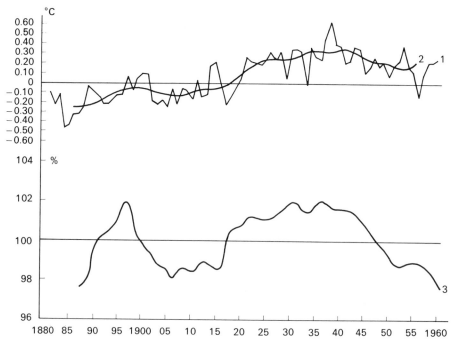

Figure 14. Secular variation of temperature and direct radiation. (After Budyko.[21])

time the lowering tendency of direct solar radiation measured at the ground. This has led Budyko[21] to develop a theory of climatic change based upon the change of solar radiation falling on the ground. Thus, the main cause of climatic change may be a change of solar radiation itself, or a change in the transparency of the atmosphere. Budyko's theory is characterized by taking the effect of latitudinal differences two-dimensionally, and by introducing a feedback mechanism of polar ice development.

Budyko's theory resides in principle on the empirical relationships between energy balances of different latitudinal zones, based on monthly averages of radiation measurement (both short- and long-wavelength ranges) at 290 stations all over the world.

The excess or deficit of radiation balance in the earth–atmosphere system at each latitude thus calculated is considered as the energy compensated for by the sensible or latent heat flowing into or out of the adjacent latitude.

Budyko obtained, again empirically, a linear relationship between this compensating energy and the temperature deviation of each latitude from the global average value. In this way he was successful in reproducing the observed latitudinal temperature distribution from the radiation distribution. His results are shown in Fig. 15.

The effect of the ice cover is now considered. After introducing empirical data on the albedo, Budyko adds the effect of ice cover to the equation connecting temperature to radiation. The term is nonlinear and this causes a self-amplifying effect.

The conclusion is that an apparent decrease of the solar constant is the origin of the decrease in temperature at the earth's surface. At high latitudes the effect is magnified as seen in Fig. 16. A 1.6% decrease of solar radiation

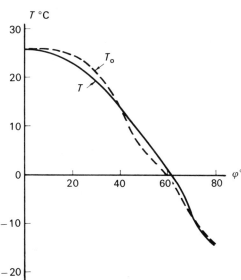

Figure 15. The average latitudinal temperature distribution. Solid line (observed); dashed line (calculated). (After Budyko.[21])

would cause a catastrophic glaciation of the whole earth. The decrease of radiation may be caused by a decrease of the atmospheric transparency of the earth, which in turn may result from the presence of dust particles in the atmosphere.

The problem of dust particles appears as a competitor to the theory of CO_2 climatic change. This was raised directly by Rasool and Schneider.[22]

Before describing this theory, I must mention the idea of the greenhouse runaway effect, first proposed by Rasool and DeBergh[23] with respect to the theory of the evolution of planetary atmospheres. The problem is why there is such a distinct difference between the atmospheres surrounding Venus, Earth, and Mars. The decisive condition is the equilibrium temperature of the planet, which is determined by its distance from the sun. We consider the status of each planet as starting without an atmosphere, and then beginning to produce an atmosphere composed of carbon dioxide and water vapor. During the production of these surrounding atmospheres, the vapor pressure of constituent gases increases gradually. Then, at a certain temperature, depending upon the equilibrium temperature of the planet, water vapor freezes to make frost on Mars and condenses to make oceans on the Earth; on Venus, because the equilibrium temperature is so high, water vapor does not condense but rather continues to increase. The increased water vapor on Venus causes an increased greenhouse effect, thus continuing to increase the temperature. The limiting factor is the beginning of chemical reactions between the planet's crust surface and the atmospheric gases. As a result, the temperature of the Venusian surface is as high as 700°K. These situations are depicted in Fig. 17. This is the greenhouse runaway effect, which is one of the background concepts of Rasool and Schneider's paper. Thus, one might suspect that an anthropogenic increase

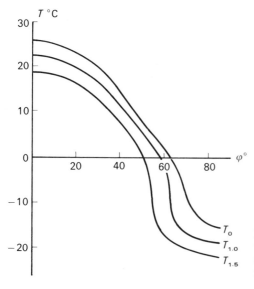

Figure 16. The dependence of latitudinal temperature distribution with varying increment (%) of solar constant. (After Budyko.[21])

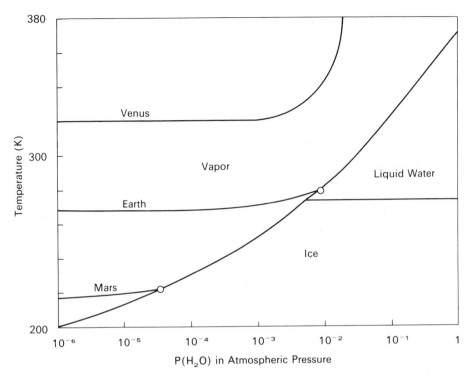

Figure 17. The runaway greenhouse effect. During the processes of the atmospheric evolution of planets accumulation of water vapor continues until freezing occurs (Mars), the condensation point (Earth), and a runaway toward very high temperatures (Venus); this depends upon the equilibrium temperature of each planet. (After Rasool and DeBergh.[23])

of CO_2 may cause the continuously increasing temperature mentioned above. Another question was whether the increase of dust particles might give rise to cooling or heating. Although Rasool and DeBergh's paper is unique in treating effects of CO_2 and dust simultaneously, in respect to the CO_2 calculations Manabe and Weatherald's paper[18] seems to be more sound, because they took into account the effect of near-infrared absorption of solar radiation as well as the temperature change of the stratosphere. With respect to the aerosol calculations, Yamamoto and Tanaka's paper[25] seems to be more rigorous and gives a clear conclusion, as will be described below. Rasool and Schneider[22] made a radiation calculation essentially similar to that of Plass,[12] but taking into account almost the same conditions as Manabe and Weatherald, and investigated the effect of an increase in CO_2 between two and eight times. They found that the CO_2 concentration did not run away, but saturated to converge at a limiting temperature, suggesting that the temperature increase for the coming several decades may be insignificant. The aerosol effect, however, turned out to give a more significant effect than that of CO_2. The method of computation is more complicated because we have to take into account scattering and absorption in both solar and terrestrial radiation on some unknown amount of aerosol, an aerosol particle-size distribution, and its refractive indices. Using the method of two-stream approximation in solving the multiple scattering problem, Rasool and Schneider reached the conclusion that assuming the present optical thickness of a dust layer about 0.1, its increase by a factor of 2–10 causes a serious extinction effect on solar radiation, whereas the emitted infrared radiation is not affected seriously by this increase, and the resulting effect is a significant cooling of the earth's atmosphere. According to this calculation, an increase of atmospheric opacity by a factor of 4 will decrease the surface temperature by 3.5°C, *which may be sufficient*, according to Budyko's paper,[21] *to trigger an ice age.*

But it seems that the situation is not so simple as that discussed by Rasool and Schneider.[22] Yamamoto and Tanaka[24] made an accurate calculation of the radiative transfer, focusing on the effect of varying values of the imaginary part of the refractive index of aerosols. As was mentioned before, the solution of equations of radiative transfer in both scattering and absorbing media of nonhomogeneous stratification is one of the most difficult problems in this field so far. The work of Chandrasekhar, an astrophysicist, done during and after World War II, and culminating in a book called *Radiative Transfer*,[25] has had a deep influence on the field of geophysics as well. Thus, Chandrasekhar was successful in analytically solving the problem of radiative transfer in plane-parallel Rayleigh-scattering atmospheres, and many geophysicists' efforts have relied on his work, so as to extend and develop the problem into a more generalized case. Yamamoto and Tanaka's work is along these same lines. They came back to Chandrasekhar's approach in solving the integral equation of radiative transfer by using the principle of invariance. And, moreover, as far as numerical solutions are concerned, Tanaka[26,27] was successful in generalizing the problem to nonhomogeneous, stratified medium. The result of such calculations is shown in Fig. 18. Here, the temperature change of the earth's surface is plotted against the change of turbidity β, with a variable parameter of the imaginary part n_i of the refractive indices. We can see in this figure that the surface temperature will increase or decrease as the turbidity increases, depending on whether the value of n_i is more or less than 0.05. According to their estimation, the present value of β is 0.05–0.075, and, if we assume the value of n_i to be about 0.01,* the doubling of turbidity may cause the surface temperature decrease by 1.3–1.8°C. This decrease may be compared to the effect of CO_2 doubling, about 2°C, as computed by Manabe and Weatherald,[18] which is considered to be the most reliable value. Thus, so far, we cannot say with certainty whether CO_2 or dust will play a major role in affecting temperature. In particular, it should be noted that the measurement of atmospheric turbidity is still in an early stage as compared with that of CO_2 measurement, and shows a large variability in different places.

---

*Very recently, Yamamoto and Tanaka determined the value of n_i experimentally, and it has indeed turned out to be 0.01 (presented at the Meteorological Society of Japan, October, 1974).

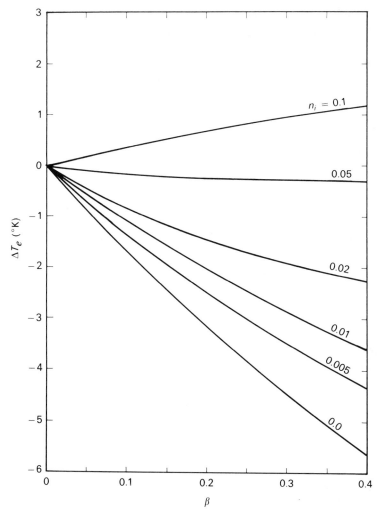

Figure 18. Effective blackbody temperature increase, ΔT_e, as a function of turbidity factor, β, on different imaginary refractive index, n_i. (After Yamamoto and Tanaka.[24])

5. The Limitations of Global-Average Models and the Appearances of the General Circulation Theory*

So far, we have discussed the radiation calculations of Plass,[12] Kaplan,[14] and Kondratyev and Niilisk,[15] in which it is assumed that a radiative equilibrium at the earth's surface occurs; that of Möller,[17] which took the humidity feedback effect into consideration; that of Manabe and Weatherald,[18] where

*The author is indebted to Dr. S. Manabe, who has kindly released his unpublished manuscript for writing this section.

thermal equilibrium in the vertical structure of the atmosphere, arising from convection, condensation, and radiation was considered; and theories of competing effects, e.g., that of dust particles, as considered by Rasool and Schneider[22] and Yamamoto and Tanaka.[24]

All of these are global-averaging models. However, the structure of the atmosphere is not globally homogeneous, having north–south gradients of temperature, and as a result giving rise to many kinds of meteorological phenomena which are compared to a heat engine driven by solar energy. Budyko's paper[21] may be said to be a first step toward a modeling of the complexity of these atmospheric processes, although in a simple style. Budyko treated the energy balance of different latitudinal belts and introduced a feedback mechanism of the spreading of the polar ice cap.

Using these considerations, Manabe and Weatherald[28] made for the first time a calculation of CO_2 climatic effect by means of a three-dimensional model originally developed by Manabe.[29,30] It is essentially the extension of the usual general circulation model now used by meteorologists. It is the time integration of the primitive equations of the atmosphere defined by simulated boundary conditions in which an energy-conserving form of the finite difference formulae is used. Topography and geometry are adopted as shown in Fig. 19. Vertical nine-layer models and horizontal 500-km mesh are used. The ocean is treated as an infinite source of soil-moisture evaporation, but heat transport by ocean currents was not considered. Radiation calculations were done in a manner similar to those of Manabe and Weatherald,[18] but on the condition of balancing the solar and net longwave radiations at the top of the atmosphere as *integrated from the equator to the pole.* Distribution of cloudiness

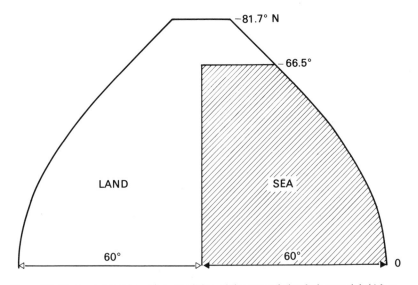

Figure 19. Topographical boundary condition of the general circulation model. (After Manabe.[29])

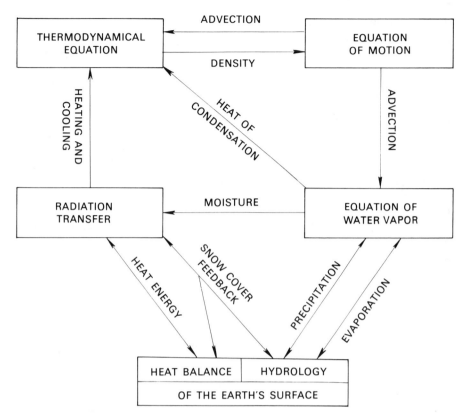

Figure 20. A block diagram showing the flow chart of the computing system of a three-dimensional general circulation model of climate. (After Manabe and Weatherald.[28])

is specified from annual mean observations and is a function of latitude and height only. The distribution of water vapor is determined by equations which take into account three-dimensional convection, evaporation, vertical mixing, nonconvective condensation, and an idealized moisture-convective adjustment. Thus, the radiative computation is coupled to a hydrologic cycle. Surface temperatures of each location are determined to balance the fluxes of solar and terrestrial radiation, and the local turbulent fluxes of sensible and latent heat. To take into account the snow- and ice-feedback mechanism, pointed out by Budyko,[21] snowfall, rainfall, and a freezing effect to form sea ice are introduced. Then, the radiation effect of ice and snow, the albedo, is introduced, as previously considered by Budyko.[21]

The model is shown in Fig. 20. The CO_2 mixing ratios of constant, 0.456×10^{-3} g/kg and 0.912×10^{-3} g/kg, are considered, i.e., the standard one and that of doubling the concentration.

The time of integration was taken as far as 800 days, to obtain a well-converging average of at least 100 days. The result of the standard case gave a satisfactory temperature profile resembling the actual one. The deviation of the profiles for the case of a doubling of CO_2 from the standard, the main result of the paper, is shown in Fig. 21. In this figure remarkable temperature increases at high latitude and near the ground are predicted. An overall temperature increase throughout the troposphere of middle and low latitudes, and a temperature decrease toward the higher level throughout the stratosphere, are projected. The large warming in the lower troposphere in high latitudes is associated with a decrease in the area of snow. The temperature increase in high latitudes is confined to lower levels because the stability of the atmosphere is better than in the other regions, so that the vertical mixing of the warming effect is suppressed. There is also an effect pointed out by Revelle,[8] i.e., the tendency of a larger contribution of CO_2 heating in a dry atmosphere at high latitudes than in a wet atmosphere at low latitudes, which plays a role to some extent. The latitudinal mean temperature distributions are drawn in Fig. 22, where the greater difference between dotted and solid lines in high latitudes can be seen.

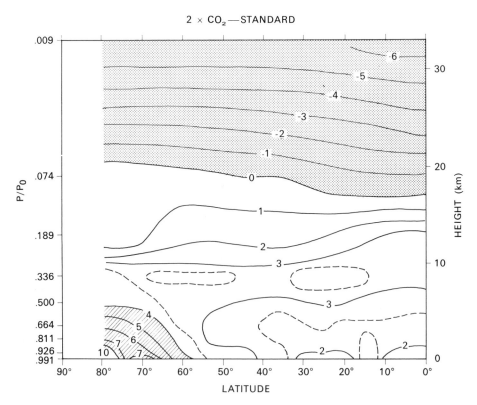

Figure 21. Distribution of temperature difference as a result of computing the general circulation model between the case of doubling CO_2 concentration and the standard one. (After Manabe and Weatherald.[28])

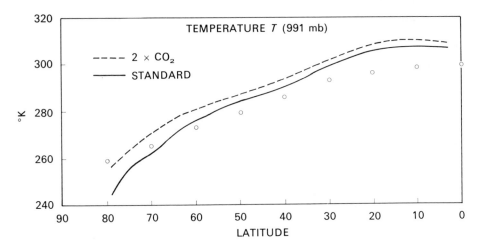

Figure 22. Mean latitudinal temperature distribution computed by the general circulation model in cases both of standard (solid line) and of doubled CO_2 (dashed line). (After Manabe and Weatherald.[28])

As to the calculation of the temperature increase caused by a doubling of the CO_2, the previous model by Manabe and Weatherald[18] was recalculated by incorporating the radiation calculations developed by Rodgers and Walshow,[16] and the results are presented in Table 3, together with the averaged value of a general circulation model.

The R–W model shows a temperature increase which is a little lower than the previous M–W model. If this revised method of radiation calculation is used in a general circulation model, the value for the temperature increase may be lowered a little more. But, in any case, we can notice that, because of the polar ice cap feedback effect, the general circulation model gives a higher value than the global-averaging model.

In addition to the temperature distributions, Manabe and Weatherald were able to deduce several interesting side results, such as the hydrological cycle, the distribution of eddy kinetic energy, and the detailed energy balances.

Table 3. Difference of Global Mean Surface Temperature between the $2 \times CO_2$ and the Standard Cases $(°C)^a$

Change of CO_2 content (ppm)	R–W Model[b]	M–W Model[c]	G.C. Model[d]
300 → 600	+1.95	+2.36	+2.93

[a] After Manabe and Weatherald.[28]
[b] Rogers–Walshow radiative convective equilibrium model.
[c] Manabe–Weatherald radiative convective equilibrium model.
[d] General circulation model.

Successful as this paper is in giving reasonable results, the model leaves much to be achieved, as the authors admit. Thus, (1) cloud formations are not considered, (2) large-scale dynamics are not considered, and (3) the effects of ocean currents are not taken into account. The importance of the first item was pointed out by Möller,[17] and this gives a negative feedback, i.e., a stabilizing effect which will have to be considered in future calculations as the item of vital importance. The importance of the third item is considered by many authors in discussions of climate. About one-fifth as much heat is transported from the equator to the pole in the oceans as in the atmosphere.[31] With respect to the second item, the dynamic effect is expressed as a local convective mixing only, and any of the effects of the usual meteorological phenomena, such as the development of cyclone and anticyclone and frontal activities, have not been explicitly incorporated. Thus, the model described is still idealized, in spite of the effort to include as many physical processes as possible. Manabe and Weatherald admit these points and indicate the extraordinarily large amount of calculation and computer time which will be needed to improve the calculation in the future.

There are other important points to be mentioned. One is the incorporation of the effect of dust particles, which have been discussed as the most important competitor to CO_2. This effect is confined to the lower layer of the atmosphere, and may not last long because of scavenging effects. On the other hand, according to Budyko,[21] Yamamoto and Tanaka,[24] and others, the influence of volcanic activity lasts several years, so that dust particles injected to the stratosphere may have an important role. According to recent advances in the chemistry of the gaseous constituents, such as sulfur dioxide[32] injected into the stratosphere by volcanic activity or from industrial activity, these may be converted to aerosols continuously, causing a long-lasting radiative effect.

Outstanding as the achievements of recent simulation experiments have been so far, and will be in the future, simulation and monitoring are not enough to attain the final objective of understanding the nature of climate. We must first compute what each physical process is doing to the climate quantitatively, as Mitchell[33] has described.

References

1. M. P. Thekaekara, *NASA Space Vehicle Design Criteria* (Environment), NASA SP-8005 (1971).
2. J. London and T. Sasamori, in *Man's Impact on Climate*, W. H. Mathews, W. W. Kellogg, and G. D. Robinson, eds., The MIT Press, 1971, p. 141.
3. R. M. Goody, *Atmospheric Radiation. I. Theoretical Basis*, Oxford at the Clarendon Press, 1964, p. 4.
4. G. S. Callender, *Tellus*, **10** : 243 (1958).
5. C. L. Wilson and W. H. Mathews, eds., *Man's Impact on the Global Environment* (*SCEP*), The MIT Press, 1970, p. 47.
6. C. L. Wilson and W. H. Mathews, eds., *Inadvertent Climatic Modification* (*SMIC*), The MIT Press, 1971, p. 232.

7. L. Machta, *Bull. Am. Met. Soc.*, **53** : 402 (1972).
8. R. Revelle, in *Man's Impact on Climate*, W. H. Mathews, W. W. Kellogg, and G. D. Robinson, eds., The MIT Press, 1971, p. 281.
9. Y. Miyake, Y. Sugiura, and K. Saruhashi, *Rec. Ocean. Work. JAPAN*, **12** : 45 (1974).
10. S. Ben-Yaakov, E. Ruth, and I. R. Kaplan, *Science*, **184** : 982 (1974).
11. G. D. Robinson, in *Man's Impact on Climate*, W. H. Mathews, W. W. Kellogg, and G. D. Robinson, eds., The MIT Press, 1971, p. 157.
12. G. N. Plass, *Tellus*, **8** : 140 (1956).
13. W. H. Cloud, *ONR Progress Report*, Johns Hopkins Univ., 1952.
14. L. D. Kaplan, *Tellus*, **12** : 204 (1960).
15. K. Y. Kondratyev and H. I. Niilisk, *Geof. Pur. e Appl.*, **46** : 216 (1960).
16. C. D. Rodgers and C. D. Walshow, *Quart. J. Roy. Met. Soc.*, **92** : 67 (1966).
17. F. Möller, *J. Geoph. Res.*, **68** : 3877 (1963).
18. S. Manabe and R. T. Weatherald, *J. Atm. Sci.*, **24** : 241 (1967).
19. S. Manabe and R. F. Strickler, *J. Atm. Sci.*, **21** : 361 (1964).
20. K. Telegadas and J. London, *Sci. Rep. No. 1, Cont. AF 19 (122)–165*, New York Univ., 1954, p. 55.
21. M. I. Budyko, *Tellus*, **21** : 611 (1969).
22. S. I. Rasool and S. H. Schneider, *Science*, **173** : 137 (1971).
23. S. I. Rasool and C. DeBergh, *Nature*, **226** : 1037 (1970).
24. G. Yamamoto and M. Tanaka, *J. Atm. Sci.*, **29** : 1405 (1972).
25. S. Chandrasekhar, *Radiative Transfer*, Oxford Univ. Press, 1950, reprinted 1960 by Dover Publ. Inc.
26. M. Tanaka, *J. Met. Soc. Jap.*, **49** : 296 (1971).
27. M. Tanaka, *J. Met. Soc. Jap.*, **49** : 321 (1971).
28. S. Manabe and R. T. Weatherald, *J. Atm. Sci.*, **32** : 3 (1975).*
29. S. Manabe, *Month. Weath. Rev.*, **97** : 739 (1969).
30. S. Manabe, in *Man's Impact on Climate*, W. H. Mathews, W. W. Kellogg, and G. D. Robinson, eds., The MIT Press, 1971, p. 249.
31. W. W. Kellogg, in *Man's Impact on Climate*, W. H. Mathews, W. W. Kellogg, and G. D. Robinson, eds., The MIT Press, 1971, p. 123.
32. R. D. Cadle, in *Physics and Chemistry of the Lower Atmosphere*, S. I. Rasool, ed., Plenum Press, 1973, p. 69.
33. J. M. Mitchell, Jr., in *Man's Impact on Climate*, W. H. Mathews, W. W. Kellogg, and G. D. Robinson, eds., The MIT Press, 1971, p. 133.

*This article appeared after the completion of the manuscript for this chapter.

The Climatic Effect of Particles in the Atmosphere

The predictions of the CO_2 "greenhouse effect" theory worked for the first 50 years of this century, but during the last 15–20 years world temperature has tended to decrease. *It is conceivable that such a decrease is associated with the great increase in aerosols in the atmosphere during that time. The study of the stability of such bodies is important, for it may allow us to influence an effect which could lead to a very undesirable, if temporary, drop in world temperatures.*

11

Aerosol Production in the Atmosphere

J. Bricard

1. *Introduction: Particle Size Distribution*

The atmosphere contains particles in suspension, near the ground as well as in the troposphere and stratosphere.* Although complex in shape, they are simulated by spherical formations with a radius R between 10^{-7} cm and 10^{-2} cm. Particles of less than 10^{-1} μ† in radius are known as "Aitken cores." They are the most numerous.

Some particles penetrate fully shaped into the atmosphere and are wind-driven, and others are spewed by volcanoes and meteorites. Others are

*The troposphere is the atmospheric layer located between ground level and a mean altitude of about 11 km (about 6 km at the poles and 17 km at the equator); the stratosphere is above that. Stratosphere and troposphere are separated by the tropopause.

†$1\ \mu = 10^{-6}$ m.

J. Bricard • Laboratoire de Physique des Aerosols, Université de Paris VI, 11, Quai Saint-Bernard, Paris V^e, France

produced *in situ* through reactive action between air-suspended gaseous impurities, between themselves, or in contact with water vapor.

In this chapter, we propose to summarize the *in situ* formation processes, as well as subsequent changes undergone by particles whatever their origin after they are in their suspended state, and by situating ourselves first in the lower troposphere and then in the stratosphere. One of the most important characteristics regarding atmospheric particles is their size distribution. On a global scale, we can differentiate between three general types of particle distribution in the troposphere: "background," oceanic, and continental.[1]

1. The most important particle distribution, from a global point of view, is that called "background." It is roughly equal in both the mid and upper troposphere. The particle population background is that which corresponds to the purest air, uninfluenced by local particle sources. It may be defined as the total number N of particles per cubic centimeter, and 700 particles/cm^3 is about the top concentration limit. This size distribution is shown in Fig. 1, and corresponds to about 200 particles/cm^3. There are few indications regarding the size distribution for particles with radii less than 0.1 μ. All the findings

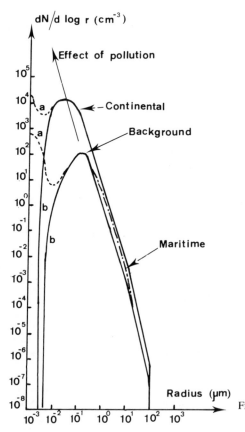

Figure 1. Size distribution of atmospheric aerosol.

obtained by optical means,[2] electrical measurements,[3] and the only measurements available at present for areas of the Atlantic[4,5] suggest that there is a maximum concentration corresponding to a radius of 0.2 μ, as indicated in Fig. 1. Since small particles attach themselves to larger particles through thermal coagulation (Brownian motion), it seems probable that the concentration decreases rapidly in proportion to the decreasing radius (full lines b in Fig. 1), except that there appears to be a continuous production of very small particles. From recent studies, it has been demonstrated that sunlight may be the cause of this production, and results obtained by Junge and Jaenicke[4] and by Renoux *et al.*[6] suggest the distribution depicted by the curve in Fig. 1. Both curves, a and b, indicate the distribution with and without the production of small particles.

2. Lower tropospheric regions above oceans (2 km in depth) contain a population of particles of oceanic origin. The oceanic-size distribution appears to differ from background-size distribution only insofar as concerns an interval of radii between 0.5 and 20 μ, within which sea-spray particles superpose on a background-size distribution, shown in Fig. 1.

3. Within the lower troposphere above continents, and in particles above regions affected by pollution, we discover a different, continental-size distribution. The concentration in number is about $10^4/cm^3$ over rural areas, and exceeds $10^5/cm^3$ over polluted large cities. The distribution shown on Fig. 1 corresponds to about $10^4/cm^3$. For particles with radii less than 0.1 μ, the size distribution has not yet been clearly defined.

2. In Situ Particle Formation

Many, if not most, air-suspended particles may be considered as having been created from a gaseous state of evolved matter that has been transformed into particles at a later stage. This is the case with SO_2 and hydrocarbons.

These gases remain gases for only a limited time in the air—a few days to two weeks maximum—and possibly for only a few hours in a heavily polluted atmosphere. Hence, balanced conditions are quickly reached even if the emission rate increases.

The process which occurs most often, which will be discussed below, is oxidation in the gas phase through productions resulting from the always-present NO_2 photolysis. Simplified, it is

$$NO_2 + h\nu \rightarrow NO + O \tag{1}$$

The atmosphere over Paris, for instance, has a concentration of 30 $\mu g/m^3$,* and this is five times more than the average for United States cities.[7] When other products are not present, atomic oxygen reacts with molecular oxygen to give ozone:

$$O + O_2 + M \rightarrow O_3 + M \tag{2}$$

*1 $\mu g = 10^{-6}$ g.

oxidizing the NO from (1) to give again NO_2

$$NO + O_3 \rightarrow NO_2 + O_2 \tag{3}$$

2.1. *Atmospheric SO_2 Oxidation*

SO_2 is a primary air pollutant, either directly (e.g., from volcanoes, domestic fuel, and coal combustion), or through H_2S oxidation from organic matter decomposition. SO_2 quantities of up to some 10's of $\mu g/m^3$ are found in pure air (rural areas),[8-10] between 0 and 4 $\mu g/m^3$ in air above oceans.[1,11] Concentrations may exceed values that are 100 or even 500 times higher in polluted air (cities).[7] H_2S does not substantially react with O_2 or O_3 in gaseous phase, and does not undergo photolysis, but it is oxidized by atomic oxygen[12] according to the following reaction[1]:

$$H_2S + O \rightarrow OH + SH \tag{4}$$

with a reaction constant of 4×10^{-14} cm^3/mol sec, at 300°K.[12] Reaction (4) is followed by a chemical chain reaction, involving products such as SO_2, SO_3, H_2SO_4, and H_2O, that takes place during photochemical smogs. Most air particles are sulfates from SO_2 oxidation. A mean figure of some $\mu g/m^3$ of sulfate in rural and ocean atmospheres[11,13] is found, as well as 7–20 $\mu g/m^3$ in cities with average pollution.[13,14]

2.1.1. *SO_2 Photochemical Oxidation*

A process which gives sulfate particle formation is direct SO_2 photochemical oxidation, schematically shown as

$$2SO_2 + O_2 + h\nu \rightarrow 2SO_3 \tag{5}$$

This reaction can be broken down to

$$SO_2 + h\nu \rightarrow SO_2^{(+)}$$
$$SO_2^{(+)} + O_2 \rightarrow SO_4$$
$$SO_4 + O_2 \rightarrow SO_3 + O_3$$

This pattern shows how low ozone levels can exist in the lower atmosphere as a photochemical by-product. Other mechanisms involve a photochemical action on various pollutants, particularly NO_2 and hydrocarbons, as these would generate radicals or atoms able to oxidize SO_2. In any event, the primary oxidation product is SO_3, then transformed into H_2SO_4:

$$SO_3 + H_2O + M \rightarrow H_2SO_4 + M \tag{6}$$

Figure 2, from Meszaros,[14] represents a concentration ratio of SO_4^{2-} and SO_2, in air, thus $[SO_4^{2-}]/SO_2$ as a function of ground solar radiation intensity within the spectrum band between 0.29 and 0.5 μ. A straight path corresponds to the computed regression. The relation between the ratio $[SO_4^{2-}]/SO_2$, solar

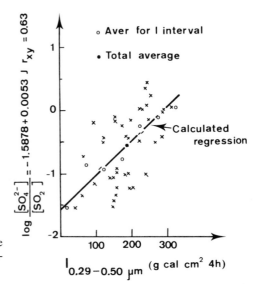

Figure 2. Relation between the mole ratios of SO_4^{2-} to SO_2, and the solar radiation intensity in the band 0.29–0.50 μ.

radiation intensity, and temperature, corresponds to a multiple correlative coefficient +0.73, and confirms the close relationship between solar radiation intensity and SO_2 oxidation. These findings agree with those obtained by Bricard *et al.*[15]

Sulfate particles exist even over oceans. Cadle *et al.*[16] above the Atlantic Ocean, and Frank *et al.*[17] over Catalina Island in the Pacific near Los Angeles, have found particles which are made of H_2SO_4 and sulfates (cf. Meszaros and Vissy[18]) that represent about one-half the total particle weight. Moreover, the ratio between the day and overnight counts shows an excess of 50% for the day, which confirms the photochemical origin, corroborating the findings of Vohra *et al.*[54]

Reactive velocity can be expressed by the following relation:

$$\frac{d\bar{SO}_2}{dt} = -k(SO_2) \tag{7}$$

where k is the photolytic rate found.

The parameter k relates to solar radiation intensity and spectral repartition, i.e., sun height and atmospheric transparency, the latter being linked to weather conditions and pollution. A discussion of the nature of the photochemical oxidation rate is given by Leighton,[19] Cadle,[20] Chovin and Roussel,[7] and in the discussions by Cox[21] and Buffalini.[22]

Table 1 summarizes some of the relevant photolytic values obtained by various authors.

Differences between the various values can be explained as resulting from the various methods used: SO_2 dosage as a function of time in some cases, dosage of formed products in other cases; different solar height above the

Table 1. Some Photochemical Rate Constants in SO_2 Reactions

Reference	Initial SO_2	Relative humidity	Rate of SO_2 consumption % per min
Gerhard and Johnstone[55]	5–30 ppm	32–91	$(1.7$–$3.3) \times 10^{-3}$
Renzetti and Doyle[56]	0.2–0.6 ppm	50	0.11
Urone and Schroder[57]	10–20 ppm	50	1.4×10^{-3}
Hall[58]	$SO_2 = 56$–230 mmHg $O_2 = 5$–200 mmHg	0	8×10^{-4}
Urone and Schroder[57]	1000 ppm	50	4.6×10^{-4}

horizon and different atmospheric transparencies, from one case to another, if the sun is used as a source; and poorly defined spectral spreads and intensities if an artificial source is used. Generally speaking, the reactions were carried out with a higher SO_2 concentration than that found in nature.

The aerosol mass produced in a given time is low in relation to the mass of SO_2 present. Let us assume that we have a concentration of 20 $\mu g/m^3$ of SO_2 with a photolytic rate of 10^{-3} per minute. Then, production on the order of 10^6 particles of radius 10^{-7} cm/cm^3 of air per second would result.

We need to involve a quantum efficiency Φ which can be defined from the photolysis rate k by the relation

$$k = k_a \Phi \qquad (8)$$

where k_a is the absorption constant equal to the product of the irradiation intensity in photons cm^3 S^{-1} times the coefficient of absorption for a given wavelength, and Φ is the ratio of the number of transformed molecules to that of the absorbed photons.

In so doing, we eliminate some of the objections which may be raised above, particularly regarding the light source intensity. The values given would suggest quantum efficiencies of 10^{-2} to 10^{-3}. More recent measurements have indicated that within the spectral region between 2900 and 3400 Å (Allen[21]; Cox[23]), and for low SO_2 concentrations in air, the maximum Φ may be as low as 3.10^{-4}.

In fact, recent experiments[24] have proved that the formation of particles within a product mixture of $SO_2 + O_2 + H_2O$ would be inappreciable and would occur only if traces of NO_2 were added, thereby involving reaction (1). This may lead to some addition complexes (NO—HSO$_4$) as has been recently suggested.[25]

Cox[21] interprets the quick oxidation of SO_2 in the presence of olefins and NO_x as an intermediate oxidizer formed by the action of ozone on hydrocarbons, itself emerging from NO_x photochemical action on these hydrocarbons. It should be noted that atomic oxygen produced by NO_2 photochemical dissociation [reaction (1)] reacts on water vapor to give rise to OH radicals according to

$$O + H_2O \rightarrow 2OH$$

The following reactions involve carbon monoxide (CO), nitrogen oxides, ozone, hydrocarbons, etc., a balance occurring between OH radicals and HO_2.[26] Now, these radicals react with SO_2 according to

$$HO_2 + SO_2 \rightarrow SO_3 + OH$$

Hence, Cox's interpretation[21] regarding the oxidation of SO_2 in the presence of olefins, i.e., that it is due to the appearance of an intermediary oxidant, which may involve the radicals shown.

The oxidation rate, depending on the degree of air pollution, is between 0.1% per hour (low air pollution), and 1–10% per hour (heavy pollution).

2.1.2. Addition Reactions: Catalytic Oxidation

Some authors[27] have indicated that the NH_3/SO_2 reactions may occur and result in an $(NH_3)(SO_2)_n$ polymer. However, recent experiments[24,28] show that the presence of ammonia does not lead to the formation of particles. But it has been realized[24] that air photolytic particles, enriched in SO_2, are of larger size when NH_3 is added. SO_2 dissolves in atmospheric water droplets to form H_2SO_3 and is then oxidized by oxygen, dissolved in droplets:

$$H_2SO_3 + O_2 \rightarrow 2H_2SO_4 \qquad (9)$$

The oxidizing action diminishes in proportion to increase acidity. Friend[28] states that the NH_4HSO_4 particle embryos will fix NH_3 to form embryos:

$$NH_3 + H_2SO_4 n H_2O \rightarrow NH_4HSO_4 n H_2O \qquad (10)$$

$$NH_3 + NH_4HSO_4 n H_2O \rightarrow (NH_4)_2SO_4 n H_2O \qquad (11)$$

$$2SO_2 + 2H_2O + O \xrightarrow[\text{NH}_4]{\text{embryo}} 2H_2SO_4 \qquad (12)$$

The products of the first two reactions are saline embryos, a medium resulting in SO_2 catalytic oxidation. Particle development continues as long as NH_3 is fixed and is able to neutralize the acid resulting from catalytic oxidation, and while SO_2 is fixed by the particle. From what Junge and Ryan[13] have said, SO_2 oxidation relates to the presence of pertinent catalyzers (ions of heavy metals) and reaches a high level only in polluted air. Such transformation ends when a certain pH value is reached in the solution. This mechanism could lead to the forming of mist particles and drops in clouds.

Urone *et al.*[29] state that, even in the absence of light, SO_2 in air is oxidized in the presence of several metal oxides (aluminum, calcium, and iron).

Atmospheric particles could have the same effect, but since the study is only qualitative the importance of oxidation cannot be appraised. It does not, however, appear to be too significant.[30] SO_2 oxidation in sulfuric acid within an urban environment was researched by Benarie *et al.*[31]

Some authors[5,32] have brought to light the formation of particles when SO_2 is present in ionized air, but the required ion concentrations are too high for such mechanisms to become sensitive to the lower atmosphere. This may take place in the stratosphere.

2.2. Hydrocarbons

Quantities reaching a mean value in the region of a few parts per million of hydrocarbons have been found in air. They emerge from human activities, such as car exhausts and fuel- or coal-burning residuals. They activate photochemical smogs in cities.

Among possible causes of hydrocarbon oxidizing, the most probable is NO_2 photolysis, shown by reactions (1), (2), and (3). However, in the case of smogs, the quantity of ozone after such reactions is lower than that which can be observed. This may be attributed to the action of atomic oxygen on hydrocarbons, leading to ozone and hydrocarbon by-products such as formaldehyde, acrolein, and peracetyl nitrate, which are characteristic of photochemical smogs.

These reactions may be shown by the simplified scheme given below

$$O + olefin \rightarrow R'O \text{ or } O \rightarrow R + OH$$

$$R \underline{\qquad} R'$$

$$O_3 + olefin \rightarrow product$$

$$R + O_2 \rightarrow RO_2$$

$$RO_2 + O_2 \rightarrow RO + O_3$$

This reaction is responsible for the production of ozone. Direct excitation of other, different organic molecules by sunlight could lead to other chain reactions.

Cadle[12] has suggested that only a small fraction of organic vapors intervenes in photochemical smogs (about 5%) and is transformed in particles, but the particle–vapor transformation mechanism (nucleation) does not seem to have been thoroughly studied, as in the case of sulfuric acid.

2.3. Natural Mists

Plant-emitted natural organic vapors (terpinols, α-pinenes) in low quantities in the air ($\neq 1 \ \mu g/m^3$ of air) were conceivably responsible for the birth of particles through similar mechanisms. Nitrogen oxides occur naturally during forest and plant fires, and, moreover, numerous hydrocarbons absorb solar radiations as electronic excitation leads to chain reactions, resulting in particles.[20,33]

Some of these organic compounds, particularly unsaturated ones, quickly oxidize even when the sun is hidden, through oxygen fastening onto carbon atoms in α positions, in relation to a double link:

$$-C-C=C-C-+O_2 \rightarrow -C-\underset{\underset{O-O}{|}}{C}=C-$$

while the peroxide radical thus formed could bring about reactions leading to the formation of particles.

2.4. Nucleation

The result of all these transformations is the formation of sulfuric acid vapors (μg/m^3) or of oxidized hydrocarbon vapors, in the presence of water vapor excesses (tens of grams per cubic meter of air). The transformation of these vapors into droplets, i.e., nuclei, is discussed in particular by Zetlemoyer.[34]

2.4.1. Homogeneous Nucleation

First, let us situate ourselves in a pure vapor environment, either mixed or not mixed with a foreign gas which does not react on the vapor and does not intervene in its condensation, and with a total absence of foreign particles in suspension in the medium (homogeneous nucleation). Let us assume that vapor molecules cluster together to form a spherical droplet embryo of radius R, volume $V = \frac{4}{3}\pi R^3$ and surface $A = 4\pi R^2$.

The first step in an approximation for homogeneous nucleation would consist in determining the variation of the free enthalpy ΔG, corresponding to the formation of an embryo, and written as

$$\Delta G = n_L(\mu_L - \mu_V)V + \sigma_{LV}A \tag{13}$$

where μ_L and μ_V represent the free enthalpy of a molecule in liquid phase and vapor phase, respectively, n_L is the number of molecules per unit of liquid volume, and σ_{LV} is the interfacial free energy per surface unit between the liquid and vapor. The quantity σ_{LV} is in fact the surface tension for a liquid.

When μ_V is higher than μ_L, i.e., when the vapor is supersaturated, the function $\Delta G(R)$ goes through a maximum, $\Delta G(R^*)$, for a value R^* of the radius, called the critical radius, and the droplet of radius R^* is in unstable balance with the vapor. The critical radius R^* is expressed as

$$R^* = \frac{2\sigma}{n_L kT \ln (p/p_\infty)} \tag{14}$$

where k is the Boltzmann constant, p is the vapor pressure, and p_∞ is the saturated vapor pressure at the same temperature.

The study of the variation of ΔG as a function of R shows that, when the embryo radius is larger than the critical radius, embryos develop through condensation, whereas embryos of shorter radius are unstable. Therefore, there are no stable droplets of radius less than the critical radius; they therefore represent the lower size limit of particles formed by homogeneous nucleation.

The problem consists of defining the number of nuclei exceeding the critical size formed per unit of time and unit of vapor volume. This is the nucleation frequency. Such nucleation frequency, represented by I, brings into play the collisional frequency between free molecules and critical nuclei. It assumes the form

$$I = KN(1) \exp\left(-\frac{\Delta G(R^*)}{kT}\right) \qquad (15)$$

where $N(1)$ represents the concentration of vapor molecules, K is the factor bringing into play the probability of collision between a vapor molecule and a critical embryo, as well as corrective factors corresponding to the reevaporation of smaller-size, unstable nuclei, toward a critical size during their growth in reaching the critical size.

Taking $\sigma = 75 \text{ dyn cm}^{-1}$ (the value for water) and $n_L = 3.3 \times 10^{22}$ molecules per cm^3, one finds with reactions (14) and (15) a nucleation frequency of 1 critical core per cm^3 per sec (the lower limit of significant frequency) if p/p_∞ is higher than about 4.5. The nucleation mechanism described therefore does not occur in nature, where p/p_∞ does not exceed 1.05 in the case of water vapor. In particular, it does not take part in the formation of cloud droplets. For water, the critical radius would be on the order of 5–10 Å under normal conditions.

Of course the reasoning here is approximate and one may entertain doubts about the significance of the concept of surface tension for particles containing only hundreds of molecules. However, the findings are on the whole verified by experience. The method of reasoning has been generalized by Reiss[35] for a system with several components. The application of such a generalization to a sulfuric acid water system was carried out by Doyle,[36] and in a simplified form by Kiang et al.,[37] who wrote the nucleation frequency formula in the form

$$I = K \exp\left(-\frac{\Delta G(R^*)}{kT}\right) \qquad (16)$$

where values of $\Delta G(RI)$ are different[37] than those of previous authors.

Calculations show that at atmospheric pressure and 10% relative humidity, above 3×10^9 molecules of H_2SO_4/cm^3 are sufficient to produce a frequency of nucleation of 1 particle/(cm^3/sec). For a given content of H_2SO_4, the nucleation frequency is linked to humidity and increases with it.

Thus, particles with a radius of about 10 Å are liable to appear. Particles coagulate among themselves and with larger particles while undergoing other transformations which we will study further below.

Lastly, there is a dynamic equilibrium between the primary particles thus formed and those which disappear through coagulation and further transformation, and this can be shown as a distribution in Fig. 1.

2.4.2. Heterogeneous Nucleation

The above computations apply only when foreign particles are present. Whenever a vapor (or a mixture thereof) contains foreign suspended particles, and conditions are propitious, we can see vapor condensation on these particles. The latter react therefore as condensation nuclei, and we therefore have heterogeneous nucleation.

Goetz and Peuscal[38] have demonstrated that when latex particles $(0.35 \ \mu\phi)$ are in a suspended state in a photochemical system (hydrocarbons, SO_2, NO_2), their sizes increase during reaction, and their mass increases with particle concentration. Moreover, the presence of particles substantially reduces homogeneous nucleation. The formation of fog by means of water condensation seems to belong to a different mechanism, because the development of nuclei, for the same oversaturation, is inversely proportional to concentration.

In the absence of the initial nuclei, the nucleation frequency relates to the concentration of reacting products. Thus, if the concentration is strong and nuclei concentration weak, homogeneous nucleation will compete with the condensation of products formed on nuclei. The latter process will be favored by a low concentration of photochemical products or a high concentration in nuclei. Working on Los Angeles smog, Husar *et al.*[39] have shown the importance of vapor condensation processes over higher-dimensioned particles $(0.1 \ \mu < D < 1 \ \mu)$, the development of which is due to condensation.

3. Transformation of Air-Suspended Particles

Gravity pulls on suspended particles in air. Their drop velocity determines their time in the air, as the sedimentation process affects only particles with a radius above about $1 \ \mu$. For particles with a smaller radius, drop velocity becomes negligible. Sedimentation accounts for about 20% of particle disappearance, the main cause being atmospheric washout precipitation and the formation of fogs.

This process of disappearance determines particular residence times in the atmosphere. It was calculated by Junge[1] from findings obtained in an atmospheric radioactive environment to be between 3 and 22 days under temperate climates, and its average is about 1 week.

Once in suspension, directly or through the above reactions, particles react with gaseous air impurities. Moreover, due to their longevity in a suspended state, particles coagulate under the action of Brownian motion. In any event, they lose their original individuality to some extent.[18]

3.1. Particle–Gas Reactions

A multitude of gaseous impurities may react on an aerosol particle, e.g., ammonia with H_2SO_4, nitric acid vapors with NaCl, ozone with organic particles, etc. The reaction rate in a gas mixture will be controlled both by gas diffusion reaction rate to the particle surface, and by diffusion rate in a particle of the gas which reacts, and thereby forms resulting products. The upper limit of the initial reaction rate can be calculated by evaluating the frequency of collisions per particle per surface unit, and by assuming that each collision leads to a reaction. Few quantitative studies have been fulfilled regarding possible chemical reactions between gas and particles in the atmosphere.

3.1.1. Formation of Ammonium Salts

Ammonium sulfate is an important component of atmospheric aerosol composition and is formed by ammonia action on sulfuric acid drops appearing during the process described above. In polluted atmospheres, we may encounter a few micrograms of ammonia per cubic meter of air and a close value of sulfate.[14] The reaction kinetics for drop formation of H_2SO_4—NH_3 were studied by Cadle and Robbins[40] and by Robbins and Cadle.[41] They found that when drops are concentrated in H_2SO_4 the reaction rate is given by the empirical relation

$$\frac{d[A]}{dt} = k\frac{3[A_o]}{r}\left[\frac{0.18(1-X)}{0.18+0.82X}\right][NH_3]$$

where $[A]$ is the acid concentration in the aerosol, expressed in units of weight or volume; $[A_o]$ is the original concentration; r is the particle radius; X is the fraction of acid reacted; and k is the velocity constant.

However, drops of sulfuric acid in a smog or a mist, such as we have discussed, are diluted. Cadle and Robbins[40] have found that this reaction is too quick to be measured by any technique which they have employed, and they suggest that every collision of an ammonia molecule with a drop leads to reaction.

3.1.2. Nitrate Formation

When NO_2 and an NaCl aerosol are found in damp air, a balanced mixture of $NaNO_3$ and HCl gas appears too quickly for the formation rates to be determined with present techniques. The first stage relates to the reaction

$$3NO_2 + H_2O \rightarrow 2NO_3H + NO$$

the equilibrium constant of which is 0.004 atm^{-1} at 300°K.

The second stage is the adsorption of nitric acid vapor on dry NaCl particles, or, when the relative humidity exceeds 75%, with the dissolving of water droplets containing NaCl. This is followed by a nitric acid reaction with NaCl and the desorption of HCl, which takes place immediately afterward or during droplet evaporation.

3.2. Particle Coagulation: Formation of Alkaline Sulfate

We have mentioned collisions resulting from Brownian motion and the progressive coagulation taking place between particles, particularly between those of different sizes.

The atmosphere contains particles of different chemical compositions and these react between themselves when they come into contact. In this case, coalescing may result in the freeing of hydrochloric acid:

$$2NaCl + H_2SO_4 \rightarrow Na_2SO_4 + 2HCl$$

The concentration of NaCl was found to be 7 $\mu g/m^3$ in the atmosphere over Frankfurt.[42] We have reason to believe that a portion of this atmospheric chlorine may be in the form of hydrochloric acid. Sulfuric acid and carbonates from the ground can react in the same manner.

4. Origin of Tropospheric Particles

Because of their suspended longevity in the air, particles evolve up to the moment when they disappear through sedimentation or washout. Particle existence in the atmosphere is a dynamic process during which there is a state of equilibrium between matter production under an initial form and disappearance under a different form. Table 2 depicts the production of various chemical

Table 2. Production of Various Chemical Species of Particles with a Radius of Less Than 20 μ

NATURAL			
Soil and rock debris		100–500	(ref. 59)
Forest fires and slash burning			
debris	(ref. 60)	3–150	(ref. 61)
Sea salt		300	(ref. 62)
Volcanic debris	(ref. 63)	25–150	(ref. 59)
PARTICLES FORMED FROM GASEOUS EMISSIONS			
Sulfate from H_2S	(ref. 59)	130–200	(ref. 60)
Ammonium salts from NH_3		80–270	(refs. 60, 63)
Nitrate from $(NO)_x$	(ref. 63)	60–430	(ref. 60)
Hydrocarbon from plant exudations		75–200	(ref. 33)
Subtotal		773–2200	
MANMADE			
Particles direction emission		10–90	(ref. 60)
PARTICLES FORMED FROM GASEOUS EMISSIONS			
Sulfate from SO_2	(ref. 59)	130–200	(ref. 63)
Nitrate from $(NO)_x$		30–35	(refs. 60, 63)
Hydrocarbons		15–90	(ref. 63)
Subtotal		185–415	
TOTAL		958–2615	

species of particles with a radius of less than 20 μ which appear directly or form in the atmosphere through the mechanisms we have described, expressed in megatons per year (for the whole earth).

The outer limits for the ratio between human activities and total emission is in the range 5–45%, and it is not presently possible to give a more accurate value. The highest contribution involving human activity is the emission of sulfate particles: about 50%.

5. *Stratospheric Particles*

Several kilometers above the troposphere, there is a specifically well-delineated layer of particles, the thickness of which is a few kilometers. Its existence has been established from studies on the ground[43] with light diffused by twilight glow, and by means of rockets,[44] and has been checked by observing the light diffused at pertinent altitudes, using searchlight beams or through the "altitude laser echo probing" method (LIDAR).

During 1960 and 1961, Junge *et al.*[45] were among the first to obtain direct samples of stratospheric particles during *in situ* probes with balloons. These particles have a maximum concentration at 18–20 km in altitude. Bricard and Vigla[46] have elaborated further on stratospheric particles in a recent paper.

5.1. *Concentrations: Properties*

Observations subsequent to those of Junge,[47] either indirectly at dawn,[43,48] or by LIDAR, have shown that particle concentration in this lower stratospheric layer can vary greatly over a period of some years.

After the volcanic eruption of Mount Agung, in 1963, an increase in particle concentration was observed. For a few years the twilight became more brilliant, but recent LIDAR observations have shown a progressive downgrading toward values that are similar to those found before the eruption. In any event, dust spewed out by the Agung eruption, added to other eruptions, did generate an increase in the mass of dust in the stratosphere by a factor of about 30.

The quantity of matter sampled between 1968 and 1970, during flights scheduled by the National Center on Atmospheric Research of the United States, was $30\text{–}40 \times 10^{-4}$ ppm by mass.

Some of these particles were in liquid form, and these were of many shapes.[49] We can group them into the following three categories:

1. Central particle, surrounded by an approximately circular group of small, separate satellites
2. Central particle, surrounded by a circular ring, without discrete satellites
3. Single particles, void of satellites

Because of these complications, the ascertaining of a size-distribution pattern becomes an uncertain task.

By proceeding with a dialysis of particles picked up in the stratosphere, Mossop[51] has shown that they contain one or several insoluble cores, which may be of cosmic origin. Regarding count concentrations, we must differentiate between particles of radius R for which $R < 0.1 \mu$ (Aitken cores) and particles with a larger radius. While we are able to define the geometry of the latter, it is only possible at the present time to determine the concentration of the former which varies depending on the given author and measuring method. While Junge[45] reports one Aitken core per cubic centimeter of ambient air by means of direct counts, it is possible to evaluate a concentration at 10^2 or 10^3 per cubic centimeter by using indirect methods.[44,50] When dealing with larger sizes than 0.1μ, the values given by Friend[52] are within a range of $0.012–0.069$ particles/cm^3, and these corroborate Mossop's findings: $0.017–0.042$ particles/cm^3.

Samplings by Junge, as indicated above, show that stratospheric particles are mainly sulfates. They show that the ratio of concentrations $[NH_4^+]/[SO_4^{2-}]$ is between 1 and 2, so that the main constituent element of stratospheric aerosol is ammonium sulfate. The mean $[SO_4^{2-}]$ concentration may be assumed as $0.1 \mu g/m^3$ of ambient air.

5.2. *On the Origin of Stratospheric Particles*

We should recall the characteristics of the environment in which stratospheric particles appear before discussing their origin, i.e., the practical lack of clouds and thermal stability in the stratosphere which enables pollutants to remain there for what has been evaluated as some years, while their stay in the troposphere only amounts to a few days or weeks, as we have seen. This evaluation of about 2 years in the stratosphere is derived from the rate of disintegration of radioactive debris injected into it by nuclear blasts. Thus, the equilibrium concentration for a source of the same intensity, can be 10 or 100 times larger in the stratosphere than in the troposphere. The accepted explanation of the origin of stratospheric particles is that they are generated from SO_2 gas borne at high altitude from the troposphere or injected upward by volcanic eruptions, and oxidized to SO_3 in the stratosphere through processes that are similar to those which we have assumed for the troposphere.

The emerging SO_3 hydrolyzes to form sulfuric acid and samplings have shown that the particles are droplets with radii from one-tenth up to several microns. However, we should note that as the concentration of stratospheric water vapor is in the order of a few parts per million, it seems difficult to account for the homogeneous nucleation process of the $H_2SO_4 - H_2O$ mixture which we have discussed for the troposphere. Indeed, we could assume that small insoluble particles, brought to light by Mossop[51] in stratospheric particle concentrations and mentioned above, are playing the role of condensation cores and would therefore explain particle formation.

Lastly, under the influence of ionizing radiations, and more particularly that of cosmic rays, some stratospheric gaseous molecules generate a separation of electric charges with a production of ionized molecules as positive ions, bearing a single positive charge while released electrons fasten on present neutral gaseous molecules to form negative ions. These ions set neutral molecules, and in particular water molecules that are present as traces, and become Aitken nuclei which can play an important role in the formation of stratospheric particles.[64,65]

In any event, during their stay in the stratosphere, that is, over one to several years, these particles develop through coagulation and drop through sedimentation. On the other hand, the Aitken cores (particularly those with radii less than 0.1 μ) emanate from earth surroundings.

Qualitatively, this general scheme explains the increase in particle concentration at the level of the layer concerned, as particles develop in the same proportion as that in which they are carried upward. But this does not explain the noted thinness (sometimes only a few kilometers) of the layer.

One proposed explanation[53] is that *in situ* coagulation of tropospheric Aitken cores is involved, but this explanation could possibly be rejected because the core concentration seems insufficient in the lower troposphere and stratosphere to be significant. However, the data are still unreliable, and more definite results will be required before this hypothesis can be definitely rejected.

There are still only few accurate data on the origin of stratospheric particles. It is not yet feasible to go through a proper selection between potential hypotheses as long as more precise information on the stratospheric aerosol size distribution, and on the respective concentrations of the various vapors present are not available.

References

1. "Inadvertent Climate Modification," Report on the Study of Man's Impact on Climate (SMIC), MIT Press, Cambridge, Massachusetts and London, 1971.
2. H. Quenzel, *J. of Geophysical Res.*, **75** : 2915 (1970).
3. C. Junge, *Meteorological J.*, **12** : 13 (1955).
4. C. Junge and R. Jaenicke, *J. of Aeronautical Sciences*, **2** : 305 (1971).
5. P. E. Coffey and V. A. Mohnen, *Bulletin of American Physical Sciences*, **17** : 392 (1972).
6. A. Renoux, J. P. Butor, and G. Madelaine, *Chemosphere*, to be published.
7. P. Chovin and A. Roussel, *Physiocochimie et Physiopathologie des Pollutants Atmospheriques*, Masson, Paris, 1973.
8. J. P. Lodge and J. B. Pate, *Science*, 408 (1953).
9. H. W. Georgii, *Experimentra Supplementum*, **13** : 14 (1967).
10. E. Erikson, *Technical Note 106*, World Meteorological Organization, Geneva, 1970, p. 31.
11. H. Rodhe, *J. of Geophysical Res.*, **77** : 4 494 (1972).
12. R. D. Cadle, *Formation and Chemical Reactions of Aerosols and Atmospheric Particles*, G. M. Hidy, ed., Academic Press, New York and London, 1972.
13. C. Junge, *Atmospheric Chemistry and Radioactivity*, Academic Press, New York, 1963.
14. E. Meszaros, *J. of Aerosol Science*, **4** : 429 (1973).

15. J. Bricard, F. Billard, M. Cabane, and G. Madelaine, *J. of Geophysical Res.*, **73** : 4 487 (1968).
16. R. D. Cadle, W. H. Fisher, E. R. Frank, and J. P. Lodge, *J. of Atmospheric Sciences*, **25** : 100 (1968).
17. E. R. Frank, J. P. Lodge, and A. Goetz, *J. Geophysical Res.*, **77** : 5 147 (1972).
18. A. Meszaros and K. Vissy, *J. of Aerosol Science*, **5** : 1 (1974).
19. P. A. Leighton, *Photochemistry of Air Pollution*, Academic Press, New York, 1961.
20. R. D. Cadle, *Particles in the Atmosphere and Space*, Reinhold Publishing Corporation, New York, 1966.
21. R. A. Cox, *Aerosol Science*, **4** : 473 (1973).
22. M. Buffalini, *Environment Science and Technology*, **5** : 687 (1971).
23. E. R. Allen, R. D. McQuincy, and R. D. Cadle, *Chemosphere*, **1** : 25 (1972).
24. J. Bricard, M. Cabane, G. Madelaine, and D. Vigla, in *Aerosol and Atmospheric Chemistry*, G. M. Hidy, ed., Academic Press, New York and London, 1972.
25. Y. Bourbogot, J. Bricard, G. Madelaine, and D. Vigla, *C.R. Académie des Sciences, Paris, Série C*, **T.276** : 547 (1973).
26. J. Bricard, P. Cazes, P. Reiss, and P. Y. Turpin, *C.R. Académie des Sciences, Paris, Série B*, **T.275** : 263 (1972).
27. V. A. Mohen, Atmospheric Sciences Research Center, *Pub. No. 165*, 1972.
28. J. F. Friend, *Proceedings of the 1st Conference on Climate Impact* (February 1972) 71.
29. P. Urone, H. Lutsep, C. M. Noyes, and J. F. Parcher, *Environmental Science and Technology*, **2** : 611 (1968).
30. A. Boulaud, Mainz Conference, 1973.
31. M. Benarie, T. Menard, and A. Nonat, *Atmospheric Environment*, **7** : 403 (1973).
32. P. E. Coffey, Thesis, State University of New York, Albany, New York, 1972.
33. F. W. Went, *Scientific American*, **192** (5) : 63 (1955); *Proceedings of the National Academy of Sciences, US*, **46** : 212 (1960).
34. A. C. Zettlemoyer, *Nucleation*, Marcel Dekker, 1969.
35. H. Reiss, *J. of Chem. Phys.*, **18** : 840 (1950).
36. G. Doyle, *J. of Chem. Phys.*, **35** : 795 (1951).
37. C. S. Kiang, D. Stauffer, V. A. Mohnen, J. Bricard, and D. Vigla, *Atmospheric Environment*, **7** : 1279 (1973).
38. A. Goetz and R. F. Peuscal, *Atmospheric Environment*, **1** : 287 (1968).
39. R. B. Husar, K. T. Whitby, and B. Y. Liu, in *Aerosol and Atmospheric Chemistry*, H. M. Hidy, ed., Academic Press, New York and London, 1972.
40. R. D. Cadle and R. C. Robbins, *Discussion Far. Soc.*, **39** : 155 (1968).
41. R. C. Robbins and R. D. Cadle, *J. of Phys. Chem.*, **62** : 469 (1959).
42. F. H. Ludlam, "Atmospheric Aerosol," *Science Progress* 65 (1954).
43. F. E. Volz, *J. of Geophysical Research*, **75** : 7185 (1970).
44. E. de Barry and F. Rossler, *J. of Geophysical Res.*, **71** : 1011 (1966).
45. C. E. Junge and J. E. Manson, *J. of Geophysical Res.*, **66** : 2163 (1961).
46. J. Bricard and D. Vigla, *Can. J. of Chem.*, **52** : 1479 (1974).
47. R. D. Cadle, A. L. Lazarus, W. H. Pollok, and J. P. Sheldovshy, *Proceedings of the American Meteorological Society*, Ramage, Honolulu, Hawaii, Institute of Geophysics, 1970.
48. A. B. Meinel and M. P. Meinel, *Science*, **155** : 180 (1967).
49. E. K. Bigg and W. H. Thompson, *Tellus*, **22** : 550 (1970).
50. S. Philipowski, J. A. Weinman, D. R. Clemska, G. S. Kent, and D. R. Wright, *J. of Geophysical Res.*, **73** : 7553 (1968).
51. J. C. Mossop, *Geochemistry/Cosmochemistry Acta*, **29** : 201 (1965).
52. J. P. Friend, *Tellus*, **18** : 465 (1966).
53. E. A. Martell, *Tellus*, **18** : 486 (1966).
54. K. G. Vohra, K. N. Vasudevan, and P. V. N. Nair, *J. of Geophysical Res.*, **75** : 2 951 (1970).
55. E. R. Garhard and H. F. Jonstone, *Industrial and Engineering Chemistry*, **47** : 972 (1955).
56. N. A. Renzetti and G. T. Doyle, *International J. of Air Pollution*, **2** : 237 (1970).
57. P. Urone and W. H. Schroder, *Environmental Science and Technology*, **3** : 436 (1969).

58. T. C. Hall, Ph.D. Thesis, University of California, Los Angeles, 1953.

59. E. Goldbert, *Atmospheric Dust, Sedimentation Cycle and Man, Geophysics* (Earth Science), (in press).

60. E. Robinson and R. E. Robins, *Emissions, Concentrations and Rates of Particulate Atmospheric Pollutants*, Final SRI Report, Project SCC 8507, 1971.

61. G. M. Hidy and J. R. Brock, *An Assessment of Global Resources of Tropospheric Aerosol* (Washington DC Second International).

62. E. Erikson, *Tellus*, **11** : 375 (1959).

63. S. T. Peterson and G. G. Junge, in *Sources of Particulate Matter in the Atmosphere, Man's Impact on Climate*, W. H. Matthews, W. W. Kellogg, and G. D. Robinson, eds., The MIT Press, Cambridge, Massachusetts, 1971 p. 310.

64. V. A. Mohnen, *Fifth International Conference on Atmospheric Electricity*, Garmisch Partenkirchen, September, 1974. To be published.

65. A. W. Castelman, *Space Science Review*, November 1973, p. 1.

There is little need to introduce a chapter on the chemistry of water pollution. The state of the rivers and lakes in industrial countries is well known, and has introduced many students to the word "eutrophication," the process whereby water effectively "dies" in respect to its bacterial response to the introduction of pollutants. The chemistry of water pollution treatment before this stage is reached, or the reversal of it after it is reached, is a central part of the entire subject of environmental science.

12

The Chemistry of Water Pollution

T. Mullins

1. *Historical*

Ever since man progressed from a hunting to an agricultural society, with the corresponding development of stable communities, the phenomenon of water pollution has been his constant companion. As agricultural methods improved, a smaller percentage of the population produced all the food needed; larger communities and diverse secondary industries developed and grew into the present modern society. Concurrent with this growth, however, was the increasing percentage of waste materials and the problems of disposal. When the total volume of waste from a community was relatively small, the easiest method of disposal was to "throw it away," usually into the nearest receptacle. Since man cannot exist without water, community development and city growth centered in areas where the water supplies were adequate and continuous. Initially this meant development in river valleys, and thus the nearest receptacle for wastes was the river.

T. Mullins • New South Wales Institute of Technology, Broadway, New South Wales, Australia
2007

The term "pollution" has been variously defined by many people, but if it may be described here as "the detrimental effects on a localized ecological structure by the addition of the waste products of a society," then it is apparent that the first noticeable pollution problems should have involved the supply of drinking water. As long ago as seventeenth-century England it was a capital offense to pollute the water supply. The same problems were also evident in such major systems as the Rhine and Seine rivers. In many parts of Europe during the middle ages the water was normally unfit to drink and ale became a popular and at times an essential beverage. This disregard for the importance of water in areas where supply was no problem was exemplified in Australia by the fate of the "tank stream." This stream was the major water supply for settlers after the First Fleet landed, but in less than 40 years it was polluted beyond use. In more recent times potable water supplies, which constitute less than 1% of the total water volume of the earth, have assumed noticeable importance, and major efforts are under way in many countries to protect them.

As civilizations progressed and trade became more important, cities were developed on trade routes and navigable waterways. The result is that, at present, narrow coastal areas contain the major proportion of population and industry. This development of course immediately opened up another disposal area. What better disposal system could be found than estuaries, bays, and the vast wide ocean? The major fallacy here of course is that "disposal to the ocean" is a misnomer, in that oceanic conditions are not met in inshore waters. What is met, however, is a coastal water system that is completely unique to each coastal section with characteristics that depend in part on the given topography and wind conditions.

It is in this particular area that the question of pollution takes on a new meaning. Is a body of water polluted when it directly affects man, or should it be classified as polluted when the ecological structure is first upset? The hydrosphere is a dynamic system containing physiochemical and biological equilibria, and there is no doubt that a normally active waterway has a large capacity to assimilate wastes. However, in many areas this capacity is now being reached or exceeded so that many waterways are becoming increasingly contaminated. Before this contamination becomes readily noticeable however, equilibria are changed and the ecological structure may be seriously affected. Some examples of water systems where the effects of pollution have become or are becoming increasingly apparent are the Adriatic, Baltic, and Mediterranean seas; the Thames, Rhine, and Seine rivers; and the Great Lakes in America and Canada. But dynamic systems have a remarkable capacity for regeneration, and with care and intelligent planning even the most seriously polluted waterways may be brought back into full use. An example of river regeneration on a large scale is the successful attempt to restore the Thames estuary.

In the specific case of the coastal system, one of the major factors involved in future planning is the detrimental effect of pollution on marine life. It is true that the vast majority of fish and other aquatic organisms caught or harvested on a commercial basis spawn and live the adolescent part of their lives in shallow water in estuaries, bays, and in the coastal water system. Some deep water fish such as the Atlantic salmon will actually migrate from salt water to fresh water streams to spawn. Many crustaceans and similar organisms will spawn in intertidal areas and live out their lives in shallow water on the Continental Shelf. Thus the greatest effect of indiscriminate dumping has been on the productivity of a vitally important area.

This area is relatively small and is related to the presence of the geological structures called Continental Shelves. These shelves underly 7.5% of the entire ocean, total approximately 18% of the earth's land area, and hold some 0.2% of the total water volume. Throughout the world, fewer than 15% of the shelf areas have been investigated in any great detail, and, in addition, fewer than 10% of the detailed coastal water circulation patterns have been investigated. Despite this lack of knowledge, however, these areas are used extensively for waste discharges. Frequently the components of these discharges are not known accurately, so that extrapolations of effects—both physiochemical and biological—and possible damage to the ecosystem are impossible. Without too much trouble it is always possible to hear the catch cry of those who are uninformed but eternally hopeful, "Don't worry about that, the sea can take it." The sea may be able to take it, but whether the sea can handle it or not is another question. It is to be hoped that the above statement never reaches the category of "famous last words."

Another form of pollution that is causing concern with respect to coastal water systems is physical in nature. Marshlands, deltas, swamps, estuaries, bays, etc., are vital components of any localized area and are of utmost importance in maintaining its productivity. Anything that seriously interferes with these systems has the overall effect of reducing this productivity. In this case, the reduced productivity is directly related to a reduction in spawning grounds.

There are numerous ways in which spawning grounds may be depleted, encompassing both deliberate and accidental actions based on ignorance or disregard of the consequences. Deliberate actions are usually taken under the guise of reclamation projects, where frequently the advantages and disadvantages on both sides are not fully appreciated. Other actions taken in a worthwhile cause sometimes have unfortunate side effects. An example of the latter was the construction of the Aswan High Dam in Egypt. This dam was designed to control the annual flooding of the Nile river, produce electricity, and enable large areas of land to be put under cultivation. The side effect is that the Nile Delta is now not subject to a yearly influx of fresh water that carries with it valuable top soil to be used as fertilizer, etc. The result is a change in the ecological structure of the Delta, with corresponding effects being felt in the Mediterranean Sea. A large proportion of commercial fish in the Mediterranean spawn in the Nile Delta and environs and so the dam may have a continuing detrimental effect on some of this commercial fishing.

2. *Pollutants*

Whether or not pollutants are physical or chemical in nature, a common ground for both is their mode of entry into the system. Broadly speaking, waterways are subject to entry of pollutants by

1. Direct discharge into the system
2. Run off and/or seepage with subsequent transport
3. River flow transport
4. Reactions and transport across the air–water interface
5. Reactions and transport across the water–sediment interface

Under all conditions of entry, however, the contaminant will be in either dissolved or particulate form, or may be converted to these forms by reactions in the hydrosphere. The fate of either form depends to a large extent on its reactivity and on the availability of reactive sites, whether physical, chemical, or biological. B. H. Ketchum has devised a chart (Fig. 1)[1] which shows in general terms the fate and distribution of a pollutant with respect to the marine environment. The chart can be extrapolated to estuarine and fresh water systems. But the individual steps are still far from completely understood, except in a broad qualitative sense, and much work is necessary before the acute and chronic effects of pollution in the hydrosphere can be evaluated.

A major feature of the chart which should be noted is the consideration given to a grossly overlooked attribute of a dynamic system. This is the phenomenon of concentration. Very few of the materials added to a water system are found to be incapable of being taken into the biological cycle or food chain. The major effect of this is felt in the higher orders of predators, where concentrations of pollutants may be so high that ecological damage becomes evident. Some examples of this are the problems arising from the accumulation in tissues of mercury, D.D.T., and polychlorinated biphenyls.

However, it should be understood that not all materials added to the hydrosphere constitute a threat to the system. Water has been classified as the universal solvent, and, as such, a water system will contain a wide variety of substances both in solution and in suspension. Table 1, taken in part from R. A. Horne[2] and E. D. Goldberg,[3,4] indicates to some extent the solvating power of water and gives some indication of the complexity of the inorganic species present in a dynamic system. This table represents an average view of seawater and it must be understood that each separate body of water contains unique species and concentrations.

It is only when a particular constituent attains a high-enough concentration, which may then affect the system, that it may be termed a pollutant. Nature has an enormous reserve and capacity to handle or consume additional levels of materials, but there is a point where accumulation exceeds consumption. At this point, where a constituent becomes a pollutant, a level is designated as the "threshold level." Below this level natural processes can, if given time, accommodate a substance. Above this level the natural processes

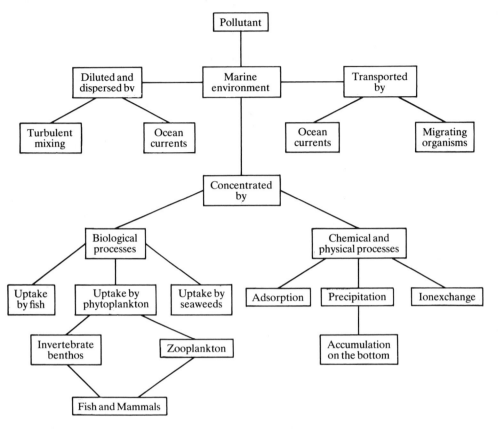

Figure 1. The various processes which determine the fate and distribution of a pollutant added to the marine environment. From B. H. Ketchum.[1]

are overloaded and the ecological structure will be upset to an extent depending on the relative concentrations of the substance.

In order to calculate threshold levels a system must be considered as a whole. This means that a particular pollutant cannot be considered in isolation. In order to do this, the individual steps in Fig. 1 must be taken into account, which requires a thorough appreciation of the rates and mechanisms of the reactions mentioned. Added to this must be an appreciation of the physiochemical reactions between pollutant and pollutant. Examples of these would be various synergistic and antagonistic reactions between metals; reactions between metals and organics; reactions between metals and complexing inorganic anions; and reactions between nonmetals. This latter requirement is the most frequently neglected since qualitative and quantitative analyses of many discharges are conspicuous by their absence. From the foregoing it should be obvious that many allowable levels and limits of pollutants are based more on guesswork than on hard experimental data.

Table 1. Metals in Seawater

Element	Abundance, mg/liter	Principal species	Residence time, years
C	28	HCO_3^-; H_2CO_3; CO_3^{2-}; organic compound	—
N	0.5	NO_3^-; NO_2^-; NH_4^+; $N_2(g)$ organic compounds	—
O	857,000	H_2O; $O_2(g)$; SO_4^{2-} and other anions	—
F	1.3	F^-	—
Ne	0.0001	$Ne(g)$	—
Na	10,500	Na^+	2.6×10^8
Mg	1350	Mg^{2+}; $MgSO_4$	4.5×10^7
Al	0.01	—	1.0×10^2
Si	3	$Si(OH)_4$; $Si(OH)_3O^-$	8.0×10^3
P	0.07	HPO_4^{2-}; $H_2PO_4^-$; PO_4^{3-}'H_3PO_4	—
S	885	SO_4^{2-}	—
Cl	19,000	Cl^-	—
K	380	K^+	1.1×10^7
Ca	400	Ca^{2+}; $CaSO_4$	8.0×10^6
V	0.002	$VO_2(OH)_3^{2-}$	1.0×10^4
Cr	0.00005	—	3.5×10^2
Mn	0.002	Mn^{2+}; $MnSO_4$	1.4×10^3
Fe	0.01	$Fe(OH)_3(s)$	1.4×10^2
Co	0.0005	Co^{2+}; $CoSO_4$	1.6×10^4
Ni	0.002	Ni^{2+}; $NiSO_4$	1.8×10^4
Cu	0.003	Cu^{2+}'$CuSO_4$	5.0×10^4
Cn	0.01	Zn^2; $ZnSO_4$	1.8×10^5
As	0.003	$HAsO_4^{2-}$; $H_2AsO_4^-$; H_3AsO_4; H_3AsO_3	—
Br	65	Br^-	—
Rb	0.12	Rb^+	2.7×10^5
Sr	8	Sr^{2+}; $SrSO_4$	1.9×10^7
Ag	0.00004	$AgCl_2^-$; $AgCl_3^{2-}$	2.1×10^6
Cd	0.00011	Cd^{2+}; $CdCl^+$	5.0×10^5
Sn	0.0008	—	1.0×10^5
Ba	0.03	Ba^{2+}; $BaSO_4$	8.4×10^4
Au	0.000004	$AuCl_4^-$	5.6×10^5
Hg	0.00003	$HgCl_3^-$; $HgCl_4^{2-}$	4.2×10^4
Pb	0.00003	Pb^{2+}; $PbSO_4$	2.0×10^3

A further problem concerning threshold levels revolves around the question of nondegradable materials. It has been mentioned previously that nature has a capacity to consume pollutants to a certain extent. But where these consumption processes are resisted or where they do not exist, for example with some metals and synthetic organics, the question of a threshold level becomes academic. It has been frequently noted that such nondegradable materials can be biologically concentrated in the food chain, and can react with other chemicals present to form chronic end products. It should also be remembered here that once a pollutant enters the hydrosphere the only feasible method of removal is a dependence on natural systems. But nature cannot handle nondegradable materials, and so it becomes imperative that possible effects be investigated before any discharge is allowed at all.

At the present time, water systems are subjected to many types of pollutants from many different sources. However, some major effects of pollution may be grouped into the following categories:

1. *Oxygen.* A great many compounds and organisms require oxygen for oxidation and decomposition. When significant quantities of dissolved oxygen in the water are removed, the environmental conditions are changed and oxygen-breathing organisms are restricted by reduced respiration. By international agreement, a mass of water should contain approximately 5 ppm dissolved oxygen to sustain a viable community. Below this level, species restriction and decreased activity become apparent.

2. *Nutrients.* Some elements and compounds which are inorganic in nature form the basic food supply for autotrophic organisms, triggering the biological food chain. Serious problems arise when the nutrient supply is hindered or enhanced to an appreciable degree.

3. *Biologically active materials.* Some chemicals, when taken into the biological chain, can have both chronic (long-term) and acute (short-term) effects. These range from interference with physiological processes to selective and wide-spectrum poisons. The former are usually very difficult to assess and control, owing in large part to lack of information, whereas the latter become readily noticeable.

4. *Suspended Solids.* These are particles capable of temporary or long-term suspension and are comprised of both inorganic and organic components. In sufficient concentrations light penetration is reduced, photosynthetic activity is affected, visual perception is reduced, and gill action may be affected. As the solid material settles to the bottom, benthic organisms are affected, resulting in reduced activity and changes in the ecosystem.

5. *Heat.* When discharges of a different temperature from the receiving water are allowed, physiological effects related to the extent of the temperature difference are noticeable. There is no simple correlation however between these, and Vernberg[5] points out that a number of variables must be considered when assessing thermal data.

6. *Radioactivity.* Radioactivity concerns specific materials particularly associated with atomic reactors. Chronic and acute effects related to atomic wastes in the physiochemical and biological spheres are variable, and information sufficient to draw adequate conclusions is not available.

These categories will be discussed further under more specific headings.

Numerous parameters have been chosen to indicate the natural or normal environmental characteristics and the extent of any changes caused by a polluting influence. Some of these parameters are useful, some are unrelated to a specific problem at hand, and some are misleading because of poor analytical and sampling techniques. There are, however, no parameters as yet in general use to indicate changes in mechanisms and reaction rates, or equilibrium states (if any), of inorganic and organic structures. What generally happens is that gross answers are postulated, so that assessments of changes in the environmental structure are mostly empirical in nature.

The physical parameters most commonly used generally encompass some or all of the following:

1. Color
2. Odor
3. Turbidity
4. Particulate matter
5. Sedimentation rate
6. Transparency
7. Temperature
8. Density
9. Currents   (position, direction, and speed)
10. Mixing   (horizontal and vertical)
11. Diffusion rate

Some or all of the following chemical parameters may be used:

12. Hydrogen ion concentration (pH)
13. Salt content (salinity)
14. Dissolved oxygen
15. Biological oxygen demand (BOD)
16. Chemical oxygen demand (COD)
17. Nitrates
18. Phosphates
19. Metals (copper, lead, iron, zinc, nickel, cadmium, mercury)
20. Oil derivatives
21. Organic residues (pesticides, fertilizers, detergents)
22. Total carbon

Despite this wide range of parameters, however, too many pollution surveys are conducted so as to render suspect any conclusions drawn. Frequently these conclusions are based on insufficient data, and there are few things worse than decisions based on inadequate work by inept personnel.

3. *Water Structures*

3.1. *Introduction*

When the question of pollution arises, even modern man is still sometimes blinded by his inability to see what he is looking at. The question here is pollution in the hydrosphere. What is pollution? What is the hydrosphere? The former is being answered in many ways by many people, but the core description remains the same. The latter question, however, is still not completely answered. There are many theories about water structure, all of which seem to work well, but a total understanding is still not fully apparent.

Water is so common throughout much of the world, and plays such a vital role in the total biological system, that it would seem to deserve much greater attention. However, because it is so common, many of its unique properties are generally overlooked. These properties are associated with the structure of the water molecule, which consists of two atoms of hydrogen combined with one atom of oxygen so as to form a triangle with the angle at the apex (the oxygen atom) equal to 105° (Fig. 2). This configuration results in an imbalance of the positive and negative charges, producing a "polar" molecule, i.e., there are residual charges on the oxygen side and on the hydrogen arms. This "polarity" results in a further unique characteristic of the molecular structure. There is sufficient attraction between opposite charges on adjacent molecules to form a weak additional bond. This is called the "hydrogen bond" (Fig. 2), and causes the molecules to clump together in groups. Because of these characteristics water has several very important anomalous physical properties. Table 2 lists some of these properties and their significance.

Two of the most important properties of water in relation to life on earth are its boiling and freezing points. Due to the presence of hydrogen bonds, which require additional energy to break them, the boiling point of water is 180°C higher than it theoretically should be. This figure is arrived at by comparing the group boiling points: hydrogen sulfide: −60.75°C, hydrogen selenide: −41.5°C, and hydrogen telluride: −1.8°C. So if water consisted of uncombined molecules it would boil at −80°C and liquid water would not exist.

As water cools its volume decreases, but the reduced kinetic energy of the molecules encourages molecular combinations, resulting in the formation of larger groups which have the effect of causing the water volume to expand. The volume shrinkage, due to cooling, and the expansion, due to formation of the molecular groups, results in an equilibrium condition at 4°C. At this temperature water attains its greatest density. Below 4°C the ratio of combined to uncombined molecules causes the water volume to expand until ice formation begins at 0°C. The end result of this phenomenon is that ice is less dense than

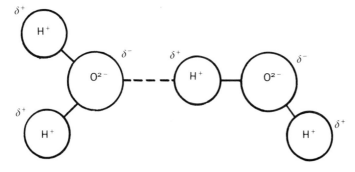

Figure 2. Structure of the water molecule and the hydrogen bond.

Table 2. Some Anomalous Properties of Water[a]

Property	Comparison with other substances	Importance in physical–biological environment
Heat capacity	Highest of all solids and liquids except liquid NH_3	Prevents extreme ranges in temperature. Heat transfer by water movement is very large. Tends to maintain uniform body temperatures.
Latent heat of fusion	Highest except NH_3	Thermostatic effect at freezing point owing to absorption or release of latent heat.
Latent heat of evaporation	Highest of all substrates	Large latent heat of evaporation extremely important in heat and water transfer of atmosphere.
Thermal expansion	Temperature of maximum density decreases with increasing salinity. For pure water is at 4°C	Fresh water and dilute seawater have their maximum density at temperatures above the freezing point. This property plays an important part in controlling temperature distribution and vertical circulation in lakes.
Surface tension	Highest of all liquids	Important in physiology of the cell. Controls certain surface phenomena and drop formation and behavior.
Dissolving power	In general dissolves more substances and in greater quantities than any other liquid	Obvious implications in both physical and biological phenomena.
Dielectric constant	Pure water has the highest of all liquids	Of utmost importance in behavior of inorganic dissolved substances because of resulting high dissociation.
Electrolytic dissociation	Very small	A neutral substance, yet contains both H^+ and OH^- ions.
Transparency	Relatively great	Absorption of radiant energy is large in infrared and ultraviolet. In visible portion of energy spectrum there is relatively little selective absorption, hence is "colorless." Characteristic absorption important in physical and biological phenomena.

[a] Source: Sverdrup, Johnson, and Fleming.[6]

water and consequently will float. This means that a body of water will freeze from the top down. When surface water cools it becomes more dense and is replaced by lighter, warmer water from below. This replacement continues until the entire body of water reaches 4°C. Upon further cooling the surface water expands and remains lighter than the deeper water so that the replacement or overturn of water ceases. On further cooling the surface water freezes, and a layer of ice forms. This ice layer becomes a heat barrier and prevents all but the shallowest of lakes from freezing solid. One effect of this replacement of cold surface water by lighter, warmer deep water is the transport of nutrients from the bottom to the surface, which ceases when the water temperature falls below 4°C.

While a number of the physical properties of water depend on the two usual parameters, i.e., temperature and pressure, there are other variables that frequently must be taken into account. The two most important of these are:

1. *Concentration of solute.* Colligative properties such as boiling point elevation and freezing point depression are dependent on solute concentrations. Frequently this means a direct relationship with salt content, introduced by infusions of sea water. Other properties, such as the solubility of gases, the solubility of metals, and metal ion interactions, are also directly related to salt content and indeed the water mass may be considered as an electrolytic solution.

2. *Water movement.* Properties such as heat conduction, chemical diffusion, and particle transport are dependent on the rate of motion, direction, and boundary mixing conditions of individual water masses. Given the proper impetus water will mix both vertically and horizontally, but without specific conditions water masses will remain discrete and mixing will occur only at the conflicting boundaries. An example of gross separation of water systems is the presence of the Gulf Stream in the western North Atlantic Ocean off the East coast of North America.

From the above it may be seen that water masses may be divided into groups and subgroups depending on a wide variety of variables. The primary and simplest division, however, is related to the concentration of dissolved salt. Bodies of water may thus be classified as being either fresh, estuarine, or marine.

3.2. *Fresh Water Systems*

The general term related to the study of fresh water systems is "limnography," which covers all aspects of fresh water characteristics (including botanical and biological systems). These structures can be subdivided and sub-subdivided, but for the purposes of this chapter a brief introduction will suffice.

3.2.1. Lentic (Standing Water)

This includes lakes, ponds, and swamps, even though some large lakes may have definite current patterns. All of these systems are in the process of changing and aging, as Hutchinson[7] states, and may be classified in a number of ways.

A simple universal classification is difficult, however, due to the wide range of boundary conditions found throughout the world. In other words, characteristics can change dramatically from one water system to another irrespective of distance.

Under these circumstances it would be correct to state that no two lakes or ponds are identical, although definite similarities may exist and may be used to extrapolate conditions and effects. Differences may be apparent in a number of ways. For example, there are characteristic differences between lakes found in tropical, subtropical, and temperate regions; under desert and semiarid conditions; and those found in the plains, foothills, or mountainous areas. Some lakes are totally unusual in that conditions and characteristics are completely unique. For example, Lake Baikal is the deepest fresh water ancient lake in the world, and more than 80% of its fauna is endemic,[36] i.e., found nowhere else. Major differences also occur in chemical content, rates of mixing, stratification, and temperature ranges. All in all it must be accepted that each lake, pond, or swamp is unique, and until sufficient information is obtained about its individual characteristics no assumptions or conclusions based on data from other systems should be accepted.

3.2.2. Lotic (Running Water)

The most basic characteristic of this type of system is the presence of a current, which plays a dominant role in habitat conditions, distribution of constituents, and horizontal movement of water masses. Once again, various characteristics may be used to classify these systems. These may be based on rate of flow, total transport of water, width, length, or point of source, but in all cases only the similarities are catalogued. This is because, similar to lentic systems, it can be truly stated that no two lotic systems are identical. In addition to this, changing boundary conditions throughout the length of a river system can totally change the characteristics from one section to another, so that no two river systems are identical in all aspects in all areas. It now becomes apparent that each system is unique in many respects and should be investigated accordingly.

3.3. Estuarine Systems

There have been numerous descriptions and definitions of estuaries, but the most accepted was proposed by Pritchard.[8] That is, that an estuary is "a semienclosed coastal body of water having free connection to the sea and having a measurable quantity of sea-salt." Large bodies of water with eventual access to the open sea, such as the Mediterranean, are excluded from this definition. Because of the access to the sea, estuaries are strongly influenced by tidal action and are usually found at the mouths of rivers debouching directly into the coastal system, as in coastal bays, marshes, deltas, and as in bodies of water behind so-called barrier islands.

In terms of structure, Pritchard classifies the following types of estuary:

1. *Coastal plain.* Formed by the flooding of a river valley by either land subsidence or a rise in sea level. Examples of this are Chesapeake Bay, on the Atlantic coast of the United States, and Port Jackson, on the West coast of Australia.

2. *Deep basin.* Usually formed by glacial action resulting in deep gouges in the rock face. These are generally restricted to one type and are normally referred to as fjords. One feature of this type of structure is the presence of a sill at the mouth which is shallower than the basin and the sea outside, resulting in a restriction of the free flow of deep water.

3. *Bar built.* Usually shallow basins between the shore and bars or barrier inlands which have been built up by sedimentation and wave action.

Estuaries may be looked upon as transition areas between the land and sea with a wide variety of unique characteristics. As mentioned earlier, estuaries form an important and crucial link in the overall productivity of a region and, as such, individual local characteristics should be the basis for a full investigation if a better understanding of these systems is to be obtained.

Another method of classification of estuaries, more pertinent to this chapter, considers one of the basic chemical characteristics of the system. This classification[8] depends on the concentration and distribution of dissolved salt, which determines the density of the water to a much greater extent than do temperature variations.

The following classifications are diagrammed in Fig. 3:

4. *Vertically mixed.* Usually shallow, with tidal action the dominating feature. The water is well mixed from top to bottom, with the salinity increasing, and with the overall flow directed toward the mouth. This type of water motion has a number of interesting characteristics that result in some pronounced pollution problems which Stommel[9] has investigated to some extent. The Severn River in England is an example of this type.

5. *Slightly stratified.* Also usually shallow, with the salinity increasing toward the mouth at all depths. The water is essentially in two layers, with the fresher overlying the more saline. The surface layer has a net seaward flow, the deeper layer has a net inflow, with a mixing layer between them. In this type of estuary the inflow of fresh water from runoff and small rivers and streams is comparable to the tidal inflow. Chesapeake Bay is an example of this type.

6. *Highly stratified.* Similar to the slightly stratified type, but with a much more pronounced halocline between the surface and bottom layers, which becomes less pronounced as the mouth is approached. The surface, fresher

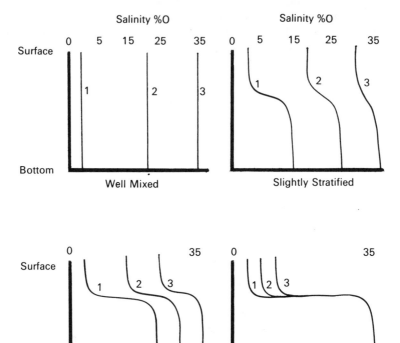

Figure 3. Examples of salinity–depth profiles for various types of estuaries. Graphs (1)–(3) show the variations as the mouth of the estuary is approached.

layer has a net seaward flow and the deeper, more saline layer has a net inflow in a manner similar to but more definite than (5) above. One major difference between these two types, however, is the nearly constant salinity of the deeper layer in this type, caused by the predominance of the freshwater inflow over the tidal contribution. Fjords are good examples of highly stratified systems.

7. *Salt wedge.* Occurs when the fresh water inflow is the dominating feature, and is usually associated with rivers with a large transport volume, such as the Mississippi. The fresh water totally overlies the salt water and flows out until it debouches into the coastal water system at the mouth of the estuary. This type of estuary has many similar characteristics to the highly stratified type, but, in addition, because of the volume and flow of the fresh water involved, Coriolis force must be taken into account. This simply means that a moving body of water will be deflected to the right in the Northern Hemisphere and to the left in the Southern Hemisphere.

In all estuaries, except the well-mixed type, the dominating feature is the circulation pattern. This is referred to as "estuarine circulation," and is apparent where there is a net seaward flow of the surface layer and a net inflow of the deeper layer, separated by a mixing layer. The mixing layer involves a vertical component which results in a net upward movement of salt water from the deeper to the surface layer. The movement of fresher water downward is virtually negligible, so that the overall effect is a predominant unidirectional mixing action. Under these circumstances estuaries have the ability, or the property, to trap and sometimes concentrate constituents. This is often referred to in the context of a "nutrient trap," where the nutrients are retained and recirculated within the estuary. This same property will obviously apply to some pollutants, so that an estuary rather than disposing of pollutants may simply recirculate them.

The estuarine circulation can be complicated to a large extent, however, when conditions are such that the estuarine structure is not consistent. This may occur when the major component of fresh water inflow is from runoff after rain. During a long dry spell the tidal action predominates and the estuary is well mixed. After rain the inflow of fresh water predominates, and the estuary may develop through a slightly stratified to a transient highly stratified system. These changes obviously cause severe complications, and such systems should be thoroughly investigated and understood before indiscriminate discharge is allowed. An example of this type of changing system is the George's River–Botany Bay system in Australia.

A considerable amount of research has been carried out on estuarine systems in general, and much information may be obtained from Pritchard,[8,10] Preddy,[11] and Ketchum.[12,37]

3.4. Coastal Systems

It would seem facetious to describe a coastal water system as a body of water with specific characteristics that is adjacent to a coast, but for many reasons such a description is necessary. In many parts of the world coastal discharges into adjacent waters are erroneously referred to as "ocean disposal," and many calculations and conclusions are based on this false premise.

Oceanographic conditions in coastal waters are for the most part uniquely different from oceanic conditions and are specific for specific areas. The only general feature found in all areas is the presence of a coastline, which sets up a primary boundary condition by limiting the possible directions of water movement. In conjunction with this are a number of factors that contribute to the formation of a specific coastal system. Some of the most important of these are outlined briefly as follows.

1. *Coastline topography.* This refers to the structure and shape of the coast. Current patterns and water characteristics are dependent to a large extent on factors ranging from the prevalence of gently sloping sandy beaches to abrupt changes associated with cliffs and the presence of sand bars or barrier islands.

2. *Bottom topography.* In this case, the presence of a continental shelf is a major factor. In shallow areas the density of the water is determined more by

salinity than by temperature, and heat penetration can be more effective. The presence, size, and shape of submarine canyons will also have a marked effect on the physical structure of the system. On the chemical and biological side, the geological structure and form of the benthic community will also have a contributing influence.

3. *Tides and tidal currents.* Depending on the position, shape, and structure of the coast, volume changes and gross water movements will vary from very large to negligible, and will occur from once to twice a day. In addition, tidal currents may induce vertical mixing, which will break down or prevent thermal and density stratifications.

4. *Seasonal variations.* In areas where the seasons are distinctive, variations in rainfall and river flow can be of major importance. When the major source of fresh water is runoff after rain and small rivers and creeks, the seasonal salinity variations in coastal water closely follow the precipitation pattern. When larger rivers are involved, however, the direct effect is to reduce the salinity of the surface layer at least, and, if sufficient mixing is induced, in the deeper layers as well. If the rivers are also fed by melting snow, the volume of fresh water debouching into the coastal stream may be sufficient to induce a specific current along the coast of low density, low salinity water. These are referred to as geostrophic currents and flow parallel to the coast.

5. *Wind.* The major force generating currents and water mass movements is wind. The water motion directed by the wind in coastal areas is modified by the coastal characteristics in such a way that distinctive and unique patterns are developed.

When all the above features are considered it seems reasonable to state that, like fresh water and estuarine systems, no two coastal systems are identical. Similarities may be apparent and various properties may be compared, but it is impossible to transpose vital characteristics from one area to another. This simply means that because a specific discharge causes minimum damage in one area, it does not follow that the same type of discharge will not cause any trouble in another coastal area.

3.5. *Ocean Systems*

Since the world ocean covers more than 70% of the earth's surface, it has been considered for some time to be the ideal waste disposal area. It may already be considered as nature's ultimate sink, but its capacity to consume the waste products from present civilization is totally unknown. In many areas, much emphasis is placed on extending submarine discharges beyond the Continental Shelf in order to safeguard beaches and other coastal areas. The assumption is that the deep ocean covers such a large area that these added materials will have a negligible effect.

Prior to this century, it was believed that no life existed in the deep ocean below about 500 m, that it was a lifeless zone, and that the bottom was a flat unrelieved plain. These assumptions or beliefs had no basis in fact and have since been proven to be completely wrong. In recent years many marine organisms have been found to live at depths in excess of 5000 m, and rather than the bottom being flat, it has been found to be as varied as surface land forms and to contain a mountain range that encircles the globe, as well as, in most areas, a very rugged terrain.

Research in the deep ocean is still very sparse, but, in addition to the above phenomena, a further characteristic has been noted with reference to the water itself. It has been shown that there are water movements at all depths and that current patterns are developed with large transport volumes. Sverdrup *et al.*[6] and Dietrich[13] discuss the formation, movement, and distribution of distinctive water masses in the ocean from top to bottom, and present a picture of the oceans as being a complex mixture containing numerous individual features. In addition, the pressures and temperatures encountered in deeper water have a pronounced effect on water structure and properties, and, as yet, no valid conclusions are possible with respect to the fate of any additions to the deep water systems. Horne[14] sums up the position at present and states that "the deep sea represents the most remarkable, the most mysterious, environment in the cosmos."

For further information on coastal and oceanic systems, the reader is directed to Sverdrup *et al.*[6] and to Dietrich.[13]

4. Reactive Sites

One of the most difficult problems associated with the determination of the state and rate of water pollution is the choice of a sampling site or sites. In order to achieve a reasonable knowledge of a water system, it is imperative that the water samples taken, or *in situ* analyses conducted, be indicative and representative of the problems at hand. All too frequently the choice of sampling sites becomes secondary in importance to accessibility and/or analytical methods. Under these circumstances results and conclusions will bear little relationship to the problems under investigation.

In all sections of a waterway, dynamic equilibria are present and in many cases are specific for the region involved. If we direct our attention to a vertical cross section of any water body, the following reaction sites can be identified.

4.1. Surface Film

This particular area generally encompasses the top 50–500 μ of a body of water. The importance of this region is reflected in the fact that all equilibria between air and water are affected by the transport kinetics contained in this fraction. A functional look at the comparative size of this area was presented by MacIntyre[15] by plotting, on a logarithmic scale, a cross section of the ocean from the diameter of a molecule of water (10^{-10} m) lying on the surface to a maximum depth of approximately 10 km. The midpoint on this scale corresponds to 10^{-3} m, i.e., 1 mm, and the contention is that the top millimeter corresponds exactly to the top functional half of the ocean. Garrett,[16] Williams,[17] and Duce *et al.*[18] have turned their attention to this layer and have found that within the top 200 μ surprisingly high proportions of virtually all classes of pollutants are concentrated. In this air–water interface, with many unique chemical and physical properties, the actual chemistry of the reactions is complex and is poorly understood. The distribution of some pollutants in this layer will be discussed further under specific headings.

4.2. Main Water Body

Depending on the depth and stratification of the water body, a series of subsections may be identified. The stratification is usually based on temperature variations with depth, and the following layers that may be generated exhibit specific characteristics and properties.

1. *Surface water.* This layer is generally related to penetration of sunlight and contains the majority of biological activities related to photosynthesis and food production.

2. *Intermediate and/or deep water.* The penetration of light and heat radiation is minimal and water characteristics depend to a large extent on individual movements of water masses.

3. *Bottom water.* This area is where transfer reactions assume added importance. In this case the transfer is from water to sediment and vice versa. If vertical diffusion in the water column is hindered, large variations in concentration gradients between bottom water and the overlying column may become apparent.

4.3. Sediment

The sediment is the ultimate receptacle for all nonsoluble and partially soluble materials, and as such is an ever-present reservoir of pollutants. Under changing conditions, such as pH and oxidation–reduction parameters, various types of compounds may be reintroduced to the water phase or precipitated out. In addition, the biological activity of bottom-dwellers may convert pollutants to various forms that may show characteristics totally different than the pollutants originally introduced.

In some instances, pollutant concentrations in sediments can be extremely useful guides, as is pointed out by McKee[43]: "Owing to the integrating effects of sedimentation and biological accumulation, the impact of potential pollutants is more easily assessed (in the sediment) than in the overlying waters."

Each of the above sections is characterized by varying chemical, biological, temperature, and pressure effects, which are reflected in the reaction kinetics of the species involved. For example, in regions of intense biological activity pollutants may be dispersed and absorbed into the biological chain at a much faster rate than in regions of poor activity. Similarly, changes in temperature will affect solubility and reaction equilibria in relation to the magnitude of change. These changes can be either a constant factor, seasonal, or, in some areas, negligible. It is an accepted rule of thumb that reaction rates will double for each 10°C (18°F) rise in temperature, which means that temperature variations can be of major importance when assessing both the acute (short-term) and chronic (long-term) effects of specific pollutants.

The role of the sediments as a storehouse and reaction site for normal constituents, as well as for pollutants, is in need of much more research, since some cherished notions of sediment stability and sterility have recently been badly shaken. The best example of the biological and nonbiological reactivity of sediments and their subecological structures is the conversion of inorganic mercury to the highly lethal methyl mercury form. Sediments are normally sites of considerable biological activity, and can vary from oxidizing to reducing environments, so that it is necessary to fully appreciate the role of a specific sediment type in conjunction with the column of water overlying it.

From the foregoing it can be appreciated that knowledge of a total system is a primary requisite before it can be definitely stated that a specific pollutant or a level of pollutants will cause no harm to the environment. All additions to a dynamic system will cause changes related to the total reactions involved. The predominant question, however, is whether or not these changes are significant in both the short and long term.

5. *Metals*

5.1. *Introduction*

It has been known for many years that most metals are normally present in all natural water systems, but, beyond determinations of some ambient concentrations (Table 1), very little systematic research has been conducted into metal–metal and metal–nonmetal reactions and mechanisms. As an example of the need to know and understand these mechanisms, one need look no further than the disastrous effects of mercury in the environment at Minamata in Japan in the middle 1950s. Inorganic and various organic mercury compounds discharged into the water system were converted to the highly lethal methyl mercury form by microorganisms in the biota. This methyl mercury was readily assimilated into the biological food chain, and thus became available to the local population through consumption of contaminated fish. This resulted in permanent physiological damage to many and in death to others.

In natural water systems, the constituents exhibit properties that may be classified as either conservative or nonconservative. The former relates to concentrations of species that remain relatively constant or exhibit a relatively constant ratio to each other throughout the system. In the case of metals this applies mostly to the more abundant lighter elements such as sodium, potassium, and calcium. The latter describes for the most part the transition metals such as iron, copper, and zinc, which show concentration variations depending on variables such as position, time, season, temperature, salinity, and, most importantly, biological activity. The mercury episode described above is a good example of biological activity. Goldberg et al.[38] have summed up this biological activity of many metals in the following way: "for any given chemical element, there will be found at least one planktonic species capable of spectacularly concentrating it." For example, plankton will concentrate copper by a factor of 90,000, lead by 12,000, and cobalt by 16,000.

These heavier nonconservative metals are generally the more reactive and are normally present in concentrations not exceeding one part per million, with most being in the micro- and submicroranges down to 10^{-10} M or lower. Because of this activity they play a critical role in the chemistry of a water system. In many instances, such as copper, zinc, cadmium, and nickel, metals present in trace amounts are vital to the biological system and may be classified as essential micronutrients. In higher concentrations, however, these same metals can act as inhibitors in enzyme reactions and in some cases can actually show a high degree of toxicity to specific organisms.

Biological activity and chemical reactivity depend in part on the forms in which the metals are present. There is a functional difference between the analytical concentration of metal ions in solutions and the concentration of metal ions available to the aquatic organisms. A gross initial separation of form that is universally accepted is the distinction between dissolved and particulate matter. If the species can be filtered through a filter of pore-size 0.45 μ, they may be considered as dissolved. Any species retained by this size filter may be considered as particulate. Under these circumstances the chemical structure is relegated for secondary investigation. It should be remembered, however, that a knowledge of species concentration and chemical composition is necessary if an understanding of the properties and reactions inherent in a water system is to be gained. The actual forms in which the metal ions are present depend to a large extent on numerous variables, such as oxidation state and complexing ability of the specific element, and will be discussed further below.

It should also be pointed out here that equilibrium conditions are not always attained by the species in question. Chemical equilibria in water systems are only reached for reactions that are fast in relation to geological and biological considerations, and so it is frequently more productive to consider waterways as "steady-state" systems rather than as equilibrium systems.

In all instances, however, metals are "nondegradable" and are not consumed, i.e., they do not disappear. They may be transferred or stored throughout a system and should be considered as "available" given specific conditions.

The introduction of additional metals to the hydrosphere is generally accomplished either by atmospheric transport or by direct discharge. Of these, atmospheric transport and subsequent transfer to the hydrosphere by gravity or rainfall is mostly overlooked. In many cases when calculations concerning the capacity of a waterway to absorb or survive extra metal concentrations are made, the only concern is with the effects of the direct discharge of the metal-containing effluent to the system. This presupposes that the contribution from the atmosphere is negligible, which is frequently a false premise. The importance of atmospheric transport is emphasized by calculations (Goldberg[44]) which show that the concentration of lead in the surface waters of the North Atlantic Ocean has risen from 0.01 to 0.07 μg/liter during the last few decades since the introduction of lead in gasoline as an antiknock compound. Regardless of the mode of entry, however, the distinction between dissolved and particulate forms is still valid, so that in the hydrosphere the characteristics of reactions and transport may be considered *in toto*.

5.1.1. Reactions

In aqueous solutions, metal ions will react with any available species present under appropriate conditions at specific rates and along specific pathways. As the conditions change, however, i.e., the environment of the reactions, the rates and equilibria will also change. This means that metal reactions in the effluent, in the receiving water, in the effluent–receiving water mix at the point of discharge, and in the effluent–receiving water mix as dilution and dispersion take place, are controlled by the overall environments of the reaction spheres. Under these circumstances it follows that the properties and characteristics of the effluent, the receiving water, and the resulting mix must be known before any valid conclusions can be drawn as to possible reactions and reaction products.

When the solvent is pure water it is common to explain reactions in dilute solutions by considering the concentrations of species involved. For example, in the general equation

$$mA + nB \rightleftharpoons xC + yD$$

the equilibrium condition is expressed by

$$K = \frac{[C]^x[D]^y}{[A]^m[B]^n} \tag{1}$$

where the square brackets denote concentrations in moles per liter.

The more correct thermodynamic expression, however, is related to the actual activity of the species present, thus

$$K = \frac{a_C^x a_D^y}{a_A^m a_B^n} \tag{2}$$

where K is the thermodynamic equilibrium constant.

These two forms of equilibrium constants can be correlated rather simply by the introduction of a specific correction factor called the molar activity coefficient, which relates the solute concentration to solute activity by the following definition:

$$a = \gamma C \tag{3}$$

where a is the solute activity, C the solute concentration, and γ the molar activity coefficient.

By substitution, equation (2) may be rewritten in the following form:

$$K = \frac{[C]^x[D]^y}{[A]^m[B]^n} \cdot \frac{\gamma_C\gamma_D}{\gamma_A\gamma_B} \qquad (4)$$

At infinite dilution, γ approaches unity, so that in sufficiently dilute solution the common practice is to substitute only concentration factors for activities in equilibrium calculations as shown in equation (1).

However, when this is not possible a further constant may be considered by defining

$$K^1 = \frac{[C]^x[D]^y}{[A]^m[B]^n} \qquad (5)$$

where K^1 may be expressed as the "apparent equilibrium constant." Thus by combining equations (4) and (5) a relationship between the apparent and the thermodynamic equilibrium constants may be obtained:

$$K^1 = K\frac{\gamma_A\gamma_B}{\gamma_C\gamma_D} \qquad (6)$$

where the correction factor is the ratio of activity coefficients. At infinite dilution this ratio is by definition unity, and $K = K^1$.

When reactions involve ions or charged species, the calculation of activity coefficients is simplified somewhat by the use of the Debye–Huckel Theory.[19] This theory states that each ion in solution is surrounded by an ionic atmosphere of opposite sign by electrostatic attraction and that in sufficiently dilute solutions the mean activity coefficient varies directly with the square root of the total ionic strength of the solution. Because of this relationship, the apparent equilibrium constant will only have a constant value if the reactions are carried out in dilute solutions of constant ionic strength. In fluctuating or in more concentrated solutions, ionic strength and activity coefficient measurements are increasingly more difficult to make.

The ionic strength μ of a solution may be equated to the concentration of each ion C_i and the net charge on the ion Z_i by

$$\mu = \tfrac{1}{2}\sum C_i Z_i^2 \qquad (7)$$

where the summation includes all the charged species in solution. In the specific cases of uncharged or neutral molecules, the contribution to the ionic strength of the solution is zero. It should also be noted that only strong electrolytes contribute appreciably to the ionic strength.

For example, the ionic strength of a 0.02 M solution of KCl is

$$\mu = \tfrac{1}{2}(0.02 \times 1^2 + 0.02 \times 1^2) = 0.02$$

or for a mixture of 0.02 M $(NH_4)_2SO_4$ and 0.05 M $AlCl_3$,

$$\mu = \tfrac{1}{2}[(0.04 \times 1^2) + (0.02 \times 2^2) + (0.05 \times 3^2) + (0.15 \times 1^2)] = 0.36$$
$$\quad\quad NH_4^+ \quad\quad\quad SO_4^{2-} \quad\quad\quad AL^{3+} \quad\quad\quad Cl^-$$

In simple systems, when the ionic strength of a solution changes, caused by changes in concentrations and charges of added species, the overall effect, from equation (6), is a a change in the apparent equilibrium constant. In more complicated systems, however, when complex formation and other competing reactions are involved, the theory begins to break down and reaction rates and apparent equilibrium constants are seriously affected.

Some equilibria and competing reactions that have been identified in water systems are

$$Ag^+ + Cl^- \rightleftharpoons AgCl \tag{8}$$

but in the presence of excess Cl^- and excess CN^-

$$AgCl + 2Cl^- \rightleftharpoons [AgCl_3]^{2-} \text{ soluble} \tag{9}$$

$$AgCl + 2CN^- \rightleftharpoons [Ag(CN)_2]^- + Cl^- \tag{10}$$

Similarly

$$Cd^{2+} + 2CN^- \rightleftharpoons Cd(CN)_2 \tag{11}$$

$$Cd(CN)_2 + 2CN^- \rightleftharpoons [Cd(CN)_4]^{2-} \text{ soluble} \tag{12}$$

In the case of copper, an oxidation reduction reaction is possible where copper (II) is reduced and some cyanide ion is oxidized:

$$Cu^{2+} + 3CN^- \rightleftharpoons [Cu(CN)_3]^{2-} \tag{13}$$

$$2CN^- \rightleftharpoons (CN)_2 \tag{14}$$

Some cyanogen may be hydrolyzed to re-form cyanide ions:

$$(CN)_2 + 2OH^- \rightleftharpoons CNO^- + CN^- + H_2O \tag{15}$$

In natural water systems, however, simple one-step reactions are generally the exception rather than the rule, and to fully understand the quantitative effects of changing conditions on an equilibrium all the intermediate species must be taken into account. For example, in the reaction involving cadmium and chloride ions, the following equilibria are involved and must be considered:

$$Cd^{2+} + Cl^- \rightleftharpoons CdCl^+ \quad\quad [CdCl^+] = K_1[Cd^{2+}][Cl^-] \tag{16}$$

$$CdCl^+ + Cl^- \rightleftharpoons CdCl_2 \quad\quad [CdCl_2] = K_2[CdCl^+][Cl^-] \tag{17}$$

$$CdCl_2 + Cl^- \rightleftharpoons CdCl_3^- \quad\quad [CdCl_3^-] = K_3[CdCl_2][Cl^-] \tag{18}$$

$$CdCl_3^- + Cl^- \rightleftharpoons CdCl_4^{2-} \quad\quad [CdCl_4^{2-}] = K_4[CdCl_3^-][Cl^-] \tag{20}$$

Under normal conditions one of these species would predominate at any given time, but all species could be present to some extent. A further discussion of this will be given below.

All aqueous reactions are dependent for the most part on physical properties such as temperature and pressure, and on chemical properties such as competing and simultaneous chemical equilibria. Another factor which also contributes to variations in chemical reactions is the presence of large concentrations of ions which affect the ionic strength and virtually cause the resulting solution to be considered as an electrolytic solution. This is the case with seawater and may be generally found in estuarine waters with normal intrusions of salt water.

For example, the second apparent dissociation constant (K_2) of carbonic acid changes from 4.2×10^{-11} in fresh water to 4.0×10^{-10} in water with a chlorinity value of 9%, and to 7.1×10^{-10} in seawater with a chlorinity value of 20%, an overall change by a factor of approximately 20 (Horne[45]).

5.1.2. *Distribution*

Metals permeate all aspects of the hydrological cycle, ranging from physical solution to participation in the biological food chain. In the physical environment metals have been found to be distributed as the soluble inorganic and organic species, as the particulate inorganic and organic species, and in all phases of the food chain ranging from plankton to benthic to free-swimming organisms. The distribution in all cases depends on factors such as temperature, salt concentration, solubility, chemical form, biological activity, and other specific localized factors. Because of this, the distribution features may show wide variations depending on locale and season. For example, in the Corpus Christie bay and harbor system, Holmes *et al.*[46] have shown that a considerable change occurs in the zinc and cadmium concentrations in the water from winter to summer, incorporating marked spatial variations from harbor to bay.

In summer, zinc concentrations range from 6–500 μg/liter depending on locale, and cadmium ranges from 3–80 μg/liter depending on locale. In winter, however, zinc concentrations range from 4–182 μg/liter and cadmium from 2–10 μg/liter. The highest concentrations were found in the surface water and secondary concentrations were found at the sediment–water interface. The distribution in the sediments also showed a marked spatial variation. From the inner harbor to the outer bay zinc concentrations ranged from 11,000–6 mg/liter and the cadmium range was 130–0.1 mg/liter.

The explanation proposed is that because of stagnant conditions during summer the metal sulfides are formed and precipitate in the harbor:

$$Zn^{2+} + HS^- \rightleftharpoons ZnS\downarrow + H^+ \tag{20}$$

$$CdCl^+ + HS^- \rightleftharpoons CdS\downarrow + Cl^- \tag{21}$$

but during winter, because of the increased interchange of water between the harbor and the bay, some of the metals are desorbed and transferred into the bay.

It should be noted in the above that the metal concentrations in the sediment, which acts as a storehouse, are many orders of magnitude greater than the concentrations found in the water. This is by no means an uncommon occurrence. Klein and Goldberg[20] reported a distribution of mercury in water and sediment from 0.1–3.6 μg/liter in the former to 80–800 μg/liter in the latter. This indicates the predominance of metals in sediments over the dissolved species in solution.

Duce *et al.*[18] have reported a further instance of metallic enrichment, as mentioned previously. They found that almost all classes of pollutants were more concentrated in the surface film (the top 100–150 μ) than in the water 20 cm below the surface. They report enrichment factors of 5.8 for particulate lead, 29 for particulate iron, 36 for particulate copper, and 50 for particulate nickel. The organic and inorganic forms of the above metals are also enriched to varying degrees.

These two areas of enrichment, the sediment and surface film, are of direct concern in our understanding of the aquatic environment. However, of more direct concern to man is the enrichment or concentration of metals in the food chain. Many heavy metals are known to be concentrated by each succeeding tropic level, but the mechanism is not well understood and knowledge of the conditions necessary for maximum concentration is very rudimentary. O'Hara[21] has determined that the cadmium uptake by fiddler crabs is at a maximum under conditions of high temperature and low salinity, but the overall uptake of cadmium and other metals by the species involved in a localized food chain is still very much unknown. Klein and Goldberg[20] have reported mercury distribution in the biosphere with a concentration factor of 1000–10,000 in fish, but also state that the data are very limited.

From the above it can be seen that metals in the hydrosphere are concentrated mainly in the surface film, sediment, and biota, and that only relatively small concentrations remain in the water itself. These concentrations in the water and the chemical forms, however, are vital to the overall chemistry of the system in question and must be known before predictions can be made.

5.1.3. Transport

The transport of metallic species, excluding the biological fraction, depends on the chemical form, solubility, and physical parameters such as water mass and sediment movements. Gibbs[47] has reported transport phenomena for some metals in river systems and has shown that the bulk of metal transported is in particulate form. In estuarine systems, however, as the salt content increases there appears to be a reversal and the percentage of the dissolved species rises. This is apparently due in part to some of the particulate material sinking to the bottom so that the total metal concentration falls, but there is no simple relationship since in many instances the concentration as well as the percentage of the dissolved fraction increases. Gibbs also reports that the particulate fraction can be subdivided and that metals will preferentially form specific structures. For example, copper and chromium are transported mainly as crystalline solids and manganese as a metallic coating. These would appear to depend in part on environmental conditions, and changes in the short term should be expected. It has been frequently found that concentrations can change dramatically in a matter of days, although there is little evidence so far of corresponding changes in chemical structures.

The predominance of the particulate form tends to hold in most areas, except the open sea, where a complete reversal takes place and the dissolved species take precedence.

When dissolved species are considered, the total transport appears to depend on water mass, tides, and current movements and mixing phenomena. For particulate materials, however, further aspects such as size, shape, and density of the forms must also be taken into account. For the dissolved species the total transport is generally from areas of higher concentration to areas of lower concentration and corresponds to a net movement downstream. In the case of the particulate species, however, a more complicated picture is involved. Very frequently particulates are trapped and accumulated in localized areas not necessarily downstream so that serious problems result. The transport of particulates depends on the properties mentioned which determine the settling velocity: the distance over which the material will travel; the deposition velocity: the minimum current velocity required to keep the material in suspension; and the scour velocity: the minimum velocity required to bring the material back into suspension after deposition. For further information on the transport of particulate matter, the reader is referred to Postma.[22]

In order to predict transport phenomena or to predict where specific pollutants will finally end up, it is apparent that the physical characteristics of a specific water system must be known. If the features mentioned, such as water mass and current movements, are not known, it is useless to make any prediction regarding transport; all that is left is to rely on divine providence and hope that nature can take care of us.

5.2. *Solubility*

When the solubility of various compounds in natural water systems is calculated, errors are frequently encountered because of a lack of appreciation of the limitations of the solubility product principle. In a solvent, the solvent molecules attract the surface molecules of the solid, competing with the attractive forces between the molecules. If the intermolecular attractive forces are overcome, the solid will dissolve until a dynamic equilibrium is reached:

$$\text{Solute (solid)} \rightleftharpoons \text{Solute (dissolved)}$$

The solubility product is related to the solubility of the salt by use of the equilibrium condition from

$$A_m B_n \rightleftharpoons mA^{n+} + nB^{m-} \tag{22}$$

where

$$K_{sp} = [A^{n+}]^m [B^{m-}]^n \tag{23}$$

Under these conditions the solubility product principle has a rather limited application and applies only to slightly soluble, completely ionized solutes, where the dissolution reaction can be explained by a single equation. But even when a salt is completely ionizable, the dissolution reaction in natural water systems may involve multistep reactions with water molecules or other substances that may be present.

In general, the solubility of a solute in a natural water system will depend on:

1. *Temperature.* For most substances an increase in temperature will result in an increase in solubility. However, there are some common salts where the reverse is true. Calcium sulfate and sodium sulfate, for example, will show a decrease in solubility with an increase in temperature.

2. *Electrolyte effect.* Similar to the equilibrium constants previously mentioned, the solubility product constant K_s is correctly related to the activities of the ionic species and is a constant at any given temperature. Equation (23) should thus be expressed as

$$K_2 = a_{A^m} \cdot a_{B^n} \tag{24}$$

or

$$K_s = [A^{n+}]^m [B^{m-}]^n \gamma_A \cdot \gamma_B \tag{25}$$

But when activity coefficients are neglected, equation (23) is a valid approximation and K_{sp} could be referred to as the concentration solubility product. The relationship of K_s to K_{sp} would thus be

$$K_{sp} = \frac{K_s}{\gamma_A \gamma_B} \tag{26}$$

Since the activity coefficients are related to the ionic strength, it follows that the solubility of the solute in question is also related to the ionic strength of the solvent, i.e., the total salt content of the water system.

3. *Time.* Time may be an important factor in natural water systems since the system itself is dynamic rather than static. This will normally imply changes in environmental conditions in addition to the fact that some substances dissolve very slowly at the best of times.

After considering the above, there are still some factors that will have an effect on solubility calculations.

4. *Particle size.* From a consideration of surface phenomena it can be generally stated that a small particle of a given solute is more soluble than a large particle.

5. *Impurities.* When solutes consist of mixed crystals or of solid solutions their properties may be completely different than the "pure" solute. This introduces errors and changes in all phases of solute properties and reactions.

6. *Chemical reactions.* In aqueous solutions the most common reaction on solution of a solute is hydrolysis, where the solvent water molecules form the reaction sphere. Most metal ions will hydrolyze, and anions of weak acids also show considerable hydrolysis effects.

For example, if the solubility of lead sulfide is calculated neglecting hydrolysis in a natural water system

$$K_{sp} = [Pb^{2+}][S^{2-}] = 8.4 \times 10^{-28} \tag{27}$$

and

$$\text{Molar Solubility} = 2.9 \times 10^{-14} M$$

But when the hydrolysis reaction is considered, the following equilibria must also be considered:

$$PbS \rightleftharpoons Pb^{2+} + S^{2-} \tag{28}$$

$$S^{2-} + H_2O \rightleftharpoons HS^- + OH^- \tag{29}$$

and

$$K_b = \frac{[HS^-][OH^-]}{[S^{2-}]} = 8.3 \times 10^{-2} \tag{30}$$

Now, from a consideration of the above,

$$[Pb^{2+}] = [S^{2-}] + [HS^-] \tag{31}$$

and

$$[HS^-] = [OH^-] \tag{32}$$

The above equations may be solved to give the molar solubility of PbS as 4.1×10^{-10} M which is a considerable increase over the solubility expressed above.

When the hydrolysis of the metal ion is added, an accurate quantitative analysis of a system containing multiple species becomes difficult.

In addition, all of the hydrolytic equilibria considered will involve either hydrogen ions or hydroxide ions, so that the solubility of a hydrolyzable salt is dependent on the pH of the solution. In natural water systems, the pH range encountered is generally from 6–8.5, which is sufficient to cause appreciable changes in solubility calculations.

For example, Moore[48] has shown that the solubility of lead in water increases appreciably from a pH of 6 to a pH of 8, and that outside this range, in both directions, the increase in solubility is considerable.

In addition to the above factors, precipitation reactions with other anions such as the hydroxide ion and complexometric reactions with other species in solution must be taken into account. The complex reactions have to be considered in the same context as the multistep reactions mentioned earlier. For example, mercuric ions in saline water would involve the following equilibria:

$$\begin{array}{c} HgCl_2 \rightleftharpoons HgCl^+ + Cl^- \\ \updownarrow \\ Hg^{2+} + Cl^- \end{array} \tag{33}$$

It has been found that, with a few exceptions, such as calcium and barium in very deep water, the concentrations of metal ions cannot be controlled by simple solubility equilibria models.

5.3. Species

As mentioned previously, the actual forms, or specific chemical species of metal ions in solution are of major importance in determining the chemical, biochemical, and biological structures of a specific waterway. Biological activity and chemical reactivity are also controlled in large part by both inorganic and organic species present in a natural water system with which the metals ions will primarily react. Most metals in solution are reasonably reactive and will form compounds or species based on the complexing ability of the individual metal. This complexing ability is a measure of the power of a metal ion to attract other ions or molecules to itself, in addition to its normal ionic bonding capability. For example, if ammonia is added to a solution of copper(II) sulfate, an intense blue color is generated according to

$$Cu^{2+} + SO_4^{2-} + 4NH_3 \rightleftharpoons [Cu(NH_3)_4]^{2+} + SO_4^{2-} \qquad (34)$$

The blue color is due to the formation of the copper tetramine ion which is a stable cation. The ammonia molecules are occupying "coordination sites" around the central metal ion, and are referred to as "ligands." In this case the ligands are all of the same species, but this is not a necessary condition for the formation of a complex ion. In many cases ligands may be mixed, and, depending on reaction conditions, they may be replaced by other ligands capable of forming the dative bonds necessary for complex formation. This ability of a complex compound to engage in a reaction when ligands are replaced is referred to as the "lability" of the complex. Where replacements are rapid, i.e., occurring within the time of mixing, the complex is "labile," but where replacements are slow or do not occur, the complex is "inert." These terms refer only to the rates of reactions, and have a direct bearing on the further availability of the metal ion concerned to other species in solution, including the biological system.

A generalized approach to the equilibrium conditions involved in complex formation can be diagrammed by considering a metal ion, M, and a monodentate ligand, L, as follows:

$$M + L \overset{K_1}{\rightleftharpoons} ML \qquad (35)$$

$$ML + L \overset{K_2}{\rightleftharpoons} ML_2 \qquad (36)$$

$$ML_2 + L \overset{K_3}{\rightleftharpoons} ML_3 \qquad (37)$$

$$ML_{(n-1)} + L \overset{K_n}{\rightleftharpoons} ML_n \qquad (38)$$

Equilibrium or formation constants can be expressed individually as

$$K_N = \frac{[ML_n]}{[ML_{(n-1)}][L]} \qquad (39)$$

with each stepwise constant being considered as a separate equilibrium condition that contributes to the whole system. The value of N represents the maximum coordination number of the central metal ion, M, for the specific ligand, L.

The most common ligands found in water systems are H_2O, OH^-, Cl^-, SO_4^{2-}, HCO_3^-, and various organic molecules, a high percentage of which are so far unidentified. The presence of these competing ligands and various reaction states of the metals has led to some confusion in the description of many metal species. However, in aqueous solution three main types of species are found to predominate.

5.3.1. Hydrated

All metal ions in solution undergo a hydration reaction where the ligand concerned is water:

$$M^{n+} + xH_2O \rightleftharpoons [M(H_2O)_x]^{n+} \qquad (40)$$

If other ligands such as Cl^-, SO_4^{2-}, or organic molecules are present, substitution reactions involving the ligands are possible, depending on the relative stabilities of the metal–ligand species formed.

If the aquo-ions are preferentially formed, the species are ionic and tend to move through the solution independently of other species present. For example, sodium chloride in water is dissociated into ions:

$$NaCl \rightarrow Na^+ + Cl^- \qquad (41)$$

which are nonassociated, and although there is some electrostatic attraction the movements are independent. A more correct description of the sodium ion formation above would be $[Na(H_2O)_4]^+$, which is sometimes neglected in the interests of speed and ease of writing.

From the Bronsted–Lowry theory, all aquo-ions are more or less acidic and exist in solution as the hydroxy form,

$$[M(H_2O)_x]^{n+} \overset{K_a}{\rightleftharpoons} [M(H_2O)_{x-1}OH]^{(n-1)+} + H^+ \qquad (42)$$

Coordinated water molecules in mixed ligand complexes dissociate in the same way. The strengths of these acids vary, as would be expected, from weak to moderately strong, depending on the properties of the coordination sphere. For example, $|Al(H_2O)_6|^{3+}$ is a moderately strong acid with a K_a value of 1.2×10^{-5}, which is similar to acetic acid ($K_a = 1.8 \times 10^{-5}$). However, a common stronger acid is $[Fe(H_2O)_6]^{3+}$, where $K_a = 6.3 \times 10^{-3}$.

5.3.2. *Ion Pairs*

Similar to the above, ion pairs are electrostatic in nature. In this case, however, the attraction is more powerful and the ions are not free to move independently, but instead form a dipolar molecular type with some water molecules trapped between the ions:

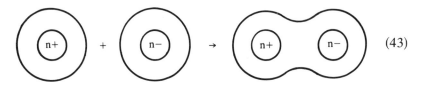

$$(43)$$

Ion pair formation

The ion pairs are capable of temporary existence only, and although they behave as molecules they are not true molecules in the normal sense. The ions are associated with each other and equilibrium conditions may be expressed as:

$$M + A \rightleftharpoons MA \qquad (44)$$

and

$$K = \frac{[MA]}{[M][A]} \qquad (45)$$

where K is the stability constant of the ion pair formation.

In normal water systems there is usually a competition for the metal ion by various anions, and Sillén[49] has pointed out that, by a comparison of respective stability constants related to the activity coefficients of the species involved, the association ability of various anions can be determined. In the competition between chloride ion and hydroxide ion, Goldberg[4] has calculated that the chloride association will dominate when

$$\log K_{Cl} - \log K_{OH} \geq -5.4 \qquad (46)$$

Metal ions such as Ag^+, Cd^{2+}, and Zn^{2+} will preferentially associate with Cl^-, but Pb^{2+} and Cu^{2+} will associate with the OH^-. As long as the stability constants are known, however, comparisons can be made and predominant species can be determined. For example, in seawater the predominant cadmium species is the ion pair with chloride $CdCl^+$.

The ion pairs, because of their polar properties, may be considered as electrolytes that are completely ionized in solution but not completely dissociated. Their formation depends on the ionic strength of the solution and on the properties of the ions in question. The competition by the anions for the metal ions thus depends on the characteristics of the localized water system in question. In their investigation, Pagenkopf *et al.*[23] showed that copper toxicity, for example, depended to some extent on the formation of $CuCO_3$, $Cu(CO_3)_2^{2-}$, $CuOH^+$, and $Cu_2(OH)_2^{2+}$.

5.3.3. Complex Ion Formation

As mentioned previously, a complex ion is formed when ions or molecules take up the coordination sites around a central metal ion, in many cases by substitution of the water molecules previously held in the coordination sphere. The inorganic anion species such as chloride ion, as above, are capable of ligand substitution and so the distinction between ion pairs and complex ions can become very blurred. When polydentate ligands (with more than one coordination site) are considered, however, the complex ion formation becomes increasingly distinctive and characteristic properties become apparent. Ligands occupying two, three, and six coordination sites are common in nature and occur frequently with humic acids. Humus is difficult to characterize chemically, however, and a wide variety of mixtures of ligands are possible. Some possible examples of ligand complexation in a six coordination sphere are:

| Bidentate | Tridentate | Sexidentate | Mixed mono-, bi-, tridentate complex |

In the formation of these complexes, the ligands may coordinate through atoms such as nitrogen where where is no hydrogen ion replacement:

$$M^{n+} + \quad \longrightarrow \quad \left[\begin{array}{c} N \quad N \\ M \\ N \quad N \end{array} \right]^{n+} \qquad (47)$$

or through the oxygen of an acid radical (or hydroxy) where there is replacement of one or more hydrogen ions:

$$M^{n+} + x \cdot H_2N - (CH_2)x - C \begin{array}{c} O \\ OH \end{array} \longrightarrow \left[\begin{array}{c} N \\ O - M/_x \end{array} \right]_x^{(n-x)+} \qquad (48)$$

amino acid

The former results in charged complexes which are generally soluble in water and which are not pH dependent. The latter may result in charged or neutral complexes which are pH dependent and, in the case of the neutral complexes, relatively insoluble in water. These are generally referred to as "salt-forming groups" and can play a major role in the biochemistry of a specific system.

This complex formation of metals and ligands plays an important part in the overall chemistry of a water system and has, in many instances, a direct bearing on the toxic effects of the complexed metal. Copper, for example, is an essential micronutrient for algae but is toxic in higher concentrations. The toxic form is the free copper ion which is also the nutrient form. Manahan and Smith[24] have shown that the complexed form of copper acts as a reservoir for the metal and replaces the free ionic form in solution when this is depleted by biological action. Complex formation can also affect the oxidation reduction potential of the aqueous system, which can have far reaching effects on various metallic properties such as solubility equilibria and biological availability. Theis and Singer,[25] for example, have shown that various humic acids (such as tannic and gallic acids), can completely retard the oxidation of iron(II) for several days.

The complexing agents or ligands necessary for complex formation have numerous sources and take various forms. Humic acids have been mentioned, but these form only a part of the humus materials which are the end products of natural vegetative decay, many of which are capable of complex formation. Other naturally occurring organic compounds, such as amino acids and vitamins, are now being identified in all natural water systems. In addition to these are species introduced into the system as waste products. These are both organic and inorganic in nature and can produce marked changes in the overall chemistry of a receiving system. Bender *et al.*,[26] for example, have identified metal complexing agents of two distinct molecular weight fractions in secondary-treated sewage, one of which is very similar to the synthetic complexers used to stimulate algal growth.

There is still very little information on the specific forms and reactions of specific metals in the environment, but light has been shed recently on some metals in particular. One example is with copper, as previously mentioned, and another is with iron. In the work reported by Theis and Singer,[25] the schematic diagram below is proposed to describe the behavior of iron in the presence of organic matter and oxygen (Fig. 4). Both forms of iron can be stabilized by dissolved organic matter with the ferrous iron in complex being resistant to oxidation. The dominant species in the ionic form is ferric hydroxide.

Kester and Byrne,[50] however, report that, in the absence of strong competition by organic compounds, the iron in solution will selectively combine with the hydroxide available at $pH > 7$ and alternate between the forms $Fe(OH)_2^+$ and $Fe(OH)_4^-$. Although HCO_3^- is relatively abundant in natural water systems at $pH\ 6.5-8.5$, its complex formation with iron is not well documented but should be significant.

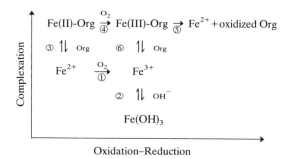

Figure 4. Behavior of iron in the presence of organic matter and oxygen.

The chemical forms of ferrous and ferric iron in solution reflect the status of most metals in that their forms are dependent on competition between complexing species, and until these species are adequately identified the reactions involved can only remain unknown or at best be crudely approximated. The biological sphere is also concerned with the ultimate form that a metal compound takes in that assimilation or absorption of the metal by an organism will depend to a large extent on this form. A situation may arise where an essential micronutrient is effectively removed from the biological cycle by maintaining a form that is not assimilated by organisms. Copper, for example, must be in the dissolved copper(II) state, but in the case of iron some organisms can absorb and utilize the colloidal form. So it is not enough to analyze a water mass and decide that a metal is present in sufficient quantity to supply the nutrient requirements of the system. A decision must be made as to the "availability" of the metal, and to do this a thorough knowledge must be gained of the chemical species finally formed by competition between all the constituents present.

5.4. *Toxicity*

In the past, toxicity studies on metal ions have too often been conducted in isolation. This means that statements can be made about the toxicity of lead, copper, or zinc, for example, in regard to a few specific species or organisms, but even in this regard adequate scientific data on heavy metal toxicity are scarce. In addition to this, a high proportion of the data available refer to the toxic effects on adult organisms only. Concentration levels and their resultant toxic effects on different stages of development in the same species, controlled in part by changing patterns of food requirements, are still far from being satisfactorily documented and understood.

It is of course necessary to understand the mechanism and role of a single metal ion in its toxic effects, but in nature metal ions rarely occur singly and so toxicity studies can sometimes be misleading. When more than one metal ion is present, which is generally the case, many combinations are possible and overall toxicity may change. If the separate metals make their presence felt independently or by joint action, i.e., additively, overall toxicity can be predicted, but the toxicity can also be greater (synergism) or less (antagonism) than predicted. Thus it is important to know what metals are present and how they react before discharging metallic wastes. For example, a synergistic reaction commonly quoted refers to the effect of a mixture of zinc and copper which is up to five times the toxicity anticipated from the combined effects (Wilber[27]). Wilber also reports that zinc and cadmium act additively and that zinc and nickel act synergistically. In the case of copper, Pagenkopf *et al.*[23] report that copper(II) is the actual chemical species that is toxic to fish.

More research is now being conducted on synergistic effects in multimetal systems and a clearer picture is beginning to emerge, but much more needs to be done in this field. Gray and Ventilla[28] have reported the synergistic effects in

Table 3. Effects of Adding Various Concentrations of Heavy Metal Ions on Cristigera Growth Rates[a]

	No ZnSO$_4$			0.125 ppm ZnSO$_4$		
	No Pb(NO$_3$)$_2$	0.15 Pb(NO$_3$)$_2$	0.3 Pb(NO$_3$)$_2$	No Pb(NO$_3$)$_2$	0.15 Pb(NO$_3$)$_2$	0.3 Pb(NO$_3$)$_2$
No HgCl$_2$	0	6.09	11.22	10.05	13.34	17.39
	0.01	10.82	12.36	6.52	15.43	20.15
0.0025 ppm HgCl$_2$	8.47	9.81	13.79	13.37	15.72	26.65
	10.59	11.49	15.19	14.48	16.74	18.54
0.005 ppm HgCl$_2$	12.24	17.65	18.54	20.89	22.65	25.96
	11.88	19.66	25.01	16.83	19.91	20.49

	0.25 ppm ZnSO$_4$		
	No Pb(NO$_3$)$_2$	0.15 Pb(NO$_3$)$_2$	0.3 Pb(NO$_3$)$_2$
No HgCl$_2$	15.76	19.88	23.82
	12.73	16.78	28.06
0.0025 ppm HgCl$_2$	16.69	26.04	47.33
	20.82	32.00	55.34
0.005 ppm HgCl$_2$	33.73	37.90	70.54[b]
	37.17	35.19	65.02

[a] J. S. Gray and R. J. Ventilla, *Ambio*, **2**(4) : 118–121 (1973).[28]
[b] Percent reduction in growth rate constant K.

a three-component system containing zinc, lead, and mercury. Table 3 summarizes the effects of various mixtures of various concentrations of metals by tabulating the percentage reduction of the growth rate of Cristigera, a sediment-living ciliate protozoan.

The authors reported significant supplemental synergistic effects in all two- and three-component combinations. From Table 3, for example, the purely additive effect of a mixture of 0.005 ppm Hg^{2+}, 0.3 ppm Pb^{2+}, and 0.25 ppm Zn^{2+} would be predicted as a percentage reduction in growth rate of 37.4%. In fact, however, the growth rate was reduced by more than 67%.

In addition to the above, other system parameters can have a marked effect on growth rates, uptake of specific metals by specific organisms, and toxicity. Salinity and temperature variations, for example, can exhibit a controlling effect on the growth rates and survival of many species of organisms. Oxygen levels can also control certain parameters aside from the respiration of the species in question. In waterways where the dissolved oxygen concentration decreases below saturation values, an elevated toxicity to fishes of zinc, lead, copper, and monohydric phenols has been reported (Wilber[27]).

Most toxicity studies are concerned with the amounts or concentrations of metals necessary to cause acute problems, i.e., those noticeable in the short term. Of more importance, however, in regard to chronic or long-term effects, are the sublethal concentrations that will cause changes in the environment and affect the uptake and assimilation of required nutrients. Many fish are sensitive to metal concentrations and will not remain in an area where the concentration exceeds the "avoidance level" for that species. This means that an area can be depopulated by metal concentrations much lower than necessary for apparent toxic effects. For some species the avoidance levels are high and they will survive even though they assimilate relatively high concentrations of toxins. Of particular interest are metals such as lead, cadmium, and mercury, which are classified as accumulative poisons. All three are known to cause severe physiological damage if consumed in sufficient quantity, and this renders contaminated fish or other species inedible.

Under these circumstances the effect is the same as if no species were present at all. By considering all the sublethal effects, a question of major importance arises. When should a waterway be declared polluted, or when does a waterway become polluted? Criteria for a definitive answer to this question are extremely difficult to promulgate since each water system is unique in its own way. This of course means that each system must be adequately examined before heavy metal concentrations are allowed to increase to an extent where significant chemical and biochemical changes are initiated.

5.5. Anoxic Conditions

In water systems, the dissolved oxygen is distributed according to various factors such as temperature, salinity, and biological activity. When consumption of oxygen exceeds supply or rate of replenishment, a condition is reached where the waterway becomes devoid of oxygen. This phenomenon is referred to as an "anoxic" condition and, aside from respiration problems, marked chemical charges occur.

Under normal conditions organic and other oxidizable materials are oxidized by the oxygen available in the water. If these materials are characterized as $(CH_2O)_n(NH_3)_m$ (Horne[45]) then the normal oxidation reaction is

$$(CH_2O)_n(NH_3)_m \xrightarrow{O_2} CO_2 + H_2O + HNO_3 \qquad (49)$$

Under anoxic conditions, however, the next best oxidizing chemical species (NO_3^-) becomes available:

$$(CH_2O)_n(NH_3)_m \xrightarrow{HNO_3} CO_2 + H_2O + N_2 + NH_3 \qquad (50)$$

and

$$NH_3 \xrightarrow{HNO_3} N_2 + H_2O \qquad (51)$$

So that concurrent with the disappearance of oxygen there is a net removal of the nutrient nitrate species by "denitrification."

After all of the oxygen and nitrates have been consumed, the normal oxygen photosynthetic cycle is replaced by the sulfur cycle

$$2SO_4^{2-} + 2H_2O \rightarrow 2H_2S + 5O_2 \tag{52}$$

$$CO_2 + H_2O + H_2S + h\nu \xrightarrow[\text{bacteria}]{\text{purple}} H_2SO_4 \tag{53}$$

or

$$CO_2 + H_2O + H_2S + h\nu \xrightarrow[\text{bacteria}]{\text{green}} S \tag{54}$$

Under these conditions the oxygen requirements are supplied by the reduction of sulfate ions. The overall reaction occurs as

$$(CH_2O)_n (NH_3)_m \xrightarrow{SO_4^{2-}} CO_2 + H_2O + H_2S + NH_3 \tag{55}$$

with the principal end product being hydrogen sulfide.

With this new set of chemical parameters numerous changes occur, among which is the lowering of the oxidation reduction potential. This results in a general reduction of the oxidation state of metal ions to a lower stable state. For example, iron occurs in the +2 state instead of the normal +3 state. This, together with the hydrolysis of the sulfide ion, equation (29), has a considerable bearing on the solubilities of various metals. Metals that normally precipitate as the hydroxide are usually the most seriously affected, with changes in solubility being quite considerable. For example, the solubility of iron(III) hydroxide is approximately 1.6×10^{-10} mol/liter, whereas the iron(II) form has a value of 5.8×10^{-6} mol/liter; also, the cobalt(III) hydroxide solubility is 5×10^{-12} mol/liter and the cobalt(II) form is 8×10^{-6} mol/liter.

6. Gases

All of the gases normally found in the atmosphere are found in solution in the hydrosphere. The interchange of these gases takes place between the atmosphere and the surface water through the barrier of the surface film and they are distributed in the water by advection and diffusion, which are affected by turbulence and mixing processes. The biologically active gases such as oxygen and carbon dioxide are also affected by photosynthesis and by the respiration of organisms. Hydrogen sulfide may also be present, but in strictly localized areas where the oxygen concentration is very low and the oxygen requirements are met by the reduction of the sulfate ion.

Under normal conditions surface waters are saturated with a gas in relation to its atmospheric concentration and to pressure, and in most cases the water will hold the maximum allowable concentration of the gas, i.e., 100% saturated. The total dissolved concentration of a gas in a liquid is given by Henry's law, which relates the mass of a gas per unit volume of solvent at constant pressure, to the pressure of the gas in equilibrium with the solvent:

$$m = Kp \qquad (56)$$

where m is the mass of gas per unit volume, p is the partial pressure of the gas in the atmosphere, and K is the proportionality constant. Most gases obey Henry's law, provided that no extremes of temperature and pressure are involved, and that the gas is only moderately soluble. The law fails to hold, however, if the gas in question is capable of a chemical reaction with the solvent. In this case only the proportion of the gas that is free or uncombined in the solution is found to obey the law. If the solubility is expressed as a volume rather than as a weight, then the Bunsen coefficient may be used, which relates the volume of gas dissolved at a partial pressure of one atmosphere to the temperature of the liquid. In most cases the relationship is inverse, the solubility of the gas decreasing with an increase in temperature.

At equilibrium, the partial pressure of a gas in solution is equal to the partial pressure of the gas in the atmosphere, and gas molecules are interchanged at the same rate. But, for a system not in equilibrium, there is a net transport of gas molecules from one medium to the other. The rate of gas transfer depends on the boundary layer of the liquid, which under these circumstances may be referred to as the "surface film." This film has a definite thickness, and the rate-determining step for gaseous exchange depends on the molecular diffusion of the gas through the film. Numerous calculations have been made of the exchange rates of various gases—especially oxygen and carbon dioxide—through the film, but in a practical situation the only agreement is that the gas transfer is not instantaneous. Since the film is the reactive site for many constituents, as previously mentioned, which may affect molecular diffusion, and since the slightest disturbance or turbulence affects the thickness of the film, transfer rates are, in a practical sense, extremely localized phenomena. It has also been shown that the solubility of gases depends as well on the ionic strength of a solution, which in natural water systems is frequently controlled by the salt content. Green and Carritt[29] have shown that the relationship is not exactly linear, but deviations from linearity may be considered negligible under normal conditions.

It can thus be generally stated that the solubility of a gas in a natural water system depends on (1) temperature, (2) pressure, (3) salinity, and (4) the thickness of the surface film.

In much of the research carried out on gaseous solubilities, equilibria, and reactions in natural water systems, the major emphasis has been on oxygen and carbon dioxide. For the purposes of this chapter further discussion will be limited to these two gases.

6.1. *Oxygen*

Because of its ease of analysis and importance in the life cycle, oxygen is the gas most investigated in the environment. However, there are still many misconceptions and disagreements regarding its properties and distribution in a water system. The overall concentrations of oxygen ranging from 6–12 ppm found in the hydrosphere are not in dispute, but the rates of replacement or replenishment are still very much unknown quantities in the short term. In the long term, over a period of a year, for example, in uncontaminated systems the enhanced capacity of cold water to dissolve oxygen during autumn and winter is balanced by photosynthetic activity and the lower solubility of oxygen in warmer water during spring and summer. This means that even though concentrations may change, the percentage saturation remains at a relatively constant level. The percent saturation, measured by

$$\frac{C_{O_2}}{C_{O_2^1}} \times 100 \tag{57}$$

where C_{O_2} is the concentration observed and $C_{O_2^1}$ is the equilibrium concentration at the temperature and salinity of the observed sample, is generally found to be between 97–105% for surface water. This tends to decrease as the depth increases, but local conditions have a dominating effect on temperature distribution.

But even in this aspect significant variations are possible. Tables for the calculation of oxygen saturation values in water and seawater are not in very good agreement, which prompted Green and Carritt[20] to publish a revised recalculated set which appears below as Table 4. As can be seen, oxygen solubility decreases with temperature and salinity so that cold, fresh water can accommodate more oxygen than warm, salty water. Since most legislation regarding oxygen levels is based on percentage saturation, it seems apparent that a standardized set of tables should be formulated and adhered to by all concerned.

The concentration of oxygen in surface water depends mostly on the gaseous exhange between the water and the atmosphere, but in deeper water that is out of contact with the atmosphere, this exchange depends on consumption by biological processes and replenishment by mixing with more oxygenated water. In both cases the dominating feature, with respect to a viable system, is the rate of renewal. A major mistake, made frequently by the uninformed, is to assume that the supply of oxygen in a depleted water system can be instantly renewed or transferred. That this is not true is evidenced by the distribution of oxygen in any normal water system. For example, in the ocean an oxygen minimum is maintained at a depth of about 1000 m. In more shallow waters wide variations can also be observed. In Botany Bay, Australia, for example, stratifications in excess of 50% of surface concentrations have been found at depths of 6–10 m. Another example of oxygen stratification is the presence of anoxic water underlying normally oxygenated water in fjord

Table 4. Oxygen Solubility (ml/liter) in Fresh Water and Seawater from a Saturated Atmosphere of which Oxygen is a 0.2094-mol Fraction Excluding Water Vapor[a]

T, °C	Chlorinity ‰							
	0	2	4	6	8	10	12	14
0	10.30	10.04	9:79	9.54	9.30	9.06	8.83	8.61
1	10.02	9.77	9.53	9.29	9.06	8.83	8.61	8.39
2	9.75	9.51	9.28	9.05	8.82	8.61	8.39	8.19
3	9.49	9.26	9.03	8.81	8.60	8.39	8.19	7.99
4	9.24	9.02	8.80	8.59	8.38	8.18	7.99	7.79
5	9.01	8.79	8.58	8.38	8.18	7.98	7.79	7.61
6	8.78	8.57	8.37	8.17	7.98	7.79	7.61	7.43
7	8.56	8.36	8.17	7.97	7.79	7.61	7.43	7.26
8	8.35	8.16	7.97	7.78	7.61	7.43	7.26	7.09
9	8.15	7.96	7.78	7.60	7.43	7.26	7.09	6.93
10	7.95	7.77	7.60	7.43	7.26	7.09	6.93	6.78
11	7.77	7.59	7.42	7.26	7.09	6.94	6.78	6.63
12	7.59	7.42	7.26	7.09	6.94	6.78	6.63	6.49
13	7.42	7.25	7.09	6.94	6.78	6.64	6.49	6.35
14	7.25	7.09	6.94	6.79	6.64	6.49	6.35	6.21
15	7.09	6.94	6.79	6.64	6.50	6.36	6.22	6.08
16	6.94	6.79	6.64	6.50	6.36	6.22	6.09	5.96
17	6.79	6.65	6.50	6.36	6.23	6.10	5.97	5.84
18	6.65	6.51	6.37	6.23	6.10	5.97	5.84	5.72
19	6.51	6.37	6.24	6.11	5.98	5.85	5.73	5.61
20	6.38	6.24	6.11	5.98	5.86	5.74	5.62	5.50
21	6.25	6.12	5.99	5.87	5.74	5.62	5.51	5.39
22	6.12	6.00	5.87	5.75	5.63	5.52	5.40	5.29
23	6.01	5.88	5.76	5.64	5.52	5.41	5.30	5.19
24	5.89	5.77	5.65	5.53	5.42	5.31	5.20	5.09
25	5.78	5.66	5.54	5.43	5.32	5.21	5.10	5.00
26	5.67	5.55	5.44	5.33	5.22	5.11	5.01	4.91
27	5.56	5.45	5.34	5.23	5.13	5.02	4.92	4.82
28	5.46	5.35	5.24	5.14	5.03	4.93	4.83	4.73
29	5.36	5.26	5.15	5.05	4.94	4.84	4.74	4.65
30	5.27	5.16	5.06	4.96	4.86	4.76	4.66	4.57
31	5.18	5.07	4.97	4.87	4.77	4.67	4.58	4.49
32	5.09	4.98	4.88	4.78	4.69	4.59	4.50	4.41
33	5.00	4.90	4.80	4.70	4.61	4.52	4.42	4.33
34	4.91	4.82	4.72	4.62	4.53	4.44	4.35	4.26
35	4.83	4.73	4.64	4.55	4.45	4.36	4.28	4.19

[a] Source: Green and Carritt.[29]

Table 4—continued

T, °C	Chlorinity ‰							
	16	18	20	22	24	26	28	30
0	8.39	8.18	7.97	7.77	7.58	7.38	7.20	7.02
1	8.18	7.98	7.78	7.59	7.40	7.21	7.03	6.86
2	7.98	7.79	7.60	7.41	7.23	7.05	6.87	6.70
3	7.79	7.60	7.42	7.24	7.06	6.89	6.72	6.56
4	7.61	7.42	7.24	7.07	6.90	6.73	6.57	6.41
5	7.43	7.25	7.08	6.91	6.74	6.58	6.43	6.27
6	7.25	7.08	6.92	6.75	6.59	6.44	6.29	6.14
7	7.09	6.92	6.76	6.60	6.45	6.30	6.15	6.01
8	6.93	6.77	6.61	6.46	6.31	6.16	6.02	5.88
9	6.77	6.62	6.47	6.32	6.18	6.03	5.90	5.76
10	6.62	6.47	6.33	6.18	6.04	5.91	5.77	5.64
11	6.48	6.33	6.19	6.05	5.92	5.79	5.66	5.53
12	6.34	6.20	6.06	5.93	5.80	5.67	5.54	5.42
13	6.21	6.07	5.94	5.81	5.68	5.55	5.43	5.31
14	6.08	5.94	5.81	5.69	5.56	5.44	5.32	5.21
15	5.95	5.82	5.70	5.57	5.45	5.33	5.22	5.11
16	5.83	5.70	5.58	5.46	5.34	5.23	5.12	5.01
17	5.71	5.59	5.47	5.35	5.24	5.13	5.02	4.91
18	5.60	5.48	5.36	5.25	5.14	5.03	4.92	4.82
19	5.49	5.37	5.26	5.15	5.04	4.93	4.83	4.73
20	5.38	5.27	5.16	5.05	4.95	4.84	4.74	4.64
21	5.28	5.17	5.06	4.96	4.85	4.75	4.65	4.56
22	5.18	5.07	4.97	4.86	4.76	4.66	4.57	4.47
23	5.08	4.98	4.87	4.77	4.68	4.58	4.48	4.39
24	4.90	4.89	4.70	4.69	4.59	4.50	4.40	4.31
25	4.90	4.80	4.70	4.60	4.51	4.41	4.32	4.24
26	4.81	4.71	4.61	4.52	4.43	4.34	4.25	4.16
27	4.72	4.2	4.53	4.44	4.35	4.26	4.17	4.09
28	4.64	4.54	4.45	4.36	4.27	4.18	4.10	4.02
29	4.55	4.46	4.37	4.28	4.20	4.11	4.03	3.95
30	4.47	4.38	4.30	4.21	4.12	4.04	3.96	3.88
31	4.40	4.31	4.22	4.14	4.05	3.97	3.89	3.81
32	4.32	4.23	4.15	4.06	3.98	3.90	3.82	3.75
33	4.25	4.16	4.08	4.00	3.92	3.84	3.76	3.68
34	4.18	4.09	4.01	3.93	3.85	3.77	3.70	3.62
35	4.10	4.02	3.94	3.86	3.78	3.71	3.63	3.56

systems. Many more examples of oxygen stratification could be given, but it is only necessary here to emphasize that the rate of diffusion of dissolved oxygen is not instantaneous and does not take place uniformly throughout the hydrosphere. Even in a relatively small localized area uniform distribution is the exception rather than the rule.

When wastes with a high BOD_5 are discharged, the dissolved oxygen in the receiving water is consumed at a specific rate and will not be replaced for some finite length of time. If the resulting mix remains on the surface the rate of renewal depends largely on the surface film (area and thickness), but if the mix is denser and sinks the rate depends on an additional number of variables. The mix at a specific depth, depending on its density, is in contact with water masses on six sides and mixing at all points of contact is possible. However, the rate of mixing and diffusion depends on differences in temperature, salinity, density, and dissolved oxygen, and on the rate and direction of movement of the mix in relation to the surrounding water. The rate of diffusion of dissolved oxygen to a depleted water mass is thus highly variable, decreases with time, depends on the oxygen differential, and is a local phenomenon. Needless to say, rates of oxygen diffusion within a water system are for the most part virtually unknown. If receiving waters are allowed to regenerate, the problem may not be serious, but for continuous discharges of high BOD_5 wastes, the problem can quite frequently attain massive proportions.

Wherever consumption of oxygen exceeds renewal a steadily declining state is achieved which has a marked effect on various organisms in the path of this moving mass of oxygen-depleted water–waste mixture. The magnitude of this effect depends on the relative rates of consumption, renewal, and water movement, and can only be totally assessed *after* the damage has been done. If a prediction is to be made regarding the extent of any damage to any ecosystem by an oxygen-demanding discharge, the variables or parameters mentioned above must be thoroughly understood and appreciated *before* an uncontrolled discharge is permitted.

When the dissolved oxygen concentration falls far enough—to less than about 2 ppm—various types of bacteria will begin obtaining their oxygen requirements by reducing the sulfate ion (SO_4^{2-}) and the water body may be considered as anoxic. Under these circumstances a fundamentally different chemistry is involved in the system than that of normally oxygenated water. This induces many changes and effects, some of which have been discussed above in the section on metals. Other effects of a more general nature, in addition to respiration problems, must still be evaluated.

In anoxic or stagnant waters the condition of a nearly lifeless zone is approached. However, life is possible and does exist even under these drastic conditions. Numerous forms of micro- and macroscopic life have been identified in even the most depleted water systems, but the life cycle is foreign to what is generally accepted as being normal in that sulfur has taken over the role of oxygen. These systems do exist naturally in various areas, as in the fjords of New Zealand and Norway, and a considerable amount of research is being conducted into their properties and characteristics. Bottom water that is in contact with the sediment, for example, appears to be more viable under these conditions than surface or intermediate waters, which are more mobile.

This type of water, however, represents only one extreme condition and can be generated rather easily. But there is another equally important condition which is much more difficult to duplicate and which has received very little attention to date. This is a body of water that is intermediate between the two extremes and can fluctuate, generally over a narrow range, depending on conditions. This water is very low in oxygen, can barely sustain a viable community, and can be driven into an anoxic condition rather easily. This change, going from just oxygenated to just anoxic and vice versa, involves a complicated set of equilibria that affects all the constituents present, such as metals and organic compounds, as well as the biological structure of the system. The stress imparted by these changes is reflected in the forms of biota present and in the chemical characteristics of the water involved. These depend to a large extent on the frequency rate and degree of change, and can fluctuate in a most confusing manner. Metallic reactions, such as solubility and complex ion or ion pair formation, are most susceptible to variations and can give confusing results if the character of the system is not fully appreciated.

Thus the question of dissolved oxygen concentration, percentage saturation, distribution, diffusion, and associated effects is not a simple one, and adequate training in the field of water chemistry is an absolute necessity before one can safely assume the responsibility of allowing oxygen-consuming discharges.

6.2. Carbon Dioxide

The carbon dioxide–carbonate system is one of the most complex and important systems in the hydrosphere. It is involved in the atmosphere–surface water interchange, the chemistry of the water system, the biological structure of organisms, and the deposition of carbonaceous sediments. In the short term it controls the pH which directly affects some of the chemical equilibria in any local system, and the ions can act as ligands for a number of complexing metals. It is also of primary concern in the biological cycle of all organisms with a carbonaceous skeletal structure, where carbon is utilized in the growth, death, and decay pattern.

Its distribution is not uniform and depends in part on the biological activity in any area. Similar to oxygen, the carbon dioxide concentration in surface waters is a function of the atmospheric concentration and partial pressure of the gas, and percentage saturation values can be determined. However, the pattern of distribution differs markedly from oxygen. For example, the surface water in the Pacific Ocean is distinctly undersaturated, whereas the surface water of the Indian Ocean near the equator is supersaturated. This points out an anomaly in the solubility equilibria of carbon dioxide: in the Indian Ocean the dissolved carbon dioxide concentration increases with an increase in temperature. The distribution of the total carbon with depth, however, seems to present a more uniform picture in that the concentration tends to increase with depth due to the settling of the decay fragments of dead organisms from the biologically rich surface water, or euphotic zone.

The available carbon is an essential nutrient and so is of primary importance in the overall ecological structure of the hydrosphere in general, and particularly in localized areas. There is at the moment a controversy concerning eutrophication in specific areas as to whether or not the available carbon is the prime activator of an algal bloom or whether excess phosphate is the culprit. In addition, carbon dioxide is a functional part of photosynthesis and it would appear that its concentration could be correlated to light penetration or to energy. This is not always true, however, since the physical chemical processes may dominate in specific areas, forcing the photosynthetic reaction to be relegated to a secondary effect.

In spite of the importance and variability of the total carbon system, the solubility of carbon dioxide in various classes of water systems and the thermodynamic equilibria involved in its reactions are not accurately known. The solubility question, for example, is much more complicated than in the oxygen system because carbon dioxide will react with water so that its total solubility does not obey Henry's law. Estimates of uncombined carbon dioxide can be correlated to atmospheric conditions by a consideration of Henry's law, but the most salient point here is that the concentration of gas shows remarkable deviations from what is predicted.

In solution, carbon dioxide will theoretically form carbonic acid at a low pH,

$$CO_2 + H_2O \rightleftharpoons H_2CO_3 \tag{58}$$

which will then dissociate in two pH-dependent steps:

$$H_2CO_3 \overset{K_1}{\rightleftharpoons} H^+ + HCO_3^- \tag{59}$$

$$HCO_3^- \overset{K_2}{\rightleftharpoons} H^+ + CO_3^{2-} \tag{60}$$

But the dissolution reaction (58) is slower than the first dissociation step, rendering carbonic acid and dissolved carbon dioxide indistinguishable, so that these two are generally combined to give the equilibrium condition

$$CO_2 + H_2O \overset{K_1}{\rightleftharpoons} H^+ + HCO_3^- \overset{K_2}{\rightleftharpoons} H^+ + CO_3^{2-} \tag{61}$$

where

$$K_1 = \frac{a_{H^+} a_{HCO_3^-}}{a_{CO_2}} \tag{62}$$

and

$$K_2 = \frac{a_{H^+} a_{CO_3^{2-}}}{a_{HCO_3^-}}$$ (63)

which results in the formation of two complementary buffer systems:

$$a_{H^+} = K_1 \frac{a_{CO_2}}{a_{HCO_3^-}}$$ (64)

and

$$a_{H^+} = K_2 \frac{a_{HCO_3^-}}{a_{CO_3^{2-}}}$$ (65)

The equilibria involving the reaction of the carbonate minerals with water and carbon dioxide,

$$CaCO_3 + H_2O + CO_2 \rightleftharpoons Ca^{2+} + 2HCO_3^-$$ (66)

can be neglected in the short term (in the order of a few hundred years) but should be considered in long-term equilibrium calculations.

From equations (62) and (63) it can be seen that the distribution of the species CO_2, HCO_3^-, and CO_3^{2-} depends on the pH of the solution and can be expressed graphically, as in Fig. 5. It may also be considered however that the pH depends on the buffering capacity of the system, i.e., the distribution of the species.

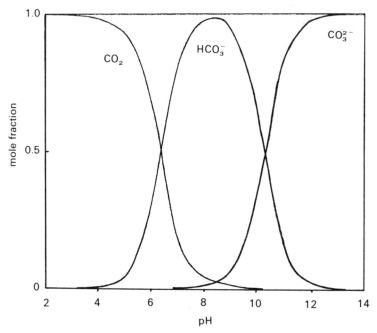

Figure 5. Distribution of $CO_2 - HCO_3^- - CO_3^{2-}$ as a function of pH.

Under normal circumstances the pH of natural systems is found to range between 7.0–8.5, and in this area the main species in solution is the bicarbonate ion. In this form the carbon is most easily assimilated by aquatic organisms and fluctuations in its availability are kept to a minimum. If the pH of the system is radically changed, however, the available carbon can be seriously depleted, resulting in a mass movement of free-swimming organisms away from the area and in the gradual depletion of benthic organisms due to disruptions in the reproductive cycle.

The dissociation constants K_1 and K_2 are in the same form as the equilibrium constants previously mentioned and are inversely proportional to the ionic strength of the solution, which controls the activity of the ions in question. At the same time, however, the constants are directly proportional to the temperature. This means that as the ionic strength increases the constants decrease (for example from fresh to salt water), and the curves in Fig. 5 are shifted to the left. As the temperature increases the curves are shifted to the right, but the total amount of carbon decreases since the solubility of carbon dioxide is expected to decrease in the warmer water. The magnitude of the shift due to both considerations is once again a local phenomenon, but the variations in most instances are not sufficient to appreciably alter the availability of the total carbon.

In addition to the disastrous effects of drastic pH changes on the carbon system, the solubility of the gas and the capacity of the hydrosphere to accommodate it with its accompanying reactions is attracting a great deal of attention. This is because of recent overall increases in carbon dioxide atmospheric concentration due to the combustion of fossil fuels and other industrial processes. Over the past 70 years the average concentration has increased by about 10%, from below 300 ppm to more than 320 ppm. It is estimated that with the expected rate of increase the average concentration by the end of this century will rise to approximately 700 ppm. Since respiration problems do not occur until the concentration of the gas is about 5000 ppm, the only real question is the effect on the environment of this expected increase.

At the present time the accepted value for the total concentration of carbon dioxide in the hydrosphere is about 4×10^{-3} M, with more than 80% of this in the HCO_3^- form. A fully saturated solution, however, will sustain a concentration of the order of 5×10^{-2} M, which is an increase by a factor of more than 10. So it appears that the increase in atmospheric concentrations can be adequately handled by the hydrosphere. But the situation is not as simple as it seems. The solubility of the gas depends on temperature, salinity, and pressure, and in the short term only the surface water will be the receptacle for the excess concentration. This means that the absorption of the gas will not be uniform, but instead will be a highly localized phenomenon. Under these circumstances some areas can become saturated or supersaturated relatively quickly while other areas will remain unsaturated, to varying degrees, for some time. At the same time, the amount of gas transferred to the water will depend on the rate of molecular diffusion through the surface film, and at present it is impossible to predict this rate. It is apparent, however, that total gas transfer is a function of the characteristics of the surface film so that the rates will vary and once again localized effects will predominate.

A further complication arises in that the extra atmospheric concentration will increase the partial pressure of the carbon dioxide, which will in turn increase the mean temperature of the earth's surface, and, finally, the mean temperature of the hydrosphere. This should of course result in a reduced capacity of the water to absorb the gas, which will eventually further increase the atmospheric concentration. Thus a cycle can be set in motion: as the water is warmed, the rate of heating increases. This cycle can be slowed down to some extent if the heating effect is sufficient to melt the ice caps. This would mean that more water would become available to absorb the gas, which would tend to suppress the cycle until the water is warmed sufficiently to begin losing the gas again, which in turn would obviously revitalize the cycle.

The reactions induced by these changes would tend to be minimal in that the bicarbonate ion (HCO_3^-) concentration would increase, but because of the buffering capacity of the water the pH change could be negligible. In this case the carbon availability would increase, which could result in increased populations of various organisms. This of course could mean significant blooms of phytoplankton and algae which would result in eutrophication on a grand scale, depending on the availability of other nutrients. On the other hand, if the pH change is considerable, drastic changes in the chemistry of the hydrosphere would follow with completely unpredictable results. The carbon equilibria would also change and the availability of the carbon in the biological sphere would be reduced with more predictable results. With less carbon for skeletal building fewer organisms would mature, and eventually aquatic life would virtually disappear.

These drastic predictions could of course be carried on *ad infinitum* until all the seas and oceans were theoretically dead and all the present coastal areas were flooded. But until a great deal more information is obtained about the carbon dioxide–carbonate cycle, no really valid predictions can be made.

The only conclusion that can possibly arise is that the carbon dioxide cycle is immediately important enough to demand significant increases in research and investigation. With more adequate data, the effects of added carbon dioxide in the atmosphere and hydrosphere can be sensibly evaluated, and decisions can be made as to whether or not these effects will be significant enough to cause a major shift in the present policy of producing energy by the combustion of fossil fuels.

7. Nutrients

7.1. Introduction

The word "nutrient" covers a number of chemical elements that are essential for the proper health and growth of all organisms. Some of these are required only in trace amounts but if any fraction is constrained or absent, the growth patterns of the subject organism are vitally affected. The most common elements utilized on the smaller scale are the transition metals such as copper, iron, cobalt, and manganese, and they are usually referred to as trace metals. The better-known nutrients, on the other hand, are the compounds obtained from the chemical reactions of nitrogen, phosphorus, oxygen, and sulfur.

In most cases the element is involved in a nutrient cycle throughout the entire biosphere, involving the atmosphere, the hydrosphere, and the lithosphere. In other cases, such as phosphorus, the cycle is short-circuited and the element is restricted in its movements, usually to two of the above. In all cases, however, the hydrosphere is never removed from the circuit, and the necessary elements are normally present there in varying concentrations.

Fluctuations in these concentrations and the distribution of the nutrients both depend on a number of variables, the most important of which is the biological activity of the area, and both can show seasonal and other effects. These short-term variations are normally restricted to surface waters (euphotic zone) where the penetration of sunlight is sufficient for photosynthetic activity. This implies that the primary consumption of the major nutrients is by plants which form the fundamental or lowest step of the food chain, and are in fact referred to as the plant nutrients.

These nutrients are initially present in the inorganic form and are converted by the metabolic action of chemotrophic organisms—which obtain energy from the consumption of inorganic compounds—to the organic form. These are the basis of the food chain and consumption by each succeeding trophic level transfers the element through the chain to the highest level available. After the death of an organism at any level, the phenomenon of decay involves further metabolic reactions which return the elements to the inorganic form, and in this way a cycle of nutrient usage is set up. When consumption exceeds rate of supply, as is the case in spring and summer when biological activity is at its peak, a steady decrease in concentration occurs. In this way the biota is restricted by the available food supply. Similarly, in autumn and winter, when plant activity is decreased, the nutrient concentrations will build up and be available for the following spring and summer.

Similar to land forms, the plants in the hydrosphere (phytoplankton) can be induced to grow more prolifically by the use of extra nutrients in the form of fertilizer. In the specific cases of nitrogen and phosphorus, as the nutrients the most easily assimilated, forms are the nitrate ion, NO_3^-, and the phosphate ion, HPO_4^{2-}. The use or presence of excess concentrations of one or both of these species will lead to an increased production of phytoplankton with the accompanying consumption of other essential nutrients, such as the transition metals. Under these circumstances the trace nutrients are exhausted first and may not be resupplied as readily as the nitrates and phosphates, so that the area in question may become less able to sustain peak growths. If, on the other hand, all nutrients are present in quantity, a condition similar to a population explosion can occur that is referred to as a plant, or algal, bloom, and which can have very serious consequences. When the bloom is sufficient to affect the overall oxygen levels of a water system it is said to induce a eutrophic effect, which can be described as an acceleration of the life cycle of the water body in question. If the acceleration is fast enough the water body will "die" in that all the oxygen will be consumed by the requirements of the decomposing algae. This latter effect is referred to as "eutrophication" and has already been noted in various areas, such as sections of the Great Lakes in the United States and Canada.

There is still much to be learned about nutrients and the biological requirements of elements for healthy growth. Some elements could be termed nonspecific in that all organisms require them, such as nitrogen for amino acids or phosphorus and sulfur for the production of protein. Others tend to be more specific, such as the copper requirements of oysters or magnesium as an integral part of chlorophyll. It seems that metals form most of the specific elements, and it is known that various organisms will absorb and concentrate specific metals peculiar to them.

It is not the intention in this chapter to investigate any further the role of metals as nutrients, but rather to have a better look at the roles of the two nonspecific elements, nitrogen and phosphorus.

7.2. Nitrogen

In one form or another nitrogen is found throughout the biosphere. In its gaseous form, N_2, which is rather inert, it forms 79% of the atmosphere and forms the major reservoir for the total cycle. In the hydrosphere its total chemical forms have a concentration of approximately 5×10^{-5} M. This latter value is a generally accepted average, but it must be remembered that it is subject to local variations to a marked degree.

Before the nitrogen can be utilized it has to be "fixed" by a biological system, which in the hydrosphere appears to be the blue green algae. The reactions involved are

$$N_2 \rightarrow 2N \quad \text{(fixation)} \tag{67}$$

followed by hydrogenation

$$2N + 3H_2 \rightarrow 2NH_3 \tag{68}$$

This ammonia produced by the fixation incorporates the form NH_4^+ in water and is next subjected to a "nitrification" reaction in two steps. The reactions in both steps are oxidations and are carried out by two species of autotrophic bacteria, nitrosomonas and nitrobacter:

$$4NH_4^+ + 7O_2 \rightarrow 4H^+ + 4NO_2^- + 6H_2O \tag{69}$$

$$2NH_3 + 3O_2 \rightarrow 2H^+ + 2NO_2^- + 2H_2O \tag{70}$$

nitrosomonas

$$2NO_2^- + O_2 \rightarrow 2NO_3^- \quad \text{nitrobacter} \tag{71}$$

so that the overall reaction is

$$NH_3 \text{ or } NH_4^+ \rightarrow NO_2^- \rightarrow NO_3^- \tag{72}$$

and the end product of the oxidation is the nitrate ion, which is the form in which the nitrogen is assimilated to form amino acids and proteins.

On the death of an organism other types of bacteria can convert the nitrogen in the amino acids to ammonia by a process called "ammonification":

$$R-NH_2-(CH_2)_x-C{\overset{\displaystyle O}{\underset{\displaystyle OH}{}}} \rightarrow NH_3 + CO_2 + H_2O \tag{73}$$

amino acid

which reintroduces ammonia or the ammonium ion back into the cycle and helps to restore the balance of nature. To complete the cycle, nitrate and nitrite ions are converted through bacterial action by "denitrification" to the gaseous states as either N_2 or N_2O. In this way, under normal circumstances, the total amount of nitrogen fixed equals the total amount returned to the environment as the gas.

As can be seen, the major role in the conversion of inorganic nitrogen to a form usable in the biological scheme is played by various types of bacteria. These are subject to environmental conditions in much the same way as other organisms, and disruptions in these conditions can result in changes in bacterial activity. Landner and Larsson[30] have investigated unusually high concentrations of nitrite ion in the bottom water in the Bay of Koping in Sweden, and have concluded that there is an overabundance of the bacteria nitrosomonas as compared with the nitrobacter. This has apparently resulted from excess fertilization of the water by runoff from fertilized land areas, thus stimulating reaction (69). But selective inhibition of nitrobacter and reaction (70) by pollution cannot be ignored, since it is known that excess nitrates, cyanates, some heavy metals, and other chemical species will inhibit the efficiency of the nitrobacter.

It now becomes possible for some environmentally dangerous conditions to arise. It is known that excess concentrations of nitrite ions (NO_2^-) are toxic to many varieties of aquatic organisms, and that excess nitrate ions (NO_3^-) can constitute a human health hazard, as well as stimulate excess algal growth. Initially the nitrogen cycle was relatively stable, and the concentrations of all the species of nitrogen compounds were such that the denitrifying reaction was capable of returning all the nitrogen fixed to the cycle. But because of the increased use of fertilizers and systems producing nitrogenous wastes this situation no longer exists. The organic wastes of a steadily increasing population also contribute to the problem. The end result is an inability of the denitrifying bacteria to keep up with the pace. This should make it theoretically possible for high concentrations of nitrates and nitrites to coexist in the same area, producing corresponding deleterious effects.

problem. It is only when these concentrations increase to above threshold levels that a problem ensues, the magnitude of which is related to the level of concentration. In the specific case of organic compounds, however, the above statements are no longer necessarily true. In some instances, such as the introduction of oil, seepages have been common in many water systems for many years, and the oil only becomes a problem in areas where its final concentration exceeds threshold levels. For other organics, however, their introduction into a water system is a new phenomenon, and for many the necessary degradative processes are either conspicuous by their absence or relatively slow on a biological time scale. Both phenomena of course raise long-term interrelated biological problems.

The organic compounds, both in solution and as particulate matter, include a wide variety of species generated from a large number of sources. The total concentration is relatively low, and rarely exceeds 2 ppm in uncontaminated waters. In polluted systems, however, the concentrations depend on the character of the discharge and on variables such as the physical and chemical properties of the receiving water, so that wide localized variations may be prevalant. The variety of species also depends on locale and over the last few years many types have been identified. They include:

1. Insecticides
2. Pesticides
3. Detergents
4. Phenolics
5. Carbohydrates
6. Protein fragments
7. Amino acids
8. Carboxylic acids
9. Humic acids
10. Vitamins

all of which have a detectable effect on any viable water community. The species in the latter half of the group constitute necessary requirements in any system and can frequently play the role of limiting nutrients. The first three, however, are foreign bodies, and frequently have an effect out of all proportion to their concentrations.

The transport of the organic compounds through the biosphere depends to a large degree on the species involved and the uses to which they are put. Sewage, for example, contains large concentrations of carbohydrates, oils, and fatty acids, and is usually discharged directly into the aquatic environment. Pesticides and industrial organics, on the other hand, are usually introduced indirectly through runoff from agricultural areas and through precipitation from the atmosphere. In many areas rain frequently contains detectable levels of polychlorinated biphenyls (PCB) and organochloro insecticides. Residues from organic industrial processes may also be discharged directly into a system as waste materials.

Irrespective of the mode of entry, the compounds are distributed in a water system in a pattern that depends on the local characteristics of the receiving water. The distribution is similar to that previously discussed for metals, and all areas contain detectable concentrations of the compounds and exhibit reactions peculiar to the area in question. Briefly, these areas are the surface film, the main body of water, the sediment, and the biological cycle. In all of these areas reactions occur between the available species present, but most of these reactions and species are not very well understood. It is known, for example, that metals such as iron, copper, and zinc are complexed readily, giving the corresponding organometallic compounds, but the rates of formation, dissociation, and transfer between species have yet to be comprehensively determined. There is also of course competition between the metals and the organics for preferential formation of the organometallic compounds so that in virtually all instances the presence of specific compounds is poorly documented. The degradation and oxidation rates of various compounds also have yet to be determined, together with the forms of the degradative products and the effects that these will have in the area in question.

The distribution of the organic materials is four-dimensional, as with all other constituents in the system, and can involve large areas, depending on individual characteristics. Oil, for example, will float and spread out on the surface of the water until constrained or until a very thin film is obtained. This film greatly increases the surface area for reactions and changes some characteristics of the oil in bulk. Other compounds will dissolve and be transported by water movements, while some will absorb on particulate species and be sedimented. Further fractions will remain as particulates to be sedimented and others will be taken quickly into the biological cycle and transported by various organisms. The distribution and degree of transport in all cases depend on local conditions and may exhibit characteristics ranging from concentration phenomena to maximum dispersion. These in turn depend on the rate and character of water and sediment movements and the chemical and biochemical properties of the compounds in question.

But, in addition to the rate and direction of movement, and regardless of whether the overall effect is concentration or dispersion, an important consideration that is frequently overlooked is the residence time of the specific compound in question. This generally refers to the length of time necessary for the compound to be degraded or changed in any way so that it can no longer be considered a potential or real problem. It may also be classified as the length of time a compound or element will remain in the environment before it is removed by one process or another. These times are highly dependent on variables such as water characteristics and biological activity and can show marked variations for the same compound in different locales. The residence times of most materials classified as pollutants have not been calculated with any degree of accuracy, and at the moment most accepted times could only be called educated guesses.

For this reason some compounds, such as pesticides and polychlorinated biphenyls, are cause for concern because they have extremely long residence times and are biologically active. Many compounds such as these are actually classified as nondegradable and were previously assumed to show no activity after disposal. This is of course a naive assumption, similar to the assumption that inorganic mercury discharged in a water system is harmless, and generally reflects sloppy thinking on the part of those would should have known better. An insecticide or pesticide, for example, does not lose its potency simply because the locale has been changed, but will continue to play the role designed for it until it is effectively removed from the system. Hindsight is always very easy, of course, but enough lessons and examples have been given by nature to indicate that much more foresight is needed in the years ahead.

In addition to these specialized attributes, further consideration must be given to the chemical and biochemical reactions entered into by the specific compounds during their periods of residence in a system. As mentioned above, it is wrong to assume that nothing will happen, since reactions reflect the chemical natures of the compounds and all are capable of some reaction, depending on many variables. These reactions determine the overall effect on the hydrosphere in both the short and the long term, and must be understood or appreciated if water systems are to be assisted in their roles as waste disposal units. But many of these reactions are unknown because the constituents of many discharges are also unknown. This raises a problem that still must be resolved, that all authorities responsible for any discharge should be fully aware of what the discharge contains and how it reacts. Some examples will be discussed below.

8.1. Sewage

The question of sewage disposal is a prime example of the phenomena mentioned above. In most cases the sewage systems are used as receptacles for industrial discharge and normal drainage water after rainfall, so that an eventual discharge of "pure" sewage is unusual in built-up and industrial areas. Even in the case of pure discharges, however, the actual composition of the effluent is mostly unknown. Broadly speaking, these constituents are generally lumped into the categories of organic and inorganic elements and compounds. The latter consist of the nitrates, phosphates, and metals which cause problems both in the biological treatment of the sewage and in the resulting mixture of the effluent and receiving waters. The former, in most cases, are not differentiated any further and are simply considered as oxygen-consuming fractions that should be treated to reduce this particular property to manageable proportions.

The scarcity of information in this regard is reflected in the parameters that are usually considered when the effects of sewage effluents are questioned. These are BOD_5, the biological oxygen demand; total solids, including suspended and dissolved fractions; and coliform, an overall indicator of bacterial pollution. On a gross scale, these parameters will give an indication of visual and superficial pollution but do not indicate in any way the chemical, biochemical, and biological reactions involved in the eventual assimilation of the discharge in the environment. In the many instances when appreciable quantities of industrial wastes and drainage water are added to the system these same parameters are correspondingly less useful in describing what actually happens in the receiving water after discharge. An added complication under these circumstances is the almost certain possibility of reactions within the effluent itself which alter the effectiveness of the biological treatment and drastically change later reactions in the water–waste mixture. These are seldom beneficial, and frequently appear to raise the toxicity of the overall discharge, which is not generally considered as a parameter under these circumstances. In any case, it is a fair assumption to state that organic pollution will change the characteristics of any natural water system to a degree depending on the quantity and type of the organic compounds present.

In the specific case of sewage there is very little information available regarding the identity of all the constituents present. Research is currently underway to reduce this problem and many constituents have been identified. However, as in many other situations, sewage reflects the habits of specific populations and so its constituents will vary both in character and concentration, depending on locale. Because of this, generalizations may be possible, but adequate analysis in each area is a necessity. In addition to this total approach, a further characteristic of sewage has to be considered which involves the actual distribution of the constituents of the effluent into clearly separate fractions. These are the soluble, nonsoluble floatable, and sludge fractions, which contain various groups of compounds in varying concentrations. These groups have yet to be adequately determined, but it appears as if most components are distributed in all three fractions to varying degrees. Some organic compounds, or classes of compounds, that have been reported (Rebhun and Manka;[39] Kolattukudy and Purdy;[40] Wallen and Davis[41]) include polysaccharides, saturated and unsaturated fatty acids, cutin (a lipid biopolymer), humic substances, fats and oils, carbohydrates, proteins, vitamins, detergents, and various classes of hydrocarbons. Numerous reactions involving these species are possible, but few have been positively identified, although the chelating effect—the metal complexing power of some of the species—has been investigated. The humic acids, for example, are readily capable of forming organometallic complexes of various molecular weight fractions which may effectively inhibit or enhance the biological uptake of some specific metals. As mentioned previously, some organometallic complexes may also act as storage systems that release the metal into the waterway as needed so as to maintain a specific concentration of the dissolved ionic species.

Interactions between the sewage compounds and the natural constituents of a water system are also being investigated, but once again information is still very sparse. Sludge, for example, because of its depositional properties, reacts to a large extent with the natural sediment, and in extreme cases totally overlies this sediment and becomes, in effect, a new bottom material. Both of these phenomena reflect the sedimentation processes of the area in question and the reaction mechanisms between the sludge and the type of natural sediment present. Some of the lipids mentioned, and fatty acids, for example, will show varying degrees of association with different types of sediment. The organisms that are then most affected by sludge disposal are benthic or bottom dwellers, and these can show drastic changes in relation to the amount and frequency of sludge deposited. This latter observation also reflects the oxygen-demanding characteristics of the sludge, which can seriously affect the normal biological system as well as cause marked changes in the chemistry of the physical system. This is done by the alteration of such parameters as the oxidation–reduction potential and the oxygen and sulfur cycles.

In the consideration of floatable materials more subtle effects are present. By their very nature these will float to the surface after discharge where they are accumulated and concentrated. This means that dilution factors based only on BOD_5 values can be totally misleading. Under the best of conditions the BOD_5 value for untreated pure sewage is normally between 200–300 ppm. When industrial wastes are also present, however, the value can exceed 600 ppm. Dilution factors calculated using these figures, which are usually obtained from the final mixture, refer to the total amount of oxygen necessary to completely oxidize the waste materials. But all too frequently this is done by a consideration of the total amount of oxygen in the water. The more correct approach is to consider the amount of "available" oxygen present, which allows for the natural consumption as well. In any viable system there is a natural BOD_5 of approximately 1–2 ppm, and by international agreement the amount of oxygen necessary to sustain a viable community is 5 ppm. Therefore in all dilution factor calculations a minimum of 6 ppm dissolved oxygen should be subtracted from the total oxygen present. This will give a residual figure which represents the "available" oxygen present. One reason why this is not done is the belief that the oxygen will be resupplied anyway by transfer from the atmosphere and from the surrounding water masses.

This excuse is not valid, however, unless the rates of oxygen transfer and diffusion and the temperature gradients are known. Frequently, and especially in the higher latitudes, the temperature of the sewage discharge is higher than that of the receiving water, which further complicates the problem. Under these circumstances the oxidation reactions can take a long time, and the sewage discharge becomes a long, persistent problem related to the rate and volume of the continual supply of fresh, uncontaminated water. This feature must also be taken into account when dilution factors are calculated. And since a concentration effect is apparent in the top few centimeters of the water body, the actual dilution of the total effluent requires an increased factor in relation to the amount of the floatable material present. In addition to this, the concentration occurs in an area of intense activity specifically referred to as the surface film.

This film at the air–water interface controls to a large extent many chemical and biochemical cycles, such as the carbon dioxide–carbonate cycle. It also affects many physical properties such as oxygen transfer, light penetration, and evaporation. It is formed naturally by surface-active lipid materials such as the fatty acids, oils, etc., which can act as reactive sites for metal-complexing reactions and for simple absorption reactions of lipophilic materials such as chlorinated hydrocarbons. In fact, it has been found that heavy metal and organic pollutants may have enrichment factors in the surface film in excess of 10^4. This means that localized high pollution concentrations are possible but may be overlooked because of poor sampling procedures. The floatable fraction contains both organics and inorganics in varying concentrations and these react under the catalytic effect of the surface film with all possible species present. In most cases the structure of the film is enhanced and "thickened," with resulting effects on transfer phenomena between the water and the atmosphere. Reactions initiated in the film also affect numerous transfer rates and mechanisms.

In many instances pollution concentrations can be a function of distance and will decrease with increasing distance from the point of discharge. In this way several zones can be identified in a moving water system under the general headings of polluted water, recovery zone, and normal water. In effect the receiving water is absorbing the polluting effluent and theoretically rendering it harmless. But this observation is mostly applied to visual and short-term effects and does not take into account the longer-term chemical and biochemical effects. Many of these are accumulative and only make their presence felt when the situation has progressed so far as to present a long-lasting problem. The zones are theoretically related to the residence times of various polluting species, but one type in particular is frequently used to give a gross approximation of the total polluting effect. This is the coliform count, which indicates the possible presence of pathogenic bacteria. It does not under any circumstances give an indication of chemical parameters, such as metals and organic compounds or complexes.

In most areas the predominant design characteristics of a coastal outfall are the coliform concentrations in the specific area concerned. This parameter is usually referred to as the decay, or dieaway rate, which is based on the length of time necessary for 90% of the bacteria to be removed from the system, and which obviously varies with the type of receiving water. But very few of these rates have been calculated under field conditions in specific areas and usually only very poor approximations are used. Calculations of the rate have also been lifted from one area, and with no further investigation, applied to another area with totally different water characteristics. It has been pointed out in some texts, and deserves added emphasis here, that decay rates calculated for one area can frequently be totally wrong when applied to another area, and this practice should be very carefully considered.

In any kind of discharge, and especially in the case of sewage, the characteristics of both the effluent and the receiving water must be taken into account. As noted above, the characteristics of sewage are frequently unknown, even under the best of conditions, and so, obviously, the resulting reactions are also unknown. As regards the receiving water, the characteristics that must be taken into account in order to calculate such factors as diffusion rates and dilution volume are:

1. Water movement relating to both direction and speed
2. Total water transport
3. Mixing rate and depth
4. Diffusion rate of specific constituents
5. Decay rate of the pollutant in the specific water in question

If these are determined, a better appreciation of the possible effects of a sewage discharge can be obtained.

8.2. *Oil*

Oil, like sewage, contains various organic and inorganic fractions, is characteristic of specific areas, and has multiple properties. In excess, it causes considerable damage to any area, but only the short-term visual effects are generally considered. As with sewage, even though oil is a mixture of chemicals, chemists and biochemists usually play the role of spectators when these polluting effects are considered. Unlike sewage, however, the introduction of oil into the hydrosphere is for the most part accidental rather than deliberate. An accidental spillage can introduce a massive overloading in a specific area, and accounts for the major series of problems encountered with this type of pollution. These may occur with accidents in shipping, such as the "Torrey Canyon disaster," and to shore facilities, underwater pipelines, and offshore oil wells, such as the "Santa Barbara leak." Deliberate discharges, on the other hand, are usually more constant, but contribute only a small percentage of the overall total to a system. These will occur when ships flush and clean fuel and oil tanks, as wastes from refineries, and with sewage discharges containing industrial wastes.

A large proportion of the above generally involves coastal systems, so that major problems seem to arise most frequently in bays, harbors, estuaries, and in the restricted coastal system itself. The high proportion of aquatic organisms spawning in these areas adds to the problem, so that the effect of oil in the hydrosphere can be magnified many times over. If only the visual effect is considered, as is mostly the case, the problem may be considered solved as soon as the oil is cleaned up or dispersed. This is analogous to cleaning the blood off an accident victim and declaring him cured. But the long-term effects, involving the reactions of various fractions in the water, can be the most disastrous and must at all times be a part of any investigation related to oil spillages. In the past, oil was frequently dispersed by sinking with carbonized sand or by adding various dispersing agents which broke the oil down into minute drops. This of course removed the visual problem but only postponed the real problem to a later time and possibly a different place. In many cases problems were magnified because the dispersant was frequently more toxic than the oil itself. But, in any event, it has been found that dispersed oil is more toxic to aquatic life than oil present as a slick on the surface.

All this seems to indicate that the emphasis placed on the physical clean-up of oil is wrong, but this is not the case. The oil must be removed from the system as quickly as possible, but the operative word is "removed," not "dispersed." The visual effect acts as an indicator of a problem which is not necessarily solved simply because the indicating effect is removed. It must always be remembered that oil is a mixture of chemicals and will react characteristically as long as any part of a spill remains in the system.

Some of the fractions contained in oil are toxic in themselves and their toxicity increases with concentration as these fractions are absorbed or dissolved in the water system. Low-boiling saturated hydrocarbons and aromatic compounds such as benzene and xylene are toxic and show varying degrees of water solubility. Some carcinogenic compounds are present in the higher-boiling fractions and appear to be related to the polycyclic compounds. The oil itself as a mixture is also toxic, but there is still only a small amount of data on the toxic effects of ingested oil. The oil will also emulsify, forming various types and concentrations of emulsions which can be toxic and can physically affect the aquatic organisms by suffocation.

As an oil slick spreads out on the surface of the water it forms a multimolecular layer which can cover very large areas. For example, 15 tons of fuel oil in a period of six to seven days will spread out to cover an area of at least eight square miles. This results in a number of effects related to reactions involving surface phenomena. It also presents a large surface area for reactions such as the solution of the water-soluble fraction found in any type of oil. As the slick expands under the influence of wind and water movements, it reacts with and forms a quasi-equilibrium with the naturally occurring surface film. The surface-active organics react and many lipophilic compounds in the hydrocarbon fractions are concentrated and stabilized. These have a direct effect on the biological and biochemical reactions within the film, and as such are reflected in the biological life in the area. The physical characteristics of such a combination include the effective thickening of the surface film, which will affect the penetration of light, inhibiting photosynthesis, and which will also seriously reduce the rate and total transport of oxygen and carbon dioxide through the film. If the slick is small in relation to the area of the water surface the oil will eventually degrade, forming tar residues and emulsions that will continue to make their presence felt.

As the oil degrades, a number of chemical reactions are involved that release some of the fractions to the water *in toto*, and which are very toxic, as mentioned above. In addition to this, a further reaction involving the actual degradation process itself becomes evident. Various microorganisms such as bacteria are needed to degrade the oil, and the rate of degradation depends on the numbers and types of bacteria present, the availability of oxygen, and the temperature. The oxidation of the oil generally proceeds more slowly in cold water, but the overall effect in any water mass is the requirement and removal of oxygen. This requirement can reach considerable proportions if the oil is allowed sufficient time to be completely oxidized. Zobell,[42] for example, has calculated that the ratio of oxygen requirement to oil volume in seawater is approximately 400,000 : 1. This means that for every gallon of oil, the total oxygen in 400,000 gallons of seawater is needed for complete oxidation. Corresponding calculations for fresh and estuarine waters depend to a large extent on locale, temperature, and other specific factors, but the resulting figures remain relatively comparable. At the same time, the rate of oxygen uptake in the surface water is reduced by the presence of the thickened surface film which tends to slow down the rate of oxidation to some extent. Other factors such as nutrient requirements and temperature will also limit the rate of oxidation. In this way, it is possible for the oil to remain in the system and to react for long periods of time. In bays and estuaries, for example, a large proportion of the oil would eventually sink to the bottom, react with the sediment, and reduce the available oxygen in the vicinity. This would reduce the viability of the system and would also reduce the rate of oxidation of the oil so that it would again remain in the system longer. In extreme cases it may be possible for a water system to become saturated with oil and to "die," remaining in this condition for some considerable time.

Another characteristic of oil pollution is the capacity of oil to attract and concentrate items such as heavy metals and pesticides such as DDT. When the oil is dispersed or spread out over a large area, the available surface area of the oil is greatly increased, which presents a very large number of reactive sites for a considerable number of reactions. Oil-soluble materials thus have ample opportunity for reaction and concentration. In the case of slick formation, the concentration is on the surface and eventually in the surface film. Dieldrin, for example, has been found to be concentrated by a factor of 10,000 in the film of a natural slick in Biscayne Bay, Florida (Seba and Cochran[34]). The concentration of metals, besides varying their toxicity, has the added effect of inducing metal organic reactions in the surface film which further complicates the molecular transport properties of the film. The reactions and concentration features of the oil slick can also short-circuit the concentration phenomenon of the biological food chain by simulating the effect of some of the lower orders of organisms. There can thus be a quicker introduction of contaminants into edible sea foods.

If the physical attributes of oil pollution, such as direct ingestion by oysters, for example, are added to the above, it can be appreciated that the problems caused by oil in the hydrosphere are frequently much greater and longer lasting than is often postulated. If sewage effects are also taken into account, areas exposed to both problems can be rendered virtually untenable for any kind of aquatic organism. Since both sewage and oil share the characteristic of being chemically unknown or underated, the long-term chemical and biochemical effects of such a combination enter the realm of uneducated speculation, and no answers are yet possible.

8.3. *Synthetic Compounds*

Pesticides, herbicides, and related compounds are among the world's most widely distributed synthetic organic chemicals. They have been identified in all phases of the biosphere in significant concentrations and their numerous biological effects have been documented. These effects cover a wide range of organisms in addition to the primary targets, and were first brought to public attention by Rachel Carson in her book *Silent Spring*. At the time, the chlorinated hydrocarbons such as DDT, lindane, aldrin, and dieldrin were in common use, although little was known about their chemical behavior in the environment.

It was known that insecticides sprayed from the air could be carried over large distances by the prevailing winds and air currents, but transport from the land itself was generally not considered. It was appreciated that some runoff could occur after rain or irrigation, but the scope of this means of transport was grossly underestimated and at times completely ignored. There was very little consideration given to the residence times of these compounds either in the soil or in adjacent water systems, and frequently whole areas were saturated with the belief that if a little is good more must be better. All of this resulted in the biosphere as a whole becoming contaminated, with interchanges of the compounds between all sections. These interchanges occurred at all boundaries, but the rates and mechanisms of the transfers were not then, and still are not fully appreciated or completely understood.

In more recent times, these insecticides and their physical effects have been scrutinized relatively carefully with the result that a number of them, such as DDT, have been banned from use in various areas. In most cases these were nondegradable, long-persisting, broad-spectrum poisons which needed to be replaced by degradable, short-term, specific poisons. These replacements, however, usually in the form of organophosphorus compounds such as parathion and malathion, are unfortunately many times more toxic, particularly to the operator who may have to come in contact with them. Parathion, for example, is about 30 times more toxic than arsenic. TEPP, one of the original "OP" pesticides, is so toxic that a lethal dose is cataloged as one drop taken orally or in contact with the skin. But the features that are as important as these lethal levels are the continuous exposure, long-term effects and the reactions which they undergo in the environment.

Even now, after more than two decades of use, the chemical reactions involving the pesticides in general use are still virtually unknown. It is appreciated, because of obvious physical evidence, that pesticides as a class are important water pollutants that affect the biota as a whole. But there is still very little information concerning their persistence, their breakdown products, and their reactions in viable water systems. It is known that some pesticides are lipophilic and will absorb on organic matter. They also destroy the larval stages of numerous organisms in the food chain, reduce some photosynthetic activity of phytoplankton, and reduce the eggshell thickness of fish-eating birds. These are all physical symptoms of pesticide pollution, however, and are the end products of chemical reactions and assimilation by the biotic community. Some pesticides have been investigated, but it is impossible to extrapolate conditions and reactions to other compounds in the same or different classes because the physical and chemical properties of these compounds vary dramatically. In the same way, breakdown products and their reactions are also impossible to extrapolate; each compound must be looked at separately.

In order to fully understand the role that a pesticide or any other synthetic organic chemical plays in the environment, a number of features must be known. These include the mode or mechanism of transfer from one system to another, the reactions of parent and degradative products, and the conditions under which these reactions will either proceed or be retarded. It must also be known whether or not other constituents in the system will act either as catalysts for specific reactions or as retarding agents. On the whole, pesticide chemistry that specifically involves reactions in the environment is still in its infancy, and, along with other areas discussed in this chapter, needs to be encouraged.

8.3.1. Chlorinated Hydrocarbons

Of all the synthetic organic compounds now found in the environment, the one that has received the most attention in terms of research and emotional stress is DDT [1,1,1-trichloro-2,2-bis(p-chlorophenyl)ethane]. It has a half-life of several years, resists chemical breakdown, and has been found in all sections of the biosphere and in all parts of the world. Penguins in the Antarctic, for example, have been found to have relatively high levels of it in their fatty tissues. It is known to concentrate in the food chain by factors in excess of 500,000. A simple comparison of its concentrations in water (5×10^{-5} ppm), in fish (2.0 ppm), and in fish-eating birds such as the cormorant (27.0 ppm), illustrates the concentration phenomenon and shows that there are really no safe levels. As long as DDT is capable of being stored and concentrated in the fatty tissues of numerous organisms, including man, it will continue to be a hazard and must be treated as such.

The organochlorines are normally strongly absorbed on particulate matter and in the soil, and in the hydrosphere this involves particulate organic and inorganic materials and the sediment. Because of this, some properties can be changed, depending on local conditions, such as the concentration of particulates and the structure of the sediment. Another factor that has a marked effect on these compounds, with specific regard to solubility phenomena, is the salt content of the waterway in question. All of these will help to determine the properties of the compounds in the selected environment and will control to a large extent the residence time of each compound. For example, in a vigorous biological community with ample dissolved oxygen and bacteria the organochlorines could be expected to be transferred from one sphere to another much faster than would be possible in a sterile environment.

All of these compounds, including DDT, lindane, aldrin, endrin, and dieldrin, to name a few, are sparingly soluble in water but show a definite characteristic, referred to as "salting out." This means that they are less soluble in salt water than in fresh water. Thus, in estuaries with varying salt concentrations, complex solubility equilibria are associated with the normal characteristics of these compounds in solution. Under these circumstances the nonsoluble fraction will be distributed further in the sediment and in the surface film, depending on local conditions. In areas where oil is a hazard the concentration effect in both the film and the sediment can be significant. Natural oils and fats inherent in sewage wastes will also dissolve and concentrate these compounds, leading to remarkable concentration factors.

Both in the water and in the sediment the degradative products of the chlorinated hydrocarbons are similar, but the metabolic pathways are possibly different. Aldrin, for example, is oxidized very slowly to dieldrin in both areas, resulting in a buildup of the dieldrin because it is much more stable than aldrin. In the case of DDT, a number of degradative products have been identified, the percentages of which depend on local conditions. In sediments, for example, the relative concentrations of the products DDD and DDE depend on the oxidation–reduction state of the sediment in question.

There also appears to be a change in percentages with depth of sediment, favoring the formation of DDE (Leland *et al.*[51]). Other products are formed, such as DDNS and DDOH, but the two diagrammed above are the major ones.

In this case there seems to be little difference between DDT and its products since all are relatively stable and have long residence times. Sediments can thus act as reservoirs and concentration media, for the above example in particular, and indeed for most other chlorinated hydrocarbons. The polychlorinated biphenyls (PCB), for example, which are used extensively in industry, are, in a broad sense, structurally similar to and more stable than DDT. They are also biologically and genetically active, but their degradation products are for the moment relatively unknown. This would indicate that in the event of a total ban on all chlorinated hydrocarbons, with no more entering the environment, there still remains a considerable backlog of these compounds which will remain active for some time to come.

In the hydrosphere itself, because of solubility factors, the concentrations of the various compounds are relatively small, but specific reactions are still inherent in the system. These reactions depend on conditions and are demonstrably different in rate and possibly in mechanism in the surface film, in fresh river water, in estuarine water, and in salt water. The transformations in each medium are poorly understood, although more insight is being gained.

8.3.2. Organophosphorus Compounds

These compounds are less stable and somewhat more specific than the organochlorines and do not normally suffer from the liability of concentration and storage in the physical environment. But the question of their possible concentration in the biological sphere still needs to be answered even though these compounds are extremely toxic. In most cases, such as parathion and malathion, the degradation is complete in a relatively short time, but there are some exceptions. Azodrin, for example, is possibly as stable as DDT or DDE, and others show degradation rates between these two extremes. The mechanisms of the degradation process are still virtually unknown, but there is evidence (Eichelberger[35]) that the reactions are biologically induced.

From the above it would seem that this class of compounds is the answer to the problem of synthetic compounds in the environment. Compounds can be chosen that are effective, and when they are removed to another sphere, such as a water system, their life expectancy is short enough so that there is no build-up of concentration. But this presupposes that the degradation products are known and are harmless; this is not the case. Parathion, for example, degrades to the corresponding thiophosphoric acid and p-nitrophenol, both of which can have serious effects on the system in question, and, depending on conditions, can have longer residence times. The degradation products of some other compounds have not yet been identified so that predictions of their effects are not possible. An understanding of transport phenomena and reaction mechanisms between the media is also lacking and will apparently remain so for some time to come. In nearly all cases, because of the need to replace the chlorinated hydrocarbons as pesticides, these compounds may be used with little consideration given to their effects in the water systems of the world. And until a proper understanding is obtained of the role played by all synthetic compounds various disasters must be expected.

9. Summary

The question of whether or not a body of water is polluted depends completely on an appreciation of the effects of added materials on the particular system. In the short term, these effects may be visually quite obvious, as with oil pollution or the mass destruction of aquatic life brought on by any number of factors. Or they may be invisible but capable of being determined physically or chemically, as with pH, temperature, and dissolved oxygen variations. In the long term, however, the effects are sublethal, not immediately apparent, and normally cumulative. These effects can be brought about by the periodic introduction of materials with long residence times or by the continuous introduction of both stable and unstable materials, and are a function of the chemical reactivities of these materials. The induced chemical and biochemical reactions involve the physical as well as the biological environment, and are the main basis for the continuation of a viable community.

It is an unfortunate fact that most of the reactions involved in both normal and polluted environments are largely unknown. So the question of what constitutes a dangerous concentration of any added material is usually based on an educated guess, or more frequently on an uneducated guess. In this way man has returned to an early primitive state and is once again at the mercy of the elements. Very frequently these dangerous concentrations are given as maximum allowable levels which allow for no variations of species or within a species. Sewage, for example, supplies nutrients to a system, but not all organisms can benefit from them. In this way, some organisms produce at the expense of others and the ecological balance is upset to a relative degree. Within a species, ranging from infant to adult, nutrient requirements change, as reflected in different degrees of tolerance to changes from the norm. It is thus possible for only adults to survive a certain level of pollutant, which in time effectively wipes out the species. The presence of adult fish in a water system is therefore not necessarily an indication that the water is suitable for aquatic life.

If effective threshold levels of any contaminants are to be accurately determined, it must be remembered that each compound and species plays a particular role in the overall ecological structure, and it is this role that must be determined. This includes such features as biological uptake, competition for chemical and biochemical reactions that are controlled by reaction rates and mechanisms, and competition for essential biological materials such as oxygen. In all facets, however, the emphasis is on a chemical reaction, whether it be the requirement of oxygen for simple oxidation or the coordination reaction between a metal and ligands to form complexes of varying stability and biological activity. With this in mind, it is relatively easy to assess the current state of knowledge concerning the factors that control the biosphere as a whole and the hydrosphere in particular.

As is pointed out in this chapter, we know very little about our environment and what makes it tick. The interrelationships between the physical, chemical, and biological spheres are only slowly being understood and there is still a long way to go. One wonders if the environment will be understood before it collapses, or whether or not predicted catastrophes can be postponed or, better still, prevented. We know so little about what we do that long-range forecasts are dramatically susceptible to the state of our knowledge, and it is hoped that as this knowledge is increased more optimistic views can be arrived at.

I believe that many problems can be prevented if a few simple questions are asked about a discharge before it is released into a water system.

1. *Where?* Very simply, this refers to the question of where the given pollutants will eventually end up. Tides, currents, water motions, and biological movements will transport those pollutants in distinctive distribution patterns and it is essential that these patterns be thoroughly understood.

2. *When?* This refers to a time-scale factor and concerns the length of time required for a body of water to attain an equilibrium, or a steady-state, condition. It also refers to the time necessary for maximum dispersion, distribution, and biological concentration. It may also be related to reaction rates and residence times of specific compounds.

3. *What?* This question refers to what is actually happening or will happen chemically in the environment due to a specific discharge. It is thus imperative to know the species and concentrations present in the effluent and in the receiving water, and the reactions involved in the effluent, the receiving water, and the resulting mix. If these are known, the overall effects of the discharge on the local ecological structure and on the biological community can be effectively predicted.

The current state of knowledge about the hydrosphere precludes comprehensive answers to these questions at the moment, but many answers can and should be determined. It at least some of these answers are required prior to giving permission for a discharge, the environment as a whole can be assisted and used intelligently in what is effectively a vast recycling scheme for acceptable waste materials.

References

1. B. H. Ketchum, "Man's Resources in the Marine Environment," in *Pollution and Marine Ecology*, T. A. Olsen and F. J. Burgess, eds., Interscience Publ., 1967, p. 4.
2. R. A. Horne, *Marine Chemistry*, John Wiley & Sons, New York, 1969, p. 7.
3. E. D. Goldberg, "The Oceans as a Chemical System," in *The Sea*, Vol. 2, M. N. Hill, ed., Interscience, New York, 1963, Ch. 1, pp. 4–5.
4. E. D. Goldberg, "Minor Elements in Sea Water," in *Chemical Oceanography*, J. P. Riley and G. Skirrow, eds., Academic Press, London, 1965, Ch. 5.
5. J. Vernberg, "Temperature Effects on Invertebrate Animals," in *Temperature, its Measurement and Control in Science and Industry*, Vol. 3, Reinhold, New York, 1963, Ch. 13, pp. 135–141.
6. H. V. Sverdrup, M. W. Johnson, and R. H. Fleming, *The Oceans*, Prentice Hall, New York, 1946.
7. G. E. Hutchinson, "A Treatise in Limnology," in *Geography, Physics and Chemistry*, Vol. 1, John Wiley & Sons, New York, 1957.
8. D. W. Pritchard, "Estuarine Circulation Patterns," *Proc. Amer. Sci. Civil Engrs.*, **81** : 717 (1955). Also, D. W. Pritchard, "Estuaries," *Amer. Assoc. Adv. Sci. Publ. No. 83*, G. H. Lauff, ed., Washington, D.C., pp. 3–5.
9. H. Stommel, "Computation of Pollution in a Vertically Mixed Estuary," Sewage and Industrial Wastes, **25** : 1065–1071 (1953).
10. D. W. Pritchard, "A Study of the Salt Balance in a Coastal Plain Estuary," *J. Mar. Res.*, **13** : 133–144 (1954).
11. W. S Preddy, "The Mixing and Movement of Water in the Estuary of the Thames," *J. Mar. Biol. Assoc. U.K.*, **33** : 645–662 (1954).
12. B. H. Ketchum, "Pollution of Estuaries and Coastal Waters" in *Man's Impact on Terrestrial and Oceanic Ecosystems*, W. H. Matthews, F. E. Smith, and E. D. Goldberg, eds., M.I.T. Press, 1971, p. 265.
13. G. Dietrich, *General Oceanography*, Interscience Publishers, John Wiley & Sons, New York, 1963.

14. R. A. Horne, *Marine Chemistry*, John Wiley & Sons, New York, 1969, Ch. 5, pp. 154–155.
15. F. MacIntyre, *Sci. Am.*, **230**(5) : 62–79 (1974).
16. W. D. Garrett, *Deep Sea Res.*, **14** : 221 (1967).
17. P. M. Williams, *Deep Sea Res.*, **14** : 791 (1967).
18. R. A. Duce, J. G. Quinn, C. E. Olney, S. R. Piotrowicz, B. J. Ray, and T. L. Wade, *Science*, **176** : 161–163 (1972).
19. P. Debye and E. Huckel, *Physik. Z.*, **24** : 185 (1923).
20. D. H. Klein and E. D. Goldberg, *Env. Sci. Tech.*, **4,9** : 765–767 (1970).
21. J. O'Hara, *J. Fish. Res. Board Can.*, **30** : 846–848 (1973).
22. H. Postma, "Marine Pollution and Sedimentology" in *Pollution and Marine Ecology*, T. A. Olsen and F. J. Burgess, eds., Interscience Publ., 1967, pp. 225–234.
23. G. K. Pagenkopf, R. C. Russo, and R. V. Thurston, *J. Fish. Res. Board Can.*, **31**(4) : 462–465 (1974).
24. S. E. Manahan and M. J. Smith, *Env. Sci. Tech.*, **7**(9) : 829–833 (1973).
25. T. L. Theis and P. C. Singer, *Env. Sci. Tech.*, **8**(6) : 569–573 (1974).
26. M. E. Bender, W. R. Matson, and R. A. Jordan, *Env. Sci. Tech.*, **4**(6) : 520–521 (1970).
27. C. G. Wilber, *The Biological Aspects of Water Pollution*, C. T. Thomas Publ., 1969, p. 70.
28. J. S. Gray and R. J. Ventilla, *Ambio*, **2**(4) : 118–121 (1973).
29. E. J. Green and D. E. Carritt, *J. Mar. Res.*, **25**(2) : 140–147 (1967).
30. L. Landner and T. Larsson, *Ambio*, **2**(5) : 154–157 (1973).
31. D. W. Schindler, H. Kling, R. V. Schmidt, J. Prokopowich, V. E. Frost, R. A. Reid, and M. Capel, *J. Fish. Research Bd. Can.*, **30**(10) : 1415–1440 (1973).
32. J. B. Upchurch, J. K. Edzwald, and C. R. O'Melia, *Env. Sci. Tech.*, **8**(1) : 56–58 (1974).
33. J. D. Williams, J. K. Syers, R. F. Harris, and D. E. Armstrong, *Env. Sci. Tech.*, **4**(6) : 517–519 (1970).
34. D. Seba and E. Cochran, *Pesticide Monitoring Journal*, December, 1969.
35. J. W. Eichelberger and J. J. Lichtenberg, *Env. Sci. Tech.*, **5**(6) : 541–544 (1971).
36. J. L. Brooks, "Speciation in Ancient Lakes," *Quart. Rev. Biol.*, **25** : 36–60 (1950).
37. B. H. Ketchum, "The Exchanges of Fresh and Salt Waters in Tidal Estuaries," *J. Mar. Res.*, **10** : 18–38 (1951).
38. E. D. Goldberg, G. D. Nicholls, H. Curl, and V. T. Bowen, *Limnol. and Oceanog.*, **5** : 472 (1960).
39. M. Rebhun and J. Manka, *Env. Sci. Tech.*, **5**(7) : 606–609 (1971).
40. P. E. Kolattukudy and R. E. Purdy, *Env. Sci. Tech.*, **7**(7) : 619–622 (1973).
41. L. L. Wallen and E. N. Davis, *Env. Sci. Tech.*, **6** : 161–164 (1972).
42. C. E. Zobell, *Adv. in Water Poll. Res.*, **3** : 85–109 (1964).
43. J. E. McKee, "Parameters of Marine Pollution" in *Pollution and Marine Ecology*, T. A. Olsen and F. J. Burgess, eds., Interscience Publ., 1967, pp. 259–266.
44. E. D. Goldberg, "Chemical Invasion of the Ocean by Man" in *Man's Impact on Terrestrial and Oceanic Ecosystems*, W. H. Matthews, F. E. Smith, and E. D. Goldberg, eds., M.I.T. Press, 1971, p. 265.
45. R. A. Horne, *Marine Chemistry*, John Wiley and Sons, New York, 1969, p. 208.
46. C. W. Holmes, E. A. Slade, and C. J. McLerran, *Env. Sci. Tech.*, **8**(3) : 255–259 (1974).
47. R. J. Gibbs, *Science*, **180** : 71–73 (1973).
48. M. R. Moore, *Nature*, **243** : 222–223 (1973).
49. L. G. Sillén, "The Physical Chemistry of Sea Water" in *Oceanography*, M. Sears, ed., Amer. Assoc. Advan. Sci., Publ. No. 67 (1961).
50. D. R. Kester and R. H. Byrne, "Chemical Forms of Iron in Seawater" in *Ferromanganese Deposits on the Ocean Floor*, D. R. Horn, ed., Lamont-Doherty Geological Observatory, Columbia University, Palisades, N.Y., pp. 107–116 (1972).
51. H. V. Leland, W. N. Bruce, and N. F. Shimp, *Env. Sci. Tech.*, **7**(9) : 833–838 (1973).

The development of many aspects of organochemical technology since the 1930s has been accomplished on the simple basis of the "improvement of life" without, unfortunately, considering the balancing decrease of the quality of life due to the rejected materials used in improving it. Among these are the effects of pesticides and plastics, and one of the most famous books in the subject of environmental science, Rachel Carson's Silent Spring, *about the effects of pesticides on the lives of birds, was a clarion bell in the early days of considerations of pollution. The organochemical aspects of pollution are visible in our daily environment, in the fouling of our lakes and streams and in the detergent flecks which now often dot our natural waters.*

13

Organic Chemical Pollution: Petroleum, Pesticides, and Detergents

H. Stephen Stoker and Spencer L. Seager

1. Introduction

The worldwide production and use of organic chemicals (both natural and synthetic) have increased dramatically since World War II. Much of the increase is the result of new compounds developed for consumer use rather than of an increased use of a few known compounds. The environmental impact of using these compounds has only slowly become apparent, and in many cases the findings have caused serious concern.

The environmental problems attributed to the use of petroleum hydrocarbons, synthetic pesticides, and detergents have received a great deal of attention, and will be the topic of this chapter. The chemical composition, sources, environmental fate, and effects of these pollutants will be discussed.

H. Stephen Stoker and Spencer L. Seager • Department of Chemistry, Weber State College, Ogden, Utah, 84408

2. *Petroleum*

Crude petroleum was first extracted from the earth in significant amounts in 1880. Production since that time has increased at a nearly constant exponential rate and on a worldwide basis now exceeds 20 billion barrels annually.[1] Refined petroleum products currently satisfy more than 60% of the world's energy requirements. It is almost impossible to use a material on such a large scale without experiencing some losses. The extent of such losses, whether intentional or accidental, is steadily increasing, and pollution of the marine environment with both crude oil and refined products is now a matter of serious concern.

2.1. *Chemical Composition of Petroleum*

A consideration of the chemical composition of petroleum is a logical starting point for a discussion of environmental effects, since both the fate of petroleum in the environment and the resulting effects on the biota are dependent on the nature of the individual chemical compounds involved.

Crude petroleum is a complex chemical mixture that contains hundreds of compounds. This chemical complexity is consistent with the accepted origin of petroleum. It is believed that petroleum was formed as the result of long-term bacterial, chemical, and geologic action on organic debris of both plant and animal origin. It is reasonable to expect the resulting petroleum to inherit, at least partially, the complex chemical nature of the parent materials.

More than 75% of the total petroleum composition is made up of hydrocarbons, compounds containing only carbon and hydrogen. The most common elements found in addition to carbon and hydrogen are sulfur, nitrogen, and oxygen. The concentration of these elements ranges from traces to about 4% sulfur, 1% nitrogen, and lesser amounts of oxygen.[2] These additional elements are usually found incorporated into hydrocarbon-like molecules rather than in the free state.

The primary difference between petroleum from different geographical areas is usually not in the compounds making up the mixture, but rather in the percentage of each compound found. Such compositional differences influence both the physical and chemical characteristics of crude oil. Some oils are almost colorless, while others are pitch black, amber, brown, or green in color. Some have pleasant odors, described as ethereal, while others smell sweet like turpentine or camphor. Still others have very unpleasant odors, usually caused by the presence of certain sulfur-containing compounds. The biological and chemical effects of the different hydrocarbons vary significantly, and it is therefore essential to consider the composition of each particular crude oil when its environmental impact is being estimated.

Petroleum composition is usually described in terms of the amounts of contained hydrocarbons which are classified into the following categories: (1) paraffins, (2) cycloparaffins, (3) aromatics, and (4) naphtheno-aromatics.

Paraffins, or alkanes, usually comprise up to 25% of the crude petroleum; they are predominantly found in the low boiling (40–230°C) fractions. The paraffins are often classified into two subclasses: normal paraffins and isoparaffins. The former contain linear chains of carbon atoms, while the latter have one or more carbon branches attached to a linear chain. Representative examples of paraffin hydrocarbons are given in Fig. 1.

The normal paraffins (*n*-paraffins) found and identified in various crude oils include methane (CH_4) and substances as complex as *n*-tritetracontane, which contains 43 carbon atoms.[3] There is considerable variety in the paraffin content of crude petroleums. Some contain large quantities of *n*-paraffins, but relatively small quantities of isoalkanes, while others contain only traces of the *n*-paraffins.

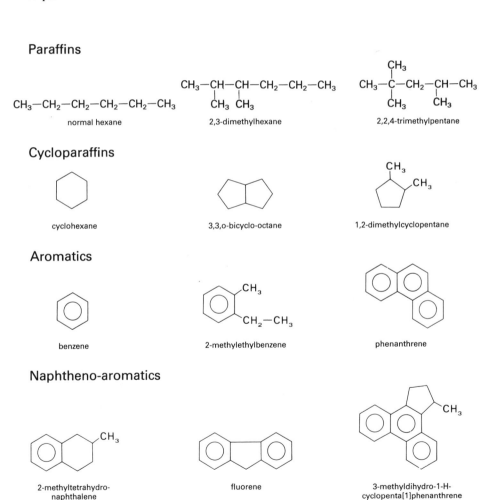

Figure 1. Common types of hydrocarbons found in petroleum.

The prevalent isoparaffins of crude petroleum are generally those with the fewest and simplest branches. Thus, more are found that contain one to four methyl groups than those with larger groups such as ethyl, propyl, and butyl groups. All isoparaffins with eight or fewer carbon atoms have been identified in petroleum, as well as 30 of the 35 C_9 isomers.[3] Much research remains to be done in order to identify isoparaffins containing 10 or more carbon atoms. These compounds will be found in the higher-boiling fractions of petroleum which have yet to be studied in as much detail as have the lower-boiling fractions.

Cycloparaffins, also called cycloalkanes or naphthenes, are ring compounds with structures represented by the examples given in Fig. 1. These simple cyclic hydrocarbons make up 30–60% of the total composition of crude petroleum. Most petroleum cycloparaffins identified at this time are monocyclic, that is, they contain one ring. However, compounds containing up to six rings are not unusual and even ten-ring examples are found in high-boiling fractions. The two most commonly found cycloparaffin ring systems are those involving 5 and 6 carbon atoms: cyclopentane and cyclohexane. Most cycloparaffins have alkyl side-chains attached to the ring.

Aromatic hydrocarbons, the third major class of petroleum-contained compounds, are also cyclic in structure, but their properties are quite different than those of cycloparaffins. The differences result from bonding characteristics which make possible the presence of delocalized electrons in aromatic rings. Benzene, the simplest aromatic hydrocarbon, and its derivatives are the predominant aromatic compounds found in low-boiling petroleum fractions. The higher-boiling fractions contain polycyclic aromatics with structures represented by fusing two or more benzene rings. Aromatic hydrocarbons are less prevalent than either paraffins or cycloparaffins in most petroleums.

A large percentage of the hydrocarbon molecules found in high-boiling petroleum fractions (above 300°C) contain structures with the combined characteristics of paraffins, cycloparaffins, and aromatics. These compounds, known as naphtheno-aromatics or cycloalkanoaromatics, generally have structures in which aromatic and cycloparaffin rings are fused. Paraffins are usually present as alkyl branches on the rings (see Fig. 1). Investigations concerning these compounds have been limited by the high boiling points involved and associated analytical difficulties.

One recent estimate of the overall composition of crude petroleum in terms of the four hydrocarbon classes discussed above is given in Table 1.[4] Some of the classes have been subdivided on the basis of the number of carbon atoms. In addition, boiling point ranges and solubility data are given. These two parameters are important in understanding the environmental fate of oil.

The residual fraction, listed at the bottom of Table 1, contains high-boiling hydrocarbons of all types. Even though their composition is not completely known the following facts have been determined: (1) they contain oxygen, sulfur, nitrogen, and trace metals; (2) their molecular structures consist of layers of condensed heterocyclic rings connected by short *n*-paraffin chains;

Table 1. Oil Composition Model[a]

Fraction description	Crude oil (% by weight)	Boiling point range (°C)	Distilled water solubility range ppm (by weight)
Paraffins			
C_6–C_{12}	0.1–20	69–230	9.5–0.1
C_{13}–C_{25}	0^+–10	230–450	0.01–0.004
Cycloparaffins			
C_6–C_{12}	5–30	70–230	55–1.0
C_{13}–C_{23}	5–30	230–405	1.0–0
Aromatic			
Mono- and dicyclic C_6–C_{11}	0–5	80–240	1780–0
Poly-cyclic C_{12}–C_{18}	0^+–5	240–400	12.5–0
Naphthenoaromatic C_9–C_{25}	5–30	180–400	1.0–0
Residual			
Includes heterocycles	10–70	>400	0

[a]Source: S. F. Moore, R. L. Dwyer, and A. M. Katz.[4]

and (3) the heterocyclic rings are of both the cycloparaffin and aromatic types.

2.2. Environmental Sources of Petroleum Hydrocarbons

During the last few years a great deal of attention has been attracted by oil pollution of the oceans resulting from oil tanker accidents or offshore drilling "blowouts." Such events, although serious, represent the source of only a small percentage of the total petroleum hydrocarbons that enter the marine environment. Most petroleum pollution of the oceans does not result from accidents or from sources large enough to attract much publicity. Various reports place the total annual influx of petroleum hydrocarbons into the oceanic environment at 5–10 million metric tons (1 metric ton = 2200 lb). Detailed statistics from one such report, that of the National Academy of Sciences, Washington, D.C., are given in Table 2.[5]

The data in Table 2 indicate that the largest influx comes from transportation-related activities. This is not surprising, since most crude oil production occurs at some distance from processing and marketing areas and consequently must be transported in ocean-going tankers. "Normal" tanker operations, rather than accidents, account for almost all transportation-related oil losses. Such normal operations include ballasting and tank cleaning (deballasting), dry docking, and terminal activities.

Over half (51%) of the petroleum losses from normal transportation operations involve ballasting and tank cleaning. A tanker, after discharging its

Table 2. Estimated Amounts of Petroleum Hydrocarbons Entering the Oceans from Various Sources[a]

Source	Metric tons/year	Percent of total
Transportation	2.13	34.9
(Normal operations)	(1.83)	(30.0)
(Accidents)	(0.3)	(4.9)
River and urban runoff	1.9	31.1
Atmospheric fallout	0.6	9.8
Natural seeps	0.6	9.8
Industrial wastes	0.3	4.9
Municipal wastes	0.3	4.9
Coastal refineries	0.2	3.3
Offshore production	0.08	1.3
(Normal operations)	(0.02)	(0.3)
(Accidents)	(0.06)	(1.0)
Total	6.11	100

[a]Source: *Petroleum in the Marine Environment.*[5]

cargo, takes seawater into its empty tanks to serve as stabilizing ballast for the return voyage. The seawater forms an emulsion with residual oil in the tanks. The oil-containing ballast is discharged into the sea when the tanker nears its destination. Other types of vessels also add to the deballasting problem by filling empty fuel tanks with seawater ballast which is subsequently put back into the oceans.

It should be noted that 80% of the world's tanker fleet uses a control measure called LOT (load on top) to reduce the amount of oil discharged during deballasting. As a result, more than 70% of the deballasting oil pollution comes from the 20% of the tankers which do not use the LOT technique.[5] In the LOT method, water and oil ballast is allowed to settle in the tank during the return voyage. The less-dense oil rises to the top. The relatively clean seawater on the bottom is then discharged. The oil, mixed with a little seawater, remains in the tanks and the next cargo is loaded on top. Except for a few special uses in which crude oil cannot contain water contamination, the LOT technique can be adopted without major tanker modifications. Most refineries accept crude oil which contains seawater; in fact, some crude oils naturally contain salt water.

Accidental oil spills resulting from the stranding or collision of large tankers are not common, but do occur periodically. The best-known tanker accident took place in 1967, when the *Torrey Canyon* ran aground off Cornwall, England. The eventual break-up of the ship resulted in the release of most of the cargo of 118,000 tons of crude petroleum. Events such as these, although serious, represent the source of less than 15% of transportation-related oil losses. However, the importance of such accidents should not be minimized; the resulting spills are small on a percentage basis but they take place along a few well-defined shipping lanes or in relatively shallow offshore areas. Their effects are therefore concentrated in small areas of the marine environment.

The environmental impact of large-scale accidents increases with the size of tankers involved. The use of increasing large tankers, especially those known as "super-tankers," has been a controversial issue. Vessels of 500,000 dead weight tons will soon be in operation, and those with 800,000-ton capacities have been projected for the next few years. Figure 2 shows a comparison between such super-tankers and previously used vessels.[6]

Compared with smaller ships, super-tankers are difficult to maneuver and require much longer stopping distances because of draft and inertia. For example, a 200,000-ton tanker requires at least $2\frac{1}{2}$ mi in which to stop, even with the engines running in full reverse.[7] While stopping under such emergency conditions, the tanker becomes difficult to steer. The loss of only a single 200–500 thousand ton tanker under conditions which prevented unloading the cargo could add 3–8% to the total amount of petroleum discharged directly into the oceans per year (see Table 2).

However, it should be pointed out that handling difficulties do not increase in a direct proportion to tanker size. Furthermore, the use of larger ships decreases the number of ships needed, and traffic can be reduced considerably. Also, since fewer ships are required the crews can be composed of only the best-trained personnel available. The larger tankers could also afford to install highly sophisticated navigation gear which might be prohibitively expensive for the many smaller ships.[8]

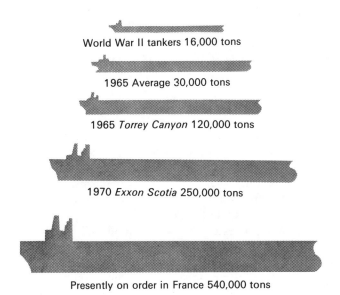

World War II tankers 16,000 tons

1965 Average 30,000 tons

1965 *Torrey Canyon* 120,000 tons

1970 *Exxon Scotia* 250,000 tons

Presently on order in France 540,000 tons

Figure 2. Increase in oil tanker capacity since World War II.[6]

The data in Table 2 show that sources other than transportation also contribute to the oil pollution problem. River and urban runoff, for example, contribute almost as much as transportation sources. Significant amounts of petroleum hydrocarbons are deposited in urban areas from a variety of sources, including oil heating systems, fallout, automobile operation, and the dumping of spent lubricants, cutting and lubricating oils, coolants, etc. Rainfall and runoff inevitably wash these materials into storm drains and then into receiving waters.

Some natural oil pollution of the marine environment also takes place now just as it has in the past. No direct measurements have been made of the quantities of oil that enter the ocean from natural seepage. However, two lines of thought indicate that it must be quite small when compared with that resulting from human activities: (1) if much oil had continually seeped into the ocean, all petroleum reserves would have disappeared long ago, and (2) we know from offshore drilling accidents that any natural oil seepage of any significant size would cause very conspicuous oil slicks to appear, and no such slicks have been observed.[8]

A small source of oil pollution in the oceans is leakage from ships and tankers sunk during World War II. As the ship hulls corrode, oil slowly escapes into the sea. The total amount of oil from this source is estimated to be approximately 4 million tons.[8]

The quantity of petroleum hydrocarbons lost directly into the oceans is small compared with the amount emitted into the atmosphere as a result of evaporation and incomplete fuel combustion. Most of these atmospheric hydrocarbons undergo photochemical reaction and are oxidized to other substances. However, a fraction exists as liquid droplets or becomes adsorbed onto small atmospheric particulates. The liquid and solid particles eventually settle or become washed out of the atmosphere, and the large amount that settles on ocean surfaces contributes significantly to marine oil pollution.

2.3. The Fate of Oil in the Marine Environment

Upon entering the marine environment, oil is transported throughout a region by physical forces. Wind is a prime mover, as well as currents, waves, and tides. A variety of natural processes, including evaporation, dissolution, emulsification, oxidation, uptake by living organisms, and sedimentation, continually alter the composition of such oil by the removal and transformation of various constituent compounds. Figure 3 shows the complex interactions that result when these processes operate on an oil spill.[9]

All types of oils contain volatile components that readily evaporate. Within a few days, 25% of the volume of a typical oil spill is lost through evaporation. The evaporative process depletes the lower-boiling hydrocarbon components regardless of the structural series to which they belong: low-boiling aromatics, paraffins, and cycloparaffins are all lost. Generally, the

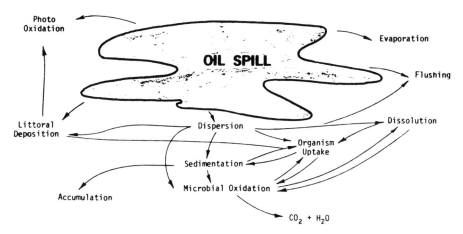

Figure 3. The interaction of factors which influence the rate of an oil spill.[9]

molecules involved are those containing 12 or fewer carbon atoms (see Table 1).

The lower molecular weight components are also preferentially removed by dissolution since they have the greatest solubilities (Table 1). Aromatic hydrocarbons have a higher solubility than n-paraffins of the same boiling point.

Biochemical (microbial) attack by bacteria, fungi, and other microorganisms occurs on compounds which have a much wider boiling range than those affected by evaporation and dissolution. No single microbial species is capable of degrading all components of a specific crude oil. Bacteria are highly selective and complete oil degradation requires the action of numerous different bacterial species. Many intermediates are produced by bacterial oxidation of hydrocarbons. Therefore, organisms are also required that will further degrade hydrocarbon decomposition products. Paraffins are most readily degraded by bacteria. Consequently, the more-resistant cycloparaffins and aromatic hydrocarbons disappear from the marine environment at a much lower rate.

Chemical and photooxidation also affect petroleum hydrocarbons, but, in an aquatic environment, the processes are not well understood. Nevertheless, photooxidation can be a significant factor under some circumstances.

Degradation rates are functions of the physical environment. As expected, temperature influences most degradation processes. The nutrient and oxygen content of the water are key factors in microbial degradation. It has been estimated that the complete oxidation of 1 gallon of crude oil requires all of the dissolved oxygen contained in 400,000 gallons of air-saturated seawater at 60°C.[8] This is equivalent to the amount of seawater in a layer 1 ft deep that covers 1.2 acres. Oxidation may be inhibited in waters with low dissolved oxygen levels caused by previous pollution. Under such conditions, bacterial degradation may cause additional damage by further depleting the dissolved oxygen. Oxygen is continually replenished in surface waters through contact with the atmosphere. However, below about 10 m this replenishment takes place very slowly.

Heavy oil residues that are not degraded or deposited in sediments are often found in the form of floating tarry globules known as tar lumps or tar balls. These are the lumps that frequently wash up on beaches. Water-in-oil emulsions, formed when water droplets become enclosed in sheaths of oil, play an important role in the formation of such tarry masses.

2.4. Biological and Physical Effects of Marine Oil Pollution

The effects of marine oil pollution on the oceans and on the life they contain are somewhat controversial topics. It is generally agreed that the effects are undesirable, but little other consensus has emerged from the numerous studies, reviews, and conferences dedicated to the topic. This is understandable for a number of reasons: (1) prespill baseline data with which to compare postspill measurements are usually unavailable; (2) uncertainties exist about the field applicability of laboratory toxicity studies; and (3) information is generally lacking about oil spill parameters such as oil composition, magnitude and duration of a spill, etc. Current conclusions must therefore be considered to be only tentative in most cases and data must be considered to be qualitative rather than quantitative.

The overall effects of oil on organisms of the marine environment may be classified into five categories:

1. Direct lethal toxicity
2. Sublethal disruption of physiological or behavioral activities
3. The effects of direct oil coating
4. Tainting caused by the incorporation of hydrocarbons into organisms
5. Changes in biological habitats

Lethal toxicity (death) can occur as a result of direct hydrocarbon interference with essential cellular and subcellular processes, especially membrane activities. The soluble aromatic fractions of oil pose the greatest environmental threat in this category. Exposure to low molecular weight (10 carbons or less) paraffins can cause narcosis, but the necessary concentrations are extremely high and would not occur from an oil spill. Available data indicate that death occurs in most adult marine organisms within a few hours after exposure to 1–100 ppm of total soluble aromatic hydrocarbons. The lethal concentrations of such compounds for larvae may be as low as 0.1 ppm. Thus, larvae appear to be 10–100 times more sensitive than adults.[4] Aromatic concentrations in the lethal range can result from unweathered oil slicks. Oil is said to be weathered after it has been in the water long enough to lose many of its volatile and soluble components. Table 3 gives estimated acute toxicity sensitivities for various forms of marine life.[4]

The death of some marine life has been found associated with every oil spill studied. The toxic effects have been generally localized, and mortality was

Table 3. Estimated Acute Toxicity Sensitivity for Various Classes of Organisms to Soluble Aromatic Hydrocarbons[a]

Class of organism	Estimated concentration (ppm) of soluble aromatics causing toxicity
Plants	10–1000
Finfish	5–50
Larvae (all species)	0.1–1.0
Pelagic crustaceans (shrimp, etc.)	1–10
Gastropods (snails, etc.)	10–100
Bivalves (oysters, clams, etc.)	5–50
Benthic crustaceans (lobsters, crabs, etc.)	1–10
Other benthic invertebrates (worms, etc.)	1–10

[a]Source: S. F. Moore, R. L. Dwyer, and A. M. Katz.[4]

greatest where the spills were confined to inshore areas possessing abundant biota. Most major spills have occurred offshore in relatively deep water, and most of the toxic oil fractions may have evaporated or have been dispersed to subtoxic levels before the oil drifted into the shallow waters near shore.

The same compounds that exert lethal effects at higher concentrations can cause problems in sublethal amounts. These problems involve interference with the chemical sensing and communications systems of marine organisms, and although they do not produce immediate death, can ultimately affect the survival of individual organisms.

Chemical communications play an important role in the behavioral patterns of many marine organisms. Marine predators, for example, are attracted to their prey by organic compounds present in seawater at concentrations in the ppb range. Similar chemical attractions and repulsions play important roles in such processes as escaping from predators, locating habitats, and sexual attractions. Sufficient information is available to draw tentative conclusions regarding the possible effects of oil on chemical communication. There is reason to believe that some compounds in polluting oil, mainly the soluble aromatics, interfere with chemical communication processes by blocking the receptors of organisms or by mimicking natural stimuli. The full implication of such communicative disruptions remains uncertain. However, concentrations of soluble aromatics in the range of 10–100 ppb may cause significant problems.[4]

Coating or smothering effects are the most damaging indisputable adverse effects of petroleum pollution. Oil- and tar-coated beaches and oil-soaked birds have been favorite topics of the visual mass media. Coating effects, which are principally associated with the higher-boiling fractions of oil (from weathered oil), are primarily a problem for intertidal low-growing plants, plankton, and birds. Mobile organisms can normally avoid exposure. Birds are particularly vulnerable to damage from coating effects for several reasons. When their inner feathers are coated with weathered or unweathered oil, insulation is lost: the insulative compounds dissolve in the oil. As a result, an afflicted bird can literally freeze to death in any season. In addition, birds whose feathers become coated with crude oil attempt to clean themselves by preening, and ingestion of hydrocarbons then becomes a problem.

Tainting, as a result of hydrocarbon accumulation in organism tissues, occurs in many, if not all, marine species. Any aquatic organism can be expected to maintain chemical equilibrium with its surroundings. If the water contains even low (ppb) hydrocarbon concentrations, these substances may be ingested and accumulate in various tissues. Such hydrocarbon incorporation is of interest for two reasons: (1) some of the potential accumulation consists of polycyclic aromatic hydrocarbons, which are known to be carcinogenic; and (2) the flavor of edible organisms is changed.

If the hydrocarbon contamination is short lived and concentrations do not get too high, self-cleansing of the organism may be nearly complete. However, the continuation of undesirable water conditions over long periods of time may result in permanent organism contamination. For example, fish and lobsters have been shown to metabolize most hydrocarbons within 2 weeks. However, the metabolism in lower organisms takes place much more slowly and is not well understood. There is no evidence for food chain magnification of petroleum hydrocarbons in marine organisms.[5]

Oil pollution can affect habitats and prevent species normally present in or on a substrate from inhabiting the area. A substrate is the material or surface from which a plant or animal obtains support. The amount and composition of oil necessary to prevent a species from utilizing a substrate is largely unknown. However, available data indicate that the presence of low to high molecular weight hydrocarbons in concentrations as low as 10–100 ppb may chemically insulate a substrate from virtually all species.[4] The effect of higher-boiling insoluble hydrocarbons depends on the degree to which an organism relies on a particular substrate. Species depending on a substrate only for passive support—those simply lying on the substrate—may be affected very little, but those living within the substrate or otherwise actively depending on it are more vulnerable.

3. Pesticides

Pesticides are chemicals used to control plants and animals considered to be undesirable. Such uses of chemicals did not originate in the twentieth century, but have been practiced on a small scale for hundreds of years. Early pesticides included arsenic compounds, lime–sulfur mixtures, and copper salts. However, the large-scale use of pesticides is characteristic only of the last 2 decades. During this time, the character of the pesticides used has changed. The earlier inorganic chemicals have been almost totally replaced by synthetic organic compounds, most of which were specifically developed for their toxicological properties.

The heavy use of these newer pesticides initially met with great enthusiasm. The synthetic formulations were easy to use, fast acting, and effective against a wide range of pests. The use of these chemicals made possible the prevention of human diseases such as typhus and malaria, and the increased production of food. In general, their use greatly improved the quality of human life. It is now recognized that these benefits have not been obtained without some penalties. Pesticides are now found distributed throughout the environment, particularly in the aquatic environment, and problems exist as a result of their presence.

3.1. *Types of Pesticides*

At the present time there are over 10,000 different pesticide formulations. The compounds are frequently classified according to their intended target, as shown by the following categories:

1. Insecticides: as the name implies, these chemicals are designed to destroy insects
2. Herbicides: these chemicals kill weeds or other unwanted vegetation
3. Fungicides: these compounds are toxic to molds (fungi) and act to help prevent plant disease
4. Other specific pesticides: these include such materials as rodenticides (effective against rats, mice, gophers, etc.), molluscicides (for use against snails, etc.), and nematocides (to control microscopic worms)

The amount of each type of pesticide used decreases in the order given in the list. Within the last 10 years the use of herbicides has increased dramatically. However, total insecticide use still exceeds herbicide use.

According to the latest estimates, 3 million different species of insects are now in existence. This is far more than the number of all other animal and plant species combined. The number of individual insects alive at any one time is thought to be about 1 billion billion (10^{18}). Of this vast number, 99.9% are, from the human point of view, either harmless or, in fact, helpful. A few, such as pollen-transferring bees, are considered indispensible. The troublesome species comprise 0.1% of the total, or about 3000 species. Most of these are agricultural pests or carriers of human or animal diseases. If humans are to have food, remain relatively free from disease, and prosper, the freedom of these selected few insects to pursue their natural activities must be eliminated. At our present stage of technological development, synthetic organic chemical insecticides are the most significant means of accomplishing this task.

Weed control, like insect control, is necessary for efficient agricultural activities. The high cost associated with using people to perform such tasks as hoeing weeds and cutting brush makes the widespread use of chemical herbicides unavoidable. In addition, herbicides have also been used by industry and government to control unwanted vegetation along highways and rights-of-way and in forests. A controversial use of them has been made in modern warfare, where herbicides were employed as defoliating agents.

3.2. Chemical Classes of Pesticides

Chemical compounds usually affect living systems or other compounds in a manner that is related to their elemental composition and molecular structure. Therefore, it is convenient to classify compounds accordingly. Such a classification system for pesticides offers an alternative to the target-classification system previously discussed. In practice both systems are in common use. Four major structural classes of synthetic organic pesticides are: (1) chlorinated hydrocarbons, (2) organophosphates, (3) carbamates, and (4) chlorophenoxy acids.

The trade names, chemical names, and molecular structures for selected members of each of these four classes are given in Table 4.[10,11] Also listed in the table are the major uses for each of the selected compounds. Note that chlorinated hydrocarbons and organophosphates function predominantly as insecticides, carbamates have multiple uses, and chlorophenoxy acids are used mainly as herbicides.

Chlorinated hydrocarbons, as the name implies are carbon–hydrogen compounds to which chlorine atoms are attached. This pesticide class is presently the most prevalent in the environment. The reasons for this include: (1) their total cumulative use exceeds that of any other class, and (2) the compounds, in general, are more persistent (resistant to degradation) in the environment than compounds of other classes. DDT, the most well known and widely used of all pesticides, belongs to this class. The heavy use of these compounds began in the early 1950s. In the early 1970s many countries began restricting the use of them due to their effect on various nontarget organisms.

The mechanism by which chlorinated hydrocarbons exert their toxic effects on organisms is not well understood. However, it is thought that they dissolve in the fatty membrane surrounding nerve fibers and interfere with the transport of ions in and out of the fiber. This latter process is essential for nerve impulse transmission. Serious interference with the ion movement results in tremors, convulsions, and death.

Organophosphates came into general use as insecticides during the late 1950s and early 1960s. The growth in use of these compounds resulted in part from the resistance (immunity) certain insects were developing toward chlorinated hydrocarbons. As a class, organophosphates have the following common structural features:

$$\begin{array}{c} R \\ \diagdown \\ \end{array} \overset{\displaystyle Y}{\underset{\displaystyle R}{\overset{\|}{\underset{\diagup}{P}}}} - X - Z$$

where X and Y are either S or O, R is a hydrocarbon or oxyhydrocarbon group, and Z is a complex organic group. Parathion and methyl parathion are the most widely used members of this class.

The site of toxic activity for organophosphates is the synaptic gap of nerves. The phosphorus-containing pesticide deactivates the enzyme acetyl-cholinesterase (ACHE) and prevents it from performing its usual function. Normally, ACHE breaks down the chemical acetylcholine after it has carried an impulse from one nerve fiber to another (across the synaptic gap). The result of interfering with this normal activity is a build-up of acetylcholine and a barrage of extraneous nerve impulses which disrupt normal function. Tremors, convulsion, paralysis, and death follow.

Carbamates are the most recently developed of the pesticide classes discussed in this section. Chemically, they may be thought of as derivatives of the parent molecule carbamic acid, which has the following molecular structure:

$$H-O-\overset{\overset{\textstyle O}{\|}}{C}-NH_2$$

Carbamate pesticides are esters of carbamic acid: the hydrogen attached to the oxygen in the parent acid has been replaced by a complex organic group. The hydrogens attached to nitrogen in the parent acid may or may not have been replaced. Carbamates are quite versatile pesticides; some function as insecticides, others as fungicides, and still others as molluscicides. Their toxic action is similar to that of the organophosphates: they deactivate ACHE.

The chlorophenoxy acids are derivatives of the simple organic acids: acetic, propionic, and butyric. They are effective herbicides whose toxic action results from their ability to mimic natural plant-growth hormones. As a result, abnormally rapid growth takes place which drains the plant's stored energy. The plant literally grows itself to death. For ease of application the chlorophenoxyacids—which are white crystalline solids—are often converted into liquid derivatives. Amine salts and esters are the forms most often used.

3.3. *Pesticide Persistence in the Environment*

Pesticide properties differ widely, both among and within chemical classes. Accordingly, their potential as environmental contaminants can vary. Among the important properties related to environmental pollution are: (1) the tendency to vaporize, (2) the tendency to dissolve in water and other solvents, and (3) the resistance to various degradation processes.

The third property is of particular importance in determining the effects of a given pesticide on the environment. This property is related to persistence, which is defined as the time required for a pesticide to lose at least 95% of its activity under normal environmental conditions and rates of application.[12] Activity loss is complete when the pesticide has been totally decomposed

Table 4. Structures, Nomenclature, and Uses for Selected Pesticides[a]

Chemical classification and trade name	Uses	Structural formula and chemical name
Chlorinated Hydrocarbons		
DDT	Broad spectrum– cotton, soybean, and peanut pests; also timber, industrial and mosquitoes.	1,1,1-trichloro-2,2-bis (*p*-chlorophenyl)ethane
Lindane	Control of cotton insects and rice stem borer; also on wood-infesting pests.	1,2,3,4,5,6-hexachlorocyclo-hexane (γ isomer)
Aldrin	Soil insecticide for control of ants, cutworms, grubs, beetles, and cotton pests.	1,2,3,4,10,10-hexachloro-1,4,4a,5,8,8a-hexahydro-1,4-endo-exo-5,8-dimethanonaphthalene
Mirex	Fire ant control, stomach poison in bait.	1,2,3,4,5,5,6,7,8,9,10,10-dodecachlorooctahydro-1,3,4-methano-2H-cyclobuta[c,d]-pentalene
Organophosphates		
Malathion	Control certain pests of fruits, vegetables, and ornamentals; also for public mosquito control.	O,O-dimethyl-*S*-(1,2-dicarbethoxyethyl) phosphorodithioate
Parathion	Larvicide for mosquito control; also broad spectrum insecticide for fruit and vegetable pests.	O,O-diethyl O-*p*-nitrophenyl phosphorothionate

Table 4—continued

Chemical classification and trade name	Uses	Structural formula and chemical name
Dichlorvos	Fumigant for household pests, active ingredient in "pest strips."	O,O-dimethyl-2,2-dichlorovinyl phosphate
Diazinon	Effective against many fruit and vegetable pests.	O,O-diethyl-O-(2-isopropyl-4-methyl-6-pyrimidinyl) phosphorothioate
Carbamates		
Sevin	On crops, especially cotton, forage, fruits, and vegetables.	1-naphthyl-N-methyl-carbamate
Baygon	For flies, mosquitoes, cockroaches, ants.	2-isopropoxyphenyl N-methylcarbamate
Zectran	Broad spectrum—against snails, slugs, moth larvae.	4-dimethylamino-3,5-xylenyl N-methylcarbamate
Dimetilan	Bait for house and fruit flies.	2-(N,N-dimethylcarbaryl)-3-(methylpyrazolyl)-(5)-N,N-dimethylcarbamate

Table 4—continued

Chemical classification and trade name	Uses	Structural formula and chemical name
Chlorophenoxy acids		
2,4-D	Control aquatic vegetation in water systems, military defoliant.	Cl—⟨C₆H₃⟩(Cl)—O—CH₂—C(=O)—OH 2,4-dichlorophenoxyacetic acid
2,4-DB	Brush control on rangelands, rights-of-way and aquatic weeds.	Cl—⟨C₆H₃⟩(Cl)—O—CH₂—CH₂—CH₂—C(=O)—OH 4-(2,4-dichlorophenoxy)butyric acid
2,4,5-T	Woody plant species and weed control, military defoliant.	Cl,Cl—⟨C₆H₂⟩(Cl)—O—CH₂—C(=O)—OH 2,4,5-trichlorophenoxyacetic acid
Silvex	Effective against woody plants.	Cl,Cl—⟨C₆H₂⟩(Cl)—O—CH(CH₃)—C(=O)—OH 2-(2,4,5-trichlorophenoxy)propionic acid

[a]Sources: *Cleaning Our Environment—The Chemical Basis for Action* and *Pesticides in the Aquatic Environment.*[10,11]

(degraded) or otherwise inactivated by chemical or biological processes. Non-persistent pesticides remain in the environment for 1–12 weeks; moderately persistent ones remain 1–18 months; and persistent pesticides endure 2 years or longer. Obviously, if degradation is rapid there is little potential for accumulation or movement of the toxic species within the environment.

Of the pesticide classes previously discussed, only chlorinated hydrocarbons fall into the persistent category. As can be seen from the data in Fig. 4, degradation times for these pesticides are measured in terms of years.[13] Most other pesticides, such as organophosphates and carbamates, decompose rapidly in soil, and their persistence is measured in terms of weeks.

There is significant variation in the time required for pesticide degradation, as determined by field studies, for single compounds. Although structure is the major factor involved in degradation rates, environmental conditions also play an important role. For pesticide residues in soils, factors such as soil type, amount of organic matter present, and extent of cultivation have all been shown to have significance.[10] In aquatic environments sorption properties are

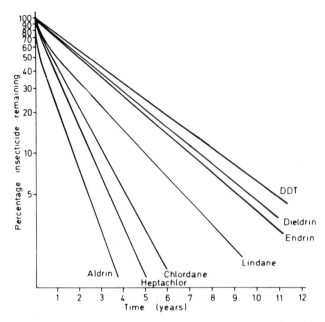

Figure 4. Degradation time for chlorinated hydrocarbons in soil.[13]

extremely important, since sorption onto sediment allows pesticides to settle to the bottom of the water body where microbial activity can take place.[11]

Although the complete chemistry involved in the degradation of various pesticides is not yet known, some generalizations can be made. Organophosphorus and carbamate compounds usually degrade rapidly as a result of hydrolysis reactions. Chlorinated hydrocarbons, on the other hand, do not hydrolyze but are broken down by slow microbial action.

3.4. *Persistent Pesticide Movement Within the Environment*

Persistent pesticides have been found in diverse parts of the environment in recent years. Such widespread detection indicates that mechanisms exist for the movement of these compounds throughout the environment.

Pesticides applied as aerial sprays, dusts to foliage, or directly to soil, ultimately reach the soil which then acts as a reservoir. From this land base, they move into air and water or are degraded. It appears that water is the ultimate depository for persistent pesticides. In the case of DDT, the most widely studied example, it is estimated that as much as 25% of the total produced to date may have been transferred to the oceans.[14]

Sufficient information is not presently available to allow a quantitative estimate to be made of the amounts of pesticides entering the aquatic environment from various sources. However, a general appraisal of the entry routes can be made. Persistent pesticides enter the world's waterways by both direct and indirect routes. Pesticide applications to water for purposes of pest control, and the discharge of pesticides as industrial wastes, are considered to be direct routes. Many organic pesticides are added directly to water to control aquatic insects, trash fish, and aquatic plants. In most of these applications the amount of pesticide added is carefully controlled. However, careful control might be lacking in mass applications such as those used for emergency mosquito control. Wastes from pesticide manufacturing and formulating plants, unless very closely controlled, contain pesticides. In addition, effluents from plants that use pesticides in their processes (e.g., moth-proofing of clothes) may contain variable amounts of pesticides.

Agricultural and urban land drainage is believed to represent the major indirect route by which pesticides enter waterways. In such drainage the pesticides may be dissolved directly in the water or sorbed (attached) to particular matter (sediments) carried along with the drainage. Pesticide sorption on sediments is considered to be the primary pathway by which pesticides move from land to water.[11] All pesticides are soluble in water to an extent that varies from very slight to great. In general, organophosphates are more soluble than chlorinated hydrocarbon insecticides, and various herbicides are highly soluble. The greatest danger from the runoff of soluble pesticides is in the period immediately following their application and prior to the time required for them to become sorbed onto soil particles. Even a very soluble pesticide, if strongly sorbed to soil, will not usually be leached from the soil. Thus, erosion is a major mechanism for surface water contamination from pesticides applied to soil.

Since pesticides are now universally present in air, the atmospheric processes of "wash-out" (precipitation) and "fall-out" (gravitational settling of particles) represent an indirect source of pesticides for the aquatic environment. The deposition of air-borne pesticides into the aquatic environment, although of lower magnitude than that from land runoff, is nevertheless considered significant. Pesticidal compounds may enter the atmosphere in several ways, including: (1) drift during spray applications, (2) volatilization from plant and soil surfaces, and (3) wind-blown soil particles. Drift results during spraying when wind-blown spray is carried away from the target area. Such spray particles may remain in the air for extended periods. Pesticide volatilization involves the loss of pesticides from treated soils and plant surfaces by evaporation. Wind can displace surface soil to which pesticides are sorbed. High winds have created dust clouds from which precipitation has deposited unusually large amounts of contaminated soil.

3.5. Effects of Persistent Pesticides in Aquatic Environments

Despite a large influx of persistent pesticides, particularly DDT, into the marine environment, the concentrations of pesticides in natural waters are quite low as a result of dilution. Pesticide concentrations are in the ppb range for fresh water and the parts-per-trillion range for ocean water. However, even these low concentrations are matters of concern because of the ability of many plants and animals to concentrate the pesticide molecules within their tissues. Chlorinated hydrocarbons, although relatively insoluble in water (1.2 ppb for DDT) are quite soluble in fats and oils, which are found in all living organisms.

The chief known hazard of persistent pesticides in the marine environment is the concentration process mentioned above. This process, often called biological amplification, produces striking results in some cases. DDT and other chlorinated hydrocarbons, for example, may be taken up selectively (due to their fat and oil solubility) by plankton and small fish. These in turn are eaten by larger organisms. When the process is repeated through several levels in a food chain, extremely high pesticide concentrations result in species at the top of the chain. Typical values for a DDT food chain concentration pattern are given in Fig. 5.[13]

One result of the biological amplification of persistent pesticides has been the reproductive failure of several birds of prey, including the brown pelican, the Bermuda petrel, the osprey, and the bald eagle. The high pesticide levels found in these birds are thought to inhibit the activity of a vital enzyme that controls calcium metabolism. As a result, eggs with thin, weak shells are produced. Such eggs are easily broken when the birds attempt to incubate them, and as a result reproduction rates decrease. There appears to be a correlation between diet and the amounts of pesticides that accumulate. Predatory aquatic birds have been found to have the highest levels.

Even though the effects of chlorinated hydrocarbons on birds have received the most publicity, several other results have been noted in other organisms, particularly fish. Different species of fish have widely varying susceptibilities to organochlorine pesticides. It is certain that large numbers have been directly killed by pesticides which have accumulated to toxic levels in

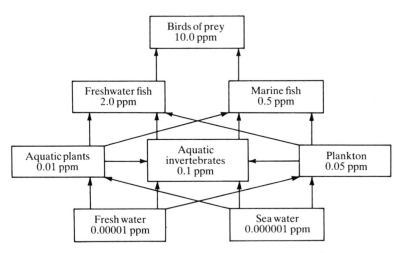

Figure 5. A typical food-chain concentration of DDT.[13]

their bodies. In addition, many indirect sublethal effects have also been noted in laboratory studies.[13] These include lowered resistance to disease, thickening of gill membranes, lower blood counts, loss of thermal acclimatization, increased respiration, and poor egg hatching ability.

Humans stand at the top of every food chain in which they are involved, and accumulation of persistent pesticides should therefore occur in them. Collected data indicate that this assumption is correct. The average amount of DDT—the pesticide present in greatest amounts—is estimated to be between 5 ppm and 10 ppm in human tissues.[13] Food is believed to be the main source, but home use and dust-laden air have been implicated as well. It appears that present exposure levels to DDT and other chlorinated insecticides among the general public have not produced any observable adverse effect. Concentrations of up to 600 ppm have been found in workers at a factory that produced the chemical. None of the workers involved have shown any demonstrable ill effects.[13]

4. Detergents

Detergents have attracted a lot of attention in the United States and in other countries because of a variety of water pollution problems caused by their constituents. In its broadest sense, the word "detergent" means anything that behaves as a cleaning agent. A more restricted definition has come into use since World War II. Accordingly, a detergent is now thought of as a cleaning agent that includes as a part of its formulation a petrochemical or other synthetically derived surfactant. The surfactants, to be discussed in the next section, provide much of the cleaning power found in detergents. We will use this limited definition of "detergent" in our discussion.

4.1. Chemical Composition of Detergents

Commercial detergent formulations usually contain one or more surfactants, or surface-active agents, and a number of builders. The surfactant serves at least two functions. It lowers the surface tension of the liquid in which it is dissolved, and it forms a stabilized emulsion or suspension with the soil particles that are to be removed. The builders sequester hard water ions such as calcium or magnesium and, in addition, react with water to form an alkaline solution in which the surfactants are most effective as cleaners.

In addition to surfactants and builders, a detergent may also contain bleaching and brightening ingredients. Bleaches oxidize colored substances into colorless chemical forms. The oxidized material is sometimes more soluble or less strongly absorbed, and may therefore be more easily removed during washing. Bleaches thus aid in the cleaning process and improve the appearance of washed items. Brighteners are fluorescent dyes which adhere to textile fibers and convert ultraviolet into visible light. The reflectance of both white and

Anionic Surfactants

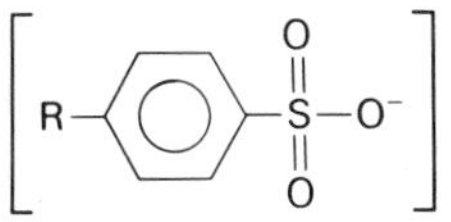

Alkyl benzene sulfonate (ABS) 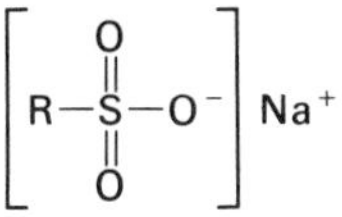   Linear alkyl sulfonate (LAS)

Cationic Surfactants

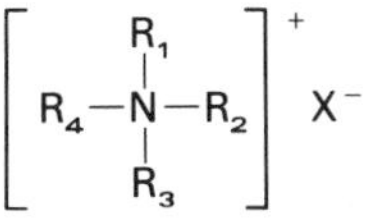

Quaternary ammonium salt

Nonionic Surfactants

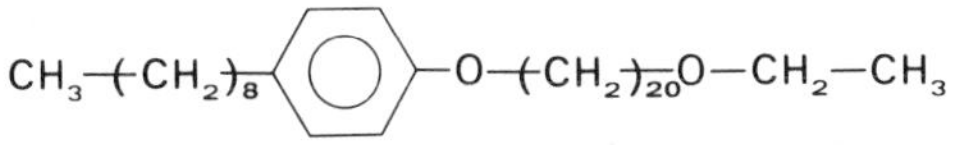

Figure 6. Structures of selected types of surfactants. R is a straight hydrocarbon chain with 12–18 carbons; R_1, R_2, and R_3 are short hydrocarbon chains; R_4 is the aromatic; and X^- is the halogen or acid group.

colored textiles is thereby improved. Other constituents (enzymes, corrosion inhibitors, perfume, etc.) may also be included to provide additional desirable properties.

A variety of different surfactants are marketed for use in different detergent formulations, including nonaqueous solutions. The surfactants generally fall into three categories: anionic, cationic, and nonionic. The chemical structures for typical representatives of the three classes are given in Fig. 6.

The surfactant to be used in a detergent formulation is determined on the basis of the conditions of use and the type of fabric being washed. Hydrophilic (water-attracting) fibers (cotton, wool, and silk) are compatible with anionic surfactants. "Permanent press" polyamide and polyester fibers are hydrophobic (water-repelling) and are therefore cleaned more easily in the presence of a nonionic surfactant.

Anionic surfactants were the first type to be used, and they still remain the principal type included in most detergent formulations. Until the early 1960s, the anionics were of the ABS type; they were replaced in the mid-1960s by the LAS type for reasons discussed in the next section. Cationic surfactants are generally more expensive to produce that the anionic types, but they may have other desirable properties such as being germicidal. Some cationic formulations are used as fabric softeners.

Detergents must contain builders because calcium and magnesium ions are present as salts in most water used in the washing process (water hardness) and are also present in soil (soil hardness) and in textiles (especially in the case of cotton). Soil hardness is a factor that is often overlooked, but it is not negligible. The term "soil" includes human waste products (perspiration, etc.), skin debris, food deposits, soot, dust, and other solid or liquid contaminants.

The most common builders now used are polyphosphates, with sodium tripolyphosphate, $Na_5P_3O_{10}$, being the most popular. The active sequestering agent in the latter is the tripolyphosphate ion, $P_3O_{10}^{-5}$, which has the structure

$$\left[O-\overset{\overset{\displaystyle O}{\|}}{\underset{\underset{\displaystyle O}{|}}{P}}-O-\overset{\overset{\displaystyle O}{\|}}{\underset{\underset{\displaystyle O}{|}}{P}}-O-\overset{\overset{\displaystyle O}{\|}}{\underset{\underset{\displaystyle O}{|}}{P}}-O \right]^{-5}$$

Detergent formulations differ from country to country as a result of factors which prevent the development of internationally useful formulations. Two of these factors are: (1) differences in washing machine design, and (2) the composition of laundry in terms of textile type.

In most European countries, washing techniques and equipment differ markedly from those used in North American nations such as Canada and the United States. Except in the United Kingdom and Norway, most European washing is done in drum-type tumbler machines that have built-in water-heating elements. Normal washing temperatures range from 60°C up to boiling. The machines are designed to minimize water consumption per load during the washing cycle. The amount of phosphate required for the treatment of water hardness is thus reduced to a minimum.

In North America, most washing machines are of the nonheated agitator or pulsator type. Similar machines, but with heating elements, are common in the United Kingdom and Norway. The agitator machine requires about four times as much water per washing cycle as the drum-type machine. One obvious effect of this is an increased requirement for phosphate builders. This is a major reason why North American detergents contain relatively higher phosphate levels than the corresponding European products.

The type of textile washed is a second important factor that influences detergent formulations. In some countries, a typical wash may contain up to 50% synthetic textiles (polyesters, polyamides) or mixtures of synthetic and natural materials. These newer fibers or mixtures are usually treated with resins or other materials to introduce properties which make them easier to handle. Permanent press treatments are examples. A property common to all the new textiles is an increased hydrophobic character which requires the use of less polar surfactants of the nonionic type. The use of nonionic surfactants is therefore increasing steadily.

4.2. Water Pollution Problems Involving Detergents

Both the surfactants and builders of detergent formulations have been implicated in water pollution. Surfactant problems were the first to be noted

8. *Man's Impact on Terrestrial and Oceanic Ecosystems,* W. H. Matthews, F. E. Smith, and E. D. Goldberg, eds., The MIT Press, Cambridge, Massachusetts, 1971.

9. "Oil and Gas Development and Coastal Zone Management," U.S. Senate Commerce Committee Hearings, 93rd Congress, 2nd Session, 1974 (Serial 93–99).

10. *Cleaning Our Environment—The Chemical Basis for Action,* American Chemical Society, Washington, D.C., 1969.

11. *Pesticides in the Aquatic Environment,* U.S. Environmental Protection Agency, Washington, D.C., 1972.

12. G. Moriber, *Environmental Science,* Allyn and Bacon, 1974.

13. C. A. Edwards, *Persistent Pesticides in the Environment,* 2nd ed., CRC Press, 1973.

14. *Chlorinated Hydrocarbons in the Marine Environment,* National Academy of Sciences, Washington, D.C., 1971.

15. *Report of the Expert Group on Detergents,* Organization for Economic Co-operation and Development, Paris, 1973.

16. H. S. Stoker and S. L. Seager, *Environmental Chemistry: Air and Water Pollution,* 2nd ed., Scott, Foresman & Company, 1976.

17. P. S. Thayer and C. J. Kensler, "Current Status of the Environmental and Human Safety Aspects of Nitrilotriacetic Acid (NTA)," in *CRC Critical Reviews of Environmental Control,* **3**, Issue 4 (1973).

The Effects of Mercury, Lead, etc.

Although much pollution is gross and macro, there are a number of materials—the most famous of which is the mercury found in some fish—that have negative influences in very low concentrations. This subject is related both to water pollution and to the extreme problem of the disposal of radioactive waste products.

14

Pollution through Trace Elements

R. R. Brooks

1. The Trace Elements

1.1. Introduction

There is an enormous literature on the chemistry, geochemistry, geology, toxicology, and epidemiology of the so-called "trace elements." Of the approximately 90 elements found in the earth's crust, a mere nine (aluminum, calcium, iron, magnesium, oxygen, potassium, silicon, sodium, and titanium) account for over 99% by weight. The other 80 or so elements, which together total less than 0.14%, are known as the trace elements and play an important part in animal and plant nutrition. These elements have an influence that is completely out of proportion to their low abundances.

The abundances of the trace elements in various geological formations are summarized in Table 1.

R. R. Brooks • Department of Chemistry, Biochemistry, and Biophysics, Massey University, Palmerston North, New Zealand

Table 1. Average Abundances of Trace Elements in Soils, the Earth's Crust, Sediments, and Igneous Rocks[a]

Element	Soil (ppm)	Crust (ppm)	Sediments (ppm)	Igneous rocks (ppm)
Mn	850	975	760	1000
Ba	500	450	690	640
Zr	300	190	200	170
Sr	300	385	450	350
Cr	200	150	130	117
V	100	145	130	90
Rb	80	165	270	280
Zn	50	125	80	80
Ce	50	46	50	40
Ni	40	95	95	100
Li	30	45	60	50
Ga	20	15	30	20
Cu	20	75	57	70
Nb	15	20	20	20
Pb	10	15	20	16
Sn	10	40	16	32
B	10	10	56	13
Co	10	35	22	18
Th	13	10	11	13
Be	6	4.5	5	4.2
Ge	5	4	4.5	2
As	5	3.4	6.6	2
Cs	5	1	10	10
Mo	2.5	2.3	2	1.7
U	1	2.4	3.2	2.6
Ag	1	0.06	0.5	0.2
Cd	0.5	0.2	0.5	0.13
Se	0.5	0.07	0.6	0.01
W	—	1.2	2	2
Sb	—	0.6	1.25	0.3
Hg	0.01	0.07	0.04	0.06
Bi	—	0.34	0.01	0.1
Au	—	0.001	—	0.001

[a] Reprinted from Andrew-Jones.[121] By permission of the Colorado School of Mines. Copyright 1968 by the Colorado School of Mines.

In order to understand more fully the role played by trace elements in the environment, we must first study the mechanisms whereby they have achieved their present distribution within the earth. Under *natural conditions* the trace elements are mobilized in two ways. First, the elements are mobilized in the *primary environment* by a process known as *hypogene mobility*, in which the liquid magma crystallizes and distributes the elements among the various rock types in a definite pattern. Thereafter, the weathering process takes place after deposition of the rocks, and the elements are mobilized in the *secondary environment* by *supergene mobility*.

1.2. *Mobilization of Elements in the Primary Environment*

Hypogene mobility occurs at depth under conditions of high temperature and pressure. The major constituents of the earth's crust form a sequence of minerals dependent on the prevailing temperature and pressure.[1] The minor elements usually occupy spaces in the lattices of these minerals according to the rules of diadochic substitution.

The residual fluids deposited as pegmatites or hydrothermal veins are usually extremely rich in trace elements. The same sequence of mobility is found in metamorphism when the last-formed minerals are the first to become liquid again as the temperature rises. The mobilization and transport of the elements in the primary environment is known as *primary dispersion.*

In the course of *primary dispersion*, the elements become concentrated in certain types of geological formations, hence leading to the formation of localized concentrations known as *ores.* There are three main geochemical classifications of the elements. *Siderophile* elements are those that are concentrated in iron deposits and in the iron–nickel core of the earth. These elements include iron, cobalt, nickel, chromium, and the platinum metals. *Chalcophile* elements are concentrated in sulfide deposits and include antimony, arsenic, cadmium, copper, lead, mercury, selenium, silver, and zinc. These chalcophile elements, though rare in overall abundance, are easy to obtain commercially because they tend to accumulate in easily accessible sulfide deposits. *Lithophile* elements, such as the alkali metals, magnesium, calcium, chromium, and vanadium, have an affinity for silicates.

1.3. *Mobilization of Elements in the Secondary Environment*

Supergene mobility in *secondary dispersion* within the surface environment is of great importance in elemental differentiation within soils and takes place under conditions of low temperature and pressure. Mobilization is strongly influenced by Eh, pH, and the stability of the minerals which have to be decomposed. The factors which ultimately cause the breakdown of the minerals are mechanical, physical, chemical, and biological. Garrels[2] has calculated the theoretical mobilities of trace elements in drainage waters from consideration of the physical chemistry of the associated mineral and ionic species. Equilibrium between mobile and immobile phases in the natural environment is seldom encountered, however, due to interferences from a wide variety of factors, such as absorption onto clay minerals and humic material.

Perel'man[3] has calculated a factor known as the "coefficient of aqueous migration" (K) which is a measure of the mobility of a particular element and is defined by the expression $K = 100\, M/aN$, where M is the concentration of the element in drainage waters (ppm), N is the concentration of the element in rocks (%), and a is the mineral residue (%) contained in the water. This treatment takes into account the Eh and pH of the water. Table 2 illustrates the relative mobilities of the elements in the supergene environment, and also shows the pH dependence of the mobility of some elements, particularly the

Table 2. Relative Mobilities of the Elements in the Supergene Environment[a]

Relative mobilities	Environmental conditions			
	Oxidizing	Acid	Neutral to alkaline	Reducing
Very high	Cl, I, Br, S, B	Cl, I, Br, S, B	Cl, I, Br, S, B, Mo, V, U, Se, Re	Cl, I, Br
High	Mo, V, U, Se, Re, Ca, Na, Mg, F, Sr, Ra, Zn	Mo, V, U, Se, Re, Ca, Na, Mg, F, Sr, Ra, Zn, Cu, Co, Ni, Hg, Ag, Au	Ca, Na, Mg, F, Sr, Ra	Ca, Na, Mg, F, Sr, Ra
Medium	Cu, Co, Ni, Hg, Ag, Au, As, Cd	As, Cd	As, Cd	
Low	Si, P, K, Pb, Rb, Ba, Be, Bi, Sb, Ge, Cs, Tl, Li	Si, P, K, Pb, Li, Rb, Ba, Be, Bi, Sb, Ge, Cs, Tl, Fe, Mn	Si, P, K, Pb, Li, Rb, Ba, Be, Bi, Sb, Ge, Cs, Tl, Fe, Mn	Si, P, K, Fe, Mn
Very low to immobile	Fe, Mn, Al, Ti, Ta, Pt, Cr, Zr, Th, Rare earths	Al, Ti, Sn, Pt, Cr, Zr, Th, Rare earths	Al, Ti, Sn, Te, Cr, Zr, Th, Rare earths, Zn, Cu, Co, Ni, Hg, Ag, Au	Al, T, Sn, Ta, Pt, Cr, Zr, Th, Rare earths, S, B, Mo, V, U, Se, Re, Zn, Co, Cu, Ni, Hg, Ag, Au, As, Cd, Pb, Li, Rb, Ba, Be, Bi, Sb, Ge, Cs, Tl

[a] Reprinted from Andrew-Jones.[121] By permission of the Colorado School of Mines. Copyright 1968 by the Colorado School of Mines.

group cobalt, copper, mercury, nickel, silver, and uranium. The group molybdenum, selenium, uranium, and vanadium mobilizes readily under oxidizing conditions, since in all cases the higher oxidation states of these elements are much more mobile.

Unlike the major elements, where concentrations scarcely vary by more than a factor of two over a wide range of geological formations, the concentrations of the trace elements vary very considerably from area to area. This is illustrated in Fig. 1, which gives the range of abundance values for some trace elements in soils.

Where trace elements have abnormally low- or high-level concentrations, adverse effects on animals and plants in the area can be recognized, and the

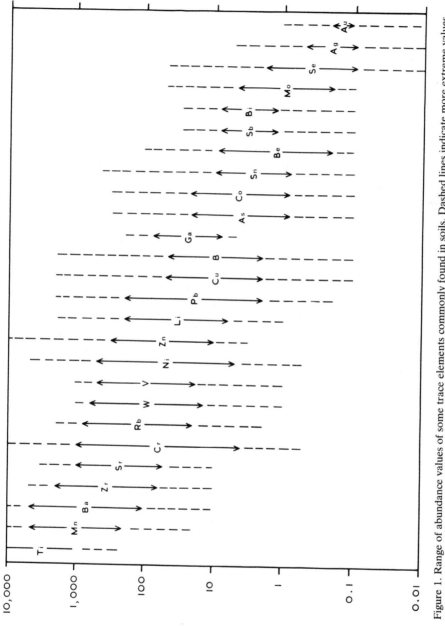

Figure 1. Range of abundance values of some trace elements commonly found in soils. Dashed lines indicate more extreme values. (Reprinted from Andrew-Jones.[121] By permission of the Colorado School of Mines. Copyright 1968 by the Colorado School of Mines.)

region is said to be a biogeochemical province.[4] Examples of this are the seleniferous soils of the western Unites States, and the areas of iodine deficiency in Switzerland and other countries. Sometimes, if a particular element is present in excess, this can result in reduced uptake of essential elements and the presence of deficiency diseases, such as hypocuprosis in cattle living in areas with a high molybdenum content in the soil.[114]

1.4. Tertiary Mobilization of the Elements

Although the above discussion refers essentially to naturally high levels of trace elements, pollution from man-made sources can easily create its own localized biogeochemical provinces, with disastrous effects on animals and humans. It may be argued that, in causing pollution, humans are only redistributing elements which are already present in the earth's crust, because, after all, no element can be created or destroyed. However what man is really doing is extracting a metal that is present in high concentrations in very localized areas and redistributing it throughout the whole world to an extent which can cause adverse effects upon the environment. A classical case of this is the mining of the locally abundant but environmentally rare element lead, and dispersing it universally via heavy industry and the internal combustion engine.

In causing pollution, man is creating new pathways for the dispersion of the chemical elements in a manner which could with justice be named *tertiary dispersion*. It is this tertiary dispersion which will be examined in this chapter in the hope that it will bring alive to many readers the potential hazards of unrestricted and unpoliced exploitation of the world's mineral resources.

2. Essential and Nonessential Trace Elements in Nature

2.1. Introduction

Most of the trace elements are found in minute amounts in all living organisms. Sometimes these elements fulfill a specific physiological function within the organism, but in the majority of cases a specific role is not evident and the element is accumulated for no apparent reason.

At present it is known that, apart from hydrogen and oxygen, about fifteen elements have a universal role in animal and plant nutrition. These are: calcium, copper, chlorine, fluorine, iodine, iron, magnesium, manganese, molybdenum, nitrogen, phosphorus, potassium, sodium, sulfur, and zinc. Cobalt and selenium have a specific role in animals, whereas boron is specific for all plants. Apart from these elements, there are others which have a specific function in the nutrition of one particular species or genus. An example of this is the utilization of vanadium in marine tunicates, which form the blood complex hemovanadin, and the specific accumulation of selenium by *Astragalus* plants which utilize this element to form seleno-amino acids.

2.2. The Physiological Role of the Chemical Elements

The physiological role of the chemical elements is summarized in Table 3. From this table it will be noted that many of the metallic trace elements have a specific function in acting as enzyme activators. It is this function which carries with it the greatest danger to the well-being of living organisms, because other nonessential elements such as cadmium, nickel, or silver can also act as enzyme activators, replace an essential element, and cause toxicity symptoms or death in the organism. It is also possible for some essential elements present in excess to displace other essential elements from their rightful function and hence produce toxic symptoms.

Interactions between essential elements which produce nutritional imbalances can take place at a number of sites, as follows[5]:

1. In the soil, preventing uptake by plants.
2. In the gastrointestinal tracts of animals.
3. At the site of metabolic action.
4. At some other site within the body by promoting excess storage in a nonmetabolic form, by promoting release from storage, combining together to form a nonmetabolizable structure, or by altering the binding sites carrying proteins.
5. At the excretory pathway by promoting abnormal excretion (reduced or excessive).

Table 3. The Physiological Role of Essential Elements in Plants and Animals[a]

Element	Physiological role
Calcium	Component of calcium pectate in cell walls and required for cell membrane stability.
Chromium	Phosphoglytamase.
Cobalt	Methymalonyl GA mutase in livers and constituent of vitamin B12.
Copper	Activator for several enzymes.
Iron	Found in prosthetic group of respiratory enzymes.
Magnesium	Metallic constituent of chlorophyll.
Manganese	Activator for pyruvate carboxylase and arginase in livers.
Molybdenum	Found in metalloflavin enzymes and in xanthine oxidase in livers.
Phosphorus	Found in nucleic acids, phospholipids, and coenzymes.
Potassium	Osmotic regulator in cell vacuoles.
Zinc	Activator for at least ten mammalian enzymes and required physiologically by all plants.

[a] Sources: R. R. Brooks and I. J. T. Davies.[14,5]

2.3. Identification of Essential Elements

The realization of the physiological role of trace elements has been a slow and gradual process and dates back to 1869 for zinc, 1917 for copper, 1928 for manganese, 1935 for cobalt, and 1942 for molybdenum. At present there is some belief that chromium may also be an essential element, but the evidence is not yet complete.

An interesting difference between the essential and nonessential elements lies in their mode of distribution in nature. In general, essential elements tend to be *normally* distributed in plants and animals, whereas the nonessential elements tend to be *lognormally* distributed. Liebscher and Smith[6] have observed this rule for humans and Timperley *et al.*[7] have established it for plants.

By a *normal* distribution, we imply that if a histogram is plotted of the frequency of occurrence of each elemental abundance value, the plot will be a perfectly symmetrical Gaussian curve. In the case of *lognormal* distributions,[8] the data plot will be asymmetric with a positive or negative *skew*. If the logarithm of the concentration data is taken, the resultant histogram will again follow a symmetrical Gaussian curve.

One useful way of deciding whether an element is normally or lognormally distributed (apart, of course, from plotting histograms) is to compare the arithmetic and geometric means with the medians. When the median and geometric mean are similar, the distribution is lognormal. A normal distribution results in the arithmetic mean and median having similar values.

The reason for the normal distribution of abundance data for the essential elements is that concentrations of these elements in organisms are *internally* controlled and tend to be constant irrespective of the amount in the environment. Nonessential elements tend to be *externally* controlled, with the result that the organism will accumulate the element to an extent proportional to its concentration in the environment.

The tendency of organisms to accumulate nonessential elements to an unlimited extent is particularly serious where pollution is involved, because toxic levels can soon be reached. For this reason, excesses of the nonessential elements tend to be more serious than excesses of essential elements. However, if essential elements are supplied to the organism in too great an excess, the regulatory mechanism can break down and toxicity symptoms or even death can result.

2.4. Essential Trace Elements in Man

Table 4 gives abundance data for six essential elements in Man.[9] The data are expressed on an ash weight basis and should be divided by about 20 to give an idea of levels expressed on a dry weight basis.

It is evident from the table that concentrations differ considerably among different organs, but there are generally high levels of most elements in the liver, which tends to act as a storehouse of trace elements.

Table 4. Concentrations of Essential Elements in Human Tissue (ppm in ashed material)[a]

Tissue	Cu	Zn	Mn	Cr	Mo	Co
Adrenal	210	1600	36	10	14–15	4–5
Aorta	97	1900	11	4.5	0–4	2–4
Brain	370	820	20	0.8	0–4	0–2
Diaphragm	150	5000	17	3.7	0–4	3–5
Heart	350	2800	23	3.4	0–4	2–3
Esophagus	140	3000	17	5.1	0–4	0–2
Duodenum	300	2500	70	3.4	1–4	1–3
Jejunum	250	2300	68	4.1	0–4	4–6
Ileum	280	3200	110	6.6	3–6	5–6
Caecum	220	3300	180	7.3	1–4	37
Sigmoid colon	230	2700	76	6.5	0–4	22
Rectum	180	3500	82	5.4	0–4	5–6
Kidney	270	4900	91	2.2	33	4–5
Larynx	59	1300	8	2.1	0–4	0–2
Liver	680	3800	130	1.5	81	4–5
Lung	130	1400	24	20	0–4	3–5
Muscle	85	4800	6	2.3	0–4	3–5
Ovary	130	1800	18	49	0–4	0–2
Omentum	190	1700	48	14	1–4	7–8
Pancreas	150	2400	110	3.7	0–4	1–3
Prostate	110	9200	19	2.2	0–4	1–3
Spleen	93	1400	11	1.3	0–4	1–3
Skin	120	1000	22	41	1–5	3–5
Stomach	230	2600	47	4.1	1–4	2–3
Testis	95	2900	19	2.4	0–4	2–4
Thyroid	100	2900	19	2.5	0–4	2–4
Trachea	65	980	14	4.7	0–4	1–3
Urinary bladder	120	3200	18	10	1–5	3–5
Uterus	110	2500	12	16	0–4	2–4

[a]Source: I. H. Tipton and M. J. Cooke.[9]

3. The Biogeochemistry of Trace Elements

3.1. Introduction

It has already been observed that trace elements are distributed in the environment via primary dispersion (igneous activity) and secondary dispersion (weathering). A third process (tertiary dispersion) results from human activities in which the elements have been rearranged in such a manner that pollution has occurred.

Trace elements find their way into humans either by direct absorption via the air or drinking water, or via the food chain. An indispensable link in the food chain is plant life (marine or terrestrial), from which humans receive their allocation of trace elements either directly, or indirectly by feeding upon herbivorous animals which depend on plants for their nutrition.

3.2. Absorption of Trace Elements by Organisms

The extent to which pollution can have an adverse effect on humans depends to some extent on the ease with which the pollutant elements can be taken up by plants. Plants absorb trace elements either via the root system or by foliar absorption. Uptake via the roots is usually the more important pathway of absorption and must be appreciated in order to understand the effects of pollutants upon the environment.

3.3. The Biogeochemical Cycle

The uptake of trace elements by plants is illustrated by the so-called *biogeochemical cycle*, which is depicted in Fig. 2. The extent of this absorption is usually a function of the *ionic potential* (μ) of the element concerned. The ionic potential is measured by the ratio of the charge and the ionic radius (Z/r). Elements of low ionic potential form easily soluble cations such as Na^{2+} and Ca^{2+}, whereas elements of high ionic potential, such as P^{5-} and N^{5-}, form soluble anions. Elements with intermediate ionic potentials, such as aluminum and ferric ions, form insoluble hydrolyzates.

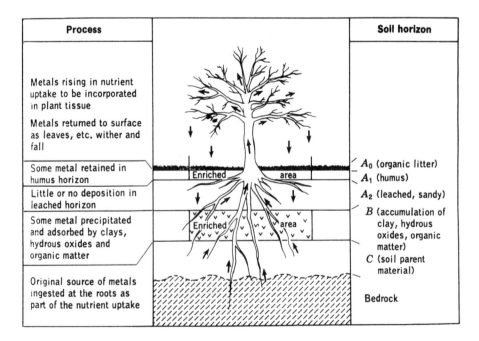

Figure 2. The biogeochemical cycle. (From Hawkes and Webb.[122])

3.4. *The Ionic Potential as a Means of Predicting Migration of Pollutants in the Environment*

The relationship between ionic potential and uptake of trace elements in plants has been studied by Hutchinson.[10] His theories are illustrated in Fig. 3, which shows the relationship between ionic potential and enrichment coefficients (i.e., concentration in the plant divided by the concentration of the same element in the substrate) for various elements in the earth's crust. Elements in the top left of the figures have low ionic potentials, whereas those in the lower right portion have high values.

Figure 3 is of great importance in understanding the probable effects of pollutants upon the environment. For example, cadmium has an ionic potential of 2.0 and is situated in the upper left hand portion of Fig. 3. Beryllium, in the intermediate zone of the figure, has a value for μ of 5.9 and is therefore easily

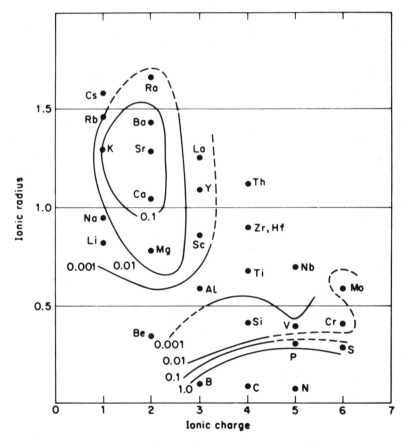

Figure 3. Enrichment coefficients of terrestrial plants (values shown in contours) shown on an ionic potential diagram. (Courtesy of G. E. Hutchinson[10] and the American Society of Naturalists.)

precipitated and has a lower tendency to migrate than has cadmium. Cadmium also is accumulated by vegetation to a greater extent and is more likely to be leached from industrial wastes. By studying the ionic potential diagrams, it is therefore possible to predict the migration of a pollutant in the environment and to assess its probable danger to humans.

3.5. *The Biogeochemistry of Trace Elements in the Marine Environment*

It is in the marine environment that some of the most spectacular accumulations of trace elements can occur. These accumulations form an extremely serious pollution problem because whereas many terrestrial organisms cannot enrich pollutants to a degree sufficient to present serious threats to public health, many marine organisms can concentrate pollutants to an alarming degree. For example, Brooks and Rumsby[11] found levels of copper and cadmium in New Zealand shellfish that were respectively 3000 and 2 million times higher in scallops (*Pecten novae-zelandiae*), and 14,000 and 300,000 times higher in oysters (*Ostrea sinuata*), than in the environmental seawater.

Goldberg[12] has discussed the enrichment of trace elements in the marine biosphere and has shown that spectacular enrichment coefficients, of the order of 10^5 or more, can occur. He showed that enrichment coefficients for metals in the marine biosphere closely parallel the order of their stability with organic ligands. Stabilities and enrichment coefficients increase with increasing basicity of divalent metal ions in the order $Cu > Ni > Pb > Co > Zn > Fe > Cd > Mn > Mg > Ca > Sr > Ba > Ra$. This order is particularly disturbing from the pollution standpoint because the most strongly absorbed ions such as cadmium, lead, and nickel are also among the most toxic to living organisms. The situation is rendered still more difficult by the fact that the organisms studied by Goldberg[12] were also the lower numbers of the food chain. Progress through the food chain successively increases the concentrations of some elements such as mercury to such an extent that *natural* levels in some commercial fish are close to, or exceed, the lowest levels now set by health authorities in many countries. It is therefore obvious that a small additional "pollution" component can be sufficient to cause a public health hazard under certain circumstances. This situation has already been reached for mercury and may soon apply to cadmium.

4. *Lead in the Environment*

4.1. *Introduction*

It can be safely said that no trace element pollutant has been studied more extensively in recent years than lead. The volume of literature on the biological aspects is now so great that it is becoming increasingly difficult to give a comprehensive review of the field. Examination of Chemical Abstracts and other review material reveals that in the period 1950–73, nearly 10,000 scientific papers were published concerning the biological effects of lead and lead pollution. Most of these papers were on the effects of lead on humans, and more than half of them were published in the four-year period 1970–73.

Most of the increased interest in lead has arisen from concern about pollution of the environment, but much of it has also arisen from the development of the carbon rod atomizer[13] which has greatly facilitated the analysis of blood samples.

4.2. The Natural Occurrence of Lead

The average lead content of igneous rocks is about 15 ppm, which places this element in the category of a rare element. Fortunately, it is concentrated in sulfide deposits which occur commonly throughout the world and it is hence easily mined and smelted. In its natural state, lead is found mainly as galena (PbS).

The lead in the earth's crust is slowly released by the process of weathering of igneous rocks and most of the lead ultimately finds its way into the oceans. The Pb^{2+} ion is fairly unstable and the lead content of the oceans is only about 0.1 ppb (0.0001 ppm) because this element is readily scavenged by ocean sediments where it accumulates as the sulfide or sulfate. Fresh waters usually contain much more lead and can have up to about 20 ppb (0.02 ppm).

The lead content of soils is of the order of its crustal abundance (15 ppm) because of the stability of this element in the geochemical cycle. Natural levels can range from 1 to 100 ppm depending on the nature of the substrate from which the soil was formed.

Vegetation (expressed on a dry weight basis) averages around 2–3 ppm lead, although this figure approaches 70 ppm when the data are expressed on an ash weight basis.[14]

It is extremely difficult to determine the natural level of lead in the air because of worldwide pollution from motor vehicle exhausts, but it would seem that the lowest values that can be found anywhere in the world are about $0.8\ \mu g\,m^{-3}$.[15]

4.3. Industrial Uses of Lead

Lead ores average about 2–20% lead, from which a concentrate averaging 60–80% is produced by a flotation process. The concentrate is roasted to remove sulfur and the lead is then smelted.

There are two main types of smelter. *Primary smelters* process the ore material and are usually large but few in number, whereas *secondary smelters* process scrap from old batteries, cable sheathing, etc.

The annual world consumption of lead is of the order of 3 million tons, of which 40% is used in storage batteries, 20% is used in petroleum as lead alkyl additives, 12% in building construction, 6% in cable coatings, 5% in ammunition, and 17% in other usages.

4.4. *Sources of Lead Pollution*

It is estimated that about 180,000 tonnes of lead are mobilized naturally each year throughout the world as a result of the weathering process.[16] In contrast to this, about 2 million tonnes of lead are mined yearly throughout the world. Probably 10% of this total is lost in treatment of the ore to produce the concentrate, and a further 10% is lost in making pig lead. Even at this stage, therefore, the amount of lead discharged into the environment is equal to the amount weathered from igneous rocks.

The storage battery industry, which uses about half the world's lead supplies, has a relatively low impact on the environment because about 80% of all batteries are recycled.

Although lead smelters are extremely serious producers of pollution, their effect is usually localized, although air and water pollution can extend up to 100 km away from these plants.[17]

4.5. *Lead Pollution from Motor Vehicle Exhausts*

Emission from motor car exhausts is by far the most serious source of lead pollution facing us today.

Antiknock lead alkyls in the form of tetraethyl lead (TEL) and tetramethyl lead (TML) have been added to most gasoline supplies since 1923. The amount at present added to petroleum spirit ranges from 2–4 g of TEL per gallon. During driving, 25–75% of this lead is emitted into the atmosphere, depending on driving conditions. Although most of this lead is ultimately deposited on the ground, the environmental air can often contain appreciable amounts of this pollutant. Cities like Los Angeles have a lead value of about 5 μg m^{-3} and this value is increasing by about 5% per annum.[18] Week-long averages of 8 μg m^{-3} can occur in San Diego.

Perhaps the most interesting revelation about lead build-up in the environment via air pollution is in the findings of Murozumi *et al.*,[19] who reported on the lead levels of dated snow strata in the Greenland ice cap. The data are shown in Fig. 4.

Up to 1750, the lead content of the ice was 20 μg per ton. As a result of the Industrial Revolution, this value had increased to 50 μg per ton by 1860. Since World War II, with the explosion of automobile ownership, the level climbed spectacularly to 120 μg per ton in 1950 and to 210 μg per ton by 1965. Despite some arguments about the validity of the data,[16,20,21] there can be little doubt that the lead content of the atmosphere has now built up to a level which can hardly fail to constitute a health hazard in many areas.

Lead pollution of the atmosphere has had some interesting distribution patterns. Cholak *et al.*[22] sampled air in the Cincinnati area from 1964 onward and found that the average lead level was only one-third that of 1946 although the number of motor vehicles had risen by 200%. The main causes of this reduction were thought to be: the decrease of air-borne particulate matter, the reduction in the use of coal for heating purposes, the dispersion of population

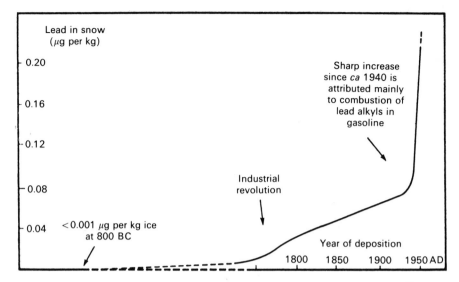

Figure 4. Lead content of snow layers in N. Greenland. (From Murozumi *et al.*[19])

from the inner city areas to the suburbs, and, finally, the channeling of traffic into newly built motorways. The experiments did not particularly indicate an overall reduction of lead levels in the air, but rather a rearrangement of its distribution. It seems likely, however, that domestic coal fires must be a significant source of lead pollution, because enhanced lead values were found in Greenland ice (Fig. 4) well before the advent of the internal combustion engine.

One of the most striking patterns of lead from automobile exhausts is its concentration in soils and vegetation along narrow strips on each side of major highways. Strangely enough, the accumulation of lead in vegetation was not noticed until Warren and Delavault,[23] in the course of biogeochemical prospecting for lead, noticed that anomalous values were always obtained near highways. They reported values as high as 3100 ppm lead in the ash of the Douglas fir, whereas background values were only 40 ppm. Following their work, Cannon and Bowles[24] determined lead in the ash of grasses bordering two major highways in Denver, Colorado. They found that the influence of lead could be detected up to 200 m from the road, and that values up to 1000 ppm were obtained for samples nearest the road. Similar studies were carried out by Lagerwerff and Specht[120] and by Page *et al.*[25]

More recently, Ward *et al.*[26] investigated the lead content of the bark of plane trees bordering a main thoroughfare in Palmerston North, New Zealand, and showed that distributions were influenced by the wind direction and by general traffic patterns along the road.

A very thorough survey of lead pollution patterns from automobile exhausts has been made by Daines *et al.*[27] They found a relationship between traffic volume, proximity to the highway, engine acceleration at constant speed,

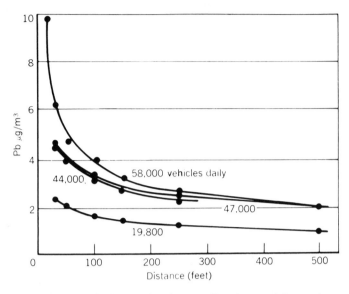

Figure 5. Air–lead values as a function of traffic column and distance from highway. (From Daines *et al.*[27] Reprinted with permission from *Environmental Science and Technology*, **4** : 318 (1970). Copyright the American Chemical Society.)

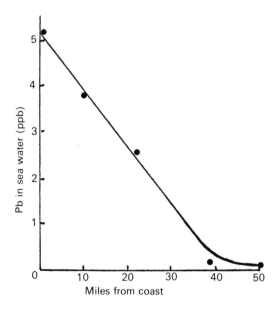

Figure 6. Lead values (ppb) in surface ocean waters expressed as a function of distance from the Los Angeles Basin. (Original data from Brooks *et al.*[28])

wind direction, and the amount of lead in the air. Particle size studies revealed that 65% of the lead between 10 m and 550 m from the roadway consisted of particles under 2 μ diameter, and 85% of particles under 4 μ diameter. They concluded that high lead pollution extended usually to about 80 m from main highways. The work of these authors is shown in Fig. 5, from which it will be seen that although lead levels tend to decrease at a distance of 80 m, there is still an appreciable effect even 200 m away.

The work of Page et al.[25] on vegetables grown near highways showed that although there was an elevated content of lead in the soils within 50 m of the roads (up to 118 ppm Pb), the relatively high values in vegetation (up to 8 ppm dry weight in strawberries) were due to absorption from aerial lead and not due to uptake from the soil via the root system.

The pollution of surface ocean water from air-borne lead is nothing short of remarkable. Brooks et al.[28] measured surface lead in ocean waters at various distances from San Pedro and Los Angeles, and found anomalous lead levels up to 25 km from land. The data are illustrated in Fig. 6.

To summarize, it appears that the effect of load pollution decreases very rapidly within a relatively short distance of major roadways. The most insidious effect of this pollution is the general raising of the lead threshold in the atmosphere as a whole throughout the world.

4.6. Lead Pollution from Mining Activities

Lead pollution from mining activities has been studied to a lesser degree than pollution from smelters or motor vehicle exhausts, but nevertheless several interesting investigations are under review in the literature.

In Derbyshire, England, extensive abandoned lead workings dating back to Roman times have resulted in a very high accumulation of lead in soils (up to 7000 ppm) near Matlock. Several investigations[29,30,31] have been carried out on grass and vegetables growing upon lead-bearing soils. Chisnall and Markland,[29] investigating farmland where several cattle and ponies had died from suspected lead poisoning, found 2–5 ppm lead in drinking water, and up to 220 ppm in grass (dry-weight basis).

In estimates on the uptake of lead by vegetables, Markland and Vallance[31] carried out pot trials on artificial composts containing up to 1000 ppm lead but concluded that uptake of this element from the substrate was relatively insignificant. The highest values (dry-weight basis) were 15.3 ppm for brussels sprouts, but six other vegetables contained an average of only about 2 ppm lead.

Crudgington et al.[30] carried out experiments on vegetables growing directly in lead-bearing soils derived largely from mining operations. Although the soils contained from 250 ppm to 6680 ppm lead, the lead content of vegetables was surprisingly low. Even potatoes growing in a soil containing 6680 ppm contained only 4 ppm lead in the peel and 0.2 ppm in the flesh.

Eighteen other different types of fruits and vegetables seldom had lead levels exceeding 1 ppm.

The remarkable capacity of vegetation to restrict uptake of lead is well known and has been reported by many workers.[14,25] Restriction of lead uptake is not a universal characteristic of all species, as demonstrated by studies on natural vegetation growing over a base metal deposit in New Zealand. Nicolas and Brooks[32] reported lead values of up to 1% in the wood ash of the tree *Beilschmiedia tawa.*

To summarize, lead pollution from mining activity does not appear to present as large a problem as motor vehicle exhausts or smelting operations because it lacks a gaseous component. Admittedly dust from opencast operations in arid areas can present a problem and can scatter pollution over a wide area, but earlier mining operations have tended to be confined underground, well away from population centers.

4.7. Pollution from Lead Smelters

In view of the obvious dangers to health of pollution from smelters, it is quite surprising that only in recent years have the true effects of pollution from this source been realized and studied to any extent. Apart from an earlier paper by Haywood,[33] most environmental studies of this nature have been carried out during the past 5 years.

The world's largest zinc smelter is situated at Avonmouth, near Bristol, England. During the past 3 years, a number of papers have appeared on heavy metal pollution from this source,[34,35,17,36] as a result of which a public outcry forced the plant to shut down for several months in 1972 so that more stringent pollution controls could be established.

Little and Martin[36] determined lead (and other metals) in leaves and soils in a grid pattern over an extensive area of several hundred square kilometers. Some of the data are shown in Fig. 7, which shows isoconcentration contours for lead in elm leaves (dry-weight basis) and in the top 5 cm of soil (expressed as ppm soluble in 2.5% acetic acid). From Fig. 7 it can be seen that measurable effects of lead pollution in soils extend to at least 10 km from the smelter and that the pattern is strongly linked with the prevailing wind direction. The lead pattern in the leaves is even more pronounced and the anomalous values can be detected up to nearly 40 km away from the source. As before, the isoconcentration contours are strongly influenced by the prevailing wind. Further work by the same authors showed that up to 85% of the lead could be removed from elm leaves by washing alone, and this gave a strong indication of the air-borne nature of the lead pollution.

A survey similar to the above work was carried out at Avonmouth by Burkitt *et al.,*[35] who detected anomalous lead values in soils and grass up to 10 km from the works. Similar work by Goodman and Roberts[37] at a Swansea (South Wales) smelter showed anomalous lead values in soils up to about

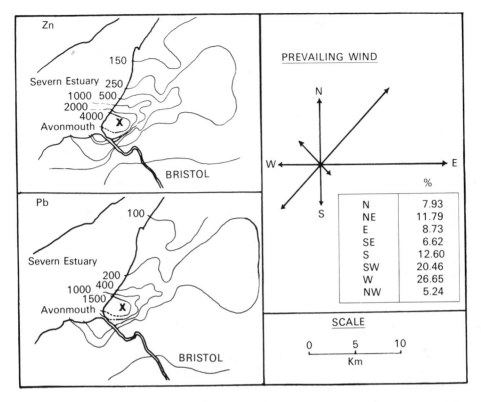

Figure 7. Lead and zinc (ppm dry weight) in unwashed elm (*Ulmus glabra*) leaves in the vicinity of the Avonmouth zinc smelter. The smelter is designated by a cross. (From Little and Martin.[36])

16 km from the plant. Enhanced values in grasses and mosses were found up to the same distance.

In Australia, lead levels were investigated in air, soil, grass, and vegetables in the vicinity of the Cockle Creek smelter, near Newcastle, New South Wales.[38] Elevated lead levels in soil and air samples were recorded at various stations near the works. Vegetables contained abnormal amounts of lead, but washing was effective in removing most of this contamination.

It is clear from the above that smelting works represent a potential source of very serious lead pollution whose effects can be felt up to 20 km away. Fortunately, smelters are few in number and the effects are much more localized than ubiquitous lead pollution from motor vehicle exhausts.

4.8. Other Sources of Lead Pollution

Two other sources of lead pollution are worthy of mention. The first of these is pollution from glazed earthenware containers. There have been many cases of lead poisoning due to the release of soluble lead from the glaze into liquids contained in the vessel. These cases extend back well into Roman times; it has even been postulated[39] that the use of lead containers for wine and other liquids was partially responsible for the decay of Roman civilization due to decimation of the ruling classes who were the more able to afford lead containers. Klein *et al.*,[40] following two cases of lead poisoning of children in Montreal, tested 264 contemporary earthenware glazed surfaces and revealed that 50% released sufficient lead for them to be unsafe for culinary use. Between 10% and 25% of the containers would have been capable of causing severe lead poisoning.

A significant source of lead poisoning, particularly in children, is lead in paint pigments. Before World War II most paints contained appreciable amounts of lead as either pigments or fillers. More recently, lead has been replaced by nontoxic titanium. In the "ghetto" areas of large cities, however, much of the prewar paint still remains and is ingested by young children who are attracted by the flakes of old paint. Hence lead poisoning from this source is essentially a factor of poverty.[41] As more and more old buildings are torn down and replaced by modern edifices, poisoning by lead paints will decrease, and ultimately will cease altogether.

4.9. The Toxicology of Lead

Perhaps no other metal, not even arsenic, has had its toxicology so extensively studied as has lead. The literature on the subject is voluminous, but it is possible in this short review to make a fair generalization. An excellent summary of the toxicology of lead has been given by Bryce-Smith.[16] Inorganic lead (Pb^{2+}) is a general metabolic poison and enzyme inhibitor (like most of the heavy metals). Young children are particularly affected and can suffer mental retardation and semipermanent brain damage.

One of the most insidious effects of inorganic lead is its ability to replace calcium in bones and remain there to form a semipermanent reservoir for long-term release well after the initial absorption. It has long been known, as mentioned above, that lead accumulates in bones, and that the high lead content of bones from the Roman period indicates a valid reason for the fall of the Roman Empire.

Organic lead, as TEL or TML, is even more poisonous than inorganic lead. The earliest symptoms of lead poisoning from this source are psychical (e.g., excitement, depression, and irritability). One of the best indicators of the degree of inorganic lead poisoning in humans is the content of this element in whole blood. Different authorities suggest safety levels in the range 0.2–0.8 ppm. Average blood levels are around 0.2 ppm for adults in the United States and the United Kingdom. The figure 0.2 ppm seems to reflect a worldwide minimum; similar values have been reported for New Guinea natives.[42]

The disturbing feature of lead levels in human blood is that natural levels (0.2 ppm) are very close to the lowest safety level (0.2 ppm) suggested by some authorities.[43,44] In a recent survey of Manchester children[45] the average value was 0.31 ppm, with 17% over 0.5 ppm and 4% over 0.8 ppm. Even if the highest safety level of 0.8 ppm[71] is recommended, it is still disturbing to find 4% of the children with levels exceeding this value.

Figure 8 shows data for blood levels in males from suburban and downtown communities of Philadelphia.[46] The figure is after Bryce-Smith.[16] As in the Manchester experiment, the figure indicates that some individuals of the downtown population had a disturbingly high lead content in their blood, exceeding even the conservative 0.5 ppm safety level of Egli *et al.*[47]

Although whole blood is the usual diagnostic agent for evaluating lead poisoning, it is also possible to use the more readily available hair or urine samples for testing purposes, although these are less reliable. There is little general agreement as to the levels in hair or urine which represent anomalous concentrations, but Kehoe[72] has established that "normal" individuals excrete urine with a lead content of about 0.30 ppm (i.e., about the same as whole blood).

Kopito *et al.*[49] have examined the lead content of human hair and concluded that "normal" values were 24 ppm, with values of up to 1000 ppm found in severe cases of lead poisoning. The same authors established normal urine level as 80 μg per 24 hr, with a maximum of 2850 μg found in a study of 17 patients displaying symptoms of lead poisoning.

Mild lead poisoning can fortunately be treated by the so-called "chelation" or "versenate" procedure, using either ethylenediamine tetraacetate (EDTA), 2,3-dimercaptopropanol, or British Anti Lewisite (BAL). The lead in the body is chelated and excreted via the urine. Unfortunately this procedure does not cure any permanent brain damage that may have occurred, but it has reduced the mortality rate from 70% to less than 5% in children admitted to the hospital.

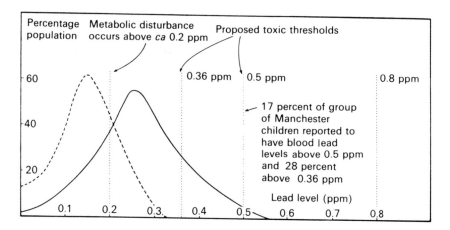

Figure 8. Lead levels in (ppm) blood of inhabitants of Philadelphia. (From Ludwig *et al.*[46])

4.10. The Control of Lead Pollution

In controlling pollution, mankind seems to lack a sense of balance. Tinker[42] has commented:

> World production of lead is 300 kt a year and of mercury only 9 kt. To get as close to the danger levels for mercury as we are for lead, we should have to eat nothing but tuna fish for months on end. Yet mercury-carrying fish are barred in more countries than have restricted lead in petrol.

It is clear therefore that the control of lead pollution is one of the most important environmental problems facing us today.

Very little can be done to control pollution from mining operations, but this is not really cause for too much concern because pollution from mining is relatively negligible as compared with pollution from other sources. Pollution from smelter operations can be reduced by improved plant control of operations, and there are encouraging signs that this is being achieved.

By far the most urgent problem is the control of lead emissions from motor vehicle exhausts. The mere reduction of the lead content of gasoline would probably be counterproductive because this is likely to lead to lower engine efficiency and an increase in the emission of other pollutants such as carbon monoxide. Many countries are seeking to reduce lead levels from the current 1 g/liter. In the United States a progressive reduction to 0.30 g/liter is proposed for 1977. In Europe, Germany hopes to reduce the level to 0.12 g/liter by 1976.

One replacement for lead in gasoline is manganese. About 10 years ago an unsuccessful attempt was made to use methylcyclopentadienylmanganese-tricarbonyl as an antiknock additive. However, replacing lead with manganese may well introduce other toxic factors into an already difficult situation, since the toxicology of manganese is not well understood at present.

Alcohols have been suggested as an alternative to lead in gasoline. Bloom[50] reported on vehicle tests with 91 octane gasoline containing 10% methanol as an additive, and found an improvement of 3-brake horsepower as compared with the lead additive.

It seems likely that the ultimate compromise solution to the lead problem in gasoline will be the removal of all additives and the design of engines which will run on lead-free gasoline. Although this may well result in a greater risk of emission of other pollutants, it will also result in greater efficiency for catalytic pollution control devices, which are readily "poisoned" by lead and other heavy metals.

4.11. The Cycle of Lead in the Environment

Figure 9 gives a representation of the cycle of lead in the environment and summarizes the facts discussed above. It is clear that man's efforts have seriously upset the natural cycle of lead, and that urgent action must be taken in the immediate future to avoid the continuing and cumulative poisoning of the environment with this ubiquitous element. As stated above, immediate action

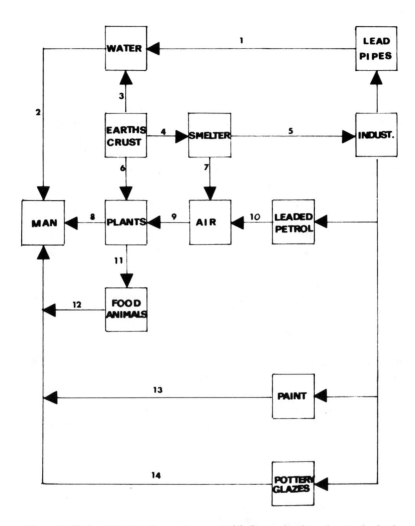

Figure 9. Cycle of lead in the environment: (1) Contamination of water by lead pipes; (2) Contaminated water drunk by humans; (3) Lead enters water via weathering process; (4) Mining operations; (5) Lead smelting produces pig lead; (6) Uptake of lead by plants via root systems; (7) Smelter fumes pollute air; (8) Contaminated plant food eaten by humans; (9) Polluted air deposits lead on vegetation; (10) Leaded gasoline pollutes air via motor vehicle exhausts; (11) Contaminated vegetables eaten by food animals; (12) Food animals eaten by humans; (13) Children ingest lead from paint flakes; (14) Lead glaze contaminates foodstuffs eaten by humans.

is imperative because in no other case are "normal" levels in human beings so close to the lowest safety limits advocated by responsible authorities.

5. Mercury in the Environment

5.1. Introduction

With the possible exception of lead, mercury as a pollutant has been studied more extensively than any other trace element during the past decade. This interest in mercury has resulted from the development of the extremely sensitive analytical technique of flameless atomic absorption[51,52] which opened a veritable "Pandora's box" of fresh information, much of it disturbing.

Before the advent of flameless atomic absorption, very little was known about mercury in the environment because levels were usually far below the detection limits of the then-existing methods of analysis. Although it had been known for many centuries that mercury is poisonous to animals and humans, it was not until the late 1950s that its extreme toxicity to humans was appreciated. In the year 1953, 52 persons living in fishing villages along Minamata Bay, Japan, died of a mysterious disease. It was some time before the ailment was correctly identified as resulting from severe mercury poisoning.[53] Further investigations revealed that the victims had eaten shellfish contaminated with mercury effluent from a nearby plastics factory. The ailment became known thereafter as "Minamata disease." The same disease caused the death of six people at Niigata in 1964.[54] The discovery of "Minamata disease" led to greatly increased research on mercury as a pollutant, spurred on not only by the new simple and sensitive methods of analysis, but also by the emergence of fresh environmental crises, as, for example, the death of 450 Iraki villagers in 1972 after eating grain that had been dusted with a mercury-containing pesticide.

5.2. The Form of Mercury in the Environment

Mercury has an extremely low abundance in the earth's crust ($\sim$0.1 ppm); however, it is readily obtained by mining operations because of its tendency to be concentrated in sulfide deposits where it exists mainly as cinnabar (HgS). In this form, mercury is relatively innocuous, but weathering, submarine vulcanism, and human activity have resulted in the accumulation of about 50 million tons of this element in the world's oceans. Natural addition of mercury to the oceans via the process of weathering is about 5000 tons per annum,[55] and a further 5000 tons is added via human activities.[56] Although mercury may initially enter the hydrosphere as the mercuric ion (Hg^{2+}), it is readily scavenged by organic matter and converted by anaerobes to produce toxic methylmercury ($CH_3Hg)^+$ and dimethylmercury ($CH_3-Hg-CH_3$).[57,57,59]

Dimethylmercury is stable at high pH values, but dissociates to CH_3Hg^+ in the low pH range. These conditions of low pH often exist in the bottom muds of streams and lakes. Since methylmercury is soluble, it is readily incorporated into organisms in the aquatic environment and ultimately finds its way into higher members of the food chain such as man.

Although mercury in the hydrosphere is the main environmental problem, it is also present in gaseous form in the atmosphere because of its relatively high vapor pressure. Natural levels are on the order of 0.003–0.009 $\mu g/m^3$, except over sulfide ore deposits.[60] Over sulfide ores and mercury deposits the concentration of this element in air increases by several orders of magnitude. For example, the air of the Ord Mine, Mazatzal Mountain, Arizona, contained 20 $\mu g/m^3$. This fact can be used for prospecting for mercury by measuring the levels of this element in the ambient air near the ground.

5.3. *Sources of Mercury Pollution*

Although natural sources are predominant suppliers of mercury to the environment, the proportion contributed by man is also greatly increasing. The natural sources of mercury are mainly from the weathering process and from terrestrial and submarine vulcanism.[61] Hoggins and Brooks[62] have studied the natural dispersion of mercury from a sulfide deposit at Puhipuhi, in the North Island of New Zealand. The data are shown in Fig. 10, which shows mercury levels in waters and stream sediments sampled at increasing distances along the course of the Wairua River, which rises in the Puhipuhi mercury deposits and enters the sea about 100 km further away. It is clear that the influence of this natural source of mercury can be traced for a distance of about 50 km in stream sediments, but only for about 5 km in the waters, as a result of the instability of the mercuric ion. Mercury has a low residence time in stream waters and is readily scavenged by sediments and in particular by the organic matter in them.[63,64] Once mercury has been absorbed on sediments, it is slowly released to the water and forms a reservoir capable of causing chronic pollution long after the original source of mercury has been removed.

Vulcanism as a source of mercury has been demonstrated by Weissberg and Zobel,[65] who showed that the Wairakei geothermal area of New Zealand is an important source of natural "pollution" by this element. As a further confirmation of this finding, White *et al.*[66] have reported mercury levels of 0.05–0.31 ppb (0.00005–0.00031 ppm) in thermal waters from Yellowstone National Park.

The world's annual consumption of mercury averages about 10,000 tons at present,[63] and about half of this total is used in the production of chlorine for bleaching paper pulp. Table 5 shows the typical percentage usage of mercury in the United Kingdom in 1969.[67]

In chlorine manufacture, mercury electrodes are used, and extensive losses of this element into the environment have occurred. It is estimated that up to 1 million lb have been lost throughout the world. One of the most disturbing examples is the case of a chlorine plant in Ontario which had been losing mercury at the rate of 15 kg/day for at least 20 years.[63] The cumulative

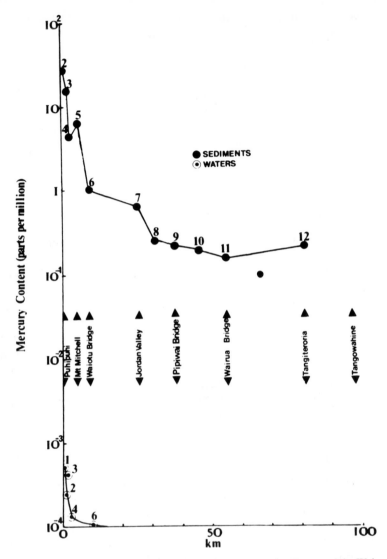

Figure 10. Mercury concentration (ppm) in waters and sediments of the Wairua River, shown as a function of distance from the mercury deposits at Puhipuhi. (From Hoggins and Brooks.[62])

total of "lost" mercury must therefore have been over 100 tonnes. River sediments near the outfall of effluent from this particular plant contained up to 0.17% mercury (practically economic-grade mercury ore).

Apart from obvious sources of pollution listed in Table 5, mercury can also be released into the air by the burning of fossil fuels.[68]

Table 5. Percentage of Mercury Used in
Various British Industries[a]

Industry	Usage of mercury (%)
Chlorine production	49.0
Switchgear/batteries	35.0
Paints	7.0
Agriculture	3.5
Pharmaceuticals	1.8
Plastics	1.8
Other	1.9

[a]Source: G. Grimstone.[67]

Although bituminous coal has a relatively low mercury content (1–25 ppb), many anthracites contain 1100–2700 ppb and can represent an appreciable pollution problem. Mercury concentrations in crude oil and petroleum are even higher and are usually in the range of 1900–21,000 ppb (1.9–21.0 ppm). Residual tars from which most of the volatile hydrocarbons have been lost can contain as much as 520,000 ppb (0.05%) of mercury.[63]

The analysis of samples from the Greenland ice cap has been useful in determining the extent of mercury pollution from industrialization.[9] This work has revealed that mercury levels have remained constant from 800 BC until the 1950s. Since then the amount seems to have doubled. The fate of mercury in the environment is summarized in Fig. 11, which shows the mercury cycle in nature.

The physiological effects of mercury poisoning have been well documented in work sponsored by the Swedish government.[69,70]

Methylmercury is particularly toxic to animals because it can readily pass the blood–brain barrier, causing injury to the cerebellum and cortex.[71] The clinical symptoms of this damage are numbness, awkwardness of gait, and blurring of vision.[53,72]

A particularly disturbing feature of mercury poisoning is that the effects are not immediately obvious. Since the brain already contains an excess of cells for any particular purpose, clinical symptoms usually do not appear until this excess has been used up. Another very unpleasant effect of methyl mercury poisoning is that mercury can enter the placenta and high levels can build up in the fetus[73] without the mother necessarily showing any signs of illness.

Clinical tests to determine mercury poisoning are based mainly on the levels of this element in whole blood. Berglund and Berlin,[74] Berglund et al.,[75] and Hammond[76] have shown that identifiable symptoms of mercury poisoning occur with levels of 0.20–0.60 ppm. Such a level would be reached by a daily intake of 0.3–1.0 mg of mercury by a healthy man. Human hair may also be used diagnostically, and neurological symptoms can appear at levels exceeding 0.05 ppm in human hair. The World Health Organization has proposed an acceptable daily intake (ADI) of 0.3 mg of mercury, of which not more than

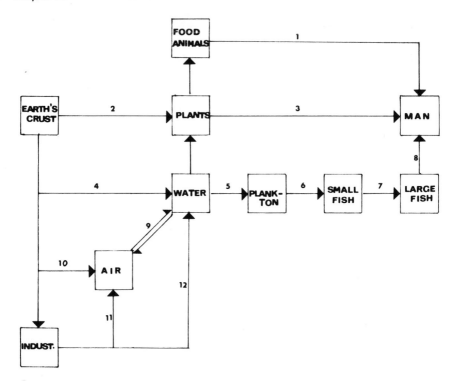

Figure 11. The cycle of mercury in the environment: (1) Food animals consumed by humans. (2) Mercury absorbed in plants via the root system. (3) Contaminated plant foods consumed by humans. (4) Mercury discharged into waters by vulcanism. (5) Plankton concentrates mercury from seawater. (6) Small fish consume plankton and concentrate mercury. (7) Large fish consume small fish with further concentration of mercury. (8) Contaminated large fish consumed by humans. (9) Mercury in air and water in equilibrium. (10) Mercury discharged into the air by vulcanism. (11) Industry discharges mercury into the air. (12) Industry discharges mercury into natural waters.

0.2 mg should be in the form of methyl mercury. An ADI is calculated by finding the lowest level which will cause toxic symptoms and then dividing by an appropriate factor to obtain a safety level which must not be exceeded. In Sweden, a factor of 10 has been recommended.[69]

5.4. *Mercury in Fish*

Although mercury is a pollutant in the atmosphere and the lithosphere, it is in the hydrosphere that its effects are most significant. As demonstrated by the Minamata disaster, this is mainly because of the tendency of the biosphere to concentrate trace elements by several orders of magnitude, as compared with concentrations in the aquatic environment. This is particularly true of shellfish and fish at the top of the food chain. In recent years many surveys have been carried out on the concentration of mercury in various edible fishes,[65,67,77] and have shown that natural levels are usually not far below the usually accepted safety limit of 0.5 ppm (net weight). Grimstone,[67] in a survey carried out in the British Isles, showed that the mercury content of pelagic fishes averaged values very close to 0.5 ppm in areas adjacent to extensive heavy industry. His data are shown in Fig. 12.

The fact that natural levels of mercury in fish are so close to the maximum permitted levels makes this element unique among other pollutants. The present high levels in pelagic fishes are not necessarily a result of human interference with nature. Indeed, it seems, from studies carried out on museum specimens of tuna caught between 1878 and 1909,[9] that these fish averaged 0.95 ppm mercury, as compared with 0.91 ppm for a random selection of recently caught fish.

It seems that fish at the top of the food chain, such as tuna and shark, contain the highest levels of mercury. This has resulted in the collapse of the school shark industry in Australia and has had a severe effect on the North American tuna industry, since specimens of tuna and shark usually average 0.5 ppm or higher.

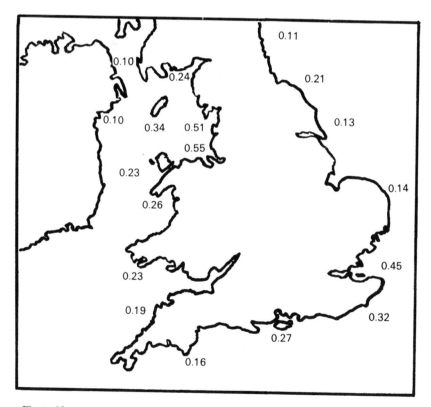

Figure 12. Average concentration (ppm) of mercury in fish caught in coastal waters of the United Kindom. (From Grimstone[67] and Anon.[79])

Among freshwater fish, the problem is hardly less serious. Weissberg and Zobel[65] recently examined the mercury content of rainbow trout in the North Island, New Zealand. They found that specimens commonly contained over 1.5 ppm mercury in areas adjacent to and downstream from the geothermal areas (see Figs. 13 and 14). At Lake Maraetai, high ratios were directly attributable to an adjacent paper mill. Figure 14 shows the relationship between the mercury levels in lake and stream sediments and the mercury content of trout caught in the area. It is clear from this figure that man-made pollution is not the only source of mercury in edible fish, and that even if all industrial pollution were to be removed immediately the basic problem would still remain in some parts of the world.

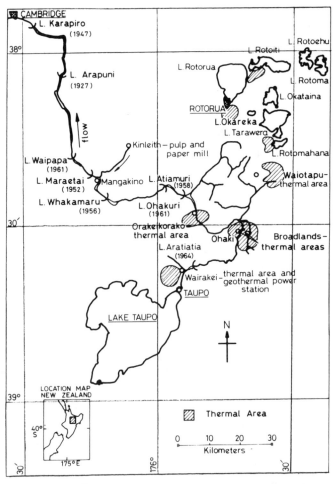

Figure 13. Map of New Zealand geothermal areas showing collection areas for rainbow trout analyzed in Fig. 14. (From Weissberg and Zobel.[65])

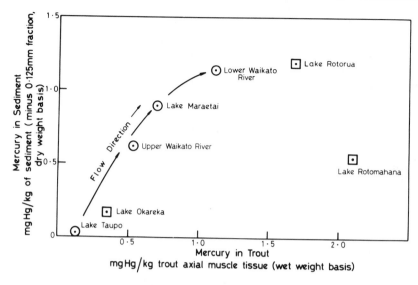

Figure 14. Mercury levels in New Zealand rainbow trout expressed as a function of mercury concentrations in lake sediments of the collection locality. (From Weissberg and Zobel.[65])

The problem of the contamination of fish by mercury is a function not only of the mercury content of the fish but also of the dietary habits of the people consuming the fish. Assuming that the 0.5-ppm safety level represents mercury completely in the methylated form, the ADI of 30 mg/day is reached by eating 420 g (about 1 lb) of fish per week. In Britain, with an average weekly consumption per person of 150 g of fish,[78] this is scarcely a serious problem, but in countries like Japan, where per capita consumption is very much higher and where the problem is aggravated by a safety level of 1.0 ppm, it is clear that a serious problem arises.

In general, benthic (bottom-dwelling) fishes tend to have a higher mercury content than pelagic (free-swimming) species. This has been demonstrated by studies of the Thames Estuary,[79] where values of 0.36 ppm and 0.45 ppm mercury were obtained for the free-swimming cod and whiting, respectively, whereas the bottom-dwellers, flounder and skate, had respectively levels of 0.74 ppm and 0.80 ppm.

Mercury levels in shellfish have been determined in Tasmania by Thrower and Eustace,[77] who found anomalous levels of this element in the soft parts of the Pacific oyster (*Crassostrea gigas*). Hoggins and Brooks,[62] in New Zealand, have assessed background levels of mercury at 0.03 ppm in the cockle, 0.06 ppm in the rock oyster, and 0.02 ppm in the green-lipped mussel. Unless samples are taken from areas near heavy industry or natural sources of mercury, shellfish normally have relatively low natural levels of mercury and can usually be eaten with safety.

5.5. *Control of Mercury Pollution*

The outcry raised against pollution of the environment has resulted in some encouraging control measures, particularly in chlor-alkali plants. Within 2 months of government suits against some of the more flagrant offenders in the United States, a survey of 50 industrial users of mercury showed that pollution of waters had been reduced by 86%.[80]

Pollution from mercury may also be reduced by outright banning of various usages. For example, in 1972, the U.S. Environmental Protection Agency moved to ban all mercury pesticides used in paint, and there is no doubt that other strictures will soon follow. Some plants have been closed down completely. For example, an ICI dyestuff factory at Dighton, Kansas, in the United States, has been shut down partially because of effluent problems.[67]

5.6. *The Future*

There is no doubt that mercury contamination will always remain a problem. Whereas it will always be possible to substitute mercury for other materials in some industries, this element will continue to be used in the fluorescent lamp industry, in chlor-alkali plants, in dentistry, and in the electronics industry. Nevertheless, with new stringent control legislation now being formulated in so many countries, there is no doubt that pollution from man-made sources can be reduced to very low levels.

Mankind will obviously never be able to control mercury "pollution" from natural sources with its particularly serious problems in the fishing industry. In a world suffering increasingly from hunger, can we afford to neglect tuna and shark as food sources, merely because their mercury levels are currently too high? The tuna problem is being tackled with some success by not taking fish from areas where individual specimens have a consistently high concentration of mercury. Fishermen are being told to ignore the larger tuna which usually have a higher mercury content. Presumably similar action could be taken for the school shark industry.

As stated previously, mercury pollution is a fact of life which we will have to live with, because, unlike lead, most of the environmental mercury is derived from natural sources. In time, the current safety limits of 0.5–1.0 ppm may have to be raised in some countries if the alternative is widespread hunger or famine.

6. *Cadmium and Zinc as Pollutants*

6.1. *Introduction*

There are groups of trace elements which are known somewhat arbitrarily as "heavy metals." This term is misleading because they are not all "heavy" in terms of atomic weight, density, or atomic number, and they are not even entirely metallic in character, as, for example, arsenic. As a rough generalization, the "heavy metals" may be said to include all metals of the periodic table except those in groups I and II. Apart from lead and mercury, which have already been dealt with, cadmium, copper, and zinc are perhaps the most important heavy metals as regards environmental problems, since they are toxic and ubiquitous in distribution throughout the world. Cadmium and zinc will be discussed together since they have a similarity in geochemistry, biogeochemistry, and industrial usage, and are usually derived from the same sources of pollution.

6.2. Natural Occurrence of Cadmium and Zinc

Cadmium and zinc are (like lead and mercury) known as chalcophile elements: i.e., they tend to be found in sulfide deposits. Igneous rocks average about 0.13 ppm cadmium and 80 ppm zinc. In the process of weathering these elements are easily mobilized and find their way into the oceans, where they have respective concentrations of about 0.1 ppb and 2.0 ppb. Cadmium and zinc, together with lead, are sometimes known as base metals and are found in the concentrates derived from the treatment of sulfide deposits. A typical zinc concentrate would have about 50–60% zinc, 0.2% cadmium, and perhaps 3% lead.

Soils contain about 4.5 ppm cadmium and 50 ppm zinc. The relatively high value for cadmium is because of its biogenic enrichment in humic material.

Vegetation contains variable amounts of both elements, but zinc concentrations (1400 ppm in plant ash) are relatively high because this element is essential to plant nutrition.

Natural cadmium levels in air are thought to be about $0.002 \ \mu g/m^3$ though very much higher values, up to $0.3 \ \mu g/m^3$ are found near zinc smelters.[81] Direct data for zinc are not available, probably because this element is less toxic than cadmium.

6.3. Industrial Uses of Cadmium and Zinc

The industrial uses of both elements are extremely widespread. The world's annual production of cadmium is around 20,000 tonnes and that of zinc is about 7 million tonnes. Zinc is used very extensively throughout industry, though mainly in galvanizing and in the manufacture of brass and other alloys. The uses of cadmium are less well known, but this metal finds an ever-increasing use in electroplating, paint pigments, plastics, and silver–cadmium batteries. If long-life silver–cadmium batteries ever replace conventional lead accumulators as a power source for electric cars, the use of cadmium is likely to increase very considerably, with important and possibly grave implications for pollution of the environment.

6.4. Sources of Cadmium and Zinc Pollution

It has been calculated that about 1 million kg of cadmium are released each year into the atmosphere from smelters and factories processing cadmium, representing about 45% of the total pollution from this element. A further 52% comes from the incineration or disposal of cadmium-containing products.[82] The tendency of cadmium to be distributed in the atmosphere is partly due to its relatively high volatility. The melting point of the pure metal is only 765°C, compared with 1744°C for lead and 907°C for zinc. In general, sources for atmospheric pollution by zinc are similar to those for cadmium.

Discharge of cadmium into natural waters is derived partly from the electroplating industry, which accounts for about 50% of the annual cadmium consumption in the United States. Another source of water pollution is the nickel–cadmium battery industry. One plant discharging into a stream that feeds into the Hudson River discharged so much cadmium that the mud carried 16.2% cadmium.[82] The greatest sources of zinc pollution in waters are smelters and the galvanizing industry.

Fertilizers, which can contain up to 8 ppm cadmium, are also a source of pollution, as cadmium builds up in crop plants and is ultimately transmitted to man.

Fungicides containing zinc and cadmium are also a potential source of pollution. This may account for elevated contents of cadmium in tobacco (about 1.4 μg per cigarette) which represents a serious hazard to smokers because cadmium is more readily absorbed by inhalation than by ingestion.

6.5. Air Pollution from Smelters

Little and Martin[36] have examined the extent of windborne cadmium and zinc pollution emanating from the Avonmouth smelter and have shown that unwashed leaves of the elm (*Ulmus glabra*) had anomalous amounts of zinc up to 12 km from the works. Isoconcentration contours for zinc are shown in Fig. 7. The patterns correspond very well with the direction of the prevailing wind. The same authors also observed elevated amounts of cadmium for considerable distances from the smelter. Typical values (dry weight basis) for unwashed elm leaves on the side of the tree exposed to the smelter were: zinc, 1650 ppm; lead, 910 ppm; and cadmium, 10 ppm. Oak leaves at the same site gave values of 6200 ppm zinc, 6800 ppm lead, and 52.5 ppm cadmium. Washing removed up to 85% of the lead but only 45% of the zinc and 28% of the cadmium, thus indicating some penetration of the leaf by cadmium and zinc.

A survey similar to the above was carried out at the Avonmouth smelter by Burkitt *et al.*[35] They found anomalous amounts of zinc and cadmium in grasses, lichens, mosses, and soils up to 13 km from the smelter. Even at this distance anomalous values were obtained, and were as follows: soils: 130 ppm zinc and 4 ppm cadmium; grasses (*Lolium perenne*): 78 ppm zinc and 4 ppm cadmium; lichens: 386 ppm zinc and 20 ppm cadmium; and mosses: 876 ppm zinc and 49 ppm cadmium.

Goodman and Roberts,[37] working in the vicinity of a smelter in Swansea, South Wales, found anomalous amounts of cadmium and zinc up to 25 km downwind from the works. At this distance, values obtained were: 60 ppm zinc and 1.3 ppm cadmium in leaves of *Festuca* sp., and 140 ppm zinc and 1.0 ppm cadmium in the moss *Hypnum* sp. At a distance of 1.5 km downwind, values obtained were: 1190 ppm zinc and 40 ppm cadmium in *Festuca*, and 543 ppm zinc and 26 ppm cadmium in soils.

6.6. Water Pollution from Smelters and Industry

Heavy metals can readily build up in seawater near sources of industrial pollution or mineralization. Abdullah *et al.*[34] have carried out a thorough survey of heavy metals in seawater at various points along the west coast of England and Wales. They reported anomalously high cadmium, copper, lead, and zinc values at various points. By far the highest cadmium concentrations were found in the eastern Bristol Channel near the Avonmouth smelter. Values ranged from 0.28–4.20 ppb, with a mean of 1.13 ppb, or about ten times the levels for open-ocean seawater. Zinc values ranged from 3.57–21.42 ppb, with a mean of 9.98 ppb. These concentrations compare with a figure of about 1 ppb for zinc in open-ocean seawater.[28] Anomalous cadmium and zinc layers were also found in Liverpool Bay and in the nonindustrialized Cardigan Bay area. High levels in Cardigan Bay were ascribed to base metal mineralization from Snowdonia.

A consequence of water pollution is the build-up of heavy metals in the marine biosphere, particularly in shellfish, which have the capacity of enriching elements by several orders of magnitude over their levels in seawater.[11] Brooks and Rumsey[83] have surveyed trace element levels in commercial fish caught from presumably nonpolluted waters off the coast of New Zealand. Table 6 gives values for cadmium and zinc in several species. Natural levels of zinc are quite high, and this element accumulates readily in certain organs, particularly the liver and kidneys. Cadmium concentrations are satisfactorily low in edible portions of the fish (i.e., muscle) but are extremely high in organs such as the kidney and liver. Levels in the livers of some species are so high that they would be likely to cause a public health problem if livers were used selectively for foodstuffs (as, for example, in the shark liver industry in parts of Europe).

Work on zinc and cadmium levels in shellfish has been carried out in Tasmania.[77] These workers found zinc levels of nearly 10,000 ppm (wet weight) in the Pacific oyster *Crassostrea gigas*, and up to 31.7 ppm cadmium. These high values were found in specimens taken from Ralph's Bay a few kilometers downstream from a large zinc smelter. These results confirmed the reasons for nausea experienced by consumers of the shellfish, since only six specimens were sufficient to produce an emetic dose of zinc. As a result of these investigations, oyster farming in Ralph's Bay ceased. The oysters from this area also had levels of cadmium 20–30 times higher than the commonly accepted safety level of 1 ppm. Experimental data are given in Table 7. Even when grown in nonpolluted areas, shellfish such as oysters usually have higher levels of zinc and cadmium than the normally recommended safety levels, and it is obvious that a special case will have to be made for these shellfish if the industry is not to be destroyed by setting unattainable standards.

Cadmium and zinc levels in seawater and marine organisms have also been investigated in the vicinity of the Avonmouth smelter. Butterworth *et al.*[17] found that anomalously high levels of both elements (>10 ppb for zinc and >0.1 ppb for cadmium) in seawater were found at distances up to 140 km from the smelter. Seaweed (*Fucus vesiculolus*) contained 800 ppm zinc and 220 ppm cadmium (dry weight basis) in estuary water near the smelter, and these levels decreased to 120 ppm and 15 ppm respectively at Hartland Point (140 km from the pollution source). The dog whelk (*Thais lapillus*) contained 3100 ppm zinc and 425 ppm cadmium at Brean (15 km from the smelter). Samples taken from other stations showed a progressive decline in concentrations of both

Table 6. Mean Cadmium and Zinc Content (ppm) of Organs of New Zealand Sea Fish[a]

Species		Organ	Cd	Zn
Trigla kumu	(gurnard)	liver	4.23	55
		kidney	0.12	14
		muscle	0.015	5.5
Polyprion oxygeneios	(hapuku)	liver	12.15	57
		kidney	5.35	22
		muscle	0.005	5.2
Arripis trutta	(kahawai)	liver	0.40	29
		kidney	2.27	67
		muscle	0.002	21.0
Seriola grandis	(kingfish)	liver	4.63	27
		kidney	0.30	201
		muscle	0.006	9.5
Latridopsis ciliaris	(moki)	liver	1.73	62
		kidney	0.50	14
		muscle	0.005	8.5
Chrysophrys auratus	(snapper)	liver	0.10	50
		kidney	0.71	49
		muscle	0.016	3.7
Cheilodactylis macropterus	(tarakihi)	liver	14.70	319
		kidney	1.67	18
		muscle	0.006	2.8
Caranx lutescens	(trevally)	liver	1.05	38
		kidney	0.85	235
		muscle	0.010	11.2

[a] Source: R. R. Brooks and D. Rumsey.[83]

Table 7. Heavy Metals in Tasmanian Oysters (Crassostrea gigas)[a]

Location	Map. Ref. (Fig. 16)	Mean concentrations (ppm wet weight)		
		Cd	Cu	Zn
Ralphs Bay	16	31.7	113	7303
		16.9	127	9122
		10.1	73	6244
		17.5	75	6301
		11.6	81	6558
		19.1	110	6838
		17.3	93	6564
Supply River	19	14.6	99	1602
Ruffins Bay	18	7.2	124	1354
Middle Island	17	5.0	136	1146
		2.8	104	526
		5.0	110	797
		4.2	192	1160
		2.3	87	451
Ulverstone	2	3.3	18	429
Devonport	3	3.2	29	572
Port Sorell	4	2.8	29	333
St. Helens	6	<2.0	10	371
Triabunna	7	<2.0	61	376
Sandspit	8	<2.0	17	359
Blackman Bay	9	<2.0	9	480
Dunalley	10	3.5	13	653
Pittwater	11	2.0	69	1038
Pipe Clay Lagoon	12	3.7	84	3449
		2.8	46	1913
		5.7	61	2966
Simmonds Bay	13	2.2	19	1136
Cygnet	14	<2.0	13	338
Long Bay	15	3.7	26	620

[a]Source: S. J. Thrower and I. J. Eustace.[77]

elements, with anomalous values still to be found at Hartland Point. It is clear from this work that cadmium and zinc pollution from the Avonmouth smelter is a serious problem off the southwest coast of England. Preston *et al.*,[84] working in the same general area, found extremely high values for cadmium in the seaweed *Fucus vesiculolus.*

6.7. The Toxicology of Cadmium and Zinc

Zinc is perhaps the least toxic of all the heavy metals discussed in this chapter. It is, in fact, an essential element in animal and plant nutrition; nevertheless, all elements are toxic if absorbed in excess, and zinc is no exception. The main physiological action of zinc is to act as an enzyme activator. In large doses it also acts as an emetic, as already discussed above in relation to oysters in Tasmania. An emetic dose is around 150 mg for a human adult male.

Compared with zinc, cadmium is a much more toxic element. It builds up in the human body over a prolonged period and the average American male is said to accumulate about 30 mg of cadmium,[85] of which 33% accumulates in the kidneys, 14% in the liver, 2% in the lungs, and 0.3% in the pancreas. The rest of the cadmium is distributed in other tissues.

Chronic cadmium poisoning produces proteinuria and causes the formation of kidney stones.[86] Brooks and Rumsey[83] have also examined the accumulation of cadmium in the kidneys of commercial sea fish in New Zealand. Schroeder *et al.*[87] have associated hypertension in man with a high cadmium/zinc ratio in the kidneys. Further evidence for the link between cadmium and hypertension has been furnished by work by Caroll[88] and Hickey *et al.*[89]

A disease specifically associated with cadmium poisoning has been recognized in Japan. This disease, known as "itai-itai" ("ouch-ouch"), is related to pollution from a mining complex and resulted in multiple fractures arising from osteomalacia.[90]

In the United States, the safety level for cadmium in drinking water has been set at 0.010 ppm. A survey of surface waters in the United States[91] showed that 4% of the 720 samples had cadmium concentrations above the safety limits. All these samples were taken from the vicinity of major industrial complexes.

The main problem with cadmium in human nutrition appears to be that the body seldom excretes as much cadmium as is absorbed. An overall daily intake of about 2 μg among humans is compounded in the case of heavy smokers, where consumption of 20 cigarettes per day can result in an additional net intake of 1–4 μg.

There is little general agreement about acceptable safety limits for cadmium in foodstuffs. In Japan, a figure of 0.9 ppm has been set for polished rice, but in other countries studies are being carried out on cadmium in foodstuffs, particularly fish, before setting arbitrary safety limits which could ruin rich fishing industries. It is possible that most countries will ultimately decide on a limit of 1 ppm, except perhaps for shellfish, where a special case will almost certainly have to be made.

6.8. *The Future*

It is probable that in the near future cadmium will join lead and mercury in forming the "Big Three" of trace elements with the greatest potential hazard to humans and the environment.

Much greater efforts will be needed in the future to reduce pollution from cadmium, and to a lesser extent, from zinc. The first steps which can be taken will be stringent control of emissions from smelters and other industrial plants. Even if this were achieved tomorrow, there still remains the problem of pollution from high concentrations of cadmium which have already built up in the environment, and it is likely to be several decades before these levels are reduced to an acceptable degree.

7. Other Pollutants

7.1. Introduction

Apart from the elements already discussed (lead, mercury, cadmium, and zinc) there are many other toxic elements whose dispersion into the environment by human activities is a cause for growing concern. One of the best sources for a study of pollution from these elements is a report by the World Health Organization,[55] from which most of the material in this section is derived. Each element will be discussed separately.

7.2. Antimony

Antimony is found along with arsenic in sulfide deposits. It is not a very significant pollutant of the environment because of its limited use by man. It is used as a constituent of alloys, such as type metal, and is also used in the match making industry.

The toxicology of antimony is similar to that of arsenic. Large doses of antimony have an emetic effect, but chronic poisoning results in disturbances of the gastrointestinal tract with vomiting, diarrhea, and lowered temperature.[92]

7.3. Arsenic

Arsenic is another of the chalcophile elements which is accumulated in sulfide deposits. It is found in concentrations of up to 1% in lead–zinc concentrates used in smelters, and because of its volatility (sublimes at 650°C) it can be readily distributed into the atmosphere. Perhaps the greatest sources of pollution from this element are agricultural chemicals such as weed killers, fungicides, rat poisons, and insecticides.

In its toxicology, arsenic is a cumulative poison which inhibits SH groups in enzymes. Chronic poisoning leads to loss of weight and loss of appetite, with accompanied gastrointestinal disorders similar to gastroenteritis.

In 1900, 70 deaths in Manchester and Liverpool resulted from the consumption of beer containing more than 15 ppm arsenic.[93]

It is not considered that arsenic is a universal health hazard as it seems to be largely an occupational risk. For example, arsenic poisoning is often found among wine growers.

Arsenic is also believed to cause lung and skin cancer in humans.[94,95,96]

7.4. *Beryllium*

Beryllium resembles aluminum in its geochemistry and is present to the extent of about 4 ppm in rocks and soils. Symptoms of beryllium poisoning are acute pneumonitis and berylliosis, which is a progressive disease affecting the heart. When inhaled as a dust it can also cause lung cancer. Although beryllium at present has limited commercial use, its inherent dangers must not be taken lightly. Early complacency about its effects led to several industrial deaths in the late 1960s.[97]

An experience in the United States[98] showed that there is a public health aspect of beryllium poisoning, as opposed to a mere occupational hazard. In 1949, the United States Atomic Energy Commission discovered that one of its beryllium plants was contaminating the surrounding air to the extent of 0.1 $\mu g/m^3$ up to 1 km away from the plant. Ten cases of beryllium poisoning resulted and one person died.

7.5. *Cobalt*

Cobalt is not normally considered to be particularly hazardous, but there have been cases of poisoning due to this element. The ADI for cobalt is said to be 0.25 mg, but ingestion of higher doses can adversely affect the hemoglobin content of the patient's blood, and produce polycythemia and hyper-lipemia.[99,100] Arena[99] also considered that cobalt causes goiters in humans. Cobalt is used mainly in the steel industry and poisoning is likely to be occupational rather than a public health risk.

7.6. *Copper*

Copper is another chalcophile element which is found in sulfide deposits along with lead, cadmium, and zinc. It is usually present in small quantities in zinc concentrates and in smelters and can be carried long distances by air and water.[34]

In work carried out near a smelter in Swansea, Goodman and Roberts[37] found anomalous levels of copper in plants and soils at distances of up to 8 km from the source.

Copper concentrations in vegetation tend to be fairly constant irrespective of the amount of copper in the soil, because copper is an essential element in plant nutrition.[7] The same observation is true for most animals. However, certain species, such as the oyster, have a physiological requirement for copper in their blood due to the formation of hemocyanin. For this reason, elevated copper levels can be expected in oysters and other shellfish. Brooks and Rumsby[11] have reported averages of 50 ppm (dry weight) for the oyster *Ostrea sinuata*. Thrower and Eustace[77] have reported average copper levels of up to 136 ppm (wet weight) in specimens of the oyster *Crassostrea gigas* growing in polluted estuaries in Tasmania (Table 7). Elsewhere, in less polluted waters, the copper levels only averaged 30 ppm with a maximum of 84 ppm.

Although copper is one of the most common of the industrial metals, it does not appear to have had its toxicity studied as much as many other elements. Although there is certainly a high incidence of lung cancer among coppersmiths,[101] there is no direct evidence that this element is carcinogenic.

Absorption of excess copper by man results in "Wilson's Disease," in which excess copper is deposited in the brain, skin, liver, pancreas, and myocardium.

7.7. Fluorine

The element fluorine is found mainly in fluorspar (CaF_2) and is physiologically very active. An extensive review of this element has been carried out by The World Health Organization.[101] Fluorine pollution is a serious problem, particularly in the vicinity of aluminum smelters, which use cryolite (Na_3AlF_6) as a flux. Fluorine tends to be accumulated in human bones progressively with age and with fluorine intake.[103] Small doses of fluoride (1 ppm in drinking water) reduce the incidence of dental caries, but an excess can cause mottling of the enamel in the teeth of humans and of domestic animals.

High doses of fluorine can be toxic to humans, with depression of collagen formation.[97]

7.8. Manganese

Although manganese is an essential element for plants and animals, chronic poisoning is sometimes observed in the mining industry. Manganese in the air may have adverse effects on humans.[105] Poisoning takes the form of progressive deterioration in the central nervous system.[106] Introduction of manganese into the environment by human activity is not really a serious problem at present, because natural levels of this element in plants, animals, and soils are so high that it would be difficult to raise them significantly by human industrial activity. Nevertheless, localized effects are sometimes found.

If at some time in the future manganese were to replace lead, as was attempted some years ago in experiments with methylcyclopentadienyl manganese tricarbonyl as an anti-knock agent, then an entirely new situation could arise that would merit serious study of the potentialities of manganese as a pollutant.

7.9. Nickel

Nickel has a very large number of industrial usages, ranging from electroplating to the manufacture of long-life batteries. It is also a common constituent of coinage. There is a fair amount of direct evidence that this element is carcinogenic.[107,108]

One of the most serious sources of nickel pollution is airborne effluent from plants using the nickel carbonyl method of manufacture. Nickel carbonyl is considered to be several times more toxic than carbon monoxide.[98] Within the lung the complex breaks down and deposits finely divided nickel, which is the main cause of cancer in that organ.

Apart from loss from nickel carbonyl plants, which are few in number, a much greater source of environmental pollution is the consumption of fossil fuels, which are estimated to release 70,000 tonnes of nickel each year into the atmosphere.[98]

7.10. Selenium

Selenium is a chalcophile element which has been studied extensively in recent years because of its association with animal health. Selenium deficiency in cattle has been recognized in many areas (e.g., "staggers" in New Zealand). At the same time, selenium is extremely toxic, and an excess of this element in diet can lead to serious poisoning effects. One of the best-known examples of this latter effect is the poisoning of cattle in the Colorado Plateau because of their consumption of selenium-accumulating plants such as the *Astragalus* species.[109] These plants frequently contain over 1% selenium (dry weight basis).

In its toxicology, selenium is considerably more poisonous than arsenic. Selenium has limited commercial uses except in the electronic and electrical industries. Because of its extreme volatility it is easily dissipated into the atmosphere. The element is so toxic that United States health authorities have set a maximum safety level of 0.01 ppm in drinking water. Copeland *et al.*[41] have reported selenium levels of up to 1.2 ppm in zooplankton in Lake Michigan downwind from Chicago. Tucker[98] has suggested that an upper limit of 0.25 ppm (wet weight) would be a realistic safety level for fish.

7.11. Tin

Although used extensively in the tinplate industry, tin is not a particularly serious health hazard. The toxic dose in man is about 2–3 g. This causes severe headaches, vomiting, and abdominal pains.[110]

7.12. Vanadium

Vanadium is a common constituent of petroleum, and by far the greatest source of pollution by this element is the burning of gasoline and oil. Mountain *et al.*[111] have studied the toxicology of this element and have found that it causes conjunctivitis, pharyngitis, and persistent coughing. Vanadium is also known to interfere with cholesterol synthesis.[112]

8. Trace Element Imbalances

One consequence of the build-up of abnormal concentrations of a particular element in the environment is that humans and animals may be adversely affected not because of the toxic effects of the element concerned but because it may lessen the availability of an essential element, and hence produce in the victim symptoms of a deficiency disease. This subject has been discussed by Furst[81] in relation to cancer studies.

A typical case of elemental imbalance is found in the effect of excess zinc in animals. The symptom of zinc toxicity is iron deficiency anemia.[113] The resulting anemia causes depression of iron-containing enzymes.

Another example of an elemental imbalance is the classic case of copper deficiency in cattle (hypocuprosis) caused by inordinately high molybdenum levels in soils.[114]

Zinc and copper are the two most important essential elements whose metabolism and absorption by plants[115] and animals are frequently upset by excess of other elements. A particularly interesting case is the cadmium–zinc content of human kidneys, which is an index of cadmium pollution.[87] The unfavorable balance can be restored to a limited extent by supplying the patient with an excess of zinc and hence relying on the mutual antagonism of cadmium and zinc to reduce the cadmium level in the organ concerned.

The study of the relationship between trace element levels in the environment and the incidence of human diseases is known as epidemiology. There is some evidence that diseases such as multiple sclerosis may be linked to trace element excesses or deficiencies. Most of the excesses are due to "natural pollution," but some may be due to human activities. The evidence has been reviewed by Armstrong.[116] Warren[117] showed that communities in Sweden and Norway with a low prevalence of the disease were living in areas with low environmental levels of trace elements.

Some epidemiological studies of deaths from malignant neoplasms have been carried out in England. It has been observed that certain parishes in West Devon have cancer mortalities well above the natural average. Warren and Delavault[118] studied the copper, zinc, lead, and molybdenum content of lettuce grown in the West Devon area, and compared these levels with data for lettuce grown in the Stoke-on-Trent region, where the incidence of cancer was low. They observed lead values which averaged five times higher in samples from the Devon area than in those from Stoke-on-Trent. The corresponding zinc concentrations were twice as high. This work by Warren and Delavault[118] involved too few samples to be statistically significant, but it did at least point out suitable avenues of future research in this field.

9. *The Future*

There are many viewpoints about the future trend of pollution in our planet and these opinions range from the downright pessimistic through to the overoptimistic. Among the wide gamut of divergent opinions there do appear to be some which are realistic and point the way to a solution or partial solution of the pollution problems which seem to threaten the future of the world, if not now at least in the not-too-distant future.

In order to seek a solution to the pollution problem, it is first necessary to identify the problem and to place it in its proper perspective. At present, pollution of the environment is a tertiary dispersion of the chemical elements from localized concentrations in ore bodies into worldwide distribution in the environment. The consequences of this dispersion are extremely serious, quite apart from health problems associated with pollution: if important elements become too widely dispersed throughout the world they will be impossible to retrieve because of the vastly greater amount of energy needed to obtain a given unit of an element from a material of low concentration as compared with one of high concentration, such as an ore body. The world will ultimately run out of these elements and will face economic collapse.

In order to place the pollution problem further into perspective, it must be appreciated exactly why it is that human activities upon this planet are harmful to the environment. In his pretechnological days, man presented no threat to the environment because he was an integral part of it. Commoner,[119] in a thoughtful paper, has used the term *environmental impact*, which is a measure of the extent to which an agency *external* to the ecosystem is able to degrade that system.

A human population will have zero environmental impact only as long as it remains part of the ecosystem. If the population is concentrated in a city it will be outside the system and will degrade it by pollution. For example, a human population within the ecosystem returns its sewage and effluent to the soil. This obviates the loss of trace elements to the hydrosphere and eliminates the need for the ecologically harmful manufacture of artificial fertilizers.

The environmental impact index (I), as proposed by Commoner,[119] is defined as

$$I = \text{population} \times \frac{\text{economic good}}{\text{population}} \times \frac{\text{pollutant}}{\text{economic good}}$$

The first term is the size of the population; the second term is a measure of the consumption or production per capita (i.e., affluence); and the final term is the amount of pollutant generated per unit of production or consumption, as represented by the quantity used to measure "economic good" in the second term.

Commoner[119] has studied growth rates of consumption or production in the United States taken as a mean over about 20 years. During that period the population increased at an annual rate of about 2%; some productions or consumptions lagged behind population growth (e.g., railroad horsepower and soap consumption); those such as steel and new copper production followed the population increase; and the growth rate of some (such as mercury consumption in chlor-alkali plants) greatly exceeded the rate of population increase. The latter group includes many items which have replaced those

Table 8. Environmental Impact for Tetraethyl Lead[a]

Index factors	1946	1967	1967/1946	% Increase
(a) Population	140.7×10^6	197.9×10^6	1.41	41
(b) Vehicle km/population (km per capita)	3171	6339	2.00	100
(c) Tetraethyl lead/vehicle km (kg per vehicle km)	85×10^{-6}	178×10^{-6}	2.10	110
Total index (a×b×c) (kg)	37.9×10^6	223.3×10^6	5.89	489

[a]Source: After B. Commoner.[119]

which are older and were less energy intensive. A good example of this is the replacement of railroad horsepower (−4%) by automotive horsepower (+7%).

Commoner[119] followed up these observations by computing environmental impact indices for a number of items, including tetraethyl lead. His findings for lead are summarized in Table 8. The environmental impact index increased by 485% between 1946 and 1967.

Similar calculations for other units of production or consumption reveal an alarming picture; but at the same time, a way out of the present predicament presents itself. Commoner[119] cites the example of gold. Because gold is considered to be valuable, it is carefully controlled and very little of it is lost to the environment. Silver is a similar case, and most commercial photographers have a procedure to recover this metal from used film. These two examples show that given *sufficient motivation*, elements can be recycled and not lost to the environment with the consequent pollution problems.

It is true that, in time, most metals will become so rare and expensive that it will be economically worthwhile to recover them, but action will have to be taken long before that time. The impact of humans upon the environment can be reduced by developing alternatives to ecologically harmful activities.

There are some signs that the move to recycle important elements is gaining momentum, and that stringent government action is beginning to reduce pollution problems in some areas, particularly in the loss of chlor-alkali mercury from paper mills. Unfortunately, there is still the reactionary view that it is cheaper to close down an industry than to attempt to control pollution from it (e.g., closing down some shark fishing and tuna fishing industries, or shutting down zinc smelters), and until this view is changed it will be difficult to bring pollution from base metals under control.

I will close this chapter with a quotation from Commoner's paper, which I think sums up the problem with admirable clarity:

It has been pointed out often enough that environmental pollution represents a long-unpaid debt to nature. Is it possible that the U.S. economy has grown since 1946 by deriving much of its new wealth through the enlargement of that debt? If this should turn out to be the case, what strains will develop in the economy if, for the sake of the survival of our society, that debt should now be called? How will these strains affect our ability to pay the debt—to survive?

References

1. N. L. Bowen, *The Evolution of the Igneous Rocks*, Princeton Univ. Press, Princeton, 1972.
2. R. M. Garrels, *Mineral Equilibria at Low Temperature and Pressure*, Harper and Row, New York, 1960.
3. A. I. Perel'man, *Geochemistry of Epigenesis*, Plenum, New York, 1967.
4. A. P. Vinogradov, *Int. Monog. Earth. Sci.*, **15** : 317 (1964).
5. I. J. T. Davies, *The Clinical Significance of the Essential Biological Metals*, Heinemann, London, 1972.
6. K. Liebscher and H. Smith, *Arch. Environ. Hlth.*, **17** : 881 (1968).
7. M. H. Timperley, R. R. Brooks, and P. J. Peterson, *J. Appl. Ecol.*, **7** : 429 (1970).
8. L. H. Ahrens, *Geochim. Cosmochim. Acta*, **5** : 39 (1954).
9. I. H. Tipton and M. J. Cooke, *Hlth. Phys.*, **9** : 103 (1963).
10. G. E. Hutchinson, *Quart. Rev. Biology*, **18** : 1 (1943).
11. R. R. Brooks and M. J. Rumsby, *Limnol. Oceanog.*, **10** : 521 (1965).
12. E. D. Goldberg, *Geol. Soc. Amer. Mem.*, **67** : 345 (1957).
13. M. D. Amos, P. A. Bennett, K. G. Brodie, P. W. Y. Kung, and J. P. Matousek, *Anal. Chem.*, **43** : 211 (1971).
14. R. R. Brooks, *Geobotany and Biogeochemistry in Mineral Exploration*, Harper and Row, New York, 1972.
15. J. Cholak, L. J. Schafer, and T. D. Sterling, *J. Air Poll. Center Assoc.*, **11** : 281 (1961).
16. D. Bryce-Smith, *Chem. Brit.*, **7** : 54 (1971).
17. J. Butterworth, P. Lester, and G. Nickless, *Mar. Poll. Bull.*, **3** : 72 (1972).
18. S. K. Hall, *Environ. Sci. Technol.*, **6** : 31 (1972).
19. M. Murozumi, T. J. Chow, and C. C. Patterson, *Geochim. Cosmochim. Acta*, **33** : 1247 (1969).
20. D. Bryce-Smith, *Chem. Brit.*, **7** : 284 (1971).
21. A. L. Mills, *Chem. Brit.*, **7** : 160 (1971).
22. J. Cholak, L. J. Schafer, and D. Yeager, *Am. Ind. Hyg. Assoc. J.*, **29** : 562 (1968).
23. H. V. Warren and R. E. Delavault, *Trans. Roy. Soc. Canada*, **54** : 11 (1960).
24. H. L. Cannon and J. M. Bowles, *Science*, **137** : 765 (1962).
25. A. L. Page, T. J. Ganje, and M. S. Joshi, *Hilgardia*, **41** : 1 (1971).
26. N. I. Ward, R. R. Brooks, and R. D. Reeves, *Environ. Pollut.*, **6** : 149 (1974).
27. R. H. Daines, H. Molto, and D. M. Chilko, *Environ. Sci. Technol.*, **4** : 318 (1970).
28. R. R. Brooks, B. J. Presley, and I. R. Kaplan, *Talanta*, **14** : 809 (1967).
29. K. T. Chisnall and J. Markland, *J. Ass. Pub. Analysts*, **9** : 116 (1971).
30. D. R. Crudgington, J. Markland, and J. Vallance, *J. Assoc. Pub. Analysts*, **11** : 120 (1973).
31. J. Markland and J. Vallance, *J. Ass. Pub. Analysts*, **9** : 119 (1971).
32. D. J. Nicolas and R. R. Brooks, *Proc. Aust. Inst. Min. Metal.*, **231** : 59 (1969).
33. J. K. Haywood, *J. Am. Chem. Soc.*, **29** : 998 (1907).
34. M. L. Abdullah, L. G. Royle, and A. W. Morris, *Nature*, **235** : 158 (1972).
35. A. Burkitt, P. Lester, and G. Nickless, *Nature*, **238** : 327 (1972).
36. P. Little and M. H. Martin, *Environ. Pollut.*, **3** : 241 (1972).
37. G. T. Goodman and T. M. Roberts, *Nature*, **231** : 287 (1971).
38. J. Mayman, *Sunday Telegraph* (Sydney), 17 September 1972, p. 35.
39. S. C. Gilfillan, *J. Occup. Medicine*, **7** : 53 (1965).
40. M. Klein, R. Namer, E. Harpur, and R. Corbin, *New Eng. J. Med.*, **283** : 669 (1970).
41. R. A. Copeland *et al.*, *Proc. 161st Ann. Meeting Am. Chem. Soc.*, 1971.
42. J. Tinker, *New Scientist*, **50** : 497 (1971).
43. S. Hernberg and J. Nikkanen, *Lancet*, **I** : 63 (1970).
44. J. A. Miller, V. Battistini, R. L. C. Cumming, F. Carswell, and A. Goldberg, *Lancet*, **II** : 695 (1970).
45. N. Gordon, E. King, and R. I. Mackay, *Br. Med. J.*, **2** : 480 (1967).

46. J. H. Ludwig, D. R. Diggs, H. E. Hesselberg, and J. A. Maga, *Am. Ind. Hyg. Assoc. J.*, **26** : 270 (1965).
47. R. Egli, E. Grandjean, J. Marmet, and H. Kapp, *Schweiz. Med. Wschr.*, **87** : 1971 (1957).
48. R. A. Kehoe, *J. Air Poll. Center Assoc.*, **19** : 690 (1969).
49. L. Kopito, R. K. Byers, and H. Schwachman, *New Eng. J. Med.*, **276** : 949 (1967).
50. H. Bloom, Alcohols as a Replacement for Lead Antiknock in Petrols, paper read at Vehicle Emission Committee of State Transport Ministry, Sydney, 1972.
51. A. E. Ballard and C. D. W. Thornton, *Ind. Eng. Chem. Anal. Ed.*, **13** : 893 (1941).
52. W. W. Vaughn, *U.S. Geol. Surv. Circ.*, **540** : 1 (1967).
53. L. T. Kurland, W. N. Faro, and H. S. Siedler, *Wld. Neurologist*, **1** : 320 (1960).
54. K. Tschuchiya, *Keio J. Med.*, **18** : 213 (1969).
55. Anon., *Health Hazards of the Human Environment*, W.H.O., Geneva, 1972.
56. E. D. Goldberg, in *Global Effects of Environmental Pollution*, Reidel Dordrecht, 1970.
57. S. Jensen and A. Jernelöv, *Nature*, **223** : 753 (1969).
58. A. Jernelöv, in *Chemical Fallout*, Thomas, Springfield, Ill., 1969.
59. J. M. Wood, F. Scott-Kennedy, and C. G. Rosen, *Nature*, **220** : 173 (1968).
60. J. H. McCarthy Jr., J. L. Meuschke, W. H. Ficklin, and R. E. Learned, *U.S. Geol. Surv. Prof. Pap.*, **713** : 37 (1970).
61. A. Eschlerman, S. M. Siegel, and B. Z. Siegel, *Nature*, **233** : 471 (1971).
62. F. E. Hoggins and R. R. Brooks, *N.Z. J. Mar. Freshw. Res.*, **7** : 125 (1973).
63. Anon., *U.S. Geol. Surv. Prof. Paper*, **713** : 1 (1970).
64. R. E. Cranston and D. E. Buckley, *Environ. Sci. Technol.*, **6** : 274 (1972).
65. B. G. Weissberg and M. G. R. Zobel, *Bull. Environmental Contam. Toxicol.*, **9** : 148 (1973).
66. D. E. White H. E. Hinkle, and I. Barnes, *U.S. Geol. Surv. Prof. Pap.*, **713** : 25 (1970).
67. G. Grimstone, *Chem. Brit.*, **7** : 244 (1971).
68. N. Grant, *Environment*, **11** : 18, 43 (1969).
69. Anon., *Nord. Hyg. Tidskr.*, Suppl. **4** (1971).
70. G. Löfroth, *Methylmercury*, Swedish Natl. Sci. Res. Council Bull. No. 4, Stockholm, 1970.
71. D. Hunter and D. S. Russell, *J. Neurol. Neurosurg. Psychiat.*, **17** : 235 (1954).
72. T. Takeuchi et al., *Acta Neuropathol.*, **2** : 40 (1962).
73. H. Matsumoto, G. Koya, and T. Takeuchi, *J. Neuropath. Exp. Neurol.*, **24** : 563 (1965).
74. F. Berglund and M. Berlin, in *Chemical Fallout*, Thomas, Springfield, Ill., 1964.
75. F. Berglund et al., *Nord. Hyg. Tidskr.*, Suppl. **4** : 1 (1971).
76. A. L. Hammond, *Science*, **171** : 788 (1971).
77. S. J. Thrower and I. J. Eustace, *Food Technol. Australia*, **25** : 546 (1973).
78. Anon., *Household Food Consumptions and Expenditure*, H.M.S.O., London, 1971.
79. Anon., *Survey of Mercury in food*, H.M.S.O., London, 1971.
80. J. T. Putnam, *Nat. Geographic*, **142** : 507 (1972).
81. A. Furst, in *Environmental Geochemistry in Health and Disease*, Geol. Soc. Amer., Boulder, 1971, pp. 109–130.
82. J. McCaull, *Environment*, **13** : 3 (1971).
83. R. R. Brooks and D. Rumsey, *N.Z. J. Mar. Freshw. Res.*, **8** : 155 (1973).
84. A. Preston, D. F. Jefferies, J. R. W. Dutton, B. R. Harvey, and A. K. Steele, *Environ. Poll.*, **3** : 69 (1972).
85. L. Friberg, M. Piscator, and G. Nordberg, *Cadmium in the Environment: A Toxicological and Epidemiological Appraisal*, Karoliska Inst., Stockholm, 1971.
86. G. Kazantzis, F. V. Flynn, J. S. Spowage, and D. G. Trott, *Quart. J. Med.*, **32** : 165 (1963).
87. H. A. Schroeder, A. P. Nason, I. H. Tyson, and J. J. Balassa, *J. Chron. Diseases*, **20** : 179 (1967).
88. R. E. Caroll, *J. Amer. Med. Assoc.*, **198** : 267 (1966).
89. R. G. Hickey, E. P. Schoff, and R. C. Clelland, *Arch. Environ. Hlth.*, **15** : 728 (1967).
90. I. Murata, T. Hirano, Y. Saeki, and S. Nakagawa, *Bull. Soc. Int. Chir.*, **1** : 34 (1970).
91. Anon., *U.S. Geol. Surv. Circ.*, **643** : 1 (1971).
92. E. Browning, *Toxicity of Industrial Metals*, Butterworths, London, 1969.

93. T. N. Kelynack, W. Kirkby, and S. Delepine, *Lancet,* **II** : 1600 (1900).
94. A. M. Lee and J. F. Fraumeni Jr., *J. Nat. Cancer Inst.,* **42** : 1045 (1969).
95. L S. Snegireff and L. M. Lombard, *Am. Med. Ass. Arch. Ind. Hyg.,* **4** : 199 (1951).
96. W. P. Tseng, H. M. Chu, S. W. How, J. M. Fong, C. S. Lin, and Shuh Yeh, *J. Nat. Cancer Inst.,* **40** : 453 (1968).
97. D. Hunter, *The Diseases of Occupation,* 4th ed., Eng. Univ. Press, London, 1969.
98. A. Tucker, *The Toxic Metals,* Earth Islands, London, 1972.
99. J. M. Arena, *Poisoning,* 2nd ed., Thomas, Springfield, Ill., 1970.
100. J. E. Davis and J. P. Fields, *Fed. Proc.,* **14** : 331 (1955).
101. T. Agnese, B. DeVeris, and B. Santolini, *Igiene Med.,* **52** : 149 (1959).
102. Anon., *Fluorides and Human Health,* W.H.O., Geneva, 1970.
103. F. A. Smith, D. E. Gardner, N. C. Leone, and H. C. Hodge, *Am. Med. Ass. Arch. Ind. Hlth.,* **21** : 330 (1960).
104. G. Nicols Jr. and B. Flanagan, *Fed. Proc.* **25** : 922 (1966).
105. D. H. Lee, *Metallic Contaminants and Human Health,* Acad. Press, New York, 1972.
106. R. H. Dreisbach, *Handbook of Poisoning,* 7th ed., Lange Medical, Los Altos, 1971.
107. E. Mastromatteo, *J. Occup. Med.,* **9** : 127 (1967).
108. W. W. Payne, *Proc. Am. Ass. Cancer Res.,* **5** : 50 (1964).
109. H. L. Cannon, *U.S. Geol. Survey Bull.,* **1085**A : 1 (1960).
110. T. Alajouanine, L. Derobert, and S. Thieffry, *Rev. Neurologique,* **98** : 85 (1958).
111. J. T. Mountain, F. R. Stockell Jr., and H. E. Stokinger, *Am. Med. Ass. Arch. Ind. Hlth.,* **12** : 494 (1955).
112. C. E. Lewis, *Arch. Ind. Hyg.,* **19** : 419 (1959).
113. S. E. Smith and E. J. Larson, *J. Biol. Chem.,* **163** : 29 (1946).
114. I. Thornton, W. J. Atkinson, and J. S. Webb, *Irish J. Agric. Res.,* **5** : 280 (1966).
115. C. A. Price, in *Zinc Metabolism,* Thomas, Springfield, Ill., pp. 69–89.
116. R. W. Armstrong, in *Environmental Geochemistry in Health and Disease,* Geol. Soc. Amer., Boulder, 1971, pp. 211–219.
117. H. V. Warren, *J. Coll. Gen. Practitioners,* **6** : 517 (1963).
118. H. V. Warren and R. E. Delavault, in *Environmental Geochemistry in Health and Disease,* Geol. Soc. Amer., Boulder, 1971, pp. 97–108.
119. B. Commoner, *Chem. Brit.,* **8** : 52 (1972).
120. J. V. Lagerwerff and A. W. Specht, *Environ. Sci. Technol.,* **4** : 583 (1970).
121. D. A. Andrew-Jones, *Mineral. Ind. Bull.,* **11** : 1 (1968).
122. H. E. Hawkes and J. S. Webb, *Geochemistry in Mineral Exploration,* Harper and Row, New York, 1962.

Pollution from Reactors

Those who wonder why our energy worries have come upon us so suddenly will have considerable interest in this chapter on low-intensity radiation. The difficulty of developing safe nuclear energy sources, according to the account given by Professor Sternglass, involves not only the difficulty of disposing of radioactive waste products, but also the untoward effects arising from changes in the background radiation. This is a new concept, born about 1972, which may have a significant effect on the future planning of our energy sources.

15

Radioactivity

E. J. Sternglass

1. Introduction

Man's awareness of the existence of natural radioactivity in the environment dates back more than three-quarters of a century to Becquerel's discovery that radiation is continuously released from uranium-containing minerals, just a few months after Roentgen's discovery of X rays in November, 1895. But only within the last 30 years has concern over man-made radioactivity arisen, as a result of the discovery of fission and its use in weapons that release large amounts of radioactive chemicals into the atmosphere.

Although the existence of serious health effects due to high levels of radiation was recognized within the first few years of its discovery, until very recently it was widely believed that there exists a safe level below which there are no detectable effects on human health. Thus it seemed that the very small doses from natural radiation sources and the still smaller man-made contributions to the background levels would have no significant effects on man or other living systems.

E. J. Sternglass • Department of Radiology, University of Pittsburgh, Pittsburgh, Pennsylvania

The strongest argument was that life apparently evolved successfully in the face of a continuous exposure to low levels of environmental radiation. And this argument seemed to be supported by the long experience with comparable levels of diagnostic X rays, for which there appeared to be very little risk of any detectable health effects.

It is the purpose of the present article to summarize the recently discovered evidence that we were misled by the relatively benign nature of medical X rays, and by the difficulty of detecting the subtle effects of environmental radiation, into regarding small continuous doses from background sources as relatively harmless.

As will be shown, the evidence gathered by a number of different investigators within the last few years now suggests that we may have underestimated the health risk of small trace quantities of radioactive chemicals in our air and water by 100–1000 times. And this has happened at the very moment when mankind has counted on being able to meet its enormous needs for energy by using uranium fission to replace the rapidly dwindling sources of fossil fuel.

2. *The Nature of Radiation and Its Biological Action*

In order to be able to understand how the magnitude of the biological effect of environmental radiation could have escaped detection in the face of a vast body of experience with medical radiation and many decades of laboratory and animal research, it is necessary to understand the nature of the various forms of radiation and their different modes of action on biological systems.[1–3]

2.1. *Forms of Nuclear Radiation*

As already recognized by the early pioneers at the beginning of the century, atomic radiation occurs in two basic forms; waves and particles. The wave type is essentially like ordinary light, but of a much shorter wavelength and therefore of greater energy per photon, or "bundle," of energy. This is the type of radiation discovered by Roentgen, and since then produced in X ray tubes for medical purposes. It is also the form of the very penetrating high-energy component of the radiation, emitted from radioactive chemicals such as radium, called gamma rays, where each photon or "quantum" carries typically the equivalent of 1-million-volt X ray photons.

The other form of energy emitted by radioactive nuclei consists of very rapidly moving particles of matter, such as the so-called beta rays, which are simply electrons, and alpha rays, which are helium nuclei. Since the discovery of these particles at the turn of the century, many other nuclear particles have been found to exist, such as the proton and a whole family of very short-lived mesons; only the neutron is of importance in the radiation effects of naturally occurring chemicals, being spontaneously emitted by some of the heavier elements, such as uranium.

One other form of corpuscular radiation is of great biological importance, and these are the so-called fission fragments, or the pieces of fast-moving nuclei created when uranium or plutonium nuclei break apart after having absorbed a neutron. These fission products are in fact isotopes of normal chemical elements such as barium, strontium, and iodine, differing only in mass from the stable chemical forms, or trace elements, in the environment. Because they are generally unstable, they in turn emit radiations in the form of gamma rays, beta rays, and X rays, often transforming themselves into other chemical elements and giving rise to a whole series of so-called "daughter products" that are taken up by biological systems along with the stable chemical trace elements in the environment.

Both wave and corpuscular forms of radiation are ultimately absorbed in matter through collisions with the atomic electrons, leading to fast-moving electrons that in turn are slowed down in a series of *ionization* and *excitation* processes. The biological damage in living cells is therefore ultimately due to the action of rapidly moving electrons, regardless of whether the original radiation was of the wave or corpuscular type, but the *spatial distribution*, or *density of energy deposition per unit distance*, will be very different on a microscopic scale.

Thus, heavy particles, such as alpha particles, produce a zone of extremely high ionization, or excitation density, in their wake. On the other hand, low-mass beta particles or atomic electrons ejected by X rays and gamma rays produce a very low density of ionizations and excitations on a microscopic scale, with very different biological effects, as will now be discussed.

2.2. Modes of Biological Action

There are basically two different ways in which the ionization and excitation of molecules in a biological system consisting of individual cells results in functional damage, and the two basic forms of radiation discussed above have very different efficiencies in producing this damage.

The first and most widely studied type of damage is of a physical, or "bullet-like" type, in which molecular bonds are broken directly in the target structure by the ejected electrons, producing ionizations or excitations. This is the type of fast, direct action that has been observed to be the principal cause of DNA damage in the nuclei of cells, leading to what is generally referred to as genetic damage, which is transmitted to future generations of cells if it is not lethal or if it is not repaired before the cell reproduces itself.

The second type of damage is of an indirect type, in which the ultimate harm to the critical target is initiated by the production of *highly reactive chemical species* which must diffuse some distance from their point of production to the critical target site. This type of *indirect, or chemical, action* is, for instance, involved when dissolved oxygen in the cell fluid captures a free electron to become the highly toxic O_2^-, or superoxide radical. This active form of oxygen is in turn able to initiate various chemical reactions that can lead to the oxidation of phospholipid cell membranes, very much in the way that ordinary air pollutants produce cell membrane damage. Although the total number of ionization processes per unit mass of tissue, which is at the basis of the unit of absorbed radiation called the rad,* can be exactly the same in the two processes, the resultant biological damage can be vastly different, depending on the *instantaneous microscopic density of the energy deposited.*

Thus, in the case of the direct type of action, leading to the rupture of bonds in the double-stranded DNA molecule, it is obvious that a massive, very densely ionizing particle, such as an alpha particle plowing through the nucleus of a cell, is much more likely to break both strands of the DNA molecule at the same time than is a low mass, sparsely ionizing particle such as a beta ray, as illustrated schematically in Figs. 1a and 1b.

It follows that repair of the DNA damage is much more difficult or actually impossible when an alpha ray has passed through a critical DNA molecule that contains vital coded information than when a beta ray or X ray has passed through the cell nucleus. Thus, in terms of direct damage to the genes, a large number of experiments have established that alpha particles are some 10–30 times as serious biologically for the same total energy absorbed on the same total dose in rads as X rays, gamma rays, or beta rays.† In fact, it is for this reason that, in the scale of biological toxicity, the present guidelines set up by the International Commission on Radiological Protection (ICRP)[4] regard the alpha-emitting elements such as uranium, radium, and thorium as by far the most toxic of all radioactive trace elements.[5]

However, paradoxical as it seems, the reverse is true for the indirect, chemically mediated type of biological damage. This is because the ultimate effect on the critical cellular structures must be conveyed by the diffusion of an activated molecule, and thus the functional damage depends critically on the chance of this molecule reaching the target structure before it is deactivated. But the chance of being able to reach the critical target structure is affected by the local concentration of excited molecules, being less when the local densities are very high, as in the case of a densely ionizing alpha particle, than when they are very low, as in the case of a fast electron. (See Figs. 1c and 1d.)

Thus, it is well known that collisions between excited molecules lead to deexcitation, so that many of the highly toxic O_2^- molecules are converted back into harmless O_2, thereby greatly reducing the overall efficiency with which absorbed energy is converted into biologically toxic molecular species when the ionization density is high. These are the so-called "recombination" processes, and they have been extensively studied for many molecular species in both gases and liquids subjected to ionizing radiation.

---

*The rad is a measure of the energy absorbed per gram of matter, defined as an energy absorption of 100 ergs/g. It applies to all types of radiation, and for X rays and gamma rays it is effectively the same as the unit of exposure, the roentgen (1 rad = 1000 mrads).

†This fact led to the introduction of the "rem" as a unit of equivalent dose absorbed, such that rem = $(DE) \times$ rad, (DE) being a factor of 10–30 for heavy particles. For X rays, beta rays, and gamma rays under 3 MeV, $DE = 1$, and 1 rad = 1 rem.

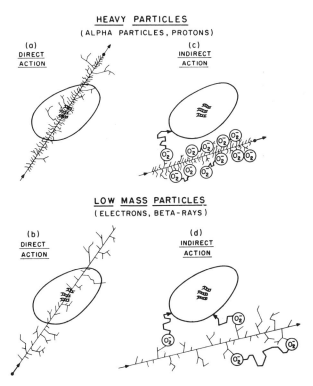

Figure 1. Basic modes of biological action of ionizing particles on living cells. (a) and (b) represent the direct action of massive and light-weight particles, primarily on the genes. (c) and (d) represent the indirect chemical action via the production of free radicals such as O_2^-. Note that the higher ionization density leads to a greater concentration of free radicals in (c) than for the lower ionization density in (d), resulting in a larger fraction of O_2^- molecules able to reach the cell membrane in (d) than in (c).

However, not only can the toxic free radicals be deactivated by collisions with each other, but their chance of reaching a vital cell component can be greatly reduced if "scavenger" molecules are present that are able to capture and deactivate the toxic molecular species. Thus, the biological effect of a given dose of radiation can be greatly influenced by the concentration of so-called "radioprotective" molecules,[1] which destroy the excitation and thus break the chain necessary for biological damage to result, as, for instance, in the case of mercaptoethanol.

Still another mechanism has recently been studied that causes the amount of biological damage to cell membranes to depend inversely on the density of ionization, and this has to do with the particular nature of phospholipid membranes. It turns out that such membranes are micelles, consisting of parallel arrays of long-chain molecules which produce a net electric field near the membrane surface, due to the fact that positively charged molecules generally project outward to a greater distance than negatively charged ones.

As a result, it has been found that cell membranes produce an electric field in the cell fluid that tends to attract negatively charged molecules, such as the highly toxic O_2^-. Furthermore, it was found by detailed computer calculations that the greater the concentration of free ions, the weaker the electric field becomes that attracts the O_2^-.[6] Thus, when ion concentrations are high, as from the passage of an alpha particle or a very high dose rate produced by an intense X ray beam, O_2^- molecules are not able to reach the sensitive cell structures as efficiently as when the instantaneous ion concentration is very low.

One therefore obtains the paradoxical result that for a given total dose, densely ionizing radiation—either from an alpha particle or from a short burst of X rays, as produced by medical diagnostic equipment—*can be biologically less damaging to cell membranes* than the protracted, low-level background radiation produced by an occasional fast cosmic ray particle or electron ejected by gamma rays from radioactive elements in the air, on the ground, or in the body.

The basic theory of recombination has been known for decades.[1] But the fact that it actually takes a much smaller absorbed dose to rupture phospholipid membranes of the type occurring in living cells when the radiation is protracted over long periods of time than when it it is given in a brief burst was only discovered within the last few years; as so often in the history of science, it happened quite accidentally, in the course of radiochemical studies of the action of radiation on synthetic phospholipid membranes immersed in water that were being carried out by Petkau, a Canadian biophysicist working in a laboratory of the Canadian Atomic Energy Establishment, in Manitoba.[7]

While carrying out measurements of the pH of an aqueous solution surrounding a small phospholipid membrane in the presence of radiation, Petkau observed that the membranes would rupture more quickly than when the X rays were turned off. He therefore decided to investigate this phenomenon further, and carried out a series of experiments to determine just how much of a dose was required to break the membranes at different radiation intensities.

Using radiation from a diagnostic X ray machine operating at 26 rads/min, he found that it took the enormous dose of 3500 rads to break the membrane, or some 35,000 times the annual dose from normal background radiation, which is close to 0.1 rad, or 100 mrads in most areas of the world. This was certainly reassuring, indicating that the membranes of living cells are not likely to be damaged by typical diagnostic X ray exposures, whose doses to the skin are generally in the range of 0.1–1.0 rad.[8]

But when he substituted a very small amount of radioactive sodium salt ($Na^{22}Cl$) in the water for the external X ray beam, such that the dose rate was reduced to only 1 mrad/min, he found that it took only about 0.7 rad to rupture the membrane, or some 5000 times less than with the high-intensity beam of medical X rays.

Subsequent studies showed that only a factor of 10–20 could be attributed to the difference between the types of radiation produced by the two different radiation sources. The remainder was found to be due to the difference in the rate at which the radiation was emitted, the dose required to rupture the membrane declining slowly as the dose rate was lowered, as shown in the plot of Fig. 2, taken from Petkau's original paper.[7]

Detailed measurements of pH in the presence and absence of the membrane confirmed that it was indeed the production of the O_2^- radical which was involved.[9] Subsequent studies on the membranes of living microorganisms in the presence of chemicals known to capture O_2^- decreased the sensitivity to damage, exactly as required by an indirect chemical action of radiation in which the free radicals reaching the membrane initiate an oxidative chain reaction that eventually destroys the membranes.[10,11] Since then, experimental studies on mice exposed to various levels of radiation have shown that enzymes known to deactivate excited forms of oxygen molecules provide a significant degree of protection to the white blood cells, which form an essential part of the defense against infectious diseases.[12]

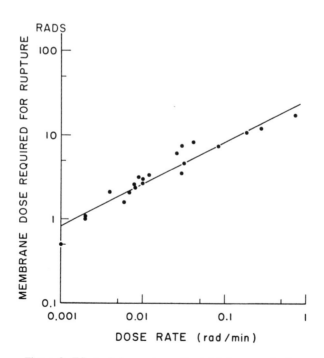

Figure 2. Effect of dose rate on the total dose required to rupture a cell membrane. Note the strong decrease in the absorbed dose measured in rads as the dose rate is reduced. (Data from Petkau.[7])

2.3. *Implications for the Risk of Environmental Radiation*

It is clear that the discovery that cell membranes require a smaller total dose for rupture as the dose rate is reduced has very serious potential implications for the likely effects of protracted background radiation, as compared with the known effects of high-dose-rate radiation. For this reason, it is essential to investigate whether this decrease in the dose needed to produce biological damage holds all the way down to the small dose rates of 10–100 mrads/year produced by global fallout,[3,11] and the normal releases from nuclear facilities.[12,13] Furthermore, it must be established whether this holds not only for isolated membranes in a test tube, but also for membranes in complex living systems where natural protective mechanisms can operate, not only in lower animals, but also in the case of man.

That this appears to be the case has recently been shown[14] by an analysis of two separate studies using very low total doses and dose rates. The first involved the precursors of blood cells formed in the bone marrow of rats,[15] and the other dealt with a study of the permeability of human red cells (erythrocytes) for individuals occupationally exposed to X rays at only a few times the normal background levels.[16]

The animal study was carried out by a group of investigators at the Oslo Cancer Hospital,[15] using very small trace quantities of strontium-90, of the order of picocuries* per gram of body weight, comparable to the concentrations observed in the bodies of newborn children during the peak of nuclear weapons testing.[17]

Using a series of matched pairs of animals given different amounts of strontium-90, these investigators found that the number of bone marrow cells was more strongly depressed for a given concentration of strontium-90, and therefore for each unit of radiation dose, the lower the concentration. Typically, the percentage decrease in the number of bone marrow cells ranged from 3.1%/mrad at an average body concentration of 1.2 pCi/g to 0.01%/mrad at a concentration of 2700 pCi/g. But this was exactly the type of behavior noted by Petkau for the case of cell membrane rupture, the sensitivity to damage becoming greater the lower the concentration of radioisotope in the liquid surrounding the cell membrane.[7]

Since the study of bone marrow cells by Stokke and his co-workers in Oslo was done before Petkau discovered the membrane rupture phenomenon, no ready explanation of the results was available at the time they were published, in 1968. But in the light of Petkau's subsequent findings, confirmed more recently by the observation of white cell depression in mice that could be protected against by an enzyme specific for the excited states of oxygen,[12] it is now clear why the lower concentrations of Sr^{90} showed a greater efficiency in depressing bone marrow cellularity than the higher concentrations.

---

* A picocurie (pCi) is 10^{-12} curies, the curie (Ci) being a unit of radioactivity corresponding to the number of disintegrations per second of 1 gram of radium, or 3.7×10^{10} disintegrations per second.

It appears that the smaller the number of radioactive atoms per unit volume, the more efficiently are the activated oxygen molecules produced able to reach the critical sites in the developing cells. Since a single molecule is able to initiate a self-propagating oxidative chain reaction over a period of hours or days, increasing the concentration has only little further effect.

In fact, one has a situation quite analogous to the case of pesticide molecules acting on critical membranes of glands controlling the thickness of eggshells in birds.[18] There, too, the effect per unit dose is greatest at the lowest concentrations, a single molecule of DDE apparently being able to affect a critical membrane site.

Thus, one is led to a dose–response function which rises very rapidly at low concentrations and then saturates rapidly for higher doses, exactly in the manner of a logarithmic type of response familiar in many sensory responses in complex living systems, such as the sense of smell (see Fig. 3[19]).

The question remains whether a similarly increased efficiency of small amounts of radiation can be demonstrated in human cell membranes. Such evidence was in fact supplied by a recent study of erythrocyte permeability carried out by Scott and his associates at San Francisco Hospital.[16]

Using groups of individuals exposed to low levels of radiation in the course of their normal work, such as radiologists and X ray technicians whose

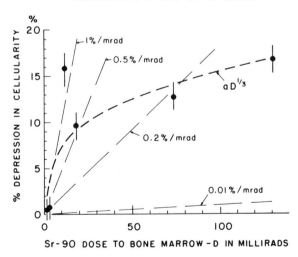

% DEPRESSION IN BONE MARROW CELLULARITY AT LOW Sr-90 DOSES

Figure 3. Form of dose-response relationship for the action of very small dose of radiation on bone marrow cells. (Data from Stokke *et al.*[15]) Note the large rise at very low doses, followed by a much more gradual rise in the effect for larger doses. Also shown is a curve of the form $aD^{1/3}$, expected on the basis of an indirect chemical action as observed by Petkau.[7]

exposure was known from their personal dosimeter readings, samples of their blood were analyzed for the permeability of the erythrocyte membranes to the passage of monovalent cations using a tracer technique that employed Rb.[86] The permeability of the erythrocyte membranes of the occupationally exposed group was compared with that of other individuals in the hospital with no X ray exposure, and a significant excess permeability was detected for the exposed individuals.

Furthermore, a plot of the percentage increase in permeability showed the same rapid rise for the lowest exposure and a leveling for the higher exposures, also observed in the animal studies of Stokke,[15] and consistent with the logarithmic, or fractional power, type of response measured by Petkau for cell membranes.[7]

Even with regard to the absolute magnitude of the effect per millirad, the result of Scott for human red-cell membranes agrees closely with the result of Stokke for the depression of the precursor cells in the bone marrow of animals given tracer doses of Sr.[90] Thus, Scott found a permeability change of some 20% for a monthly dose of about 8 mrads, which, for a typical lifespan of red cells of 3 months, works out to 0.8%/mrad. This may be compared with an effect of 0.6%/mrad at a comparable dose rate of 6 mrad/month, observed by Stokke.[15]

Clearly, all this evidence suggests that at very low dose rates, where the repair of direct damage to the DNA is apparently very effective, it is the indirect, chemically mediated damage to cell structures, such as the phospholipid membranes that dominates, owing to the lower total dose needed to produce functional damage when the ion concentrations are very low.

That the nature of this difference is dependent on the ion density, or dose rate, can be seen from an examination of Fig. 4, taken from Sternglass.[14] In this plot, the total dose D_o needed to produce a given effect, such as membrane rupture, or doubling of the normal rate of genetic defects, has been plotted against the dose rate in rads per minute.

It is seen in the figure that the data points involving some form of membrane damage fall along a broad band that declines as the dose rate is lowered from very high rates to those encountered in the natural environment. In particular, the data on isolated membranes obtained by Petkau are seen to fall along a line roughly parallel to that found by Stokke for bone marrow cellularity, but with a lower absolute value corresponding to the absence of natural protective chemicals in the *in vitro* laboratory study.

By contrast, the data observed for genetic mutations in mice[20] and for cancer risk in both animals[21] and man[22] show the opposite tendency at high dose rates, the amount of dose needed for permanent genetic damage of increased cancer risk declining slowly as the intensity of the radiation is increased. This fits the hypothesis that genetic damage is primarily due to a direct-hit type of process, where the likelihood of damage to both strands of the DNA before repair can take place increases directly as the instantaneous density of ionization is increased.

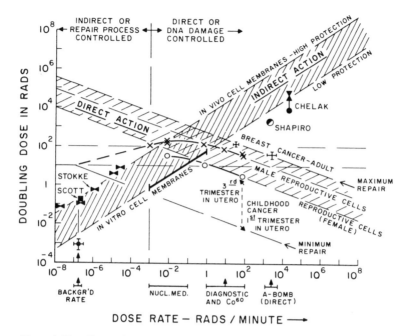

Figure 4. The effect of dose rate on the total dose needed to produce a given amount of biological damage for both the direct and indirect mechanisms.[14] Note that the indirect action causes the critical dose D_0 required to decrease as the dose rate declines toward those of background radiation, opposite to the trend for the direct action, where D_0 tends to increase as the rate of radiation absorbed decreases, leading to very large differences between expected and observed effects when the direct action on the genes is assumed to be the dominant mechanism at low dose rates.

In terms of a complex organism consisting of many cells, it is clear that whichever mechanism requires the lowest dose will dominate at a given dose rate. At high doses and dose rates, it will be the DNA damage, leading to cancer through the triggering of uncontrolled growth, for which the "doubling dose" is of the order of 10–100 rads according to a series of recent reviews of the subject.[21,22,23] On the other hand, below a dose rate of about 10^{-3} rad/min, which is 10,000 times the natural background dose rate of 10^{-7} rad/min, the indirect, chemically mediated damage has the lower doubling dose. This leads to such functional effects as a depletion of white blood cells involved in the immune process, thereby decreasing the resistance to infectious diseases,[24] a phenomenon previously associated only with very high doses given at high dose rates, such as those that result from the flash of a nuclear bomb.[25]

Other conditions associated with cell membrane damage dominating at background rates would be expected to occur in large populations exposed to very low radiation levels. These would involve developmental defects produced by failures of proper cell division or cell differentiation during embryonic development, leading to increased rates of birth defects, spontaneous abortions, and deaths in early infancy as a result of immaturity, hormonal deficiencies, and accompanying defects in the immune system.

The same damage to the immune system would also be expected to lead to a decreased ability to detect and destroy cancer cells,[26] thus leading to a rise in cancer mortality associated with a decreased resistance to the spread and growth of tumor cells.

In fact, all the diseases normally associated with aging, such as those of the lung, the heart, and the circulatory system, would be expected to increase as the production of free radicals leads to an increased efficiency of free radical damage to cell membranes. Such generalized aging effects of radiation have long been observed for animals exposed to radiation at the high dose rates used in all early laboratory studies,[27] but because of the large dose required for membrane or DNA damage at high dose rates, the small dose from background radiation had been discounted as a significant factor in the normal aging process.

However, it is now apparent that the new evidence of Stokke, Scott, and Petkau on the indirect chemical effect of radiation completely alters this expectation. Since the dose required to double the normal changes of membrane-type damage appears to be of the order of 100–200 mrads at background rates, as compared with 100–200 rads at the dose rates of the medical X rays on which both our human data and animal studies were largely based, it now seems that the role of background radiation on human health, as distinct from its genetic effect, was underestimated by some 100–1000 times.

The new findings of strong dose-rate effects indicate that the assumption of a constant risk per rad independent of dose rate—and underlying all existing estimates of the likely health effects of small radiation exposures arrived at by the various national and international bodies who had set the existing permissible doses prior to the discoveries of Petkau, Scott, and Stokke—can no longer be regarded as valid. The assumption of strict linearity for the relation between dose and risk used in these estimates to extrapolate from high doses to low doses, accepted as the more conservative hypothesis when compared with the threshold hypothesis, now turns out not to have been conservative enough. The studies of Petkau, Stokke, and Scott, carried out at low dose rates for which no previous experimental information had been available, show that the dose response obeys a logarithmic or fractional power law that is much steeper at low doses than at high doses, as illustrated in Fig. 5 taken from a paper by Baum.[28]

Since the slope of the dose response curve is proportional to $(1/D_o)$, it means that the slope at very low doses, found by drawing a straight line from the high-dose data obtained at high dose rates, is some 100–1000 times smaller than the actual slope for small increments of radiation above background.[14] Instead of the hoped-for "safe threshold" or even the "conservative" linear relation between dose and response, the new data show that the biological effects on cell membranes are "superlinear" near the origin, rising very rapidly

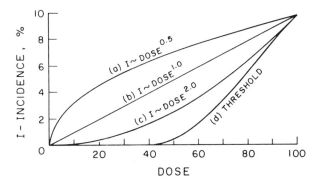

Figure 5. Various theoretically possible forms of the dose-response relationship for the effect of radiation on living organisms. (a) is the type of fractional exponent low predicted by an indirect chemical mechanism of the type found to hold for phospholipid cell membranes by Petkau.[7] (b) is the direct or linear dose response observed whenever the dose rate is constant and the range of total doses is not too large, previously believed to represent the most "conservative" possible assumption. (c) represents a quadratic type of law found to hold for cancer incidence at high total absorbed doses. (d) represents the "threshold" type of response observed for some effects such as immediate death at high whole-body doses. Note the large differences in risk predicted for low doses using the various assumptions.

at low levels, and then saturating for higher doses absorbed in a given period of time.

It remains to be seen whether the effects on cell membranes observed in the laboratory have resulted in detectable increases in the effects on human health.

3. *Low-Level Radiation and Human Health*

The discovery that protracted radiation is more damaging to cell membranes than to the genes of living cells completely changes our expectations as to the type and magnitude of health effects that we would expect to find for large human populations exposed to small amounts of radiation over long periods of time.

The dominance of somatic damage to cell membranes by free radicals over genetic damage to the reproductive cells by direct action means that birth defects among the newborn would be developmental rather than genetic in origin. Similarly, any increases in cancer would reflect damage to the body's defenses against the proliferation of cancer cells, rather than the triggering effects of damage to the genes. But most important would be the fact that conditions never previously associated with small radiation doses should be detectable, namely, infectious diseases such as influenza and pneumonia, as well as chronic diseases associated with aging such as emphysema, heart disease, diabetes, kidney disease, and stroke.

As to the magnitude of the effects to be expected, instead of the changes of the order of a few percent per rad expected on the basis of all earlier genetic and high dose rate somatic effect studies,[22] the results of Petkau, Stokke, and Scott would lead one to expect effects of the order of 0.1–1.0%/mrad, or about 100–1000 times greater.

Added doses up to a few times the normal background of 50–120 mrads/year are known to have been experienced by many population groups. These fall into the following major categories:

1. Those exposed to abnormally high concentrations of naturally occurring radioisotopes
2. Those exposed to low levels of diagnostic X rays and radioisotopes
3. Those exposed to fallout from distant weapons tests
4. Those exposed to the releases from nuclear reactors and associated chemical fuel reprocessing facilities.

3.1. *Exposures to High Levels of Natural Radiation*

3.1.1. *Uranium Miners*

The earliest health effects of naturally occurring isotopes were observed among the men who worked in the European uranium mines of Joachimsthal and Schneeberg.[8,29] The miners were known to suffer from a mysterious lung disease that was eventually identified as lung cancer. The agents responsible were found to be the radioactivity decay products of radon, a chemically inert but radioactive gas which was itself a daughter product of uranium. Subsequent detailed studies of American miners showed that the doses received varied from as little as 0.5 rads/year for well-ventilated mines, to 100 rads/year in those with no ventilation. The presence of dust and the effects of smoking were found to increase the biological risk by 5–10 times, a finding subsequently confirmed in animal studies.[30]

Since the dominant type of radiation producing the exposure of the lung membranes are alpha particles,[29] one would expect an effect nearly independent of dose rate.

This is indeed found to be the case, the risk increasing directly with the calculated exposure up to rather large accumulated doses of 2000 rads and an estimated doubling dose D_0 of the order of 50–100 rads.

However, it was found that the risk per rad tends to be somewhat higher for individuals in the low exposure groups.[29] This observation is consistent with the fact that a portion of the exposure derives from the formation of free radicals, especially by the beta rays and gamma rays emitted by some of the radon daughter products whose role becomes relatively more important as the dose, and thus the dose rate, approaches normal background values.

3.1.2. *Radium Dial Painters*

Another group of individuals exposed to high doses of naturally occurring radioisotopes were the radium dial painters.[8] Here the exposure resulted primarily from radium ingested accidentally when the workers licked the brushes used to paint the luminous dials. Since the primary type of radiation exposure was again due to alpha particles, this time to the bone, where radium concentrates because of its chemical similarity to calcium, one would expect a relatively high D_o. This is in fact found to be the case, typical doubling dose estimates being on the order of 140 rads, characteristic of all such measurements involving high instantaneous ion densities in the more radiation-resistant organs.

Unfortunately, the relatively high dose required for the densely ionizing alpha particles emitted by radium to produce bone cancer misled the early investigators, who were attempting to set permissible levels of bone-seeking radioisotopes. As compared with the much lower doubling doses for the beta rays from Sr^{90} producing bone-marrow cell damage and discovered by Stokke many years later, the relatively low biological effect of radium in bone caused the permissible limit for beta ray emitting isotopes—arrived at during the time of the Manhattan Project, in 1941[8]—to be set at too high a value.

3.1.3. *Developmental Defects in Newborn Children*

A third population group exposed to abnormally high background radiation are children born in areas with high uranium and thorium content in the local rocks or soil. In this case, it is primarily the externally produced gamma rays that are the source of the exposure, rather than the very short range alpha particles that dissipate their energy typically in distances of less than 0.1 mm in solids, or in a few inches of air.

A study of the number of children born with some form of visible birth defect, in areas of New York State that differed widely in their concentrations of naturally radioactive isotopes in the local rock, did in fact show a significantly higher incidence in areas of high granitic rock and shale concentrations than in basaltic sandstone and limestone areas, corresponding generally to their uranium and thorium content.[31] The observed rates of congenital defect mortality in areas of rock possessing high activity were typically some 20–40% greater than in areas of rock with lowest activity. Although no detailed population–dose measurements were carried out in this study, subsequent surveys of variations in terrestrial gamma activity indicated that the ground gamma doses per year after shielding corrections vary typically between 30 mrads and 60 mrads in the northeastern United States.[32]

These variations must be compared with the total dose from all natural sources, including cosmic rays, which add about 50–60 mrads/year in this area, and another 18 mrads from internal emitters, mainly K^{40}. When shielding by building materials as well as by the body itself is taken into account, Oakley[32] arrives at total bone marrow doses of about 87 mrads/year for the average United States population.

Thus, the percentage increase of the effect per millirad of about 30% for 30 mrads, or close to 1% per mrad, is of the same order as observed many years later by Stokke in studies of reduction in the number of bone marrow cells in animals.[15] Likewise, it fits the permeability changes in the red cells of occupationally exposed individuals observed by Scott.[16] However, because the effects of background were so much larger than expected, on the basis of genetic studies carried out with X rays.and gamma rays at high dose rates as well as with individuals who accidentally ingested radium, these and similar findings by Wesley[33] were unfortunately discounted and regarded as spurious.[34]

In particular, the very strong correlation of the mortality rate due to congenital malformations with cosmic ray intensity all over the world—from about 1.8 per 1000 births near the equator to more than 5 beyond 50°N, as demonstrated by Wesley[32]—could not be reconciled with the known low sensitivity of the genes. But it clearly fits the sensitivity for indirect chemical action on the cells of the developing embryo, according to which almost the entire variation in the rate of malformation can be explained by the background levels, as concluded by Wesley.[32]

3.1.4. *Chromosome Abnormalities*

More recently, studies have been reported of the incidence of chromosome abnormalities in individuals who live in areas of high background radiation from thorium-bearing sands (e.g., monazite sands in Brazil),[35] and in individuals who work in factories where monazite sand is processed commercially.[36]

In this case, the exposure is due primarily to the airborne activity of Pb^{212} and the external gamma radiation from the ground, both of which give rise to relatively low ionization densities since the Pb^{212} emits beta but no alpha particles.[36]

Both the population living in the high monazite areas and the workers in the monazite plant had a higher incidence of chromosome abnormalities, as compared with control groups of the same population. However, the most significant finding was that the dose response for workers known to be exposed to different levels of radiation showed the same logarithmic type of response that was observed by Scott[16] and by Stokke[15] (Fig. 3), rising very rapidly at low exposures and leveling off for the more heavily exposed individuals.

Thus, in the case of defects involving deletions, the frequency increased from 0.90% to 2.00% as the average air concentration of Pb^{212} increased nearly tenfold, from 0.007 to 0.09 pCi/liter. But a further tenfold increase to 0.90 pCi/liter for the most heavily exposed group only increased the frequency of deletions to 2.57%, consistent with the logarithmic type of response observed by Petkau and Scott for indirect, chemically mediated effects on critical cell membranes.

In summary, both the dose-rate independent sensitivity to the densely ionizing radiation produced by alpha particles and the rapidly changing sensitivity per millirad at very low doses of beta and gamma rays are now seen to be consistent with the most carefully studied effects of natural sources of background radiation on man available at the present time.

3.2. Low-Dose Medical Exposures

Although most diagnostic X ray exposures, involving typical total doses of 10–1000 mrads take place in very brief periods, or at dose rates of 1–100 rads/min, which are millions of times greater than the exposure rates for environmental sources (Fig. 4), it is nevertheless of interest to briefly summarize our experience with health effects for this type of source, especially for those cases where the total doses accumulated in a few hours or days are very low.

3.2.1. X Ray Exposure of Radiologists

Perhaps of greatest relevance are the extensive epidemiological studies comparing radiologists with physicians in other specialities, carried out most recently by Seltser and Sartwell.[36] Because the precise doses and dose rates were not available, these investigators compared the leukemia rate, heart disease mortality, and life expectancy of radiologists with the rates for groups of specialists in orthopedic surgery, internal medicine, ophthalmology, and pathology, known to use X rays in decreasing order.

Furthermore, they were able to compare the incidence of mortality rate for those radiologists who were more heavily exposed in the period before 1945, when awareness of the hazard was less, with the rates for those who began their practice in later years.

The results indicated that the mortality rates were indeed ordered in the expected manner, with radiologists showing the highest rates in the early years prior to about 1950, when the excess mortality ranged from about 60% for heart disease to 600% for leukemia. Furthermore, again as expected, in the most recent period the excess risk for radiologists disappeared as both awareness of the risk and the use of dose-reducing advanced technology increased.

Rough estimates of the accumulated yearly doses and the rate at which the dose was received may be obtained as follows. The present maximum permissible dose for occupationally exposed individuals is 5 rads/year. The majority of radiologists today receive no more than one-fifth to one-tenth this dose, mainly in the course of fluoroscopic examinations, or some 5–10 times the normal background dose of 0.1 rad in any given year.

For this total dose, the instantaneous dose rate is roughly 10–100 times 0.5–1.0 rad/yr, or at a rate 50–1000 times the background rate of 2×10^{-7} rad/min, since fluoroscopic procedures typically occupy less than 20 hr/week and the X ray beam is turned on only about 10% of the total time needed for the entire procedure.

Examination of Fig. 4 indicates that for dose rates of 10^{-5}–10^{-4} rad/min the doubling dose D_o for membrane-type effects on bone marrow cells, as measured by Stokke,[15] is 3–10 rads. Thus, at the present total yearly doses of 0.5–1 rad, one would not expect to find more than a 5–10% increase in mortality rates, below the sensitivity of the epidemiological technique employed.

However, prior to the 1950s, when annual exposures were 10–100 times larger, or 5–100 rads, the situation was different. For these annual doses, the dose rate was also 10–100 times greater, increasing the doubling dose to 10–30 rads according to Fig. 4. Thus, doses of 5–100 rads accumulated per year would be expected to have resulted in mortality rate increases of the order of 50–500%, and this was in fact the range of health effects observed for the early radiologists.

Therefore, the most detailed study of a large population group exposed to X rays at dose rates below 10^{-3} rad/min appears to be consistent with the type of dependence on dose rate discovered since then for cell membrane damage by Petkau,[7] observed for bone marrow cells in animals by Stokke,[15] and reported for red cells in occupationally exposed individuals by Scott.[16]

The reason that these studies aroused no particular concern at the time they were published is also clear in retrospect. Since the doubling doses deduced for the more heavily exposed early radiologists were in the range of 10–30 rads (or higher, depending on the years of dose and assumed accumulation) they were not very different than what was measured in animal studies, or for the survivors of Hiroshima and Nagasaki at much higher doses. In fact, they reinforced the belief that the existing animal and human data were indeed an adequate basis for predicting the effects of small doses from environmental radiation, in the absence of the recognition that the 1000-fold higher dose rate of the X ray exposures was a crucial factor.

3.2.2. *Intrauterine Exposure of the Developing Infant*

The other large human population group for which the health effects of very low total doses have been studied extensively are children who were exposed during their early development in the course of abdominal X rays taken of their mothers during pregnancy.

The first of these studies was carried out by Dr. Alice Stewart of Oxford University in 1958.[38] This pioneering study represents the first quantitative evidence that total doses comparable to those received from background radiation in one year can produce serious health effects in man. Stewart discovered, in the course of interviews with mothers whose children had died of leukemia, that those mothers who had been exposed to a typical series of about three X ray films had nearly twice the likelihood that their children would develop leukemia before age ten than those who had reported no X rays during pregnancy.

These initial findings were seriously questioned because the typical dose per film to the infant was only about 0.3–0.5 rad, giving an average doubling dose of only 1–2 rads, much lower than had ever been observed in animal studies, and far below what was then widely believed to be a safe threshold dose. However, these results have since been fully confirmed by the comprehensive studies of MacMahon,[39] who used the hospital records of some 800,000 children born in New York and New England, and by a still larger study by Stewart involving a follow-up of more than 10 million children born in England and Wales between 1943 and 1965, published in 1970.[40]

Not only did the latter studies confirm an average increase in risk of childhood cancer and leukemia of about 572 extra cases for a population of 1 million children exposed to 1 rad, corresponding to a dose of only 1.2 rads needed to double the spontaneous rate of about 700 cases per million children born, but it also disclosed a direct, linear relation between the risk and the number of films taken during the examination (see Fig. 6).

Such a linear dose–response curve is, of course, a very strong argument for a cause-and-effect relationship since there is no connection between the number of films taken and any possible prior condition of the mother that might influence the risk of the child developing some form of cancer. It is also an extremely strong piece of evidence that there is no safe threshold for cancer in man, at least down to total doses of a few hundred millirads, below the presently permissible population dose of 500 mrads in any given year.

However, the evidence for a linear dose response was misleading as far as the true hazard of environmental radiation is concerned. It did not give any indication of the fact that when indirect, membrane-type damage is involved, for which dose rate becomes a crucial factor, one gets a "superlinear" curve where different doses are administered over the same period of time, that is, at different instantaneous dose rates.[14]

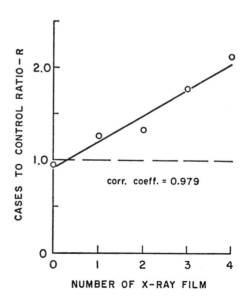

Figure 6. The relative risk of childhood cancer as a function of the number of X-ray films during pregnancy (after Stewart and Kneale[4]). These data are obtained at a constant dose rate, and show no evidence for a safe threshold.

Unlike environmental exposures, in the case of exposures from a series of X ray films each exposure is essentially of the same length, and the dose per exposure is very nearly constant. As a result, no matter how large the total number of films or the total dose received, each point on the curve of Fig. 6 is taken at the same instantaneous dose rate. Typically, the rates are 1–5 rads/sec, or 60–300 rads/min in an exposure lasting only $\frac{1}{10}$–$\frac{1}{5}$ sec, for which the total dose is typically 100–500 mrads per film.

This is very different than the situation in an environmental exposure, where, for instance, a given population group receives 100 mrads in one year and 200 mrads in another year, as a result, for example, of localized fallout. In this case, the lower dose is given at half the dose *rate* of the higher exposure, for which the studies of Petkau revealed a 40% greater sensitivity per millirad absorbed, thus leading to a steeper than linear dose response as the dose declines.

But there is still another reason why the plot of Fig. 6 seriously underestimates the true magnitude of the hazard of small environmental radiation doses as compared with the same dose from medical X rays. It turns out that the overwhelming majority of pelvic X ray exposures take place in the last 3 months of pregnancy, when the infant is far advanced in its development, including its ability to repair DNA damage and to recognize bacteria, viruses, and cancer cells. Thus, for the full-term infant at the time of birth, the risk of cancer is already quite low, within a factor of 10–20 of that for an adolescent or young adult.

However, for the case of radiation received from environmental sources, the dose is given throughout the period of development. This includes the time of greatest sensitivity in the first 3 months, when cells are dividing most rapidly, cell differentiation is proceeding, and organs are still being formed. During this period, the enzymes involved in DNA repair are still incompletely produced, and the ability to recognize and destroy cancer cells, viruses, and bacteria is still in its early stages. Thus one would expect to find a greatly increased sensitivity to any environmental agent at this time, and this has, of course, been most tragically demonstrated in the case of thalidomide.[41,42]

Such an enhanced sensitivity is exactly what Dr. Stewart found when she examined separately the small fraction of all cases where the mother was accidentally exposed to X rays in the first 3 months of pregnancy.[39,40] For these cases, the risk of subsequent development of cancer was 15 times greater, corresponding to a doubling dose of a mere 80 mrads, an amount of absorbed radiation equal to less than one-sixth of the presently allowed maximum permissible level for the general population by the Federal Radiation Council, and one-sixtieth of the annual dose presently permitted for those occupationally exposed.

In still another respect, the data on childhood leukemia and other cancers obtained by Stewart and MacMahon fail to reveal the full nature of the hazard of small amounts of protracted radiation. Childhood cancers are relatively rare, occurring only in about 1–2 children out of every 1000. Other causes of death in childhood are some 10–20 times more frequent, and it was only in the last few years that a study was published showing that not only the risk of cancer, but also the risk of mortality from other diseases, such as infectious diseases associated with damage to the immune system, is increased by a comparable amount.

This detailed study, involving a carefully planned follow-up of thousands of children whose mothers were exposed to X rays during pregnancy for various medical reasons, was carried out by a group of physicians and statisticians at Johns Hopkins University over a period of 10 years.[43] As reported by Dr. Lilienfeld and his co-workers, the most clear-cut evidence was found in the white population group, although increased risk of death from infectious diseases was observed for both white and nonwhite children.

Thus, among the 5264 white children who had received intrauterine X ray exposures consisting typically of 3–5 films, the leukemia rate was 18.2 per 100,000, compared with a rate of 6.1 for the control group, or a 200% increase in risk. For infectious diseases, the rates were 14.2, versus 7.2 for the exposed and control groups, and for diseases of the respiratory system, 50.7 versus 16.4. For the typical dose of 1000–2000 mrads, these results represent a sensitivity of the order of 0.1%/mrad, or about 10 times less than for infants irradiated in the first trimester, as found by Stewart.

There was almost no increase in the rate of congenital anomalies (24.3, versus 21.5 expected), in accordance with the fact that most of the X rays were given late in the course of pregnancy, just before birth, when organ development had already been completed.

Once again, these results illustrate the point that data obtained in studies of effects due to medical X rays tend to underestimate the seriousness of the problem of environmental exposures, when radiation is delivered not only at a much lower rate but also continuously throughout the developmental process. One should therefore not be surprised that radiation doses received from radioactive chemicals in the environment will be more serious in their biological effects on the developing young infant than one would be led to believe either from data observed for brief, high-dose-rate exposures of human and animal populations, even when such factors as biological concentrations in the food chain and various organs have been taken into account.

3.3. Health Effects of Low-Level Fallout

Concern about the possible health effects of fallout from nuclear weapons was initially confined to the rather heavy local fallout of fission products and radioisotopes produced by the absorption of neutrons in the air, the soil, and the bomb casing which reaches the ground in a matter of only minutes to hours, particularly for weapons detonated near the ground.[12,44] For this situation, radiation doses can be many tens to thousands of rads, and therefore there was never any doubt as to the great biological hazard of massive nuclear fallout.

However, the situation was very different for the so-called "distant" or "global" fallout, produced by the gradual descent of small fallout particles that were carried to very high altitudes above 30,000–40,000 ft, where the radioactive dust would be carried around the globe in a matter of 10–14 days by the circulating jet stream. It was expected that the particles would spend many months or even years in decaying before they reached the ground, so that the radiation doses would be small compared with the natural background levels of some 80–100 mrads/year.

Since about 90% of the high-altitude fallout is brought down by rain or snow,[45] it was found that the ground deposition, and thus the amount of radioactivity in water, milk, and general diet, was closely proportional to the annual rainfall in areas of the same latitude (see Fig. 7).

As a result, the primary concern centered mainly around the contamination of milk by isotopes that were known to be taken up by cows and secreted with the milk, particularly the elements iodine, cesium, strontium, and barium, whose radioactive isotopes were particularly strongly represented among the fission fragments of uranium and plutonium nuclei.[12,45]

Iodine was early recognized to present a special hazard because it concentrates chemically in the thyroid gland and produces energetic beta rays, therefore resulting in a very high dose to this critical organ, typically some 100 times greater than the gamma-ray dose due to the iodine in the body or on the ground.

The isotope I^{131}, with a half-life of 8 days, is of particular concern, since a short half-life means that a given number of atoms produce many disintegra-

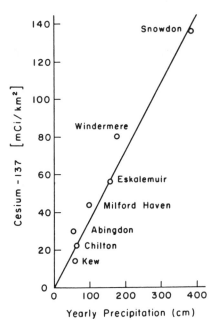

Figure 7. Typical relation between fallout deposition and yearly precipitation. (From P. Weish and E. Gruber, *Radioactivity and Environment*, Fischer, Stuttgart, 1975.)

tions per unit time, making it a much more intense source than other isotopes of longer half-life, such as I^{129}. Furthermore, the time during which one-half of the radioactive atoms created decay is long enough to allow passage of I^{131} through the milk pathway, from the production of the isotope to its deposition on the grass and its consumption in the milk, all taking place within a matter of 1–2 weeks.[12]

The isotopes of cesium are also of great biological concern since they are readily taken up by all animals, having the same chemistry as the essential trace elements Na and K. Because Cs^{137} has a very long physical half-life (30 years), it tends to accumulate in the environment, even though the typical time it spends in the human body, or the "biological half-life," is only about 100 days. Since it resembles Na and K, it goes to all cells of the body and gives a fairly uniform whole-body dose of about 1 mrad/year for every 100 pCi per kilogram of body weight.[46]

It has been found that if milk containing 3600 pCi/liter and weighing about 1 kg is consumed continuously, the internal concentration of Cs^{137} due to the milk alone leads to a maximum whole-body dose of about 500 mrads, or to an average dose of about 170 mrads in a large population, presently regarded as the maximum permissible dose for the general population in the United States. Since the external dose from cesium on the ground is about twice the internal dose, and since other, shorter-lived fission products give an external dose roughly comparable to the Cs^{137} alone, it turns out that, to a very rough approximation, the yearly dose to the whole body from all sources is about 0.2–0.3 mrad whenever the milk contains 1 pCi/liter of Cs^{137}.

Thus, since the levels of Cs^{137} in milk were in the range 10–150 pCi/liter during the height of nuclear testing in the Northern Hemisphere,[47] whole-body doses ranged up to 45 mrads/year in the early 1960s, comparable to the magnitude of the typical annual doses in the United States from cosmic rays (40–50 mrads/year) and natural radioactivity in the ground (35–45 mrads/year), as recently summarized by Oakley.[48]

Comparable doses are received by the bone marrow as a result of Sr^{90}, a daily consumption of 200 pCi/liter of milk leading to an average of about 170 mrads/year.[49] This works out to about 1 mrad/year when the milk concentration is 1 pCi/liter, leading to the rule of thumb that for almost any of the most important isotopes, 1 pCi/liter consumed during a whole year adds about 1–2 mrads to the natural radiation dose for organs that selectively concentrate certain isotopes such as Sr^{90} and I^{131}.

However, because Sr^{90} accumulates in bone and is only slowly released with time, with a typical biological half-life of 3–5 years depending upon age,[50] it builds up in the human body and thus represents a greater long-term hazard than Cs^{137} with regard to leukemia and cancers of those organs where Sr^{90} and its radioactive daughter Y^{90} concentrate.

These considerations make it clear that if the levels of natural radioactivity are indeed able to produce detectable differences in disease at the sensitivity of 0.1–1.0% per mrad typical of membrane damage at low dose rates, then one would expect to find comparable effects of fallout from distant nuclear detonations. These would be most easily detected not for direct effects on the genes, but for somatic damage on cell membranes and therefore for such conditions as developmental disturbances in the early embryo leading to birth defects.

3.3.1. *Fallout and Congenital Anomalies*

The first evidence suggesting the possibility that relatively low levels of fallout from distant nuclear tests were capable of increasing the incidence of developmental defects in man was discovered by Dr. L. J. le Vann, in the course of a survey requested by the Canadian Minister of Health for the Province of Alberta, in March, 1960.

As described by Dr. leVann,[51] the initial survey, for children born in Alberta in 1959, showed a relatively low rate of anatomically recognizable birth defects of only 7.76 children per 1000 births, corresponding to 257 children out of 37,996 born, 89% of which could be traced.

However, a second survey, carried out in 1962 for children born the previous year, and undertaken because of the rising concern over malformations attributable to certain drugs such as thalidomide that had just been introduced in Canada, showed a very much greater incidence. In this survey, out of 38,762 births recorded, 99% of which could be traced, there were 528 children with 631 separately identified birth defects. This was a rate of 13.8 per 1000 children born, representing an increase of 271 cases, or close to a 78% rise in only 2 years.

There were a number of reasons why it did not seem possible to attribute this sudden and statistically highly significant increase to the introduction of new drugs such as thalidomide. Out of the 528 children born with birth defects, only two were found to have been born with the characteristic limb deformities to mothers who had taken this drug. In fact, it is known that thalidomide acts primarily during organ development in the first 2 months of pregnancy, and since the drug was not introduced into Canada before 1961, only children born in the last 4 months of 1961 could have been affected.[40]

The second reason that led Dr. leVann to conclude that drugs were not likely to account for the sharp rise was the fact that the greatest increase took place in the most northerly part of the province closest to the Arctic Circle, where the precipitation was greater than in the southern part of Alberta during 1960–1961. Thus the long-lived, high-altitude fallout from the large series of 22 nuclear tests carried out by Russia at Novaya Semlya, north of the Arctic Circle, in 1958, would affect children born in northern Alberta much more than it would those in southern Alberta, some 850 miles to the south. In fact, the ratio of the rates for the northern and southern regions was 18.6 to 14.0, or 1.32, similar to the ratio of the rainfall, namely, 17.62 to 12.33 inches, or 1.42.

Furthermore, if there was any variation in the use of drugs between the southern, more densely populated urbanized area and the sparsely populated, rural northern area, one would clearly expect the greater use of newly introduced drugs to be in the south and not in the north.

Nevertheless, because of the absence of any figures as to what fraction of the mothers in the different regions had taken some drug during pregnancy, it was possible to raise doubts as to the conclusiveness of the evidence that linked the rise in congenital defects to fallout rather than to drugs. Furthermore, the lack of any direct measurements of fallout levels for the different areas in Alberta represented a serious drawback.

These problems can be overcome, however, in the light of our more recent knowledge, on the basis of which it is now possible to arrive at the necessary information on the percentage of children exposed to drugs, as well as the approximate levels of radiation involved.

With regard to the fraction of the mothers who used drugs in each region, this can be obtained from an analysis of the detailed figures given by le Vann for the number of congenital malformations in each region among children whose mothers had taken drugs and for those whose mothers had not, as summarized in Table 1.

The analysis is based on the assumption that for small increases in radiation levels close to background, the effect is very nearly linearly dependent on dose when the dose rate is nearly the same. Furthermore, it is now known, from numerous studies carried out since 1963, that the effect of a small increment of radiation on different segments of a population with different "spontaneous" rates is to increase each by very nearly the same factor.[21] Thus, a dose that increases the incidence of cancer in one segment of a population with a spontaneous rate of 100 per 100,000 by 50%, to a rate of 150 per 100,000, will in general produce the same 50% increase in another group that has a spontaneous incidence rate of 200 per 100,000, causing it to increase to 300 per 100,000.

As applied to the data gathered by le Vann, this means that the children, both those exposed to drugs and those not exposed, can be assumed to have had their risk of developing a congenital malformation increased by essentially the same factor for a given fallout dose. Using the data on the number of malformations supplied by le Vann for the two regions with the greatest difference in rainfall, namely, the southern and northern areas, as summarized

Table 1. Number of Congenital Malformations in Alberta for 1961 With and Without Drugs Taken by Mother During Pregnancy

	Northern area (18,072 births)	Southern area (14,240 births)
No drugs	192	78
With drugs	145	122
Total malformations	337	200

in Table 1, one can write the following two pairs of equations for the number of malformations for the drug-exposed children and those not exposed to drugs for each of the two different regions:

$$N_{ds} = [c \times n_s \times f_{ds}][a_d \times D_{rs}] \tag{1}$$

$$N_{os} = [c \times n_s \times (1 - f_{ds})][a_0 \times D_{rs}] \tag{2}$$

In these equations, N_{ds} is the number of malformations among children exposed to drugs in the southern area, and N_{os} is the corresponding number for those not exposed to drugs, n_s is the total number of children born in that region, c is the average number of different malformations per child (equal to 1.19 from leVann's data), f_{ds} is the fraction of all children exposed to drugs during their development, a_d is the risk per unit of radiation for children also exposed to drugs, a_0 is the risk per unit radiation for those who were not exposed to drugs, and D_{rs} is the radiation dose in the southern region.

The second set of equations contains the same quantities for the northern region:

$$N_{dn} = [c \times n_n \times f_{dn}][a_d \times D_{rn}] \tag{3}$$

$$N_{on} = [c \times n_n \times (1 - f_{dn})][a_0 \times D_{rn}] \tag{4}$$

By taking the ratios of the equations (1) to (3) and (2) to (4), the actual values of the doses drop out, and it is possible to solve for those values of f_{ds} and f_{dn} which give the same ratio of (a_d/a_0) for both north and south.

The result is $f_{ds} = 0.42$ and $f_{dn} = 0.26$ for a common value of $(a_d/a_0) = 2.15$. This means that 42% of the mothers in the more urban southern area probably took drugs during their pregnancy, as compared with only 26% of the mothers in the more rural and remote northern region, a result that agrees with what one would expect from the difference in the availability of drugs in the two areas, as well as what has been reported for various populations in different western countries, including nearby Manitoba.[52]

Most interesting is the fact that the ratio of the risk per unit of environmental radiation for children of mothers who took drugs, a_d, to that for those who did not, a_0, is very nearly 2 to 1. This clearly indicates that the risk of malformation in the presence of either normal background radiation or fallout is essentially twice as large when drugs are taken during pregnancy as it is normally, thus fully justifying the present medical practice of avoiding all types of medication as much as possible, especially during the first 3 months of pregnancy.

The absolute values of a_d and a_0 can be obtained from knowledge of the normal background radiation levels in southern Alberta, D_{rs}, which is very close to the United States, for which detailed studies were carried out by Oakley.[31] According to Oakley, the cosmic ray levels for the high plateau just east of the Rocky Mountains are 50–60 mrads/year. For the radiation from terrestrial sources for this type of area, Oakley gives a best estimate of 45.6 mrads/year, so that the total external dose for southern Alberta may be estimated at 95–105 mrads/year to within ±15%.

Using $D_{rs} = 100$ mrads/year so as to include a small contribution of 5–10 mrads due to fallout, equations (1) and (2) give $a_0 = 0.79\%$/mrad and $a_d = 1.70\%$/mrad. These values can now be checked for the northern region, making use of the fact that the ratio in rainfall as given by leVann for the two regions is 1.42. The dose from the combined effect of fallout and background radiation in northern Alberta for 1960–1961 may thus be estimated as close to 142 ± 20 mrads/year.

This estimate, based on natural background and rainfall, is further substantiated by direct measurements of Cs^{137} in milk for Canada and other areas of the northern hemisphere for 1960, given in the United Nations Reports on the Effects of Radiation,[53] and more detailed measurements of radioactivity in whole milk for northern and southern Alberta reported by the Canadian Ministry of Health and published by the U.S. Department of Health, Education and Welfare.[54]

For 1960, the year that most of the children studied in 1961 were conceived, the average concentration of Cs^{137} in the Canadian milk was 55 pCi/liter.[53] This corresponds to an average whole-body dose of about 10–20 mrads/year from all fallout sources consistent with a range of 5–40 mrads/year in going from the southern to the northern part of Alberta. This pattern is consistent with the lower average of 32 pCi/liter for the United States to the south and the sharply higher value of 150 pCi/liter for Norway, whose latitude is just north of that of Alberta.[53]

The existence of an increase in fallout levels with distance toward the Arctic Circle following large-scale tests at Novaya Semlya is further confirmed by the direct measurements of the activity in milk for the northern part of Alberta that includes Edmonton, and for the southern part that includes Calgary.[54] Thus, following the last atmospheric tests in the arctic, the milk analyzed by the Canadian Ministry of Health in the spring of 1963 for the months of May and June showed 117 and 211 pCi/liter of Cs^{137} for the northern zone, while for the southern zone the values were 92 and 199 pCi/liter, again confirming the fact that northern Alberta received more fallout following large atmospheric tests in the arctic than the southern part, a pattern continued southward into the United States.

With the dose estimate of 142 mrads/year, equations (3) and (4) lead to $a_0 = 0.85\%$/mrad, and $a_d = 1.82\%$/mrad, in good agreement with the values obtained for the southern region. Thus, the sensitivity of the fetus to environmental radiation is essentially the same as that of cell membranes as found by Stokke[15] and Scott,[16] corresponding to an increase in risk of about 1% for each mrad.

The data are summarized in the plot of Fig. 8. They indicate that for the population of the southern region not exposed to drugs, the number of congenital malformations was only about 8 per 1000 births, not very different than the rate of 9.2 separate malformations per 1000 births for the total population of Alberta in 1959, when the accumulated Sr^{90} and Cs^{137} was much lower and thalidomide had not yet been introduced. Those exposed to the

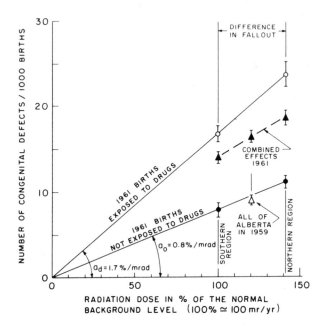

Figure 8. The effect of radiation and drug exposure during pregnancy on the incidence of congenital malformations for infants born in Alberta, Canada, in 1959 and 1961. The lower curve represents those not exposed to drugs, the upper curve represents those exposed to drugs in addition to fallout for children born in 1961. The dotted curve represents the data when those who were exposed to drugs were counted together with those not so exposed. In all cases, those who received the heavier fallout radiation in the northern region are seen to show a greater incidence of birth defects. Also, for all of Alberta, the total incidence rose between 1959 ($\triangle$) and 1961 ($\blacktriangle$).

heavier fallout in the northern part in 1960–1961, but not exposed to drugs, had their risk increased to 11.4 per 1000 births.

Contrary to the misleading impression gained by leVann that drug use had little influence based on a comparable frequency of reported drug use by mothers of normal and congenitally defective children, the plot clearly shows the now well-established effect of drugs during pregnancy.[41,42] For the 42% exposed to drugs in the southern region that had low fallout, the rate of malformations doubled from 8 to 17 per 1000 births. For the 26% of all infants in the north who had received exposures to both drugs and heavy fallout, the data indicate that the rate of malformations actually tripled, from 8 to 24 per 1000 infants born. Thus it was the mistaken implicit assumption that about half the mothers everywhere in Alberta had taken drugs which led leVann to interpret the data as showing no significant effect of medication during pregnancy. However, the detailed analysis indicates that only about one-third of all mothers appear to have taken some form of medication during the critical period of early pregnancy, so that if drugs were not a factor, only one-third of those born malformed in all of Alberta should have been found to have been born to mothers who took some form of medication.

These findings indicate that Wesley was probably correct in his conclusion that the broad pattern of the geographic distribution of congenital defects with latitude, other than those due to local differences in drugs, diet, or fallout, is connected with the increasing natural background radiation levels toward the poles. However, radiation from such other natural sources as uranium and thorium daughter products released from rock or in the fly ash of coal would also be a factor in the heavily industrialized northern countries.

3.3.2. The Relative Toxicity of Drugs and Environmental Radiation

The conclusion that drugs typically involving doses of 1–100 mg/day can produce detectable effects seems easier to accept than the idea that very much smaller amounts of radioactive chemicals in the diet should be able to produce comparable biological damage. However, the reason for the much larger toxicity of radioactive elements can be understood in terms of the indirect chemical action of the radiation emitted by these substances, whose total energy per decay process is millions of times greater than the energy needed to produce an excited molecule.

In the case of Cs^{137}, each gram has an activity of about 90 Ci, or a rate of disintegration equal to that of 90 g of radium. Thus, the typical daily intake of about 100 pCi, or 100×10^{-12} Ci, for individuals in northern Alberta in 1960 from all dietary sources represents only 1.1×10^{-12} g. This is a mere one-millionth of one microgram, a billion times less than even a small 1 mg daily dose of a drug.

Due to the finite biological half-life of 100 days for Cs^{137}, however, there is a build-up in the body that accounts for a factor of about 100 in relative toxicity, leaving a factor of 10^7 to be explained. Now each disintegration of a typical fission product such as Cs^{137}, Sr^{90}, or Y^{90} leads to the release of one or more beta rays whose energies are generally between 0.2 and 2 MeV. This may be compared with the typical energy needed to excite a molecule, which is of the order of 0.1–1.0 eV.

It follows that a single radioactive atom can give rise to as many as 10 million excited molecules, spread throughout a volume dictated by the range of the ejected electron of many millimeters, typically some 10,000–100,000 cells, each 10^{-4}–10^{-3} cm in diameter. Thus, whereas a single nonradioactive atom or molecule might lead to the excitation of a single critical molecule in a single cell, a radioactive atom emits enough energy to lead to as many as 10 million excitation processes spread over some 100,000 neighboring cells. And when the instantaneous density of excitations is small, the chance that some of the activated molecules created will diffuse to a vital part of the cell structure before being deexcited is much greater than when the density is high. This accounts for the very great toxicity of trace quantities of radioactive chemicals in our environment, especially for rapidly dividing and differentiating cells.

That the effect of low-dose-rate background radiation leads primarily to somatic rather than to genetic damage is perhaps most clearly evident from an examination of the effect of fallout on the mortality rate due to infectious diseases. Such increases in mortality rates result from a lowered ability to recognize and destroy viruses and bacteria, an effect known to follow radiation damage to the bone marrow cells involved in the production of white cells, as studied by Stokke[15] for low doses of Sr^{90} in laboratory animals.

Figure 9 shows a plot of the mortality rate of newborn infants in the United States in the first year of life due to influenza and pneumonia, as given in the U.S. Vital Statistics[35] for the period 1938–1974. It is seen that there was a very rapid decline in the mortality until shortly after the end of World War II, the rate dropping from about 8 to only 3.4 between 1938 and 1946, presumably due to the improvement in medical care and the introduction of sulfa drugs and antibiotics.

However, beginning in about 1951, at the same time as the beginning of the first atmospheric tests of nuclear weapons in the continental United States which caused Sr^{90} levels to rise sharply, there was a complete and sudden halt in this downward trend. In fact, the trend began to completely reverse itself in coincidence with the largest nuclear test series in Nevada, which took place in

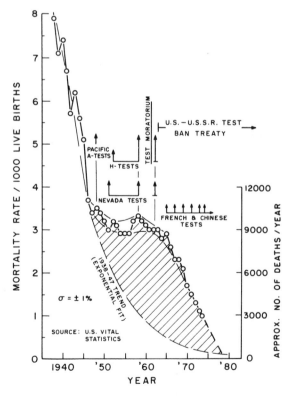

Figure 9. Effect of low levels of fallout radiation on the resistance to infectious diseases among the newborn. The data plotted is that for influenza and pneumonia mortality rates in the United States for infants 0–1 year old before, during, and after the period of heaviest nuclear weapons testing in the atmosphere (1951–63).

1957,[44] beginning a renewed decline during the period of the test moratorium which ended again with the resumption of testing by the U.S.S.R in 1961.

Only after the large amount of Sr^{90} in milk from the renewed tests had begun to decline, following the signing of the test ban treaty in 1963, did a renewed drop in mortality rate take place. The rate of decline was not as rapid as prior to the onset of nuclear weapon detonations, in accordance with the fact that some atmospheric tests by France and China continued to add Sr^{90} and other radioactive isotopes to the world's air and water even after 1963.

Nevertheless, the sudden renewal of a decline in mortality rates, in coincidence with the measured decline of fallout in milk,[56] means that the effect could not have been of a genetic type, affecting the reproductive cells of the parents rather than the cell membranes of the developing infant. The sharp decline from 3 deaths per 1000 births in 1965 to only about 1 per 1000 in early 1974 thus confirms the importance of the indirect, somatic action of radiation, with a risk increase per millirad consistent both with the results for bone marrow cells damaged by Sr^{90} as measured by Stokke,[15] and with the effect of fallout on developmental defects as observed by leVann.[51]

The data on influenza and pneumonia mortality rates for the newborn also point out that it is not only the obvious anatomical defects that one must be concerned about, but the more subtle, and often more numerous biochemical disturbances that lead to mental retardation and a host of metabolic diseases that are only discovered later in life. It is apparently these chemical effects on cell membranes and other vital cell structures that contribute to the lowering of the immune system's ability to detect and destroy viruses and bacteria, a subtle form of damage that may also be an important factor in the body's ability to control the proliferation of cancer cells.[57,58]

The possibility that the dose-rate-dependent indirect form of radiation damage may be the dominant factor in the body's ability to prevent the proliferation of cancer cells at the low doses encountered for environmental radiation is strongly supported by a number of recently published investigations.

Thus, the first study of cancer induction in animals by the protracted action of inhaled plutonium at total doses well below 50 rads, reported by Sanders and his co-workers in 1973,[59] indicated a form of dose–response relationship that rises much more rapidly at small than at large doses (Fig. 10). This is exactly the type of dose–response relation observed by Stokke[15] for small doses of Sr^{90} affecting the number of cells in the bone marrow involved in the immune process, where again the effect per rad increased as the dose was reduced (see Fig. 3).

Using a logarithmic plot (Fig. 11), it is seen that the cancer incidence of 4.3% for the control group that received only background radiation fits on a straight line with the data points at 9, 32, and 375 rads.

This result is consistent with the hypothesis that a significant fraction of the so-called spontaneous cancers are in fact induced by the normal background radiation during the 2-year life-span of these animals. Figure 11 shows that a

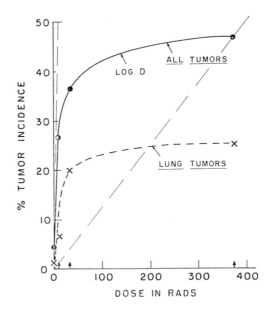

Figure 10. Form of the dose-response relation for cancer induction by small doses of inhaled plutonium-238 in rats (after Sanders[59]). Solid curve shows the trend of incidence for tumors of all organs other than the mammary gland; the dotted curve shows the data for lung cancer alone. Note the sharp rise at very low doses, similar to that expected for an indirect chemical mechanism, giving an incidence much above that expected by a linear extrapolation from high doses (dashed line).

dose of only about 180 mrads is needed at background levels to double the spontaneous incidence rate of cancer. This value is some 200 times smaller than the 34 rads which would have been obtained on the basis of an assumed "conservative" linear dose–response relationship from the point at 375 rads, typical of the values of 10–100 rads that have until now been widely accepted as the doubling dose for estimating the risk of cancer from environmental sources.

The results of Sanders and his co-workers have since been confirmed by an independent study in other animals by Little *et al.* at the Harvard School of Public Health using very small amounts of polonium instilled into the lungs.[60] Furthermore, the more rapid rise in the incidence of lung cancer for the smallest doses occurred both when the polonium was given attached to small insoluble particles such as are found in cigarette smoke and air-pollution, and without such particles. But the rate of lung cancer was greater when insoluble dust particles acted as carriers, in agreement with observations for uranium miners.[29]

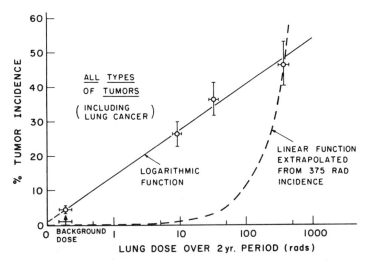

Figure 11. Data of Fig. 10 for tumors in the rat following low doses of plutonium-238 inhalation replotted on a semilogarithmic scale. Note fit to a straight line on this plot, passing through the point for the controls exposed only to background radiation. The dashed line represents the prediction of the linear relation of Fig. 10. From this plot, it is possible to determine that the dose required to double the incidence from its normal value of 4.3% is only 180 mrads when given over a 2-year period.

These results, together with the discovery that radioactive isotopes such as naturally occurring Pb^{210} and Po^{210} are trapped by the fine hairs of tobacco leaves[61] has now led to the suggestion that the very small amounts of radioactivity in cigarette smoke may be the single most important factor in the large increase in lung cancer throughout the world.[62]

Thus, environmental radioactivity, acting synergistically with dust and chemical air pollutants, may play a far greater role in all types of cancers and other diseases than has been realized in the past. It is consistent with the otherwise unexplained sudden rises in certain types of cancers a few years after nuclear weapons fallout began to appear in the atmosphere, as illustrated for the cases of pancreatic and lung cancers in Japan by Figs. 12 and 13.[63]

4. Summary and Conclusion

The discovery that the indirect chemical effect of radiation on cell membranes dominates over the direct physical action on the genes at very low dose rates has many far-reaching implications.

First of all, it explains the totally unexpected large effect of very small doses of natural and man-made background radiation as compared with

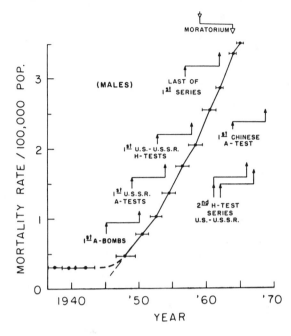

Figure 12. Mortality rate for pancreatic cancer in Japan before and after the onset of nuclear fallout. Note the sharp rise of 1200% in 20 years for male Japanese beginning in 1948, preceded by an essentially level trend during the preceding 10 years.

similarly small doses from medical X-ray sources in terms of the great difference in the instantaneous ion concentrations on the scale of single cells.

Thus, the discoveries of Petkau, Stokke, Scott, Sanders, and Little, provide an explanation for the origin of the radiation controversy,[64] brought about by the fact that the actual number of infant and fetal leukemia deaths as correlated with the levels of radioisotopes in milk during the period of nuclear testing was some 100–1000 times larger than expected, on the basis of the known genetic damage or the carcinogenic effect of high-dose rate medical and direct bomb radiation.

Only through the dose-rate dependence of the efficiency with which free radicals such as O_2^- can initiate oxidative chain reactions in phospholipid membranes can the unexpectedly high degree of toxicity of background radiation be explained. As the data of leVann[51] show, it now appears that for the most sensitive stages in the development of multicellular organisms such as man, each radioactive atom is some 10–100 million times more toxic than a molecule of the most potent teratogenic substances such as thalidomide.

The physical reason lies in the large number of excited molecular states of oxygen that can be created by a single disintegration process, turning a molecule essential for life and found in large concentrations in the cell fluid of all aerobic cells into a highly toxic substance, such that a single excited oxygen molecule can initiate the destruction of an entire cell. Thus, oxygen, on which all higher forms of life depend, turns out also to be the source of greatest vulnerability when combined with low densities of ionizing radiation compara-

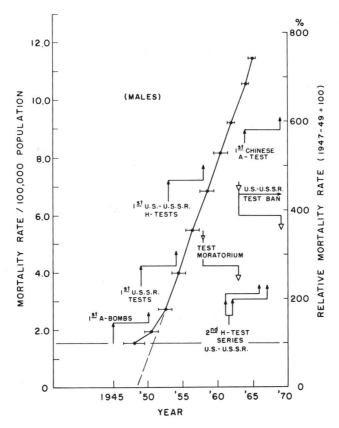

Figure 13. The mortality rate for lung cancer in Japan following the onset of nuclear weapons fallout. Note the sudden acceleration of the mortality rate some 5–10 years after the first nuclear detonations in 1945, the same delay observed for the uranium miners after they began to work in the mines.

ble to those of the natural background levels now existing in the earth's environment.

Thus, the extreme efficiency of the oxidative processes initiated by low levels of radiation not only explains the statistical findings of fetal deaths,[65] infant mortality,[65] prematurity,[65] leukemia,[63,65] and other cancers associated with low levels of fallout,[65] and releases from nuclear facilities that have typically been no greater than 10–100 mrads, well below the present permissible level of 500 mrads for the general population.[67,68] They also explain why total mortality rates from all causes for adults as well as the newborn[66] can be affected so much more strongly when the radiation is spread out over days or weeks than when it is received in short bursts, as in the case of diagnostic X rays.

From the point of view of evolution, the primary emphasis had to be given to the protection of the genes through the development of extremely efficient chemical repair mechanisms. Only this way could the gene pool be protected against the effect of the ever-present natural radiation background, and the relatively high stability of species over periods of millions of years be ensured.

But the same consideration does not hold either for the individual cell of a multicellular organism, or for the member of a large species. Both the individual cell and the individual member of a whole species are not vital; in fact, their continual death and replacement through reproduction is an essential feature of the whole evolutionary process. Once the individual cell or member of a species has performed its evolutionary function of reproduction, it is no longer needed, so that extension of its life span beyond the age of reproduction is not necessary.

Thus, premature aging, which can be produced by the action of free radicals,[69] is not a serious threat to the survival of a species, even though it is a major problem for human society, where each individual hopes for a long and healthy life far beyond the age when the individual has performed his evolutionary task of reproduction.

The implications are therefore clear: if man wants to enjoy a long and healthful life, as free as possible from the debilitating effects of birth defects, cancer, heart disease, and other chronic conditions normally associated with aging, he cannot afford to increase to any significant degree the levels of unavoidable background radiation already existing in his environment, anymore than he can afford to add carcinogenic or teratogenic chemicals to his air, water, or diet.

References

1. D. E. Lea, *Actions of Radiations on Living Cells*. London : Cambridge University Press, 1962.
2. P. Alexander, *Atomic Radiation and Life*. London: Penguin Books, Inc., 1965.
3. *Report of the United Nations Scientific Committee on the Effects of Atomic Radiation*. United Nations, New York, 1962.
4. *ICRP Report of Committee II on Permissible Dose for Internal Radiation*. London: Pergamon Press, 1959.
5. A. Brodsky, "Determining Industrial Hygiene Requirements for Installations Using Radioactive Materials." *Am. Ind. Hygiene Assoc. J.*, **26** : 294, 1965.
6. A. Petkau, "Radiation Effects with a Model Lipid Membrane." *Can. J. Chem.*, **49** : 1187, 1971.
7. A. Petkau, "Effect of $^{22}Na^+$ on a Phospholipid Membrane." *Health Physics*, **22** : 239, 1972.
8. *Report of the United Nations Scientific Committee on the Effects of Atomic Radiation*. Annex G. United Nations, New York, 1962.
9. A. Petkau and W. S. Chelack, "Radioprotective Effects of Cysteine." *Intl. J. Radiobiology*, **25** : 321, 1974.
10. W. S. Chelack, M. P. Forsyth, and A. Petkau, "Radiobiological Properties of Acholeplasma Laidlawii B." *Can. J. Microbiol.*, **20** : 307, 1974.
11. A. Petkau, W. S. Chelack, S. D. Pleskach, and T. P. Copps, *Radioprotection of Hematopeitic and Mature Blood Cells by Superoxide Dismutase*. Paper presented at the Annual Meeting of the Biophysical Society, Philadelphia, Pa., 1975.

12. Virginia Brodine, *Radioactive Contamination*. Harcourt Brace Jovanovich, New York, 1975.
13. *Radioactive Waste Discharges to the Environment from Nuclear Power Facilities*. U.S. Environmental Protection Agency, Washington, D.C. (BRH/DER 70-2) and Addendum (October, 1971) Nt'l. Tech. Inf. Service, Springfield, Va., 22151 (ORP/SID 71-1): See also Ref. 25, Vol. I, Annex A, Chap. II for releases and doses from the entire nuclear fuel cycle.
14. E. J. Sternglass, *Implications of Dose-Rate Dependent Cell-Membrane Damage for the Biological Effect of Medical and Environmental Radiation*. Proceedings of Symposium on Population Exposure. Knoxville, Tenn. Oct., 1974 (Conf-741018) Nt'l. Tech. Inf. Service, Springfield, Va., 22151.
15. T. Stokke, P. Oftedal, and A. Pappas, "Effects of Small Doses of Radioactive Strontium on the Rat Bone Marrow." *Acta Radiologica*, **7** : 321, 1968.
16. K. G. Scott, E. T. Stewart, C. D. Porter, and E. Sirafinejad, "Occupational X-Ray Exposure Associated with Increased Uptake of Rubidium by Cells." *Arch. of Envir. Health*, **26** : 64, 1973.
17. *Report of the United Nations Scientific Committee on the Effects of Radiation*, 24th Session, Suppl. #13 (A/7613) Annex A, Chapt. II. United Nations, New York, 1969.
18. R. W. Risebrough, *Effects of Environmental Pollutants Upon Animals Other than Man*, in "Effects of Pollution on Health," Vol. 6 of Proceedings of the 6th Berkeley Symposium on Mathematical Statistics and Probability, edited by L. M. LeCam, J. Neyman, and E. J. Scott. Univ. of California Press, Berkeley, 1972.
19. D. Schneider, "The Sex-Attractant Receptor of Moths." *Scient. Am.*, **231** : 28, 1974.
20. W. L. Russell, L. B. Russell, and E. M. Kelly, "Radiation Dose Rate and Mutation Frequency." *Science*, **128** : 1546, 1958.
21. BEIR Report, National Academy of Sciences, Washington, D.C. *The Effects on Populations of Exposure to Low Levels of Ionizing Radiation*. November, 1972.
22. C. W. Mays, R. D. Lloyd, and J. H. Marshall, *Malignancy Risk to Humans from Total Body Gamma Rays*. Proc. 3rd. Int'l. Congr. Int'l. Rad. Protect. Assoc. (IRPA), Sept. 9–14, 1973.
23. *Report of the United Nations Scientific Committee on the Effects of Radiation*. 19th Session, Suppl. #14 (A/5814) Annex B, United Nations, New York, 1964.
24. E. J. Sternglass, *The Role of Indirect Radiation Effects on Cell Membranes in the Immune Response*, in "Radiation and the Immune Process," Proceedings of the 1974 Hanford Radiobiology Symposium. Division of Technical Information, ERDA, Oak Ridge, Tenn. 1976. (Conf-740930.)
25. *Report of the United Nations Scientific Committee on the Effects of Atomic Radiation*, 27th Session, Suppl. #25 (A/8725) Vol. II, Annex F, United Nations, New York, 1972.
26. G. W. Casarett, *Pathogenesis of Radionuclide Induced Tumors*, in "Radionuclide Carcinogenesis," ed. by C. L. Sanders *et al.*, AEC Symposium Series Vol. 29. U.S. Atomic Energy Commission, Off. of Inf. Services, 1973. (Conf-720505.)
27. *Report of the United Nations Scientific Committee on the Effects of Atomic Radiation*. Annex D. Chapt. III, Section 122ff. United Nations, New York, 1962.
28. J. W. Baum, "Population Heterogeneity Hypothesis of Radiation Induced Cancer." *Health Physics*, **25** : 97, 1973.
29. F. Lundin, J. K. Wagoner, and V. E. Archer, "Radon Daughter Exposure and Respiratory Cancer," Joint Monography #1, National Institute for Occupational Safety and Health, and National Institute of Environmental Health Sciences. U.S. Dept. of Health, Education and Welfare, 1971 (PB-204-871), National Tech. Infor. Service, Springfield, Va., 22151.
30. B. O. Stuart, D. H. Willard, and E. B. Howard, *Uranium Mine Air Contaminants in Dogs and Hamsters*, in "Inhalation Carcinogenesis," ed. by M. G. Hanna, Jr., *et al.* AEC Symp. Series, Vol. 18. U.S. Atomic Energy Commission, Division of Tech. Infor., 1970, Springfield, Va., 22151, National Tech. Inf. Serv. (Conf-691001.)
31. Gentry, J. T., E. Parkhurst, and G. V. Bulin, "An Epidemiological Study of Congenital Malformations in New York State." *Am. J. Public Health*, **49** : 497, 1959.
32. D. T. Oakley, *Natural Radiation Exposure in the United States*. U.S. Envir. Protection Agency, June, 1972 (ORP/SID 72-1).

33. J. P. Wesley, "Background Radiation as the Cause of Fetal Congenital Malformation." *Int'l. J. Rad. Biol.*, **2** : 297, 1960.
34. W. J. Schull, *Radiation and Human Genetics in Radiation Biology and Cancer.* University of Texas Press, Austin, p. 423, 1959.
35. M. Barcinski *et al. Cytogenetic Studies in Brazilian Populations Exposed to Natural and Industrial Radioactive Contamination.* World Health Organization Meeting, Mol, Belgium, 1972, also *Am. J. Human Genetics,* **27** : 802, 1975.
36. C. Costa-Ribeiro *et al.* "Radiobiological Aspects and Radiation Levels Associated with the Milling of Monazite Sands." *Health Physics,* **28** : 225, 1975.
37. R. Seltser and P. E. Sartwell, "The Influence of Occupational Exposure to Radiation on the Mortality of American Radiologists and other Medical Specialists." *Am. J. Epidermiol.*, **81** : 2, 1965.
38. A. Stewart, J. Webb, and D. Hewitt, "A Survey of Childhood Malignancies." *Brit. Med. J.*, **1** : 1495, 1958.
39. B. MacMahon, "Pre-natal X-Ray Exposure and Childhood Cancer." *J. Nt'l. Cancer Instit.*, **28** : 1173, 1962.
40. A. Stewart and G. W. Kneale, "Radiation Dose Effects in Relation to Obstetric X-Rays and Childhood Cancers." *Lancet,* **1** : 1185, 1970.
41. W. Lenz, *Chemicals as a Cause of Human Malformations*, in "Against Pollution and Hunger," ed. by A. M. Hilton, Halsted Press, New York, 1974.
42. See J. Warkany, "Congenital Malformation through the Ages," in *Drugs and Fetal Development* ed. by M. A. Klingberg, A. Abramovici, and J. Chemke, Plenum Press, New York, 1972, p. 7.
43. E. I. Diamond, H. Schmerler, and A. M. Lillienfeld, "The Relationship of Intrauterine Radiation to Subsequent Mortality and Development of Leukemia in Children." *Am. J. Epid.*, **97** : 283, 1973.
44. *The Effects of Nuclear Weapons,* ed. by S. Glasstone. U. S. Atomic Energy Commission, 1962, U.S. Government Printing Office, Washington, D.C., Chapt. 9.
45. *Report of the United Nations Scientific Committee on the Effects of Atomic Radiation.* Annex F, Part I.
46. Ibid., Annex F, Part III, Sect. 38.
47. *Report of the United Nations Scientific Committee on the Effects of Atomic Radiation,* 27th Session, Vol. I, Annex A, Sects. 222–239, 261–264, 267–268 and earlier U.N. Reports.
48. Oakley, op. cit., Sect. 3.3.2 and Fig. II.
49. *Report of the United Nations Scientific Committee on the Effects of Atomic Radiation,* 27th Session, Vol. I. Annex A, Sects. 178–205.
50. Ibid., Vol. I. Annex A, Sects. 195–204.
51. L. J. leVann, "Congenital Abnormalities in Children Born in Alberta During 1961." *Can. Med. Assoc. J.*, **89** : 120, 1963.
52. See article by N. W. Choi *et al.*, in *Drugs and Fetal Development*, ed. by M. A. Klingberg, A. Abramovici, and J. Chemke, Plenum Press, New York, 1972, p. 511 and other articles in the same volume.
53. *Report of the United Nations Scientific Committee on the Effects of Atomic Radiation.* Annex F, Part II, Table XXIX.
54. *Radiation Health Data and Reports.* U.S. Dept. of H.E.W., Vol. 4, #10, p. 505 (Oct. 1963).
55. *U.S. Vital Statistics,* Summary Reports and Annual Volumes.
56. *Report of the United Nations Scientific Committee on the Effects of Atomic Radiation,* 27th Session, Vol. I, Table 29.
57. K. Irie, R. F. Irie, and D. L. Morton, "Evidence for in Vivo Reaction of Antibody and Complement to Surface Antigens of Human Cancer Cells," *Science,* **186** : 454, 1974.
58. E. L. Felix, B. Lloyd, and M. Cohen, "Inhibition of the Growth and Development of a Transplantable Murine Melanoma by Vitamin A." *Science,* **189** : 886, 1975.
59. C. L. Sanders, "Carcinogenicity of Inhaled Plutonium–238 in the Rat." *Radiation Res.*, **56** : 540, 1973.

60. J. B. Little, A. R. Kennedy, and R. B. McGandy, "Plutonium–238 Exposure and Lung Cancer in Hamsters." *Science,* **188** : 737, 1975.
61. E. A. Martell, "Radioactivity of Tobacco Trichomes and Insoluble Cigarette Smoke Particles." *Nature,* **249** : 215, 1974.
62. E. A. Martell, "Tobacco Radioactivity and Cancer in Smokers." *American Scientist,* **63** : 404, 1975.
63. E. J. Sternglass, *Epidemiological Studies of Fallout and Cancer Mortality,* in "Radionuclide Carcinogenesis," C. L. Sanders, R. H. Busch, J. E. Ballou, and D. D. Mahlum, eds., ERDA Symposium Series, Conf–72050, pp. 1–14, 1973.
64. See, for instance, "The Environmental Revolution," a collection of articles from the *Bulletin of the Atomic Scientists,* Educational Foundation for Nuclear Science, 1020–24 East 58th Street, Chicago, Ill. 60637. Also, "Poisoned Power" by J. W. Gofman and A. R. Tamplin, Rodale Press, Inc., Emmaus, Pa., 18049, and *Low-Level Radiation,* by E. J. Sternglass, Ballantine Books, New York, 1972, reprinted by the Friends of the Earth Foundation, San Francisco, Cal., 1976.
65. E. J. Sternglass, *Evidence for Low-Level Radiation Effects on the Human Embryo and Fetus,* in "Radiation Biology of the Fetal and Juvenile Mammal," Richland, Wash., May 5–8, 1969, M. R. Sikov and D. D. Mahlum, eds., ERDA Symposium Series, Conf–690501, pp. 693–718, 1969.
66. L. B. Lave, S. Leinhardt, and M. B. Kaye, *Low-Level Radiation and U.S. Mortality,* Working Paper 19-70-1, Graduate School of Industrial Administration, Carnegie–Mellon University, July, 1971.
67. E. J. Sternglass, *Environmental Radiation and Human Health,* in "Effects of Pollution on Health," Vol. 6, in Proceedings of the Sixth Berkeley Symposium on Mathematical Statistics and Probability, L. M. LeCam, J. Neyman, and E. L. Scott, eds., pp. 145–221, University of California Press, Berkeley, Ca., 1972.
68. M. H. DeGroot, *Statistical Studies of the Effect of Low-Level Radiation from Nuclear Reactors on Human Health,* in "Effects of Pollution on Health," Vol. 6, in Proceedings of the Sixth Berkeley Symposium on Mathematical Statistics and Probability, L. M. LeCam, J. Neyman, and E. L. Scott, eds., pp. 223–234, University of California Press, Berkeley, Ca., 1972.
69. D. Harman, Aging: "A Theory Based on Free Radical and Radiation Chemistry." *J. Gerontol.,* **11** : 298–300, 1956.

Energy

The plans for the energy future formed during the 1960s took negligible account of the problem of pollution. We now have a double problem: how to reduce the pollution of our present energy sources, the fossil fuels, while being threatened by their exhaustion in the absence of a satisfactory situation regarding replacement. The natural reaction we have is to reduce our concern for pollution from the fossil fuels, and, for example, to burn coal with an increasing amount of sulfur or to reduce our standards for safety precautions in building nuclear reactors. It is a classic example of the situation described by the saying "out of the frying pan into the fire."

16

Abundant, Clean Energy

J. O'M. Bockris

1. The Primacy of the Energy Supply

The amount of energy we have per capita of population affects the level of material wealth at which we live. Household heating, the ease and speed of travel, the production of goods from factories are all very energy dependent and their availability to average members of the population is clearly a function of the ratio of the price per unit of energy to the price of a day's work. In the material sense, there were no radical changes in the way people lived in cities during the millenia that preceded the eighteenth century. From that century onward, man began to harness energy, almost all of it in the form of heat energy from the burning of fossil fuels, and the world has since undergone some very striking changes.

One of the striking things which impresses itself upon those who travel widely is the immense difference which still exists between the material standards of average populations in different countries. When the countries are in the "overdeveloped" group (e.g., the European countries), the differences of

J. O'M. Bockris • Department of Chemistry, School of Physical Sciences, Flinders University of South Australia, Bedford Park, South Australia 5042

standards among countries are not great, e.g., the probability that a given citizen is the owner of a car may vary between 0.1 and 0.5 among countries in this group. But at the other end of the spectrum in a less developed, if not primitive country, such as India, one sees, for example, that the cooking utensils could be the same as those used 1000 or more years ago. Clothing among the villagers is not much different than that of much earlier times. Transportation is largely by bicycle or on foot. Manufactured goods are rarer, more expensive, and, for about three-quarters of the population (that of the villages), nonexistent. The possession of a car is a rare and special thing, a sign of great material superiority.

This can be illustrated by a plot of income per person as compared with the energy per person for various countries. Such a plot is shown in Fig. 1. It contains a clear message: *the standard of living (income per capita) is linearly connected to the energy per capita.*

One must think about this when considering the cutting back of energy supplies which now threatens citizens of countries in the affluent world, due to the fact that we have entered an exhaustion phase with two fossil fuels (oil, and natural gas) which has come upon us more quickly than we had expected;

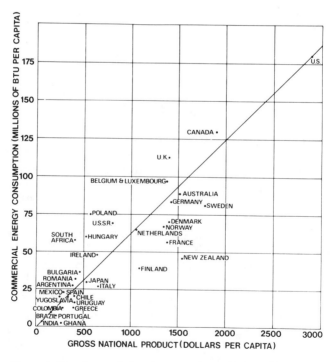

Figure 1. Rough correlation between per capita consumption of energy and gross national product is seen when the two are plotted together.[1]

these fuels may exhaust themselves before their successors, the atomic and solar energy sources, are ready.

The amount of energy per capita governs the standard of living. The price of a unit of fuel governs the energy per capita. Hence, the fuel price has great influence on the living standard: the latter will decline linearly with the reciprocal of the price of a unit of energy at constant income level.

2. *The Classical Concepts of the 1960s in Respect to the Energy Future*

When one considers the disarray in energy matters which characterizes the western nations in the 1970s, and the threat of an energy disaster which is applicable within the lifetimes of most citizens living, most people find it difficult to understand why this was not foreseen at an earlier time. It is interesting to see what considerations were current some 5–10 years ago, in the 1960s, when there was no consciousness that there would be any difficulty with the supply of energy for hundreds of years.

Thus, in the 1960s, the following two assumptions were unquestioned keystones in discussions of energy.

1. *The supply of fossil fuels (coal, oil, and natural gas) is effectively "infinite."* It was realized that this statement was not to be taken literally, that fossil fuels would eventually run out, but their exhaustion was always put at some vague point "hundreds of years in the future," and under these circumstances no interest could be shown in attempting to replace them with some other system. Thus, fossil fuel exhaustion had been discussed and predicted for many decades. However, it was found for many decades that the discoveries each year exceeded the fraction of the known reserves used up in one year. "They will always find some more" was the general impression. Corresponding to this viewpoint, there were, in United States universities in 1960, no institutions for the study of new energy sources, though there were, for example, about a dozen institutes for the study of materials science.

2. *Atomic power would be available "soon" in massive quantities.* This assumption began to be made in about 1950, after the successful running of early atomic reactors. It was thought that atomic energy could be made so readily available that the price of energy could be decreased to perhaps one-tenth. It seemed, in 1950, that atomic energy would be liberally available during the 1960s (as it turned out, by 1969 less than 1% of United States electricity was atomic in origin).

It is interesting to trace the path by which the optimistic assumptions of the 1950s broke down. The first, that concerning fossil fuels, turned out to be wrong in respect to oil. A lesson which had to be learned was that it was not the total supply which was of importance, but rather ease of recovery. This affected the price. There might be much undiscovered oil in off-shore positions, for example, but its price by the time it had been extracted was often too great for the product to be of use. When this factor was taken into account, the only large cache of oil known to exist from which the oil is easy to extract, and therefore extremely cheap (<$1 per barrel at the resource base), is in the Middle Eastern countries, above all in Saudi Arabia. The fact that so much attention has been given to the findings near Prudhoe Bay in Alaska, one of the most inhospitable places in the world, is a reminder of the present difficulty of finding new oil under conditions which make its recovery useful. More will *not* always be found (see Figs. 2, 3, and 4, and Table 1). Somewhat similar statements apply to natural gas.

In the case of coal, there was, and is, a great deal of coal, as long as needs are considered in terms of "the present." Thus, if one takes a country's energy need at a given time, and divides it into the coal reserves in similar units, one comes up with a comfortable "several hundred years" of available coal.

However, the situation with coal demonstrates the falling away of the comfortable concepts of yesteryear. There is, first, the difficulty of availability, as with oil. How much of the world's known reserves of coal can be recovered? The answer depends on the available technology and on the acceptable price of the product. It is only safe to assume the technology of the present. If it is then assumed that a price greater than double that of 1975 (in constant dollars!) is

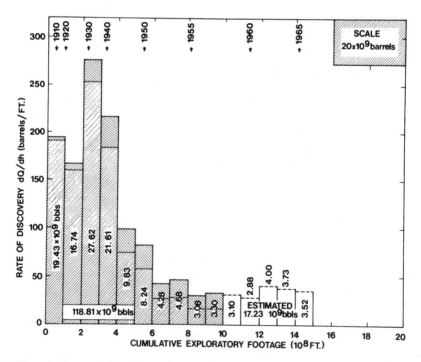

Figure 2. Average United States discoveries of crude oil per foot of exploratory drilling for each 10^8-ft interval of exploratory drilling.[2]

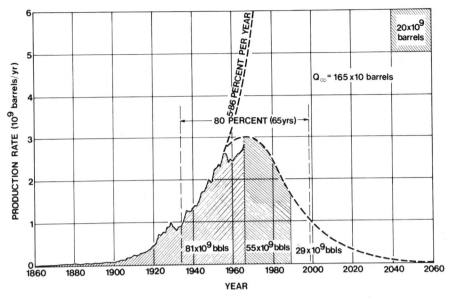

Figure 3. Complete cycle of United States crude oil production.[3]

not acceptable, then only about one-quarter of the world's coal can be considered available.

Another factor which radically changes the projections from those of the 1960s is that associated with the growth of need, which arises from the population growth and the growth of living standards. This factor was radically

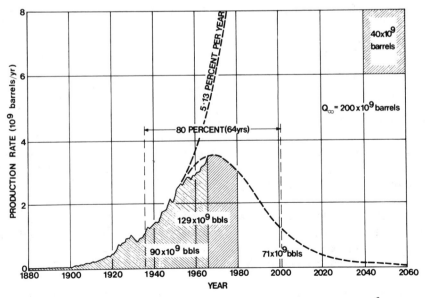

Figure 4. Complete cycle of United States production of petroleum liquids.[3]

Table 1. Energy Contents of the World's Initial Supply of Recoverable Fossil Fuels[a]

Fuel	Quantity	Energy content		Percent
		$(10^{21}$ J(th))	$(10^{15}$ kWh(th))	
Coal and lignite	7.6×10^{12} tonnes	201	55.9	88.8
Petroleum liquids	2000×10^{9} barrels $(272\times10^{9}$ tonnes)	11.7	3.25	5.2
Natural gas	$10,000\times10^{12}$ ft^3 $(283\times10^{12}$ m^3)	10.6	2.94	4.7
Tar-sand oil	300×10^{9} barrels $(41\times10^{9}$ tonnes)	1.8	0.51	0.8
Shale oil	190×10^{9} barrels $(26\times10^{9}$ tonnes)	1.2	0.32	0.5
Total		226.3	62.9	100.0

[a]Source: M. K. Hubbert.[2]

underestimated until the 1970s, and is a major source of delay in our waking up to the poor energy outlook. It seems to have been beyond the comprehension of those few scientists who contemplated the energy future in the 1960s to believe that growth would continue and become exponential. A typical attitude would be: "We have had a phenomenal increase in demand in recent times, but of course this kind of thing cannot continue." On this basis, the projection of demand was always made by dividing the rate of use at the time into the total available amount of the resource. Even now, a trace of the same attitude exists. The demand for energy in affluent countries, extrapolated to 2000, will be about three times greater than at present. However, no commentators seem to extrapolate past this date to the far greater demands of, for example, 2050. Yet, if the exponentially increasing demand stops at 2000, it will be because the world population has stopped growing at this point (an exceedingly optimistic assumption), or because the living standard is falling (a very unsettling possibility).

The breakdown of optimistic assumptions about atomic energy is more complicated. There are two reasons. First, the practical achievement of suitable reactors has turned out to be more difficult from the technical and engineering point of view than has been projected.* Thus, things which it was assumed

---

*The poor achievements of atomic energy, in conflict and competition with the international oil companies, is interesting to contrast with the fabulous, much more unlikely, achievements of the U.S. Space Agency, which had no competitor.

would be ready by 1960 were not ready by 1970, and do not seem likely to be ready for another generation, if at all. At the same time, the earlier cost assumptions proved to be particularly optimistic. In particular, the breeder reactor, a reactor which would be capable of making more *active* atomic fuel from the action of the rare, active U^{238} on the plentiful U^{235}, is not yet technologically satisfactory, although the first breeder reactor was built in 1953.

But another point which recently has begun to play a large part in the public attitude toward atomic energy is the question of pollution. When atomic reactors work, they produce waste products. These cannot be rejected into rivers and seas because they are radioactive and would be very dangerous to the populace. They cause cancerous growths due to radioactive particles interfering with the reproduction mechanisms of cells within an organism. The way to dispose of these waste products with safety is not clear, and, at present, they are stored uncertainly in steel cannisters beneath the earth (the lifetime of the radioactive pollutants is much longer than that of the steel cannisters).

A last factor is psychosocial. The dangers of atomic reactors are real, and probably worse than has been admitted by their designers. However, there is an element of panic and exaggeration in the community concerning the massive use of atomic energy and this feeds back upon the possibility of building atomic reactors because citizens' groups frequently obtain injunctions to cease building until an investigation has been made as to safety.

In summary, we face a bleak energy future because the two great assumptions of the 1960s have turned out to be false. We do not have enough fossil fuels—when the population and demand growth is taken into account—to last into a time when atomic energy will be safe, ready, and extensively built. Use of what we have pollutes the air. The Great White Hope of the energy future—atomic energy—has proved to contain unexpected difficulties in its safe development, and its dangers have frightened many responsible public groups into restrictive reaction, probably with justification.*

3. *The Effect of the Arab–Israeli War, October, 1973*

During the first years of the 1970s, the attitude toward energy was ambiguous. On the one hand, a realization was growing that shortages might occur in the realizable future, and that we needed to build atomic plants at a greatly reduced rate. On the other hand, the era in which environmental

---

*One of the difficulties in judging the correctness of assertions made by the opponents of atomic energy concerns the fact that it will take a long time before we know if they are right. People die from cancer all the time, and we will have to wait until statistics are available to see whether there are extra deaths in regions of the world into which atomic reactors are introduced. We will have to wait several generations before birth statistics show whether the fear of genetic damage is justified. But our energy situation is such that we cannot wait 50 years before we decide whether or not to build enormous atomic plants.

considerations were considered as important was burgeoning. These considerations seemed to go against the alternatives. Thus, environmentalists were
savage in attacks against the internal combustion engine and the use of
gasoline; but they were also afraid about prospects for atomic reactors. Some of
this negativity arises particularly because of proof of the enhanced sensitivity of
the fetus to low-intensity radiation. The increased use of coal to replace oil and
natural gas, with sulfur-containing components, is looked on by the influential
environmentalists with disfavor (see Chapter 15).

When the Arabs crossed the Suez Canal in October, 1973, the effect on the
fuel situation was dramatic. Used for many years to defeat at the hands of the
Israelis, the Arab world gained immense confidence by its partial victory.
Shortly afterward, probably as a result of the victory, the oil-rich Arab states
felt united enough to announce that they would introduce a partial embargo
upon oil deliveries to the western world. In view of the small reserve (1–2
months) carried in most countries, such an embargo was tantamount to a threat
of extinction, and led to great anxiety in Europe and Japan, and to some
perturbation in respect of fear of cut-backs in the United States. The result was
easy to predict. Frightened western statesmen were pleased to accept a price
rise of 400% for crude oil, and to pass this on gradually to retail customers.

Japan was more affected than any other country because it depends very
greatly upon Middle Eastern oil. European countries were also heavily
affected. The United States was affected less, because a large (though declining) fraction of its oil comes from sources within the country, predominantly
from the oil wells in the states of Texas and Oklahoma.

The Arab price increase of 1974 provided an accelerator in the consideration of the energy situation. Leaders in the oil-producing countries, particularly
the Shah of Iran, drew attention to the fact that even the Middle Eastern
supplies will exhaust in 20–30 years.

The political–economic happenings of 1973–1974, were, then, the shock
which broke the suspended muse of earlier years, and gave impetus to turn
away from the doubtful and exhausting fossil fuels, with their pollutive
difficulties and threat of politically based embargo, to the new sources of
energy.

4. The Time Scale in the Exhaustion of the Fossil Fuels

The investigations to which the oil price crisis of 1973–1974 gave rise
showed a worrisome situation. The assumptions of earlier times that atomic
energy would take over from oil and coal have proved to be wrong, not only
because of the difficulties explained above, but because it was seen to be
impossible to build enough atomic plants before the worldwide exhaustion of
the present sources of oil and natural gas.

This left only coal, and it is important to examine this as a separate
resource, to estimate whether products from it could be made to replace oil and
natural gas.

5. *The Unacceptability of Coal as a Long-Term Replacement for Oil and Natural Gas*

If one takes into account the expansion of the population, and the fact that oil will be exhausted, worldwide, by the end of the century, it is possible to impose on previous estimates the requirements for coal to take over from oil and natural gas and to supply more people. The result is a graph such as that shown in Fig. 5.

It is worthwhile explaining the nature of this graph. It is based upon some assumptions and one of these is the resource base, i.e., how much total coal there is. The figure shows the effects of three different assumptions. Second, it involves the assumption of an increase of 8% per year in need, and it has been pointed out that this arises not necessarily because of a continued increase in living standards, which seems unlikely at present, but because the populations

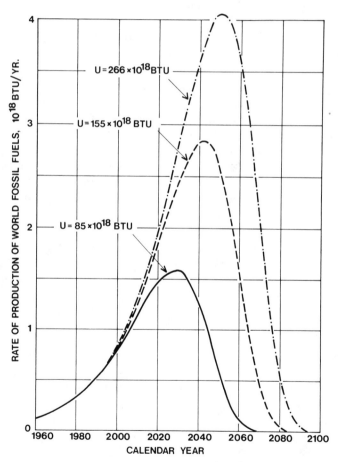

Figure 5. Effect of type of function used on the projected rate of world
fossil yields.[4]

of most countries are increasing rapidly. *The maximum use rate of coal will occur as early as* 2030, *and not later than* 2050.

However, there are difficulties in the use of coal which may make the situation less favorable even than that indicated by Elliott and Turner. The first is pollutive. We are saddled with air pollution from gasoline. A hydrogen economy (see Chapter 17) would cure these troubles. Suppose, then, we start to replace gasoline by coal. Two new pollutive hazards would hit us. First, coal contains sulfur, generally varying between about 0.3% and 6%. There is no economical way to remove this sulfur on a large scale. Burning it to produce electricity, or converting it to methane or liquid fuels, brings about the problem of what to do with the sulfur. It if enters the atmosphere, we will have an increased pollution problem. Another pollution problem will occur with further fossil fuel use: the effect of the aggregation of CO_2 in the atmosphere (see Chapter 10). The amounts which could enter the atmosphere if our energy were supplied by coal could have damaging climatic effects.

But there is another kind of reason why coal may not be a solution to the exhaustion of oil and natural gas. This concerns the difficulty of getting so much coal out of the ground in the appropriate time scale without using, for example, atomic explosives. Coal presently supplies less than 20% of the energy of the United States. Were it to have to supply near to 100%, now, the amount we would need would be increased several times. However, we will need more energy in the future, some three times more by 2000. We should therefore need some ten times more coal per year by 2000 than we produce now. This means that, in some 25 years, we would have to build ten times more coal mines than we have now.[5] Were these coal mines ten times bigger than the coal mines of today, it can be shown that, in the United States, we would have to open a new giant coal mine every second day until the year 2000! This is obviously an entirely impractical demand, and neglects the question of whether we would find about 2 million people in the United States willing to become miners, and how we would manufacture the equipment in the time period concerned.

Coal, then, seems to be an unlikely source of energy to replace the oil and natural gas, which, even in the Middle East, will have exhausted (assuming the continuation of growth) by the end of the century.

6. *The Vital Question of Time*

We are in a difficult situation, and all of the difficulty arises from the time scale. We have only one generation of people to go before we will be suffering from a great energy shortfall; and perhaps only 10 years to go before our standard of life will be noticeably affected by the shortage of energy, i.e., its steep extrainflational price rise.

Now, if we could, as in a fantasy, switch on one of the new sources of energy which are *theoretically* available, the situation would not be worrisome. However, it takes very long (decades) to research and develop the conversion equipment, and further decades to build it (see Fig. 6 and Table 2).

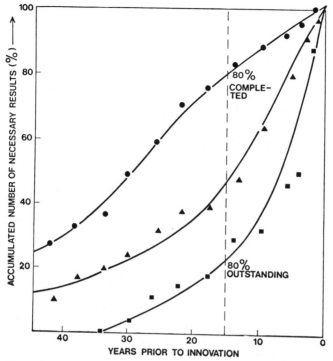

Figure 6. Number of years of basic research (●), applied research (▲), and development (■) necessary for the realization of innovations.[6]

Table 2. Time Development of Some Energy Sources[a]

Source	Electricity	Steam engine	Fission	Fast breeder	Fusion	Solar
Scientific feasibility established	Faraday (1831)	Newcomen (1712) Watt (1765)	1942	1950	1975?	1954
Useful power	Sturgeon (1836)	Many developments	1955	1960	?	?
Economic power	Siemens (1856)	1785	1965	1980–90?	?	?
Time	25 yrs	20 yrs	23 yrs	35 yrs+	—	—

[a]Source: C. N. Watson-Munro.[7]

7. The Position of Atomic Energy

It is usual to present the sources of abundant, clean energy for they are few and are easily enumerated. First, there is the atom. Here, we have three possibilities. We are using one of them, called fission, which works by processes such as those shown in Fig. 7.

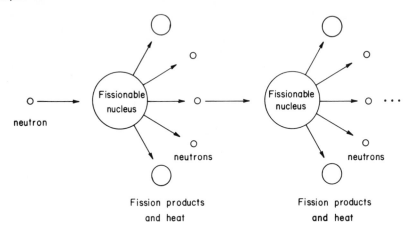

Figure 7.   Chain reaction of atomic fission.[8]

Fission is no answer to our longer-term energy problems, because what it uses is the tiny fraction of uranium oxide which is radioactive. But this fraction (that associated with the isotope called U^{235}) is only 0.7% of the total uranium content. Thus, in extracting U^{235} from the total uranium ore and burning it in fission reactors, we are using up material which, as we shall see below, can be better used in another process. In fact, extrapolating our use of uranium, there will not be much left by the end of the century (when oil and natural gas, and perhaps coal,* will also be near exhaustion at acceptable prices) to use for breeding from the residual inactive U^{238}.

The second way we use atomic energy is in the breeder process. In this, we utilize the initial active U^{235} to activate the atoms of the negative U^{238}. The processes are shown in Fig. 8.

Thus "breeder reactors" *would* be more valuable than the fission reactors, since they will convert a substantial quantity of inactive U^{238} into a radioactive material. If we could build working breeder reactors we might we able to use this process for producing energy from uranium for several hundred years. There are, however, the points already outlined above: we are not yet ready technologically, and may not be until after oil and natural gas have become very rare; and, even when we are, there is the pollution problem of the radioactive wastes and what to do with them. One idea is to put them into disused mine shafts. Another idea is to take them to Antarctica and bury them in deep ice. A more far-fetched idea is to put them into orbit around the sun. The latter idea would become feasible if there is a 1–2 order of magnitude reduction in the cost of putting material into orbit.

The third way in which we could use the atom is by fusion. This idea is different than the fission process outlined above, and works as shown in Fig. 9.

*Not because total coal could be exhausted before several more decades, but because of difficulties in getting the coal out of the ground at a sufficient rate.

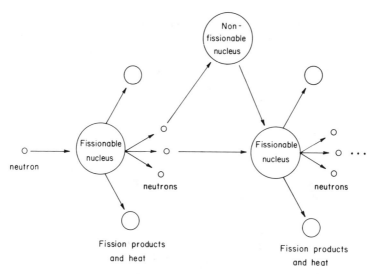

Figure 8. The breeder. Some of the neutrons produced can be used to create new fuel.[8]

If it can ever be made to work, this would be a utopic process, because it would produce no radioactive wastes, and the necessary raw material, deuterium, is available in the sea to such an extent that we would not have to worry about running out for millions of years.

The snag here is that the fusion process has not been maintained since research began some 20 years ago. The trouble is, essentially, that it needs immense temperatures to start it off, temperatures in the region of 10^{8}°C, and it is difficult, even in the laboratory, to make the technology which allows a mass of gas to reach this immense temperature.

When do researchers in this field believe that they will be ready? Everyone is more cautious now, after the unexpected slowness in the development of the breeder. Fusion scientists talk about success in the laboratory in about 2000. (Laboratory successes with fission first occurred in 1942.) The hopes for fusion

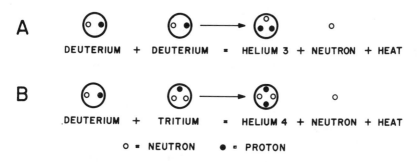

Figure 9. Some fusion reactions.[8]

are based on paper studies, and, although this criticism can also be made of other projected developments, e.g., those of the hydrogen economy, the technology needed for fusion is immensely complicated (and hence much more doubtful of achievement) compared with that for many other devices. In fact, no price projections can yet be made for fusion.

A changed attitude toward the development of atomic power has spread among the scientific community in the last 2–3 years. Although, before about 1972, the coming of massive atomic power was regarded as a certain, immensely powerful and utopic means of providing the energy basis for our future, it is now realized that it will not be ready in time in sufficient amounts, i.e., before the oil and coal run out; that (in respect to fission and breeders) it may be dangerous if it *is* developed extensively; and we may, therefore, by no means be going in the right direction in attempting to obtain inexhaustible, clean energy through the atom. Unfortunately, an inertia momentum is associated with the atomic energy establishments of many countries. These establishments involve tremendous government budgets, great buildings and plants, thousands of men in full career with high salaries, all of which has a great momentum, however unfortunate for the population. The reversal of atomic reactor development will not be easy. The reversal of research and development on fusion is not desirable, for if fusion does ever become successfully developed, it would be a splendid help, and perhaps the main source of future energy. But its achievement is so doubtful that we cannot rely on it. The point to understand is that we have taken the wrong path in energy development. We chose the breeder reactors, which are difficult to engineer, and have hitherto avoided those ways which may be less novel, less scientifically challenging, but are much more likely to bring us to energy sufficiency before the end of the fossil fuels, and with much less final danger.

Some speculative ways to represent the situations which are possible in respect to energy are shown herewith in Figs. 10 and 11.

8. *The Inexhaustible, Clean Energy Sources*

One of these sources may be fusion, but there is no doubt that fusion will be ready only after the exhaustion of oil and natural gas. Unless we are able to replace fusion with another means of getting energy on a large scale, there are significant doubts about the continuation of affluent technological civilization past the first 2–3 decades of the next century; i.e., we could be in our last 50 years of life as we know it.

9. *The Solar Sources*

A large amount of energy arrives on the planet by radiation from the sun. The figure is about 1.7×10^{17} W. The amount of energy used by man at present

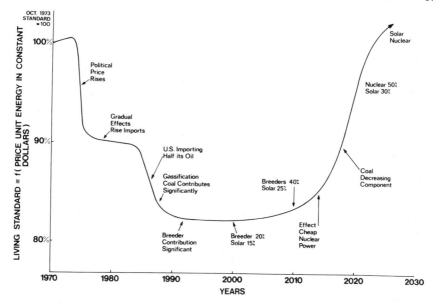

Figure 10. Speculation on a positive solution for the near energy future.[9]

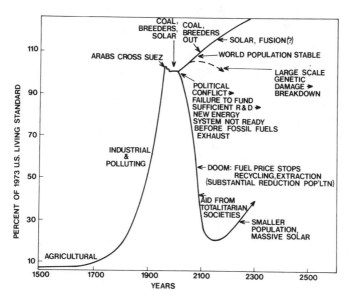

Figure 11. Speculation on triumph and disaster (United States).[9]

is in the region of 10^{10} kW, so that, if we could capture significant amounts of solar energy and convert it to electricity, all would be well. It is possible to do this in a thought experiment, assuming that we have, say, 1% of the land area of the world taken up in collecting solar energy. Let it also be assumed that we have collectors which have an efficiency of 10%; we could then collect a total amount of energy equal to 10^{11} kW.

Is this enough? This can be calculated by assuming a certain world population* and then assuming that each person consumes a certain amount of energy.

The present population of the world is about 3×10^9. If the population is increased to 5×10^9, and everyone (including those in less developed and even primitive countries) consumes at the rate of modern Americans (equivalent to 10 kW per person), which is clearly an overrequirement, even then we would have more than enough energy from solar sources collecting at 10% on 1% of the earth's surface to supply it all.

This, then, is an encouraging introduction to solar sources, but it may be said, on the other hand, that at the moment we have not yet engineered collectors which are 10% efficient, and which also work economically. The statement "solar energy is free" is a deceptive one. It may be that the fuel is free, but the fuel cost is only one component of the price of the converted usable fuel (electricity, hydrogen, methanol). The cost of this depends upon the cost of the conversion equipment. Money costs, the amount depending upon the reigning interest rates of the day. If a converter costs $500 to transduce 1 kW of solar power to electricity when the sun is overhead, and the rate of interest to be paid on the money is 10%, then one has to reckon with $50/year for 1 kW and then calculate the number of kWh which it can transduce from the solar energy impinging upon it at various angles and with the known degree of cloud for the area. Thus, because the energy from the sun arrives at the collectors free, the cost transduced to a useful form depends upon the cost of the transducers, the storage system, the transmission, the insurance, the maintenance, the profit, and the banks' rate for loans (with its counterpart, the rate of inflation).

The next question, therefore, is: what are the methods usable for transducing solar energy at this time? There are approximately five of these. Let us begin with the most conventional.

9.1. *Photovoltaic Devices*

In Fig. 12 one can see how a photovoltaic device works. A stream of photons from the sun impinges upon the upper part of what is a two-part sandwich consisting of some semiconductor, e.g., silicon. The photons activate

---

*There is no doubt that, if the world population continues to grow and exceeds a certain value, there is *nothing* we can do to avoid an energy disaster. In one sense, therefore, the limitation of the growth of the population, and indeed its *decrease*, is the most important thing of all in our attempt to obtain energy stability and constancy.

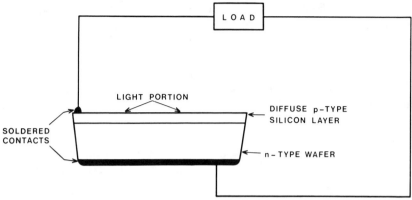

Figure 12. Mechanism of function of photovoltaic generator.[10]

electrons from the valency band of silicon and these electrons then migrate to the boundary between the two types of silicon. As they form an excess on the upper boundary there is a potential difference across this boundary, because a potential difference always occurs if two materials that have different charges on their surfaces are in contact. The potential difference that turns up at the boundary when the photovoltaic converter consists of two types of Si, is about 0.8 V. To obtain a larger potential one would have to put many silicon cells into series. One can obtain any desired amount of current by extending the area of the silicon cells.

This photovoltaic method of collecting solar energy could be built into a solar farm, and it is possible to calculate what the area of a farm of this kind would have to be.

One has to start out with some assumptions. Let us calculate the area of Australia needed to supply the energy requirements of Japan, assuming: (1) a Japanese population of 100 million, and (2) an energy requirement of 10 kW per person (the 1966 requirement per American was 8.3 kW[11]). We would need 10^9 kW. The maximum solar power falling on a surface perpendicular to the sunlight is equivalent to 2.0 cal/cm² min, or 0.14 W/cm². For 100% conversion of this maximum sunlight, the area required is therefore about 7×10^{12} cm, or 7×10^{13} for 10% conversion. Suppose we assume 8 hr of sunlight per 24 hr; then the area needed is some 2×10^{14} cm². But sunlight is not always perpendicular, even in optimal climates, and is obscured by cloud, dust, etc. Let us take a diminishing factor of three to cover these factors; then 6×10^{14} cm² is the required area—a square with a side of 250 km, or about 150 mi.

The corresponding figure for Australia, with 15×10^6 people, and the same energy consumption per person, would be, then, a square with a side of some 60 mi; for the United States, with 200 million people, about 220 mi. These very rough estimates show that patches of Central Australia, not larger than a couple of the larger Australian homesteads, could supply the total energy needs of the first two superpowers.

There are difficulties in the photovoltaic route: they are economic. Silicon sounds as though it should be easily available. The oxide, silica, is one of the more abundant materials in the earth. Sand consists of silica and there is much silica in rocks. However, what is needed is silicon, the element, so that first there must be a reduction stage to obtain the material from its plentifully available oxide. It is not this part which causes the economic trouble, although it does put a lower limit on the price of Si. The trouble lies in a different direction, and here one has to enter in a bit more to the requirements of a photovoltaic system.

One of these is that it should have great purity. When the electrons are excited by photons, they are *deactivated* (i.e., lose energy) when they strike impurities. "Deactivated" means that they lose the high energy which has shot them into the conduction band and they go back into the lower energy levels of the valency band of the material (if it is a semiconductor), and are no longer available in diffusion in the conduction band. Hence, they do not reach the boundary between the two types of silicon and, therefore, are no longer available for building up a potential difference. All this means is that we need exceedingly pure silicon, which accounts for some of the expense of making the silicon photovoltaic.[12]

Another expensive part of the manufacturing arises because the silicon must not only be pure, but largely in the form of a single crystal, or, rather, if not just one single crystal, at least a material highly single crystalline in character. Thus, most materials are "polycrystals," and consist of a large number of tiny bits, individual crystals. If one tries to make a photovoltaic crystal from polycrystalline material in the case of silicon, then, when the electrons, activated by light, pass through from one polycrystal to the next, they become deactivated in striking the side of the crystal, and thus only an insignificant number break the boundary between the positive and negative types of Si, and set up the potential difference. Only pure single crystals of silicon will give a reasonably high voltage.

Until 1973, the cost of producing the pure single crystal silicon needed for photovoltaics was prohibitive. It was estimated as being $100,000 per kW. In starting off with this figure, and making the economic calculations, one finds that the cost of electricity, or of the hydrogen fuel secondarily produced from it, would be 100 times greater than that which we pay now! For this reason, solar energy conversion was thought to be economically impractical, particularly in the 1960s, when the energy horizon was cloudless and a deeper look at the energy future was not deemed worth the time.

"Necessity is the mother of invention," and, as the energy picture began to darken (1972 onward), research to make cheaper silicon photovoltaics has increased. The situation is now better. Cheaper methods have been invented for producing silicon in a form suitable for making photovoltaics. It is not yet possible to purchase the silicon at the cheaper price, because part of the cost reduction is in having a large plant for mass production, and one has not yet been built. However, scientists now project that, were such a factory built, and utilizing the new methods of making silicon single crystals, the cost of the silicon would drop to \$250 per kW. The cost becomes meaningful if compared with the cost of the alternative means of reaching the same objective. The right one to make a comparison with is an atomic plant, at a cost of \$700–1000 per kW. The solar alternative begins to look good, but there is no great cause for optimism yet, because, on the one hand, \$250 arises from a calculation, and a systems analysis; it has not been attained in commercial practice; and, second, because when one talks about \$250 per kW in solar energy, the kW referred to would be produced only if one had the sun, without cloud, directly overhead. Of course, this means that the Si which could cost \$250 in a few years will produce less than 1 kW for most of the daylight hours. To be sure, one can to some extent compensate for this situation by using a mechanical tracker which keeps the crystals perpendicular to the sun as it appears to move. Clouds and darkness reduce the amount of time one can receive solar energy. An atomic plant can produce energy 24 hr per day, and one must take into account the fact that the solar plant produces at the best for 12 hr per day, and even less than this because of cloud effects. Probably, therefore, one has to multiply the projected \$250 by at least three, and perhaps by as much as nine, to make the price comparable with the 1 kW cost of an atomic reactor, and it will be seen that this photovoltaic method of obtaining solar energy is still more expensive, even as a projection, than the actuality of atomic energy. Conversely, the solar energy converter is *completely* clean, our ability to collect energy by it forevermore is about as certain a piece of practical knowledge that we have; and there is no danger of exhaustion, as there is with the fuels of the fission reactor.

9.2. *The Mirror Concentrator (or "Power Tower") Method*

This is a simpler method of obtaining solar energy than the photovoltaic one, and it is shown in Fig. 13, from which the nature of the method is clear. It consists of having an array of tracking mirrors which move with the sun. The object of the movement is that they track the sun as it moves across the sky during the day, but there is another objective in having the array pointed in a certain way. This is so that the mirrors focus the sun's energy all on the same small area, and this area is that of a boiler on a tower several hundred feet above the array. The boiler will contain a suitable working fluid. It could be water, but is more likely to be some other liquid, e.g., a low-boiling liquid such as ammonia. The boiler could work a conventional heat engine, or a more advanced device, a magnetohydrodynamic generator, in which heat is used to ionize gases, and the ions are passed between a magnet which then becomes a generator because its N and S poles attract particles of different charge.

The trouble with studies of solar energy is that little of it has become actual so far, because so much more research money, time, and realization has gone to atomic energy. Thus, we can only go upon projections, which are reasoned, detailed analyses, taking into account every cost which can be thought of.

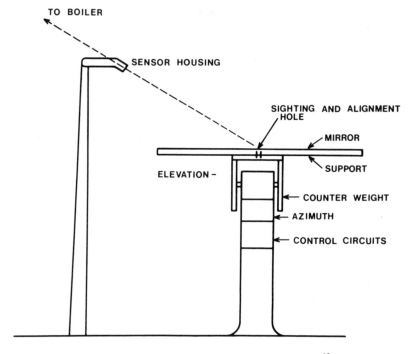

Figure 13. Diagram of a single-mirror pedestal system.[13]

According to Hildebrandt,[13] the projected cost of the electricity that could be produced from the mirror concentrator method would be in the vicinity of 0.7 cents per kWh^{-1}, and this compares well with the cost of electricity (about 1 cent) from coal plants at this time.

9.3. Satellite-in-Orbit Method

The most imaginative of the solar energy collection methods is the satellite-in-orbit method (Fig. 14). It assumes that it is possible in the near future (25 years) to orbit platforms at <$100 per pound (the present cost is ±$5000 per pound). The platforms would have placed on them photovoltaic converters, similar to those described above. The ones described above are earthbound. Were they placed upon an orbiting platform, there would be certain advantages. The first one would be that, in space, the absorption due to the earth's atmosphere is absent, so that there is a greater amount of solar energy falling on one square meter than on earth. Weather and clouding decrease are absent, too, but the major gain is the absence of night. The orbited solar collectors could constantly collect solar energy and convert it to electricity: one could produce energy constantly, as with atomic reactors.

What of returning it to earth? This would be done by converting the electricity to a frequency which could be transmitted in a beam. The microwave

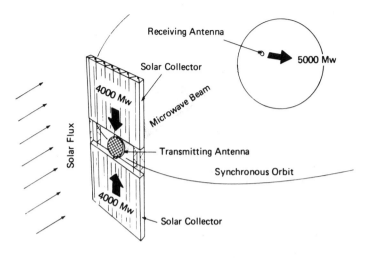

Figure 14. Design principles for a satellite solar power station.[14]

region would be the best one because a beam of microwaves coming down onto earth from the satellite could pass through cloud layer with little absorption. As the microwave beam approached the earth, collecting antennae could be put in the way, and so the electricity produced at the platforms could be converted back into electricity on the earth, and this used to produce hydrogen, which would be a store to be turned to when needed (our energy needs are cyclical— we need more during the day than during the night—so a hydrogen store would be necessary).

All this seems science fictional, but it has been systems analyzed, and the engineering seems feasible. The doubt arises from cost, and the cost is that of putting something into orbit. At present, this is about $5000 per pound, and, when one calculates back to the weight of material which would be needed on the orbited platforms, one finds that the cost per kilowatt would be far too high. However, advances are being made in space technology, and it may well be that within a decade or two the cost of orbiting material will have been reduced to about $100 per pound. Projections of this order of magnitude have been made in the publications of 1973–1974.[14] If this comes about, then the orbiting satellite might be the best way of collecting solar energy.

9.4. *Sea–Solar Power*

The surface of the tropical oceans is about 25°C in temperature. At a depth of 1–2 km, the temperature is reduced to about 5°C. Wherever there is a temperature difference it is possible to extract energy. Thus, one form of solar energy would be a sea–solar generator the essential parts of which are shown in Figs. 15, 16, and 17. One would be a boiler which would have a low-boiling liquid in it, e.g., ammonia; 25°C is enough to boil ammonia, which would work

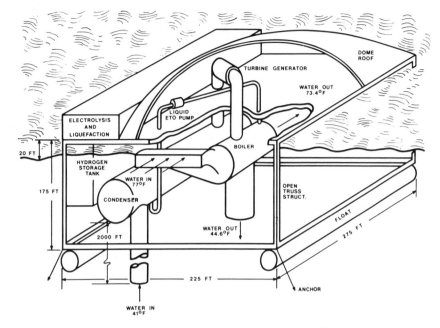

Figure 15. Solar–sea power plant schematic.[15]

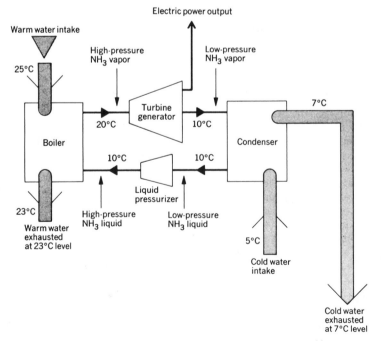

Figure 16. The basic components of a solar–sea power plant.[16]

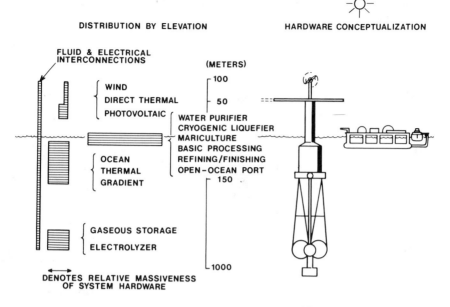

DISTRIBUTION BY ELEVATION   HARDWARE CONCEPTUALIZATION

Figure 17. Typical systems location.[17]

a heat engine or turbine. If the cycle is to continue, the ammonia has to be condensed, and this is where the cold water from the depth would come in. It would pump into a condenser, held below the boiler, and here the ammonia would be condensed, again to be evaporated, work the heat engine once more, and so on.[16] Once a continuous cycle is going, electrical generators could be worked, etc.

This is a helpful and attractive method for collecting solar energy, although, like the mirror concentrator method, it is mainly a paper study (large-scale models are being built). On the positive side, there is one particular advantage: the sea–solar power collector mines solar energy. Thus, going back to the pools of Middle Eastern oil, one is seeing there the result of solar energy collected over hundreds of millions of years in the photosynthetic productions of vegetation, and the decay of the vegetation to oil. The reason oil is so cheap (apart from political price hikes) is that the fuel is just sitting there as a result of millions of years of insolation. It is easy to extract. Its conversion to mechanical energy is direct: one does not have to make it into some intermediate fuel.

When we try to collect solar energy from the desert floors, from direct radiation coming in at a given time, the most we can obtain is the amount arriving at that moment. But the ocean thermal gradient collector is somewhat like oil; it represents not the solar energy of the moment, but that stored up from preceding insolation. Hence, in terms of energy per unit area, it can collect far more than can the photovoltaic and mirror concentrator methods.

There are two negative points on the sea–solar power collector, although they are not likely to be overwhelming. The efficiency of collection is very low. Carnot efficiency is

$$\frac{T_2 - T_1}{T_2}$$

and if $T_2 - T_1$ is about 20, and T_2 is about 300, the efficiency one can expect is in the region of 7%. In practice, it is expected to attain about 3%. This would be a disappointing efficiency were it not for the fact that the actual energy—the solar energy stored in the hot water—is free, so that the only disadvantage of poor efficiency is that we would have to pump a great deal of water through the sea–solar collector. Indeed, the amounts are immense, but one must come back to hard economics and ask simply what pumping energy is required to pump these amounts of seawater through the collector; it turns out that the cost of that pumping is far less than the value of the energy which can be produced (Table 3).

One other disadvantage of the sea–solar power plant is that either it must move or the water must be moving through it by some natural force (i.e., one without cost). If one of these does not happen, then, of course, the temperature gradient will be reduced as the collector uses up the energy, and the local source will be diminished. An ideal situation would be in the Gulf Stream, where the water runs at 4 mph, and the collector could be kept satisfactorily. Alternatively, it might be possible to have the collectors in the form of lengthy platforms, but small enough (100 m in length?) so that they could be maneuvered as ships. They could move in a circle with a radius of a few hundred miles at low speed. In this way, they would always have the required thermal gradient between them and the ocean depth.

The key point is the price of the energy that will be produced from these devices, and here again we must fall back on *estimates*, for no such devices have

Table 3. Cost Summary of Solar–Sea Power[a]

Main plant	$9,560,000
Auxiliaries	443,200
Structure and assembly	4,210,000
Assembly of cold pipe	262,000
Engineering and supervision	724,000
Contingency	1,448,000
Total	$16,647,000

Cost per kW = $166.00
Yearly owning and operating cost = $1,870,000
Rated yearly capacity = 876 million kWh
Estimated yearly output = 650 million kWh
Cost per kWh = $0.00285

[a]Source: J. Hilbert Anderson and James H. Anderson.[18]

yet been built. The estimates indicate excellent performances, and in 1975 dollars about 0.7 cents per kWh would be expected.

How may these sea solar collectors transmit power to land? This is an easy question to answer if the machine is stationary. If the land is close, cables can be used. Alternatively, one could produce hydrogen from the electrolysis of desalinated seawater.

9.5. *The Solar–Gravitational Source: Wind Energy*

The windmill was introduced into Europe in the twelfth century and was one of the earliest forms of man's attempt to utilize natural forces to give him energy (Figs. 18, 19, and 20). In the form of a wind generator producing electricity, the wind generator is familiar to people who work on farms. The massive production of wind energy has not often been discussed, except in Russia, but attempts to make it successful have not been great, in the past, usually on the grounds that the wind rotor (the giant propellor used in place of windmill sails) has hitherto always broken off when a storm has driven it too fast.

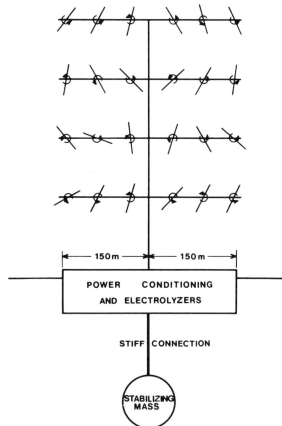

Figure 18. Small rotor alternative. Large rotor has a rotor of 25 m radius. A large number of smaller rotors could be used. A motor generator would be attached to each.[21]

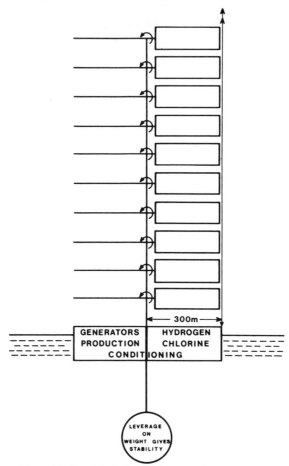

Figure 19. Possible design for large sea-borne generators:
alternative possible arrangement. Individual wind panels
(Al frames and sail-cloth) rotate central drive shaft and then
rotate themselves to minimum profile position for rotation
direction against wind.[21]

Recently, an enhanced interest has been shown in respect to wind
generators. One reason arises from the studies of Mullett in Australia.[19,20]
Mullett drew attention to two facts:

1. Although wind might seem to be a sporadic source, it is reliable if one
integrates over time spans on the order of a year or so. Thus, although at any
given day one cannot know far in advance at what velocity the wind will blow in
a given location, there is a repeatability about the winds which blow over the
span of a year. The mean of all winds for a given year is called the "mean annual
wind," and this is a reasonably constant figure for any part of the world.

2. Mullett[19,20] pointed out that there are wind belts in the world, areas
where, in terms of the mean annual wind, there is a "constant" wind which
blows at average velocities of 15–20 mph. An example of this is the wind which

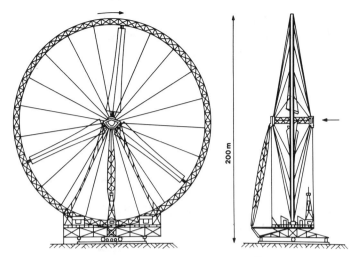

Figure 20. A proposed aerostation.[19]

blows in the Southern Pacific, below the latitude of 40°. Here, a constant wind blows around the world, touching the Cape of Good Hope, passing across the Southern Pacific, stroking the southern coasts of Australia and New Zealand, and traveling around the world to the western coast of South America.

Numerous other winds of this type exist—none perhaps so vast as the Great Southern Wind—and, with knowledge of the mean annual wind velocity it is possible to calculate how much energy they would give. To do this, it is necessary to know the formula that relates power produced from a wind generator when the velocity of the wind passing through it is v and the density of the wind is ρ.

This formula is not straightforward, for it depends upon detailed hydrodynamic arguments concerning the precise form of flow of the wind, the rotor, etc. However, a rough formula, giving the maximum possible degree of conversion is

$$\text{Power} = 16/27 \times 1/2\rho v^3$$

per unit area swept out by the rotor. The $1/2\rho v^3$ factor is the total kinetic energy which passes through the rotor with a velocity v and the $16/27$ factor arises as a maximum of what may be collectible on the rotation of the rotor.

More detailed considerations bring in other efficiency factors, which lower the power generated perhaps by another half. However, there are more subtle mathematical matters to resolve. The meteorological data consist of values of v, the mean annual wind. This is taken from instruments, called anemometers, that measure the instantaneous velocity of the wind.

The mean annual wind cubed, however, differs from the mean of the cubes of the wind, taken at individual times, and it is the latter factor which we should be using. The difference between the two factors is large, about 2.5, in favor of the mean of the cubes of the wind. Hence, if one utilizes the cube of the mean annual wind, one is shortchanging the possible energy which can be derived from wind generators, and one should multiply the power given by the above formula by about 2.

All this refers to a unit area, so that the potential possible area of a wind generator becomes important. Suppose that the wind generator has a radius r, and other efficiency factors are 0.75, then:

$$P_{elec} = 0.88 \, \rho v^3 r^3$$

It is interesting to experiment with this equation and see what sort of radius would be necessary to produce an impressive amount of power from one rotor. One has to assume some mean annual wind, and it is acceptable to assume a mean annual wind of 20 mph; this value of wind could exist, for example in the Southern Ocean and in many other places. Indeed, if one could accept a minimum velocity of 15 mph, there are many places in the world where one could collect wind, averaged over the whole year, e.g., the northwest areas of the United Kingdom, and part of the northeastern coast of the United States.

In megawatts, one obtains, for a radius of 100 m,

$$P_{elec} = 2 \times 10^{-4} \, v^3 \text{ MW}$$

Substituting the value of 20 mph into the above formula, one finds that the energy per rotor having a radius of 100 m is

$$P_{elec} = 5.4 \text{ MW}$$

This is an impressive amount of energy, and the question, of course, is: what is the feasibility of building such a gigantic rotor, and could one engineer a structure which would be stable on the sea, could live through storms, etc.? The frontier of engineering progress is in this field, and, as with so many other matters concerned with nonatomic research in energy, it is not being researched and developed. There is, for example, the requirement of deactivating the rotor so that it does not move at more than a certain velocity in the presence of a wind of more than, say, 45 mph. Even if rotors with radii as large as 100 m prove to be impractical, it would be possible to produce the same amount of energy per unit area by splitting them up into, say, ten rotors of lesser radii.

A calculation on these lines can be made to show the sort of area which would be involved in collecting enough wind energy to supply a city of 1 million people. Thus, let us take 4 MW per rotor, 10 kW per person; one would then need $10^4/4$, or 2500, rotors of 100-m radius. If the rotors were separated by 1 km, one could have them in a square whose side is about 50 km.

How would the wind energy be transduced to useful energy? The most reasonable way would be to use groups of rotors,[21] e.g., ten of them, and to connect their production of electricity to some neutral structure, e.g., the electrolysis devices producing hydrogen from the sea. It must not be forgotten, however, that this would have to be accompanied by chlorine reconversion

Table 4. Cost Estimates for a Wind-Based Hydrogen
Energy System[a]

System component	$ per kW
Wind generators	100–300
Electricity generator	50–150
Hydrogen production	100
Undersea storage facility	50
Fuel cell	150–350
Total	450–950

[a]Source: J. O'M. Bockris.[21]

plants. These plants would take off chlorine, which would accompany the oxygen gas evolved, and convert it chemically back to oxygen. An alternative would be to solar-distill the water to fresh water at the site of use.

What of cost calculations here? It must be again stressed that these are systems analyses and that no machines of this type have actually been built. However, the cost analysis might be similar to that shown in Table 4.

If one takes $500 per kW, the cost per kWh of hydrogen energy at T thousand miles from the source would be (U.S. dollars, p is the cost of money in percent per year)

$$\left[\left(\frac{500p}{100 \times 8760}\right)100 + 0.15T\right] \text{ cents}$$

plus maintenance.

If $T = 1$ and $p = 10$, and maintenance is 10%, the cost would be 0.79 cents per kWh.

10. *Consequences in Energy Research*

The general situation of the world in respect to energy is that, in the affluent free-enterprise states, we are in a serious battle between the ending of the fossil fuel supply, and the researching, developing, and building up of a new energy source. Failure to win this battle will mean a cut-back until the new (abundant, clean) source is researched, developed, and built up throughout the affluent world. During this time, the world, with its economics being so energy dependent, will become less affluent; and, because the attraction of a free enterprise society is its relative affluence, which balances its higher degree of uncertainty and anxiety, the world will become less stable.

The future of the energy base used to be thought of as atomic, but there are strong objections now being made against a future based upon atomic technology, except for the utopic concept of fusion, where research will not let us hope for practical engineered devices until well into the next century.

Apart from these complicated and less hopeful looking developments, are those in solar and solar–gravitational energy sources. Here, the prospect looks much more favorable. It is true that atomic energy looked favorable at first, and then the outlook became less favorable. Can this happen with solar energy, and solar–gravitational sources? Possibly, but it seems improbable. The engineering needed is classical, and would have a more solid basis for the assertion that the building of inexhaustible, clean energy sources based upon solar and solar–gravitational, not atomic, sources, is the likely basis of an affluent future for mankind.

Thus, the accomplishment of an era of massive atomic energy supply involved the great new ideas of fission (1939) and fusion (the 1950s), very advanced new feats in engineering, not yet accomplished in 1975, and far-out feats in the storage of exceedingly dangerous wastes which (in 1976) there are not even agreed upon as ideas. But the accomplishment of a massive supply of solar energy needs only *realization* of its availability; the economy of the less-conventional ways of collecting it, and, finally, its long-distance transmission in hydrogen. No new engineering feats are needed. To be sure, much development work is needed, but it all involves engineering which is well known and which does not require new wonders. There is no time for new wonders in the time scale of needed energy development.

The fact that these sources are entirely clean, whereas the atomic sources are extremely polluting (except for fusion), seems to put pressure upon nations to change the proportion of research money devoted to energy matters, deemphasizing fission and breeder reactors, and emphasizing solar and solar–gravitational sources. These should surely be supported at an amount, in terms of U.S. dollars, in the order of hundreds of millions per year (i.e., equivalent to the budget of breeder program).

11. *Could the Purely Gravitational Sources Be Used?*

There would be no wind energy were it not for the differences in temperature that cause the rising of air masses. Conversely, there would be no wind if it were not for gravitation, which works upon energy differences to cause shifts in air masses. Purely gravitational sources, however, are seen in hydroelectric power, and in the energy of the tides. At present, we use only a fraction (10%) of our available hydroelectric power; we do not use more because most of it is in far-off places of the world, where cables would have to be so long that the energy would cost too much to send to the place of use. However, it may be possible to use hydroelectric sources more in the future, because hydrogen could be a cheap way to transport energy around the world.

Tidal energy has been regarded as a massive source of energy, but it turns out that, upon consideration,[22] only if the tide is great—more than 40 ft— would it pay to produce tidal power stations (see Fig. 21). Nevertheless, there are some parts of the world where such tides do exist, and one is at Broome in

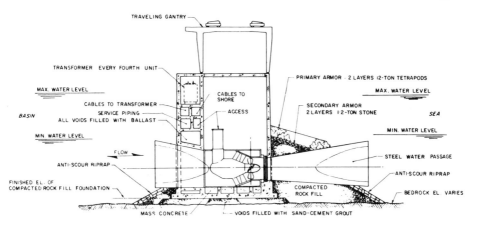

Figure 21. A tidal power house.[22]

Northwest Australia. Another is in the Bay of Fundy in Nova Scotia. A tidal station has been built in Rance, France, and is in regular working order. Here, the method of conversion would be that the flow of water, properly ducted, would drive rotors, the rotors would drive generators, etc. Again, the question of hydrogen transport comes in. Broome is more than 1000 mi from the nearest city, Perth, and once more the transportation of energy would depend upon a cheap method, e.g., the flow of hydrogen through pipes.

References

1. E. Cook, *Scientific American*, **225** : 83–91 (1971).
2. M. K. Hubbert, "Energy Resources for Power Production," in *Environmental Aspects of Nuclear Power Stations*, Vienna, International Atomic Energy Agency, 1971, pp. 13–43.
3. M. K. Hubbert, "Energy Sources," in *Resources and Man, A Study and Recommendations by the Committee on Resources and Man of the Division of Earth Sciences*, National Academy of Sciences, National Research Council, Freeman, San Francisco, 1969, pp. 157–242.
4. M. A. Elliott and N. C. Turner, presented at the American Chemical Society Meeting, Boston, Massachusetts, April 1972; M. A. Elliott and C. G. von Friedonsdorff, *Natural Gas of North American, Memo. No. 9*, **2171** : 2 (1968); M. A. Elliott, *Proc. American Power Conference*, **24** : 541 (1962).
5. J. Arthur, Chairman, Duquesne Light Company, Pittsburg, Pennsylvania, quoted in D. B. Thompson, *Industry Week* 26th November, 17 (1973).
6. G. Friborg, *Forskning och Framsteg (Stockholm)*, **3** : 2 (1969).
7. C. N. Watson-Munro, *World Energy Resource for the Next Century*, University of Sydney, 1974.
8. J. O'M. Bockris and Z. Nagy, *Electrochemistry for Ecologists*, Plenum, New York, 1974.
9. J. O'M. Bockris, *A Solar–Hydrogen Economy*, A.N.Z. Book Co., Brookvale, New South Wales, 1975.
10. J. O'M. Bockris and S. Srinivasan, *Fuel Cells: Their Electrochemistry*, McGraw-Hill, New York, 1970.
11. R. P. Hammond, in *Electrochemistry of Cleaner Environments*, J. O'M. Bockris, ed., Plenum, New York, 1972.

12. J.O'M. Bockris, *A Solar–Hydrogen Economy*, A.N.Z. Book Co., Brookvale, New South Wales, Chapter 8, 1975.

13. A. F. Hildebrandt, G. M. Haas, W. R. Jenkins, and J. P. Colaco, *E. & S. Transactions of American Geophysical Union*, **53** : 684 (1972).

14. P. Glaser, Hearing U.S. Congress, 31 October, 1973.

15. H. L. Olsen, G. L. Dugger, W. B. Shippen, and W. H. Avery, in *Proceedings of Solar–Sea Power Plant Conference and Workshop*, sponsored by NSF, Division of Advanced Technology Applications, Washington, D.C., June, 1973, p. 185.

16. C. Zener, A. Lavi, and C. Chang Wu, *Solar Sea Power*, First Quarterly Progress Report on NSF/RANN Grant No. 39114, October, 1973.

17. W. Escher, in *Proceedings of Solar–Sea Power Plant Conference and Workshop*, sponsored by NSF, Division of Advanced Technology Applications, Washington, D.C., June, 1973, p. 96.

18. J. Hilbert Anderson and James H. Anderson, *Mech. Engr.*, **43** : 41 (1966).

19. L. F. Mullett, "Wind as a Commercial Source of Energy," presented at Engineering Conference, Canberra, 1956.

20. L. F. Mullett, private communication, 1975.

21. J. O'M. Bockris, *Advanced Energy Conversion*, **14** : 87–91 (1975).

22. F. L. Lawton, in *Tidal Power*, T. J. Gray and O. K. Gashus, eds., Plenum, New York, 1972.

In considering a world in which we will be decreasingly fueled by the fossil fuels, and doubtfully by atomic energy, it seems as though our prospects may include the transportation of energy over very long distances, for it may be that the solar sources of the Southern Hemisphere will increasingly fuel the energy sinks of the Northern Hemisphere. The transfer of energy over long distances is most economically performed with hydrogen. But, if one converts energy to hydrogen for transport, it may be that hydrogen will also be an interesting general fuel for many other purposes. Such a situation has given birth in modern times to the concept of a hydrogen economy.

17

The Hydrogen Economy

J. O'M. Bockris

1. Definition of a Hydrogen Economy

During the 1970s a concept grew up: one of the better ways to reduce the spread of pollutants from the burning of fossil fuels would be to replace these with hydrogen. Thoughts concerning this were expressed in the United States in respect to a desire to avoid pollution from automobile exhausts, but they spread to the concept that hydrogen would be introduced into industry to replace natural gas as it was exhausted. Thus, one could see the possibilities of the building up of an energy system in which the medium is hydrogen. A system of industry, transportation, and household energy which depends on piped hydrogen as a fuel is called a "hydrogen economy."

The relation of the energy medium to the economy arises because there are likely circumstances (transport of fuel over long distances from large atomic or distant solar sources) whereby the use of hydrogen as a medium would lead to a diminution of energy costs, and would therefore be a basic step to overcoming the high price of fossil fuels from Middle Eastern sources since October, 1973.

J. O'M. Bockris • Department of Chemistry, School of Physical Sciences, Flinders University of South Australia, Bedford Park, South Australia 5042

2. The Cost of Energy and the Distance between Its Source and Place of Use

In the classical systems of energy production, fuel is transported from the place where it is produced to the place where it is used. For solid fuels like coal, these distances may be a few 100 km, but for oil and natural gas the distance may reach a few thousand kilometers. Thereafter, the fuel is refined and distributed over short distances to local gas stations and dispersed to industrial establishments. The electricity-producing "power stations" are local (within tens of kilometers) to centers of use. The electricity grid, which connects power stations near towns, transfers electric power over distances of up to a few hundred kilometers.

The above limitation in distance over which energy can be sent was another impetus for the hydrogen economy. The limitation in the transfer of electric power lies in the electrical resistance of the cables. This is intrinsic for a given diameter, temperature, and metal. Increase of the diameter reduces resistance, but increases the cost of cables. Thus, for a voltage at which transmission occurs, a given cable gives a certain voltage loss. With electrical transmission voltages in the region of 500,000 V, some 600 km is the maximum distance over which transmission can economically occur (see Table 1).

There are two ways, within the context of present engineering, to extend the length over which electric energy can be transferred. In the simpler method, the line voltage is raised. The current for the same energy is less and resistance losses are reduced. However, the maximum to which raising the line voltage is feasible is 700,000 V. At higher potentials, e.g., 1,000,000 V, the effect of the high electrical fields upon surrounding objects and the danger of sparking to earth both cause difficulties.

Another possibility lies in the use of superconductors. At present, these would require enclosing in liquid helium, or hydrogen, to maintain tempera-

Table 1. Relative Prices of Delivered Energy, 1972 ($/million BTU)[a]

	Electricity	Natural gas	Electrolytic hydrogen
Production	2.67[b]	0.17[c]	2.95–3.23[d]
Transmission	0.61	0.20	0.52[e]
Distribution	1.61	0.27	0.34
Total	4.89	0.64	3.81–4.09

[a] Source: J. O'M. Bockris.[1]
[b] Equivalent to 9.1 mils/kWh.
[c] 1975 costs will be about 300% greater than this.
[d] Assuming power purchased at 9.1 mils/kWh.
[e] Assuming pipeline hydrogen, at $3/million BTU used for compressor fuel in optimized pipelines, compared with natural gas at $0.25/million BTU.

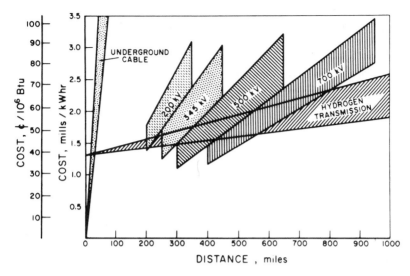

Figure 1. Relative costs of energy transmission by electricity cables and by hydrogen
pipeline.[6]

tures in the wires at which superconductivity can exist, and analysis shows that
the cost of this would be too great. Present theory makes it seem unlikely that
room-temperature superconductivity will be attained.[1]

Within this situation, therefore, a different solution was suggested, and the
essential point was mentioned for the first time by Lawaczeck, in 1933.[2] It was
pointed out that the energy to pump hydrogen through a pipe was less than the
work done to push the same amount of energy in electrical form through a wire.

Bockris, in 1962,[3] suggested the transfer of solar-produced hydrogen from
sea-borne platforms to towns, and Justi, in 1965,[4] published a diagram showing
a method by which solar energy could be collected and passed to various
hydrogen-using facilities, e.g., households, industry, transportation.

In 1969, Bockris[5] suggested that calculations be made in detail—
Lawaczeck does not seem to have made his suggestions quantitative—and
these calculations were carried out by Gregory et al., resulting in the diagram
shown in Fig. 1.

Consideration of this shows that, when a comparison is made between the
cost of pushing energy in the form of electricity through a wire with the cost of
pushing energy in the form of hydrogen through a pipe, the costs are in favor of
pushing electricity through the wire, up to a distance of a few hundred miles. At
higher distances, however, the costs are lower if the energy is first converted to
hydrogen and the hydrogen is then passed through a pipe.*

The details depend upon parameters of the situation, in particular the
transmission voltage; the higher this is, the longer the grid must be before it is
cheaper to send energy through hydrogen.

---

*At first, energy is lost by the electrolysis, so that the transfer of electricity is cheaper.

Another vital aspect is the cost of the electrolysis of water, which would have to take place before the electricity could be converted to hydrogen. Thus, the cost of electrolysis depends not only upon the thermodynamic energy needed to split water, but upon a quantity called "overpotential." Electrolysis takes place at a rate i, in A/cm^2, related to the overpotential (predominantly at the oxygen electrode), by the equation

$$\eta = \frac{RT}{\alpha F} \ln \frac{i}{i_o} \tag{1}$$

where α is a constant, about one-half, i_o is the equal and opposite current density at the reversible potential, and i is the current density for hydrogen evolution. However, this overpotential depends on not only i, the current density, e.g., what the electrodes are made of, so it is an important variable in deciding under what conditions transmission of energy in hydrogen becomes more economical than transmission of energy through wires.

Note from Fig. 1 that if the energy is transmitted in underground pipes, rather than overhead grids, the situation is more in favor of the transmission by hydrogen. Thus, the cost of laying high-voltage cables underground, in terms of energy sent per unit distance, exceeds that of electrolysis and transmission of gas through pipes after a few hundred miles. This is an important observation, less so at the moment, but likely to become more important in the future. Most countries face a rising population and, therefore, a rising energy demand. Were it to be brought to the cities in overhead wires, there would be increasing sight pollution, so that there will most likely be a trend to bring it underground. Such a consideration greatly favors the gas-in-pipe concept, and hydrogen economy.

A less favorable aspect is: What will be the use of the hydrogen fuel when it arrives at the other end? If it is to be converted to electricity, another factor to be taken into account is the efficiency of the fuel cell which would have to be used in the conversion. This will not be better than 0.7, and could be 0.6, so the distance at which a hydrogen economy would become viable, when the final goal is electricity production, would be increased. However, in most affluent communities electricity is only some 15% of the total energy demand. Hence, most of the hydrogen which is envisaged in a hydrogen economy would be used as such at the other end, e.g., in transportation, industry, and households (only a small fraction of that entering a household would be used for electricity). That used for heating, cooling, etc., would be used without reconversion to electricity, with extra energy loss (25–33%) in the fuel cell to be added to loss in electrolysis (10–25%).

One aspect must be clarified about the concept of a hydrogen economy: it is to be thought of in reference to nuclear, solar, and solar–gravitational (wind) sources. It is not to be thought of, except in terms of preliminary experiments, as to be used in conjunction with residual fossil fuels. This enhances its relevance, for the majority of energy users in the world are in the Northern Hemisphere, whereas solar energy, and probably the greatest wind sources, are in the Southern Hemisphere. Therefore, in the future, we may have to look to energy sources that are very far distant from the sites of use.

The main objection to nuclear sources (in particular, against breeders) is that they will pollute in a dangerous way. This pollution will grow worse as the size of the nuclear plant increases, so that large nuclear plants will be difficult to place near towns, not only because of dangers associated with air pollution and the disposal of wastes, but particularly in respect to thermal pollution. Were very large nuclear reactors placed a large distance away from the towns, then pollution problems would be reduced, but only in conjunction with a scheme such as hydrogen transmission, which reduces the economic penalty of a great distance between the energy source and the energy sinks.

3. *Some Consequences of a Hydrogen Economy*

Let it be supposed that we have proceeded some 2–3 decades so that a hydrogen economy is incipient. At such a time, then, hydrogen would be available in abundance. In particular, air vehicles would be running on it (see below) and it would fuel a portion of the transportation system. Cars could either run directly on hydrogen, in internal combustion engines, or on hydrogen via electrochemical fuel cells, thus avoiding all pollution.

However, this would only be the most obvious means of utilizing hydrogen. Its production on a very large scale, from nuclear and solar sources, would make it cheap compared with other forms of fuel, and entirely nonpolluting.

Let us look, then, at some of the other uses to which hydrogen could be put (Fig. 2).

3.1. *Metallurgy*[8]

Metallurgical plants produce much pollution, as one can observe by looking at the pollutive haze over a manufacturing town as one approaches it from the air. This consists not only of smoglike material associated with automobile exhausts, but also of the material which arises from factory wastes. If hydrogen could be used cheaply and abundantly, these pollutive emissions will no longer be present. Hydrogen could be used to reduce iron oxide to iron, instead of carbon, which is used at present in the form of coke. The present so-called blast furnaces eject a mixture of CO and CO_2; if hydrogen were used, then the iron oxide would be reduced

$$Fe_2O_3 + 3H_2 \rightarrow 3H_2O + 2Fe \tag{2}$$

and pure steam would be the only product ejected from the furnace. The steam could be condensed and utilized.

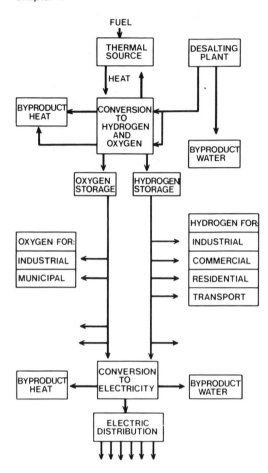

Figure 2. A hydrogen economy.[7]

There is a more attractive possibility in copper ore plants which for decades have been evolving SO_2 into the atmosphere. Thus, the sulfide ores in which copper, nickel, and other metals are found, are blasted together with air to form SO_2 and the corresponding metal oxide. The SO_2 is vented into the atmosphere, with undesirable environmental effects. If hydrogen could be used to convert sulfides to copper, the by-products would be sulfur and water, and solid sulfur could be collected and used in manufacture.

There are other concepts for obtaining sulfide minerals without polluting the atmosphere. One of them is the electrolysis of the sulfide. However, its competitive position, compared with the hydrogen-oriented process, would depend upon the price of electricity, and, if the energy source were nuclear and solar plants at substantial distances from the site of use, it would depend upon the price of hydrogen which could be delivered to the place at which the ore body was available. The direct use of hydrogen, without the efficiency loss in converting it back to electricity, would probably be economically preferable.

3.2. Chemical Technology

Hydrogen is much used even now in chemical technology, for it gives methanol and ammonia. What will tend to happen, as the price of natural gas rises and that of hydrogen falls, is replacement of most of the functions of natural gas by hydrogen. Thus, organic chemical technology at present relies largely upon natural gas, oil, or coal as its basic product. Whereas it has always been realized that these sources of supply will eventually exhaust themselves, the estimated exhaustion time has now been reduced to a few decades, the same order of time that it is likely to take to build up solar (and atomic) sources so that they can give the energy we will need in the year 2000, which at that time will not be available in sufficient quantity from natural gas and oil, perhaps not even from coal.

There is one carbonaceous substance of which we do have a great deal, to such an extent that we consider it "inexhaustible," namely, CO_2. Future CO_2 would come from one of two sources. First, it could be obtained by heating carbonate rocks. But there is also a great deal of CO_2 in the atmosphere, and this is increasing yearly with injection from the burning of present fossil fuels. This CO_2, when reacted with hydrogen to form HCHO or CH_3OH, could be the source of the organic chemicals for the future. The basic reaction would be the simple one

$$4H_2 + CO_2 \rightarrow 2HCHO + H_2O \tag{3}$$

Other compounds could be built up from this reaction.

Thus, in a nonfossil fuel-dependent future, the organic chemical industry, including that connected with the building up of plastics, is likely to depend upon hydrogen and carbon dioxide.

In particular, a proposal has recently been made by Steinberg *et al.*[9] that hydrogen from electrolysis could be combined with CO_2 for methanol, and this is a fuel which would be more convenient to use in ground transportation than hydrogen. It seems, however, that methanol will be more expensive than hydrogen, since its formation depends upon the availability of hydrogen.

3.3. Miscellaneous

There are areas in industry which at present depend on the use of natural gas. Processes involving paper, clay, and glass products, together with elements of the primary metal industry, all use natural gas, and these will, in the future, have to use other chemicals, one of which would be hydrogen. Gas turbine plants, now fueled by natural gas, could be fueled by hydrogen. One of the questions which will remain for reanalysis as the decades proceed is the competition between the thermal route toward energy, via a turbine or combustion engine, and the fuel cell route, in which hydrogen would be brought into an industrial plant and used to produce electricity locally. The resolution on this competition depends upon the efficiency in the conversion of energy, and may favor a fuel cell conversion.

3.4. Transportation

In respect to ground transportation, there are three possible fuels for the future:

3.4.1. Electricity

The first is electricity, and the much-discussed electric car. Electric car research has been relatively unproductive during the last decade because it has been misoriented toward minor improvements in electric cars (e.g., switch gears, more suitable bodies, or devices for smoothing out the load on the battery[10]) when the real problem which must be overcome is the development of a high-energy-density battery.

The lead–acid battery, the battery which we presently use in our cars, is a poor performer in respect to energy density, showing a value of only 10–15 Wh/lb.[11] In respect to this factor, it is the least efficient type of battery.* There are a number of better batteries in research programs, e.g., the zinc–air battery, or the molten salt lithium–chlorine battery. However, the funding given to research on these devices has so far been small, so that the high energy density batteries have not developed, and the electric car, because of the lack of research, still has the limitations known to it: a top speed of 40 mph and a 40-mi radius between replacement of its batteries.

This performance would make a suitable transport vehicle for city driving, where about two-thirds of the petroleum we use is burnt. However, consumers wish to have a vehicle which will do both the city driving, which is for the daily work run, and the long-distance driving, which is for enjoyment. Hitherto, therefore, the commercial prospects for selling lead–acid-driven electric cars have been nonexistent. The situation could change because of the large price increases which will befall our petroleum fuels during the next few years. Electricity costs should rise less steeply (insofar as they will be based on coal and atomic sources) and this could favor electrically driven cars. If research into high energy density batteries were funded at an effective level in organizations that would profit by success, the result could be that electric cars might have top speeds and ranges similar to those of the less efficient internal combustion cars, e.g., diesel-driven vehicles.

3.4.2. Methanol

Methanol has been proposed as an automotive fuel. Superficially, its properties are excellent, for it is an easily contained liquid, not unlike petroleum in this property. It would need a substantial modification of the present car engine, but not the radical changes in total car design needed if we went

---

*This has an undesirable effect on most people's opinion of batteries. The lead–acid battery is known to all, in fact it is equated with the word "battery." Its satisfactory function in starting internal combustion engines does not belie its low position as far as electrical energy per unit weight is concerned.

over to electric cars. There is thus much to be said—from the corporate viewpoint as well—in favor of methanol. This aspect will be reviewed in Chapter 18.

The economic aspect is the controlling one. It is difficult to estimate the future cost of methanol, were it to be produced in very large quantities. The question will only be decided over the next 1–2 decades. Thus the cost of both atomic and solar energy will help determine the matter: methanol will only be a medium of energy, never a basic resource such as oil or coal. At present, it looks as though methanol could be more expensive than hydrogen, because, eventually, it must be produced by adding something to hydrogen. Methanol could be considered as a molecule in which CO_2 carries some hydrogen around with it. In respect to pollution, methanol would be far better than gasoline. But it would still produce excess CO_2, unless it was produced by extraction of this from the atmosphere.

3.4.3. Hydrogen

The advantages of hydrogen as an automotive fuel lie in three directions. First, it would be entirely nonpollutive, because its only by-product is water. Second, it can be used in the present internal combustion engine with minor alterations, the main one being in the ignition system; and third, if the main energy supplies of the future are to come over long distances, and thus arrive in the form of hydrogen, it would be particularly convenient to use hydrogen directly without converting it to anything.

But hydrogen as a transportation fuel has a negative side. It is more difficult to carry than methanol, for it will need high-pressure cylinders, if it is gaseous, and cryogenic tanks, if it is liquid (the cost of liquefaction will also add about 25% to the cost of the fuel). A few possibilities are shown in Fig. 3.

The work of Murray and Schoeppel[13] on hydrogen engines has shown that NO impurities can be made negligible using H_2 as a fuel. NO is one of the most poisonous products of internal combustion engines.

Thus, the future transportation fuel has not yet been decided for land-based vehicles.[14]

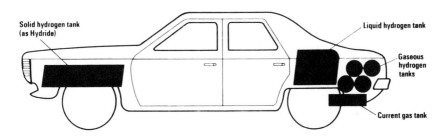

Solid hydrogen tank (as Hydride)

Liquid hydrogen tank

Gaseous hydrogen tanks

Current gas tank

Figure 3. Hydrogen storage options.[12]

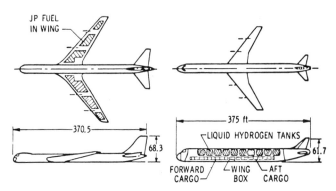

Figure 4. Subsonic cargo aircraft, payload = 265,000 lb, range = 5070 nautical miles.[15] (a) With JP fuel: take off weight = 1,500,000 lb; empty weight = 646,377 lb. (b) With liquid hydrogen fuel: take off weight = 915,000 lb; empty weight = 521,963 lb.

3.5. *Transportation in the Air*

The still-doubtful identity of the future fuel for land-based vehicles, with hydrogen the leading competitor among hydrogen, methanol, and electricity, is changed to one of certainty in the air. Hydrogen is the most likely future fuel (Fig. 4) for air transportation because of the advantages of complete cleanness, and the easy adaptation of present jet transducers* to run on hydrogen. The prospects of running aircraft on fuel cells or batteries are small.† This might be thought to be peculiar because of the use of fuel cells in space vehicles. However, in space vehicles, fuel cells are used for auxiliary power (heating, lighting), whereas blast-off power is performed by rocket jet using the hydrogen–oxygen reaction. In aircraft, however, we need power for the jet transducers. But the power-to-unit-weight ratio of the combustive type of transducer is far greater than that of the electrochemical transducer (although the latter exceeds the former in energy density).

Correspondingly, methanol would not be a suitable fuel for the advanced type of jet aircraft we expect in the future. The present supersonic transports have difficulty in being commercially viable, because they burn their fuel at such a rate that their range tends to be smaller than is commercially acceptable. The same thing would apply to SSTs running on methanol.[16,17]

These difficulties are avoided by the use of hydrogen, which is lighter than methanol in weight per unit of energy. There is no doubt that hydrogen is the best fuel for advanced aircraft and will be used in the next generation of aircraft, the hypersonic vehicles.

An important advantage in using hydrogen as a fuel for advanced aircraft would be its use as a source of cold for a refrigerant liquid to cool the airframe.

---

*A B57 aircraft of the U.S. Air Force was run on hydrogen as early as 1957.
†Short-range helicopters could be thus powered.

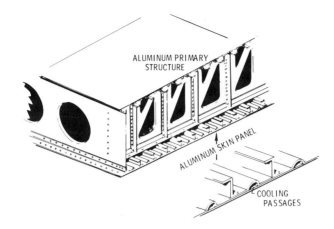

Figure 5. A proposed cooled wing structure.[18]

The weight of aircraft bodies, and hence their payloads, depends on their density, and alloys of the lighter metals, e.g., Al, Mg, are indicated. However, their strength at the higher temperatures reached by the aircraft skin is poor, and would only be adequate if the skin were cooled in flight (Fig. 5).

3.6. *Household Energy*

The use of hydrogen in households will follow the diffuse energy concept. Instead of transmitting electricity to households, as at present, the concept is to transmit hydrogen in pipes, similarly to the transmission of household cooking gas. In the house, the hydrogen will be converted to electricity for those functions of the household which need it, and the rest will be used for heating.[7]

There is an advantage, thus, in "the total energy concept." The hydrogen which produces the electricity also produces heat, both in-house. If we had electricity production at a power station tens of miles from the point of use, we would have to use cooling water to deal with this heat, and thus the heat is wasted and rejected to river or sea. If we produce the electricity *in* the house, the heat is available and can be used to heat the house.[19]

Many novel things in households could be run with hydrogen. For example, lighting need not be carried out by the usual means, but could be done by making phosphorus give out a bright light. Heating could be carried out by passing hydrogen through a ceramic containing a catalyst, so that the ceramic would become hot, without a flame. Cooking could be carried out in a similar way.[19]

An unexpected benefit in the use of hydrogen in houses could be the provision of auxiliary water supplies. There would be no problem in providing extra water in houses with hydrogen, with the condensation of the water produced in the use of hydrogen. If apparatus could be designed to utilize all the water produced, at the present use of electricity, the amount would be some 2–3 gallons per household per day. The fraction of energy used in houses which is electric may increase, and some have suggested that it will be five times higher in 2000 than it was in 1972.[20] Finally, if there were a total hydrogen economy, the amount of water produced would be about 14 gallons per person per day.

4. *Large-Scale Production of Hydrogen*

4.1. *Introduction*

The present mode of producing hydrogen is from natural gas, but this cannot be the source of future production of hydrogen because of the coming exhaustion of natural gas. One possibility would be to produce hydrogen from coal, and this can happen chemically, e.g., by the single reaction

$$C + H_2O \rightarrow CO + H_2 \qquad (4)$$

However, although coal might be the source of hydrogen for the beginning of the hydrogen economy, the decades in which we shall be able to rely on a coal-based economy would be limited (between 3 and 8 decades), so that it is more reasonable to look toward water as the source of future hydrogen and discuss the splitting of the H—O bond by chemical or other means.

4.2. *The Chemical Generation of Hydrogen from Water*

One first thinks whether it is possible, thermally, to decompose water. Thus, if one heats water directly, one obtains about 1% hydrogen in the water–hydrogen mixture at a temperature of 2000°K. This is too high a temperature for an acceptable industrial process, for the cost of the materials which would stand up to such temperatures would be impossibly high, and their availability impractically low. Further, a large-scale supply of heat at 2000°K is not in prospect.

However, the chemical production of hydrogen from water would be possible from another approach. It is not necessary to go *directly* from water to hydrogen. Rather, we could induce the reaction of water with some material, M, to produce MO + hydrogen. Thereafter, we could decompose the MO by heating it, reproducing the M, and releasing O_2.

$$M + H_2O \rightarrow MO + H_2$$
$$MO \rightarrow M + H_2O_2 \qquad (5)$$

The advantage of this path is that it would give rise to the production of hydrogen at a lower temperature than that necessary for straight decomposition.

Thus, the major determinant of the amount of hydrogen in equilibrium with water at a given temperature is not only the heat of reaction (this, by Hess' law, is independent of the path), but also the entropy, through the thermodynamic relation

$$\Delta G = \Delta H - T\Delta S \qquad (6)$$

If the entropy of the reaction is positive, we can decrease ΔG, and hence increase the H_2 in equilibrium with water at a given temperature.

Suppose, then, in a series of cyclical chemical reactions that result in H_2 production, we utilize high temperatures for those reactions which have a positive ΔS (these will decrease the net ΔG for the overall process of water decomposition), but low temperatures for those reactions which have a negative ΔS (the $T\Delta S$ component will tend to increase ΔG and decrease the hydrogen in equilibrium with water at a given temperature), then we may end up with a net numerically reduced ΔG, for $H_2O \rightarrow H_2 + \frac{1}{2}O_2$ and a reasonable amount of hydrogen without the temperature being raised to an unacceptable value.

What is an acceptable temperature value? Although this does not only depend on the thermal stability of the construction materials of the reactors, it does become difficult to find materials which are stable for long periods over 1000°C. However, there is a more important reason why 1000° should remain about the maximum acceptable temperature for the chemical production of hydrogen from water. One needs a source of heat to drive the reaction along (the ΔH value for the production of hydrogen from water is positive), so that we need to put heat into the reaction to make it go to the right. What large-scale source of heat would be available in the future for such processes? It would be nuclear heat, and the temperature at which this heat would be available in large quantities is about 1000°C.

As it turns out, there is no practical example of the production of hydrogen from water, which in reality is as simple as that used to illustrate the principle above utilizing the symbol M. One of the simplest of the indirect ways is given by the reaction sequence[21,22]

$$2CuBr_2 + 4H_2O \rightarrow 2Ca(OH)_2 + 4HBr \qquad (730°C)$$

$$4HBr + Cu_2O \rightarrow 2CuBr_2 + H_2O + H_2 \qquad (100°C)$$

$$2CuBr_2 + 2Cu(OH)_2 \rightarrow 2CuO + 2CuBr_2 + 2H_2O \qquad (100°C)$$

$$2CuO \rightarrow Cu_2O + \tfrac{1}{2}O_2 \qquad (900°C)$$

4.3. Difficulties of Chemical Generation

The difficulties of any process aimed at commercial production must always be taken comparatively. The method with which a chemical approach to the decomposition of water should be contrasted is the electrochemical method. Here, the method itself is splendidly simple: it is the electrolysis of water familiar to high school students. But the objection is that we would have to measure the value of the energy we could get from using hydrogen as a fuel at a distance, compared with the cost of the original energy which drives the method, which gives the electricity, which produces the hydrogen. Thus, the appropriate comparison to make is the efficiency of the chemical production of hydrogen from heat, as sketched out above, not directly with the electrochemical production, but with the efficiency of the conversion of heat to electricity, multiplied by the efficiency of the conversion of electricity to hydrogen.

Here, then, is a difficulty in the electrochemical method, the well-known electrolysis method for breaking down water to hydrogen and oxygen. It does not lie in the electrochemical splitting of water; it arises from the fact that one has to proceed, first of all, from some heat source, e.g., nuclear heat, to electricity. Indeed, it is not the direct manufacture of electricity that is inefficient, but rather the conversion of the heat to mechanical work in the engine which drives the generator. Typically, the conversion of heat to electricity would occur with an efficiency of 30–40%, so that up to 70% of the heat is not being converted to mechanical energy, and therefore, not being converted to electricity. (The mechanical-work-to-electricity stage is of high efficiency, and can be as much as 80%.)

This means that the electrochemical method of conversion of water to hydrogen is going to start with a kind of "down" situation in the energy needed to obtain one mole of hydrogen, for some 60% or more of the energy has been lost to heat in making the electricity.

From this point of view, then, the *direct* chemical conversion of heat to hydrogen looks good, for it would avoid the loss of energy involved in the heat-to-mechanical-energy conversion.

So far, there seems to be much in favor of the chemical approach, but closer examination shows that there are two substantial counterbalancing difficulties. The efficiency loss in the electrochemical method is real enough, but we have forgotten to ask whether there are any corresponding (if different) energy losses which may occur in the chemical approach. In the electrochemical approach, the loss is connected with heat-to-mechanical stage, and this involves the Carnot cycle, in which heat is converted to mechanical energy at

$$\frac{T_2 - T_1}{T_2}$$

where T_2 is the top temperature of the process, that at which the heat enters the conversion of energy, and T_1 is the bottom temperature of the process, that at which the heat leaves the conversion engine.

At first, the chemical process seems to have none of this, but consideration reveals that it does have something similar. Thus, in the indirect conversion one carries out the separate processes at different temperatures and pressures. Hence, there must be an *expansion* of gases as they go from a cold to a hot temperature, or a *contraction* as they are cooled down again in various parts of the cyclical process. Taking into account the work done in these processes gives energy losses like those involved in the cyclical Carnot process, covering the conversion of heat to mechanical energy.

Funk and his colleagues[23-25] have been industrious at pointing out the limitations of the chemical method, and Table 2 is reproduced from their work. It shows that the efficiency of the chemical method may be less than that of the net efficiency of the electrochemical method, including the efficiency of the electrolysis process itself (this is high, 75% or more) and the efficiency of the heat-to-mechanical-energy conversion (30–40%).

It was stated above that there were two difficulties in the chemical method. One is that there are substantial efficiency losses. What is the other? It is the difficulty connected with the complexity of the apparatus and the high temperatures involved. Thus, the electrolysis method consists of one vessel and one

Table 2. Vanadium Chloride Process Data Tabulation[a]

Parameter	Units	System				Remarks
Maximum helium temperature	F	2000	2000	1500	1500	
Minimum helium temperature	F	37	37	67	67	
Helium system pressure	atm	10	1	10	1	He pressure drop equals 10 psi in all cases
Process heat input	kcal	155	155	475	475	
Total heat rejected	kcal	312	382	698	998	Includes unrecovered shaft work
Total heat regenerated	kcal	262	262	872	872	
Helium flow	$\dfrac{\text{mol}}{\text{mol } H_2}$	113	113	383	383	
Helium pumping power	kcal	3	24	14	105	100% efficiency
Separation work (VCl_4 and He)	kcal	18	18	23	23	
Total separation work	kcal	33	33	38	38	4 stages, plus VCl_4 and He separations
Shaft work input	kcal	69	90	90	18	Separations work at 50%; pumps work at 100% efficiency
Thermal power	kcal	230	300	296	600	30% efficient (heat to work)
Total reactor thermal power	$\dfrac{\text{kcal}}{\text{mol } H_2}$	385	455	770	1075	Process heat plus shaft work
Figure of merit, η	%	18	15	9	6	$\Delta H_0/Q_t$

[a] Source: J. E. Funk.[25]

series of tubes, pipes, and connections, etc. In the chemical method, there would be a series of vessels for the chemical cycles. Moreover, some of them would be working at temperatures of 700° or more, and this would produce highly corrosive conditions. The costs of the plant would be likely to be much greater than those of the electrochemical method, where the apparatus itself can be very simple and therefore cheap.

Lastly, there is an unexpected difficulty in the cycle chemical method.[26] No cycle is 100% cyclical, but rather is cyclical only to a certain degree. Thus, with each cycle a certain amount of material will remain unconverted. Let us suppose our degree of cyclicity is as much as 99.9%. There will still be 0.1% of material left over. But large plants will be dealing with many thousands of tons of material per day and so, insofar as it is not reconverted completely, tens of tons of material will be left over, unrecycled, and this net use-up of materials would add substantially to the cost of the process.

However, these objections to the chemical method are only weights in a scale which will have to swing one way or the other in the coming decades. The chemical method uses heat directly to drive the reaction (avoiding the Carnot loss in using electricity), but the penalty is that it must go through many stages, and this introduces inefficiency of cost. Let us see whether one can do better in another way.

4.4. Electrochemical Process[26]

The electrochemical process for producing hydrogen from water is known to those who have had a course in chemistry. It involves the connection of an electron-giving electrode (cathode) to one pole of an outside power source, and the connection of an electron-accepting electrode (anode) to the other pole of another power source, and the dipping of these two electrodes into an alkaline or acid solution of water. Hydrogen is given off, separated from oxygen, at the

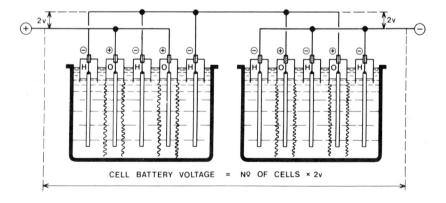

CELL BATTERY VOLTAGE = N⁰ OF CELLS × 2v

Figure 6. Unipolar cell construction.[27]

cathode, and this is the purest way to produce hydrogen. Oxygen is given off at the other electrode.

There are many examples of this electrolytic production of hydrogen today, and an example of a typical electrolysis plant is shown in Fig. 6.

4.5. The Effect of Temperature on the Efficiency of Electrolysis

The equation $\Delta G = \Delta H - T\Delta S$ has been quoted, and, in the case of the reaction for the decomposition of water to hydrogen and oxygen, the value of ΔS is positive, so that, with increase of temperature, the ΔG value will become less positive, i.e., will tend to become more negative. Thus, at higher temperatures the equilibrium of water, hydrogen, and oxygen is more to the hydrogen side, so that we ought to be able to decompose water more easily at higher temperatures than at lower ones.

Correspondingly, a graph which shows the dependence of the potential necessary to decompose water at higher temperatures shows that one may use a smaller potential upon the cell at higher temperatures to produce hydrogen (Fig. 7).

Of course, the heat to keep the cell at the high temperature costs, too, but the cost of a certain amount of energy as heat is less than the cost of electricity, because of the Carnot efficiency loss in converting heat to electricity. It will be advantageous, therefore, if one utilizes electricity to electrolyze water at a higher temperature than room temperature. Indeed, there is no need to restrict

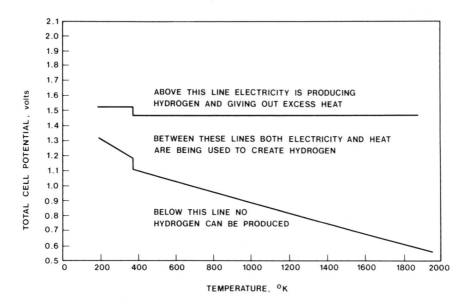

Figure 7. Electricity and heat: gain and loss in the electrosynthesis of hydrogen.[26]

oneself to temperatures at which water is liquid; one can also utilize steam and an apparatus for doing so.

4.6. The Conversion of High-Temperature Heat to Hydrogen

The use of high temperatures to facilitate the chemical reaction of the dissociation of water to hydrogen and oxygen may be associated with expensive equipment. There is one method of converting heat to hydrogen, however, which could give in the future a suitable adjustment to the difficulties inherent in the use of high temperatures.

Thus, Fig. 7 shows that, as the temperature increases, the reversible potential for the electrolysis of water to hydrogen *decreases*. Superficially, it looks as though, as long as one keeps the electrolysis vessel sufficiently hot, e.g., at 900–1000°, one could electrolyze at a thermodynamic potential less than that necessary at room temperatures. As the cost of a given amount of hydrogen is proportional not only to the cost of electricity per unit, but also to the magnitude of the potential difference applied across the cell, then there would, as pointed out earlier, apparently be economic gains by utilizing the highest temperatures at which the expense of the apparatus will not counteract the advantage.

A difficulty, in principle, of electrolysis as a method for obtaining hydrogen is that, in making the electricity from an energy source such as coal or atomic energy, we have to pass through a heat engine; in the process of converting the heat from coal or the atom to mechanical energy (and thence to electricity), some 60% of the energy is lost.

However, if we can use high-temperature electrolysis, only part of the energy to decompose water is coming via electricity obtained through a heat engine. Thus, the energy used to produce hydrogen directly from water will be cheaper than the room temperature methods, which go through "pure electricity," i.e., electricity produced from a heat engine.[28]

One would go about calculating the cost of hydrogen from the high-temperature method as follows: The cost (in cents) of the electricity, plus plant costs, of 10^6 BTU of hydrogen is[29]

$$229Ec + 48 \tag{7}$$

where E is the potential applied to the cell to make it work at a desired current density $(0.1–1 \text{ A/cm}^2)$, and c is the cost of electricity in cents per kWh. 48 is the allowance in cents per 10^6 BTU for "fixed costs," based upon a prognostication due to Gregory *et al.*[30] for the development of emerging technology electrolyzers. It represents maintenance, profits, taxes, cost of money at 10%, etc., and has been raised 20% over the 1972 projection of Gregory *et al.* (The fixed-cost figure would be about 80 cents per 10^6 BTU for older technology.) The 1974 value of c for the United States, given by Sporn,[31] is 9.4 mils. However, if low use time power were used, the costs would be 3–6 mils. We here assume that the costs are either 9 mils or 5 mils. The costs of electricity for the massive electrochemical production of hydrogen would tend to be lower than that quoted for other purposes, since the hydrogen synthesis plant could be placed at the site of the electricity production, i.e., there would be no cost of electricity distribution. Cheap nuclear electric power may eventually (after 2020) be available from nuclear sources at about 3 mils.

Nothing has been said about how the electrolysis of steam at 900–1000° could be achieved. Such a project might seem unlikely. However, it was achieved in 1966.[6] This was at a time at which much fuel cell research was being done in the American space program. One of the difficulties of fuel cells, as with uneducated electrolyzers, is the presence of overpotential. This overpotential is the departure of the electrical potential of an electrochemical cell, or electrode, from the potential which it would have according to thermodynamic reasoning. The latter refers to conditions of *equilibrium*, i.e., zero net rate, and, if one wants a reaction to actually occur with a net rate in any given direction, one has to encourage it by biasing the potential at the electrode in excess of the thermodynamic value appropriate to the temperature, concentration, etc. This excess is the overpotential, and the greater it is, the greater is the potential needed to decompose water to hydrogen at a given rate.

Overpotential is reduced with an increase in temperature, so that one way of optimizing fuel cells is to make them function at high temperatures, and thus reduce the overpotential, which, in the case of fuel cells (they produce electricity from hydrogen and oxygen), reduces the potential at which the electricity gets produced, and hence reduces the power (potential times current) produced by the cell.

One way to work a fuel cell at a high temperature is to use an anion-conducting oxide, in thin, wafer form, as the electrolyte between electrodes. Steam is brought up against one side of the zirconium oxide membrane, and both sides are connected to metallic leads.

Hydrogen gas strikes one side, dissociates to atoms and forms protons, injecting electrons into the circuit. These flow round it, doing work, and they are then attached to O_2 and form O^{2-} ions, which migrate through the ZrO_2 to meet the protons and form water. Hence, the fuel cell action is the formation of water and the production of electrical energy.

Correspondingly, as an electrolysis device, water arrives on the ZrO_2, dissociates to form $2H^+$ and O^{2-}. The current from the outside source neutralizes the protons which form H_2 on one side of the ZrO_2, and the O^{2-} diffuses and gives its electrons to the conductor on the other side, evolving O_2.

The materials problems associated with such a cell are great. The ohmic resistance of the zirconia needs reducing—the wafer needs to become thinner, or the specific conductivity higher—so that less potential drop is lost in forcing the ions through the zirconia. Of course, the thinner the membranes, the more likely they are to rupture. Thus, the future of this method of producing hydrogen, achievement of which would be a stroke in favor of cheap hydrogen as a fuel, depends much upon the mechanical *stability* of thin zirconium oxide wafers.

4.7. *The Crude Electrolyzers in Use Today*

Most electrolyzers in present use are somewhat over-simple in design (see Fig. 6). These designs are only trivially different from those that were constructed at the beginning of the century! Only in the last decade has knowledge become available which could considerably improve them. Several things are wrong with the present electrolyzers.

1. Bubbles are allowed to come out between the electrodes and cause an increase of resistance, thus taking up unnecessary voltage loss between the cells.

2. There has been no allowance for electrocatalysis, and the reduction of activation overvoltage, which (for a working electrolysis cell) increases the potential to obtain a rate of evolution above the thermodynamic minimum voltage.

3. Electrodes in fuel cells have structures different than the crude planar structures of old electrolyzers. Porous electrodes can be more effective in electrolysis than these planar electrodes, and no consideration has been given to this in the present electrolyzers.[6] Improvements in performance (lowering of voltage for a given current) from recent work are shown in Fig. 8.

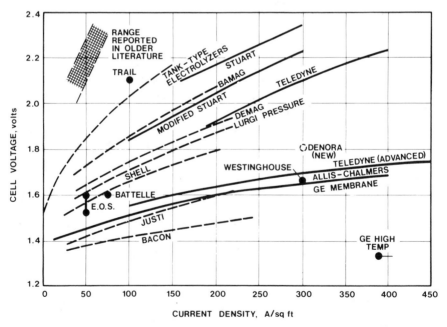

Figure 8. Cell operating performances of various advanced electrolyzers.[32]

5. Would a Hydrogen Economy Run into Materials Difficulties?

An essence of the hydrogen economy is the sending of hydrogen over long distances through pipes. Among problems which have been examined in this century, one of the most researched is the interaction of hydrogen with metals. Surprisingly, deleterious things happen.[33] When hydrogen comes into contact with some metals, it causes their embrittlement.[34] This means that, under certain conditions, the metal breaks down after contact with hydrogen, becomes less ductile, weak, and, under stress, undergoes splitting.[35]

The mechanism of this breakdown is well understood.[36] Metals contain voids, holes inside the structure, before atomic hydrogen begins diffusing there from the lattice. When hydrogen arrives on the outer surface of a metal, by dissociation of molecular hydrogen from the gas phase, or deposition from solution, it begins to diffuse into the metal. Some of the diffusing hydrogen atoms arrive on the outside of the voids and diffuse into them, arriving in the adsorbed state on the walls of the voids. There is then desorption of the hydrogen atoms to the gas phase, whereupon they become molecules. This hydrogen has its own characteristic internal pressure, and this pressure is great, on the order of tens of thousands of atmospheres. If the hydrogen internal pressure exerts a stress upon the void walls greater than the yield stress of the metal, the metal begins to yield internally, and the small voids increase in size.

When this happens in millions of voids (i.e., flow), and some of these spreading voids join, greater voids are formed. The effect is a weakening of the metal lattice. If it is under strain, it may crack.

5.1. The Effect of Pressure on Hydrogen Embrittlement

The pressure at which hydrogen will be maintained within the pipes which lead from its source to the site of use will be above atmospheric, because of the desirability of having pipes that are small in diameter. 100 atmospheres might be such a pressure.

This pressure would increase the likelihood that gaseous hydrogen will break down metals. The probability of hydrogen dissolving in the metal is proportional to the square root of the external pressure, and a greater amount of hydrogen will arrive at voids when the outer pressure is high.

These effects may be small compared with indirect ones. Local stresses may affect hydrogen solubility. It can be shown that the local stress in a metal affects solubility through the equation[37]

$$c_\sigma = c_0 e^{\bar{V}\sigma/RT} \tag{8}$$

Thus, as the local stress, σ increases, the local solubility will increase in an exponential way. When σ is a practical value of about $10\ kg/mm^2$, the solubility begins to get greater than in the absence of such local stresses.[38]

If local solubility is above that of the surroundings, the amount of hydrogen at the local stress spot will be high, and there is a greater likelihood that embrittlement damage will begin at this point.

So far, the situation has been discussed as though the surface of the pipes were uniform. However, in many metals, micro*pits* form from the surface, and the stress is larger at the bottom of these pits than at the surface. Thus, if the depth of a pit is l and the radius of curvature at the bottom is r, then the strain at the tip is related to the macrostrain in the surroundings by the equation[39]

$$\sigma_{\mathrm{tip}} = \sigma_{\mathrm{macro}} \left(\frac{2l}{r} \right)^{\frac{1}{2}} \tag{9}$$

When one takes into account the fact that the typical value of l may be 1 mm and that of r about 10^{-6} cm, and then carries through the consequences of this in the above equation for solubility as a function of stress, one finds that the solubility of hydrogen at the bottom of pits may be several orders of magnitude larger than that without stress, thus giving rise to the top of a pit as the likely point at which embrittlement effects will be seen.

5.2. The Relation between the Type of Steel and the Damage of Hydrogen

The amount of carbon which can be in iron, thus the type of steel, has been examined particularly by Nelson (Fig. 9).

Small amounts of chromium and molybdenum reduce the attack of hydrogen on iron, e.g., an alloy containing 2.25% chromium and 1% molybdenum resists attack well.

5.3. Experimental Studies of Pressure Effects upon Vessels

Studies have been made which show that some types of steel, e.g., martensite steel, are more susceptible to embrittlement than, say, perlite steel.

Up to about 5000 psi (357 atm) there is less danger of attack. Above that pressure, hydrogen in pipes would reduce in strength, probably near welds and notches. Thus the flow of hydrogen through pipes may have to be undertaken at a smaller pressure than one would use if, for example, one were flowing nitrogen through the pipe.

5.4. Hydrogen Environment Embrittlement[41]

A new type of embrittlement of metals by hydrogen has been discovered in recent times, called "hydrogen environment embrittlement," and this type of embrittlement is the most relevant to the hydrogen economy. Storage tank failures have occurred by this type of embrittlement, characterized by the formation of surface cracks. The attack may occur at 1 atm (see Table 3).

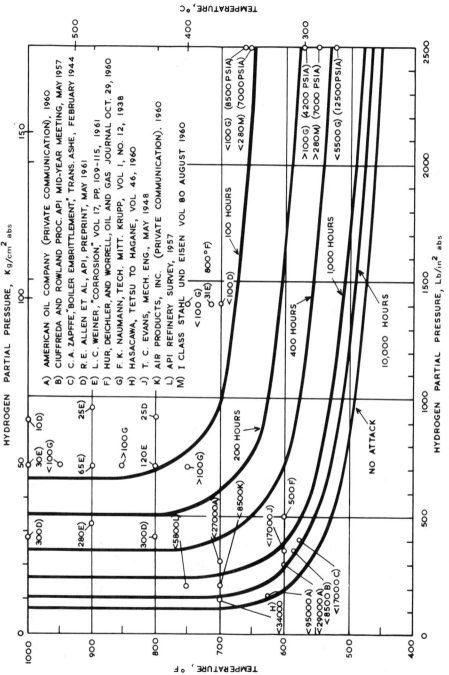

Figure 9. Time for incipient attack of carbon steel in hydrogen service.[40]

Table 3. Characteristics of the Types of Hydrogen Embrittlement[a]

| Characteristics | Types of embrittlement | | |
	Hydrogen environment embrittlement	Internal reversible hydrogen embrittlement	Hydrogen reaction damage
Usual source of hydrogen	Gaseous (H_2)	Processing ⎫ electrolysis ⎬(H) corrosion ⎭	Gaseous or atomic hydrogen from any source
Typical conditions	10^{-6} to 10^8 N/m^2 gas pressure	0.1–10 ppm average H content	Heat treatment or service in hydrogen, usually at elevated temperatures
	Most severe near room temperature	Most severe near room temperature	
	Observed $-100°$ to $700°C$	Observed $-100°$ to $100°C$	
	Gas purity is important	Strain rate is important	
	Strain rate is important		
Crack initiation	Surface or internal initiation[b]	Internal crack initiation	Usually internal initiation from bubbles or flakes
Rate-controlling step	Adsorption = transfer step $\begin{pmatrix} \text{Adsorption or} \\ \text{lattice} \\ \text{diffusion} \end{pmatrix}^{b}$ = embrittling step	Lattice diffusion to internal stress raisers	Chemical reaction to form hydrides (e.g., in titanium or gas bubbles)

[a] Source: W. B. McPherson and C. E. Cataldo.[41]
[b] Unresolved.

Hydrogen environment embrittlement can be overcome. However, alloy material, such as chromium and molybdenum, may have to be used in the steels and this will increase their price.

Nickel and nickel alloy steels are best, as are steels containing copper, aluminum, and, as mentioned above, molybdenum and chromium.

5.5. *The Effects of Traces of Oxygen in the Hydrogen*

An important contribution to the subject of hydrogen attack on metals has been made by Johnson.[43] He showed that if hydrogen contains a small amount of oxygen (e.g., 200 ppm) embrittlement does not occur.

The mechanism of this effect may be that the oxygen causes protective oxide films which hydrogen does not reduce, so that in the presence of the oxide it never gets to adsorb onto the surface.

5.6. *Rhythmic Variation of Stress*

In the period of hydrogen use as an energy medium, the pressure would rise in times of low use, when the hydrogen producer will continue to work speedily but hydrogen is not being removed from the pipe at the same rate. The maximum use periods are likely to be twice per day—one in the early morning, one toward dinner time—and these will give rise to minima in pressure in the pipe, with corresponding maxima twice a day.

Repeated stress is destructive to metals, for it causes in them a phenomenon known as fatigue, another cause of breakdown. In Fig. 10 results are shown

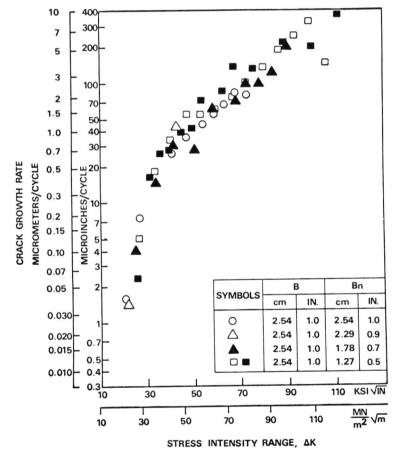

Figure 10. Cyclic load crack growth rates as a function of stress intensity range for inconel 718 TDCB specimens with various side groove depths exposed to 34.5 MN/m^2 (500 psi) hydrogen at ambient temperature R = 0.1, 1.0 cps.[44]

which suggest that, with two cycles of pressure changes per day, the lifetime of a pipe will be between 50 and 100 years.

6. *The Ways in Which Hydrogen Could Be Converted Back to Energy*

6.1. *Introduction*

The hydrogen economy presupposes that hydrogen is being taken from its source to the user sites in pipes, but at these user sites it must be reconverted to some form of energy, e.g., to mechanical energy for use in transports or to electrical energy for use in households.

There are two ways in which this can be achieved:

1. Hydrogen could be used as a gas to fuel internal combustion engines.
2. Hydrogen could be used in fuel cells to produce electricity. Thus, this electricity could be used directly (as in heating), or with electric motors to produce mechanical power.

The pros and cons of which method will be used are complex, and change with the time and the place at which the use will occur. The efficient fuel cell is the obvious transducer to use when electricity is needed from the hydrogen (e.g., for running electronics and lighting in a house), whereas the combustion engine is the one to use when mechanical power is needed. This judgment may be less valid when economics are brought in, for the fuel cell is a more efficient converter than the combustion engine, and the fuel cell–electric motor combination may prove to be the more economical choice.

6.2. *The Effect of Hydrogen upon the Choice of Transducer*

With hydrogen as a general fuel replacing gasoline, the pros and cons of which converter is best will change. For example, gasoline is an excellent fuel for an internal combustion engine, but a poor fuel for a fuel cell.[8] Hydrogen is a good fuel for both an internal combustion engine and a fuel cell.[45] Relatively, therefore, the fuel cell gains when hydrogen is introduced instead of gasoline (Figs. 11 and 12).

However, it is not only the fuel cell which is changed by the presence of hydrogen. Instead of the old-fashioned, boiler-type steam engine, hydrogen, if it could be used in combination with oxygen, would give a steam engine of great convenience. The function would differ from that of the conventional steam engine, in which the steam expands in cylinders to push a piston. In the hydrogen–oxygen steam engine, the hydrogen and oxygen would be brought into combination to produce high-pressure steam, and this would be directed onto turbine blades. Such steam engines would be small in size, 100 times smaller than their bulky counterparts. Correspondingly, rocket engines are

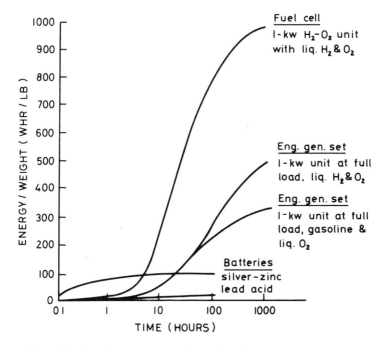

Figure 11. Plot of energy–weight ratio as a function of time for some energy converters.[46]

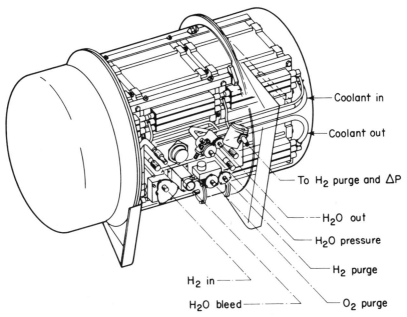

Figure 12. The General Electric ion exchange membrane hydrogen–oxygen fuel cell battery. Layout of a three-stack fuel cell.[47]

used in space vehicles and could be used to power, e.g., electric conversion equipment in power stations, as long as both hydrogen and oxygen could be brought together to power them.[48]

6.3. Pollution in the Use of Hydrogen in Engines

Where the hydrogen is used in one of the combustion-type engines, or in a fuel cell, the pollution position is excellent. In a fuel cell, the production from the use of hydrogen will be exclusively water. In one of the combustion engines, there may be some NO produced, and this is a pollutant. However, the amount produced will be so much less than that presently produced by internal combustion engines using gasoline that it can be regarded as negligible.

7. A Research Recommendation

Research on internal combustion engines is a far more nurtured subject than research into electrochemical fuel cells, which has recently been carried out intensely only in the brief period, 1960–1968. Hence, fuel cells have a way to go before they have had the research attention which internal combustion engines have had. The introduction of a hydrogen economy would tend to further fuel cell research, for, as stated, fuel cells act well with hydrogen as a fuel and tend to give more efficient conversion to electricity than hydrogen. Research gains in fuel cell technology could be helpful to lowering the cost of energy.

8. More Details on Ways in Which Hydrogen Would Be Used in the House

The entire household energy system could be run on hydrogen. One could feed electricity produced from fuel cells into the electric lines, and hydrogen directly into heating and cooking systems. However, this would not be the best way to use hydrogen, and, if we are going to introduce a new medium of energy, it is best to introduce corresponding advances.

8.1. Lighting, Heating, and Cooking

Lighting need not be carried out electrically, but by the production of "cold light," which is formed by interacting hydrogen with phosphorus.[7]

Correspondingly, cooking stoves in a hydrogen economy would not have flames, or visible hot plates. Hydrogen would be led into a ceramic material, which would contain a catalyst, and the hydrogen–air reaction taking place within the ceramic would cause the latter to heat up and act as a hot plate.

Variations in temperature would be attained through varying the flow of hydrogen.

Such a method could be used for both space heating and for cooking. However, space heating would better be carried out by the use of heat pumps, working electrically.

8.2. Electricity in the House[49]

It is possible that less electricity would be used in a house in the hydrogen economy, e.g., cooking would be more likely to be carried out in the way described above. The minimal use of electricity would be in running the electronics, and the electricity would be produced in a fuel cell, probably housed in the basement of the house, in the adjacent garden, or on the roof.

Another method of heating the house would be to duct the heat produced in the functioning of this central fuel cell to the various rooms.

One must take into account the hump which would have to be overcome in introducing hydrogen into houses: the present piping would not be acceptable, and would have to be replaced. The burners themselves, and most equipment, would have to be installed new. The initial price difficulty would be the main one in converting an old house to the hydrogen economy.

8.3. Dealing with Sewage[50,51]

Normally, sewage has to pass through miles of piping to a central sewage works, and trash collection involves burning, which is pollutive. There are a number of ways in which trash and sewage could be dealt with in a hydrogen economy. It could be pyrolyzed in hydrogen to give methane, an odorless, colorless, nonpolluting gas which might have uses in secondary energy production.

In some situations in a hydrogen economy, oxygen could be piped to houses. Aerobic sewage treatments have been developed: the sewage is brought into contact, at temperatures near to 100°C, with pure oxygen, and conversion to CO_2 occurs.[52]

8.4. A Detailed Discussion of the Derivation of Water from Hydrogen

If a household consumes hydrogen as its main energy source, the amount of water produced can be calculated if one knows the electricity use of the household per day and calculates the Faraday law equivalent of water. Electricity consumption of a house is 35 kWh per day, but one should allow for a higher energy equivalent in a hydrogen economy, since the function of the house would depend on the hydrogen. Let it be assumed that this will be, then, 100 kWh per day. The number of Wsec is $100,000 + 1 \times 60 \times 60$, and, if one assumes that the potential of the fuel cell from which the energy is being produced is in the neighborhood of 1 V the number of coulombs (= current in amperes × time in seconds) would be $100,000 \times 60 \times 60$.

However, the Faraday equivalent, the number of coulombs which are equivalent to 1 mol of electrons, is 96,500 C, and 2 mol of electrons would be used up for 1 mol of water.

It follows that the number of moles of water produced per 100 kWh per day (and according to our present assumptions per household), would be

$$\frac{100,000 \times 60 \times 60}{2 \times 10^5}$$

To convert this to gallons, one first converts it to grams by multiplying by the molecular weight of water, or 18. Remembering that the density of water is 1, this is also the number of cubic centimeters of water. To get liters, one should divide by 1000. There are 5 liters per gallon, so that division by 5 gives the number of gallons per household per day, about 6.

This is not a great amount, but it could be significant in respect to drinking water.

If one thinks of the production of water within a hydrogen economy, not as the household-centered matter, but in a general way, one would have to take into account the entire energy consumption. Then, one gets a substantially larger figure, about 14 gallons *per person*, per day. It is possible that arrangements could be found for collecting *some* of the water produced in a hydrogen economy. There could perhaps be two water supplies, a highly pure one for drinking, and a second-class supply acceptable for many other purposes.

9. *Environmental Effects of a Hydrogen Economy*[53]

At first, when one considers the introduction of a hydrogen economy, one assumes that it will be "perfect" in respect to environmental considerations, a satisfactory long-term economy, because it converts water to hydrogen and oxygen, and recombines them to form water again.

However, air pollution is not the only type of pollution. In considering various economies, one must consider not only the air pollution aspects, but the other environmental effects.[54] By this is meant the usefulness of resources. It is important to compare the hydrogen economy, therefore, with other economies such as a coal–synthetic fuel economy, a nuclear–hydrogen economy, a nuclear–all-electric economy, etc. Such comparisons are shown in Table 4. The various hydrogen economies do indeed rank higher in an environmental sense than do the older economies using fossil fuels. It is noteworthy that, although a solar–hydrogen economy is environmentally the best, a solar–electric economy would be better because it would involve the use of less materials. However, the assumption made in the table is that the solar–hydrogen economy would work in connection with photovoltaic cells, which will consume much silicon, and, if the solar–hydrogen economy works in connection with sea–solar power units, or wind power, the effect on the environment would be smaller, and the

Table 4. Environmental Damage Comparisons[a]

System	Environmental damage index (1970 levels)	Ratio of damage to that of present system
I (coal–hydrogen)	670	1.64
I* (coal–synthetic fuel)	554	1.36
II (nuclear–hydrogen)	483	1.18
II* (nuclear–electric)	461	1.13
III (solar–hydrogen)	259	0.63
III* (solar–electric)	173	0.42
Present mixed system (calculation omitted)	408	1.00

[a] Source: H. J. Plass.[54]

solar–hydrogen economy would be environmentally better than a solar-electric one.

10. *How Would a Hydrogen Economy Be Developed?*

It is important to keep one's feet on the ground, and realize that the advantages of a hydrogen economy can only be reached by stages. A new energy economy has to be built up by research, development, pilot plant building, a small test operation, a larger test operation, and finally, a whole community working on it. The necessity to use this stage-by-stage approach is one of the reasons that the changeover to a new energy situation will be dangerously slow, for we are not certain that we will have usable fossil fuels after 2000, and after about 2050 it is unlikely that they will be available.[55] Stage 1 in a hydrogen economy might begin with coal as the source of hydrogen. The coal would be gasified by heating it with water, to form a mixture of CO and H_2. The CO would be oxidized to CO_2 and rejected into the atmosphere, and the H_2 would be piped off to a nearby plant. Here, it would be used to produce electricity, probably in a rocket engine.

It is important to give sufficient economic incentive to organize this early trial, as, for example, with a Ford Foundation or a government subsidy.[56]

In Stage 2, there could be the coupling of a hydrogen plant with a nuclear fission plant, near a smoggy area such as Los Angeles. The hydrogen would be brought to the town and utilized to produce electricity or to run a section of the transportation systems.

In Stage 3, one could take a small town (e.g., 10,000 people, in a "government town" like Los Alamos) and introduce a solar collecting station. Some experimental, large aerogenerators could be set up in a mountain area near Los Alamos.

Stage 4 would be an important stage, involving many billions of dollars, and would attempt to run a chunk of the economy on hydrogen. Thus, household, transportation, and industry would all run on hydrogen, and a suitable place for developing this would be Hawaii. There, one has an encapsulation of the entire United States economy, but isolated. Hawaii is a good place for collecting wind power, a steady wind blowing for more than 10 months of the year. A water-borne nuclear reactor could be used to aid the wind generators. The development of solar collectors could be attractive. If the Hawaii project were begun by 1990, one might expect conversion within perhaps 10–15 years.

After the suggested Hawaiian trial had been in operation for a few years, it would be desirable to try another major area before deciding on the general introduction of a hydrogen economy, and this could be, for example, Miami, because of its proximity to the Gulf Stream and the possible use of the sea–solar plants as the origin of power. One would aim at the conversion of this city to a solar–hydrogen economy by 2020.

These concepts, originating in 1974, can be compared with NASA plans of the same year. NASA proposes developing at the White Sands Missile Range in New Mexico the production of H_2 via steam reforming to natural gas; at the same time, it will build up thermochemical and electrolytic systems to produce H_2 from solar energy.[57]

References

1. J. O'M. Bockris, *A Solar–Hydrogen Economy*, A.N.Z. Book Co., Brookvale, New South Wales, 1975, Ch. 8.
2. Lawaczek (1930), quoted by R. O. Lindstrom, *ASEA Journal*, **37**(1) : (1964); also quoted by E. W. Justi, *Leitungsmechanismus und Energieumwandlung in Festkörpen*, Vandehoeck and Ruprecht, Göttingen, 1965.
3. J. O'M. Bockris, Idea Statement written to Westinghouse Company, 1962.
4. E. W. Justi, *Leitungsmechanismus und Energieumwandlung in Festkörpen*, Vandehoeck and Ruprecht, Göttingen, 1965.
5. J. O'M. Bockris, *Chemical and Engineering News*, **50**(44) : 26–35 (1972).
6. D. P. Gregory, D. Y. C. Ng, and G. M. Long, in *Electrochemistry of Cleaner Environments*, J. O'M. Bockris, ed., Plenum Press, New York, 1972.
7. C. Marchetti, *Chemical Engineering and Economic Review*, **5**(1) : 23 (1973).
8. J. O'M. Bockris, *A Solar–Hydrogen Economy*, A.N.Z. Book Co., Brookvale, New South Wales, 1975, Ch. 14.
9. M. Steinberg, M. Beller, and J. R. Powell, *A Survey of Applications of Fusion Power Technology to the Chemical and Material Processing Industry*, Brookhaven National Laboratory, New York, May, 1974.
10. D. Whitford, D. A. Atkinson, and K. Rush, "Flinders Electric Vehicle Project," Final Report, Stage I, School of Physical Sciences, Flinders University of South Australia, January, 1975.
11. J. O'M. Bockris and D. Drazic, *Electrochemical Science*, Taylor and Francis, London, 1972, Ch. 8.
12. R. J. Schoeppel, *Chem. Tech.*, **27** : 476 (1972).
13. R. G. Murray, R. J. Schoeppel, and C. L. Gray, paper presented at the 7th Intersociety Energy Conversion Engineering Conference, San Diego, September, 1972.

14. L. O. Williams, *Astronautics and Aeronautics,* **13** : 42 (1972).
15. W. J. Small, D. E. Fetterman, and T. F. Bonner, Jr., paper presented at the Intersociety Conference on Transportation, Denver, Colorado, 23–27 September, 1973.
16. W. J. D. Escher and G. D. Brewer, paper presented at the AIAA 12th Aerospace Sciences Meeting, Washington, D.C., 30 January–1 February, 1974.
17. W. J. D. Escher, "Prospects for Liquid Hydrogen Fueled Commercial Aircraft," Report PR-37, Escher Technology Associates, St. Johns, Michigan, September, 1973.
18. R. D. Witcofski, paper presented at the American Chemical Society Symposium on Non-Fossil Chemical Fuels, Boston, Massachusetts, 9–14 April, 1972.
19. G. deBeni and C. Marchetti, *Euro Spectra,* **9** : 46 (1970).
20. R. P. Hammond, in *Electrochemistry of Cleaner Environments,* J. O'M. Bockris, ed., Plenum Press, New York, 1972.
21. "Hydrogen Production from Water Using Nuclear Heat," Progress Report No. 3, EURATOM Joint Nuclear Research Center, Ispra, Italy, December, 1972.
22. G. deBeni and C. Marchetti, "Mark-1, A Chemical Process to Decompose Water Using Nuclear Heat," paper presented to the Am. Chem. Soc. Meeting, Boston, 9 April, 1972.
23. J. E. Funk and R. Reinstrom, *I. & E.C. Process Design Develop.,* **5** : 336 (1966).
24. J. E. Funk, W. L. Conger, and R. H. Carty, paper presented at the Hydrogen Economy Miami Energy (THEME) Conference, March, 1974.
25. J. E. Funk, paper presented at the *Am. Chem. Soc. Div. Fuel Chem.* (April, 1972) 79.
26. J. O'M. Bockris, *A Solar–Hydrogen Economy,* A.N.Z. Book Co., Brookvale, New South Wales, 1975, Ch. 9.
27. A. K. Stuart, paper presented at the Am. Chem. Soc. Symposium on Non-Fossil Fuels, Boston, 13 April, 1972.
28. J. O'M. Bockris, *Energy Conversion,* 1975.
29. J. O'M. Bockris, paper presented at the Hydrogen Economy Miami Energy (THEME) Conference, March, 1974.
30. D. Gregory, P. J. Anderson, R. J. Dufour, R. H. Elkins, W. J. D. Escher, R. B. Foster, G. M. Long, J. Wurm, and G. G. Yie, *A Hydrogen Energy System,* Institute of Gas Technology, Chicago, 1972, p. II-18.
31. P. Sporn, private communication, 1974.
32. D. Gregory *et al.,* op. cit., p. III-18.
33. E. C. W. Perryman, *J. Inst. Metals,* **78** : 621 (1950–51).
34. E. Heyn, *Stahl and Eisen,* **20** : 837 (1900); **21** : 913 (1901); *Metallographist,* **6** : 39 (1903).
35. R. Staehle, paper presented at the International Conference on Stress Corrosion Cracking and Hydrogen Embrittlement of Iron-base Alloys, Unieux-Firminy, France, 12–16 June, 1973.
36. J. O'M. Bockris, *A Solar–Hydrogen Economy,* A.N.Z. Book Co., Brookvale, New South Wales, 1975, Ch. 12.
37. N. J. Petch and P. Stables, *Nature,* **169** : 842 (1952).
38. P. K. Subramanyan, "Hydrogen in Metals," in *M.T.P. International Review of Science, Electrochemistry,* Series 1, Volume 6, J. O'M. Bockris, ed., Butterworths, London, 1973.
39. J. O'M. Bockris, W. Beck, M. A. Genshaw, P. K. Subramanyan, and F. S. Williams, *Acta Met.,* **19** : 1209 (1971).
40. G. A. Nelson, *Hydrocarbon Processing,* **44**(5) : 185 (1965).
41. W. B. McPherson and C. E. Cataldo, "Recent Experience in High Pressure Gaseous Hydrogen Equipment at Room Temperature," Technical Report No. 8-14.1, American Society for Metals, Metals Park, Ohio, October, 1968.
42. R. L. Savage, L. Blank, T. Cady, K. Cox, R. Murray, and R. D. Williams, ed., *A Hydrogen Energy Carrier, Vol. II: Systems Analysis,* NASA–ASEE, September, 1973, p. 110.
43. H. H. Johnson, paper presented at the Hydrogen Economy Miami Energy (THEME) Conference, March, 1974.
44. R. J. Walter and W. T Chandler, "The Influence of Hydrogen on the Fracture Mechanics Properties of Metals," Pub. 572-K-22, Rockwell International, Canoga Park, California 91304, 1974.

45. J. O'M. Bockris and D. Drazic, *Electrochemical Science*, Taylor and Francis, London, 1972.
46. W. Mitchell, Jr., *Fuel Cells*, Academic Press, New York, 1963.
47. J. L. Schanz and E. K. Bullock, paper presented at the American Rocket Society, Santa Monica, California, 25–28 September, 1962.
48. J. O'M. Bockris, *A Solar–Hydrogen Economy*, A.N.Z. Book Co., Brookvale, New South Wales, 1975, Ch. 13.
49. L. Lessing, *Fortune*, **85** : 138 (1972).
50. R. C. Corey, "Pyrolysis, Hydrogenation and Incineration of Municipal Refuse—A Progress Report," Proceedings of the 2nd Mineral Waste Utilization Symposium, ITT Research Institute, Chicago, Illinois, March, 1970.
51. W. F. Schaffer, Jr., "Cost Study of the Treatment of Sewage Sludge by the Wet-Air Oxidation Process, Using Oxygen Produced by Low-Cost Electricity from Large Nuclear Reactors," Oak Ridge National Laboratory, February, 1968.
52. M. D. Rickard and A. F. Gaudy, Jr., *Journal of Water Pollution Control*, **29** : 405 (1968).
53. J. O'M. Bockris, *A Solar–Hydrogen Economy*, A.N.Z. Book Co., Brookvale, New South Wales, 1975, Ch. 16.
54. H. J. Plass, paper presented at the Hydrogen Economy Miami Energy (THEME) Conference, University of Miami, 18–20 March, 1974.
55. J. O'M. Bockris, *A Solar–Hydrogen Economy*, A.N.Z. Book Co., Brookvale, New South Wales, 1975, Ch. 18.
56. R. L. Savage *et al.*, op. cit., p. 149.
57. K. E. Cox, private communication, 8 October, 1974.

Older considerations suggested that the replacement for the internal combustion engine would be the electric car. Indeed, this may be the shape of things to come. However, there are powerful forces—perhaps more those of the inertia due to invested capital than those due to lack of appropriate scientific advance—that point to a continued use of internal combustion, and the question then becomes: with what fuel? The main point, of course, is that the fuels must be environmentally clean. It turns out that there are very few possibilities. One of these continues to be electricity.

18

Environmentally Clean Fuels for Transportation

J. O'M. Bockris

1. Introduction

There has been a change of emphasis in energy affairs, starting in the 1970s, away from coal and atomic energy and toward solar power. The origin of this shift in emphasis was environmental. It arose from the well-known smog pall so frequently seen over Los Angeles, and the reasoning that this pall would soon appear over other cities. Earlier smog disasters, e.g., in London, 1952 (four thousand deaths) were recalled.

From discussions published between 1969 and 1973, it became clear that, although there were many sources of air pollution, the principal origin of smog was emissions from automobile exhausts.[1] Photochemical smog formation became part of the environmental scene (see Chapter 9), and pressure began to be exerted in the late 1960s about the control of exhausts from cars, and the replacement of gasoline-driven cars by cars having other transducers. The

J. O'M. Bockris • Department of Chemistry, School of Physical Sciences, Flinders University of South Australia, Bedford Park, South Australia 5042

suggestion most frequently made was (and is) the battery-driven electric car, with special attention given to the development of high-energy-density batteries.

This situation is the background to the present chapter. Two preliminary remarks should be made: .

1. The strength of the environmental movement to replace internal combustion engines and to diminish other sources of pollution has been muffled by the still greater danger of a large rise in the price of fuel, which could lead to cutbacks in the standard of life as a whole. We are between the Scylla of poisoning the air by burning fuels which pollute, and the Charybdis of a declining living standard because of the high price of clean fuels. If the energy situation is to be relieved at reasonable costs, coal will have to be used to replace the failing reserves of natural gas and oil, and if the pollution arising from the coal is to be avoided by superpurification techniques, the cost of fuels will rise more steeply than in recent experience, with corresponding effects on the standard of life.

2. The second point, the basis of this chapter, is the failure of alternatives to the internal combustion engine. These alternatives are the Sterling engine, the gas turbine, and the battery or fuel-cell-driven electric car. All have drawbacks compared with the internal combustion engine, but the net offering which such power sources have to make is negative only if one considers the picture from the viewpoint of advocates of the internal combustion engine. The aspects stressed are those of overwhelming acceleratory ability, and vast amounts of reserve power. Silence, lack of pollution, a rational fuel policy for the long term, and acceptable speed and range—none of these factors are not considered, and do not play a part in the use of the media to persuade the consumer what kind of a product he actually needs. Thus, already by the 1950s, the power-to-weight ratio of the internal combustion engine resembled size in the brontosaurus: it is something good, but overdone past all reason, and obviously slated for a cul-de-sac in the evolution of transportation.

The so-called "failure" of the alternative sources mentioned above is a failure to communicate with the public. The present system of corporate–government–people interaction is such that the net pressures are in favor of continued use of the internal combustion engine, *or of a very near relative.* Those who are aware of the advantages of alternatives do not have sufficient money to buy space in the media and let people know.

2. *The Inability to Attain a Satisfactory Long-Term Situation by Cleaning Up the Internal Combustion Engine*

In Chapter 9, the chemistry of the internal combustion engine has been described, with respect to the possibilities of satisfactorily reducing its pollutive character. The measures that can be taken—some of which have been, or are being, taken—will reduce pollution from the internal combustion engine, and may reduce net emissions for a decade or even 15 years.

However, a long-term clean-up of the internal combustion engine is not practical. The reasons are:

1. The problem is not "removing the pollutant material from the exhaust stream," but doing this with a low use of energy and chemicals, without the addition of large, extramechanical devices, under conditions which do not significantly reduce splendid acceleration, in a way which does not require new skills in the driver, at constant speed-running, rather than on a test bench, and with the start–stop driving of towns, and, above all, at an *acceptable price*.

2. The attitude taken to most problems in the pre-1970 era was that resources were unlimited. However, the satisfactory reduction of unsaturates needs a catalyst. If a catalyst such as platinum could be used, success would be more likely. But noble metal catalysts would not be available for long if used in cars.[2]

3. The price limitation. Executives of companies keep their well-remunerated positions if they maintain the companies' profits year by year. They will only amicably accept a government law if it does not reduce profits significantly. A government can force companies to do something by law, but it cannot make an unpopular law or it will be diminished by the populace, which does not understand questions such as exponential growth in pollution, the relation to catalysis, etc. If the price of cars is raised, extra-inflationally, there will be buyer's resistance and a diminishing of profits. Hence, some possibilities for eliminating pollution must be rejected.

4. Practical conditions of driving. In experimental work on car engines running at a constant speed, it is relatively easy to reduce pollutants to an acceptable amount. Two different factors arise in running cars in real situations. They are not driven at constant speeds, but accelerated, deaccelerated, and idled. They are not driven for 5000 or 10,000 mi, but for 50,000 and 100,000 mi. These factors make demands upon the development of an economically acceptable anti-pollution device which requires a test more sophisticated than many. Among other requirements, the device must stay "active" for a long time, for, if it leads to the car having to be serviced more than is acceptable, sales will fall, or drivers will have the device removed, etc.

5. A negative conclusion concerning the continued use of the internal combustion engine must be associated with an expectation for the future. However, if growth favorable to a free enterprise economy is to continue, the expansion rate of the car population will be such that a reduction of pollution by 90% from the 1968 level will not be sufficient. Pollution would start to increase again (see Fig. 1). This conclusion, like other limits-to-growth predictions, can be avoided by zero or negative growths. The difficulty is that such measures will not be taken in a society based on expansion and growth. They would result in gross unemployment. While unemployment is allowed to occur, the democratic vote will be (out of fear) entirely negative to the cutbacks in growth necessary for survival.

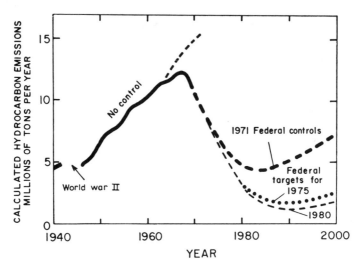

Figure 1. Estimated effect of federal controls in the United States on hydrocarbon emissions from passenger vehicles. A rising trend is expected after 1990 because of the increasing numbers of cars.[3]

6. The points concerning the pollution problems of gasoline fuels are economic and sociological, i.e., "realistic." However, there is a scientific point which militates against a good solution. The simultaneous oxidation of unsaturated hydrocarbons, and other oxidative compounds, is difficult. The heterogeneous catalyst chemist must devise a situation in which, with a limited contact time and area, oxidation at a sufficient rate of a large number of different kinds of compounds can be carried out. The task is to·raise the rate-constant for the velocity of a number of heterogeneous-catalytic reactions greatly, when the catalyst area is restricted, the flow rate is controlled within certain limits, the temperature stays effectively the same for the oxidative reaction of all the hydrocarbons emitted to CO_2, while the catalyst remains cheap, and must not deactivate over at least 6 months' running (5000–6000 mi).

As a final twist of the screw, the pollutants are likely to contain vapors of lead compounds, about the best-known catalyst poisons. And, if they are removed (as is necessary anyway because of lead poisoning), the concentration of unsaturated compounds in the exhaust will increase, with a resultant increase in smog.

Many American companies have been working intensely on the problem of exhaust purification for decades. For example, the duPont company, in the 1950s in particular, made a considerable effort over several years to find out how to remove automobile exhausts.

These facts tend toward the conclusion that the problem is not soluble under acceptable practical conditions, except by the (hitherto unacceptable) change in the type of motive unit in the automobile.

There is perhaps one possibility which could make the internal combustion engine viable for the long term. If the manufacturing organizations reduced the vehicles to small microvehicles (for one or two passengers), gave up air conditioning and heating, and were content with a very low-powered performance now regarded as unacceptable, they could pass the standards which are being regularly put up and taken down again by the United States government. But a far better solution would be to move on to a radically different fuel.

3. *Survey of Near-Future Fuels*

Energy considerations depend on the time scale. During the next 10 years we will certainly continue to use gasoline as the major fuel. In the near future, up to 25 years, we will continue to use carbonaceous fuels, although they may consist of mixes of gasoline, synthetic natural gases, or other products from coal, e.g., methanol.

Methanol does not have the status of hydrogen or electricity, and it would not be easy to couple it with solar or atomic sources. On the one hand, it has a more favorable status because its acceptability is larger: it represents—to the public—a negligible change, no safety precautions would have to be taken, and driving habits would hardly have to change.

The replacement of gasoline by methanol would certainly be an advance on the pollution front. Methanol burns cleanly to CO_2, and the effects of CO_2 pollution will probably not obtrude for some 25 years.

However, methanol has a less favorable status than hydrogen or electricity for the following reasons:

1. It is not likely to represent a permanent fuel. Thus, permanence would be possible if CO_2 could be converted to methanol, but this would need massive amounts of hydrogen, so that methanol derived from CO_2 would be more expensive than hydrogen.

2. Correspondingly, methanol cannot be compared favorably with electricity, because electricity is the product of future solar and nuclear sources.

The following factors, apart from those mentioned, will influence the choice of a clean fuel (see Fig. 2).

1. *Availability.* A satisfactory *permanent* new fuel for transportation that is both cheap and clean may not be available in sufficient quantity before oil and natural gas exhaust themselves (before 2000).* This would apply to electricity

---

*The figure 2000 is a compromise between warring considerations. On the one hand, considerations of the total fossil fuels available, predominantly coal, suggest that an extension of energy from coal could be made up to a possible date of, say, 2030 (and conceivably higher dates, see Linden[4]). However, there are many considerations which suggest that coal cannot be used as a main energy source to such times. The principal one is the difficulty of obtaining enough coal in the relatively short time available (1–2 decades). *Even Middle Eastern oil sources have a lifetime of less than 25 years*, and the practicality of the United States existing on imported oil past 1985, as examined by Hottel,[5] seems questionable.

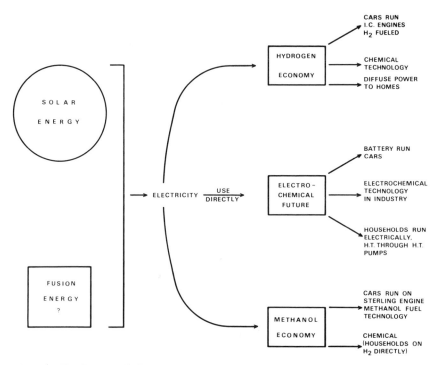

Figure 2. The three possibilities for future fuels. The sources of abundant, clean energy will increasingly be solar and fusion, although the successful development of the latter is doubtful. The present fission reactors have life of only 2–3 decades before uranium exhaustion. The so-called breeder reactors are technologically uncertain, but the main reasons they are decreasingly seen as a part of the future is that they produce plutonium, an extremely dangerous pollutant. The figure shows the possibilities which become increasingly clear. It is likely that a mixed use of all three fuels will evolve, and probably become a mixture of electricity and liquid H_2 during the middle decades of the twenty-first century. An acceptable, affluent, clean, stable planet is still possible but two changes of magnitude are essential and urgent. (1) The world population growth *must be reduced at once*. (2) Investment of research and development money far exceeding that on space must be rapidly made in developing a solar–hydrogen economy.

from nuclear power which would be an acceptable, clean fuel (although see Chapter 16 for the difficulties of breeder reactors). No country will have sufficient nuclear reactors to provide the transportation energy by 2000, in terms of electrolytic hydrogen or hydrogen-based methanol.*

2. *Economics.* Predictions and discussions of economic quantities are made more difficult because of the inflational characteristics of free enterprise economies. That the price of fuel will rise is an accepted commonplace. But the rise originates in the monopoly of the OPEC countries. American consumers cannot do much about it. It is a different matter when it comes to an increase in the costs of fuel that result from making it clean. Then, the considerations become more tenuous. Each year the situation becomes less favorable, but only

---

*However, energy from an electrified car fleet could be derived by utilizing the plant during its low-use time, or at night; see Hietbrink et al.[6]

by a small amount, i.e., the public will resist the real extra-inflational price rises necessary to bring about clean fuel.

3. *Compatibility with sociological considerations.* On one side is the resistance to change from the internal combustion engine that arises from the inertia of capital invested in the means of production. The automotive manufacturing capacity of the main western countries is a primary component of their industry. Its dislocation from a too rapid change to the manufacture of a new kind of car would cause damage to the country's economy (in the case of the United States, and therefore to the world economy). Thus, it is not realistic to press for a short-term (5-year) change from one source of power to another. Such a change must occur over a longer time (10–20 years), although there may be resource-based reasons (e.g., a climb in the price of carbonaceous fuels due to their exhaustion) which will give rise to enhanced rate of change.

4. *Compatibility with longer-term considerations.* Although we must take into account the present situation, the strongest element must be compatibility with future sources of energy. This is an important, and little-discussed, pressure upon the choice of a new fuel. These sources (Chapter 16) will be either nuclear, and thus produce heat or electricity, or solar, the latter taken in the wider sense to include not only heat and electric energy, but also indirect solar energy, namely the solar-gravitational source, wind.

Thus, the following choices for new fuels are time dependent.

3.1. *Synthetic Natural Gas*

One acceptable fuel would be the synthetic natural gas methane. This could be used with little modification of the combustion engine. The combustion of low-weight hydrocarbons in methane occurs more completely than that of the high-weight hydrocarbons in oil. Methane burns clean, except for CO_2.

Supplies of real natural gas are subject to exhaustion and are hardly worth consideration. However, the conversion of coal to methane is a known process, and a simple one. "Upgrading" is produced by the reaction of hydrogen with CO_2 to form methane

$$C + H_2O \rightarrow CO + H_2$$

$$2H_2 + CO \rightarrow CH_4 + H_2O$$

Predictions of the cost of synthetic natural gas are proportional to the price of coal, but the 1975 prediction would suggest a price of $2.00 per MBTU, and this is less than the 1975 price of gasoline in the United States, which, direct from the refinery, without taxes, is in the vicinity of $2.60 per MBTU (1975 dollars).[7]

The continuance of synthetic natural gas depends upon the price of coal, and difficulties of coal have already been mentioned: building sufficient mines in time, training miners in time, and the difficulty of the sulfur in coal, and its evolution to polluting SO_2.

How long could natural gas be a fuel for transportation? Much would depend upon the mining situation and the potential research advances in sulfur removal. An advance in underground gasification would enhance the possibilities. Atomic explosives *might* be utilizable to increase access to coal fields, but more than a decade has passed since atomic engineering began to be evaluated in specific cases, and nothing has been accomplished.

3.2. *Methanol*

The relatively favorable pollutive properties of methanol have been stressed. The price of methanol (see Table 1) after exhaustion of available coal, depends principally upon the price of water-related hydrogen.

One possibility with methanol would be its extraction from atmospheric CO_2,[8] which could be combined with the CO_2 to give CH_3OH. Were the CO_2 used for methanol manufacture obtained from the atmosphere, then the difficulties with a future CO_2 effect would be overcome because the CO_2 would be removed and replaced at an equal rate.

The competition between hydrogen, methanol, and electricity will be influenced by the following factors. What would be the price of the water? Water, in the amounts needed for methanol production for automobiles, could not be regarded as cheap. It must be fresh, and this would have to come from river or lake sources, or by desalination of seawater. Much depends upon the availability of cheap energy. If it is available, cheap desalination is possible. Desalination on a large scale is cheaper by atomic rather than by solar methods. If hydrogen for transportation is only acceptable as a liquid, and the cost of liquefying hydrogen on a large scale is greater than the cost of extracting CO_2 from the atmosphere and the cost of combining it with hydrogen to form methanol, methanol will be cheaper than hydrogen. At present, it seems that liquefaction of hydrogen will be the cheaper of these two possibilities.

3.3. *Hydrazine*

The favorable side of the case for hydrazine as a fuel for transportation arises from its properties. It is liquid at room temperature, and, in the pure

Table 1. Projected Costs of Large-Scale Methanol Production (Cost per MBTU)[a]

Year	Coal gasification to methanol	Anaerobic digestion of vegetable matter	Nuclear electrolysis to hydrogen, CO_2 + hydrogen, etc.
1973	2.25[b]	3.00	4.50[c]
1974	4.50[d]	3.45	5.17[e]

[a] Source: J. O'M. Bockris.[7]
[b] Coal at $5.00 per ton.
[c] 1973 dollars, 10% return on investment.
[d] Coal at $10.00 per ton.
[e] 1974 dollars, 15% return on investment.

state, it is odorless and colorless. It forms an excellent fuel for fuel cells, and has been used satisfactorily as a fuel for an electric truck developed by the U.S. Army. Further, it has been used by the U.S Army in fuel-cell-based backpacks for soldiers, for operating radio devices, etc. Its use to provide electric power to villages in Vietnam has been reported.

Hydrazine has a final excellent property: It is made from resources which will remain plentiful, nitrogen from the air and hydrogen from water.

The negative side of hydrazine arises from its price. It is several times greater than the price of rival fuels. However, this is not the real price for small amounts, but rather the predicted price for very large amounts. This should become five times less than that quoted. It is unlikely that the price difficulty is substantial. In the present process for the manufacture of hydrazine, the principal cost element is the extraction of hydrazine hydrate from aqueous solution. If this could be avoided by using the material in aqueous solution, then prospects for cheap hydrazine would be increased.[9]

One of the negative features of hydrazine is its reported carcinogenic property.

Further, N_2H_4 dissociates to NH_3. Thus, hydrazine-driven vehicles have been accompanied by an unpleasant ammonia odor. The dissociation is metal based and may be controllable.

3.4. Hydrogen

Hydrogen as a fuel for transportation has been suggested from early times. Jules Verne[10] makes Captain Nemo tell his passengers aboard the Nautilus that the fuel of the future will be water, the products therefrom, etc. J. B. S. Haldane,[11] in a lecture at Cambridge University in 1923, predicted the use of hydrogen as a universal fuel. Lawaczek,[12] in the 1930s, pointed out that hydrogen could be passed through pipes more cheaply than electricity could be passed through wires. Erren,[13] in the 1930s, collaborated with Zeppelin engineers to feed the hydrogen from lift balloons into the diesel motors of Zeppelins during descending phases of flight. King *et al.*[14] and Erren[13] both did work in the 1930s on the use of hydrogen in internal combustion engines. Schoeppel has been the leader of such work in modern times, in the 1970s in the United States.[15-20] He, and, in separate work, Adt,[21] have both shown that there is no doubt about the possibility of using hydrogen as an automotive fuel, but the significant factor, and the overwhelming advantage to its use, is that it would be possible *with little modification of the present automobile engine.*

Thus, the main alterations needed would be in the mode of ignition, fuel injection being more suitable than carburetion.

The possibility of using hydrogen as a fuel for automotive transportation is consistent with the use of hydrogen for other purposes. Thus, hydrogen could be used to give household power, as the medium of the long-distance transmission of energy, and as a motive power in industry. In these cases, the positive effect of the use of hydrogen would be to give a clean alternative to the present use of natural gas or gasoline.

The availability of hydrogen fuel would encourage the use of fuel cells which function on hydrogen with high efficiency. One possibility would be cars in which hydrogen was the medium of energy for long-distance running (fuel cell drive in electric motors), with accelerative modes based on a temporary switching in batteries.[22] The batteries would be charged from stored liquid hydrogen, via fuel cells.

With these things in favor of hydrogen, what is the negative side? This consists of three areas:

1. Doubts concerning the price of the fuel. However, the price of hydrogen underlies the future price of methanol and hydrazine, so that, after the exhaustion of practical coal resources (between 2000 and 2050), all depends on solar or nuclear sources producing cheap hydrogen, even if the automotive fuel is methanol or hydrazine.

2. The difficulty of storage and carrying the hydrogen. This is a significant difficulty which could give rise to preference for methanol or for hydrazine. Conversely, the technology of storing liquid hydrogen has improved in recent years, and it seems likely that a cryogenic storer at a price of $300–400 may be manufacturable.

3. A liquid gas tank which would have to be 3.8 times larger than the gasoline tank.

4. Lastly, there are the psychological difficulties, e.g., the memory of laboratory hydrogen bangs from school, the image of the Hindenberg accident, etc. Hydrogen can explode if it is mixed in a ratio with air between 18 and 59 vol percent. It is thus more dangerous than methane, whose explosion limits are much narrower. However, it is not easy to find, outside the laboratory, a situation in which the explosion ranges are likely to be met. It is impossible to reach them in the open air. In a closed space, the hydrogen always rises, so that, as regards small leaks, a "closed" area with plenty of ventilation in the ceiling, properly arranged to prevent the gathering of pockets of hydrogen, would make an explosive ratio difficult to achieve from a slow leak.

Some properties of hydrogen counteract the more dangerous properties of other materials. The comparative explosive power of equal volumes of air and hydrogen, and air and natural gas, is three times less with hydrogen; in an accident which pierced the gas tanks, liquid hydrogen would spill on the roadway and would tend to evaporate rapidly, not spreading at a low level. The heavier petroleum vapor tends to spread near the ground, increasing the probability of its contacting hot bodies and therefore causing an explosion.

3.5. *Electricity*

The classical concepts of transportation technology in the year 2000 looked to the electric car. Scientifically, technologically, and socially, the advantages of electricity as an energy medium are considerable. It is completely clean, and the product of nuclear heat or solar energy. A nuclear or solar–electrical economy is a possibility, with household energy being electric, and many of the functions of factory and industrial use arising in an electrical way. Electrochemical technology would have to grow to have an increase in breadth.

The negative side of electricity arises thus:

1. Its introduction as an automotive fuel would cause a rebuilding of much of the means of producing cars. It would have to take place over, say, 20 years, to cushion this dislocation.

2. Hitherto, electric power sources do not give electric cars a performance which makes them viable with the present internal combustion engines running on gasoline.[22] This may be an overaffluent criterion, and future cars may function largely as intraurban transports, with interurban transportation occurring in an advanced kind of tracked vehicle whose cost of energy per passenger–mile is less than that of other kinds of transportation. Hence, if electric cars reach 40 mph and have a range of 50 mi, this might be acceptable. The 300-hp, 250-mi range, 120-mph car, will perhaps be regarded in the future much as the huge, powerful, fast, steam locomotives of the 1930s are now regarded. But electric cars running on lead–acid batteries would have difficulty in providing heat, and air conditioning would be out of the question. Will the U.S. Congress, so interrelated to the interests of the corporations, fund research for the high energy density batteries which would overcome such difficulties and make electric cars satisfactory, but damage the great automotive companies, not forgetting the large amount of the Economy they affect?

4. Economic Comparisons of Available Fuels

Such comparisons must be speculative, for the coming price rises in gasoline are uncertain, and price developments in alternative fuels are very uncertain. With these difficulties of price comparisons in mind, the following may be stated.

It is important to make the comparison on a uniform basis. The untaxed cost before profits at the site of production is the appropriate price for comparison. Thus, the cost of gasoline at the pump is not to be compared with the cost predicted for fuels which have not been commercialized, for these predicted costs are the sum of the cost of the materials, the amortization of the plant, maintenance, insurance, etc., and neglect factors such as profit, tax, etc.

It is important to compare meaningful quantities. Comparisons of cost of fuel per unit weight or volume are meaningless. Rather, comparisons of costs for a given unit of energy should be made.

Lastly, the environmental characteristics of fuels must be compared. It is useless to claim that fuel *A* is cheap and fuel *B* expensive, if fuel *A* is environmentally damaging and fuel *B* is ecologically sound. Herein lies one of the greater difficulties of prediction. How much damage is too damaging? How much will the public pay to have a cleaner fuel if the damage is invisible? How many of the facts about damage are relevant? How much is the damage realized?

On top of these difficulties of comparison is the effect of inflation, although this is the lesser of the difficulties because price rise is uniform among competitive products.

It would not be of great use to record here the projected prices of electricity, hydrogen, and methanol. First, the prices must necessarily be projections, and the degree of uncertainty of these overlaps any significance in the projected price differences. Also, much depends on the distance between source and sink. Electricity is likely to be cheaper than hydrogen with a few hundred miles of the energy source, and more expensive at distances of 500 mi or more. Methanol will have to be made from H_2 and so will be more expensive than H_2 gas, but will it be more expensive than liquid H_2, which is obviously the fuel one should compare with methanol? And, what of accounting for the cost of cryogenic storage? On top of all these factors will be inflational price changes.

For the unit of 10^6 MBTU, a price in the region of \$2 (1975 dollars) should be attainable for gaseous H_2, with improved electrolyzer technology and using low use–time power. This is a few tenths of one percent less than 1975 gasoline, ex refinery, and without tax. There are possibilities that even lower prices could be obtained from the hypothetical and unresearched direct conversion of water to hydrogen, using solar light via enzymes or electrochemical cells.

Experience shows that although predictions made for a technology are often overoptimistic, really new technology can produce great price falls. One of the more certain aspects of the future is that the fuel which looks cheapest at present, natural gas, will be expensive in a 25-year time scale because of its exhaustion.

Correspondingly, the price of solar electricity is, on a large scale, impossibly high with present technology, but it seems likely that emerging technologies (e.g., sea–solar power) will bring about a cost reduction of 10^2–10^3 within 10 years (although this statement is based on systems analysis).

Nevertheless, it is unlikely that we can expect a further era of cheap energy, compared with the price of energy in the mid-1960s. There was a special element about the price of energy in the first half of the nineteenth century. Essentially, a happy, spendthrift era was made possible by the unrealizing owners of the oil of the time. Thus, seen in retrospect, the fossil fuels were a special cache of solar energy collected over millions of years (Fig. 3). Regarded realistically, and with omniscience, man should have carefully rationed his fossil fuels by price or ordinance, keeping plenty in reserve until he had developed and proved a new source of abundant, clean energy. Thus, in view of their vital character to the technology then built up, *and their limited nature*, the proper attitude would have been to evaluate oil, natural gas, and coal as "deposits" of great value, and price them as such. Instead, until 1973, man has been spending them with abandon, selling them with what now looks

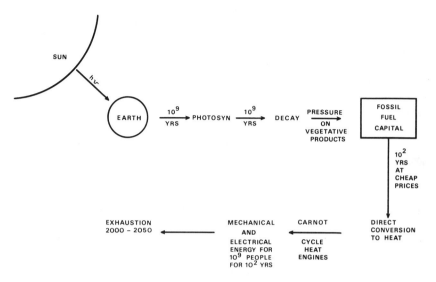

Figure 3. The price of energy or the level of affluence: fossil fuels were a free gift. Fossil fuels (coal, oil, natural gas) are stored solar energy. The sun insolated the earth for most of its life (*billions* of years) and the decayed vegetation from this gigantic amount of energy input did not all go back to CO_2, but partly, in processes that occurred very long ago and over a very long time, became converted to a variety of hydrocarbons. These products of photosynthesis are the fossil fuels, and those available to us are distributed over the earth. Some are little available (those where the earth is covered with sea) and some (e.g., in the great Middle Eastern fields) are almost like lakes of oil, just beneath the surface. The price at which they could be made available to the public (and on which the price until 1974 was based) is the relatively small cost of getting them out of the ground, transporting them to the refinery, and delivering them to the pump. The cost of this, plus profit, maintenance, insurance, etc., is the "real" (i.e., normal or "fair") price, and that which is now charged by the OPEC countries is a price that results from a monopoly situation. However, an argument can be made that this (or an even higher price) is "right." Thus, man did not realize he was spending capital, a once-only gift and distinctly exhaustible. The huge expansion and prosperity of the world in the twentieth century was all oil based. But (see D. H. Meadows *et al.*, *The Limits to Growth*) it will all turn to dust and ashes if abundant, clean energy is not achieved in time (i.e., before fossil-fuel exhaustion). Hence, a brake upon expansion and oil (i.e., basic capital) spending is essential, even though it will lead to depression, unemployment, and will put great strains on the maintenance of the western free enterprise system. The living standards of the western nations in the twentieth century have been based on spending capital, and resemble that of a family which received a legacy and decided to go on a free-spending binge, instead of investing the money to get income. The extremely serious question is: Do the western nations have enough time left to build abundant, clean energy before the capital (OPEC oil and western coal) is exhausted?

like incredible cheapness, and exhausting his only sure energy (i.e., technologically life-giving) capital. In this respect, the 400% price-raising action of the OPEC countries is rational. It may have to be greater if solar energy is not developed quickly.

Does present knowledge suggest a prospect of renewed cheap energy in the 100-year future (Fig. 4)? One can define "cheap" as one-tenth the price of energy in 1970. The answer is in the negative. Thus, let us suppose that nuclear fuel were free, as is, in a sense, energy from solar sources. The situation is different than that of large oil fields. There (e.g., in Saudi Arabia), the crude oil simply must be raised from the ground by derricks and stored. *It does not need expensive collecting equipment.* The collecting of solar energy by mirrors, heating, and working fluid, conversion to electricity, and finally its conversion to some storable fuel, perhaps hydrogen, all requires the purchase and construction of equipment. The cost has to be amortized, or paid for, at, for example, 12% per year. Cost estimates suggest that there is, at best, a prospect of a price equal to the pre-1974 price of energy. The best prospect in respect to price is energy from fusion. However, the machines that are envisaged at present as making hydrogen fuse in a practical way are very complicated and very large engineering structures. It is principally the capital cost of these machines, and of those for the conversion of their electricity to hydrogen,

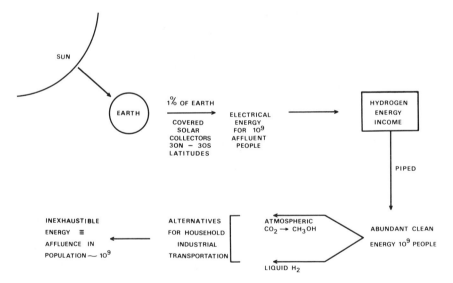

Figure 4. The price of energy or the level of affluence: abundant, clean energy. It will be possible (*if* there is sufficient research and development investment *in time*) to support about the same number of people as we have now with the income from the sun, i.e., by covering 1% of the earth's surface with solar collectors that are 10% efficient. But we will not be spending the gratuitous inherited capital of the fossil fuels, but the income our technologists can bring to us from the sun. Will we have energy at the fossil fuel price? It seems improbable, although possible. How will that affect our living standards, our social systems?

which form the basis of the price. Rough estimates have been made of a price as low as 0.3 cents (1973 dollars) per kWh.[23] However, we are at present so far from any realization of fusion power that a predicted price here has less meaning than predictions concerning, for example, wind power, where the converters involve entirely classical engineering.

These somber remarks are not meant to be taken as the final word. Finality in science is a presumed attribute only of the great negative laws of physics and chemistry, e.g., that the speed of light is the maximum possible speed, or that heat does not transfer spontaneously from a colder to a warmer body. It is conceivable that future researchers might come up with some means of obtaining energy at costs even cheaper than those connected with the predictions of the AEC in respect to electricity from fusion. Perhaps large orbiting satellites, beaming solar power to earth, might provide such very cheap energy.[24] At present, these seem to be one of the most expensive of the future forms of energy, but, were it possible to undergo considerable advances in space technology, and to come to a price of $100/lb for this lift-off (now $5000/lb), the prospect of cheap solar power could be within our grasp.

5. *Compatibility with Present Technology*

Methanol and synthetic natural gas are the most compatible fuels for the future. They could be integrated into the present technology with little difficulty. Their acceptability from an environmental–chemical viewpoint depends upon the source of the energy from which they come. If they are from coal, they will not be well accepted because they would continue to offer some pollutional problems. However, their acceptability must be limited after 2000 because of the increased danger of CO_2 and the greenhouse effect (see Chapter 10).

In developments that are time dependent, it is necessary to take into account amortization. No organization pays cash for the machines it buys; the money is borrowed. But it must be repaid, and the money to repay comes from the earnings of the device built. Hence, a question which would concern those corporations whose directors decide to build machines to interface with synthetic natural gas or methanol would be: How long will these fuels be part of the accepted technology? One can conceive of a smooth transition between a petroleum-based economy and a methanol-based economy. The question would be: How long will methanol be used before there is a cheaper source of energy (e.g., hydrogen from fusion)?

Will coal be available in very large amounts after, say, 2000? If not, the amortization proposition for a methanol-based automotive industry could be doubtful (since it could not begin until, say, after 1980). Eventually, such an industry will have to run on hydrogen-based electricity, or, more doubtfully, on methanol from atmospheric CO_2 and hydrogen from nuclear electricity.[7] The question must then be asked: Why have a short-term intervening phase of only

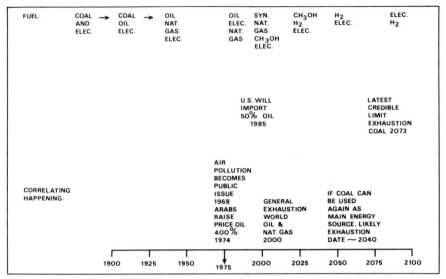

FUEL:	COAL AND ELEC.	→	COAL OIL ELEC.	→	OIL NAT. GAS ELEC.	OIL ELEC. NAT. GAS	SYN. NAT. GAS CH₃OH ELEC.	CH₃OH H₂ ELEC.	H₂ ELEC.		ELEC. H₂

Figure 5. Time of dominance of various fuels (partly projected). The figure shows the likely variation of fuels with time. Only the position of methanol seems uncertain.

some 2 decades with methanol? Those positive to the development of a methanol economy will assert that methanol will be made from atmospheric-produced CO_2 and nuclear-based hydrogen (see above). But here the economics would seem to favor hydrogen.

The position in respect to the compatibility of future fuels with present and future sources is balanced between hydrogen, electricity, and methanol (Fig. 5). It is not possible to resolve the issue except to say that, with present technological knowledge, the best prospect seems at present to be with hydrogen. However, to obtain this conclusion, one has to take into account not only the advantages of hydrogen in the transportation situation for sea, ground, and air, but also the situation with respect to the widespread potential uses of hydrogen in industry, long-distance transportation from probably distant sources, and the advantages of its distribution to private households. It is the net picture which looks better with hydrogen, with which one could do all of these things.

6. Methods of Storing Electricity for Use in Transportation: The New Batteries

There is an illusion about batteries. Most people only know the lead–acid battery, that usually used in starting the internal combustion engines of most cars. The functioning of this battery is as follows:

at the cathode

$$Pb^{2+} + 2e \rightarrow Pb$$

and at anode

$$Pb^{2+} + 6H_2O \rightarrow PbO_2 + 4H_3O^+ + 2e$$

The battery works well for starting the internal combustion engine; for this purpose, a large amount of power is needed per unit weight of battery, but only a small amount of energy, since the battery functions for only a few seconds before recharging. Hence, a battery with a large power-to-weight ratio (which the lead–acid battery has)[25] is needed, but one with a large energy-per-unit-weight ratio (which the lead–acid battery does *not* have) is not needed. The reverse is needed, however, when one considers an electrochemical motive source for cars. Here, a large energy per unit weight is needed (as well as a large power per unit weight).

Many new batteries have recently been developed to the laboratory and testing stage. Some of these are shown in Table 2. A more favorable energy-per-unit-weight ratio is available with these than with the lead–acid battery. Insufficient developmental work has been carried out to bring most of these new batteries to a commercial level.

Typical of these batteries are the hydrogen–air battery and the zinc–air battery. The hydrogen–air battery acts like a fuel cell. The difference from a fuel cell is that, instead of having outside sources of hydrogen enter the battery that are then electrochemically burnt to hydrogen ions, while at the other electrode oxygen molecules form which are electrochemically reacted to OH^- ions (so that the net reaction is water formation), the hydrogen is obtained back from the water by electricity. The oxygen produced when the water within the cell is electrolyzed can be vented, because of the oxygen available from the surrounding atmosphere.

The hydrogen–air storage battery has an energy-to-weight ratio of >1000 Wh/lb, and a power-to-weight ratio of some 10 W/lb. This is not satisfactory for the power aspects of an automotive source, but it is satisfactory in respect to its energy per unit weight, and it could be combined with a high

Table 2. Battery Parameters[a]

	W/kg	Wh/kg
Zinc–air	100	100
Ni–H$_2$ (compressed)	300	80
Ni–Fe	40	50
Zn–O$_2$ (compressed)	300	160
Aluminum–air	150	240
Sodium–sulfur	100	100
Sodium–air	200	400
Li–Cl$_2$ (liquid storage)	300	500
H$_2$–air	100	2000

[a] Source: "Technology of Efficient Energy Utilization."[26]

Table 3. Possible Hybrid Drive Trains for Wheeled Vehicles

Type of fuel	Accelerator move	Steady running	Special example	Good features	Remarks
Chemical, H_2	Flywheel	Chemical motor drives flywheel.	Synthetic natural gas drives Sterling engine—flywheel—fluid drive.	Low pollution; good performance from high power density of flywheel; motor runs constant velocity, hence pollution reduced.	Flywheel could break loose in accident; super flywheel?
Electricity	High power density battery.	High energy density battery.	Ni-Zn cell in conjunction Zn-air cell; light-weight electric motor.	Zero pollution; would couple well with coal economy; coal used → electricity.	Performance still significantly less than gasoline internal combustion system.
Chemical fuel, H_2	High power density battery.	High energy density battery.	Na-S battery; methanol fuel cell; light-weight electric motor.	Low pollution; uses methanol from coal.	Interim system? No noise pollution.
Chemical fuel	High power density battery.	Chemically-driven motor works generators, charges battery.	$Fe-O_2$ battery; synthetic natural gas-driven motor; light-weight generator, electric motor.	No new technology.	Too heavy?
Chemical fuel	High power density battery drives wheels.	Chemically-fueled external combustion engine drives wheels, charges battery.	Fe-Ni battery,; methanol-driven Sterling engine.	Should → excellent performance.	Cost methanol? Sufficient Ni?

Electricity	Turbo-hydraulic drive.	High energy density battery drives constant speed motor which drives hydraulic train.	$Zn-O_2$, turbo-hydraulic.	Battery delivers current constant speed, hence long life.	Complexity of system → poor economics?
H_2	Internal combustion engine.	Fuel cell drives motor.	H_2 internal combustion engine to 50 kph and fuel cell drives light-weight motor, torque applied 5th wheel.	Economy long distance running, very low pollution.	Best proposition? Weight?
H_2	Air motor, compressed air storage of energy.	Fuel cell drives motor, compresses air.	Air turbine; H_2 fuel cell; printed circuit motor.	Zero pollution, excellent acceleration.	Best proposition? Weight?
H_2	Flywheel gives high power per unit weight.	Fuel cell drives motor which tops up flywheel.	Super-flywheel design (high energy density); H_2 fuel cell gives 65% efficiency.	Zero pollution, excellent acceleration.	One drive train; best proposition? Super-flywheels do not consist of high inertia solid structures but consist of largeness of components, so that break-loose less fatal.

Notes: The various hazards that the evolvers of a new system for driving cars must avoid are: First, economics. The cost per passenger–mile must be be minimal and not too greatly increase that of pollution-giving and exhausting petroleum. Second, the fuel has to give minimal pollution. Third, it must be acceptable in performance. The numerous possibilities will be attempts to compromise and optimize in respect to the degree to which cost per passenger–mile will be increased, and performance sacrificed, in order to run on fuels which will be clean and inexhaustible.

On the whole, the obvious alternative of a battery-driven electric car demands rather large sacrifices in performance, although it gives a good economic prospect.

Two hybrids attempt to increase performance while maintaining *nearly* the same excellent nonpollutive situation which the simple electric car offers; and attempting to keep costs minimal.

In the table, chemical fuels are synthetic natural gas (methane), methanol, and hydrazine.

power density battery to provide a hybrid system with good acceleration and long range.

The zinc–air battery is a relatively new development in its modern form. It utilizes the air electrode, developed by NASA in the 1960s. It utilizes zinc electrodes that can be recharged according to the equation

$$ZnO + H_2O + 2e \rightarrow Zn + 2OH^-$$

This battery has a power-to-weight ratio of about 100 W/kg and an energy-per-weight ratio of about 100 Whr/kg. It is not satisfactory as yet because the number of deep discharges and recharges which can be suffered by a battery before it must be rebuilt has an important economic influence upon its viability. About 1000 cycles is the going rate for "viability," and the zinc–air battery runs about 250 cycles. However, with less-deep discharge the battery would remain active for far more than 250 cycles.

7. Possible Developments in Electrochemical Transportation

Electrochemical transportation will be reviewed in Chapter 19. Two points are outstanding:

1. If the lead–acid battery is to be used, one difficulty is its lifespan. Thus, a motive force for a vehicle must withstand sudden accelerative demands, i.e., very high currents for short times. This means that the battery is asked to pass a high current across its plates. This overheats the plates and leads to the formation of nonuniform crystals and an increased size of crystals, so that the number of times it can be charged and recharged is decreased. (Thus, an important property of a battery is charge acceptance. But this is proportional to the area in contact with the solution.) In economic terms, this may mean that the owner must buy a battery more often than he would consider economic, e.g., every 6 months.* This difficulty can be overcome by a device developed for the first time at the Flinders University of South Australia, by Darryl Whitford and others. The battery is made to discharge at a constant rate through a constant-speed motor (while the car is in use in any given period) and the acceleration and braking is done through a separate hydraulic system which is motivated by the electric motor. This system removes the violent variations in current across the battery, associated with acceleration and braking, that normally damages plates, so that the latter *should* last for several years. This is an attractive solution, for the lessened cost of the fuel for electric cars means nothing if new batteries would have to be bought yearly or bi-yearly.

2. If we are to have satisfactory electrically fueled transportation, a high energy density battery must be developed. Little research in the free enterprise countries is being done on this.

*This difficulty does not apply to the slower, older electric cars, but to the recent attempts to make (relatively) high-performance vehicles still having lead–acid drive.

8. Hydrogen Hybrid Cars

Many possibilities arise from the use of hydrogen in transportation. The simplest is with hydrogen as a fuel for an internal combustion engine. However, hydrogen could be used in some interesting hybrids. One could build a hydrogen–electric car with fuel cell drive for the town, where high speeds are not necessary. Acceleration could be added by having iron–air batteries of small capacitance—therefore light weight—which would allow rapid acceleration on starting, and which would be charged from the fuel cells on board. But this hybrid could be replaced by another in which a small internal combustive motor (hydrogen fueled) would be responsible for charging a battery. Electric drive would be used for acceleration and running below 50 kph. Above 50 kph, the combustive motor would pull. All these systems would provide environmentally clean systems, although the all-electric systems would be the cleanest (see Table 3).

9. Concluding Remarks

The question of a clean fuel for future transportation is only partly a technological question. Hydrogen and electricity are possible clean fuels. Methanol is another possibility. However, the time at which these fuels will be available, the price, etc., depends upon research, the intensity of that research, and this research depends upon sufficient funding. However, heavy research funding of a new technology for mass use is preeminently a *political* matter, because the success of the research could upset the situation of the great capital groups. The main questions are: How much research funding? Will the work be done in companies which would profit and not lose by its success?

References

1. E. S. Forest, "Pollution," *Commercial Car Journal,* **87** (1970).
2. J. O'M. Bockris and S. Srinivasan, *Fuel Cells: Their Electrochemistry,* McGraw-Hill, New York, 1969, p. 633.
3. J. O'M. Bockris and D. Drazic, *Electrochemical Science,* Taylor and Francis, London, 1972.
4. H. R. Linden, World Congress on Gas, Chicago, 1973.
5. H. C. Hottel, "Challenges in Production of Fossil Fuels," *Chemical Engineering Progress* **21** : 311 (1973).
6. E. H. Hietbrink, J. McBreen, S. M. Selis, S. B. Tricklebank, and R. R. Witherspoon, in *Electrochemistry of Cleaner Environments,* J. O'M. Bockris, ed., Plenum Press, New York, 1972.
7. J. O'M. Bockris, *A Solar–Hydrogen Economy,* A.N.Z. Book Co., Brookvale, New South Wales, 1975.
8. M. Steinberg, M. Beller, and J. R. Powell, *A Survey of Applications of Fusion Power Technology to the Chemical and Material Processing Industry,* Brookhaven National Laboratory, New York, 1974.

9. J. O'M. Bockris and D. Drazic, *Electrochemical Science*, Taylor and Francis, London, p. 225, 1972.
10. J. Verne, *Twenty Thousand Leagues Under the Sea*, Paris, 1869.
11. J. B. S. Haldane, quoted in R. A. Erren and W. A. Hastings-Campbell, *J. Inst. Fuel*, **VI** : 277 (1933).
12. J. Lawaczeck, quoted by R. O. Lindstrom, *ASEA Journal*, **37** (1) : (1964).
13. R. A. Erren and W. A. Hastings-Campbell, *J. Inst. Fuel*, **VI** : 277 (1933).
14. R. O. King, W. A. Wallace, and B. Mahapatra, *Can. J. Res.*, **26F** : 264 (1948).
15. R. G. Murray, R. J. Schoeppel, and C. L. Gray, paper presented at the 7th Intersociety Energy Conversion Engineering Conference, San Diego, September, 1972.
16. R. J. Schoeppel and S. Sadiq, paper presented at the Frontiers of Power Technology Conference, Stillwater, Oklahoma, 15–16 October, 1970.
17. R. J. Schoeppel, *Chemtech*, **27** : 476 (1972).
18. R. J. Schoeppel, paper presented at the Gas Symposium of the Society of Petroleum Engineers of AIME, Omaha, Nebraska, 21–22 May, 1970.
19. R. G. Murray and R. J. Schoeppel, paper presented at the Intersociety Energy Conversion Engineering Proceedings, Boston, Massachusetts, 3–6 August, 1971.
20. R. J. Schoeppel and R. G. Murray, paper presented at the Frontiers of Power Technology Conference, Stillwater, Oklahoma, 28–29 October, 1968.
21. M. R. Swain and R. R. Adt, Jr., paper presented at the 7th Intersociety Energy Conversion Engineering Conference, San Diego, California, 25–29 September, 1972.
22. J. O'M. Bockris, *Search*, **4** (5) : 144 (1973).
23. A. M. Weinburg and R. P. Hammond, *Am. Sci.*, **58** : 412 (1970).
24. P. E. Glaser, *Science*, **162** : 857 (1968).
25. J. O'M. Bockris and D. Drazic, *Electrochemical Science*, Taylor and Francis, London, p. 228, 1972.
26. "Technology of Efficient Energy Utilization," Report of a NATO Science Committee Conference, Les Arcs, France, 8–12 October, 1973.

We have seen that transportation is the greatest polluter. In this book we discuss transportation solutions in two ways: in terms of fuels which may be used in the future and which will be clean, and in terms of electric cars. "Electrochemical transportation" would be the appropriate designation, because the essence of this system does not concern the propulsion system, brakes, and mechanical features as much as it concerns the power source. In the following chapter the situation is very briefly expressed, with an estimate of the relative roles to be played by the fuel cell and the battery.

19

Notes on Electrochemical Transportation

J. O'M. Bockris

1. General Background

Evidence has been given in this book that air and noise pollution are connected with our use of internal combustion engines in transportation. There are three possible solutions to the dilemma we are in, i.e., that of wishing to have good transportation in order that our lives may continue in the same form as hitherto, and knowing that our present mode of transportation is not possible in the long term. The first two solutions are to use methanol or liquid hydrogen to replace gasoline, but the third way is to use electrochemical power sources to run our cars. Regarded as fuels, these three methods have been already reviewed in Chapter 18, so that this chapter will be brief.

Electric cars are the populist solution. Due to the growth of the electrical supply, low use period electric power would be sufficient to provide enough electricity to charge the batteries—or produce hydrogen—for all cars in the

J. O'M. Bockris • Department of Chemistry, School of Physical Sciences, Flinders University of South Australia, Bedford Park, South Australia 5042

Table 1. Projected Electrical Energy Requirements for Electric Passenger Cars[a]

Year	Total passenger cars, million	Total electric passenger cars, million	Estimated electric energy required for electric cars, billion kWh	Estimated electric power generation United States, billion kWh	Estimated power available by leveling diurnal load (20% increase), billion kWh[b]
1980	110	1.1	9	2830	566
	110	10.0	88	2830	566
2000	150	15.0	120	8250	
	150	150.0	1200	8250	1650

[a]Source: E. H. Hietbrink et al.[1]
[b]This assumes that charging will occur during off-peak hours.

United States by 2000. Thus, no special new means of producing energy for cars would be necessary if we took the electric car solution, whereas if we took the methanol production or the hydrogen solution, a very large new industry for fuel production would become necessary as the petroleum industry was wound down.

The basis for above statement is shown in Table 1.

2. Characteristics of Present Electric Cars

Some characteristics of present electric cars are shown in Figs. 1 and 2.

The typical performance of a lead–acid-driven electric car is referred to as 40–40, meaning a top speed of 40 mph and a range of 40 mi. The actual range depends upon driving conditions—how many starts and stops—and the 40 mi quoted is reduced to about 25 mi in bad driving conditions.

Nevertheless, an average electric car, lead–acid battery driven, would be practical for urban traffic, particularly for buses, shorthaul trucks, etc.

3. The Lead–Acid Battery

It is central to the position of electric cars to explain the pros and cons of lead–acid batteries. Those not familiar with the battery area may imagine that the lead–acid battery is the only feasible battery for driving electric cars. This is not the case. Any battery consists essentially of two plates. In the lead battery, they are of lead and lead oxide (which conducts). During discharge and

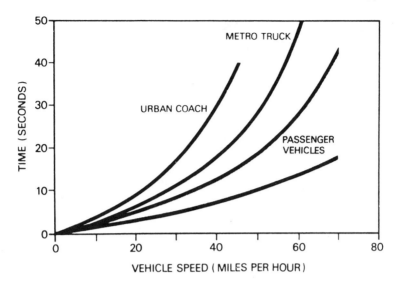

Figure 1. Typical acceleration requirements for selected vehicles.[1]

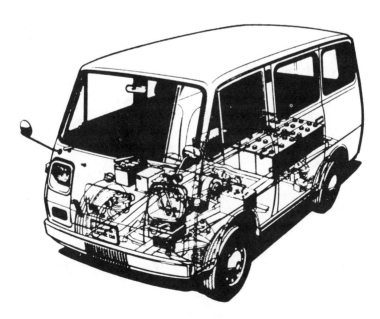

Figure 2. Daihatsu van, 40 mph, 50-mi range.[2]

electricity production, the lead oxide plate has the reactions

$$PbO_2 + 4H^+ + 2e \rightarrow Pb^{2+} + 2H_2O$$

i.e., it produces Pb^{2+} in solution. The Pb plate has the reaction

$$Pb \rightarrow Pb^{2+} + 2e$$

Essentially this is an historical accident, because of the fact that the properties which lead–acid batteries have fit their use at this time, that is, being *auxiliary* power sources used for starting gasoline engines in cars.

Lead batteries would be heavy per unit of energy stored, in which characteristic the lead–acid battery is the poorest battery known. But it does have good properties as far as power per unit weight is concerned. When one uses a lead–acid battery to start a gasoline engine, one demands from it a burst of high power. The amount of energy used is small, because one utilizes it only for a short time. Therefore, it does not matter at all, in its use as a starter battery, that the lead–acid battery is heavy per unit of energy stored.

Looked at from the point of view of the power source to run a car, the situation is different. The lead–acid battery, then, has three great negative features:

1. Its energy density is low, so that cars run upon lead–acid batteries will be heavy, slow, and will not have a good range. This is only a characteristic of the fact that they are being run on this particular battery, the lead–acid battery. *It would not be a feature of electric cars if they were run on other batteries.*

2. The lead–acid battery has an Achilles heel: it undergoes side reactions which produce, gradually, as it is charged and recharged, a layer of lead sulfate on an electrode, and the battery loses its ability to be recharged.

3. Were we to go, during the next 10–20 years, entirely to an electric car fleet, at least in the West, where cars are so much more numerous per head of population than in the East, we would need a great amount of lead. Talking about the period of time in which a resource will last is a complex business, because the answer is affected by many factors: new technology for discovering resources, degree of recycling, etc. However, neglecting the use of lead in electric cars, Meadows *et al.*[3] gave the time of exhaustion of lead as 21 years, or 64 years if the assumed lead resources are multiplied by five (some lead producers are suspected of hiding the extent of their resources). Were we to use lead for batteries in cars, the use-up rate would be greatly increased, so that it would seem improbable that we could supply cars with lead for more than a few years.

Why have the high energy density batteries (see below) not been developed earlier?

The answer is a commercial one. People do not support applied research unless they see a market. A market for high-energy-density batteries has not been present, and until some major movement is made to tax polluting cars off the roads, it may not be there, and therefore the pulse toward research on high-energy-density batteries may still be lacking.

4. The Possibility of Introducing Electric Transportation Now

Would it be possible in the present state of knowledge about batteries to introduce electrochemical transportation?

This would necessarily have to be done slowly. First, even were there to be a government regulation, it would be physically impossible to produce a changeover to an electric car fleet in less than about a decade. Within 3–4 years of the resolve to do it, it would be possible to replace many of our trucklike vehicles, post office vans, shorthaul delivery vehicles, etc., and to then introduce low-cost, low-mileage cars in near-city transportation, probably municipally owned, and involving a system of credit card or magnetic tape payment for shorthaul use. Such a system was begun in Amsterdam in 1974.

The expansion of an electric car fleet to private owners, for passenger wagons, may not come until a battery has been provided which would not be subject to the exhaustion of its principal material, and to other difficulties of the lead–acid battery.

5. Electric Cars Now Available

In the United States, two electric cars are available at the moment. They are the Sebring-Vanguard (made in the United States) and the Elcar (made in Italy). They both run on lead–acid batteries, and the range is up to 50 miles and 25–50 miles, respectively. Top speeds are said to be 28–50 mph and 25–35 mph, respectively.

6. Brief Notes on High-Energy-Density Batteries under Research

In stating what high-energy-density batteries are being researched, the situation varies, depending upon which battery is considered. Exotic high-energy-density batteries, those involving alkali metals, need more research than do the down-to-earth replacements for the lead–acid battery, e.g., the iron–air battery under development by the Westinghouse Corporation, etc.

Some idea of the progress in battery research during the last few years is given in Fig. 3.

The energy densities of a few batteries are shown in Chapter 18, Table 2.

What degree of research is necessary to commercialize these batteries? First, the problem of material availability must be considered. Batteries that use sodium and sulfur, lithium and chlorine, iron and air, perhaps zinc and air,

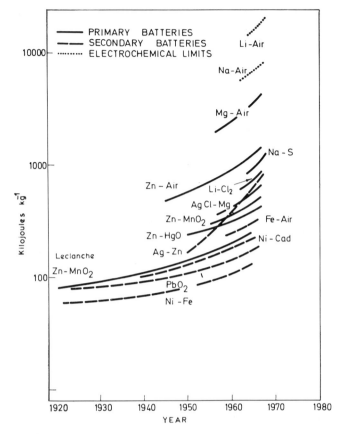

Figure 3. The improvement of energy density for various types of
battery during the last 50 years. Metal–air batteries seem to have
the steepest curves, promising high capacities in the 1980s.[4]

etc., are those which should be considered. Those which use less-available
materials are to be less considered.

It is difficult to state the amount of research which must be done to
commercialize these batteries. However, it should be compared with the vast
gains which would arise were we to be able to run at least an intraurban car fleet
upon low use time electric power, and thus clear our cities of the major source
of air pollution.

7. Hybrids

These have been mentioned elsewhere (Chapter 18). Batteries tend to
have high power densities, while fuel cells have high energy densities. Conversely, batteries have poor energy densities and fuel cells have poor power

Figure 4. Electric car powered by a hydrogen–air lead battery power system.[5]

densities. It follows, obviously, that a union of these two power sources within the car, with circuitry such that the battery would provide the power for acceleration and the fuel cell the energy for steady running, would be the development needed.

Other hybrids are referred to in Chapter 18.

An electric car which utilizes this principle has been developed by Kordesch, as shown in Fig. 4.

8. The Electric Motor

If one refers to the weight of electric motors in ordinary textbooks of engineering, one finds that it is in the region of 10 lb per horsepower. This is a negative feature for electric cars, because the overall weights of petroleum engines are about 2 lb per horsepower. In an electric car, one would have to put the weight of the power sources together with the weight of the motor, and this would then be more than ten times the weight of the corresponding present system. Hence, research is needed to reduce the weight of electric motors. There is some evidence that lightweight electric motors, with weights on the order of 1 lb per horsepower, can be made.

9. The Public Electric Car

This system stresses the idea that, for the most part, cars are needed by individual persons for only a small part of the day. Hence, there is a waste in the concept of a personally owned car standing in a parking lot while its owner does his work, using it, say, for half an hour at the beginning and end of each day. It would be more reasonable to have cars publicly owned, and for them to be available to individuals for the journey from the suburbs to the towns and back again at the beginning and end of the day, being in use in the town during the day. The capital financing would have to be carried out by a central organization, probably the state, and the use costs would be billed to the consumer.

The object that such a system would reduce "freedom," would not allow personal use of cars for pleasure, could be met by the availability of other types of cars on hire for out-of-town pleasure trips.

10. *The Flinders Electric Car*

An advanced electric car is that developed within the Flinders University by Whitford, Atkinson, and Rush. So-called "high-performance" electric cars, running on lead batteries, have the difficulty that the battery plates last only 6–12 months, because, in sharp acceleration, currents of many thousands of amperes pass across the plates.

The concept developed in the Flinders electric car is that the batteries drive a motor at constant velocity. The torque for acceleration and driving is taken hydraulically and transmitted to the wheels (see Fig. 5). Thus, the batteries should last 5 years.

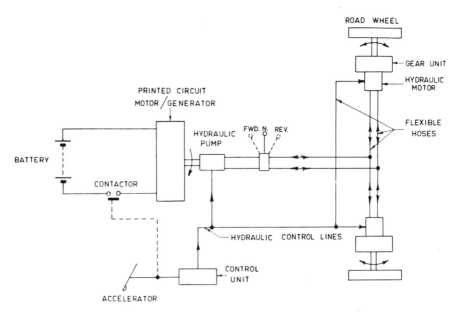

Figure 5. Flinders electric vehicle propulsion and control.[6]

11. *A Hypothetical Electric Car*

Among the options which could be developed for an electric car, one choice would be a car with the following specifications:

1. *Power source.* This would be a hybrid, for reasons stated. It would be best to use a hydrogen-driven fuel cell, with hydrogen kept in cylinders and using liquid hydrogen. Batteries would be iron–air batteries, but they would be of small capacitance because the main power source of the car would be the fuel cell.

Another possibility would be to have a methanol-fueled fuel cell, but this possibility depends upon the development of such fuel cells, which is behind the development of hydrogen fuel cells.

2. *Drive.* The drive for the electric car could be a fluid drive which would be developed as in the Flinders electric car. This would reduce the strain on the batteries during acceleration. Electric braking and freewheeling would be available. Electrically regenerative charging of the batteries would be an advantage.

3. *Battery recharge.* This would happen partly from the regenerative charging, but also from the fuel cell. Circuitry would have to be developed so that the fuel cell charged the battery when needed.

4. *Range.* Fuel cells have a large energy density, and a considerable range would be possible. Thus, energy densities of 1000 Whr/lb are possible for fuel cells, and this is more than ten times the energy density of lead–acid batteries. Thus, the range on a fuel-cell-driven car would be more than ten times that of lead–acid battery car, or more than 400 mi with one fueling.

5. *Acceleration and speed.* In respect to the top speed of electric cars, any speed would be obtainable, but what would be gained in speed would be lost in range.

In a fuel-cell-backed, battery-driven car, speeds of 75 mph should be available, but the ability of the car to maintain such speeds for long distances is more doubtful; it would be subject to the poor power density of the fuel cell.

6. *Heating and cooling of interior.* One of the problems of electric cars is that, because they are short of energy, cooling and heating is more difficult than with gasoline-driven cars. With a fuel-cell-backed, battery-driven car, it should be possible to develop a heat-pump-type of reserve cycle heating, because, under these conditions, the efficiency of the use of energy for heating is more than 100%. Such a device could provide cooling, but there is a possibility that one might have to derive heating and cooling from solid-state devices, namely, those involving storage of energy in solids. The solid would be cooled to a certain temperature, below room temperature, during the night, and inserted into the car for use during the day. As it melted it would remain cool, so that air blown over it could be used to cool the car.

12. *Why Do We Not Yet Have Electric Cars?*

It is an interesting sociological problem that the pollutional contributions of our present cars have been in the public domain for 5 years, but no mass movement to replace them with electric cars has taken place.

The position of the automotive concerns is to heap criticism upon the performance of electric cars. Referring to those which existed in 1910, they are subject to ridicule, just of course as the 1910 internal combustion car could be ridiculed.

However, there is no escaping the fact that lead–acid-driven cars would have a poor performance compared with modern petroleum-run cars. This is clearly counterbalanced, and surely overcome, by two facts:

1. Were electrochemical power source development to be pursued, there would be no essential difference between the performance of internal combustion and electrically driven cars.
2. More important, the question of performance (top speed and acceleration) has been exaggerated in its importance to the public. What the public wants is reliable transportation that does not pollute. The strength of the media has been used to avoid a confrontation between the wish of the automotive concerns to continue to produce the same product, and the public interest.

A change to electric cars would be against the interest of the large automobile corporations. The reason for this is not clear, for a reasonable rejoinder would be: "Can't they make profits with electric cars, too?" Doubtless, the answer to this question in the long run is affirmative, but, from the corporate point of view, the transfer to the electric car would have negative consequences in the short term, since it would mean, over a period of perhaps 10 years, much reconstruction of the manufacturing plants. Corporations work on a short-term viewpoint—the profits each year—so that reorganization to build electric cars would not be acceptable to the directors as regards the short-term profit picture.

Thus, the question of electric cars is primarily a political one. The political situation must be solved before the research funds are given and the remaining technological problems can be solved. But massive research funding of electric cars in the United States (in institutions not under the control of the corporations) does not seem a likely proposition.

References

1. E. H. Hietbrink, J. McBreen, S. M. Selis, S. B. Tricklebank, and R. R. Witherspoon, in *The Electrochemistry of Cleaner Environments*, J. O'M. Bockris, ed., Plenum, New York, 1972.
2. J. O'M. Bockris, "Electrochemical Transportation," *Search*, **4**(5) : 144 (1973).
3. D. H. Meadows, D. L. Meadows, J. Randers, and W. W. Rehrens III, *The Limits to Growth*, Potomac Associates, Washington, D.C., 1972.

4. J. O'M. Bockris and D. Drazic, *Electrochemical Science*, Taylor and Francis, London, 1972.
5. K. Kordesch, in *Modern Aspects of Electrochemistry*, Vol. 10, J. O'M. Bockris and B. E. Conway, eds., Plenum, New York, 1975.
6. D. Whitford, D. A. Atkinson, and K. Rush, "Flinders Electric Vehicle Project," Final Report, Stage I, Flinders University of South Australia, January, 1975.

Fresh Water from the Sea

The limitation to the opening up of many areas of the world, as in Australia, is the absence of fresh water. This is needed not only for the simple purpose of drinking, but in vast quantities for numerous other purposes to support life, e.g., in agriculture, and perhaps as the source of hydrogen fuel. A sufficiently cheap process for the reclaiming of fresh water from the sea would be one of the great contributions to a satisfactory future, and would give us enough time to set up a stable population and avoid large-scale disaster.

20

The Desalination of Water

E. D. Howe

1. Introduction

The desalination of water is the removal of all or part of the dissolved salts from seawater or from other saline water supplies. It is presently used for the production of high-purity water for certain industrial and chemical processes and, on a limited scale, for potable water to be supplied to people living in arid areas, such as the lands bordering the Arabian Gulf and some of the islands in the Caribbean. Population pressures in the United States have also led to demands for fresh water which cannot be met economically with natural fresh water if it must be brought from considerable distances. Therefore desalination is becoming increasingly attractive as a method for freshening local supplies of saline water. This chapter presents the fundamentals of the various desalination processes as they are applied to, or proposed for, the production of "fresh" water.

Desalination as a means of producing potable water supplies has been considered feasible only in very recent years. The earliest installations were on the islands of Curacao and Aruba in the Caribbean, where the development of

E. D. Howe • Department of Mechanical Engineering, University of California, Berkeley, California 94720

oil refineries and the attendant population increase made necessary the furnishing of high-quality potable water regardless of the cost. This was followed by similar demands in the oil-producing areas around the Arabian Gulf, notably Saudi Arabia and Kuwait. Still later, the growing concern for adequate fresh water supplies in Europe and in the United States in the face of periods of drought became sufficiently great that serious research and development efforts were undertaken. This work began about 1951, and has continued to the present time. The major problem is one of cost, since most of the processes of desalination, like other chemical refining processes, require relatively expensive equipment and appreciable amounts of energy. The cost of desalted water in 1950 was about $1.30/m^3$, and in 1974 this figure had been reduced to about $0.23/m^3$ for distilled water, and to a lesser cost for water produced from low-salinity well water by electrodialysis and reverse osmosis plants.

One of the reasons for continuing research and development efforts is the great diversity of possible methods for desalting water. The salts dissolved in saline water supplies are nonvolatile, so that both evaporation and freezing will separate pure water from saline water. Also, dissolved salts are ionized, so that electrical methods are feasible. The phenomena of capillarity and osmosis suggest that interactions with certain properties of membranes and other solids have possibilities for separatory schemes. As a result, the following processes have been considered for use in the production of fresh water, either from the oceans or from the many saline sources on continents: distillation, electrodialysis, ion exchange, solvent extraction, freeze separation, hydrate separation, and reverse osmosis.

Of the above-mentioned processes, distillation, electrodialysis, and reverse osmosis have been developed to the extent that they can now be considered commercial for relatively large-scale production. Distillation is capable of treating waters of all salinities, including seawater, whereas the two latter processes, electrodialysis and reverse osmosis, are limited to waters of comparatively low salinity. Ion exchange is presently used chiefly for "polishing" high-purity water for use in nuclear reactors and other demanding industrial situations. Its use for softening fresh water, i.e., exchanging objectionable ions for acceptable ions, continues, but its application to large-scale desalting operations has been hindered by the cost of regenerants. Freeze separation was developed in one prototype plant at Eilath, Israel, where it was operated for a brief period, but was finally "placed in mothballs," due to the construction of a larger distillation plant. Solvent extraction and hydrate separation methods have not progressed beyond the laboratory stage. Missing from the above list is simple precipitation, due to the fact that there is no known "cheap" chemical for causing the precipitation of one of the most soluble chemicals, sodium chloride, which constitutes the major portion of the materials dissolved in saline waters.

In order to compare and assess the several processes for desalination, it is essential to recognize the place of water in the environment, and to consider the degree of purity required for water supplies, as well as those characteristics of water which affect the application of the various processes to desalination.

2. The Nature of Water

2.1. Environmental Aspects of Water

The basic essentials of the earth's environment are water, land, and air. Because of the earth's rotational speed, its size relative to the sun, and its distance therefrom, water can exist as a liquid on the surface of the earth and as a vapor in the atmosphere. These conditions, which are different from those on other planets in our solar system, make possible the existence of organic life as we know it. Water is the chief component of the fluids within animals and plants, and thus serves as the medium for conveying nutrients to growing cells, for carrying away waste materials, and for eliminating heat by evaporation. Since the latter two functions remove water from the organism, there must be periodic replacement of the water. Such water is absorbed through the root membranes of plants and through the membranes lining internal organs in animals and man. This absorption is the result of osmotic pressure differences between the internal fluids and the water, and can occur properly only when the concentration of dissolved salts is less in the water than in the internal fluids.

When the dissolved salt concentration in water is higher than that in the internal fluids, the osmotic transfer of water takes place in the reverse direction and the plant or animal becomes dehydrated. Observations to determine the maximum acceptable concentration of dissolved salts in the water indicate that this quantity depends to some extent on the plant or animal. However, drinking water for people and most animals should contain no more than 500 ppm dissolved salts. For most plants the limit is slightly higher, with many plants thriving on waters that contain as much as 1200 ppm. Since most of the water on the planet is in the oceans, with an average salt concentration of about 35,000 ppm, the question arises as to the source of the relatively pure water required by man, animals, and plants.

2.2. The Hydrologic Cycle

Fresh water is furnished by the natural processes of the hydrologic cycle, by which water is evaporated from the oceans and precipitated over the land. The cycle begins with the absorption of solar energy by the water in the oceans, thus heating the water and causing part of it to evaporate and become humidity in the air. This humidified air is carried by winds overland and, when it comes into contact with cold masses of air, the humidity condenses and is precipitated as rain or snow. Exposure to the sun melts the snow, and the meltwater joins with the rainwater to form rivers, or to percolate into the ground and proceed slowly through the soil back toward the sea. The outflow of rivers and the slow seepage of underground water into the sea completes the hydrologic cycle. During rainfall, plants are watered and small streams collect the water as it drains into the low portions of the terrain. Animals and men obtain their supply of fresh water from flowing streams, lakes, or dug wells, causing only slight delays in the return of the water to the oceans.

The problems of water supply arise from the fact that the distribution of precipitation is far from uniform, in either time or location. As a result, many of the land areas between 40°N and 40°S, where the climate is most habitable, are either arid or semiarid. In these regions of low rainfall, groundwater is apt to be saline, and hence unsuitable for drinking purposes. In such cases, either low-salinity fresh water must be imported, often from great distances, or local saline supplies must be desalted.

2.3. Chemical Properties of Water

In order to appreciate the significance of the several processes for desalination, it is essential to understand the chemical properties of water. Water is a very unique and irregular chemical material, for its properties are radically different than those of other chemicals. The crystalline state of watei, ice, is less dense than its liquid state, so that ice floats on liquid water. If ice were heavier than liquid water, marine life would be impossible, for the oceans and lakes would freeze from the bottom upward. Under present conditions, ice floating on the surface of the oceans and lakes prevents the water below from freezing, and thus permits the marine biota to survive the winter weather.

A second peculiarity of water is the large span of temperature between its freezing and boiling points, and the elevation of these temperatures. If one were to plot freezing and boiling points against molecular weights of the chemicals in the same column of the periodic table as oxygen, the data for all compounds with hydrogen would fall on smooth curves, except for water. From such curves it would seem that the freezing point, when correlated to the molecular weight of water, should be $-100°C$ and its boiling point $-91°C$. These are radically different than the actual properties of water, namely, $0°C$ and $100°C$ for the freezing and boiling points, respectively. The temperature interval between freezing and boiling is actually $100°$ as compared to the projected interval of $9°$ based on the molecular weight correlation. A related anomalous property is the latent heat of water evaporation, which is much larger than its molecular weight would indicate on a comparative basis with other H_2 compounds.

While the water referred to in the above paragraph is simple H_2O, with atomic weights of hydrogen and oxygen taken as 1.007 and 16, respectively, normal water supplies from precipitation contain a mixture of isotopic compounds. This is to be expected, since hydrogen has two known isotopes, and oxygen five, besides the normal forms. Of the many possible combinations of these isotopes, not more than five have been detected in waters from precipitation. These are $H_2{}^{16}O$, $H_2{}^{18}O$, $H_2{}^{17}O$, HDO, and D_2O, where D indicates the isotope of hydrogen with an atomic weight of 2, and the superscripts show the atomic weights of the oxygen isotopes. The concentration of most isotopic compounds other than $H_2{}^{16}O$ is very small, the largest being that of $H_2{}^{18}O$, which amounts to about 2000 ppm. Heavy water, D_2O, amounts to only about 200 ppm. In most of the calculations made for desalination it is sufficient to use the properties of normal water, $H_2{}^{16}O$, since it represents over 99% of the water present.

Two qualitative properties of water should be noted, namely, its very great stability as a chemical material and its solvent capacity. Water is at once the most plentiful material on the surface of the earth and the most stable. It can be evaporated, condensed, frozen, combined with other materials, and still return to its pure liquid form. The amount of water present on the earth and in the atmosphere is essentially the same today as it was millions of years ago. This may be contrasted with other common materials used by man, such as iron and other metals which corrode, fuels which are oxidized, and fissionable substances which split into two or more substances. These latter materials can be returned to their original states only by the expenditure of considerable energy and careful manipulation. The great stability of water makes it possible to use a great variety of processes for its desalination.

The second qualitative property is the solvent capacity of water. Water is capable of dissolving almost all inorganic chemicals to some extent, and especially so in its pure form. As a result, there is great difficulty in maintaining a high degree of water purity. For example, when very pure distilled water is distributed to work stations in a laboratory, there is noticeable reduction in its purity at the work station, due to the picking up of ions from the walls of the piping. This solvent capability is both the source of the need for desalination, and the means by which the unwanted chemical ions may be carried away from the desalting plant, using only a very small amount of water. For example, if half of the seawater entering a distillation plant is recovered as distilled water, the other half can absorb and carry away the unwanted salts without problems of precipitation.

2.4. Chemical Composition of Various Water Supplies

The composition of natural water supplies varies from that of almost pure water from rain or melting snow, through the salinities of water that has percolated through soluble components of soils to wells, and finally to that of the seawater into which all rivers flow. Table 1 shows the composition of a few types of water supply.

The entries in Table 1 for seawater refer to "standard" seawater,[1] which is recognized as the basis for investigations of scaling tendencies and other problems. Waters from various oceans differ slightly in both composition and total dissolved salts; that from inland seas and gulfs varies from a low of 7000 ppm in the Baltic Sea to 43,000 ppm in the Arabian Gulf. Also, the entries show only those ions which have concentrations of 1 ppm or more.

Table 1. Ionic Composition of Water Supplies
(in ppm)

| Dissolved ions | Seawater[a] | Saline water | | | Fresh water |
		Webster South Dakota wells	Coalinga California wells	Utah wells	Los Angeles California aqueduct
Cations					
Na^+, Sodium	10,561	92	500	685	40
Mg^{2+}, Magnesium	1272	102	72	139	6
Ca^{2+}, Calcium	400	220	120	266	26
K^+, Potassium	380		9		5
Fe, Iron		28			0.03
Mn^{2+}, Manganese		28			0.005
B, Boron			3		0.58
Total	12,613	470	704	1090	77.62
Anions					
Cl^-, Chloride	18,980	19	252	1940	19
SO_4^{2-}, Sulfate	2758	856	1170	33	23
HCO_3^-, Bicarbonate		331		67	
Br^-, Bromide	65				
NO_4^-, Nitrate			0.8		0.2
F^-, Fluoride			0.2		0.7
Total	21,803	1206	1423	2040	43
Total Dissolved Ions	34,416	1676	2127	3130	121

[a]Source: M. W. Kellogg Co.[1]

Some 40 ions have been identified in seawater, but those having concentrations less than 1 ppm are not listed in the table.

Three of the entries for saline water show the composition of waters in places where electrodialysis plants have been operated, and are representative of the water analyses with which this type of desalination plant has been most successful. The last column gives a water analysis for an existing natural potable supply of fresh water. The value of about 200 ppm for total dissolved salts is very nearly an average for acceptable potable water supplies, although some individual water sources show 50 ppm or less, while others approach the 500-ppm limit. It is noted that the ions of the major ionic quantities are the same in all the waters listed. This would be true of most water analyses, for the main variations in the compositions of saline waters are chiefly in the ions of very small concentrations, such as iron, manganese, boron, nitrates, and fluorides.

2.5. Scaling Problems in Saline Waters

In addition to the problem of separating fresh water from these solutions, which is the subject of this chapter, the chemical constituents and their concentrations are particularly significant as they affect the tendency for scale formation.[2] It is noted from Table 1 that seawater contains large quantities of ions that, upon heating, tend to produce three objectionable scale-forming compounds, namely, $CaCO_3$, $Mg(OH)_2$, and $CaSO_4$, in various forms. This occurs because these compounds have inverted solubilities, i.e., solubility decreases as temperature increases. Precipitation of these salts in distillation plants generally results in the formation of scale which adheres to heat transfer surfaces and increases resistance to heat flow.

The first two materials, $CaCO_3$ and $Mg(OH)_2$, result from the breakdown of bicarbonate ions in the saline solutions, according to the reactions:

$$2HCO_3^- \rightleftharpoons CO_3^{2-} + CO_2 + H_2O \rightarrow \tag{1}$$

Part of the carbonate ions may further break down as follows:

$$CO_3^{2-} + H_2O \rightleftharpoons 2OH^- + CO_2 \tag{2}$$

The Ca^{2+} and the Mg^{2+} ions combine with the negative ions to form

$$Ca^{2+} + CO_3^{2-} \rightleftharpoons CaCO_3 \tag{3}$$

$$Mg^{2+} + 2(OH)^- \rightleftharpoons Mg(OH)_2 \tag{4}$$

Reactions (3) and (4) tend to be in the direction of left to right at high temperatures, thereby enhancing the problem of scale formation. Prevention or elimination of the scale components can be accomplished by the addition of acid to the saline water, the reaction with the bicarbonate ion being as follows:

$$H^+ + HCO_3^- \rightleftharpoons CO_2(gas) + H_2O \tag{5}$$

The CO_2 gas can be eliminated by use of a degassing process.

The last material, $CaSO_4$, is more difficult to control, since this compound is not soluble in acids. The solubility of $CaSO_4$ in seawater is shown in Fig. 1. In this figure it will be noted that three crystalline forms of $CaSO_4$ can be distinguished under different conditions of temperature and concentration (the anhydrite form, $CaSO_4$; the dihydrate form $CaSO_4 \cdot 2H_2O$; and the hemihydrate form, $CaSO_4 \cdot \frac{1}{2}H_2O$), and that the anhydrite limits the maximum temperature allowable for standard seawater in a distillation plant. Since the concentration of the brine increases as it moves through the plant, it is important to keep the temperature–concentration relations such that all state points lie in the nonscaling parts of the diagram. The operating line for the 12-effect Freeport distillation plant[3] is shown in Fig. 1, roughly parallel to and slightly below the solubility limit line for the anhydrite. It must be noted that in plants of this type, where vapor is formed at the heat transfer surface, there may be very local areas where evaporation causes concentration to be much greater than average. To avoid scale formation under such conditions it is important to provide for forced circulation of the water across the heat transfer surface, thereby diffusing the locally concentrated liquid into the surrounding, more dilute solution.

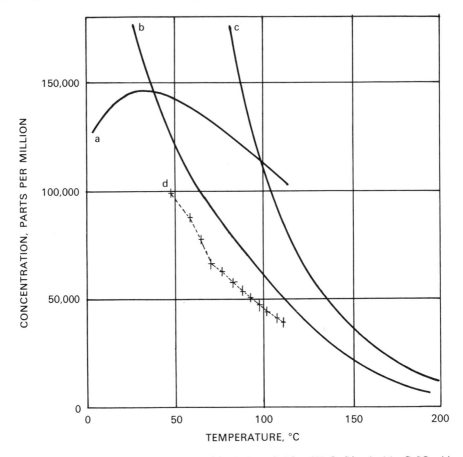

Figure 1. Solubility of $CaSO_4$ in seawater: (a) dihydrate $CaSO_4 \cdot 2H_2O$; (b) anhydrite $CaSO_4$; (c) hemihydrate $CaSO_4 \cdot \frac{1}{2}H_2O$; (d) operating line Freeport plant.

2.6. Elevation of Boiling Point due to Concentration

Besides contributing to the scaling tendency, the presence of dissolved salts in the water increases the boiling point, i.e., the temperature of vapor formation. That elevation of temperature is usually referred to as the boiling point rise (BPR). Because most of the solutes in saline water are ionized, the laws of ideal (nonionized) solutions do not apply, so it is necessary to rely on empirical data. The BPR for seawater has been determined by Fabuss and Korosi[4] for a wide range of temperatures and concentrations. When the BPR values are plotted against temperature with concentration as a parameter, the results show a linear relationship described by the following equation:

$$BPR = a + bT \qquad (6)$$

where a and b are the intercept and slope, respectively, of the lines for given

Table 2. Constants for Calculating Boiling Point Rise of
Seawater Solutions[a]

Concentration factor	(a)	(b)
0.4	0.0840	0.001193
1.0	0.226	0.003092
1.4	0.331	0.004442
1.8	0.440	0.00578
2.0	0.504	0.006579
2.4	0.628	0.008083
3.0	0.883	0.01045
3.6	1.055	0.01297

[a]Concentration factor is the ratio of concentration in ppm to 34,416, the ppm of standard seawater. (a) and (b) were calculated from data of Fabuss and Korosi,[4] for use in equation (6), with T in °C.

concentrations. Values of a and b are given in Table 2 for several concentrations, thereby providing a means for determining the BPR.

3. Economic Considerations

The primary obstacle to the desalting of water is its excessive cost. Fresh water has traditionally been one of the cheapest commodities, selling for only a few cents per ton. The cost of water from each of the desalination processes used or contemplated so far, is several times the price presently charged for natural fresh water in urban areas. Research and development efforts in the 1950s and 1960s have reduced the cost of distilled water per cubic meter from about $1.30, in 1950, to slightly less than $0.23, in 1974. By comparison, the cost of natural fresh water in United States cities is about $0.05. Continuing research and development efforts should further reduce this gap.

While the very low cost of natural fresh water cited above is typical of many urban areas in the United States, there are exceptions for which costs are much higher and desalination plants have been justified. One example is the city of Coalinga, California, where local wells furnished water of about 2100 ppm salinity. For potable supplies, fresh water was transported some 40 miles by railroad tank cars and distributed in a separate piping system, with the cost to the consumer being $1.85/m³. Installation of an electrodialysis plant in 1955 made it possible to satisfactorily desalt the local well water at a cost of only about $0.37/m³.

The situation in Key West, Florida illustrates the comparison between the cost of desalination and that of water delivered by a long pipeline. Key West, a small island at the end of a chain of islands, received its fresh water from a pipeline approximately 100 mi long. Population growth increased the demand far beyond the capacity of the pipeline. A study of the alternatives of either

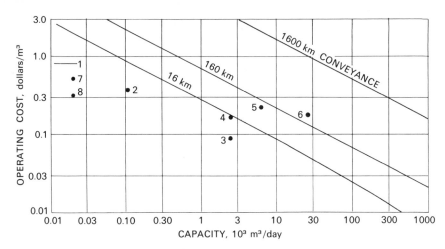

Figure 2. Operating costs of water supplies. (1) Basin type solar stills; (2) electrodialysis, Coalinga; (3) electrodialysis, Buckeye @ 98% load factor; (4) electrodialysis, Buckeye @ 38% load factor (actual); (5) distiller, Key West; (6) distiller, Rosarito Beach; (7) reverse osmosis, Coalinga, experimental plant; (8) reverse osmosis, Coalinga, production plant estimate.

constructing a new pipeline or installing a distillation plant concluded that the cost of distilled water would be less than that of fresh water delivered by a new pipeline. A primary reason for the high cost of the proposed fresh water pipeline was the comparatively small quantity of water involved, about 9000 m^3 per day.

The relation between plant capacity and the cost of fresh water has been studied by Koenig,[5] and is illustrated by the chart in Fig. 2, adapted from his report. In this figure, the cost per cubic meter of natural fresh water is plotted logarithmically as a function of plant capacity in cubic meters per day, with the distance over which the fresh water must be transported as a parameter. It is noted that the costs of fresh water are greater in smaller plants than in larger ones. Also shown on the chart are costs achieved at a few desalination plants. From this chart the conclusion can be drawn that desalination plants, in spite of their relatively high production costs, have the possibility of being competitive with natural water plants of small sizes, especially when fresh water must be brought from long distances.

4. *Cost Components*

The two major components of the cost of desalted water are the cost of energy and the cost associated with capital investments in the plant. These two components are often interrelated; for example, in multiple distillation plants any reduction in the energy required is accompanied by an increase in the complexity and physical size of the plant, thus increasing the capital cost. An economical plant design must therefore represent an optimization in which the costs of capital investment are balanced against those of energy. Since the cost of fuel varies from one place to another, the optimized design of a given size of plant is likely to be different for various locations.

In addition to reuse of heat in the distiller itself, there have been other efforts to increase the effectiveness of heat usage, and thus decrease the cost of energy by taking advantage of such means as the use of waste heat and the dual-purpose concept. In this latter concept, a distillation plant is combined with a steam–electrical generating plant, the objective being to use the exhaust steam from the turbogenerator as the source of heat for the distiller. This scheme must recognize the fact that distillers are operated with maximum temperatures of about 120–150°C, whereas steam turbines are operated with steam temperatures of 650°C or more. As a result, the amount of steam required by the turbine will be increased by this process, since the steam cannot expand in the turbine below the 120°C noted above; thus the hoped-for economy is partially offset. One method for evaluating the cost of the steam to be charged to the distiller is to measure the increase in steam necessary to produce in the dual plant the amount of the electricity desired, over that which would have been needed to produce the equivalent electricity in a normal condensing turbine operation. This quantity is obviously much less than the amount of steam received by the distiller, and so the cost will be less than for steam supplied to a separately operated distillation plant.

This scheme, while very attractive from the fuel aspect, has not been widely adopted because of its limited potential contribution to the water supply. In the first place, a dual-purpose plant, sized to meet the average electrical demand and operated continuously at maximum load, would produce only a portion of the average water demand. Second, the demand for electrical power on most systems fluctuates considerably, so that the average diurnal load may be only 50% of the turbine capacity. In contrast, distillers operate best at constant load, so that a distiller would have to be sized to correspond with the minimum electrical load on the power plant, thus reducing the water production still further. For these reasons it can be expected that only a small fraction of the total water demand will be produced by dual-purpose plants. However, from an environmental point of view, the dual-purpose concept should be used as fully as possible, since it decreases the total amount of fuel burned and decreases the amount of heat which must be disposed of at the end of the process.

Other schemes have been proposed for decreasing energy costs by using nonfuel sources of energy, such as the use of solar radiation, the use of geothermal heat, and the use of temperature differences betwen surface water in the oceans and the water at depths of 400 m or more below the surface. It is sufficient to note here that each of these has only limited potential and, of the three, only solar distillation has been used commercially for desalination.

5. *General Technological Problems*

As previously indicated, the economics of desalination plants are strongly influenced by the amount of energy needed and by the annual fixed charges for the capital investment. In attempting to reduce these costs, it is essential to inquire what limits there may be to such reductions. The limit for the energy required can be derived mathematically, and is the minimum energy of the separation of water from a saline solution. The limit for capital investment has to do with the estimated useful life of the equipment, and is influenced by items such as the resistance to corrosion of the materials used. Operational costs, other than those for energy, depend to some extent on the avoidance of scale deposition. All of these technological problems have been studied, so that a few summary comments can be made.

The very least amount of energy that an ideal desalination plant would require is the amount necessary to separate isothermally a very small amount of pure water from a large body of saline water. Figure 3 shows schematically a

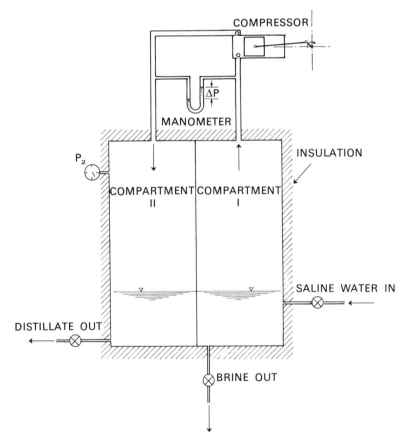

Figure 3. Ideal device for separating pure water from saline water.

device for accomplishing such a separation, assuming mechanical energy. Saline and fresh water in compartments I and II, respectively, are confined in a box which is perfectly insulated from the surroundings. The partition between the two is thermally conducting so that the temperature is the same in both compartments. All other gases are removed so that water vapor alone fills the volumes above the liquid. A manometer connecting the two compartments shows the pressure over the fresh water, in II, to be greater than that over the saline water, in I. A compressor, jacketed to carry away the heat of compression, is operated very slowly to compress the vapor isothermally and reversibly, from the low pressure in I to the saturated vapor pressure in II. Assuming the water valves to be closed, so that no liquid enters or leaves the system, the work required to compress and transfer a very small amount of water vapor from I to II can be computed from the equation

$$\text{work} = -RT\ln(p_1/p_2) \tag{7}$$

where p is the vapor pressure, T is the temperature in °K, R is the gas constant, subscript 1 is the saline water side, and subscript 2 is the fresh water side. The values for p_2 can be obtained from the pure water saturated vapor tables at temperature T, and for p_1 from the BPR tables or curves for the particular salinity and temperature.

Figure 4 shows values of the minimum work of separation for seawater solutions at 20°C. It will be noted that the minimum work required for separation increases as the concentration increases, i.e., the vapor pressure of I decreases so that the compressor must do more work in pumping vapor from I to II. It should also be noted that the differential pressure between I and II increases with temperature, so that there is a curve similar to that in Fig. 4 for each temperature. While Fig. 4 has been drawn for seawater solutions, it can be used as a first approximation for other saline waters.

The minimum work described above refers to a static situation, in which there is no flow of liquid and the amount of water vapor transferred from I to II is so small compared with the amounts of water in each of the compartments that there is no change in the salinity of the water in I. If pipe connections to the two compartments are made as indicated in the figure, and saline water is pumped in, brine is pumped out, and fresh water is removed, all at steady rates proportioned to maintain constant levels in the two compartments, it is possible to explore the effects of varying the recovery ratio, W_2/W_1, where W_2 is the rate of fresh water flow out of II and W_1 is the rate of flow of saline water into I. When W_2/W_1 is very small, the concentration of salts in I is very nearly the same as that in the incoming saline water, and the energy of separation is the same as that shown on Fig. 4 for the particular salinity. As W_2/W_1 is increased, the salinity of the water in I will be increased. Assuming water in this compartment to be thoroughly mixed, the compressor must do the work corresponding to the salinity in I, and which is given on Fig. 4. By using a concentration of seawater for the incoming saline water, W_1, values of minimum work have been determined for several values of W_2/W_1, the results

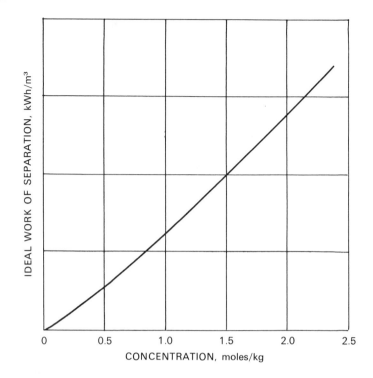

Figure 4. Ideal energy of separation of fresh water from saline water, showing effects of concentration.

of which, when plotted, form the curve labeled "internal mixing" on Fig. 5. This curve shows a rapidly increasing energy demand accompanying an increased recovery ratio, and represents the minimum work required when the entire separation takes place under conditions of a specific recovery ratio.

Another suggestion to further reduce the theoretical limit of minimum work of separation is to conceive a situation in which several sets of compartments, such as shown in Fig. 3, are placed in series, the brine effluent from one being used as the saline feed to the next one. The quantity W_2/W_1 is then the ratio of the sum of the pure water increments produced by all of the devices, divided by the saline water input to the first one in the series. When the number of devices is so large that each one produces an infinitesimal quantity of pure water, the total energy required will be minimized and will be equal to the average ordinate under the "internal mixing" curve between zero and the recovery ratio in question. The lower curve on Fig. 5 shows values of this quantity, and is labeled "differential separation." These two curves of Fig. 5 represent extremes between which would lie the idealized systems most nearly resembling practical desalting systems, since all practical systems have some internal mixing and none of them completely approaches the differential separation condition.

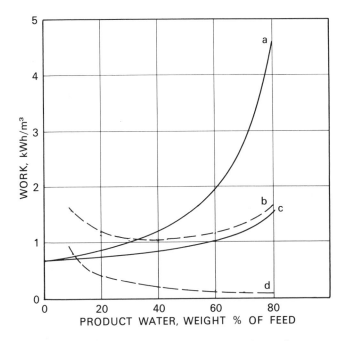

Figure 5. Ideal energy of separation of fresh water from saline water, showing effects of recovery ratio and pump work: (a) internal mixing; (b) differential + pump work; (c) differential process; (d) pump work.

In the above arguments the work of pumping the liquids into and out of the separatory systems has been ignored. As pointed out by several writers, this quantity could be very large for low values of recovery ratios, since extremely large amounts of saline water and brine would need to be pumped through the saline water compartment. Since this hydraulic work would be greater than the minimum work of separation, and decreases as the recovery ratio is increased, the sum of pumping work and separation work will have a minimum value at some specific recovery ratio. To illustrate this point, a curve of pumping work has been added to Fig. 5, and another curve plotted to show the sum of pump work and differential separation work. The minimum on this latter curve falls at about a 50% recovery ratio. The curve for pumping work in Fig. 5 is based on a head loss of 30 m and a pump with 100% efficiency. Other assumptions of head loss and efficiency would shift the minimum point slightly.

While the foregoing discussion has been based on a mechanical system shown in Fig. 3, the conclusions drawn can be generalized by noting that the pressure ratio in equation (7) is based on the use of the perfect gas law. For substances to which this law applies, the pressure ratio indicated is also the ratio of the activities of the water component in the two states. By convention, the activity of water vapor at saturated conditions is unity, so that equation (7) becomes:

$$\text{work} = -RT \ln a_1 \tag{8}$$

Table 3. Comparison of Energy Required by Desalination Plants with Ideal Minimum

Plant description	Actual energy / Ideal minimum
3-effect distiller, Kuwait (1954)	81
4-stage flash distiller, Kuwait (1958)	57
6-effect distiller, Aruba (1962)	48
Vapor compression distiller	29
36-stage flash distiller, San Diego (1962)	37
12-effect distiller, Freeport (1965)	29
24-stage flash distiller, Malta (1967)	22
Electrodialysis, Coalinga, California—2127 ppm (1956)	39[a]
Electrodialysis, Buckeye, Arizona—2101 ppm (1962)	55[b]

[a]Based on stack power only, assumed as half of total power used.
[b]Based on stack power only, not including energy for pumping.

where a_1 is the activity of water vapor in equilibrium with the salt solution. This equation, it is recognized, is a familiar expression for minimum energy, and applies regardless of the type of the energy involved. Thus it is relevant to systems using electrical, chemical, or heat energy as well as to the mechanical system already described.

The application of equation (7) to systems using heat energy can be achieved by using the temperature limits of the system for determining the efficiency of a Carnot (reversible) cycle between those temperatures. The required heat input for a distiller, then, is the minimum work from Fig. 5 divided by the Carnot efficiency.

The degree to which a few actual desalination plants approach the idealized systems to which Fig. 5 applies is indicated in Table 3. From this table it is observed that the ratio of actual input energy to the minimum energy of separation for an ideal separatory system is considerable, even in the most favorable cases, thus justifying the need for further research and development efforts.

6. Processes for Desalination

The desalination of water could presumably be accomplished by utilizing one or more of the many properties of saline solutions. Of the numerous possible processes that have been considered, a few presently have commercial status, and a number of others have been the subject of experimental work. As previously mentioned, those having commercial status include distillation in several forms, electrodialysis, and reverse osmosis. Ion exchange has certain special applications, such as the production of high-purity water for industrial purposes. Solvent extraction, freeze separation, and hydrate separation have been the subject of experimental investigation, but have not been applied in production plants.

The possible processes for desalination can be grouped under several different classifications. One classification would group them according to the phenomena involved, such as those utilizing a phase change of water as in distillation, freeze separation, and hydrate separation; those utilizing surface properties of membranes as in electrodialysis and reverse osmosis; and those using ion-selective properties of solids and liquids as in ion exchange and solvent extraction. Another possible classification is one specifying the type of energy used; for example, heat energy is used in several forms of distillation; mechanical energy is used in vapor compression distillation and reverse osmosis; electrical energy is used in electrodialysis; and chemical energy is used in ion exchange, solvent extraction, and precipitation. This classification has a direct relation to the cost of energy, heat energy being the least expensive and chemical energy the most expensive.

There is also another classification that distinguishes between the processes for which the energy required is the same regardless of the initial salinity, and those for which the energy required is strongly dependent upon the initial salinity. Grouped according to this classification are those processes which basically withdraw pure water from the saline feed and which require essentially the same energy regardless of the initial salinity; these include distillation, freeze separation, hydrate separation, and solvent extraction. The second category includes those processes that basically withdraw ions from the saline feed, such as ion exchange and electrodialysis; these require energy, proportional to the ions removed, which is directly related to the initial concentration of the saline feed. It has been found that when the quantity of ions involved is a small fraction of the product water, the energy required for their removal can be less than that required for processes in the first category above. Reverse osmosis is in a category lying between the other two, because it both withdraws water from the saline solution, and requires less energy for low initial salinities than for high ones. The presentation of the various desalination processes in this chapter will follow this latter classification.

6.1. Chemical Precipitation Process

As previously stated, there are no large-scale applications where desalination can be accomplished by precipitation at the present time, due to the very great solubility of most sodium and chloride compounds. A review of the previous emergency applications will illustrate the problem. In 1885, Kays reported experiments for the production of fresh water from seawater in life boats. It was proposed to store emergency chemicals in lifeboats in rubber-covered bottles, and the formulation recommended was four grains of free citric acid plus 960 grains of silver citrate for each pint of seawater (equivalent to 0.55 g of citric acid plus 132 g of silver citrate per liter of seawater). Kays reported that addition of these chemicals rendered the seawater "drinkable."

Recent analysis of such a chemical reaction shows that the addition of these chemicals precipitated the chloride ions and calcium ions, thereby reducing the total dissolved ions from over 34,000 ppm to about 15,000 ppm, still far above the acceptable 500 ppm limit.

A more modern formulation for such an emergency was that used on life rafts during World War II. Pellets containing silver zeolite and barium hydroxide were stored dry on the rafts. When required, these pellets were crushed and dropped into seawater contained in special flasks. After shaking and allowing time for the chemical reactions to take place, it was possible to suck the water out of the flask through a built-in filter. The chemical reactions that took place were the exchange of sodium ions with silver ions in the zeolite (thus withdrawing sodium ions), and the precipitation of silver chloride (withdrawing chloride ions) and barium sulfate (withdrawing sulfate ions). These reactions were capable of reducing the salinity from over 34,000 ppm to about 2000 ppm, a salinity regarded as low enough to prevent serious dehydration of the body.

As is noted, both of the foregoing formulations used silver ions, and thus were extremely expensive. While there are a few other chlorides of low solubility and a few relatively insoluble sodium compounds, none of these is sufficiently cheap to be used on a large scale for desalination. The case of an emergency on a life raft is one in which the cost could be justified in terms of human life.

6.2. Ion Exchange

The use of silver zeolite in the foregoing formula for precipitation of the chloride ions and removal of the sodium ions suggests the possibility of using an ion exchange process with some material other than silver as the exchangeable ion. On a small scale such a process has been developed for the "polishing" of high-purity water for industrial processes, such as boiler feed in nuclear reactor power plants. In these cases the exchangeable ions are hydrogen ions for positive ion replacement, and hydroxide ions for negative ion replacement. After ion exchangers become exhausted in this application, acids and bases are used for regeneration. In spite of the high cost of these regenerants, their cost can be justified because the need is great and the amount of ions to be removed is small. However, for large-scale potable water supplies such high regenerative costs are prohibitive. For example, the cost of the cheapest acid, sulfuric acid, is about $1.00/m^3, and the cost for the regenerant base at least as much. Because of the high cost of acids and bases for regeneration, there have been numerous efforts to reduce such costs with schemes by which the regenerant can be removed from the waste brine.

As indicated above, the principle of ion exchange depends upon the fabrication of ion exchange beads that have positive and negative sites which attach to the undesirable ions in the solution to be desalted. The nature of these beads is more fully discussed by Helfferich[6] and Kunin,[7] but it is sufficient here to understand the problem of keeping the beads active for the required ion exchange. There has not yet been satisfactory application of the regeneration of resin beads for the production of high-purity water. Of the several proposals, one was suggested by Gilliland,[8] who presented the idea of regenerating the exchanger with ammonium bicarbonate in a mixed bed operation. The unwanted ions in the saline solution would be left on the exchanger, leaving the product water containing the ammonium bicarbonate in solution. Upon heating the product water, carbon dioxide and ammonia gas would be released, collected, and redissolved, to be used over again for regeneration. This idea was tested by Stewart[9] on a laboratory scale, who concluded that the removal of the dissolved gases would require as much heat as would have been required to distill the product water in the first place. Another proposal was to use other regenerants which could be recovered by heat, but all have been found to use as much heat as would have been necessary for the distillation of the product water initially.

A more successful application of ion exchange has been proposed for the softening of seawater, particularly to prevent scale deposition as part of the pretreatment of feed water in a distiller plant, thereby permitting higher brine temperatures in those plants and reducing the energy required. The idea of this application is to remove calcium ions by regeneration with sodium ions from the effluent concentrated brine. Klein[10] studied this process and concluded that regeneration could be satisfactorily accomplished with a brine concentrated to four times the salinity of seawater. Klein's studies were based on the fixed bed concept of ion exchange. Kunz[11] studied the same type of softening process, with fluidized beds in which the "softened" effluent of the first bed is discharged as distiller feed. The exchange material of this bed, in fluidized form, is passed on to the second bed and is regenerated by the stream of concentrated brine running countercurrent to it. In this manner the inventory and cost of ion exchange materials was reduced. His studies showed that the brine temperature in seawater distillation plants treated in this manner could be as high as 160°C without scale formation. This temperature should be compared with the 120°C limit imposed by scaling tendencies in distillers without descaling treatment.

6.3. Electrodialysis

Electrodialysis is the desalination process in which the ions are withdrawn from the saline feed by the use of electrically charged plastic membranes. These membranes contain the equivalent of ion exchange sites and are made in pairs, with one membrane having positively charged sites and the other negatively charged sites. Under the influence of an electrical potential, positive ions can pass from site to site through the negatively charged membrane, but will be repelled from the positively charged membrane. Negative ions are repelled from the negatively charged membrane but can pass from site to site through the positively charged membrane. The structure of those membranes is tight enough that water transfer through the membrane by osmosis is kept to a minimum.

6.3.1. *Electrodialysis Membranes*

The membranes are the significant feature of this process.[12] The first electrodialysis membranes were made by grinding ion exchange beads into a powder, and dispersing this powder into an inert plastic binder which could be solidified in sheet form. This resulted in a heterogeneous membrane with sites of a given electrical polarity distributed through it more or less uniformly. Membranes are also made by polymerization techniques similar to those used in preparing ion exchange beads. Sheets of plastic film are produced by polymerization and the charged sites are added by chemical treatment of the film, resulting in a homogeneous structure with charged sites distributed throughout the membrane. Some films of this type have been produced in continuous strips with widths up to 1 m, while others have been cast between glass plates of sizes about 0.5 m by 1.0 m. A wide variety of substances have been used for the plastic film, including a copolymer of styrene and divinyl benzene, a polymer of phenol-formaldehyde, a copolymer of polystyrene and polyethylene, and cellophane. For the production of charged sites the plastic films have been subjected to processes such as sulfonation for cation membranes, and chloromethylation followed by amination for anion membranes. Since electrodialysis membranes may be subjected to considerable water pressure, the films must be quite strong. Some manufacturers provide the required strength by using relatively thick membranes, while others use fabric inserts made of glass cloth or of Dynel fabric. The useful life of electrodialysis membranes is about 5 years maximum, and may be appreciably less under unfavorable conditions.

6.3.2. *Principles of Electrodialysis*

The basic idea of electrodialysis is to use an electrical potential across a stream of saline water to cause the ions to migrate across the stream as carriers of the electrical current. By placing a pair of oppositely charged membranes in the stream parallel to the direction of water flow and perpendicular to the direction of the electrical potential, the ions can be removed from the portion of the stream between the membranes, due to the selective properties of the membranes. As the ions pass through the membranes they enter the portion of the stream adjacent to the electrodes. The electric current is carried by the migration of the ions through the water and membranes, but it is carried by electrons in the external circuit. Therefore it becomes necessary that a transfer from ions to electrons take place at the electrode surfaces by means of oxidation and reduction reactions. The reactions most generally accepted usually involve the production of gases to accomplish this transfer.

The usual reduction reaction at the cathodic electrode produces hydrogen gas according to

$$2H^+ + 2e^- = H_2(gas)$$
$$E_0 = 0.00 \text{ V} \qquad (9)$$

in which e^- is the electron and E_0 is the electrode voltage. So that the hydrogen ion will not need to come from the breakdown of water, it is usual to furnish it by acidification of the feed water entering the electrode compartment. At the anodic electrode several oxidation reactions are possible. Because corrosion of the electrode by oxidation is a prevalent problem, the anodic electrode is frequently made of highly corrosion resistant materials, such as tantalum plated with platinum, to minimize this tendency. When corrosion is minimized and the feed water contains chloride ions together with a low pH, the oxidation reaction releases chlorine gas according to

$$Cl_2 + 2e^- = 2Cl^-$$
$$E_0 = 1.36 \text{ V} \qquad (10)$$

When the amount of chloride ion is small, the pH low and a number of other negative ions present, the water tends to break down to produce oxygen gas according to the equation:

$$O_2 + 4H^+ + 4e^- = 2H_2O$$
$$E_0 = 1.23 \text{ V} \qquad (11)$$

Each of the gases noted above must be removed from the electrode compartments to prevent the formation of gas pockets. This is accomplished either by the venting of the compartments themselves, or by degassing the effluent brines, the former method being more certain.

6.3.3. Electrodialysis Apparatus

The apparatus to accomplish this electrodialysis separation of water is known as an electrodialysis stack. Since the magnitude of the electrode voltages are, individually, at least as great as the voltage drop across the two membranes of the pair and the water between, a single pair of membranes between electrodes would utilize approximately one-third or less of the total voltage in transporting ions out of the product stream. To offset this inefficient use of electrical energy, electrodialysis plants consist of stacks containing 275–500 pairs of membranes between a single set of electrodes; for by so doing the voltage required for the oxidation and reduction reactions at the electrodes can be made a very small fraction of the total voltage.

Figure 6 shows a schematic cross section of an electrodialysis stack, containing six pairs of membranes between electrodes. Saline water is shown entering all of the compartments at the bottom and flowing upward over the membrane surfaces, and through all compartments in parallel. Electrodes at

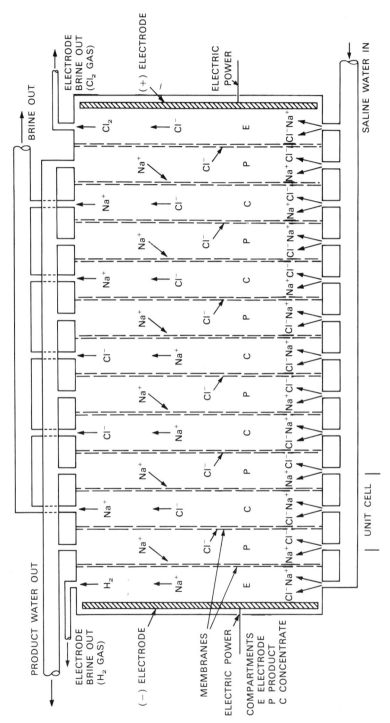

Figure 6. Electrodialysis stack—schematic only.

the ends of the stack are connected to an external power source, and distribute the electrical current over the entire active area of the membranes. Ions carrying current—exemplified by sodium and chloride ions—are shown as being repelled by one membrane and attracted to the other. Ions passing through most of the membranes are trapped in the concentrate compartments and carried out with the waste brine. The desalted feed is then removed as product water from alternate compartments. Passing through many pairs of membranes obviously saves energy, because the increased number of ions that pass out with the brine flow do not have to be oxidized or reduced and hence do not require the proportional voltage drop associated with these latter reactions. Ions passing through the end membranes and entering the electrode compartments are carried out with the electrode brines, or are oxidized or reduced. For instance, chloride ions may leave the anodic electrode compartment either as an ion or a gas, depending on whether these ions have participated in the oxidation reaction to produce electrons in the external circuit. If they have not, then water will probably break down to produce the required electrons. At the other end of the stack, hydrogen ions are shown as contacting the electrodes, thus participating in the reduction reaction and thereby absorbing electrons from the external circuit. After reduction, the ions are shown as leaving in the form of hydrogen gas.

The repeating unit in an electrodialysis stack is referred to as a cell, and consists of a pair of membranes, a concentrate compartment, and a product compartment. These are indicated in Fig. 6. While the figure shows the compartments to be much wider than the thickness of the membranes for purposes of clarity, the compartments are actually only about 1 mm thick and the membranes vary from 0.15 mm to nearly 1 mm, so that the thickness of the unit cell will be of the order of 2.5–3 mm. Gaskets fill the compartments between the membranes and are formed of sheets of polyethylene or another nonconducting plastic, continuous around the edges to prevent leakage and cut out to form the desired water passages across the membrane. Two general arrangements of water passage are in use, the sheet flow type and the sinuous path type. In the sheet flow type the polyethylene sheet serves as a gasket and has a large rectangular cutout in the center, leaving narrow strips around the periphery. Slits through two opposite edges permit water to flow into and out of the central cutout. A mesh or corrugated spacer of the same thickness as the gasket fills the central cutout area and serves to produce turbulence in the water flowing through it as well as to hold the membranes apart. The sinuous path type of spacer contains narrow cutouts which guide the water back and forth across the sheet, which has U bends at the edges that cause the water to travel as much as 4 m between inlet and outlet in some instances, even though the sheet is only 0.5 m square. Spacers across the narrow cutouts serve the same purpose as mentioned above.

The manifolds for conducting liquids into and out of the stack are shown in Fig. 6 as being external to the stack, again for the sake of clarity. However, in an actual stack, the manifolds are formed within the stack and consist of holes

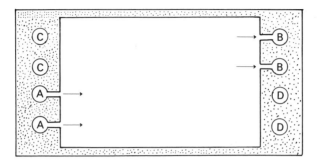

Figure 7. Sheet flow type spacer for electrodialysis—schematic
only.

punched through the membranes and spacers, as indicated in Fig. 7. This figure
shows a spacer gasket for a sheet flow type of cell. There are four circular holes
in each end of the spacer gasket, with slits connecting one pair of holes on each
end to the central cutout. The membranes each have similarly located holes but
no slits. When the slits are positioned as shown in the figure, water will flow
from A to B across the membrane surface. When the gasket is turned over, the
slits will be positioned to cause water to flow from C to D. Thus, if the product
compartments are served by the gasket in the position shown, the concentrate
compartments will be served with the gaskets turned over from the position
shown. With this scheme, the manifolds can be contained within the stack.
Outside of the end membranes and electrodes are iron or steel plates that are
squeezed together with long bolts or jacks to control or minimize leakage of
water from the edges of the gaskets. Holes in these end plates register with the
manifold holes in the membranes and connect to the piping external to the
stack.

6.3.4. *Performance in Electrodialysis Stacks*

The performance of an electrodialysis stack can be expressed in terms of
efficiency and capacity. The current efficiency, or coulomb efficiency, is the
ratio of the number of chemical equivalents of salt ions removed from the
product water to the number of Faradays of electricity used, or, in terms of
symbols,

$$\eta_c = \frac{W(C_1 - C_2)}{(I \times N/F)} \tag{12}$$

in which η_c is the current efficiency, W is the rate of flow of product water in
kg/sec, C_1 is the concentration of the saline water supply in gram-
equivalents/kg, C_2 is the concentration of the product water, I is the current in
amperes, N is the number of pairs of membranes, and F is the Faraday's
constant, 96,500 C per gram-equivalent. Values of η_c reported for existing
electrodialysis plants are over 90%, with some as high as 96%. It should be
emphasized that the current "efficiency" measures the effectiveness of the

current usage, and is not an energy efficiency because it does not include voltage.

The capacity of an electrodialysis stack can be specified in terms of the product water flow rate and the percentage of dissolved salts which can be removed in one pass through the stack. The product flow rate depends on the hydraulic resistance of the flow passages and the pressure provided. For example, one stack of 275 pairs of membranes is rated at 950 m^3 per day with a pressure drop across the stack of about 2.8 kg/m^2. The percentage of ions removed in a single pass through the stack varies from 40% to 60%, and depends for its exact value on the nature of the dissolved ions as well as on the design of the cell. With this limitation on the percentage of salts removed, it will be realized that saline water of more than 900–1000 ppm initial concentration must be passed through at least two stacks in series if the product water is to contain less than 500 ppm. For example, the Coalinga water described in Table 1, containing 2127 ppm, would require three stacks in series, and seawater, containing 35,000 ppm, would require seven stacks in series to produce water with less than 500 ppm.

6.3.5. *Limitations of Electrodialysis*

6.3.5a. *Resistance.* Since the conductivity of water is proportional to the amount of ions in solution, its reciprocal—resistance—increases as water becomes desalted, and is extremely high for desalted water. Thus, the resistance of the water in the product compartments of the stack increases as the water progresses across the membranes. Since the available voltage drop is uniform across the surface of the membrane, any large increase in resistance will decrease the current and hence the number of ions migrating through the membranes near the water outlet. By limiting the percentage of ions removed to the 40–60% mentioned, the effectiveness of ion transport through the membranes is maintained at a satisfactory level near the outlet side of the product compartments.

6.3.5b. *Polarization.* Another factor influencing the limitation of ion removal is the electrical current density. Since the amount of ion transport is proportional to the current through the stack, it is desirable to use as high a current as possible. However, the phenomenon of polarization places a limit on the maximum current which can be used under given stack conditions. "Polarization" in this situation is the substitution of hydrogen and hydroxide ions from water splitting for solute ions as carriers of electricity through the membranes. Water splitting occurs at the surfaces of the membranes when the number of solute ions arriving at the surface is insufficient to carry the current through the membrane, a condition which is related to the hydrodynamic flow conditions in the product compartment. The flow of water in these compartments is characterized by a turbulent core with viscous boundary layers adjacent to the membrane surfaces. Ions are delivered to the stream side of the boundary layer by the eddying motion of the turbulent core, but must be transported through the boundary layer by diffusion at a rate proportional to the local difference in concentration between the two sides of the boundary layer. Under given flow conditions, the transport of solute ions through the boundary layer is sufficient at low current densities. As current density is increased above a certain value, the transport of solute ions to the membrane surface becomes too small, and water molecules break down to furnish hydrogen or hydroxide ions that carry the electricity through the membranes. At the same time it is found that voltage must be increased disproportionately to achieve these high values of current density.

This phenomenon of polarization has been studied extensively by Cowan,[13] who originated a scheme of plotting which indicates quite clearly the presence of polarizing conditions. This scheme uses measured values of voltage drop and current through the stack, the vertical coordinate being voltage/ current, and the horizontal coordinate the reciprocal of current. For an electrodialysis stack this plot has the typical shape shown in Fig. 8, although this particular curve does not apply to any specific stack but is presented for illustration only. Beginning at the low-current, right-hand end of the curve, there is a negative, linear slope. As the current is increased, reading from right to left, it reaches a value of diminishing slope until the slope is zero at a minimum value of the ordinate. As the current increases further the line slopes upward toward the vertical axis. In order to seek an explanation for these

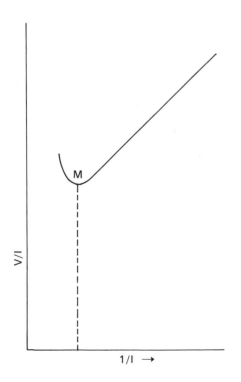

Figure 8. Cowan plot showing polarization in electrodialysis.

changes of slope, Cowan measured ionic concentrations and pH values in both product and concentrate compartments. He found that, beginning at the point *M* on the reciprocal axis and proceeding to the left, there was an increase in the hydroxide ions in the concentrate compartment; he took this as an evidence of the water splitting mentioned above. Further support of his conclusion comes from the observation that $Mg(OH)_2$ scale tends to form in the concentrate compartments of electrodialysis stacks when these are operated with certain waters at high current densities. Since the presence of hydroxide ion cannot be accounted for on the basis of the initial concentration and pH values, it is concluded that water splitting must have taken place. To avoid polarization it is standard practice to operate electrodialysis stacks at current densities well below the critical value, marked *M* on Fig. 8, and in some cases at values only 70% of *M*.

Because of the complicated nature of the flow path in the stack, it is necessary to estimate empirically the values of critical density. One of the many equations for such estimates is as follows:

$$i^*/C_2 \rightleftharpoons K e^{0.02T} v^{0.6} \tag{13}$$

where i^* is the critical current density in mA/cm^2; C_2 is the concentration of exit product stream in equiv/l; *K* is the constant, based on fraction of sodium ions; $e = 0.0677 + 0.1495$ (ppm of Na^+/ppm total); *v* is the velocity of product stream in cm/sec; and *T* is the temperature in °C.

Values computed by this equation represent the value at point *M*, and should be multiplied by 0.7 to obtain conservatively safe values of critical current density.

6.3.6. *The Buckeye, Arizona Electrodialysis Plant*

From the foregoing discussion of electrodialysis, it should be helpful to consider its practical application at the Buckeye plant. A description of the plant has been presented by Katz,[14] together with figures for water analyses before and after desalination. This plant was installed in 1962, with a rating of 650,000 gpd (2460 m^3/day) to serve as the entire city water supply. The well water contained total dissolved salts amounting to 2167 ppm, divided among the several ions as indicated in Table 4. After desalination the water contained only 398 ppm, with that ionic distribution also shown in Table 4. Since the specifications for the plant required the reduction of total dissolved salts to 500 ppm, containing not more than 250 ppm of chloride ion, the plant obviously met the conditions. The plant consisted of six Ionics Mark III stacks arranged in two stages in series, with three stacks per stage. These stacks have water flow paths of the sinuous type, in which the water crosses the approximately square membrane surface eight times in going from the inlet manifold opening to the outlet manifold opening. The manifolds are located across the center of rectangular membranes and gaskets, thus providing in each stack double squares of membrane pairs served by a single central manifold system.

Table 4. Analysis of Water at Buckeye, Arizona[a]

	Saline water		Desalted water	
	ppm	eq/kg	ppm	eq/kg
Na^+	670	0.0291	145	0.0063
Ca^{2+}	128	0.0064	9	0.0004
Mg^{2+}	16	0.0013	1	0.0001
Positive ions	814	0.0368	155	0.0068
Cl^-	1120	0.0316	207	0.0058
SO_4^{2-}	161	0.0034	12	0.0003
HCO_3^-	66	0.0011	22	0.0004
NO_3^-	4	0.0001	1	0.00002
F^-	1.84	0.0001	0.88	0.00005
Negative ions	1352.84	0.0363	242.88	0.0066
Totals	2167	0.0731	398	0.0134

[a]Values are from W. E. Katz[14] converted to the metric system.

The principal design characteristics of the standard version of these stacks is given in Table 5. It should be noted that the stacks supplied to Buckeye were rated at $820 \text{ m}^3/\text{day}$ rather than the standard $950 \text{ m}^3/\text{day}$. Therefore it is estimated that the Buckeye plant had fewer membrane pairs than the standard 275 pairs, probably not more than 240 pairs, based on the reduced flow.

Katz[14] states that the saline feed to the three stages is divided between the product and concentrate compartments, with 25.2 1/sec entering the product compartments, and 12.6 1/sec going into the recirculating concentrate stream. The demineralization indicated in Table 5 would effect a reduction of 57.2% in each of the two stages, with the resulting concentration of the product water leaving the first stage having a value of 0.0313 equivalents/kg.

Table 5. Characteristics of Standard Ionics Mark III Stack[a]

Maximum hydraulic flow rate, per stack	$950 \text{ m}^3/\text{day}$
Pressure drop at maximum flow	2.8 kg/cm^2
Number of pairs of membranes	275
Size of membranes	$45.7 \text{ cm} \times 101.6 \text{ cm}$
Total area of membranes	255 m^2
Percent of area available for ion transport	70%
Demineralization	
At full flow	48%
At low flow	57%
Current: full flow, 3000-ppm feed	100 A
Voltage: full flow, 3000-ppm feed	1000 V

[a]This information has been taken from W. E. Katz[14] and converted to the metric system.

Using the above values, the following calculations can be made:

1. *Electric current required in each stack*

$$I = W(C_1 - C_2) \times 96{,}500/(N \times \eta_c) = \text{current in amperes}$$

where W is the flow rate in kg/sec (8.4 for each stack); C_1 is the concentration of saline water entering, in equiv/kg (0.0730); C_2 is the concentration of product water leaving, in equiv/kg (0.0313); N is the number of membrane pairs (240); and η_c is the coulomb efficiency (estimated 95%). Therefore,

$$I = 8.4 \times (0.0730 - 0.0313)/2 \times 96{,}500/(240 \times 0.95) = 74.1 \text{ A}$$

It should be noted that the factor of 1/2 applied to the concentration is necessary because one Faraday of electricity will transfer one equivalent each of positive and negative ions.

2. *Current density*

$$i = I/A = \text{A/cm}^2$$

where A is the membrane area $(45.7 \times 101.6 \times 0.70)$. Therefore,

$$i = 74/(45.7 \times 101.6 \times 0.70) = 0.0228 \text{ A/cm}^2 = 22.8 \text{ mA/cm}^2$$

3. *Electric energy to stack*

$$E_s = I \times V/(W_2 \times 1000)$$

where V is the voltage (1000 V), I is the current (74.0 A), and W_2 is the product flow rate in m^3/hr $(8.4 \times 3600/1000)$. Therefore,

$$E_s = 74.0 \times 1000/(8.4 \times 3600) = 2.45 \text{ kWh/m}^3$$

4. *Electrical energy to plant*

$$E = 2E_s + E_p$$

where E_s is the energy per stage, and E_p is the pumping energy. Therefore,

$$E = 2 \times 2.45 + 0.72 = 5.62 \text{ kWh/m}^3$$

It is noted that energy required for the second stage is the same as for the first stage, because the change in concentration, expressed in equivalents, is the same in the two stages. The pumping energy was estimated on the basis of the total flow and a pressure drop of 2.8 kg/m^2 in each stage, using the usual hydraulic power calculations.

The plant at Buckeye occupies a building approximately 10 m by 20 m in area. Each stack measures about 0.5 m by 1 m in plan area, and is a little over 2 m high, since it is mounted on a pedestal with piping underneath. Besides the six stacks and several pumps, there is a filtration unit containing 10-μ fiber filter cartridges, and a rectifier unit for converting the local 60-cycle AC electrical power to the direct current required by the stacks. The product water leaving the plant is pumped to the 1000-m^3 town reservoir, and the waste brine to the town's waste treatment plant. This arrangement of equipment is typical of that found in any electrodialysis plant for desalting water.

6.4. *Reverse Osmosis*

Reverse osmosis, as a process of desalination, logically follows electrodialysis since it also uses plastic membranes. It is a process which has some of the characteristics of both of the two categories, i.e., the energy required is affected by the initial salinity and it is a process in which water is removed from the solution.

6.4.1. *Osmotic Pressure*

Osmosis is a natural process in which the water component moves through a membrane separating two solutions of different concentrations. As indicated earlier, it is the process by which water is absorbed into all living organisms. The classical demonstration of osmosis involves two containers filled with solutions of different concentrations, the container filled with the more concentrated solution being closed and fitted with a pressure gauge, and the other, filled with pure water, being open to the atmosphere. The two containers are connected by a tube across which is an osmotic membrane which holds the two liquids apart. At zero time the pressure gauge is attached and shows a reading of zero. As time goes on, the reading of the gauge increases, the water level in the open container falls, and that in the closed chamber rises. These changes continue until the pressure gauge ceases to rise and the water levels remain constant, at which point the pressure indicated is the osmotic pressure for the concentration of the solution in the closed container at that time. If a source of pressure is now connected to the closed container and the pressure is increased still further, there will be a flow of water through the osmotic membrane back into the open container. In this case the osmotic flow has been reversed, illustrating the basic scheme for desalination by reverse osmosis. Since osmosis normally causes water to flow from the less saline solution into the more saline solution, the desalting process, which withdraws water from a saline solution, must be operated at pressures greater than the osmotic pressure difference between the saline source and the product water.

Since the desalination process requires determination of the osmotic pressure of a saline solution, it is important to understand its magnitude and variations. The osmotic pressure can be computed from the equation

$$\pi \bar{V}_1 = RT \ln (p_2/p_1) \tag{14}$$

where π is the osmotic pressure, $\bar{V}_1$ is the partial molal volume of the water component, R is the gas constant, T is temperature, p_2 is the saturated vapor pressure of pure water, and p_1 is the vapor pressure over the solution. This equation is valid for dilute solutions, such as those of interest in desalination. Values for p_1 can be obtained from the BPR as in the calculations of minimum work. It will be noted that the right-hand member of equation (14) is similar to

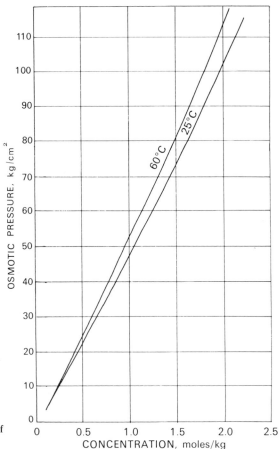

Figure 9. Osmotic pressure of NaCl and sea-salt solutions.

that in equation (7), which is a result of the fact that both equations can be derived using reversible isothermal processes.[15] Values of the osmotic pressure of seawater solutions and sodium chloride solutions have been determined and are plotted in Fig. 9, in which values of osmotic pressure are plotted against concentration. It will be seen that the osmotic pressure increases almost linearly at low concentrations and then rises more rapidly, giving the curve a concave upward shape. The curves are for two temperatures, covering the range of temperatures commonly found in reverse osmosis plants. The data for both seawater and sodium chloride fall on a single line, probably because sodium chloride is the chief component of the solutes in seawater. These curves can be used as first approximations for the osmotic pressures of most natural saline waters.

6.4.2. Membrane Fabrication

While the natural osmotic process has been understood for many years and has seemed a logical process to apply to the desalination of water, there

were no known membranes, either from organic sources or synthetic formulations, which were sufficiently strong and permeable for the purpose. In the
1950s, Reid[16] showed that cellulose acetate film would exclude most of the ions
in saline waters, but concluded that the material was not sufficiently porous to
water to be used satisfactorily as an osmotic membrane. Loeb and Sourirajan[17]
soon afterwards found that cellulose acetate membranes could be made porous
to water by including a leachable material, e.g., magnesium perchlorate, in the
casting solution when the membranes were made, and then leaching out the
material after the membrane was cast. Their disclosure of this scheme, in 1960,
set off a wave of experimentation with these membranes and led to the present
situation in which there are several commercial desalters of the reverse osmosis
type.

The Loeb–Sourirajan membranes were cast from the following solution:

Cellulose acetate	22.2% by weight
Magnesium perchlorate	1.1% by weight
Acetone	66.7% by weight
Distilled water	10.0% by weight

The membranes were cast on glass plates under carefully controlled temperature conditions, between $-8°C$ and $-11°C$, exposed to the air for 3–4 min,
soaked in ice water for periods up to 1 hr, then removed from the casting plate
and soaked in hot water, from $79°C$ to $82°C$ (for seawater), and finally stored in
water at room temperature until used. For use in waters of other salinities, the
temperatures of the hot water were varied. Studies of the structure of these
membranes disclosed the fact that they were porous to water but had a very
dense skin on the air-dried side which served to exclude the salt ions. The other
side, next to the glass plate during the casting, was quite porous and would not
exclude ions if placed on the saline water side. It was also shown that the
magnesium perchlorate was entirely leached out during the two soaking
periods.

While these membranes were used for much of the early development
work, Loeb and co-workers[18] continued exploring various possible materials
and disclosed another formulation which could be cast at room temperature
and was therefore much more satisfactory. They showed that a solution
containing

Cellulose acetate	25% by weight
Acetone	45% by weight
Formamide	30% by weight

could be cast at $23°C$ to give a very satisfactory membrane. They also showed
that the proportions of the ingredients could be varied considerably—cellulose
acetate from 20% to 30%; acetone from 35% to 60%; formamide from 10%
to 40%; thus giving membranes with a range of porosities and selectivities. This
type of membrane is now widely used. Other formulations have been originated by various inventors and are also in use.

6.4.3. Performance

The two principal performance characteristics of membranes used for desalination by reverse osmosis are the flow rate of water per unit area, generally referred to as the flux rate, and the selectivity, or the degree to which ions are excluded. Membranes made of the materials described above are generally of thicknesses between 100 μ and 250 μ, and show flux rates in the range 400–2000 liter/m^2/day, for membranes in flat sheets or in tubes 1 cm or more in diameter. It has been found that none of the membranes is completely selective, i.e., excludes all of the dissolved ions, but that they can be expected to exclude a large fraction, up to 90–95%. The flux rate of water through a membrane depends upon the exact formulation and heat treatment of the membrane as well as upon the pressure drop across the membrane. The membrane's selectivity, or its reciprocal, the flow of ions through the membrane, are also both affected by membrane preparation, as well as being strongly dependent on the difference in concentration between the two sides of the membrane. The flow rates of water and ions are found to be related, since increasing one by changing the membrane preparation causes the other to increase as well. This interdependence of flow rates places limitations on the degree to which the water flux rate can be increased by changing the formulation and heat treatment of the membrane. While a further discussion of this problem is beyond the scope of this chapter, the subject, discussed from the standpoint of coupled flows involving irreversible thermodynamics, is dealt with by Lonsdale.[19]

6.4.4. Polarization

The water flux rate through the membrane is further affected both by the formulation and heat treatment of the membrane and by the phenomenon of polarization. Polarization in reverse osmosis is the serious reduction or cessation of water flow through the membrane due to increased local concentrations of ions which produce counteracting local osmotic pressures very close to the applied pressure. This phenomenon is due to the effects of the boundary layer resistance to the diffusion of ions. The flow of water past the membrane surface is characterized by the existence of a turbulent core with a viscous boundary layer next to the membrane surface. As water is forced through the membrane, ions are left behind and must pass through the boundary layer to get to the turbulent core and be carried away. Passage through the boundary layer takes place by diffusion, is controlled by the local concentration difference between the two sides of the boundary layer, and is therefore slow. Since the saline water becomes more and more concentrated as it travels along the membrane surface, the concentration differential across the boundary layer becomes smaller, reducing the rate of diffusion of the ions and allowing the local concentration next to the membrane to be built up. This increase in concentration corresponds with an increase in osmotic pressure and a resultant decrease in local flux rate, or polarization. To minimize this effect, velocities of water flow are kept as high as possible to reduce the thickness of the boundary layer. This is in contrast to polarization in electrodialysis, in which polarization results from a scarcity of solute ions next to the membrane surface; in reverse osmosis, it results from too many solute ions next to the membrane surface.

6.4.5. *Membrane Assemblies*

While the first membranes were cast in sheets and used in devices patterned after the plate-and-frame filter press, it was soon realized that tubular construction would have advantages over flat membranes when it came to problems of sealing, etc. However, the flat membrane seemed to lend itself better to quantity production methods, so that one of the devices in use today combines the advantages of flat membrane fabrication with a tubular assembly. This device, known as the "spiral-wound blanket assembly," was developed by the Gulf General Atomics laboratory and is being used in a number of plants. As shown in Fig. 10, two long strips of membrane, with the saline water side outermost, are sealed around two sides and one end, enclosing a sheet of porous material. The unsealed end is then sealed to a pipe in which holes are drilled to lead to the porous sheet between the two membranes. This blanket of membranes and porous sheet is then wrapped around the central pipe, together with a sheet of screen of the same area, the screen serving as a spacer between the turns of the membrane blanket. The entire assembly is then slipped inside a tight-fitting larger pipe which serves as a pressure vessel. In operation, saline water is pumped into one end of the larger pipe and circulates across the membranes in the spaces of the screen sheet. Desalted water, passing through

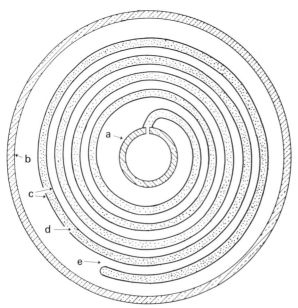

Figure 10. Reverse osmosis desalter—spiral blanket type (schematic cross section): (a) product tube; (b) casing; (c) membranes; (d) porous material for conveyance of desalted water; (e) this space kept open for flow of saline water by mesh (not shown).

the membranes, is collected by the central pipe and conveyed out of the device. Practically, several of these spiral-wound units are placed end to end in a single outer casing, so that saline water passes from one of these assemblies to the next. This type of reverse osmosis desalter has been quite successful and is represented by a number of commercial installations.

The concept of a tubular membrane, rather than the spiral-wound blanket mentioned above, has appealed to a number of manufacturers as a more direct way of making use of the inherent advantages of tubular construction. One of the first concepts was that of the porous tube, made of glass fibers and lined with a cellulose acetate membrane cast within the supporting tube after the tube had been formed. Originated by Havens,[20] this idea was developed and several commercial plants constructed. Tubes 1.25 cm in diameter were wound with glass fiber filaments and bonded with resin, the resulting tubes being quite porous and yet capable of supporting a membrane tube with pressures up to 100 kg/cm². These fiber tubes were then passed through a machine which coated the interior with a reverse osmosis membrane. The major problem of this design was the sealing of the ends of the tubes in such a manner that saline water from the insides of the tubes could not leak into the fresh water which had passed through the membrane lining and dripped from the outside of the glass fiber tube. This problem was solved by Havens by assembling shell-and-tube devices in which the saline water flowed through the membrane-lined tubes under pressure, and the fresh water dripped into the shell and was there collected.

Other manufacturers have developed schemes for casting membranes in tubular form, and then inserting these membrane tubes into metallic or plastic tubes for supporting the membrane under pressure. An example of this type of construction is that used by Loeb[21] in assembling a 20-m³/day plant at Coalinga, California, for testing membrane formulations. As shown in Fig. 11, the tubular membranes, approximately 25 mm in diameter, were wrapped in several layers of a porous nylon cloth and inserted into copper tubes, the latter having been prepared by drilling small holes through the tube walls at intervals

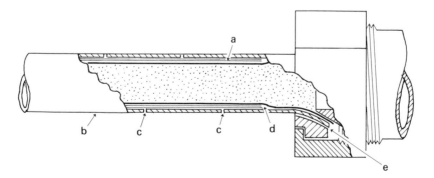

Figure 11. Tubular-type reverse desalter—schematic only: (a) porous liners; (b) copper tube; (c) drip holes; (d) membrane tube; (e) end seal by flare fitting.

of about 7.5 cm along the length. End sealing of the membranes was provided by the use of flared fittings, the membrane being clamped within the flare. In operation, saline water is pumped into the membrane tube and desalted water is forced through the membrane into the porous liner, through which it passes to the holes in the copper tube and drips into an outer casing. The Coalinga plant contained 112 tubes of 3.05 m length, giving a membrane area of approximately 37 m². This plant operated for some 5 years and produced an average 20 m³/day. Various manufacturers have produced tubular units with different schemes for end sealing and other details of construction, so that the above description is intended only to point out the problems and to represent one of several possible solutions.

Still another variation of tubular construction is the self-supporting tubular membrane. Where other designs had depended on the use of metallic or plastic tubes for strength to support the membranes, Mahon[22] pointed out that capillary tubes of cellulose triacetate would be self-supporting in sizes of from 70–100 μ outside diameter, with wall thicknesses of 5–10 μ. More recently, the DuPont Company has been producing reverse osmosis desalting devices made of hollow nylon fibers with 45 μ outside diameter and 24 μ inside diameter. As described by Cooke,[23] these devices are exemplified by an assembly some 3 m long and 36 cm in diameter, in which are 28 million fibers of the above diameters. These fibers make return bends, with the free ends cast into a block of epoxy. After casting, the epoxy is machined away on the face away from the fibers to expose the holes in the fibers. The fibers are then placed in a steel casing and the epoxy block sealed into the end of the casing. Saline water is pumped into the casing the desalted water issues from the internal holes in the fibers and is collected in a cap over the epoxy header. These devices are referred to as "permeators," and the size noted above is said to contain 7850 m² of membrane area and to produce 22.7 liters/min of desalted water when pressurized to 42 kg/cm².

6.4.6. Cost of Desalted Water

Relatively few data have been published regarding the cost of desalted water from plants using reverse osmosis. In one of the few such reports, Johnson et al.[24] studied the data from the 20-m³/day Coalinga plant after it had been operated for $3\frac{1}{2}$ years. From this study they concluded that the product water from a production plant of the same size and design should have cost about $0.31/m³, with the following component items:

Capital charges	$0.11 per m³
Membrane replacement	0.07
Maintenance	0.09
Operation, including power and filters	0.04
Total	$0.31

Due to the experimental nature of the plant, and the need for frequent gathering of data, the actual operating cost was nearly $0.51/m^3$. Further reductions in water cost from production plants could be expected as a result of an increase in size. The above report included an estimate of $0.13/m^3$ for water from a plant of the same design but having a capacity of $200 \text{ m}^3/\text{day}$, rather than $20 \text{ m}^3/\text{day}$ as at Coalinga. The development of this method of desalination is being pursued vigorously, and plants capable of producing $600–1000 \text{ m}^3/\text{day}$ are being planned, so that significant reductions of product water cost can be expected.

6.5. *Distillation*

Attractive as they may seem because they are sensitive to the initial concentration of the saline water, electrodialysis, reverse osmosis, and ion exchange have not yet displaced distillation as a method for obtaining fresh water from seawater or other highly saline sources. All of the large plants for desalting seawater use some form of distillation. In general, all forms of water distillation plants include the three basic elements of the hydrologic cycle: the production of water vapor due to heating, the transport of this vapor to a cool region, and the condensation of the vapor by the removal of heat. Since there are numerous ways in which each of the elements can be accomplished, there are several different forms of distillation, those to be discussed here being solar distillation, multiple-effect distillation, multiple-stage flash distillation, and vapor compression distillation.

6.5.1. *Solar Distillation*

While it is one of the less important schemes for distilling water from the standpoint of worldwide plant capacity, solar distillation applies the basic elements noted above in a straightforward manner, and serves as an effective introduction to distillation. The basic type of solar still consists of a shallow basin of water covered with a sloping transparent cover, as shown in Fig. 12. Solar radiation passing through the transparent cover is absorbed by the black surface of the basin liner and serves to heat the water in the basin. Vapor is formed when the water temperature is greater than the temperature of the transparent cover, and leaves the water surface as humidity of the air above the surface. This humidified air, being less dense than the cooler air near the edges of the basin, rises and comes into contact with the transparent cover. Due to its transparency, the cover is cooler than both the water and the rising humid air, and therefore condenses part of the humidity as product water and cools the rest of the air. As it cools, the air descends along the sloping cover and returns to the surface of the water, thus establishing a convection current. In this device solar radiation supplies heat for the vapor formation, convection currents provide transport, and the transparent cover serves to condense the vapor, thus accounting for the three basic elements noted above.

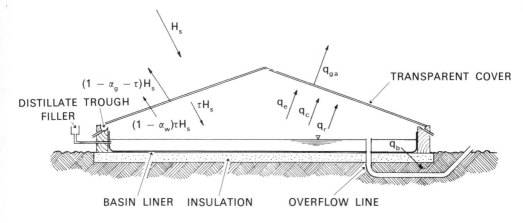

Figure 12. Basin-type solar still—schematic cross section.

This basin-type still is a single-effect device, since the heat from solar radiation is used only a single time and then dissipated from the transparent cover. The effectiveness of heat usage is further limited due to the humidification and dehumidification of the circulating air. Since this dry air and its associated uncondensed vapor are also heated and cooled, it transfers heat from the water to the transparent cover without this heat being used for humidification. This loss due to circulating air may be as low as 4% of the solar radiation collected at water temperatures near 65°C, but increases to about 20% for temperatures below 40°C.

This basin-type solar still is most advantageous in small sizes, especially in places where fuel must be imported at considerable cost, such as on sparsely populated tropical islands and isolated homes or small communities in remote arid places. The amount of water evaporated in a solar still is proportional to the amount of solar energy entering the still. Because the solar radiation intensity varies from hour to hour during the day, the solar still operates at a variable rate, with water production beginning in the morning after sufficient radiation has been absorbed to heat the water to a temperature greater than that of the transparent cover. Production of vapor continues after sunset until the water temperature again drops below that of the cover, at which times the production decreases to zero. In addition to diurnal variation, solar radiation varies with the seasons, so that a solar still has a maximum daily production rate in the summer and a minimum in the winter. If solar stills are the sole producers of fresh water, sufficient distilling and storing capacities must be provided to furnish water as needed in spite of the annual and diurnal variations in production rates.

The estimation of the diurnal production of a solar still is based on a heat balance between the incoming solar radiation and the various ways in which heat energy is dissipated from the still. Arrows have been added to the cross section of a still in Fig. 12 to show the directions of energy flux on which such a heat balance would depend. Various writers have proposed schemes for relating the several quantities involved, a good summary being that presented by Talbert *et al.*[25] Of all the schemes listed by Talbert, that originated by Morse and Read[26] seems to be the best adapted to simple calculations, and is recommended for those who wish to go into the subject in greater detail. It is sufficient for the purposes of this chapter to consider three of the seven equations involved in their analysis, which are as follows: Overall heat balance,

$$\alpha_g H_s + \alpha_w \tau H_s = q_{ga} + q_b + c_{wg}(dT_w/dt) \tag{15}$$

heat balance on the transparent cover,

$$q_{ga} = q_r + q_c + q_e + \alpha_g H_s \tag{16}$$

and heat transfer due to evaporation and condensation inside the still

$$q_e = 0.0254\left[(T_w - T_g) + \frac{P_w - P_{og}}{39 - P_w}(T_w + 460)\right]^{\frac{1}{3}} (P_w - P_{wg})h_w \tag{17}$$

The symbols used in these equations are defined as follows: H_s is the incoming solar radiation, α_g is the absorptivity of transparent cover for radiation, α_w is the absorptivity of water basin for radiation, τ is the transmissivity of transparent cover for radiation, q_{ga} is the heat transfer between transparent cover and surroundings, q_b is the heat loss through base and edges of still, c_{wg} is the equivalent heat capacity of still structure, contents, and bottom insulation or ground, T_w is the temperature of water, t is time, q_r is the heat transfer by radiation between water and cover, q_c is the heat transfer by convection between water and cover, q_e is the heat transferred from water to cover by evaporation and condensation, T_g is the temperature of transparent cover, P_w is the vapor pressure of water at the water surface, P_{wg} is the vapor pressure of water at the cover surface, and h_w is the latent heat of evaporation.

The production rate is proportional to the quantity q_e so that this quantity should be made as large as possible to achieve maximum rates. From the above equations the conditions for maximum q_e would include the following observations:

1. The transparent cover should absorb very little radiation and transmit as much as possible, thus involving the properties of the glass or other material, i.e., α_g must be small and τ large.
2. The heat loss to the ground and from the edges of the basin must be minimized, thus suggesting insulation of these parts, i.e., q_b must be small.
3. The heat capacity of the still basin and contents, c_{wg}, should be small, so that a rapid increase in temperature will occur in mornings to start production. This implies that the structure should be as light as possible, and of low specific heat materials, and further that the amount of water should be small.

From equation (17), q_e will be maximized when the temperature difference between the water and transparent cover is greatest. This temperature depends on the convection and radiant heat transfers between water and cover, as well as on conditions outside the still.

Daily production rates are plotted against solar radiation intensity in Fig. 13 for several basin-type solar stills of various designs. These curves represent clear-sky-performance data, with cloudy-day-performance data lying below the curves. Also plotted on Fig. 13 are lines of constant efficiency, the latter being defined as the ratio of the product of the distillate rate and the latent heat of the vapor, to the incident solar energy. It will be noted that the performance curves cross the efficiency lines, indicating that the still operates with increasing efficiency as solar radiation intensity increases. That this should be true follows from the fact that the temperature inside the still increases with the solar radiation intensity and causes the vapor to be formed at higher temperatures and partial pressures. As vapor pressure rises, the amount of air circulating with the vapor decreases, and hence the loss due to air circulation is decreased.

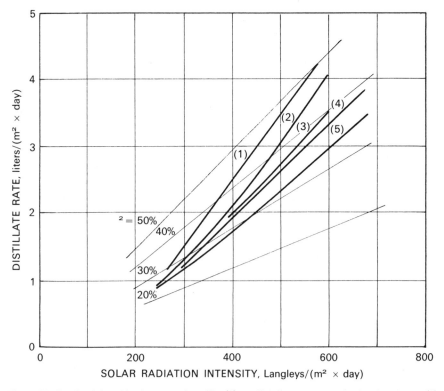

Figure 13. Productivity of basin-type solar stills: (1) small, asbestos cement basin, glass cover; (2) large, deep basin, asphalt basin liner, glass cover; (3) large, butyl basin liner, inflated plastic cover; (4) medium, plastic basin liner, glass cover; (5) large, plastic basin liner, glass cover.

Solar stills are expensive to construct, for they cover large areas. Designers have used a wide variety of materials in trying to cut down costs: glass and sheet plastics for the transparent covers; wood, concrete, and sheet metal for cover supports; asphalt, polyethylene, and butyl rubber sheets for basin liners. Another choice is that relating to the geometrical shape of the still. Most designers use the symmetrical "greenhouse" form of Fig. 12, but some prefer a single-sloped form, and others an "inverted Vee" form with the distillate collection trough running along the center line.

Other measures for increasing the efficiency of usage of solar energy have been suggested, all of which involve the collection of solar heat separately from the distillation process, in which case the solar radiation is used to heat water without evaporation, and this warmed water is then used as the source of heat in a conventional distillation plant. In this case the determination as to whether or not the process is economical will depend strongly on the cost of solar heat collection.

6.5.2. *Multiple-Effect Distillation*

Most large distillation plants use heat energy from fuels rather than from solar radiation because the product water is cheaper than in the smaller capacity solar plants, and can be depended on to produce to their designed capacities regardless of the weather. In order to keep down the costs of operation of these plants, it is essential that the cost of fuel, the major component of cost, should be minimized. Several design schemes are in use for this latter purpose, the oldest of these being known as multiple-effect distillation.

In multiple-effect distillation, the heating effect of steam produced from fuel heat is reused, thus multiplying its effectiveness several times as compared with single-effect devices. This is accomplished by placing several evaporator–condensers in series, each unit in the series being operated at a successively lower pressure. Steam from an external source condenses in the first unit and thus furnishes heat to produce a nearly equal quantity of vapor at a slightly lower temperature than the steam. This vapor is conducted to the second unit, where it condenses and produces a nearly equal quantity of vapor at a still lower temperature. The same process is repeated in the remaining evaporator–condensers in the series, except for the last one. Vapor from the last unit is conducted to the final condenser, in which a very large amount of cooling water serves to condense the final vapor and to carry away the heat energy from the plant.

In order to understand how the above processes are carried out, note the details of Fig. 14, in which a two-effect usage of heat energy is shown. Steam from an external source is condensed in the evaporator of the first effect to form the vapor which is used to heat the second effect. The vapor from the second effect is then condensed in a separate condensing unit. Each of the two effects consists of a tubular heat exchanger, i.e., a cylindrical shell with tubes connected to tube-sheets near the ends of the shell, these latter being sealed

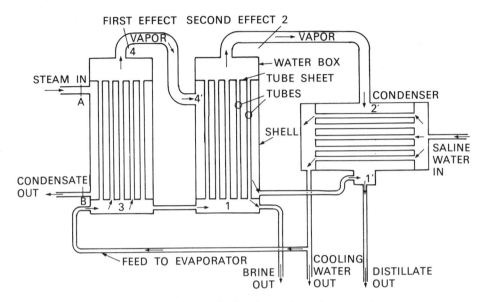

Figure 14. Two-effect distillation plant—schematic cross section.

around their peripheries so that the condensing steam on the outside of the tubes cannot mix with the evaporating liquid on the inside of the tubes. The ends of the large shells are fitted with water boxes, so that water can be conducted into the tubes and vapor or liquid can be removed from the upper ends of the tubes. While the diagram shows the condensing vapor to be on the outside of the tubes—the shell-side—some plants have been built with the vapor condensing on the inside of the tubes.

In the above example, the condensation of 1 kg of external steam produces nearly 2 kg of vapor from the saline water, so that the heating effect of the fuel, i.e., the external steam, has been multiplied by a factor of nearly two. By using more effects the heating value of the fuel would be multiplied by a number nearly equal to the number of effects used. In practice, the greatest number of effects used in a large plant has been 12, although as many as 30 have been used in an experimental plant with rotating heat-transfer surfaces. Because of the fact that the latent heat of steam increases as temperature is decreased, and further, because the cold water must be heated up to the maximum temperature in the plant, the gain by adding effects is actually less than the number of effects. A close approximation to the gain is to use a factor of 0.8 to arrive at a measure of the added efficiency due to the number of effects. For example, a 12-effect plant will produce water amounting to about 9.6 times the weight of external steam condensed in the first effect. This ratio of product water to external steam used is frequently employed for comparing the steam economies of different plants, and is known as the gained output ratio, GOR.

In order to appreciate the thermodynamics of the process, it is useful to trace the paths of water and vapor through the plant in Fig. 14. Saline water

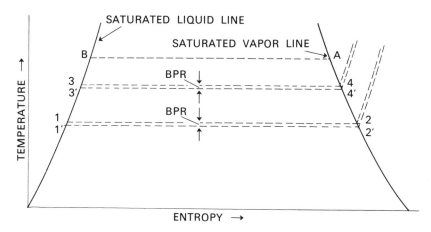

Figure 15. Changes of state of water in 2-effect distiller.

enters the condenser and is heated by condensation of the vapor therein. Leaving the condenser, most of this water is discharged as cooling water to carry away the bulk of the heat from the plant, for heat input of external steam must be accounted for in heat contained in the discharges from the plant of the cooling water, the distillate, and the brine. The balance of the water from the condenser is the feed to the first effect. Numbers in Fig. 14 and corresponding state points on the temperature–entropy diagram of Fig. 15 indicate the progress of the water and vapor through the plant. External steam is shown as entering at A as saturated vapor and returning to the boiler as saturated liquid at B. The temperature difference, $T_B - T_3$, causes heat transfer to produce vapor at (4). This vapor, passing to the shell of the second effect, condenses there at temperature $T_{4'}'$, the difference $T_4 - T_{4'}'$ being the boiling point elevation due to salinity. Condensation, in this example, is completed at (3') and causes evaporation along the line 1–2. This vapor then enters the condenser at (2') and is condensed along the line 2'–1'. Restrictions in the liquid lines are used to maintain the desired distribution of pressures in the plant. In an actual plant, liquid-to-liquid heat exchangers would be used to further heat the incoming feed water by cooling the brine and distillate. These were omitted from Fig. 14 for purposes of simplicity.

While it is clear that a maximum number of effects would be desirable from the standpoint of fuel economy, it will be realized that each added effect contributes to the capital cost. Thus to minimize the cost of distilled water, it will be necessary to balance the fuel cost against the added capital cost. The cost items can be expressed algebraically, a simplified version being that presented by Gilliland,[8] namely

$$C^* = C_1 n + C_2/n, \text{ cost of distilled water} \qquad (18)$$

in which n is the number of effects; C_1 is the annual capital cost per unit of water produced in one effect; and C_2 is the cost of steam per unit of water produced in a single-effect distiller. The cost of water would be minimized when $n = (C_2/C_1)^{\frac{1}{2}}$, determined by applying the usual differentiation procedure. Values of C_2 and C_1 will depend on the exact circumstances of a plant, their magnitudes being as follows for the conditions of a particular plant:

$$C^* = 0.00330n + 1.33/n$$

The number of effects for minimum water cost under these conditions would be 20 in this case, with a minimum cost then being $0.13/m^3$. To show the results of changing the number of effects, the three terms in the above equation have been computed for a series of values of n and plotted on Fig. 16. It will be seen that the total cost curve is very flat near the minimum point and that the advantage of increasing the number of effects from 10 to 15 is much greater than that of increasing from 15 to 20. While the constants in the equation will vary considerably, the type of curve will be similar to that in Fig. 16 regardless of the constants involved. Also, it has been found that from a practical

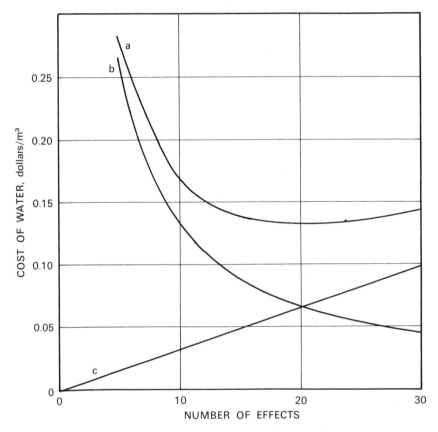

Figure 16. Cost economy of multiple-effect distiller: (a) total; (b) fuel or energy cost; (c) capital amortization cost.

viewpoint it is best to have the number of effects slightly less than that corresponding to the number for minimum cost.

While the practical considerations noted above limit the number of effects used in actual plants, there is merit in inquiring as to the maximum number of effects in an idealized situation, where no limit is placed on the sizes of heat exchangers. Consider a hypothetical plant with a great number of evaporator–condensers, each of which has a heat transfer area so large that an infinitesimal temperature difference is sufficient to accomplish the transfer of heat. If steam is supplied at 125°C and cooling water at 25°C, the number of effects possible will be determined by the average BPR, since the vapor produced in the first stage at 125°C will condense in the second stage at $(125 - BPR)°$, etc. Thus the maximum number of effects will equal the temperature interval, 100°, divided by the average BPR, 0.455°, for seawater, or 220 effects. Applying the 0.8 factor to this number gives $GOR = 176$ kg water per kg of steam. The latent heat of steam at 125°C is 535 kcal/kg, or 0.622 kWh/kg, so that this GOR corresponds to an energy requirement of 3.53 kWh/m^3 of product water, which may be compared with the ideal minimum mechanical energy of about 0.7 kWh/m^3 from Fig. 4, divided by the Carnot efficiency for the temperature used, giving a value of 2.79 kWh/m^3. Thus the conditions chosen for this hypothetical plant would require an ideal minimum amount of energy approaching that of the ideal work of separation. Although 220 stages is the theoretical maximum limit, it is very impractical from many points of view.

In addition to the number of effects, the arrangement of the heat transfer surfaces and other components has an important bearing on the operation and economy of the plant. Evaporator–condensers have been made with a variety of configurations—those with long tubes and short tubes, with horizontal or vertical tubes, with saline water inside the tubes or outside, with water flowing upward through the tubes or downward, with external piping or with piping inside a casing containing several effects, etc. The combination which seems to be preferred uses vertical tubes with saline water flowing downwards on the interior surfaces of the tubes, and with several effects enclosed within a single casing. The tubes are fluted or corrugated to maximize the effectiveness of heat transfer. Passages within the casing conduct vapor and water between effects and thus avoid the external piping implied in Fig. 14. The approximate arrangement in one effect of this type of plant is shown in Fig. 17, the same components appearing in all effects except the last, which has water-cooled condensing tubes instead of evaporating tubes.

In Fig. 17 it will be seen that the incoming saline water is shown as moving from one effect to the next without evaporation and in a direction opposite to that of the brine, distillate, and vapor, and is being heated a small amount in each effect. This scheme is preferred in order to have the saline water to enter the first effect with minimum salinity, thereby minimizing the danger of scale formation in this zone of maximum temperature. This routing of the fluids in the plant is known as "forward feed," since the evaporating brine is moving forward with the distillate and vapor from the first effect to the second, etc.

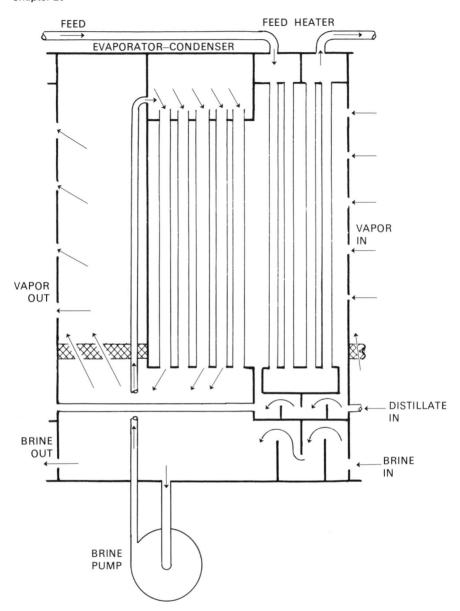

Figure 17. Multiple-effect distillation module—schematic cross section.

Since Fig. 17 shows only one effect in the series, the other components of the plant are not evident. Pumps are necessary to drive the saline water through the heaters to the first effect, to remove the distillate and brine from the last effect and condenser, as well as to circulate the water in each effect as shown. A vacuum pump is required to remove noncondensible gases and also serves to maintain the pressure gradient through the plant, which causes the distillate, brine, and vapor to progress through the several effects. Control elements are needed to maintain liquid levels within chosen limits in the several units.

6.5.3. Multiple-Stage Flash Distillation

One of the operational problems of multiple-effect plants involves the deposition of scale during vapor formation at the heat transfer surfaces in the plant. Flash distillation is a method in which evaporation takes place away from the heat transfer surface, and, along with other advantages, offers in some respects a superior process for desalination.

In flash distillation the saline water is heated to a maximum temperature under pressure and without evaporation, and then expanded through orifices to lower pressures. Each expansion is very nearly a constant enthalpy process, so that a small amount of the expanding water vaporizes, and when separated from the liquid, can be condensed as pure water. Such evaporation involves a very small amount of the expanding liquid—approximately 10%, with water expanded from 120°C to 40°C, the range common to modern plants. This indicates that the amount of water circulating through the plant will of necessity be about 10 times that of the distillate produced.

Figure 18 shows schematically a cross section of a modern multiple-stage flash distillation plant. As shown, the plant has three sections: the heat rejection section, the heat recovery section, and the brine heater. Physically, the heat rejection section and the heat recovery section consist of large casings with partitions across them. While the early designs of flash distillers used four or five stages of water expansion, each stage a separate casing, the modern design, invented in about 1960, uses 20–40 or more stages of expansion, with

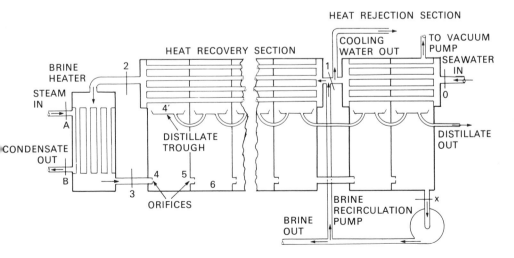

Figure 18. Multiple-stage flash distiller.

four or more partitions in one casing. The condenser tubes, grouped near the middle of the top of the casing, pass through tight-fitting gasket holes in the partitions. Condensate trays beneath the condenser tubes leave spaces near the sides of the casing through which the vapor can rise in passing from the water surface to the tubes and from which it condenses to the condensate trays. This condensate passes from stage to stage through the U tubes indicated or through submerged orifices. The third portion of the plant is the brine heater, in which steam from an external source is condensed to furnish heat energy.

By expanding the saline water in small increments rather than in one stage, the vapor will be formed and condensed at a series of successively decreasing temperatures, as shown by the steps in Fig. 19. The feed water can be pumped countercurrently through the condenser tubes in series so that its final temperature is close to the maximum value of the steam temperature in the brine heater. The resulting temperature relationships in this type of plant are shown schematically in the figure, in which temperatures are plotted as ordinates against locations in the plant as abscissae, with numbers on this diagram corresponding to locations in Fig. 18.

A large quantity of saline water is pumped through the condenser tubes of the heat reject section and is heated from temperature T_0 to T_1. Most of this water is discharged as cooling water and a small amount of it is mixed with recirculating brine, the latter at temperature T_x, the temperature at exit. This mixture is passed through the condenser tubes of the heat recovery section,

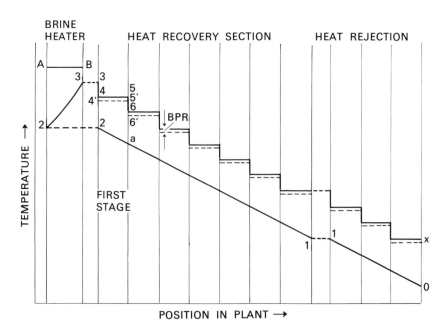

Figure 19. Temperature profile in multiple-stage flash distiller—schematic only.

where its temperature is increased from T_1 to T_2. The temperature of this water is further increased from T_2 to T_3 in the brine heater by condensation of external steam at temperature A. This hot water then expands into the brine chamber of the first stage, with a reduction in temperature from 3 to 4. The vapor produced by expansion condenses along the line 4'–5', which is below the liquid temperature 4–5 by the amount of the BPR. The heat released by condensation of this vapor causes the water in the condenser tubes to increase in temperature from a to 2. The same sequence of events is repeated in each succeeding stage, with the brine leaving the plant at T_x. While plants have been designed for a variety of temperature conditions, it may be noted that T_3 usually ranges from less than 90°C to nearly 125°C. Values of T_0 are approximately 20°C, and those of T_x are from 5° to 8° above T_0. The temperature rise in the brine heater is generally about 6°.

The effect of changing the number of stages can be evaluated from Fig. 19. Various numbers of stages could be used for a given set of values of T_0, T_2, T_3, and T_x, the largest number being equal to the temperature interval, T_3-T_x, divided by the BPR. The smallest number would result if the saline water is expanded from T_3 to $(T_2 + \text{BPR})$ at entrance to the first stage, which would result in zero temperature difference between condensing vapor and cooling water at the end of the stage, and hence in zero heat transfer. The temperature difference for heat transfer is the average vertical intercept between the horizontal condensation lines, i.e., between 4'–5', and the liquid heating line, a–2. If the number of stages is increased, the height of the "steps" will be decreased and the intercept distance will be increased. Since the rate of heat transfer is proportional to the intercept distance, an increase in the number of stages improves heat transfer within each stage, which is the most important result of such a change.

From Fig. 19 it can also be deduced that changing the number of stages will have no effect on the amount of steam used in the brine heater. This follows from the fact that the heat required in the brine heater is that necessary to heat the saline water from 2 to 3, and for given values of T_2 and T_3 will be the same regardless of the number of stages. The GOR is therefore independent of the number of stages.

Since the GOR is independent of the number of stages, it is necessary to consider its relation to other variables. By making heat balances on various portions of the plant, as well as on the entire plant, it has been shown[27] that the steam economy of an idealized plant is dependent on certain temperatures, the latent heat, and the recovery ratio, the relation being as follows:

$$\frac{W}{W_s} = \left(\frac{W}{W_f + W_r}\right)\left(\frac{h_{fg}}{C}\right)\left\{(T_3 - T_0) - \left(\frac{W}{W_f + W_r}\right)\left(\frac{h_{fg}}{C}\right)\right\}^{-1} \qquad (19)$$

where W is the weight of distillate; W_s is the weight of external steam; W_f is the weight of saline water feed; W_r is the weight of recirculated brine; h_{fg} is the latent heat of evaporation; and C is the specific heat of water. It is noted that for seawater, W_f is equal to the weight of waste brine plus the weight of the distillate and, further, that the latter two are nearly equal to each other. Since W represents only about 10% of the brine passing through the evaporators, it follows that W_r must represent a large fraction of the brine to be evaporated. The recirculation of a large share of the brine has the practical advantage that this part of the feed stream need not be re-acid-treated for scale control, which thus saves the cost of the acid.

From a mathematical viewpoint, equation (19) has the general form of

$$y = ax/(b - cx)$$

or

$$xy - by/c + ax/c = 0 \tag{19a}$$

This is recognized as the equation of a rectangular hyperbola offset from the origin. The feature of the hyperbola which is important here is that these curves are asymptotic to their axes. This means that the curve W/W_s is asymptotic to a vertical line at $W/(W_f + W_r) = C(T_3 - T_0)/h_{fg}$, so that very small changes in the recovery ratio, $W/(W_f + W_r)$, will produce large changes in the steam economy, W/W_s. From this it can be concluded that the steam economy and the GOR of a multiple-stage flash distillation plant are dependent primarily on the recovery ratio, and that a particular plant will operate most efficiently under closely controlled design conditions.

To illustrate the effect of varying the recovery ratio, it is noted that the published design figures for the 24-stage plant at Malta[28] show this plant to be designed with $W/W_s = 9.4$ at $W/(W_f + W_r) = 0.093$, when the temperatures and other variables in equation (19) are as follows: $T_3 = 89°C$; $T_0 = 28°C$; $h_{fg} = 536$ (steam); $h_{fg} = 564$ (distillate); $C = 0.931$ (brine); and $C = 0.942$ (distillate). Values of W/W_s computed for these data are shown in Table 6 for values of recovery ratio near 0.093. In comparing the Malta design figures with those from equation (19), it can be seen that there is rather good agreement between the two. It can also be noted that changes of about 1% in the recovery ratio change the steam economy by 14–17%.

Table 6. Minimum Number of Stages for Multiple-Stage Flash Plants

$W/(W_f + W_r)$ (computed)	W/W_s	n (minimum)
0.090	7.75	9.5
0.092	9.64	12.2
0.093	11.0	14.5
0.094	12.9	17.4
0.095	15.0	21.2
Malta design		
0.093	9.4	24.0

As previously mentioned, the number of stages used in a plant is somewhat arbitrary, although they must be greater than some minimum value. The minimum number of stages will depend on the temperature intervals T_3-T_2 and T_3-T_x. Noting that T_x will decrease as steam economy is increased, more stages will be accommodated in the interval T_3-T_x. Values of T_x have been computed for the idealized plant of equation (19), and the minimum number of stages tabulated in the last column of Table 6. These figures show that the Malta plant conditions would require at least 12–15 stages for its minimum number. The actual number of stages in the Malta plant was 24.

In a separate study, El Ramley *et al.*[29] made a statistical analysis of the relation between capacity and number of stages for plants constructed up to 1970, and arrived at the conclusion expressed in the equation

$$n = 19 + 1.6 \times 10^{-3} \, \text{m}^3/\text{day} \tag{20}$$

Applying this to the Malta plant capacity of 4500 m^3/day gives the number of stages as 26.2. This compares favorably with the actual 24 stages of the plant, being well above the theoretical minimum number of 12–15 stages.

Plants of the multiple-stage flash type have been constructed in sizes up to 19,000 m^3/day per unit, with the cost of water produced being reduced as the unit size is increased. There have been several design variations introduced through the ingenuity of the designers, particularly in the arrangement of the condenser tubing. Figure 18 and the accompanying analyses relate to plants with condenser tubes running continuously through several stages, but several very satisfactory plants have been constructed with sets of condenser tubes running perpendicular to the direction of brine flow and serving a single stage. These and other variations are amenable to the same principles and analyses which have been presented for the long tube arrangement.

6.5.4. *Vapor Compression*

6.5.4a. *Principles.* While multiple-effect and multiple-stage flash distillation both use heat as the major source of energy, it is also possible to use mechanical energy in a heat pump cycle, generally referred to as vapor compression distillation. The principle of operation is very similar to that of the idealized desalination device in Fig. 3, the major differences being that the compression process is essentially adiabatic rather than isothermal, and the resulting temperature difference, used to transfer heat from condensing vapor to evaporating liquid, is finite rather than infinitesimal.

Figure 20 shows, schematically, the arrangement of apparatus for this process. Saline water enters the system through the two heat exchangers, which cool the distillate and reject brine, respectively, and proceeds to be heated in the tube side of the evaporator. This heating of the water produces vapor which is withdrawn from the evaporator by a compressor and compressed to a slightly higher temperature before it is returned to the shell side of the evaporator. Here it comes into contact with the condenser tubes through which water of a

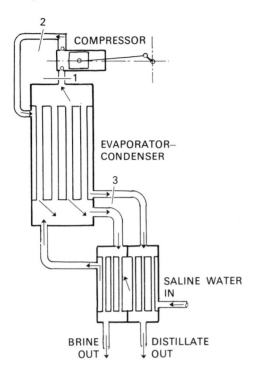

Figure 20. Vapor compression distiller.

slightly lower temperature is flowing, thus causing the vapor to condense and thereby supplying heat to the water inside the tubes and producing more vapor. Brine is withdrawn at a constant rate so as to maintain the evaporating liquid at a given concentration.

The temperature–entropy diagram in Fig. 21 shows the idealized cycle of state changes followed in this type of plant. Vapor at state point 1 is compressed

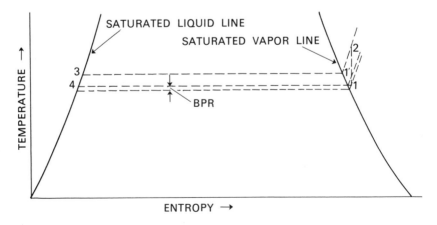

Figure 21. Changes of state of water in vapor compression distiller—schematic only.

adiabatically to point 2, and is then cooled and condensed along the constant pressure line 2–3. The temperature difference, T_3-T_4, causes heat to be transferred to the evaporating liquid, and results in vapor formation along the line 4–1. State point 1 lies in the superheat region because of the boiling point rise, so that compression from 1–1′ is necessary to increase the pressure to that of point (4). The further compression from 1′–2 provides the heat transfer differential, T_3-T_4.

The temperature at which most vapor compression devices operate is 100°C at point 1, which corresponds to atmospheric pressure, so that no vacuum equipment is needed. The differential temperature for heat transfer, T_3-T_4, is on the order of 6°C, which means that the latent heats are very nearly equal for condensation and evaporation, and each kilogram condensed produces nearly a kilogram of vapor. The small shortage of heat from this source, plus heat losses from the outside of the equipment, require the addition of a small amount of heat. If the unit is driven by an internal combustion engine, the heat increment needed is frequently taken from the jacket cooling water of the engine.

Since the operating temperature of 100°C is considerably above that of the saline water supply, which is usually around 20°C, the waste brine and distillate streams are passed through liquid-to-liquid heat exchangers to heat the incoming saline water feed, as shown in Fig. 20. Because both fluids are liquid and no evaporation is involved, the heat transfer is rather poor, requiring heat exchangers with quite large heat transfer surfaces, in some cases nearly as large as the evaporator–condenser itself. Since this increases the capital cost considerably, other means of heating the incoming water have been proposed. The most promising alternative scheme seems to be to use vapor compression in combination with a multiple-stage flash distiller.

In this combination a vapor compression plant would be placed in the brine circuit between the brine heater and the orifice at inlet to the first flashing stage. The distillate from the vapor compression plant would be admitted to the condensate tray in stage 1 through an orifice. Here it is cooled as it flashes and condenses its part of the vapor while heating the condenser tubes in this and succeeding stages. At the same time the residual brine from the vapor compression plant would be passed through the orifices at entrance to the brine chamber of stage 1, where it flashes and condenses to contribute its share of vapor for heating in stage 1. By moving from stage to stage, both of these liquids are cooled, and at the same time the brine adds more distillate by the usual flash distillation process. The heating of the brine in the condenser tubes is very nearly the same as it would be without the vapor compression unit, since all of the water heated in the brine heater is flashed in the several stages, part flashing as brine and the balance as distillate. For this reason the flash distiller may come to be preferred to the liquid-to-liquid heat exchanger for vapor compression.

6.5.4b. *Performance.* The energy required by the vapor compression process is much less than the value represented by the latent heat of the vapor produced, because of the heat pump nature of the process. A heat pump only changes the temperature level of the vapor, because it is a fact that the major share of the heat involved is furnished by condensation of the vapor which has been compressed. In this application, the work done by the compressor is equal to the heat required to change the temperature of the liquid distillate from T_4 to T_3, about 6°, requiring 6 kcal/kg, whereas the heat transferred between condensing vapor and evaporating liquid is the latent heat, or about 530 kcal/kg. In actual vapor compression plants the energy required by the compressor to recycle the latent heat over the 6° range is approximately 12 kWh/m^3.

As in other processes, the cost of water produced by a vapor compression plant is made up of two principal components, namely, the capital charges on the equipment and the energy for driving the compressor and pumps. The capital charges reflect the cost of heat exchangers, compressor, and pumps. These capital charges decrease as the temperature difference for heat transfer in the evaporator–condenser increases, because of the decrease in the heat transfer area needed. However, any increase in this temperature difference also represents an increase in the cost of energy needed by the compressor. Thus it is necessary to balance the savings in capital charges against the increase in energy costs to determine the minimum cost of water. When this is done it is generally found that the minimum cost of distilled water will coincide with the condition at which the energy cost equals the capital charges, which for simple vapor compression plants involves temperature differences in the condenser–evaporator of about 4–6°.

The attractiveness of the high efficiency of the vapor compression process has led to several attempts to build larger plants. The direct-displacement compressors that had been used previously were limited in capacity to about 200 m^3/day, but had been favored because they could be made to deliver full capacity of distillate under conditions of accumulating scale on the heat transfer surfaces, by simply increasing the delivery pressure of the vapor and the power to the compressor. With improved methods of scale control it was possible to use other types of compressor. One of the first attempts to overcome this size limitation was the construction of the 4750-m^3/day plant of the OSW at Roswell, New Mexico,[30] which used an axial flow compressor, with the water pretreated to remove the scale-forming ions. Not all of the distillate went through this compressor as vapor, since the plant combined a two-effect distiller with the vapor compression feature. The plant was so arranged that the compressor received vapor from the low-pressure effect and delivered it to the high-pressure effect, where it served as heating steam. Vapor from the high-pressure effect then served as the heating steam in the low-pressure effect. Thus the compressor was employed to compress roughly half of the vapor. This pointed the way to larger sizes, so that it is now possible to obtain compressors large enough to produce several times as much distillate as in the Roswell plant.

The other major component of the plant, the evaporator–condenser, has also been the subject of changes for increasing its capacity. The limitation in this device has been the heat transfer rate. The first improvement was that of forced circulation, which was practiced in the Roswell plant, accomplishing some reduction in the size and cost of the equipment. Later improvements came from the introduction of fluted and corrugated surfaces, which increased heat transfer rates by several times. By placing corrugated tubes vertically and arranging for the saline water to flow downward on the inside of the tube while the steam condenses on the outside of the tube, it is found that most of the outside wall of the tube remains dry, with the condensate drawn by the surface tension into the "valleys" of the corrugations. It is reasoned that this improves heat transfer by allowing condensation on the "dry" portions of the tube, thus avoiding the thermal resistance of the water film which covers the entire surface of smooth tubes. With this type of tube and the improved compressors noted above, there is opportunity for vapor compression distillation to be more widely applied to large plants for water desalination.

6.6. Freeze Separation

Freeze separation is related to vapor compression distillation, since it uses a vapor compressor in a heat pump cycle. It differs from vapor compression distillation in that most of the product water bypasses the vapor compressor, with only about $\frac{1}{8}$ of the product water passing through the vapor phase.

The basic idea of freeze separation is that crystals produced from saline water do not contain the dissolved ions, and when separated from the residual brine can be melted to give pure water. In practice, none of the freeze separation devices used for desalination has accomplished complete separation of ice crystals and brine, so that the product water has contained dissolved salts amounting to several hundred ppm.

Although several other schemes have been tried, the two most promising systems for desalination are the vacuum flash vapor compression system (VFVC) and the refrigerant flash vapor compression system (RFVC). The first system was invented by Zarchin[31] and was developed by the Fairbanks Morse Division of Colt Industries[32] to the point where a prototype production plant was installed and operated at Eilath in Israel.

The vacuum flash vapor compression system is shown schematically in Fig. 22. In this scheme, incoming saline water is deaerated and precooled before being sprayed into the freezing chamber, where the pressure is so low that the corresponding saturation vapor temperature is below the freezing temperature of the saline water, which is a little below $-2°C$ for seawater. A small amount of water flashes into vapor during the spraying process, absorbing heat and causing the formation of ice crystals in the liquid that weigh about seven times as much as the vapor. The ice crystals are removed by a slurry pump, passed through a wash column to remove the brine, and delivered to the melter. Meantime, the vapor has been compressed and sent to the melter, where it melts the ice by condensation. From the figure it will be noted that the freezer,

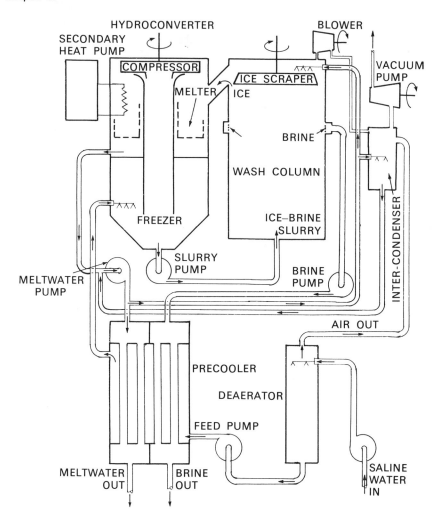

Figure 22. VFVC freeze separation distiller—schematic cross section.

compressor, and melter are contained in a single unit, called the hydroconverter. The wash column is supplied with a small amount of the product water, which washes the ice crystals just before they are scraped back into the melter. This wash water flows countercurrently to the ice crystals and mixes with the waste brine as it leaves the plant. Waste brine and product streams flow through the precoolers to exchange heat with the incoming saline water. Since the work of compression adds heat to the hydroconverter, a separate refrigerator is employed to carry away this heat, together with any heat which has leaked into the plant. While the principal energy consumption in the VFVC system is that used by the main vapor compressor, this secondary refrigerator is also a major absorber of energy, since it must pump the compressor energy and leakage losses from the 0° level in the melter to the ambient temperature of about 20°C.

The refrigerant flash vapor compression system has the advantage of avoiding the high vacuum of the VFVC system already described. In this system, a precooled stream of saline water is mixed with a precooled stream of refrigerant liquid, and the mixture of water and refrigerant is sprayed into the freezing chamber at a pressure which will cause the refrigerant to vaporize at about −4°C. The formation of refrigerant vapor absorbs heat and causes ice crystals to form. These crystals are pumped out in a slurry and washed, and the vapor is compressed and used for melting the washed crystals, as in the other process. The chief difference between the two processes lies in the fact that the refrigerant condensate is mixed with the product water from which it must be separated. One of the preferred refrigerants has been n-Butane, which can be separated by decantation since it is only slightly soluble in water. The energy requirement of this system is about the same as for the vacuum flash vapor compression system.

6.7. Hydrate Separation

A process related to the freeze separation systems is the hydrate separation system, in which crystals of one of several hydrates, known as clathrates, are used in place of ice. The clathrate structure is formed of a large number of water molecules arranged in a cagelike formation, with one molecule of the hydrating agent acting as the keystone which holds the crystal in shape. The ratio of water molecules to "guest" molecules varies with the hydrating agent, being as large as 17 for propane, but only 8 for methyl bromide. A few of the hydrating agents are listed in Table 7, from which it will be seen that some of

Table 7. Hydrating Agents for Desalination of Water[a]

Material	Reactor conditions	
	Temperature (°C)	Pressure (mm Hg)
Chlorine	28.7	4570
F–31 (CH_2ClF)	17.9	2147
Methyl bromide	14.7	1151
F-142b(CH_3CClF_2)	13.1	1743
F-12(CCl_2F_2)	12.1	3435
F-11(CCl_3F)	8.0	419
Propane	5.7	4141

[a]Adapted from Briggs and Barduhn.[33]

them form crystals at temperatures much higher than freezing, and a few of them at close to ambient conditions. Use of these crystals would reduce the problem of heat dissipation noted in the freeze separation process, in which much energy was used in pumping the heat from freezing temperature to ambient conditions.

A plant for hydrate separation would be arranged in a manner similar to the VFVC plant shown in Fig. 22. A hydrate reactor would replace the freezing chamber of the VFVC system, and both saline water and liquid hydrating agent would be sprayed into this chamber at controlled pressure and temperature conditions. Part of the hydrating agent would flash into vapor, absorbing heat and causing hydrate crystals to form. The crystals would be washed to rid them of the brine and passed to the melter, where the condensation of the compressed vapor from the reactor vessel would furnish the required heat. Separation of the product water and hydrating agent would present the same problems as in the RFVC system, with agents insoluble in water being decanted. In this system the melter would be located outside of the reactor vessel, rather than being located in a combined unit such as the hydroconverter of the VFVC system.

6.8. *Solvent Extraction*

The last of the processes for desalting water by removing the water from saline solutions is solvent extraction. The great success and wide usage of solvent extraction, in fields as diverse as pharmaceutical preparation, oil refining, petrochemical manufacturing, and food processing, has led to experimentation with it for desalination.

The most desirable solvent extraction process for desalination would be one which would extract the unwanted dissolved salts from water. However, there seemed to be no material for which this was true, so that all of the studies have dealt with solvents which absorb water preferentially to the dissolved salts contained in it. Such solvents dissolve water in the amount of 30–40% by weight at ambient temperature, but retain only 10% or less at temperatures in excess of 65°C. Certain secondary and tertiary amines have been mixed to form solvents capable of operating in the temperature range indicated.

Some experimentation has been done with a desalination process in which the following steps have been studied: (1) the extraction of water in a column where the saline water and solvent move countercurrently; (2) separation of water and solvent by heating and decantation; and (3) stripping of solvent from product water and brine to recover the solvent for reuse. The separation step requires heat energy, which could be partially furnished by heat interchange between incoming and outgoing streams. The stripping step has been the most difficult, since failure to recover all of the expensive solvents has caused excessive make-up costs. So far, the solvent extraction method has been confined to the laboratory, with little prospect of competing with distillation or other methods for the desalination of water.

References

1. M. W. Kellogg Co., *Saline Water Conversion Engineering Data Book*, U.S. Dept. of Interior, Washington, D.C., 1965.
2. J. W. McCutchan and E. N. Sieder, "Scale Control in Saline Water Evaporators": Review of Current Status, Res. Develop. Prog. Rep., No. 411, Office of Saline Water, April, 1969, 62 pp.
3. Dow Chemical Co., "An Engineering Evaluation of the Long-tube Vertical Falling-film Distillation Process," Office of Saline Water, Res. Develop. Prog. Rep., No. 139, Feb. 1965, 282 pp.
4. B. M. Fabuss and A. Korosi, Office of Saline Water, Res. Develop. Prog. Rep., No. 384, 1969, U.S. Dept. of Interior, Washington, D.C.
5. L. Koenig, The Economic Boundaries of Saline Water Conversion, *Journal of the American Water Works Assoc.*, **57**(7) : 845–862 (1959).
6. F. Helfferich, *Ion Exchange*, McGraw-Hill, New York, 1962.
7. R. Kunin and R. Myers, *Ion Exchange Resins*, John Wiley, New York, 1950.
8. E. R. Gilliland, Fresh Water for the Future, *Ind. Eng. Chem.*, **47** : 2410 (1955).
9. P. B. Stewart, Sea Water Demineralization by Ammonium Salts in Ion Exchange, *Advan. Chem. Ser.*, No. **27** : 178–191 (1960).
10. G. Klein *et al.*, "Design and Cost of Ion Exchange Softening for a 50 mgd Sea Water Evaporation Plant," Univ. of Calif., Berkeley, Calif., SWCL Rept. No. 68–2, Sept, 1968, 59 pp.
11. J. K. Kunz, "Ion Exchange Pretreatment for Sea Water Desalination Plants," *Desalination*, **3**(3) : 363–372 (1967).
12. Bureau of Reclamation, "Test Manual for Permselective Membranes," Office of Saline Water, Res. Develop. Prog. Rep., No. 77, U.S. Dept. of Interior, 1964.
13. D. A. Cowan, "Interaction of Technical and Economic Demands in the Development of Large Scale Electrodialysis Demineralizers," *Advan. Chem. Ser.*, **27** : 224–231 (1960).
14. W. E. Katz, "Some Practical 1959 Advances in Electric Membrane Demineralization," *Advan. Chem. Ser.*, **27** : 232–246 (1960).
15. E. B. Millard, *Physical Chemistry for Colleges*, McGraw-Hill, New York, 1926, p. 161.
16. C. C. Reid *et al.*, "Water and Ion Flow through Imperfect Osmotic Membranes," Office of Saline Water, Res. Develop. Prog. Rept., No. 16, U.S. Dept. of Interior, Washington, D.C.
17. S. Loeb and S. Sourirajan, "Sea Water Demineralization by Means of a Semipermeable Membrane," UCLA Engineering Rept. No. 60–60, 1960, Univ. of Calif., Los Angeles, Calif.
18. S. Manjikian, S. Loeb, and J. W. McCutchan, "Improvements in Fabrication Techniques for Reverse Osmosis Desalination Membranes," *Proc. 1st Intern. Symp. Water Desalination, Washington, D.C.*, **2** : 159 (1959).
19. H. K. Lonsdale, "Theory and Practice of Reverse Osmosis and Ultrafiltration," in *Industrial Processing with Membranes*, R. E. Lacey and S. Loeb, eds., John Wiley, New York, 1972.
20. G. G. Havens, *Desalination by Reverse Osmosis*, V. Merten, ed., The M.I.T. Press, Cambridge, Mass., 1966.
21. S. Loeb and F. Milstein, "Design, Development, and Testing of a 500-gpd Osmotic Desalination Cell," Dept. of Eng. Rept., No. 62–52, 1962, Univ. of Calif., Los Angeles, Calif.
22. H. I. Mahon, "Hollow Fibers as Membranes for Reverse Osmosis," Nat. Acad. of Science, *Nat. Research Council, Publ.* No. **942** : 345–354 (1961).
23. W. P. Cooke, "Hollow Fiber Permeators in Industrial Waste Stream Separations," *Desalination*, **7** : 31–46 (1969).
24. J. S. Johnson, J. W. McCutchan, and D. N. Bennion, "$3\frac{1}{2}$ Years' Experience with Reverse Osmosis at Coalinga, California," School of Engineering and Applied Science Rept., No. 69–45, July, 1969, Univ. of Calif., Los Angeles, Calif.
25. S. G. Talbert, J. H. Eibling, and G. O. G. Löf, "Manual on Solar Distillation of Saline Water," Res. Develop. Prog. Rept., No. 546, Office of Saline Water, U.S. Dept. of Interior, 1970.
26. R. N. Morse and W. R. W. Read, "The Development of a Solar Still for Australian Conditions," *Mech. Chem. Eng. Trans., Inst. of Engineers, Australia*, MC3, May, 71–80 (1967).

27. E. D. Howe, *Fundamentals of Water Desalination*, Marcel Dekker, 1974.

28. "Flash-Evaporator Plant in Malta," *The Engineer, London*, **224**(5834) : 643–646 (1967).

29. N. A. El Ramley, J. M. English, and J. W. McCutchan, "The Multi-stage Flash Desalting Process: Its Commercial Applications," Rept. No. UCLA Eng., 7079, Aug. 1970, Univ. of Calif., Los Angeles, Calif., 125 pp.

30. P. L. Geiringer *et al.*, "First Annual Rept: Brackish Water Conversion Demonstration Plant #4, Roswell, New Mexico," Res. Develop. Prog. Rept., No. 169, Office of Saline Water, U.S. Dept. of Interior, 1965, 87 pp.

31. A. Zarchin, Demineralizing of Sea Water by Freezing, Israeli Patent #7764, 1953.

32. R. Consie *et al.*, "Vacuum Freezing Vapor Compression Desalting Process," Res. Develop. Prog. Rept., No. 295, Office of Saline Water, U.S. Dept. of Interior, 1967, 99 pp.

33. F. A. Briggs and A. J. Barduhn, *A.C.S., Advances in Chemistry Series*, **38** : 190–199 (1963).

All of the preceding chapters assume that pollutants can be appropriately measured, and thus they rest upon analytical chemistry.

21

Analytical Chemistry of Pollutants

L. E. Smythe

1. *Sampling*

1.1. *General Considerations (Static versus Dynamic Situations)*

In chemical analysis the most careful and skilled laboratory work, often using expensive physicochemical instrumentation, results in virtually worthless data unless the sample taken is representative of the material whose composition is to be determined. In a typical situation, for example, in water quality monitoring, the main steps and considerations can all influence the overall error. These are outlined in Table 1.

It will be appreciated that a project could founder if the best judgments were not made in steps 1 and 2 (Table 1). Serious determinate errors could also arise, particularly in trace chemical analysis, in any of steps 2–4, by either contamination or loss of the analyte.

L. E. Smythe • Department of Analytical Chemistry, University of New South Wales, Sydney, New South Wales, Australia 2006

Table 1. Steps and Considerations for Water Quality Monitoring

Steps	Considerations
1. Sampling site selection	Dynamic nature of the environment (incl. physical and geographical features); season, flow, representative area
2. Sampling	Dynamic vs. static methods; containers, preservation; statistics
3. Processing	Preconcentration and separations; matrix considerations; chemical reactions
4. Measurement	Instruments or methods; quality control (incl. standards, blanks, calibrations, etc.); statistics
5. Data processing and verification	Precision and accuracy; statistics
6. Interpretation and action	Comparison with other (e.g., baseline) data; statistics

There are few, if any, reference books which provide concise, step-by-step procedures for the selection of sampling sites for gaseous, liquid, or solid samples prior to chemical analysis. For a large comprehensive project involving the chemical analysis of numerous and varied pollutants, the prescribed reading for sampling site selection might include sections of literature in the following areas: geography, geology, ecology, sociology, and engineering. This emphasizes the value of interdisciplinary project teams. For limited surveys, rapid selection may be made with a knowledge of only the general sampling procedures. The most useful general surveys of principles and methods of sampling for chemical analysis purposes are available in well-known treatises.[1-3]

Sampling is the process of extracting a small fraction of material from a large bulk, so that the fraction is representative of the bulk in every way. Some preliminary considerations, in addition to those outlined in Table 1, would therefore include: the nature of the material (physical state: gas, liquid, or solid; its heterogeneity; its stratification); the end use of the product or material; the variation to be permitted in the product or material; the relationship between the cost of chemical analysis, the cost of sampling, and the cost of the product; the expected concentration range of the analyte in relation to accepted standards or regulations; and the required precision and accuracy for the chemical analyses. The last-mentioned requirements can also greatly affect method, selection, and costs.

Despite its limitations, a "static" sample is often taken for the chemical analysis of pollutants. Continuous, or in-line, sampling and chemical analysis is often only selected if overall costs in this category are in line with the overall project–investigation costs. A further consideration involves the dynamics of analyte concentrations in a medium, i.e., the change in analyte concentration with time. Isolated samples for the examination of pollutants can be of great value provided that all the important collection data are recorded and the use and interpretation of resultant data are only made with due regard to the limitations. Unfortunately, these considerations are not always followed and data may be used out of their correct context, often by persons who have not been involved in either the sampling or the chemical analysis.

A "static" sample may either be chosen at random or according to a statistically based plan. In simple terms, the statistical plan theoretically gives every particle or portion of the analyte an equal chance of appearing in the sample. The actual statistically based sampling procedure involves the removal of portions from every section of the sample. Combination, mixing, and resampling may then be carried out until a suitable laboratory-size sample is obtained. The most suitable laboratory size is often also dictated by the expected amount of analyte present, any necessary preconcentration and separation procedures, and the limit of detection or sensitivity of the method of chemical analysis.

Random-choice sampling is often more difficult than anticipated because personal prejudices can exert an influence. Careful thought is required on the various sources of determinate error, and the relationship of these to inbuilt indeterminate error. Any sample treatment (e.g., crushing or grinding, sub-sampling, and storage) should be considered with caution.

A matter of increasing complexity for the analytical chemist, and also one of great importance for those involved in general pollution studies, concerns the specific composition and form of the analyte when it may be mixed with many extraneous substances (i.e., the matrix). It is now obvious that there are increasing demands for information on the chemical form (and sometimes physical state) of the analyte in complex matrices. Consequently, the analytical chemist may be required not only to say how much of the analyte is present, but in which form it is present. The forms may play important roles in sampling procedures, storage, chemical processing prior to chemical analysis, and the actual determination and assessment of results (including toxicity or possible corrective treatments). Table 2 lists some possible forms of two typical pollutants in environmental materials.

1.2. *Storage of Samples: Contamination, Loss, and Change*

Contamination, loss, and change in both chemical and physical forms can take place at any stage in the sampling, storage, and processing procedures. In the macro-regions these problems may not be severe or require special considerations. In the micro-trace-level regions, major problems can

Table 2. Some Possible Forms of Two Typical Pollutants in Environmental Materials

Analyte	Matrix or material	Some possible forms
Pb	Air (near highways)	As gas, aerosol or solids, Pb, Pb oxides, Pb basic carbonate, $PbClBr$, $PbSO_4$, lead phosphates, etc; Pb alkyls and other metal–organic compounds
Cd	Waters	Cd^{2+}, Cd inorganic compounds, Cd organic compounds, Cd adsorbed on solids, Cd coatings on solids (precipitation and coprecipitation), Cd incorporated in solid biological materials, Cd incorporated in crystalline structures

eventuate. The recommended approaches not only vary according to physical state (gas, solid, or liquid) but also must relate to each individual case. Unfortunately, at the present time, it is not possible to give a useful summary of the necessary approaches owing to the great complexities and continual changes in recommended procedures. Useful guidelines may be obtained in the following areas: storage and preservation of water samples[4]; ultrapurity in trace analysis (including contamination)[5]; losses during dry or wet oxidation[6]; methods of air sampling and analysis[7]; and methods of soil analysis.[8,9]

2. Chemical Analysis Methods

2.1. Selection of Methods

There are approximately 50 fairly broad classifications of methods for chemical analysis. The classification used by Meinke[10] and Pickering[11] should be studied by those who are not conversant with the broad field of chemical analysis. It is also recommended that the broad classifications used in typical modern reviews of advances in analytical chemistry[12] be studied to obtain the necessary perspective. It will be appreciated that many hundreds of different methods of chemical analysis are available for the determination of pollutants. The selection of the most suitable methods is often difficult and may not necessarily be based on objective scientific considerations. Broad experience in and an understanding of analytical chemistry constitutes the most suitable basis for the selection of the appropriate methods, and yet this experience, understanding, and training in analytical chemistry is often not possessed by those who wish to examine the environment for pollutants. It is perhaps this aspect which is responsible for the often unreliable data which are reported in this area. However, some general guidance in the selection of the most suitable methods can be given.

Step 1.
a. Select the broad area for study (e.g., air, water, or soil).
b. List the analytes or substances to be determined (e.g., inorganic, organic, particulate, biological, etc.).
c. List the major constituents of the matrices in which the analytes (b) are to be determined.
d. Read as much as possible about the possible sampling requirements (see 1).

Step 2. Consider if an official or "recognized" method is available (e.g., references 4, 7, 9). This can save much time and is important if the data are to be reported on a local governmental or international basis. Governmental or semigovernmental agencies and national standards bodies can often assist with information on recognized methods. Methods for air and water quality are also examined and recommended by the International Standards Organization (ISO) with headquarters in Geneva. This is the international specialized agency for standardization, comprising the national standards bodies of 73 countries. More than 150 technical committees of this body meet and recommend appropriate standards, which are published and which represent a global consensus of opinion. Examples of two committees are: (1) Technical Committee ISO/TC 147—Water Quality; and (2) Technical Committee ISO/TC 146—Air Quality. Both of these committees also have subcommittees with extensive experience. For example, the present subcommittee on water quality comprises: (1) terminology, (2) physical, chemical, and biochemical methods, (3) radiological methods, (4) microbiological methods, (5) biological methods, and (6) sampling. Further details can be obtained from a recent report.[13]

If Step 2 has provided a suitable method, the remaining considerations may involve those detailed in Step 1, together with: laboratory space, facilities, staff, time, cost, and the assessment and reporting of results.

Step 3.
a. If Steps 1 and 2 have not resulted in the selection of a suitable method, a rapid survey of the most popular methods should be undertaken.[10,11,14] These are also summarized in Table 3.
b. Other sections in this chapter, suitable treatises,[1,2,3] and recent reviews[12] should then be consulted for the final selection.

Step 4. Problems may result either from the lack of suitable facilities, equipment, and experience, or from difficulties in the determination of one or more analytes in difficult samples or matrices. Adaption or research and development may then be necessary. This is not an area for the novice. Time-consuming and expensive work can result and this often involves the use of intercomparison (standard) samples.

Table 3. Comparison of Methods for Chemical Analysis of Pollutants

Method	Sample	Instrumentation	Cost $	Specificity	Time hr/sample	Sensitivity	Precision %	Applications
1. Gravimetric	SLG	conventional lab.	200	good	1–2	100 mg–1 g 1–10 μg	0.005–0.01 0.1	major constituents
2. Titrimetric	SLG	conventional lab.	200	good	0.25–0.5	$10^{-2}M$ in sol'n. $10^{-5}M$ in sol'n. 10^{-6}–$10^{-7}M$ in sol'n.	0.01 0.1 0.2–1.0	major and semi-micro constituents
3. Visible spectrophotometry	SL	colorimeter; spectrophotometer	200; 800–3,000	fair	0.5–1.0	10–100 ppm in sol'n. 0.005–0.1 ppm in sol'n.	1–5 5–10 $(0.1)^a$	semi-micro and micro-constituents (trace metals)
4. Ultraviolet spectrophotometry	SLG	UV spectrophotometer	3,000	fair	0.5–1.0	10–100 ppm in sol'n. 0.005–0.1 ppm in sol'n.	1–5 5–10 $(0.1)^a$	semi-micro and micro for chromophores and organic compounds
5. Flame emission spectroscopy (FES)	SL	flame photometer; flame spectrophotometer	800; 3,000–5,000	good	0.25–0.5	0.1–10 ppm in sol'n. 0.001–0.1 ppm in sol'n.	0.5–3 5–10	micro; for alkali, akaline earth, G11A and some transition metals
6. Atomic absorption spectroscopy (AAS)	SL	AA spectrophotometer	4,000–9,000	excellent	0.25–0.5	0.1–10 ppm in sol'n. 0.001–0.1 ppm in sol'n.	0.5–3 5–10	micro; for transition and some semi-metals

Method	State	Instrument	Cost	Rating		Concentration		Application
7. Gas chromatography	LG	gas chromatography	3,500–5,000	excellent	0.25–0.5	major component 1-2% 0.1-1% 0.01-0.1% 10-100 ppm 10 ppm	0.1 0.2-0.5 0.5-1.0 1-5 5-10 ≥10	major to micro; organic and organometallic constituents
8. Anodic stripping voltammetry	L	DC-pulse polarograph	2,500	good	0.25–0.5	10-100 ppm 0.1-10 ppm 0.001 ppm (in sol'n.)	1-2 3 5	micro; selected trace metals; Ag, Bi, Cd, Cu, Fe, In, Pb, Sb, Sn, Zn
9. Microscopy	SL	polarizing microscope	800	excellent	0.25	—	—	micro; for particle size, nature of compounds, fibers, etc.
10. Spectro-fluorimetry	SL	recording spectro-fluorimeter	7,000	good	0.5-1.0	0.001-10 ppm	0.5-10	micro; for inorganic and organic constituents
11. Emission spectroscopy	SL	photographic/direct-reading spectrograph	60,000	excellent	0.1-0.25	major component 0.1-100 ppm	10-15 5-10	major and micro; many elements
12. X-ray fluorescence spectrometry	SL (intact)	XRF spectrometer	50,000	good	0.25-0.5	10-200 ppm generally	1-2	semi-micro; elements in soils
13. Liquid chromatography	SL	high-pressure liquid chromatograph	6,000–8,000	good	0.5-1	0.001-1 ppm	2-20	micro; mainly organic constituents

continued on page 684

Table 3—continued

Method	Sample	Instrumentation	Cost $	Specificity	Time hr/sample	Sensitivity	Precision %	Applications
14. Polarography	L	multifunctional polarograph	10,000	good	0.25–0.5	10–100 ppm 0.1–10 ppm	1–2 3	semi-micro and micro; organic constituents and elements
15. Infrared spectroscopy	SLG	infrared spectrometer	3,000–15,000	fair	0.5–2.0	major component 10–100 ppm	5 5–10	Major and micro; mainly gases and organic species
16. Microbiology	SL	microbiology equipment	3,000	fair	12	0.005–1 ppm	20	Micro; organisms and trace metals

[a] Using precision (differential) techniques.

Notes: A summary table comparing methods must, of necessity, suffer from limitations. This table comprises the most popular methods for the chemical analysis of pollutants as suited to the medium-sized laboratory, based on personal experience. It is obvious that major factors in this selection have been the availability and cost of facilities. There is little point in recommending, say, spark source mass spectrometry or neutron activation analysis for the determination of selected analytes, if these facilities are beyond the resources of or are otherwise unavailable to, the medium-sized laboratory. The main methods omitted from the table are: electrodeposition; coulometry; ion-selective electrodes; chronopotentiometry; amperometry; turbidimetry and nephelometry; atomic fluorescence spectroscopy; electron probe microscopy; thermogravimetric analysis; differential thermal analysis; thermal analysis; thermometric titration; electron microscopy; X-ray crystallography; polarimetry; optical rotatory dispersion; refractometry; magnetic susceptibility; electron spin resonance spectroscopy; nuclear magnetic resonance spectroscopy; mass spectrometry (including spark source and isotope dilution MS); organic microanalysis; thin-layer chromatography; ring oven chromatography; activation analysis; radioisotope dilution analysis; vacuum fusion mass spectrometry; Mossbauer spectroscopy; residual resistance ratio; photoelectron spectroscopy. Summary details of these methods are readily available,[10,11,12] and some are described below. The following headings are used for Table 3:

1. *Method.* The commonly accepted name is listed for convenience, but many variations are possible. For example, "titrimetry" can involve different processes such as: proton transfer (acid-base and nonaqueous titrations); electron transfer (redox titrations); association (complexometric and precipitation titrations).

2. *Sample.* The most usual forms of the sample suited to the method are listed: Solid (S), liquid (L), gas (G).

3. *Instrumentation.* The most usual name and cost (United States $) are listed.
   Cost $. Overall costs (instrument cost plus time per determination) can also play a vital role in the selection of a suitable method.

4. *Specificity.* As a general guide, the range used is: poor, fair, good, excellent. The intention is to indicate the general ability of the method or instrument to measure the analyte or signal in the presence of matrix interferences. The classifications are obviously rather broad and subjective; suitable separation or preconcentration methods can often be used to greatly improve specificity.

5. *Time.* These estimates are rather arbitrary but may be of some assistance. In this table the overall time is estimated for the determination of an analyte in a small number of samples including the more usual chemical processing (e.g., dissolution of sample). For example, the actual instrumental measurement in atomic absorption spectroscopy (AAS) may take less than 2 min, but dissolution of a soil sample prior to AAS may take 0.5 hr.

6. *Sensitivity.* The more usual concentration ranges are given. The attainable precision can vary with the concentration determined.

7. *Precision.* The term is used here in the broad sense of measurements which relate to the variation among test results themselves. Accuracy data (measurements which relate to the difference between the average test results and the true result when the latter is known or assumed) can only be considered for a particular case, and are not included in the table.

8. *Applications.* The most common application for the chemical analysis of pollutants is listed.

2.2. *Adaption and Development of Methods*

Step 4 in the selection of methods comprises a time-consuming and difficult area which is often avoided, unless it is an area of research interest or a pressing solution is required. Regrettably, many chemists have neither the experience nor the basic training in the "analytical chemistry approach" to undertake fruitful work in this area. However, some general points for guidance can be given:

1. A thorough understanding of the overall composition of the sample (i.e., analyte plus matrix), its preparation, storage, and any subsequent treatment.
2. A detailed understanding of the best current methods and their deficiencies. This implies a good knowledge of relevant literature and adequate experience with the necessary laboratory and instrumental techniques.
3. The known deficiencies should then be listed with particular attention to:
   a. Sample processing or pretreatment.
   b. Specificity with respect to the analyte
   c. Influence of matrix
   d. Amounts or concentration ranges for determination of the analyte
   e. Sensitivity and precision required for the specified amounts or concentration ranges
   f. Skills and facilities required; time and cost
   g. Performance with respect to standards or certified materials, and especially with respect to accuracy
4. Decision whether to undertake development of the existing method or to develop a new method.

It would take many volumes to detail the development approaches recommended for the 50 broad classifications of methods for chemical analysis. An idea of the work required can be illustrated in the case of the development of a new or improved method involving spectrophotometry. Broadly, the approach applies not only to the familiar UV–visible spectrophotometry, but also to other methods involving spectrophotometry such as atomic absorption spectroscopy (the latter with some additional considerations) or spectrofluorimetry, etc.

The basic approaches required for the development of a new method of spectrophotometry may be studied in two typical papers. Yoe[15] gave details of a rapid systematic method for the screening of organic reagents for inorganic ions, utilizing small-scale testing procedures. The reagent is dissolved in a suitable solvent and its reactions are noted with 75–80 ions on a spot plate or in microglass cells (0.5 ml) in acid, neutral, and alkaline media. Further studies may then be made with the addition of buffers, complexing agents, etc.

In the reverse manner, West[16] has outlined the procedure adopted for the qualitative examination of a range of anionic metallochromic reagents in the search for a suitable reagent for the spectrophotometric determination of a particular metal; in this case, Ag.

When one or more color reactions are obtained with a particular reagent, it is then necessary to make a qualitative assessment of the intensity of the color, its rate of development or fading, and the dependence of these factors on the pH of the medium, solvent, etc. It is also possible to assess the potential sensitivity and selectivity of the reaction at this stage by an experimental determination of the detection limit on a spot plate. Semiquantitative investigations may also be made of the effect of masking agents on the color of the reaction, in the presence or absence of other ions. Preliminary spot plate or microglass cell studies can, in this way, provide a great deal of valuable information which may assist in the formulation of the general approach to the quantitative study of the color reaction.

A further approach in the development of a new spectrophotometric reagent is to study a currently used reagent and from structural considerations to tailor more selective or sensitive reagents. There are hundreds of examples in the literature,[17,18] but an excellent example is the development by Florence[17,18] of the reagent PADEP (2'-(2-pyridylazo)-5'-diethyl aminophenol) from considerations of the use of PAR (4-(2-pyridylazo)resorcinol). Florence used the reagent PAR for the highly selective and sensitive determination of U. With a masking solution containing cyclodiamine tetraacetic acid (CyDTA), fluoride, and sulfosalicylic acid, no interference was encountered from over 30 metal ions. The introduction of the diethylamino group in place of the hydroxyl group in PAR led to PADEP, resulting in much greater sensitivity. Some data for the two reagents are

$$\text{PAR: } \varepsilon\ 38{,}700; \sigma_p = -0.37; \lambda_{max} = 530\ \text{nm}$$

$$\text{PADEP: } \varepsilon\ 76{,}000; \sigma_p = -0.70; \lambda_{max} = 564\ \text{nm}$$

where ε is the molar absorptivity; σ_p is the Hammett substitution constant[19,20] for the p-position; and λ_{max} is the wavelength for maximum absorption.

The important factors that influence a new spectrophotometric method of chemical analysis and which require study are summarized below. It is, however, instructive to read several of the many hundreds of examples in the literature. Two recent examples for the determination of traces of Mo in environmental materials are described by Haddad[21] for spectrophotometry and spectrofluorimetry based on the use of a ternary complex, and by Kim[22] for atomic absorption spectrophotometry based on the solvent extraction of the Mo-thiocyanate complex.

The factors to be examined in the development of a spectrophotometric method for chemical analysis can be summarized as follows:

1. Choice of solvent
2. Spectral characteristics of the complex
3. Choice of pH and consideration of pK of any buffers used
4. Effect of reagent concentration

5. Checks of conformity to Beer's law and the optimum concentration range
6. Stability of the color
7. Effect of temperature
8. Precision and sensitivity
9. Interferences (check for 40–50 elements and anions likely to be present, usually in 100-fold excess of analyte; check for inert salts, usually at 1 g/50 ml level)
10. Masking and separations
11. Checks with standards or standard materials

Further details of such checks are given by Kirkbright.[23]

3. Air

3.1. Introduction

The quantitative chemical analysis of gaseous pollutants may involve concentrations of the analyte varying from high (e.g., in smokestacks or chimneys) to ultra trace (e.g., in the upper atmosphere). A great difficulty confronting the analytical chemist in this area centers around the provision of adequate standards and the comparison and evaluation of "equivalent" methods. Standard mixtures of pure gases are available from standards organizations and from commercial sources, but these only account for a very small proportion of the many constituents which are required to be determined in air. In many cases, it is virtually impossible to prepare and keep standards of the particulate materials, aerosols, and many organic and inorganic compounds in air. Aggregation, deposition on container walls, and chemical changes, are but a few of the factors which prevent the preparation of useful standards. In general, the development of sensitive chemical analysis methods for pollutants in air has far surpassed the facility to calibrate them accurately. The conventional methods for the preparation of sub-ppm standard gas mixtures used bag-type containers or evacuated glass or metal sample bombs. The reliability of the standards depended on the accuracy with which the aliquot and diluent were metered and also on the general compatibilities and storage qualities of the mixture. More recent techniques for the preparation of gas mixtures are based on the permeation rates of gases or vapors through Teflon tubing.[27] Calibrated tubes are also available commercially.[28]

Reference methods associated with ambient air quality standards have been recommended for photochemical oxidants, CO, NO_2, SO_2, hydrocarbons, total suspended matter, and for asbestos, by the U.S. Environmental Protection Agency,[24] but there are difficulties with the "equivalence" of the various reference methods.

The worldwide state of analytical chemistry as applied to clean air can be evaluated in the proceedings of international congresses,[25] in reviews,[12] and in compilations of official methods.[26]

3.2. *Optical Properties*

The optical properties of air offer scope for the application of instrumental methods for the evaluation of particulate matter for air quality standards. Such standards are stated in terms of visibility in kilometers or in $\mu g/m^3$. The more simple equipment is based on measurement of decrease in opacity or reflectance (soiling) of deposits on filters.[12] More recently, microwave excitation (with wavelength isolation filters) and laser beams have been used to monitor air pollutants.[12] Laser optoacoustic spectroscopy gas analyzers now provide a single system which can automatically analyze a mixture of gases with high rejection ratios and sensitivity.[29] A discretely tuneable CO_2 laser provides a beam which is chopped and passed through a gas cell (containing the air sample), with both microphone and IR detectors linked to signal-detection electronics and a small computer. Detection limits in the range 1 ppb have been reported.[29]

3.3. *Filtration and Identification*

Many air quality standards are based on the collection of particles from aerosols or from air by filtration on a substrate to determine mass concentration and composition. Most of the international and national standard methods are based on the use of long-term, high-volume samplers using glass fiber filters or those made from inert esters of cellulose, with uniform pore sizes ranging from 8 μ to 0.01 μ. The specified times for sample collection range from 24 hr to 1 month continuous sampling. These long-term samplers give average values which obviously bear very little relationship to peak values of the particle or analyte in the air. Even in the space of 24 hr weather conditions can vary greatly. Humidity and moisture, for example, could vary within this time from very low to high, with continuous rainfall. The wind velocity could also vary from zero to very high, thereby causing greatly varied dispersion.

Increasing attention is now being given to hourly and even shorter-term sampling. Such sampling can provide valuable information about source contributions and provide peak values which may have great significance to health. Several rapid microtechniques are now available for taking air samples as small as 200 ml with the collection of only 1 or 2 μg of constituents. An example is the direct determination of lead airborne particulates by nonflame atomic absorption spectroscopy.[30] The filter disks were less than 3 mm in diameter and total sampling and chemical analysis time was well under 5 min for determination.

A great variety of alternative filter media are being studied and compared. Maximum collection efficiency is required for sub-μ particles, without interference from absorption of reactive gases, in order to obtain the total mass of the sample with minimum flow resistance. Neutral glass filters are now available to obviate the absorption of acid gases by the residual alkalinity of the more usual glass fibers.[12]

One of the most versatile approaches to the collection and examination of filter deposits is provided by commercially available apparatus, filters, and methods.[31] In all cases the filter medium can be selected with a precisely defined pore structure and the collected material is then examined by one or more of the following methods: microscopic identification (visible light techniques: oblique incident light, transmitted light, and phase contrasts); electron microscopy and electron microprobe spectroscopy; chemical spot tests (including Weisz ring Oven[32]); infrared spectroscopy; UV–visible spectrophotometry; analytical flame spectroscopy (FES, AAS, and AFS); emission spectroscopy; X-ray fluorescence spectroscopy, X-ray diffraction; and radiochemistry (including activation analysis). Of particular assistance to those undertaking trace chemical analysis of pollutants under clean conditions[5] are the notes, methods, and bibliographies[31] on such aspects as: clean room sampling; full-flow and isokinetic sampling; particle counting analysis; and microbiological examination.

3.4. *Continuous Mass Analysis*

Static methods of mass analysis for air samples involve several handling steps and are time consuming and costly. In recent years attention has been paid to continuous mass analysis. The continuous gravimetric analyzer weighs the amount collected after conditioning.[33] Other methods are based on: impaction onto adhesive-coated crystals[34]; deposition onto a piezoelectric crystal[35]; and the absorption of either α[36] or β radiation[37] by the mass. Most of these methods suffer disadvantages which preclude their present acceptance as standards.

3.5. *Inertial Size Separation*

Air quality monitoring also requires information about particulate mass and concentration as a function of particle size. Various types of impactors can be used for size-selective sampling, but many suffer from deposition and fractionation errors in the submicron region, and the comparison and standardization of the various systems involve many difficulties. Discussion of design and limitations is beyond the scope of this chapter, but the most important designs involve cascades, cylinders, impingers, centrifuges, cyclones, and impactors of various types.[12,31]

3.6. *Inorganic Composition*

The inorganic composition of air includes gases, the elements, and their compounds. The latter two classes of constituents can also be present in vapor form or as aerosols. Many standards or regulations for clean air still specify many of the allowable concentrations in terms of the total element, but increasing attention is now being paid to the form of compounds in which the pollutant is present (e.g., Pb in Table 2). This has arisen from the recognition that the form of the element is all important in absorption and ultimate toxicity. Air containing oxides of nitrogen and compounds of sulfur in the presence of ozone and moisture will obviously lead to the production of a complex and often unstable mixture. Likewise, the effects of inhalation of air containing varying compounds of mercury or lead, often in aerosol form, will be closely dependent on the forms of the element actually present, and only limited use can be made of data for total-element concentrations. It is likely that considerable development of new methods of chemical analysis will be required in the next few years to specify the amount and form of a trace constituent in air. At present very few suitable methods are available, as can be gauged from recent reviews.[12]

3.6.1. *Volumetric Methods*

The Orsat method[38] is one of the older, more simple, means for determining macroamounts of gases in air. It is mainly used for the analysis of one or two constituents in a known gas mixture. A measured volume of gas is passed through several reagents and the decrease in volume caused by each reagent is measured in a simple volumetric apparatus. Confining liquids are used to transfer the gas from one pipette to another in the Orsat apparatus. A commonly used liquid is mercury, covered with a film of water to maintain constant humidity in the gas. The main gases which can be analyzed in this way are: CO_2, SO_2, H_2S, HCl, O_2, CO, and H_2. Although some of these gases are pollutants in air, the method is not recommended for trace analysis.

3.6.2. *Spectrophotometric Methods*

The more familiar chemical methods for the analysis of gaseous pollutants in air are usually based on spectrophotometry. A useful book, which gives working details of such methods, has been written by Ruch.[39] Simple methods are given for the following inorganic constituents: NH_3, As, Br_2, CO_2, CS_2, CO, CCl_4, Cl_2, ClO_2, CrO_2Cl_2, $(CN)_2$, CNBr, CNCl, boranes, F^-, NH_2, NH_3, HCl, HCN, HF, H_2S, I_2, $Fe(CO)_5$, $Ni(CO)_4$, NO_2^-, NO_2, O_3, $COCl_2$, PH_3, SbH_3, SO_4^{2-}, SO_2, S_2Cl_2, SO_3, SO_2Cl_2. The required volumes of air may be measured by displacement, with a mechanical gas meter, or with flow meter tubes. With a reactive constituent, it may first be absorbed or reacted with the required reagent before the residual gas is measured. Absorption of the analyte is usually carried out with two fritted bubblers in series. Alternatively, the sample is passed through a filter medium of specified pore size,[31] and the material which is retained is dissolved and analyzed. Although these methods are generally simple and use inexpensive equipment, they can be costly in terms of time and operational staffing. Many of the methods are also prone to interferences, and should therefore be used with the appropriate checks and quality controls.[40,41]

Table 4. Features of Analytical Flame Spectroscopy

Name	Development	Instrument basic features	Cost $	Elements
Flame emission spectroscopy (FES)	1850–present	Grating 0.5 Å 1st ord.; UV–visible range PM detectors; mech. or elect. modul.; amplif.; 0–200-μ slit	8000–10,000 (filter flame phot. 1,000–6,000)	Alkali metals, alkaline earth elements G.111A (Al, Ga, In), lanthanides, some transition elements (e.g., Re, Ru, W) approx. 24 elements give lower det. limits (total approx. 65)[a]
Atomic absorption spectroscopy (AAS)	1952–present	As FES, plus range of hollow cathode lamps (HCL)	8000–10,000 plus HCL 1,000–2,000	Transition elements and some semi-metals, Ag, As, Au, B, Be, Bi, Cd, Co, Fe, Hg, Ir, Mg, Ni, Pb, Pt, Sb, Si, Sn, Te, Zn; approx. 21 elements give lower det. limits (total approx. 65)[a]
Atomic fluorescence spectroscopy (AFS)	1964–present	1. As FES, but special optical and electronic design[c]; high int. HCLs or microwave-excited electrodeless discharge lamps required 2. Nondispersive systems using solar blind PM detectors	21,000[b] 15,000[b] 35,000[b]	Those elements mostly having fluorescence emission between 1,800–3,500 Å: Probably sig. lower detection limits for: Ag, As, Au, B, Bi, Ce, Cd, Co, Cu, Fe, Ge, Mn, Mo, Ni, Pb, Sb, Sc, Se, Te, Tl, W, Zn (total approx. 22 elements)

[a] Analytical flame spectroscopy can be applied to a total of some 65 elements: FES and AAS give comparable detection limits for approximately 17 elements.

[b] Technicon AFS-6 commercial instrument using ordinary HCLs cost $21,000 (now discontinued); probable cost of first-generation instrument: $15,000; basic cost of UNSW computer-interfaced AFS instrument (excluding labor and development costs): $35,000.

[c] Doublebeam; integration and signal sorting; condensing mirrors; wide, nearly square 'slit'; special uv sens. PM.

3.6.3. *Analytical Flame Spectroscopy Methods*

A summary of the features of analytical flame spectroscopy is shown in Table 3. Methods 5 and 6 (Table 3) are now widely used for the determination of many inorganic constituents which are pollutants in air. A third related method, atomic fluorescence spectroscopy (AFS) was not included in Table 3, because it is in a comparatively early stage of development, and no commercial instruments are currently available.

The three methods (FES, AAS, and AFS) are complementary and each offers certain advantages which are shown in Table 4. AFS appears to offer

Table 5. Flame versus Nonflame Methods

Method	Advantages	Disadvantages	Systems and costs $
Flame	Simple; variable temp. 800–3,500°C; ox. or red. conditions; suits solutions; detection limit to 10^{-10} g; suits FES, AAS, AFS; precision ±5%	Short residence time of atoms in flames; influence of physical (sol'n.) parameters (viscosity, surface tension, etc.); influence of flame chem. processes (reaction with radicals, oxygen, anions, etc.); influence of matrix; sample processing (e.g., dissolution); 5–10 ml sample	Generally inexpensive; nebulizer, flow control, burners ≪$1500
Nonflame	Simple; drying, ashing, and atomize steps 110–3,500°C; reducing conditions (N_2 or Ar); simplified sample processing suits micro-samples (e.g., 1–20 μl soln's. or 1–5-mg solids); reduced influence of radicals, anions; significantly lowered detection limits to 10^{-12}g; suits mainly AAS and FES; precision ±5–15%	Microanalytical skills required; optical system requires more careful adjustment; matrix and non-AA effects can influence; requires fast recorder or integration; at present unsuited to commercial high-sample throughput	Generally more expensive (auto. power supply, special attachment, electrodes or ribbons req'd). 1. Perkin Elmer HGA (Massman type) >$4000 2. Ta strip (Barnes) $2400 3. Varian-Techtron CRA63 (Matousek type) $2400

Table 6. Advantages of Analytical Flame Spectroscopy

What it does
  Quantitative analysis of approximately 65 elements; applicable to most environmental materials.
What it does not do
  Provide information on state of chemical combinations (specific compound, valency, or ionic form).
Advantages
  Speed. With simple chemical processing can attain 200,000 determinations/instrument/year (AAS).
  Analysis cost. Has revolutionized chemical analysis in many areas (e.g., mining, metallurgy, water, etc.). As low as $1.50/element/sample (AAS).
  Instrument cost. Basic $10,000. Low maintenance.
  Routine operation. Most suitable.
  Chemical processing. Generally minimal and simple. Most interferences can be removed.
  Precision. Usually ±5% (overall); frequently ±1%.
  Percentage accuracy. With good standards, in correct range, usually better than ±5% (overall).
  Micro-samples. With nonflame methods 1–20 μ l or 1–2 mg.
  Limit of detection. With nonflame methods to 10^{-12} g.

considerable advantages for the trace chemical analysis of approximately 22 elements and the method is well covered in recent reviews[42–44] and in an article dealing with applications to the chemical analysis of environmental materials.[45] The use of nonflame methods for AAS and AFS is of particular interest in the determination of inorganic pollutants in very small samples of environmental materials.

Table 5 lists some of the advantages and disadvantages of flame compared with nonflame methods. Nonflame methods are currently undergoing rapid development and will find wide applications in the chemical analysis of environmental materials.

The overall advantages of analytical flame spectroscopy for the determination of inorganic pollutants are shown in Table 6.

Applications to the determination of inorganic constituents in air samples are listed in recent reviews.[12] A typical rapid procedure for short-term air sampling capable of wide application has been described by Matousek and Brodie.[30] Although this method describes a rapid nonflame (AAS) procedure for the determination of lead airborne particulates, it could be easily adapted for other elements. Sampling and chemical processing requirements, prior to determination of inorganic constituents in air samples by analytical flame spectroscopy, are essentially similar to those described above for spectrophotometric methods. Generally, the methods are more rapid and less subject to interferences from matrix constituents. The few exceptions relate to thermally stable compounds in the flame, and mainly concern certain oxides, AlO_2^-, SiO_3^{2-}, SO_4^{2-}, and PO_4^{3-} (except for certain nonflame applications[30]). Acid dissolution procedures using other than sulfuric and phosphoric acids are therefore recommended. For example, the following acids, alone or in combination, are recommended for dissolutions: HCl, HNO_3, HF, $HClO_4$. It is recommended that $HClO_4$ always be used in the presence of some HNO_3, and that care be taken with organic material and certain inorganic compounds.[6] Separation procedures may be also necessary to remove and concentrate the analyte from matrix interferences following dissolution. A separation procedure of wide application in analytical flame spectroscopy involves the use of the APCD (ammonium pyrolidinecarbodithioate)–MIBK (methyl isobutyl ketone) extraction with direct aspiration of the solvent into the flame.[46]

3.6.4. *Emission Spectroscopy Methods*

Emission spectroscopy (ES) offers many advantages (Table 3) for the simultaneous qualitative and quantitative determination of inorganic pollutants. Some 70 metallic and metalloid elements can be rapidly determined in a few milligrams of a solid sample or in an appropriate solution using the rotating disk or plasma jet techniques.

The fundamentals of analytical emission spectroscopy are well covered in a recent book by Mika and Torok.[47] The best compilation of a wide range of methods for emission spectrochemical analysis is available in an ASTM book.[48]

Instrumentation is expensive compared with that required for, say, spectrophotometry or atomic absorption spectroscopy, but the additional cost can often be easily justified by high sample throughput, extreme versatility, and excellent specificity. Consideration of criteria for the selection of ES instrumentation is beyond the scope of this chapter, but some general guidance can be given. Many industries and large laboratories have sufficient throughput of similar sample types to justify purchase of a special instrument. For example, these may be designed as direct readers for either ferrous or nonferrous samples. A direct-reading grating vacuum spectrograph can also be used for the rapid determination of nonmetallic elements, including carbon, nitrogen, oxygen, and halogens at concentrations as low as 10 ppm.[49] Direct readers designed for the rapid analysis of metallic and nonmetallic impurities in aluminum and aluminum alloys, for example, are often computer-coupled and can provide a printout of many impurity concentrations in 1–2 min. In the author's opinion, the most versatile emission spectrograph for general purpose work in a research, industrial, or environmental laboratory working with pollutants, is a research direct-reading–photographic instrument[50] that costs approximately \$60,000. Multipurpose emission spectrographs usually suffer deficiencies of design and performance in one or other mode of operation, but there are exceptions.[50] The special characteristics of this instrument include a photographic recording head which can rapidly be interchanged with direct-reading heads, with precision mounting of up to 16 exit slits and photomultipliers. A particular head can be set up, for example, for the determination of selected pollutants in soils, and another head in a similar manner for selected pollutants in waters. Changeover to suit a new matrix is then very rapid using the preset heads. Direct reading brings many advantages in terms of convenience, accuracy, and precision in ES. Precision can often be improved to $\pm 1\%$ of concentrations above 0.5%. Perhaps the most important advantages are the elimination of photographic emulsion calibrations and measurement, which can take 20–30 min. In addition, the photomultiplier tubes provide a measurement of light intensity range which is 10^3 times that of a photographic film. One of the most important recent advances in the design of such a multipurpose ES is concerned with the development of a new optical alignment servo system which eliminates the need for elaborate temperature and barometric pressure controls in the laboratory. This servo system is based on the continuous monitoring of a mercury line via a PM tube, servo amplifier, and servomotor, operating a deflection plate at the entrance slit.[50] Except for the few moments while a sample is being exposed, the servo system operates continuously and corrects for changes in position as small as one hundredth of the exit slit width. The electronic readout in such an instrument operates a conventional teletype printer unit and is fully computer compatible. Special logarithmic readout is available; this is ideally suited to ES and also provides automatic ranging and single-point standardization for maximum accuracy.

An ES method well suited to the determination of a range of inorganic pollutants in liquid effluents and in environmental samples is the rotating disk technique for solutions.[48] This is not an ultratrace technique for specified elements, but is very useful in the concentration range 0.005–10.0%. However, with suitable preconcentration procedures the range can be considerably lowered. Samples should be in solution form, in either dilute HCl or HNO_3. Indium or certain other elements may be used as an internal standard. The solution is placed in a glazed porcelain combustion boat and continuously introduced into the spark gap by means of a rotating graphite disk electrode. Excitation is between the upper edge of the disk and a graphite rod which is supplied by a controlled high-voltage, air-interrupted-type spark discharge. The rotating disk apparatus is available commercially[51] and usually consists of a graphite or tantalum spindle rotated at 10 rpm by an electric motor and having a means for attaching a graphite disk electrode onto the spindle.

The detection limits reported in the literature for ES range down to nanogram level for many elements, but the experimental limit of detection for each element depends on the nature of the sample (e.g., matrix), the excitation method, and the properties of the ES system used. Generally ES is not an ultratrace method, and with complex matrices there are very few reliable quantitative methods below the 1–10 ppm level. The numerous advantages far outweigh the few disadvantages.

3.6.5. *Gas Chromatography Methods*

Gas chromatography (GC), or, more recently, gas–liquid chromatography, is based on the volatilization of thermally stable analytes which have a vapor pressure of approximately 0.1 mm or greater at temperatures less than

400°C. It is one of the outstanding and more recent methods which have revolutionized the chemical analysis of major and minor components (analytes) for both organic and inorganic analyses (Table 3). Trace organic analysis comprises the area of greatest application for gas chromatography, but there are several GC techniques available for inorganic pollutants. Some of the inorganic constituents may be relatively involatile and may also be of fairly high molecular weight. Special sampling and processing techniques may be used in such cases, and these include pyrolysis, derivatization, and the indirect analysis of reaction products.[52] There has also been a recent review of the GC of inorganic substances by Anvayer and Drugov[53] which covers the metals and their oxides, hydrides, halides, chelates, solid nonmetals, and isotopes. A promising area for trace analysis of inorganic constituents involves the conversion of the trace element to a chelate compound or organometallic and subsequent GC determination using electron capture detection.[52] A flame photometric detector can also be used in GC for metal-containing compounds, as described by Aue and co-workers.[54] The time required for chemical analysis using GC is normally from a few minutes to half an hour (Table 3). However, for some complex samples, the time involved in sample separation, quantitative data reduction, and sample identification can extend to several hours. The accuracy of GC analysis is governed by the sampling and injection procedures, attainable resolution, the detectors and detector calibration, peak area measurements, and the availability of suitable standards for GC. The precision attainable depends greatly on the particular analytical chemist's experience and also varies for different concentration levels (Table 3).

In recent years the versatility of GC has been greatly extended by the so-called ancillary techniques.[52,55] This refers to the coupling of different instrumental or chemical methods with GC in one unified system. Examples are the coupling of GC with infrared and Raman spectroscopy, mass spectrometry, NMR spectroscopy, thin-layer chromatography, microreactor systems, and pyrolyzers.

Detailed considerations of the use of these ancillary techniques for the examination of inorganic pollutants in air are beyond the scope of this chapter; further details are available in the book by Lettre and McFadden.[55] However, the GC–MS–computer system is worthy of more detailed consideration because of its great potential in the chemical analysis of air and its pollutants. This field has been reviewed by Junk[56] and Karasek.[57] The latter author quotes a limit of detection of 10^{-11} g and a limit of identification of 10^{-10} g for GC–MS. Comparable figures for IR and UV spectroscopy are 10^{-7} g and 10^{-6} g, respectively. The most critical part of the GC–MS system is perhaps the interface device, which must remove the bulk of the carrier gas from a GC peak and transfer it into the mass spectrometer ion source. This involves a reduction in pressure from 760 Torr in the GC (peak), to about 10^{-3} Torr or less, which is required by the MS ion source. The various interfaces can be classified as direct coupled (using capillary tubing); effusive type (using molecular flow through a frit); jet orifice (using expansion and enrichment); and permeable membrane (using a thin elastomer membrane with preferential conductance to organic compounds). The choice of each interface system rests on the experience of the user and the intended use. A versatile GC–MS–computer system costs $100,000–150,000, but costs are falling with miniaturization. Applications to the determination of pollutants in air are as yet fairly limited, possibly owing to the cost of equipment and its current unsuitability for routine air monitoring. The system does, however, approach the concept of a single instrumental facility for the widest range of air pollutant analyses.

3.6.6. *Other Spectroscopy Methods*

A range of other spectroscopy methods are available for the indirect determination of inorganic pollutants in air. These include: Infrared (IR), Ultraviolet (UV), correlation, luminescence, X-ray fluorescence, X-ray diffraction, and electron microprobe spectroscopy. A detailed examination of these methods is beyond the scope of this chapter, but some general notes may be of assistance.

3.6.6a. *Infrared Spectroscopy.* IR spectroscopy is a versatile technique which is finding increasing application in the determination of a limited number of inorganic pollutants in air. The main applications are for organic constituents of air, but inorganic gases such as CO, CO_2, SO_2, NO, NO_2, and NH_3 can be determined with detection limits of 1–100 ppm using a 10-m pathlength gas cell and ordinate expansion. The detection limits (ppm) for IR gas analysis are usually stated in terms of ability to measure the absorption signal with a 2:1 or 3:1 peak-height-to-noise-level ratio. However, care should be taken not to interpret these limits as necessarily relating to air and other gas mixtures in the "field" situation. Such limits are usually quoted for calibration mixtures in the laboratory containing a known amount of ultrapure-grade gas mixed with an inert gas such as helium or nitrogen. With most air samples intense bands occur in such regions as water vapor and carbon dioxide, with consequent interferences. For many air pollution applications IR analyzers are designed for a specific application, such as motor vehicle exhaust gas analysis or process gas analysis. The instruments may be of either nondispersive or dispersive design. The former is based on broad-band IR absorption and is sensitized for the analyte of interest by means of a detector, a special cell, or a filter. Commercial analyzers of this type are usually manufactured for a particular application.

In a typical nondispersive IR analyzer for stationary-source monitoring of SO_2 radiation from two separate filaments passes through a chopper (typically 10 Hz), through optical filters to reduce background interference from other IR-absorbing components, and then through two cells. One cell is a reference cell containing a nonabsorbing reference gas such as N_2 or Ar, and the other is a cell through which the air sample (containing SO_2) is continuously drawn. The detector consists of a gas-impermeable double chamber with a partition consisting of a flexible metal diaphragm. Both chambers of the detector are filled with SO_2 and suitable windows are provided. The IR radiation passes through each cell (reference and sample), and the SO_2 in each detector is heated differentially because a portion of the IR is absorbed by SO_2 in the sample stream. The pressure rise is greater in the detector chamber on the reference side and this causes a diaphragm displacement which is electronically measured as a capacitance change. The chopping of the IR radiation causes a periodic capacitance change which modulates a radiofrequency signal from an oscillator. This signal is then demodulated and amplified and the output is fed to a meter or recorder. The following typical meter ranges are provided: 0–2000 and 0–10,000 ppm; 0–2 and 0–10%; 0–20 and 0–100%. To eliminate or minimize interference from H_2O, CO_2, and aromatics which absorb in the region of SO_2 absorbance, auxiliary absorption cells or optical filters are often incorporated in the nondispersive IR analyzer.

3.6.6b. *UV Spectroscopy.* Commercial instruments which use nondispersive absorption techniques in the UV regions for the monitoring of specific inorganic pollutant gases are also available. There are two main types of instrument for the measurement of SO_2 in gas streams. The first employs a double-beam arrangement through the reference and sample cells, with UV-sensitive photomultiplier detection. The second type of instrument is based on the measurement of the difference between absorption by the sample at the absorbing wavelength of SO_2 and at some other nonabsorbing wavelength. This type of analyzer is generally subject to fewer interferences than the IR type. For example, interference from NO_2 can be corrected for by proper selection of the reference and sample wavelengths. UV-dispersive instruments with a provision for selection of wavelengths usually suffer from the inability to select absorption regions free from interferences. Similar remarks apply to the dispersive IR instruments which have been described above.

3.6.6c. *Correlation Spectroscopy.* Correlation spectrometers are available for the determination of selected inorganic gaseous pollutants in the atmosphere. These spectrometers use either skylight or artificial light. In a typical skylight correlation spectrometer for the remote sensing of SO_2, the skylight is collected by a telescope, collimated, dispersed by a grating, and focused onto a correlation mask. The correlation mask has a pattern which is formed by depositing aluminum on glass and subsequently removing slits of aluminum that correspond to absorption lines of SO_2. If this mask is regularly shifted relative to the incident spectrum, the photomultiplier tube observes a minimum when the mask lines correlate with the SO_2 lines and a maximum when the mask is shifted away from the correlation position. The difference in light intensities measured by the photomultiplier is then a measure of the SO_2 concentration between the light source and the spectrometer. If an artificial light source (e.g., quartz–iodine or xenon) is used, the application is usually across a defined distance, such as across a chimney stack or between two buildings. Such instruments can provide long-line data employing units of concentration times distance (e.g., ppm $\times$ m). Such correlation spectrometers are usually highly specific and suitable for long-term continuous monitoring.

The artificial source also allows monitoring at night or when there is insufficient skylight.

Second derivative UV spectrometers are relatively new, but offer promise for both ambient air and stationary source monitoring.[58] Hager has listed[58] the minimum detectability (in ppb) for the following inorganic pollutants: NO (5), NO_2 (40), SO_2 (1), O_3 (40), NH_3 (1), and Hg vapor (0.5). A typical second derivative gas analyzer has also been described by Hager.[58]

3.6.6d. *Luminescence Spectroscopy.* Many optical techniques for air pollutant monitoring (e.g., IR and UV) are based on absorption spectroscopy. Measurement of a signal reduction or of a ratio is consequently required in absorption spectroscopy. The required signal often arose from an instrument incorporating a multipass optical cell (20–50 m) with a relatively large volume and internal surface area susceptible to the adsorption and reaction of reactive pollutants. In contrast, methods based on luminescence (including fluorescence and chemiluminescence) involve the measurement of an increased (emission) signal. In particular, trace chemical analysis methods based on either atomic or molecular fluorescence have the potential for greatest sensitivity, because the fluorescence intensity (or radiance) is proportional to the intensity of the exciting radiation. In effect, a signal can be "turned up" by increasing the intensity of the exciting radiation. Trace concentrations of pollutants are consequently conveniently measured by modern photomultiplier techniques in small volumes. Specificity and sensitivity are high compared with absorption techniques.

A great deal of development has been carried out on the applications of chemiluminescent reactions to the measurement of many inorganic and organic air pollutants.[59] Chemiluminescent reactions may result when a gas of interest reacts with a reagent, which may be another gas, a reagent supported on a surface, or a liquid. Interference can occur if another molecule reacts in the same way with the reagent, yielding an exothermic reaction and chemiluminescence, which may overlap with the spectral emission band of the pollutant. Interference can also occur through the presence of quenching agents in ambient air. Two of these, oxygen and nitrogen, are present in almost constant amounts, however, and any quenching effect is also fairly constant. Such interferences are generally not serious, and luminescence reactions are finding increasing applications for the measurement of air pollutants in ambient air. The main inorganic air pollutants which are currently measured by automated instrumentation employing specific chemiluminescent reactions are ozone, sulfur compounds, and oxides of nitrogen. For each of these pollutants there are a number of specific chemiluminescent reactions. An automated gas phase–chemiluminescent ozone monitor has been developed from the work of Nederbragt *et al.*[60] and is available from at least six commercial sources.[59] The monitor is based on the reaction between ozone and ethylene at atmospheric pressure. λ_{max} for the chemiluminescent reaction is $\sim$435 nm; the intensity of emission is directly proportional to ozone concentration in the range 0.003–30 ppm and there are very few interferences.

A flame photometric detection system is used in a chemiluminescent monitor for sulfur compounds in air. The system was first described in a patent by Draeger[61] and has been improved and modified by a number of other workers.[59] When air containing sulfur compounds is burned in a hydrogen-rich flame, an intense chemiluminescence occurs as a series of evenly spaced bands at 350–450 nm. This chemiluminescence results from the decay of an excited S_2^* species formed from the recombination of S atoms produced in a reducing flame environment. The flame uses an approximate 1 : 1 ratio of air to hydrogen, and the cool region is viewed with a photomultiplier behind a 394-nm interference filter. The specificity of this type of monitor has been extended by coupling in a special GC column which can differentiate between H_2S, SO_2, and CH_3SH at concentrations in the range 0.005–10 ppm.[62]

The chemiluminescent reaction between ozone and nitric oxide is used as the basis of an automated $NO–NO_2–NO_x$ air monitor. The direct chemiluminescent reaction is applicable only to the detection of NO. The now well-known reaction is

$$NO + O_3 \rightleftharpoons NO_2^* + O_2 \text{ (approximately 10\% yield)}$$

$$NO_2^* \rightleftharpoons NO + h\nu \text{ (continuum 590–2750 nm)}$$

The intensity of the emission is proportional to the mass flow rate of NO into the reaction chamber. O_3 for the reaction is generated in the monitor by passing cylinder air or oxygen through an ultraviolet source. As the O_3 and NO mix, the chemiluminescent emission is proportional to NO concentration and is measured by a photomultiplier tube. NO_x analysis is obtained by dissociating the NO_2 in the air stream by means of a thermal converter which may contain Mo metal at 200°C and has an efficiency of >99%. A typical modern monitor of this type, costing approximately \$3500, has dual channels and can provide simultaneous measurement of NO, NO_x, and NO_2, with detection limits of 2 ppb and several range settings between 0.2 ppm and 5.0 ppm full scale.[63]

Monitors for inorganic pollutants in air, based on either atomic or molecular fluorescence emission, have as yet received little attention. With molecular fluorescence, there are literally hundreds of reactions between inorganic and organic species that might be investigated as a basis for methods of determining trace pollutants in air. However, many of the reactions are sensitive to quenching processes, particularly in aqueous solution.[64]

3.6.6e. *X-ray Fluorescence Spectrometry.* The advantages of X-ray fluorescence spectrometry (XRF) for the chemical analysis of trace inorganic pollutants in air mainly reside in the nondestructive determination of a wide range of elements with good specificity, speed, and precision (Table 3). The method is readily adaptable for inorganic pollutants collected on air filters or in solution. With the most favorable cases, or where preconcentration methods are used, the sensitivity can extend to the sub-ppm level, but is more typically 10–200 ppm. Commercial instruments are available for XRF work with wavelengths shorter than 1 nm, and consequently the range of elements extends upward from sodium (atomic number 11). XRF is also well suited to automation and data evaluation, often using coupled (or dedicated) minicomputers.

The conventional wavelength-dispersive XRF spectrometer generally consists of a high-wattage X-ray tube with either a tungsten or chromium target (the tungsten tube being used for high-energy lines and the chromium tube for low energies). Samples may be placed either singly or automatically in the path of the tube radiation. The characteristic X rays induced in the sample are passed through a collimator on the periphery of the goniometer circle, diffracted by a suitable analyzing crystal, and detected by a scintillation or proportional counter. The counter is located on the goniometer circle at an angle satisfying the Bragg condition. This type of spectrometer has an overall low efficiency owing to several intensity losses through the restriction of solid angle, the low reflectivity of the analyzing crystal, and, in some cases, detector losses.

Energy-dispersive XRF is rapidly developing with solid-state detectors of high resolution used in conjunction with improved radioisotope or X-ray tube excitation. The more recently developed tubes of the transmission–target type effectively generate X rays with very little of the *bremsstrahlung* which causes degradation of background and interference in trace analysis.[66] Used in the direct-excitation mode, the tube requires only 40 W of power. Suitable small-power supplies can be constructed which do not require water cooling and special high-power AC lines. A typical energy-dispersive XRF system consists of the excitation means described above, the sample, and the detector, ranged in a fixed geometry. The overall efficiency of this system is much better than the conventional technique because of fewer, less restrictive solid angle losses, the lack of diffraction losses, and a near 100% detection efficiency over a wide energy range. The usual detector is a windowless silicon (Si) detector and spectral peaks down to and including carbon can be measured. At lower energies the conventional technique has a slight advantage over the energy-dispersive system, but the reverse is the case in the important high-energy area. One of the greatest drawbacks in the conventional technique is the necessity for relatively cumbersome spectral scanning, more difficult (related) data collection, and the complication with higher-order reflections. However, an electronic scanning, simultaneous display of all elements is available in real time with energy-dispersive XRF systems. All of the incoming signals are processed and sorted according to their energies; there remains only the choice of a range in which to look, by setting the electronic gain of the system. The memory of a coupled multichannel analyzer or minicomputer may be continuously monitored by a display oscilloscope or television, allowing the operator to inspect the entire analytical range at a single glance. With the coupling of a minicomputer total analysis is feasible, including matrix corrections and comparison with standards. The technique may also be extended by use of a secondary target system. Moreover, the adaption of the energy-dispersive X ray system to the scanning electron microscope, transmission electron microscope, and electron microprobe offers far greater potential than the older adaption to the conventional crystal spectrometer. The potential of the energy-dispersive XRF system for the determination of inorganic pollutants in air is clearly shown in composite spectra for Br and Pb in engine exhaust gases, and for Ni, Cu, Zn, Hg, As, Pb, Mo, Ag, and Cd, on a "millipore" filter.[65] With this system it is quite feasible to analyze 50–100 samples per 8-hr day.

3.6.6f. *Electron Microprobe.* Electron microprobe methods have not been extensively used for the chemical analysis of inorganic pollutants in air samples, because the technique is more suited to the distribution of elements present in a solid sample rather than for the determination of average composition. It is possible that the technique could be useful for the examination of inorganic pollutants on air filters or on other solid surfaces. The technique is suitable for investigating the distribution (on the μ scale) of elements above atomic number 5 at concentrations as low as 100 ppm with relative errors of $<3\%$ in most cases.[66] The in-depth and lateral resolutions are above 1–3 μ, and scanning techniques are available. The use of energy-dispersive systems coupled with this technique should also be noted (see above).

3.6.6g. *X-ray Diffraction.* In conventional X-ray crystallography, the sample is exposed to a beam of monochromatic X rays and the image formed by the scattered X ray is recorded on a photographic film. The characteristic diffraction patterns are usually regarded as the ultimate method of structure determination for solid pollutants which can be collected. However, more recently, the X-ray diffraction data is also supported by data obtained from IR and laser Raman, Mössbauer, nuclear magnetic resonance (NMR), electron spin resonance (ESR), and photoelectron spectroscopy.[10,12] Particles as small as 5 μ will often give an X-ray diffraction pattern with three or more lines which can aid in identifying the form of the inorganic pollutant obtained from an air sample. As mentioned earlier, the possible forms of pollutants are receiving increased attention. Further information on these specialized techniques can be obtained from recent reviews.[12]

3.6.7. Electrochemical Methods

Electrochemical methods such as polarography or coulometry are well suited to the determination of inorganic pollutants in air sample filters and other media; until recently, however, there has been a general reluctance by those studying environmental pollution problems to use them as regular laboratory methods. This has probably been due to the feeling that such methods are time consuming and require tedious manipulative procedures. This state of affairs has prompted analytical chemists to write on such subjects as: Is polarography dead?[67] and What has happened to polarography?[68] The answers are that polarography is now flourishing as a method for the trace chemical analysis of a wide range of elements and also of organic compounds. Pulse polarography and anodic stripping methods now offer considerable advantages, including lower limits of detection for nearly 25 elements as compared with analytical flame spectroscopy, molecular spectrophotometry, and neutron activation analysis.[67,68] The advantages of these methods are also listed in Table 3. The costs of modern polarographs range from $2500 for a pulse polarograph to $10,000 for a multipurpose polarograph of great versatility. A pulse polarograph is an asset to any small laboratory.

Polarography can be applied in principle to any of about 80 elements which are capable of electrooxidation or electroreduction. If the half-wave potentials are sufficiently different (usually >0.05V), several elements can be simultaneously determined in the same solution. For pulse polarography, limits of detection are in the range 0.02–20 ppb for nearly 20 elements,[67] which can be of importance in air pollution studies. Anodic stripping has limits of detection in the range 0.01–0.2 ppb for the following elements: Bi, Cd, Cu, Fe, In, Pb, Sb, Sn, and Zn.[67] Suitable procedures for the dissolution of inorganic materials prior to polarography were described above.

The determination of oxygen as an impurity or pollutant in process gases, or in applications such as food processing, can also be carried out by polarographic analyzer instruments that are commercially available.[69] The oxygen sensor electrode contains a silver anode and a gold cathode which are isolated from the sample by a thin membrane of Teflon. An aqueous solution of KCl, soaked into suitable paper, is retained in the sensor by the membrane and serves as an electrolytic agent. Oxygen from the sample diffuses through the thin Teflon membrane and takes part in the following oxidation–reduction reactions:

$$O_2 + 2H_2O + 4e \rightarrow 4OH^- \quad \text{(at cathode)}$$

$$4Ag + 4Cl^- \rightarrow AgCl + 4e \quad \text{(at anode)}$$

With a DC voltage applied across the electrodes the oxygen in the sample is reduced (cathode reaction) and a current flows which is proportional to the amount of oxygen present in the sample. The response of the system to changes in concentration is fairly rapid (90% in 20 sec) and the sensor can be used in liquid or gas streams up to pressures of $34.3 \times 10^4 \, N/m^2$, or submersion in liquid up to a 30-m depth. The monitor is useful in the range 0–25% oxygen.

Coulometry may also be applied to the determination of traces of inorganic pollutants obtained from air samples. Coulometry involves the quantitative electrochemical conversion of a constituent in solution from one initial oxidation state to another well-defined oxidation state. The product of current $\times$ time is equivalent to the quantity of electricity (coulombs) and this can be measured with inexpensive solid-state electronics with high precision and accuracy. Another way of viewing the process is to regard the electrons necessary to perform the conversion as the measured reagent. Two types of coulometry can be carried out in specially designed electrolysis cells: controlled potential coulometry and constant current coulometry. In the former type, a potential is selected and controlled so that only the desired reaction can proceed, the current decreasing with time to near zero. In the second type, a depolarizer introduced into the electrolysis system reacts at a "generator" electrode to produce a titrant for the species in question. The current is usually held constant by appropriate electronic means and the time is measured. Coulometry can be adapted in the macro- to microrange for an appreciable number of elements. Summary tables of the various applications are readily available[70] and should be consulted for specific applications.

Ion-selective electrodes are finding increasing applications for the determination of a range of cations, anions, and gases. The theory and applications of these electrodes have been described by Koryta[71] and by Moody and Thomas[72] and the recent literature has been reviewed by Buck.[73] Ion-selective electrodes are generally simple to design, rugged, and of moderate cost ($100–200 for each electrode). They can best be used with pH meters provided with special scales: 2- and 4-decade log; known increment; known decrement. The electrodes can be used over wide ranges in concentration with varying precision (in brackets): 10^{-3} M–saturation in solution (0.5–2%); 10^{-5}–10^{-3} M in solution (1–5%); 10^{-7}–10^{-5} M in solution (2–30%). About 30 cations and anions can be determined in selected matrices using ion-selective electrodes, and the electrodes are also available for gases such as oxygen, carbon dioxide, and ammonia. The limitations of these electrodes are due to interference effects termed "electrode" and "method" interferences. The former interferences are those in which the electrode cannot distinguish between the analyte ion and one or more ions in the matrix. The latter interferences occur when some characteristic of the matrix prevents the electrode from sensing the ion of interest. Most of this type of interference stems from the dependence of potential response on ionic activity rather than on ionic concentration. To help overcome ionic strength dependence, both standards and samples are treated with buffers which keep pH and ionic strength at constant levels. Caution is therefore advised in the use of these electrodes, and considerable attention should be paid to solution equilibria and electrode limitations. Ion-selective electrodes are generally used in the direct potentiometry mode (similar to pH electrodes) with correlation of the electrode electromotive force to a standard or calibration curve, but they can also be used in the potentiometric titration mode (direct or indirect) with greatly improved precision. Manufacturers' leaflets and data provide much useful information, and trials are recommended before purchase of electrodes for a specific application. Some promising ion-selective electrodes, incorporating an ion exchanger in an inert polymer matrix, which is used for coating platinum wire, have been described by Cattrall and Freiser[74] and by Moody, Oke, and Thomas.[75,76] An example is a Ca^{2+}-selective electrode using the Ca form of didecyclphosphoric acid in dioctyl-phenylphosphonate in a PVC matrix.[75,76] A simple ion-selective electrode of this type can be made by dipping a platinum wire in the active constituent mixed in a solution of polyvinyl chloride dissolved in tetrahydrofuran or carbon tetrachloride.[74] These inexpensive electrodes require some care, and there is "trial and error" involved in their preparation, but they do have good response and stability.

3.7. Organic Composition

3.7.1. Introduction

Widely varied organic substances can be present as pollutants in air. Organic smoke can arise from stationary fuel-burning sources, motor vehicles (diesel or gasoline), shipping, locomotives (diesels), aircraft, and industrial processes. Solid particles, which are mainly organic in composition, can range from the various forms of carbon through to particles containing high proportions of polynuclear aromatic hydrocarbons. Aliphatic compounds such as methane, acetylene, propane, hexane, heptane, pentane, and octane can also arise from the above-mentioned sources. Aromatic compounds which can arise in the same way include benzene, toluene, xylene, chlorobenzene, and cyclohexane. Activities by man also promote the release of appreciable quantities of more exotic organic compounds used as aerosol propellants, detergents, pesticides, herbicides, fungicides, weedicides, preservatives, and numerous organic compounds used or produced as by-products in industrial processes. In addition, the microbial constituents of air may also require evaluation as pollutants in certain medicinal operations or "clean-room" conditions. The range of methods which can be used for the chemical analysis of these organic pollutants, from ultra-trace to macro concentrations, embraces nearly all of the 50 broad classifications. The choice of a suitable method for the chemical analysis of an organic pollutant depends primarily on the constituent present and its level, the matrix, and whether continuous long-term or short-term monitoring is required. Table 3 may be of assistance in a preliminary survey of the suitability of methods.

3.7.2. Specific Compounds in Air

Reviews of the literature[12] or reference books[26,30] should be consulted in the search for a suitable method for the determination of a specific organic pollutant in air. Reference books generally give details of the older "accepted" method for such pollutants.[26,30] Katz[77] has written a guide to the selection of suitable methods for the measurement of air pollutants, but it does not include many improved methods developed during the past 5 years. In most cases, conventional laboratory equipment and instrumentation available to the smaller laboratory are described. Unless the method is familiar, operational trials will be necessary and some adaption and development may be necessary. A general procedure for the collection of a volatile organic pollutant in air involves the use of special absorbing solutions which may be either acid, alkaline, or neutral, depending on the chemical properties of the analyte (pollutant).[26,30] In other cases, the solution may be either water, alcohol, or a solution containing a reagent capable of forming a suitable reaction product. The solutions may be contained in one or more fritted bubblers in series, or midget or standard impingers may be used.[26,30] The resultant solutions may then be examined by one or more of the methods already described. Where the volatile organic constituent is adsorbed onto or contained in particulate matter in the air, it may be necessary to carry out a prior filtration through an aerosol filter medium[31] which is subsequently extracted or examined by other means.

3.7.3. *Instrumental Methods*

Instrumental methods suitable for the determination of specific organic pollutants in air include those using: optical properties (e.g., smoke detection), filtration, continuous mass analysis, inertial size separations, gas chromatography, infrared, ultraviolet, correlation, and luminescence spectroscopy, and electrochemical methods. Not all of these instrumental methods are "approved" or incorporated in standard or "official" publications. In addition, many of the more exotic methods excluded from Table 3 (see notes on Table 3), such as gas chromatography–mass spectrometry (GCMS) and ESR and NMR spectroscopy, are unlikely to be incorporated in standard, or "official," methods for another decade, even though they may be extremely versatile. It is recommended that regular reviews on fundamental analytical chemistry[12] be consulted for the possible application of these and other instrumental methods for the determination of organic pollutants in air. However, some selected instrumental methods which have not been described earlier in this chapter will be briefly described in order to further illustrate their role.

3.7.3a. *Hydrocarbon Analyzer.* Portable hydrocarbon analyzers incorporating flame ionization detection are now available commercially. Analyzers of this type are capable of routine monitoring of a number of organic pollutants, including acetylene, propane, methane, hexane, heptane, pentane, octane, acetone, toluene, xylene, benzene, ethyl acetate, ethyl alcohol, and methyl alcohol. However, unless specifically modified (e.g., by incorporation of gas chromatography), the basic analyzer is usually designed for methane-in-air determinations. This analyzer, incorporating flame ionization detection, uses the principle that when a hydrogen flame burns in "normal" air it contains very few ions. Introduction of a trace of hydrocarbon, such as methane, into the flame results in a complex ionization, producing a large number of ions. If a polarizing voltage is applied between the burner jet and a collector plate, an electrostatic field is produced in the vicinity of the flame. Ion migration results from this field effect, and positive ions are attracted to the collector and negative ions to the burner jet. The small ionization current thus established can be measured with an electrometer amplifier circuit, and is directly proportional to the hydrocarbon (methane) concentration in the flame. The amplifier circuit provides a driving voltage for an indicating meter or recorder. It should be noted that the magnitude of the signal is indicative of the number of carbon atoms passing through the flame and is also dependent on sample flow rate. A hydrocarbon analyzer calibrated at a given sample flow rate for traces of methane (CH_4) in air will give twice the signal for a C_2 hydrocarbon, 3 times for

C_3, 4 times for C_4, and so on. The instrument should therefore be calibrated for specific hydrocarbons or for mixtures. Under favorable conditions this hydrocarbon analyzer can be provided with sensitivity ranges from 1 ppm to 10% methane for full scale. A typical response time is less than 0.5 sec for 90% of final reading as methane.

3.7.3b. *Smoke Nephelometer.* A good example of optical monitors for the detection of traces of smoke in air, such as inside telephone exchanges as a fire-detection device, is provided by an instrument recently developed by Packham and Gibson of the Australian C.S.I.R.O. and Post Office.[78] The more conventional early-warning detectors are of the ionization type based on measurement of changes in the conductivity of air as induced by a small radioactive source. This conductivity is changed by the presence of either traces of smoke, or changes in humidity, temperature, or air speed. The latter interfering effects are prevalent in return-air ducts where air conditioning dilutes smoke below the limit of detection. A dual-beam nephelometer is approximately 100 times more sensitive for smoke detection in such air-conditioned enclosures than is an ionization-type detector. In the dual-beam smoke nephelometer, one beam of light is passed through a cell containing air drawn from within the building, and another beam of light is passed through a cell containing air drawn from outside the building. The difference in the amount of light scattered is proportional to the smoke particles generated within the building. Warning signals can be produced in this way with conventional electronics.[78]

3.7.3c. *GC–MS with Multiple Ion Detection.* The gas chromatography-mass spectrometer (GC–MS) system for gas analysis has been briefly described above. Recently, compact mass spectrometers capable of fast scanning and high sensitivity, and capable of accepting such new concepts as chemical ionization and multiple ion detection (or scanning) have become commercially available for approximately $40,000.[79] A typical instrument of this type is double focusing and based on a reverse Nier–Johnson geometry with high-efficiency ion source, field-regulated magnet, and high-speed pumping system. It can be coupled with a GC detector (using impact ionization) which is fitted to the ion source. The electron energy of this detector is only 20 eV, so that the helium carrier gas is not ionized. The system can also be linked with a minicomputer (8K) and a modular unit for peak jumping according to the multiple ion detection technique (MID).[81] In conventional mass spectrometry, the same amount of time is spent on small peaks and the areas between peaks as on large peaks. Consequently, an undue proportion of instrument time is spent on areas containing no useful information. This time is essentially wasted because ions of mass-to-charge (m/e) values of interest are continually being produced and are collected only when a particular m/e value is measured by the detector. MID—multiple ion scanning or mass fragmentation—is achieved by the instrument monitoring 2–16 ion signals simultaneously or in rapid succession. Consequently, a larger proportion of time is spent on important peaks, more ions are collected, and improved ion statistics are obtained.[80] The signal-to-noise ratio is improved, and with it, the limit of detection and precision for the determination of trace organic pollutants in air samples. MID is particularly useful for small samples, but a disadvantage is that it is necessary to know what peaks must be monitored before the sample is examined. GC retention times, relative peak intensities, mass peak positions, and other data may be required to take full advantage of MID.

3.7.3d. *Liquid Chromatography.* Liquid chromatography and high-pressure liquid chromatography (HPLC) are rapidly emerging as some of the most powerful of all the various chromatographic methods for the identification of organic pollutants. The method is well described in books[81,82] and in recent articles[83-85] which provide full details. HPLC can be regarded as an extension of the traditional forms of chromatography which make use of a liquid mobile phase, i.e., paper, thin-layer (TLC), and column chromatography. The basic liquid chromatograph (Table 3) is comprised of a special chromatographic column, a pump, and a suitable detector. A suitable solvent is pumped into the column and the injected sample is separated into its components in a manner similar to other kinds of chromatography. The special column in earlier forms was about 500 mm long with an inside diameter of about 2 mm packed with spheres $10-30\ \mu$ in diameter. High-pressure pumps employing pressures of up to $500 \times 10^5\ N/m^2$ were necessary to reduce the separation time to several minutes. Lower pressure pumps ($<300 \times 10^4\ N/m^2$) usually resulted in column separations taking hours or days. However, modern column packings which are now fully developed and commercially available use the pelliculars and irregularly structured microparticles for rapid separations at pressures of less than $300 \times 10^4\ N/m^2$.[84] In view of this, Taggart[84] has suggested that HPLC refer to high-performance liquid chromatography rather than to high-pressure liquid chromatography. Rapid developments in HPLC have accompanied the development of improved columns, column packings, pumps, and detectors. The general advantages of the method are shown in Table 3, but performance continues to improve with developments in instrumentation. For example, the two detectors in common use in HPLC are ultraviolet (UV) and refractive index (RI), but more recent instruments incorporate improvements in each of these systems. Earlier UV detectors were based on pen-lamp mercury sources, with simple wavelength isolation in the two principal regions using filters. Later UV detectors incorporated small monochromators and improved sources for wavelength selection and improved precision over a wider range. Earlier RI detectors had low stability and precision, but current detectors can reliably measure differences in 10^{-8} units in RI. Other types of detection systems, such as those based on molecular fluorescence, flame ionization, infrared absorption, and atomic absorption, are under investigation.[12] It should be mentioned that columns, packing materials, and column fittings are a relatively expensive part of work with HPLC. In order to carry out a reasonable range of work it is not unusual for an amount equivalent to the basic price of an HPLC instrument (approximately $4000) to be spent on necessary accessories.

3.7.4. Thin-Layer Chromatography (TLC)

TLC is an excellent method for the semiquantitative analysis of non-volatile organic or inorganic compounds and may be applied to the determination of constituents of particulate contaminants in air. The method and its applications are detailed in a book by Stahl,[86] and also in a recent book by Touchstone (ed.)[87] which deals with quantitative aspects.

In the more usual form of TLC a glass plate supports a thin adsorbent bed or film and a spot of a mixture is deposited near one edge. The edge below the spot is dipped into a selected solvent which flows by capillary action into the bed. Suitable choices of mixture–solvent–adsorbent cause the compounds of the mixture (sample) to move or develop so that the components separate. The plate is usually removed from the solvent after the solvent front has moved over 80–90% of the distance. In practice, 200–1000 μg of a mixture is applied to the plate as a spot, and typical limits of detection are in the range of 10–1000 μg of a compound. As with many other methods of chemical analysis, suitable preconcentration procedures may be used to lower the limit of detection. The main techniques for quantitative evaluation of the separated compounds on a TLC plate include visual comparison, measurement of spot areas, and visual, UV, and reflectance or fluorescence evaluations of spots. A spot may also be separated from the plate, the compound extracted from the absorbent material, and the extract then analyzed by a number of well-known methods.

Despite its many advantages, TLC also has several disadvantages. Some of these are: the spreading of TLC spots results in decreased sensitivity and possible number, which spread, as distance and time, from initial development increase; the poor definition and increased breadth of spots causes difficulties in measurement; TLC is closely dependent on operator technique; and the time required to increase the distance between TLC spots increases as the square of the distance ratio. A recent technique known as programmed multiple development (PMD)[88] is very useful in overcoming some of these difficulties in TLC. Following each development, the solvent is automatically removed from the plate by controlled evaporation with a heating element or by using an inert gas stream. The procedure is then repeated for each development. With increasing cycles, TLC spots are narrowed rather than spread and the sensitivity increases with time. Other advantages, such as the separation of a greater number of components in a mixture, are said to result from PMD.[88]

3.7.5. Organic Microanalysis

Organic microanalysis of separated or purified organic and organometallic compounds has been used for many years to aid in establishing the identity of compounds. The techniques of organic microanalysis are also commonly used in association with other physicochemical techniques such as IR or NMR spectroscopy. The principal microanalyses which are available as a routine laboratory service in organic chemistry are: C—H, N, O, S, P, As, halogens (Fl, Cl, Br, and I), O·CH$_3$, N·CH$_3$, CO·CH$_3$, C·CH$_3$, active H, carbonyl, C=O, equivalent weight, molecular weight, double bonds, saponification value, and a range of metallic elements. The techniques are rather specialized and are better carried out by a laboratory which has specialized equipment, with experienced staff, and operates on a routine basis. The available textbooks[89,90] which mainly describe the slower classical chemical procedures for organic microanalysis, should be supplemented by reading recent reviews of developments in this field.[91]

In organic microanalysis, the elements or constituents are present in small samples but in relatively high concentration. The sample size is usually 1–10 mg and the sensitivity is typically 10 μg with a precision of $\pm0.5\%$. More accurate results can be obtained with larger samples, but these take more time to analyze. Commercial CHN autoanalyzers have been on the market since 1964,[91] and these are invaluable for the laboratory which processes a larger number of samples. The better-known, older Pregl method for C and H, and the Dumas–Pregl method for N, are still used in many smaller laboratories but are becoming less economic to operate. In the former method, a weighed sample of about 10 mg is combusted in a stream of oxygen and the resulting H_2O and CO_2 are absorbed in two tubes, the first containing phosphorous pentoxide on pumice and the second soda asbestos.[89,90] The absorption tubes are weighed before and after the combustion and the necessary calculations are carried out to provide the percentage composition of the sample with respect to C and H. A typical time for a complete analysis is 40–45 min, exclusive of all weighings. In the second method, a weighed sample of about 5 mg is mixed with Co_3O_4 and heated in a pure CO_2 atmosphere. After reduction of any resulting oxides of nitrogen, the CO_2 and N_2 are passed into a gas burette filled with potassium hydroxide solution, and the unabsorbed N_2 component is measured and necessary calculations carried out to provide the percentage of N in the sample. A typical time for this analysis is 35 min, exclusive of all weighings.

A great deal of attention has been directed toward the use of CHN and other autoanalyzers for organic analysis which can speed up the determination time. Some of the more recent autoanalyzers are coupled to a dedicated minicomputer, and in this case the cost can exceed $30,000, as compared with $12–13,000 for basic instrumentation. The main advantages of a basic CHN autoanalyzer are:

1. Most of the older methods for CHN required an air-conditioned, dust-free laboratory. With the autoanalyzer, the only weighing carried out is that of the sample.
2. With autoanalyzers, the productivity of CHN analyses can be increased two or three times.
3. The required operator skill can be reduced for operation of the autoanalyzer equipment, as compared with the older methods.
4. The sample size can be reduced to 0.5–3.0 mg, depending on the design of the autoanalyzer.

Manufacturers' catalogues supply useful descriptions and data relating to autoanalyzers for CHN as well as for P, S, and other constituents of organic compounds.

3.8. *Radioactivity*

3.8.1. *Introduction*

Radioactivity levels in the air above "natural" or normal levels can arise from both natural and man-made influences. During the last 30 years a great deal of radiochemical analysis development has been directed toward the determination and regular monitoring of radionuclides in air. This may be done on a local scale, such as in the vicinity of nuclear reactors or radiochemical processing plants, inside a radiochemical laboratory, or on an international basis for surveillance purposes related to both nuclear explosions and for the monitoring of nuclear reactor and processing operations in a country of interest. Natural nuclides entering the atmosphere can arise from a variety of sources, such as radioactive ore deposits, certain beach sands, or gases issuing from volcanoes. In such cases, the main emission is a radioactive gas such as radon. However, most monitoring of radionuclides in air is directed toward measurement of man-made radionuclides over and above the natural levels.

This field is so extensive and specialized that it will not be discussed in detail in this chapter. The reader is recommended to read some recent texts[92,93] and reviews[94] for essential background information.

3.8.2. *Radioactive Particulate Contaminants in Air*

Suitable methods for the determination of radioactive particulate contamination in air range from those issued by international agencies, such as the IAEA,[95] to manuals issued by atomic energy research laboratories,[96] and the many individual papers listed in Nuclear Science Abstracts. Collection media range from high-volume air samplers, using special graded pore size filter media[31] and impingers of various types, to relatively crude collection devices on wings of high-flying aircraft, using adhesive surfaces. For the smaller laboratory, simple equipment and procedures are described using special membrane (graded pore size) filters.[31] The main requirements are a suitable aerosol open filter holder, special membrane filter papers, a calibrated vacuum pump, and either a proportional counter or scintillation counter.[31,92–95] Membrane filters are particularly suitable for sampling radioactive particulate contamination because of their high retention. The particle is deposited on or close to the front surface and corrections for absorption of the radiation are either eliminated or greatly reduced. Limiting orifices can be inserted in the filter holder to provide collection rates in the range 0.5–10.0 liters/min for up to approximately 30 min. The lower the expected radioactive contamination the greater the flow rate used for collection. Counting for α-, β-, or γ-active nuclides is then carried out on the filter and, in some cases, radiochemical separations of certain nuclides may be required.[92–96]

Contact autoradiography is also used for the examination of particulate material collected on membrane filters. This simple technique also allows microscopic examination of the material. The contact autoradiography is carried out by placing the dried filter face down and clamping in contact with either X-ray film, Kodak nuclear track plates, or Kodak autoradiographic plates. The exposure time, which is usually about 2 hr, will depend on the level of radioactivity of the contaminant, the particle size, and the type of emission. Emission from radionuclides produces stellar radiates (spots) in the emulsion. Burnett[97] has shown that the size and intensity of spots produced in this manner depend on the exposure time and radiation intensity of the individual particles. By calibrating against particulate matter produced from standards it is possible to obtain semiquantitative measurements on both overall activity of the sample and activity of individual particles. For example, Burnett showed that a 1-mm-diameter spot on an X-ray plate corresponds to approximately 10^4 disintegrations, or to a 30-min exposure to a particle of $1.5 \times 10^{-4} \mu$ Ci activity.

As radium and radon occur in the greatest concentration in radioactive minerals, radon 222 and thoron are commonly determined in air as "natural" pollutants in areas where these minerals occur. The decay products of radon become attached to particulate matter in air and are commonly determined by passing the air through a filter[98] or over a negatively charged electrode[99] at a known rate for a given length of time. The measurement can be made by either alpha-, beta-, or gamma-ray counting, but alpha activity is usually measured, and if necessary is aided by alpha spectrometry[100,101] because of the presence of other beta- and gamma-emitting nuclides present in "normal" air and resulting from fallout following nuclear explosions in the atmosphere. Most of the current methods are scintillation counting.[100,101] Other collection methods for the determination of radon 222 in air use either activated charcoal[102] or collection by freezing in a liquid air-cooled trap.[103] The radon 222 content of "normal" air is in the range 0.03–1.1 pCi/l.[104]

The methods described above are illustrative of the numerous methods for the determination of natural radioactivity, in the form of particulate contaminants, in air. However, in the past 30 years, with the development of atomic energy, many methods have been developed for the determination of man-made radionuclides and particularly the fission products in air. The most important of these fission products, as pollutants in the atmosphere, are the nuclides of relatively long half-life, which enter the atmosphere either as gases adsorbed on particulate matter, or as elements or compounds. These nuclides and their half-lives (in brackets) are: Kr 85 (10.4 years); Sr 89 (54 days); Sr 90 (28 years); Ru 106 (1 year); Tc 127m (105 days); Te 129m (34 days); I 131 (8 days); Xe 131m (12 days); Xe 133m (2.3 days); Xe 133 (53 days); Cs 137 (33 years); Ce 144 (282 days); Ba 140 (12.8 days).

Many of the methods for the determination of one or more of these nuclides involve radiochemical concentration and separation techniques,[92,93,95,96,100,101,105] and detailed consideration is beyond the scope of this chapter. Similar considerations apply to other important man-made radionuclides which are present in the atmosphere. These include tritium (12.26 years); C 14 (5.8×10^3 years); Cr 51 (27.8 days); Co 58 (71 days); Co 60 (5.27 years); Fe 59 (45 days); Mn 54 (314 days); Mo 99 (2.75 days); Zn 65 (245 days); Pu 239 (2.43×10^4 years); and other actinides. Many of these radionuclides arise from nuclear explosions, reactor operations (including corrosion products), or other nuclear processing. Suitable methods for these radionuclides are also found in the methods listed earlier for fission products and in various international[95] and national publications.[96]

3.9. *Instrumentation Developments*

Overall consideration of the preceding sections dealing with the chemical analysis of pollutants in air will indicate that the choice of physicochemical instrumentation is often dictated by the size of the laboratory coupled with the complexity and number of determinations which are required. The smaller laboratory will often prefer methods based on more simple instrumentation, e.g., spectrophotometers possibly preceded by filtration and microscopy. Alternatively, a small instrument may be purchased or built for the determination of a single pollutant in air under specified conditions.

The considerable advances in solid state and microcircuitry, coupled with improvements in the specificity and sensitivity of detection systems, has made multipurpose instruments a more attractive proposition for the larger laboratory. There is a definite trend toward continuous operation remote-sensing instruments for air pollution studies. While some of these may only be designed for the detection and measurement of a single constituent, it is often feasible to control the operation of several different types on a continuous basis with dedicated minicomputers, which are now available for less than $10,000. A multipurpose instrument with the greatest potential in this field is possibly the GC–MS–computer system. It is recommended that developments in these areas be followed in regular review series.[12]

4. *Water*

4.1. *Physical Data and Gases*

Instrumental methods are well suited to the determination of physical data and gases in various waters. The main instrumental methods for the determination of physical data and gases in water are shown in Table 7.

Certain procedures, such as the determination of threshold odor[4] and taste,[106] were not included in Table 7 because they are not based on instrumentation. Further details on the instrumentation and operating principles relating to Table 7 are readily available in the literature.[4,106,107–110] However, some selected instrumental methods will be described in further detail to illustrate their capabilities in this field.

Table 7. Instrumental Methods for the Determination of Physical Data and Gases
in Water

Method	Parameter measured
Iron-constantan thermocouple	Temperature (°C)
Epply pyranometer	Solar radiation (J/cm^2 min)
Spectrophotometric or tristimulus	Color units
Turbidimetric or nephelometric	Turbidimetry (J.C.U.)
Wheatstone bridge	Specific conductance (μS/cm)
Electrometric	H$^+$ activity (pH)
Potentiometric	Oxidation–reduction potential ORP(mV)
Gas-sensing electrode	Gases (e.g., O_2, NH_3, SO_2) (mg/liter)
Reflectance	Oil and grease (mg/liter)
Polarographic	Biological oxygen demand BOD (mg/liter)
Nuclear counter	Radioactivity (pCi/liter)
Conductometric	Salinity
Gravimetric	Density
Combustion–infrared	Biodegradability
Centrifugation–ultraviolet	Biochemical activity
Centrifugation–gravimetric	Colloids
Centrifugation	Clay minerals

4.1.1. Instrumental Methods for Dissolved Gases

Electrochemical methods based on coulometry or polarography can be used for the determination of low levels of certain gases, such as oxygen, in air, gas streams, and waters. A sensor electrode system using a silver anode and a gold cathode, which are isolated from the sample by a thin membrane of Teflon, has been described above. This system is commercially available as instruments incorporating dissolved oxygen (DO) probes capable of operating in waters up to depths of 200 m or up to pressures of about 205×10^4 N/m. The instrument is usually provided with a readout meter incorporating an oxygen scale for various ranges (dissolved oxygen ranges: 0–1 ppm, 0–10 ppm, 0–25 ppm; gaseous oxygen: 0–2.5%, 0–10%, 0–25%, and 0–100%). The submersible sensor contains separate temperature and DO sensors and temperature compensation is usually available in the range 0–60°C. The response time for dissolved oxygen is 90% in 15–30 sec (depending on variations in design of the DO sensor) and the response time for gaseous oxygen is 90% in 10 sec. A further variation in design of the DO sensor will permit ppb dissolved oxygen measurement in boiling water.

While oxygen may not be strictly regarded as a pollutant in water, its presence (particularly a deficiency) and relation to tests for biological oxygen demand (BOD), chemical oxygen demand (COD), and organic carbon (OC), is a most valuable indicator of several types of pollution in waters; but mainly for relation to gross discharges of organic material.[111] The DO probe procedure is also an alternative finish to the BOD biochemical-type procedure, using a modified Winkler with full-bottle technique, which will be described below.

Some comments may be appropriate regarding possible interferences to methods involving DO probes. While dissolved organic materials are not known to interfere, dissolved inorganic salts can cause interference.[4] As the membrane-type DO probe responds to partial pressures of oxygen, which in turn can be a function of dissolved inorganic salts, conversion factors are required for brackish water and seawater. These factors can be easily calculated from dissolved oxygen-saturation-versus-salinity data. Reactive gases such as chlorine or hydrogen sulfide will also interfere with DO probes.[4] Chlorine will depolarize the cathode, causing excessive probe outputs, while long exposures will coat the anode with the chloride of the anode metal and desensitize the probe. Hydrogen sulfide will also interfere in different ways, depending on the applied potential in relation to the half-wave potential ($E_{\frac{1}{2}}$). If the applied potential is less than $E_{\frac{1}{2}}$ coating of the anode with the sulfide of the anode metal can take place. When the applied potential is greater than $E_{\frac{1}{2}}$, interference by hydrogen sulfide can take place. It is useful to note that variations in pH of waters will not affect the performance of DO probes. However, the probes are temperature sensitive and are consequently provided with temperature compensation owing to a temperature coefficient of 4–6%/°C, depending on the characteristics of the membrane used.[4]

Other instrumental methods for the determination of gases in water are: the thallium column method for trace oxygen in the range 5–1000 $\mu g/l$[112]; gas chromatographic determination of gases such as O_2, N_2, CH_4, CO_2[113]; the spinning-disk stripper combined with coulometry for oxygen[114]; and the determination of N_2O in water by gas chromatography using a 5 Å molecular sieve trap.[115]

4.1.2. *Chemical Methods for Dissolved Gases*

Ruch[39] gives details of fairly simple chemical methods for the analysis of 176 gaseous pollutants, many of which could be applied to water using stripping, displacement, or other collection techniques. Other standard and semiofficial compilations of chemical methods should also be consulted.[4,107–110] Some of the more important of these methods will be described to illustrate typical chemical approaches.

The Winkler azide method[4,108,110] for dissolved oxygen in water, often used as a finish for the BOD method, is perhaps the most well known of the chemical methods. The sample is treated successively with $MnSO_4$, KOH-NaN_3-KI, and H_2SO_4. The precipitate $Mn(OH)_2$, which first forms, combines with the dissolved oxygen in the sample, forming a brown precipitate of manganic hydroxide, $MnO(OH)_2$. Following acidification, the $MnO(OH)_2$ forms manganic sulfate, $Mn(SO_4)_2$, which acts as an oxidizing agent to release free I_2 from the KI. This free I_2 is then titrated with $Na_2S_2O_3$, since it is stoichiometrically equivalent to the dissolved oxygen in the sample. The result may be returned as mg DO, mg/l BOD (when used in the biochemical analysis procedure), or as ml O_2 per liter at STP. It will be appreciated that corrections for O_2 solubilities are required for variations in barometric pressure. The following constituents which may be present in water can cause interference in the Winkler method: general oxidizing or reducing materials (e.g., NO_3^-, Fe^{2+}); SO_3^{2-}; $S_2O_3^{2-}$; polythionate; free Cl_2; OCl^-; organic matter which is readily oxidized in highly alkaline solution, or which is oxidized by free I_2 in acid solution; high amounts of suspended solids; domestic sewerage; biological flocs; color which may interfere with the endpoint.[4,108,110] In such cases the DO probe may be used. Spectrophotometric methods for the determination of ozone in water have been described by Czerniec *et al.*[116] and by Wierzbicki and Pieprzyk,[117] but care should be taken in respect of the presence of other strong oxidants. The first method is based on the discoloration of methyl red dye by ozone, and the second method is based on the liberation of I_2 from an acidified starch-Kl solution with a spectrophotometric finish at 570 nm

Chemical methods for the determination of carbon dioxide in water may be based on gravimetric or volumetric procedures.[118] Although not strictly regarded as a pollutant in water, CO_2 measurements are of importance in overall studies of pollution.

Ammonia may be regarded as a gaseous pollutant in water and many methods are available for its detection.[108,110,118] Examples of automated chemical methods for the determination of ammonia are: using nitroprusside as a catalyst;[119] phenol hypochloride (applicable to seawater)[120]; Nessler reagent[121]; and a spectrophotometric method using coulometric generation of hypobromite.[122]

Marks[123] has reviewed 14 chemical analysis methods for the determination of residual chlorine in water and waste water. Many of these are either simple colorimetric methods (e.g., for swimming pools) or are based on spectrophotometry using chromophoric agents.[118] Some recent examples are: syringaldazine; tyrosine; and *N,N*-diethyl-*p*-phenylenediamine (for chlorine dioxide and chlorite).

4.2. *Inorganic Composition*

4.2.1. *Introduction*

Inorganic pollutants in waters may include: dissolved gases, particulate matter (which may contain metals in the suspended form), metal cations, anions, and organometallic compounds (as solids or in solution). The basic approach may be to determine inorganic composition as dissolved metals, suspended metals, total metals, extractable metals, gases, anions, or organometallic constituents. Organometallic constituents may, of course, form part of the extractable metals component. Increasing attention is now also being paid to the chemical or physical form in which the inorganic constituents are present (see Table 2). Definitions of and treatments for the various forms of inorganic constituents have been described by Kopp and Kroner,[124] and, more recently, by Gibbs.[125] However, these considerations should be preceded by considerations of sampling schemes and sampling handling and storage.[4,108,126]

A brief consideration of sample handling and storage of waters is advantageous at this point. The water-sampling program should be designed to provide minimum time between sample collection and analysis. There is little doubt that many earlier chemical studies were of little value owing to lengthy collection times and storage. For example, the pot collection of rainfall samples on a roof or other exposed area, over a period of more than 1 week, nearly always results in algal or bacterial growth in the sample and losses of inorganic constituents via incorporation in the microorganism and adsorption on the container walls. It will be appreciated that the growth of algae or bacteria will result in the utilization of essential elements and compounds from the water. Preservation of water samples is possible,[4] but this is seldom complete and unequivocal. At best, the recommended methods of preservation are limited and they mainly reduce or retard biological action, the hydrolysis of chemical compounds, and the volatility of certain constituents. A suitable procedure for field use employs pressure filtration at the sampling site, using the commonly accepted 0.45-μ-pore-size filter,[4,108,125,126] and is carried out within 2 hr of actual sample collection. The filtered water sample is then preserved in the appropriate manner (e.g., by the addition of redistilled HCl to make the pH 1.5), and the filter medium and residue may also be preserved with a volatile organic medium consisting of 1% of a mixture of 1 part o-fluorotoluene, 2 parts n-butyl chloride, and 1 part 1,2-dichloroethane.[125] Alternative preservatives for water include other mineral acids, $HgCl_2$, or NaOH. Refrigeration or freezing is also used.

The problems of sample handling and storage may also be magnified by sample contamination, reagent contamination, and poor cleaning of apparatus.[4,5,126] Failure to observe these precautions often results in worthless data. Although detailed consideration of the necessary precautions is beyond the scope of this chapter, a brief comment will be made concerning cleaning procedures for both containers and laboratory apparatus.

Many of the commercial polymer or plastic containers are sources of inorganic and organic contamination. In most cases this is surface contamination (e.g., sodium or zinc stearate releasing agents, or dust), but in other cases, the contaminant is incorporated in the polymer (e.g., tin in polyvinyl chloride, various unpolymerized monomers, or organic plasticizers such as esters of phthalic acid). The necessary cleaning procedures, or trial storage of standards or blank solutions, should be carried out to establish the contaminant. New containers or apparatus of plastic or glass should first be cleaned with a suitable detergent solution which is nonmetallic. The detergent manufacturer's literature should be consulted or the necessary checks carried out. Plastic containers may then be immersed overnight in the appropriate-strength mineral acid (e.g., diluted HCl or HNO_3) and then rinsed quickly with filtered tap water, followed by distilled or demineralized water. Although demineralized water contains small amounts of soluble organic materials and "fines" from the ion exchange material, this is not generally a problem. "Organic-free" demineralized water is often obtained by the distillation of demineralized water containing some potassium permanganate.

New glassware may also be cleaned in a similar manner, but one of the most effective strong cleaning agents is a mixture of equal parts of concentrated nitric and sulfuric acids. Appropriate care should be taken in handling this mixture, which also increases in volume if exposed to the atmosphere for some time due to the absorption of water vapor. Chromic acid cleaning solution should be avoided, owing to contamination of the glassware from chromium. The history of all old or used glassware or plastic materials should be known before use. Problems may arise from changes in surface (e.g., by abrasion) or from "memory," following storage of stronger solutions.

To prevent losses via adsorption or hydrolysis, stock or standard solutions should be prepared in a concentrated form at levels of at least 100 mg/liter. Provided that there is no deterioration of these solutions by exposure to light, heat (evaporation), or contamination with dust, the daily working solutions are prepared by the appropriate dilutions.

Mention has been made of the type and form of constituents in water. Kopp and Kroner[124] listed four types of metals:

1. *Dissolved metals.* Those metals which pass through a 0.45-μ organic membrane filter.[31] The pH of the filtrate is then adjusted to a pH of about 2 with redistilled concentrated HNO_3.

2. *Suspended metals.* Those metals which are retained in the particulate matter on a 0.45-μ organic membrane filter. The volume of the sample is noted and the filter and retained particles are transferred to a beaker and digested by evaporation to near dryness several times with concentrated nitric acid.[124,126] Finally the suspended metals are obtained in an HCl solution.

3. *Total metals.* The sum of the concentrations of metals in both the dissolved and suspended fractions. This can also be obtained by the vigorous digestion of an unfiltered sample. In the latter case, 3 ml of redistilled concentrated HNO_3 is added per 100-ml aliquot of sample. Repeated evaporation to dryness is then carried out with HNO_3, and the solution is finally acidified with concentrated HCl.[124,126]

4. *Extractable metals.* These may be considered to be acid extractable or extractable by well-known chelation–solvent extraction procedures. In the former case, the unfiltered sample is digested with hot dilute mineral acid such as HCl, by heating for 15 min on a steam bath or hot plate. The solution is then cooled and filtered through a 0.45-μ organic membrane filter.

In the latter case, many workers have not realized that their solvent extraction procedures, applied to "raw" water samples, only extract metals in a readily available form.[125,127] Gibbs[125] has shown that studies of the chemical mechanisms in the transport of trace metals in rivers involve the following forms: ionic species and inorganic associations; organometallic complexes; constituents adsorbed on solids; metallic coatings (by precipitation and coprecipitation on solids); constituents such as elements incorporated in biological materials; and constituents incorporated within crystalline structures. Gibbs used a separation scheme to provide data on these forms in two river systems.[125] It is obvious that, with further refinements, such separation schemes will find increasing use in pollution studies. For example, an element such as Cr may have variable toxicity in water, depending upon whether it is present in ionic form in the (111) or (VI) oxidation state. It would also be expected to play a variable role if present in any or all of the forms studied by Gibbs.[125] Doolan and Smythe[127] showed that all trace Cd present in water was not present in a readily extractable form. Conversion to a readily extractable form could easily be achieved with water samples by the addition of H_2O_2 to the sample followed by irradiation with UV light. The addition of zinc as a synergistic agent promoted complete extraction of Cd, with 2-mercaptobenzothiazole in *n*-butyl acetate, at levels down to 2×10^{-5} ppm.[127]

Consideration is also required of the classes of waters which may be examined by physical, chemical, and microbiological methods. Table 8 summarizes the characteristics of the various classes.

Table 8. Characteristics of Classes of Waters for Chemical Analysis

Class	Characteristics
Drinking water	controlled quality; often filtered and chlorinated; fluoride may also be added as a trace additive; generally low salinity and dissolved solids content.
River water	variable composition often related to flow rate, geology or pollution; wide variations possible for suspended and dissolved trace constituents.
Lake water	fairly uniform composition; some variation with depth.
Seawater	high salt content, chiefly Na^+, K^+, Cl^-, and SO_4^{2-}, present in same proportion throughout oceans of the world; elements have complex form and reactions.
Bore or artesian water	high salt content; variable composition; can contain high amounts of some elements (e.g., F^-).
Effluents	extremely variable composition, depending on: location, sources of pollution, variation in discharge rates.
Sludges	semi-solid mixtures; often associated with effluents or sewerage.
Sewerage	essentially a mixture of sludge and liquid; varying degrees of treatment (primary, secondary and tertiary); can contain industrial wastes, detergents, etc.
Miscellaneous waters	rainwater (often collected from roofs), well water, estuarine water (often saline), boiler water, process water.

Detailed storage and processing procedures are available in the literature.[4,108,124,126]

4.2.2. *Separation and Preconcentration Procedures*

Many inorganic pollutants may be present in water in trace amounts, necessitating separation and preconcentration procedures. The primary aim of a separation procedure may only be to separate the analyte from interfering matrix components. This separation is often also combined with preconcentration, to improve sensitivity and limit of detection.

Separation and preconcentration procedures should be both rapid and specific. Many chemical and physical methods are available for separation and preconcentration of inorganic pollutants. These most commonly used methods include evaporation, chelation and solvent extraction, ion exchange, coprecipitation, chromatography (gas, liquid, and thin layer), and electrochemistry. The last two methods have already been discussed.

4.2.2a. *Evaporation.* Evaporation is a useful but slow method for concentrating inorganic pollutants in water prior to chemical analysis. Care should be taken with waters containing high total dissolved solids (TDS) owing to the solubility of some components being exceeded and resulting in precipitation. A twenty-times concentration of a 500-ml water sample may take many hours. Contamination during evaporation may take place via dust or by uptake of certain elements from the evaporation vessel (e.g., silica and other elements from a porcelain evaporation basin with imperfect glaze). Losses may also occur if some components are volatile (e.g., Hg, As, Sb, Se, etc.) or by adsorption on the walls of the evaporation vessel.

4.2.2b. *Chelation and Solvent Extraction.* Advantages of chelation and solvent extraction include simplicity, speed, high specificity, and also direct use of the organic extract in certain instrumental methods (e.g., AAS and spectrophotometry). A major advantage leading to high specificity is the separation of the analyte from unwanted bulk matrix components. Control of specificity may be achieved in this manner by pH control with acids, alkalis, and buffers, and sometimes with change of oxidation state. Other controls include salting out, salting in, use of mixed solvents, synergism, and stripping or back extraction. The literature should be consulted for more detailed considerations of chelation and solvent extraction.[128]

Marcus[128] has provided the following useful classification of solvent extraction systems in which the systems are classified according to the reaction involved and the nature of the solute–solvent interaction. However, it should be borne in mind that there are gradual transitions from one mode of behavior (or system) to another.

1. Systems involving the distribution of neutral covalent molecules between aqueous solutions and inert solvents, e.g., I_2 (aqueous solution): hydrocarbon solvent.

2. Halo-metallic acid extraction. These systems involve anionic complexes of metals with ligands (mainly halide anions) with solvating solvents, containing mainly O as donor atom; e.g., $PaCl_3$: hexone (methyl isobutyl ketone, MIBK).

3. Systems where the solvating solvent directly solvates the metallic cation; e.g., $UO_2(NO_3)_2$: tributyl phosphate (TBP), $(CH_3 \cdot CH_2 \cdot CH_2 \cdot CH_2O)_3PO$.

4. Liquid cation exchange systems comprising mainly organic acids and acidic phosphorus esters dissolved in diluents; e.g., Lanthanides: dibutyl phosphate (DBP); also TBP $(Ln):(CH_3 \cdot CH_2 \cdot CH_2 \cdot CH_2O)_2 \cdot PO_2H$ (some further solvation may take place, forming $Ln(Bu_2PO_4, HBu_2PO_4)_3$).

5. Chelating extraction systems in which the metals are complexed by reagents satisfying simultaneously the charge neutralization and coordination number requirements; yielding compounds which are much more soluble in inert organic solvents than in water; e.g.,

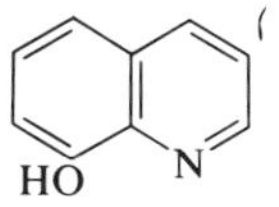

oxine (8-hydroxyquinoline)

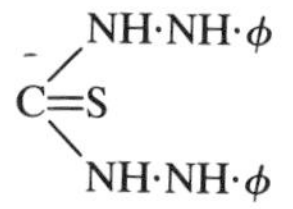

dithizone (DTz)

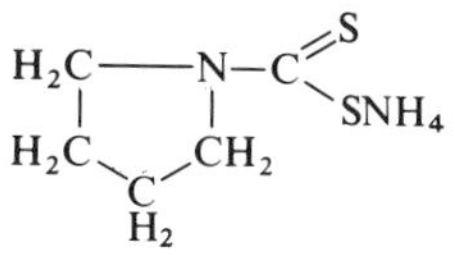

ammonium pyrrolidene carbodithioate (APCD)

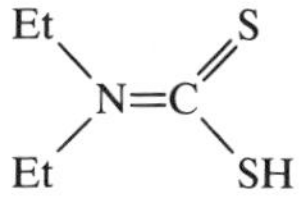

diethylcarbodithioate (DCD)
(acid form)

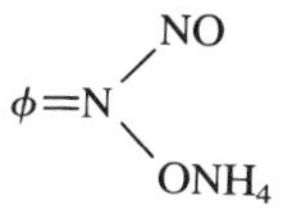

cupferron (ammonium salt of N-nitrosophenyl hydroxylamine)

The above reagents are typical of many which are available for the chelating extraction of inorganic metal pollutants. The reagents are generally dissolved in solvents such as $CHCl_3$ or CCl_4, if insoluble in water, and the solution of reagent is shaken with an aqueous solution of the metal to be extracted. A small amount of the chelating agent is considered to pass into the aqueous phase where it reacts with the metal ion, forming a neutral complex. This neutral complex, being more soluble in the organic solvent, then distributes preferentially to the organic phase. Alternatively, some of the soluble compounds (e.g., ammonium salts) are added to the aqueous solution of the metal to be extracted and the complex is extracted into the solvent phase. For example, a 1% solution of APCD in water (prepared fresh daily) can be used for the extraction of one or more of the following metals, using MIBK and with suitable adjustment of pH: As, Cd, Co, Cr, Cu, Fe, Hg, Mn, Mo, Ni, Pb, Se, Tl, U, Zn. This is a system of wide applicability for the determination of inorganic metal pollutants in water.[126]

6. Ion pair formation (liquid anion exchange) systems. In extraction with strongly basic reagents, the solvent (usually diluted by an inert diluent) is attached so strongly to protons that stable onium species are formed, such as trialkylammonium cations. The onium species extract metal complexes by ion pair formation rather than by direct solvation. Since the anion attached to the ammonium cation may be exchanged for other anions, or for complex metal cations, these solvents are often called liquid anion exchangers, because they resemble the resin exchangers in many of their properties, e.g., $UO_2(SO_4)$ in H_2SO_4 solution: triisooctylamine (TIOA) $(i\text{-}Oc_3\text{-}NH)_4UO_2(SO_4)_2$ (species extracted).

Quaternary amines extract by a similar mechanism, but are also effective in nonacid solutions.

7. Large cations and anions can form ion pairs which, as with the chelates, do not require further solvation, and behave in inert solvents as if they were covalent molecules (see also class 1); e.g., Cs extracted with tetraphenyl borate $[NaB(\phi)_4$; sodium salt] into nitrobenzene.

The most commonly used of the extraction classes 1–7 are (5), (6), and (4). Further information on the theory of solvent extraction is available in the literature.[128,129] The general aim in using solvent extraction in the laboratory is to form a stable extraction system wherein the analyte has a low solubility in the aqueous phase, but has high solubility in the organic phase. Hand-operated extractions, using extraction or separating funnels, are usually preferred in the laboratory for chemical analysis. However, automatic continuous extraction systems are also available. The analytical chemist should be well aware of all the physical and chemical aspects of solvent extraction, including the partition constant and extraction coefficient, the completeness of extraction, and the selectivity of extraction.[128,129]

4.2.2c. *Ion Exchange.* Ion exchange is a widely used, highly selective, but rather slow method for the separation and preconcentration of inorganic (and many other classes of) pollutants. Its first widespread use for chemical analysis probably originated over 30 years ago, for the separation and identification of the actinides. Some essential aspects of ion exchange will be discussed, but the literature should be consulted for further information.[130] There are three main classes of ion exchange materials which are used for chemical analysis:

1. Liquid ion exchangers (described above).
2. Solid inorganic ion exchangers (e.g., natural zeolites; hydrous TiO_2, ZrO_2; zirconium phosphate; microcrystalline ammonium molyb-

dophosphate; and potassium or ammonium hexacyanocobalt(II) ferrate(II).

3. Solid organic ion exchangers (e.g., ion exchange celluloses; anion exchange resins; cation exchange resins; chelating resins; ion retardation resins; and macroporous resins).

Full consideration of the properties and many uses of these materials in chemical analysis is beyond the scope of this chapter.

In their more usual forms in the laboratory—(2) and (3)—ion exchange resins are graded powders or beads of porous insoluble three-dimensional polymeric compounds. The polymer contains firmly bonded organic functional groups that may be associated with exchangeable cations or anions which are exchanged for ions in solution. The polymer is supplied in varying degrees of polymerization and cross linkage. Analytical-grade ion exchange resins usually contain little "fines" and soluble organic materials, in contrast to some of the commercial resins which are used on a large scale. The inorganic ion exchange materials are usually prepared synthetically and are often used with control of pH.[131] Both classes of ion exchange materials (2 and 3) are available with weak or strong exchange capabilities. The exchange capacity is usually expressed as meq/g of the moist resin. With a cation exchange resin in the H form, it is understood that the exchangeable cation of the resin is H^+, and that this is released during exchange. The Na form (sodium form) is also common. In contrast, anion resins are commonly used in the hydroxyl (OH) or chloride (Cl) forms. The ion exchange celluloses are also available in both anion and cation forms, and are prepared commercially from highly purified cotton and wool celluloses. These exchange materials are mostly used for the separation and purification of enzymes, hormones, serum components, acidic and basic proteins, etc., and are seldom used for separations of inorganic constituents.[130] The chelating resins have amine carboxylates as the functional groups, and although not used extensively for inorganic separations, are finding important uses in the chemical analysis of seawater.[130,132] The ion retardation resins are also mainly used for the separation of biochemicals. A typical ion retardation resin is a spherical bead containing paired anion and cation exchange sites.[130] The macroporous resins contain micropores, similar to those encountered in gel-type resins, and the dense and rigid matrix is permeated by macropores.[130] Separation by these resins depends on the kinetics of exchange for large molecules which enter the macropores being relatively rapid, compared to the slow entry of small ions capable of entering the micropores.

When solid organic ion exchangers are used for separation and preconcentration, it is important that the basic effect of varying degrees of crosslinkage be appreciated. The percentage of crosslinkage determines the swelling characteristics, solubility, and selectivity, as well as other chemical and physical properties. In general terms, as crosslinkage increases permeability decreases, wet volume capacity increases, and selectivity increases. As crosslinkage decreases permeability increases, wet volume capacity decreases, and selectivity decreases.

The use of ion exchange resins for the concentration of trace inorganic constituents from water results from the high affinity of the resins for ions in very dilute solution. Provided that the inorganic constituent is in the correct ionic form and oxidation state, a large volume of water can be passed through a small resin column with virtually complete retention of the ionic species. When the column is later eluted with a small quantity of acid or other appropriate eluent, quantitative recovery and concentration can often be achieved.

Ordinary ion exchange resins are generally unsuited to the concentration of traces of ionic constituents from stronger electrolytes such as seawater. In this case, the major ions such as Na^+, Ca^{2+}, Sr^{2+}, and Mg^{2+} occupy the exchange sites preferentially. Chelating resins are more useful in these cases.[130,132]

The ion exchange resins may be used in the laboratory, as in the column and batch methods. In the column method, a suitable quantity of the resin is carefully packed into a glass or plastic tube with a porous plug and tap at the bottom. The resin is carefully slurried into the column and packed and backwashed to avoid channelling or voids. The column is usually left filled with liquid after preparation. A suitable height-to-column-diameter ratio is 10–15 to 1. The exchange capacity of the resin should be checked before deciding on the size of the column. In the batch method, a suitable quantity of the resin is added to the solution, which is stirred for a time. It is more difficult to secure complete extraction using the batch method.[129]

4.2.2d. Coprecipitation. The physical phenomenon of coprecipitation has plagued analytical chemists carrying our gravimetric analyses for many years. The effects of simultaneous precipitation of similar compounds, often with occlusion or adsorption of other species, can also be used for some separations and preconcentrations of inorganic pollutants prior to chemical analysis. However, the coprecipitation steps suffer by often being complex, lengthy, and tedious. Coprecipitation on indium or aluminum carriers has long been used for emission spectroscopy, yielding concentration factors of up to 30,000 times when the ashed coprecipitate is used for arc emission spectroscopy.[133,134] Precipitation by ferric hydroxide has been used for the concentration of elements, such as Be, in marine waters.[135,136] Other similar coprecipitations include sodium carbonate (for seawater)[137]; thionalide (2-mercapto-*N*-naphthyl-acetamide)[138,139]; oxine (8-hydroxyquinoline) with thionalide and aluminum[138]; and thionalide in acetone or alcohol with pH control.[139] These methods should be consulted as a guide to suitable methods which could be used for the separation and preconcentration of a variety of inorganic pollutants.

4.2.3. *Instrumental Methods*

Phillips *et al.*[140] and Coleman[14] have prepared short surveys on the use of instrumentation for water quality monitoring. In this section reference will be made to instrumental methods which are described earlier in this chapter, and further comment will be made on aspects which particularly concern methods for the determination of inorganic constituents in water.

4.2.3a. *Microscopy.* The use of microscopy for the examination of particulate matter in air (Table 3) indicates that it would also be suitable for the examination of water. With comparatively inexpensive equipment, rapid detection of as little as 10^{-12} g of particulate matter is possible. Moreover, particles may be characterized by use of six parameters (e.g., color, shape, and birefringence) using the classification system of McCrone and Delly.[141] Useful information on suitable microscopy techniques for particulate matter is also available in commercial leaflets.[31]

4.2.3b. *Scanning Electron Microscope and Electron Microprobe.* Particulate matter of both inorganic and organic composition may also be examined in residues obtained from water samples using the scanning electron microscope. This instrument has high magnification and superior depth of field, making it very suitable for showing the shape and surface of relatively large particles. The electron microprobe can also be used to determine the inorganic composition of a single particle in a semiquantitative manner, and useful data to aid identification is listed in "The Particle Atlas."[141]

4.2.3c. *X-ray Diffraction.* X-ray diffraction is often useful for the determination of the form and structure of inorganic particles (as small as 5 μ) isolated from water. Whittig[142] has provided useful details of X-ray diffraction techniques for mineral identification and mineralogical composition applicable to soils and to particulate matter from water.

4.2.3d. *Spectrophotometry.* Instrumentation suitable for the spectrophotometric determination of inorganic constituents in water is readily available in most laboratories (see Table 3). Suitable spectrophotometers for either manual or automatic operation are available from many commercial sources.[140] Many of the instruments are provided with digital display and BCD output, and may either be linked to automatic sampling equipment or to minicomputers. Semicontinuous analyzers using spectrophotometry are often preferred for the determination of metals and anions, mainly because of simplicity in operation and low cost. A known fraction of the water is selected and the analyte is converted into a form suitable for spectrophotometric measurement by one or more chemical reactions. Proportioning pumps may be used to take the sample and measure and mix the required reagents prior to passing to a color development section, and finally to the absorption cell. The samples are taken on a regular basis, with the time interval carefully selected in relation to significant changes in the sample stream being monitored. Typical semicontinuous systems for water monitoring are described by Philips *et al.*[140] The general development and influence of autoanalyzers and data processing in analytical chemistry has been reviewed by Smythe.[143]

4.2.3e. *Analytical Flame Spectroscopy.* The related and complementary methods FES, AAS, and AFS are well suited to the determination of a wide range of inorganic constituents in water. These methods have been described earlier in this chapter (see Tables 3–6). A good summary of suitable methods for water has been prepared by Parker.[126] The following papers and reference books are also recommended:

1. FES: Pickett and Koirtyohann,[144] Mavrodineanu[145]
2. AAS: L'vov,[146] Slavin,[147] Price,[148] Christian and Feldman[149]
3. AFS: Winefordner and Elser,[42] Kirkbright and West,[43] Browner,[44] Winefordner[150]
4. General: Mavrodineanu,[145,151] Winefordner and Vickers[152]
5. Multielement flame spectroscopy: Busch and Morrison,[153] Mitchell, Jackson, and Aldous,[154] Fassel and Kniseley[155]

A more detailed consideration of the many advantages of analytical flame spectroscopy is beyond the scope of this chapter. The more recent and promising aspects involve the use of AFS and multielement flame spectroscopy, and it is likely that the whole area will play an increasingly important role in the analytical chemistry of inorganic pollutants. The undoubted success of AAS, following its discovery and development by Walsh,[156] has greatly stimulated interest in the area. The trends in AAS, its history, and its economic benefits to the economies of the world have been reviewed by Brooks and Smythe.[157]

4.2.3f. *Emission Spectroscopy.* ES is very well suited to rapid surveys of the levels of inorganic pollutants in nearly all environmental samples (see Table 3). Coupled with various rapid preconcentration procedures (see above), ES is eminently suitable for the monitoring and control of all classes of waters (see Table 8). Excitation sources which may be used with ES include direct current arcs, gas-shielded arcs ("plasma jets"), microwave and radio-frequency plasmas, and high-voltage spark. Most of these sources can be used for the direct analysis of the inorganic constituents of water, but preconcentration steps are necessary for detection in the important range 0.001–1.0 ppm. Most preconcentration steps suitable for ES involve the production of quite concentrated matrices, and these are best tolerated using the DC arc (100% solids) and HV spark (5% solids).[158] Promising preconcentration techniques for use with the above excitation sources and ES in the examination of waters appear to be evaporation to dryness, solvent extraction using APCD–MIBK at pH 4, and coprecipitation on Al (20 mg) with oxine, thioanilide, and tannic acid.[138,158]

4.2.3g. *Gas and Liquid Chromatography.* These instrumental methods have been described above (Table 3) and are generally not well suited to the determination of a wide range of inorganic pollutants in water. The main disadvantages include limited scope and range; greater requirement for time and skilled operation; necessity for changes in columns and operating conditions; and often difficult and time-consuming processing for conversion of the trace element to a stable chelate compound or organometallic. Future developments with high-speed chromatography and automation may improve this situation.

4.2.3h. *Infrared Spectroscopy.* This method, which has been described above (see Table 3) is better suited to the determination of inorganic gases and organic constituents in water than to the determination of inorganic metals. Since an appreciable number of purely inorganic compounds provide useful infrared spectra in the solid state, it would be possible to use this method for the examination of certain water residues obtained by evaporation.

4.2.3i. *Ultraviolet Spectroscopy.* Ultraviolet spectroscopy is better suited to the determination of organic constituents and certain gases in water than to the determination of inorganic constituents in aqueous solution. The method has been described earlier (see Table 3).

4.2.3j. *Luminescence Spectroscopy.* Many of the reactions with organic compounds which might serve as a basis for the determination of inorganic pollutants in water are sensitive to quenching processes.[64] Luminescence-based methods do not at present offer marked advantages over well-known methods for the determination of metals in water. However, some recent developments in the appreciation of stable ternary complexes offer promise for the determination of trace metals in water. Haddad *et al.*[21] have described a method for trace Mo, based on the reaction of molybdenum oxypentathiocyanate ion with the dyestuff Rhodamine B (RhB) to produce the ternary complex $MoO(SCN)_5(RhB)_2$. The formation of this complex is accompanied by a color change and the extinction of the fluorescence of RhB. The method has a detection limit of 0.05 μg and could be adapted for the determination of trace Mo in water.

4.2.3k. *X-ray Fluorescence Spectroscopy.* XRF has been described earlier (Table 3) and is well suited to the determination of trace inorganic constituents in water using either filtration procedures for particulate material or suitable preconcentration procedures. Its main advantages reside in the nondestructive determination of a wide range of elements with good specificity, speed, and precision. Filtration and preconcentration procedures add to the time requirements, but the method ranks highly with ES for extreme versatility.

4.2.3l. *Electrochemical Methods.* Electrochemical methods, such as AC and pulse polarography, anodic stripping voltammetry, coulometry,[159] and ion-selective electrodes, offer inexpensive, speedy, selective, and sensitive approaches to the determination of a wide range of metals in water.[67-74,159] The methods have been described above (see Table 3) and adaption to the examination of water is generally accomplished without any separation or preconcentration procedures.

4.2.3m. *Spark Source Mass Spectrometry.* While primarily suited to the trace chemical analysis of conducting and semiconducting solids, spark source mass spectrometry (SSMS) can be used for powders, insulators, and liquids such as water. A survey of some recent applications of SSMS has been published by Brown *et al.*[160] The SSMS can detect over 80 elements at concentrations as low as 1-10 ng/g, with a precision, at the lower end of the range, of $\pm$5-20% for photographic detection and $\pm$2-5% for electronic detection.[161] As regards wide cover of elements, the method can be compared to emission spectroscopy (ES). SSMS is far more sensitive than ES, but the preparation of samples in a suitable form is more time consuming and subject to possible loss in sensitivity and/or contamination. Several preconcentration techniques used in ES could be also applied in SSMS.

4.2.4. Chemical Methods

Chemical methods for the determination of inorganic pollutants are described under several other headings in this chapter. Since most of these methods are based on reactions and measurements in aqueous systems, their adaption to the examination of water presents no great difficulty. Adequate guidance and information is available in various official or standard books and in reviews.[4,107,110,118,124,126]

Some of the newer or less known chemical methods will be briefly described under this heading.

4.2.4a. *Catalytic Methods.* Catalytic methods for trace metal analysis have been reviewed by Batley.[162] The methods appear to offer promise for further development and application to the determination of pollutants in water. Catalytic methods to determine the analyte are based on the concentration dependence of the catalytic effect of a chemical species in a chemical equilibrium. The catalyst is regenerated by a cyclic process, enabling its detection in trace amounts. Measurement under nonequilibrium conditions enables significant lowering of limits of detection below those found by the conventional equilibrium techniques. Batley has compared the sensitivity of catalytic analysis with anodic stripping voltammetry, spark source mass spectrometry, and neutron activation analysis for the ppb determination of Ag, Ru, Co, Cu, Fe, Mn, and V.[162] The sensitivity and accuracy of catalytic methods are shown to compare very favorably with the other methods. The use of separation and preconcentration procedures can increase the sensitivity of catalytic methods by at least one order of magnitude.[162]

The catalytic effects of metal ions on polarographic reductions have been known for many years,[163] but only recently have developments led to the determination of nanogram amounts of certain metals such as uranium[163] and the platinum metals.[163,164] In work with Rhodium(III)–cysteine solutions, Alexander and Orth[164] have shown that catalytic hydrogen waves can be used for the trace analysis of Rh, giving a detection limit of 2×10^{-9} M. The work has now been extended to include the trace analysis of serum proteins using catalytic polarographic waves of Rh(III) and also the trace determination of Pt in other materials.[165,166]

Catalytic methods form part of an important and promising area in trace analytical chemistry; namely those using kinetic (or dynamic) techniques, in contrast to the normal equilibrium techniques. A comprehensive reference book on kinetics in analytical chemistry by Mark and Rechnitz[167] indicates a promising area for the development of new methods for the determination of trace inorganic pollutants in water and other materials. In addition, kinetic approaches may well solve many problems regarding the form or nature of various elements, such as chromium in water.[167,168]

4.2.4b. *Chemical Methods Suited to Oceanography.* Both seawater and estuarine water (which may contain fresh water) have a high salt content and contain other complex organic and inorganic materials, and therefore often preclude many chemical methods of analysis which may apply quite well to fresh water. A report by Major *et al.*[169] gives details of a number of chemical and instrumental methods well suited to oceanography. The chemical methods detailed are dissolved oxygen, dissolved orthophosphate, total phosphorus, reactive silicate, nitrate, ammonia, urea, particulate total nitrogen, particulate total phosphorus, particulate carbon, and certain organic constituents. A detailed review with 331 references on inorganic analytical methods used in oceanography has been prepared by Spencer and Brewer[170] and a general review of all aspects of water analysis is also available.[118]

Chemical methods for the determination of inorganic pollutants in seawater must be suited to a total salt concentration of 35.1 g/kg. While the major components are present in the same proportions in all oceans and are independent of the total salt content, the minor components (including pollutants) vary with locality, depth, etc. There are substantial ranges listed for minor components, and suitable tables listing the minor element concentrations in seawater[172] should be consulted before selection of a suitable chemical method. In most cases preconcentration methods are also required prior to chemical analysis. For trials of suitable methods or for the development of new methods, it is often convenient to work with a purely inorganic artificial seawater matrix. An example of an artificial seawater matrix is given by Parker.[126]

4.3. Organic Composition

4.3.1. Introduction

The organic composition of water is extremely variable and complex. The type of water (Table 8), its location, and the wide range of dynamic physical processes and chemical reactions involving organic compounds and living matter, all necessitate individual approaches to the determination of organic composition and organic pollutants.

Many of the physical, instrumental, and chemical methods which are described earlier in this chapter may be applied to the determination of organic pollutants in water. It is not possible to give details of these methods without undue repetition. The reader is recommended to consult a comprehensive review[118] before commencing a specific project.

It is necessary to broadly subdivide the areas concerned with the organic composition of water. Table 9 gives a broad classification of the organic composition of water.

Although specific organic pollutants may fall into more than one of these classifications, some useful comments, from the analytical chemistry viewpoint, may be made about each class.

Table 9. Broad Classification of Organic Composition of Water

Classification	Comments
Gases and volatile compounds	Components with high vapor pressure which may be stripped from water
Organics	Wide range of general compounds, excluding other classifications in this table
Pesticides and herbicides	Man-made pollutants
Detergents	Man-made pollutants
Living matter and associated debris	Plants (including algae), fish, micro-organisms, etc., decomposition products
Solid organic debris	Man-made polymers, other solid organic debris and decomposition products

4.3.2. Gases and Volatile Compounds

Abnormal levels of gases such as methane, which occurs naturally in some waters, may require determination as pollutants. Some man-made organic gases may enter water as by-products of refineries and other industries. Other volatile compounds which may be present in water include alcohols, esters, ketones, phenols, and aromatics. Eleven organic compounds contributing to taste and odor were detected in a United States river.[111] In finished water from a New Orleans water treatment plant, 35 organic compounds, including volatile compounds such as acetone, bromoform, chloroform, and methyl chloride, were also detected.

The main methods used for determinations of these pollutants include: displacement or stripping with an "inert" gas (e.g., N_2), which may be followed by adsorption on active carbon or other collecting agents at low temperature; thin-layer chromatography; gas–liquid chromatography; mass spectrometry; ultraviolet and infrared spectroscopy; polarography; liquid–liquid extraction; ion exchange; and NMR and ESR spectroscopy. Most of these methods have been described earlier and further information can be obtained from reviews[118] and other publications.[108,172]

4.3.3. General Organics

A very wide range of instrumental and physical methods can be used for the determination of general organics in water. Some idea of the complexity of necessary determinations for organic pollutants may be gauged from the listing of 717 products in only one area of river water in the United States.[172] It is recommended that a review[118] and references[107–110] be consulted prior to selection of a method.

4.3.4. Pesticides and Herbicides

Pesticides and herbicides are commonly determined in drinking water and in river water, from where they may reenter the food chain or other biological cycles. The extremely low levels, coupled with the complexity and similarity of the compounds involved, necessitates an individual approach to each chemical analysis problem. Preconcentration techniques are often required. These include liquid–liquid extraction; thin-layer chromatography; paper chromatography; derivatization followed by extraction and the use of absorbents or special ion exchange resins. Reviews[118] should be consulted for further guidance.

4.3.5. Detergents

A wide range of nondegradable and biodegradable detergents may be present in water at the ultra-trace level. Methods range from simple qualitative shaking and visual tests to instrumental and chemical methods. Similar considerations to those briefly outlined for pesticides and herbicides also apply to the determination of detergents. Automatic methods based on spectrophotometry, polarography, and the various types of instrumental chromatography, are well suited to the routine determination of detergents in water.[118] These methods are also widely used for smaller numbers of specific determinations.[118] Some other instrumental methods such as mass and infrared spectrometry are seldom used for the analysis of detergents in water at the ultra-trace level.

4.3.6. Living Matter and Associated Debris

The natural cycle of living matter, metabolites, and debris, in varying stages of life or decomposition, not only give rise to "natural" levels of organic substances in water, but also incorporate and metabolize other organic and inorganic pollutants. The wide range of organic and organometallic compounds in solution can range from humic acids to dimethyl mercury. Plants, fish, and microbial organisms may incorporate or metabolize both inorganic and organic pollutants; often with either concentration or discrimination. For this reason, the analytical chemist can often determine the pollutant in living matter, which may either be important in the food chain or represent a convenient "preconcentration" stage.

Filtration procedures[31] may be used to isolate algae, plankton, microbial organisms, or fine organic debris prior to chemical analysis, using methods which have been described.

Larger samples of living matter should be taken with due regard to the necessary precautions of collection and storage.

Microbial pollution in water may be measured either by comprehensive laboratory procedures which are beyond the scope of this chapter, or by simple portable field kits.[31] A recent publication[173] on the biological analysis of water and waste water gives details of sampling procedures, laboratory preparation and testing, and field testing equipment, and also has a glossary of terms and a bibliography. The most important methods are: total coliform, fecal coliform (FC), fecal streptococcus (FS), FC–FS ratio, staphylococcus aureus, and pseudomonas aeruginosa.[173]

4.3.7. *Solid Organic Debris*

Increasing quantities of man-made solid organic debris are entering rivers, lakes, and oceans. The most important solid organic debris probably consists of various synthetic polymer materials and their decomposition and degradation products. Many polymers, such as the polyolefins, have slow decomposition and degradation in water. This may be either an advantage or a disadvantage as far as the total environment is concerned. Other more reactive polymers, with incorporated modifiers and plasticizers, degrade more rapidly in water, and chemical analysis of water may then be required. It is recommended that a comprehensive review be consulted regarding suitable methods for the determination of polymers, monomers, and degradation products in water.[174] Many of these methods require adaption to trace levels in water. The main class of polymers which may be considered as possible pollutants in water are acrylics, polyacrylonitrile, polyamides, polyesters, polyethers and epoxides, polyolefins, polystyrene copolymers, and polyvinyl chloride and other vinyl polymers.

4.4. *Radioactivity, Isotopic Analysis, and Nuclear and Atomic Activation Analysis*

4.4.1. *Radioactivity*

A high proportion of natural and man-made radioactive gases and particulate matter in the atmosphere enters water systems (Table 8). While sampling, storage, processing, and preconcentration techniques are different for water as compared with the atmosphere, the final instrumental and radiochemical methods are very similar. The important classes of radionuclides present in water are radionuclides arising from "natural" sources, man-made radionuclides, and fission products. Detailed consideration of suitable methods is beyond the scope of this chapter and suitable reference books and reviews should be consulted.[92-105] Methods for measuring the trace element and radionuclide levels in natural waters and the use of these measurements for studying biogeochemical processes in water systems are also described by Perkins and Rancitelli.[176]

4.4.2. *Isotopic Analysis*

Deuterium, tritium, carbon 14, and oxygen 18 are not usually determined by analytical chemists as pollutants, although they may well be regarded as such in localized areas. These and certain other nuclides are mainly of importance in hydrological investigations, biological cycles, or age dating. Reviews[94,118] should be consulted for further information.

4.4.3. *Nuclear and Atomic Activation Analysis*

This is a rather specialized area of analytical chemistry which usually requires the facilities of nuclear research centers. Facilities range from nuclear reactors and accelerators to relatively small radionuclide sources emitting neutrons. It is possible to carry out a useful range of work in normal or low-level laboratories, but the greatest potential of the methods for the determination of traces of the elements in environmental samples is seldom achieved without special nuclear experience and facilities. While neutron activation analysis (NAA) competes with analytical flame spectroscopy, electrochemistry, luminescence methods, and spark source mass spectrometry as one of the most sensitive methods for the determination of trace elements, it often suffers from attendant lengthy radiochemical processing. As a result of this handicap, attention has been directed, during the past 15 years, to the development of automatic instrumental methods for NAA which are also nondestructive of the sample. Many of the improved methods have been based on the use of fast (14-MeV) neutrons, improved solid state detectors such as the Ge(Li) detector, and multichannel spectrometry. Activation analysis may also be carried out with protons, gamma photons, and He^3 and other ions. A useful guide or introduction to activation analysis has been edited by Lyon,[176] and this should be supplemented by more recent state-of-the-art summaries[177,178] and by recent reviews.[179]

5. *Soils*

5.1. *Introduction*

A soil may be defined as the natural weathered material in which plants grow and by which they are supported and supplied with water, elements, and compounds derived from minerals and organic matter. The main constituents of a soil are mineral particles—clay, silt, and sand—incorporated with air, water, humus, and living matter. It follows that a large proportion of the analytical chemistry associated with soils has been directed toward the evaluation of agricultural properties. The chemical composition of a soil may vary widely and the definition of a soil pollutant, whether natural or man-made, depends on soil type and location and also on the composition which is "normally" expected. Comparisons with "normally" expected constituent levels can be achieved by using either standard soil samples or similar soils from other areas. Standard soil or sediment samples are not generally available for use in pollution studies and the second alternative is usually followed. A stream sediment sample is often similar in composition to certain soils and these samples are also examined in pollution studies. Soils and sediments may contain elements and compounds incorporated in the air, water, humus, and living matter; adsorbed or exchanged onto soil colloids, clay minerals, or humus; incorporated within crystalline structures; or as metallic coatings (by precipitation and coprecipitation) on soil particles.

The physical and chemical methods which are suitable for the examination of pollution in soils and sediments are not too dissimilar from those used in biogeochemistry. Brooks[180] lists the following avenues for geochemical exploration: plant, soil, bog, water, stream sediment, overburden, and rock. With the exception of water, which has been covered above, all of these materials are capable of pollution to varying degrees and may be examined in environmental studies.

5.2. *Inorganic Composition*

The major and minor inorganic constituents of a soil are often related to the composition of the underlying rock, from which they have been derived by factors responsible for soil formation. This derivation forms the basis of geochemical prospecting by soil sampling.[180] The minor elements in soil assume importance in pollution studies because they include many elements which play a role in nutrition and disease in plants and animals. While the nature of the parent element determines the abundances of the elements in the profile as a whole, their distribution within the soil profile is determined by their mobility, together with other factors such as soil climate. Data are readily available on the major and minor constituents of rocks and the major soil types.[171,180] Brooks[180] provides convenient tables showing average abundances of trace elements (approximately 35) in soil, crust, sediments, and igneous rocks; the range of abundance values of some elements commonly found in soils; the distribution of trace elements in the soil horizons of different soils; and the relative mobilities of elements in the supergene environment. It is recommended that these data, as well as the mechanism of the adsorption of ions onto clay minerals and humus,[180] be studied prior to commencing any inorganic pollution studies involving soils.

The available physical and chemical methods of analysis for inorganic pollutants in soil embrace a large proportion of all the methods which have been described earlier for air and water. The taking and storage of soil samples should be carried out according to well-established procedures. The most valuable compilation of methods for the examination of soils has been edited by Black *et al.*[9] The following listing of chapter headings will indicate major sources of information suited to the determination of inorganic pollutants in soils: elemental analysis by X-ray emission spectrography, optical emission spectrography, flame photometry, absorption spectrophotometry, and polarography; cation exchange capacity; total exchangeable bases; exchange acidity; hydrogen ion activity; soluble salts; fusion with Na_2CO_3 for total elemental analysis; methods for the following elements: Si, Fe, Ti, Al, Ca, Mg, Mn, K, Na, P, Mo, B, Co, Cu, Zn, S, Se, Cl, Br, F, N (various forms), and carbonate.

While most of the above information relates to methods suitable for the determination of the total amount of an inorganic constituent in a soil, this does not provide all the information necessary for associated inorganic pollution studies. It has been indicated in the above introduction for soils, and in earlier remarks concerning the type and form of constituents in water, that the inorganic composition (and pollutants) in a soil may be present in different forms. Consequently, data on the total amount of element present may provide incomplete information for pollution studies, unless there are also data on the form of the element. Whether the element is present as a cation, anionic complex, organometallic complex, in adsorbed or metallic form, or incorporated in the soil mineral crystalline lattice, is very relevant to such studies. Unfortunately, there is an almost complete lack of information in this area in relation to inorganic pollution studies concerning soils. The almost complete lack of suitable methods constitutes a challenge to analytical chemists. However, studies have been carried out on the "availability" of elements in soils for agricultural purposes and more recently for some pollution studies.[181,182] Agricultural and soil scientists have for many years used various aqueous extracting media to simulate the uptake of nutrients by plant roots, giving rise to data on "available" nutrients. In inorganic pollution studies of soils, three main extracting solutions have been used[181,182]: water, ethylenediamine tetraacetic acid (EDTA), and acetic acid. It is important to appreciate, however, that all such extracting solutions constitute arbitrary, relative methods and that the results are dependent on factors such as soil pH, soil type, form of element (e.g., oxidation state, etc.), particle size of soil sample, solution strength, and the shaking or elution time. By common acceptance, and for comparative purposes, such solutions provide useful data on the inorganic pollution of soils. Typical pollution studies[181,182] have concerned the following extractable elements in soils: Zn, Cu, Pb, and Cd. However, with recent interest in other elements which play a role in health and disease,[183,184,185] it is likely that some of the following elements will also assume importance in soil pollution studies: Cr, Sn, V, F, Si, Ni, and Sr. New or modified methods will be required for determination of the form and amount of these elements in environmental materials.

5.3. *Organic Composition*

The organic nature of the upper horizons, involving humic layers of soil, is largely a function of the vegetation cover and living matter in the soil. Organic debris in varying stages of decomposition, together with humic acids, humic complexes, proteins, many other organic compounds, and living matter, may be considered to be "normal" constituents of a soil. Humic layers in particular, and the lower horizons, can become enriched in inorganic and organic constituents as a result of the biogeochemical cycle. Microorganisms play an important role in the mobilization and fixation of elements, and also in the cycles of C, N, P, and S compounds.[180]

Physical and chemical methods for the determination of organic pollutants in soils should therefore be suitable for use in the presence of a large number of naturally occurring compounds. Conventional methods[9] used by agricultural and soil scientists may be adapted for some of these purposes. The compilation by Black *et al.*[9] lists the following sections dealing with organic and microbial constituents: total carbon; organic carbon; partial extraction and gross fractionation of organic matter; inositol hexaphosphate; amino sugars; amino acids; microbial populations by direct microscopy; methods for determining microbial populations or counts; aerobic spore-forming bacteria; nitrifying bacteria; denitrifying bacteria; rhizobia; azobacter; actinomycetes; fungi; algae; protozoa; nematodes; mites and other microarthropods; enzymes and microbial respiration. The choice of suitable methods for the determination of organic pollutants in soil should be based upon consideration of the above methods for "normal" constituents, together with a consideration of Table 3 and previous sections of this chapter.

The taking and storage of soil samples for the determination of organic pollutants should be carried out according to well-established procedures.[9] However, the storage conditions may require judicious modification where the determination of volatile constituents (e.g., pesticides) is concerned. Reduced time between sampling and examination or refrigeration in sealed containers may be required.

5.4. *Radioactivity*

The conditions and principles already outlined for determination of inorganic and organic pollutants in soils may be applied, in most circumstances, for the determination of pollution by radionuclides.

Gaseous radionuclides in soil may be displaced and collected by conventional methods, using vacuum displacement with an inert gas or by heating. "Available" radionuclides may be extracted from soils by various aqueous, acid, or chelate extracting systems described earlier. It is stressed that these are arbitrary, relative methods, which may be accepted for certain pollution studies. In most circumstances, total dissolution of the soil is carried out by conventional fusion methods with fluxes such as carbonates, borates, pyrosulfates, and hydroxides of alkaline elements[9,186]; or, by conventional acid attack with an acid or mixture of acids.[187,188,189,190] The radionuclides separated from the soil may then be determined by methods described earlier.

Rapid methods have also been developed for the determination of radionuclides in whole or untreated soil samples. The soil may be examined in this way using γ spectrometry for comparatively large samples or by the measurement of thin layers of finely ground soil using other detection equipment.[95-97,100-101]

6. *Use and Interpretation of Results*

6.1. *Introduction*

In the past, many analytical chemists have not stressed or accepted their responsibilities in the use and interpretation of results. An enormous amount of virtually useless data has been produced by analytical chemists who have had no voice in the collection and storage of *representative* samples or in the interpretation and use of results. In a somewhat similar manner, scientists who have had no formal training or experience in analytical chemistry often rush into the adoption and use of ill-chosen methods. The problems of inexperience and choice are also magnified by lack of appreciation of fundamental concepts of accuracy and precision, contamination, quality control, and the correct use of intercomparison standards. Trace chemical analysis is an extremely exacting discipline. In environmental investigations the planning, sampling, chemical analysis, and interpretation and use of results is moving increasingly toward a team effort, involving scientists in several disciplines, because of the wide range of skills and experience that are required.

6.2. *Accuracy and Precision*

An analytical method can be specified in terms of accuracy, precision, and range. Each of these parameters can be defined.

6.2.1. *Accuracy*

Accuracy is defined as the deviation of the mean value of several replicate analyses of a sample of known composition from the "true" or accepted value for that sample.[41] The deviation can be positive or negative and is often called "bias." (Standard samples are usually required for comparison. Standard addition techniques are not always satisfactory, because the form of the analyte in the addition may not necessarily be the same as in the actual sample.)

6.2.2. *Precision*

The precision of a group of measurements refers to how closely the individual measurements agree. This can be determined statistically by carrying out more than one determination on a sample, and sometimes up to 30 results are required for adequate confidence. When the values of a set of measurements are close to each other, they are said to have high precision, and when scattered, low precision. High precision does not necessarily mean high accuracy because a constant error may be present that causes bias in all the results. Low precision indicates errors large enough to produce untrustworthy results, unless the central value of a large number of values is taken. The standard deviation is a measure of precision and may be calculated from the expression

$$\sigma = \pm \left(\frac{\Sigma \Delta^2}{n-1} \right)^{\frac{1}{2}}$$

where σ is the standard deviation, Δ is the difference of each individual result from the arithmetic mean, and n is the number of replicates in the group.

Statistical tables indicate that 997 results in 1000 should fall within $\pm 3\sigma$, and a curve drawn at 3σ from the mean is referred to as the 99% confidence limit. The 95% confidence limit is a curve similarly drawn 2σ from the mean. An approximate, but rapid means for calculating σ, makes use of the expression

$$\sigma = \frac{\text{Range}}{c}$$

where range is the difference between the greatest and smallest value obtained, c is a constant for different values of n, and n is the number of determinations.

The following values for c may be used, depending on the value of n: n: 2, c: 1.13; n: 3, c: 1.69; n: 4, c: 2.06; n: 5, c: 2.33; n: 8, c: 2.87; n: 10, c: 3.08.

Care should be taken when using the range of some calculations, because the derivation of σ from the range assumes that Gaussian law applies.[191]

6.2.3. Range

The word "range" used in the analytical sense should not be confused with its normal statistical meaning which is briefly given above.[191] The analytical range is the range of concentration over which the element (or analyte) can be determined within tolerable limits of accuracy and precision. This implies that all analytical methods have a limit of detection (or sensitivity) and an upper limit, and as these limits are approached the standard deviation about the mean increases rapidly.

6.2.4. Limit of Detection

The limit of detection, or sensitivity, of an analytical method may be defined as the concentration of an analyte which can be detected with 95% certainty. It is that quantity of the analyte which gives a reading, value, or signal equal to twice the standard deviation of a series of at least ten determinations at or near the blank level.

There is some difference of opinion about the definition of the limit of detection under different circumstances, as will be seen in the following example. The limit of detection for a method can be determined by recording ten values for the analyte at or near the blank value ($\bar{x}$ is the mean of these readings) and also recording ten values for the blank ($\bar{x}_b$ is the mean of these values). It can be shown that

$$(\bar{x} - \bar{x}_b)_{\min} = k\sqrt{2} \cdot \sigma_b$$

where $(\bar{x} - \bar{x}_b)_{\min}$ is the minimum detectable difference between the average analytical reading and the average blank reading, σ_b is the standard deviation of the blank reading, and k is a constant defined by the degree of confidence required. Values for k at different confidence levels are: 1.0 (68.3%); 1.96 (95.0%); 3.0 (99.7%). If we substitute 1.96 in the above expression, $(\bar{x} - \bar{x}_b)_{\min}$ is approximately equal to $3\sigma_b$ rather than 2σ. The discernibility or sensitivity of a method is proportional to the slope of the curve relating the signal magnitude to the amount of material present, and this will reflect directly on the ability to ascertain a difference between the signal and the blank at the detection limit; i.e., given adequate precision, the greater the sensitivity, the better the detectability.[192] Winefordner and co-workers[193,194] and Eckschlager[191] have further investigated the extent to which the absolute value of the blank experiment elevates the detection limit and also influences the reliability of trace analyses. Winefordner and co-workers[193,194] define the detection limit as the least signal which may be distinguished from zero noise by means of Student's t-test. This detection limit is not constant, and depends, for example, on the number of parallel determinations carried out. The problem of detection limit including the notation and terminology is still unresolved, but the first definition given above will suffice for most practical purposes.

It will be of interest to note that the limit of detection can be expressed in either of two ways.[192] The *absolute limit* is the smallest detectable weight of a substance expressed in micrograms, nanograms, picograms, etc. The *relative limit* is the lowest detectable concentration expressed as a percentage, parts per million on a weight or atomic basis, micrograms per milliliter or cubic centimeter, etc. It is usually expressed in a form related to the total sample, in units traditional to the analytical chemist and bearing in mind the type or purpose of the chemical analysis.[192] Further information on the above matters, and on other statistical methods which are applicable to the chemical analysis of pollutants, is available in the literature.[191-194]

6.3. *Quality Control*

Quality control for ensuring consistent and even quality of products is used in industries ranging from steel and foodstuffs to pharmaceuticals. Quality is usually maintained in such industries by adopting various standards or by setting upper and lower limits for a production run. The appropriate control may involve random or regular sampling with consequent adjustment, recycling, or rejection.

Comprehensive and effective quality control of laboratory operations is also a necessity for reliable chemical analyses. The quality control may involve the performance of laboratory personnel, physicochemical instruments, methods, and reagents. In general terms, the maintenance of consistent quality in the results of chemical analyses may involve any or all of such performances. The primary aims are to keep short-term bias and long-term drift to a minimum. These aims may be achieved by the introduction of quality control procedures which incorporate prior agreement on the necessary accuracy and precision for, e.g., a batch of chemical analyses, or the performance of a physicochemical instrument over a specified period.

There are several useful ways in which such quality controls can be achieved. Quality control charts provide a rapid visual means for the recognition of performance in terms of accuracy and precision without the burden of arithmetic calculations and masses of data. Timms[41] expresses this well by saying that: "In the laboratory, such calculations can be a burden to the analyst or supervisor and, unless made frequently become more a record of history than a guide to current work."

6.3.1. Control Charts for Accuracy

It is too often assumed that the overall accuracy of a method of chemical analysis can be checked in the laboratory with the appropriate standard of the analyte, without the presence of the matrix of the "real" sample. Standards of the analyte alone have severe limitations if the results are to be used for accuracy evaluations for a sample which includes analyte plus matrix. For this reason, standard materials are generally used when the accuracy of a method of chemical analysis is to be ascertained. Repeated measurements are then made of the standard, using the chosen personnel, instruments, methods, and reagents. A quality control chart can then be prepared using a central line drawn at the "accurate" value, and individual results are plotted.[40,41] Lines may be drawn at distances 2σ and 3σ above and below the center line. The 2σ control limit may be regarded as a warning limit, and the 3σ control limit may be regarded as the limit to initiate corrective action. These limits may be set, and if necessary further refined, in the course of trials or operation. The practice in an operating laboratory usually involves the injection of samples (standards or substandards) of known composition into the laboratory run, in a manner unknown to the operator (analytical chemist). In interlaboratory or "round-robin" exercises,[40] the values for the analytes are not known by the laboratory and samples are usually coded. On other occasions, a standard sample with a known value for the analyte may be used by the analytical chemist for a personal check of his method or instrument.

6.3.2. Control Charts for Precision

In this case, replication of measurements on a single item is required to evaluate variability, preferably over the whole range of concentrations being determined. The center line may be fixed at the average range, which provides an estimate of the standard deviation. Upper and lower control limits may be set as appropriate.[40,41] In general laboratory practice routine samples can be duplicated on some random basis. In other checks on precision the analytical chemist may carry out ten replicates on a known sample. A further scheme involves the injection of special samples in a laboratory to cover the entire working analytical range.[41] Other schemes are also available in the literature.[191,195]

6.3.3. Quality Control Charts for Instruments

It is not generally appreciated that a simple quality control chart may be used to check the performance of a physicochemical instrument such as an atomic absorption spectrophotometer. Such a chart can provide useful daily or weekly checks of instrument performance under chosen conditions. The chart readily enables an operator to check for drifts or changes in AAS instrument performance over specified periods, occasioned, for example, by deterioration in nebulizer efficiency, amplifier or photomultiplier performance, etc. A standard solution of an element at a concentration of 1×10^{-3} ppm (prepared freshly by dilution from a more concentrated standard stock solution) is used to record absorbance with all instrument settings in their optimum mode. The optimum absorbance is then selected as the central quality control line and regular plots, under the same conditions, show the performance of the instrument at a glance. These quality control charts are particularly useful when one instrument is used by a number of operators. A quality control check before use ensures that the instrument is functioning properly.

6.4. *Intercomparison Standards*

There are not nearly enough recognized international standards of soils, rocks, plants, and other environmental materials. This is understandable because the preparation and certification of such standards requires considerable time, large numbers of chemical analyses, and involves considerable cost. The importance of such standards cannot be overemphasized. They are not available to the individual in large amounts (usually only 50 g) and consequently it may be necessary to prepare substandards based on the standard. This is not easy.

The main sources of environmental standards are commercial organizations, standards organizations such as National Bureau of Standards (NBS), some universities, geological survey organizations, and research associations. The NBS has issued trace element standards of orchard leaves, tomato leaves, bovine liver, and coal (for mercury content). Most of the standards, such as orchard leaves, are issued with values for about 20 elements. Other values then accumulate via the publications of other research workers and laboratories, who give their own results. One of the older standards is Bowen's Standard Kale.[196] Since its issue 8 years ago, it has been analyzed by hundreds of analytical chemists, and their results are scattered through the literature. There are many accounts of intercomparison chemical analyses in the literature. Some years ago the research laboratories of the International Atomic Energy Agency issued standard intercomparison samples of uranium ore, which were analyzed by a large number of laboratories.[197] The well-known lunar rock and soil samples also formed the basis of many interesting intercomparison analyses.[198] There is certainly a growing need for similar intercomparison samples in the analytical chemistry of pollutants.

6.5. *Interpretation*

Interpretation of chemical analysis results should always involve the analytical chemist. Pollution studies also increasingly involve teamwork spanning several scientific disciplines, and, wherever possible, there should be adequate interdisciplinary consultation on the use and interpretation of data resulting from the chemical analysis of pollutants in the environment. Base-line data are an invaluable aid to interpretation. Where these are not available, comparison should be made with closely related material from other areas or with standard materials. Increasing use should also be made of environmental impact surveys[198] prior to commencement of developments which may result in changes in the environment.

Morrison says, about impact surveys,

More recently, society has recognized that in addition to these customary economic analyses and discussions of need, there should be a detailed assessment of the effect of a proposed development on the environment and thus its ecological, separate from its monetary, benefits and costs; put together, these assessments comprise an Environmental Impact Statement. The preparation of such a Standard should be done by a team of physical and social scientists and engineers; likewise reviews of statements will generally require an interdisciplinary team effort.[198]

A details consideration of various "environmental stresses" and the derivation and use of an environmental impact index has been recommended by Commoner.[199] While this index is primarily related to production and population, such considerations also form a valuable background for the interpretation of data derived from chemical analyses of pollutants.

References

1. W. W. Walton and J. I. Hoffman, "Principles and Methods of Sampling," in *Treatise on Analytical Chemistry*, I. M. Kolthoff, P. J. Elving, and E. B. Sandell, eds., Vol. 1, Wiley–Interscience, New York, 1959, Ch. 4.
2. C. R. N. Strouts, J. H. Gilfillian, and H. N. Wilson, *Analytical Chemistry: The Working Tools*, Oxford, 1955, Ch. 3.
3. R. C. Tomlinson, in *Comprehensive Analytical Chemistry*, C. L. Wilson and D. W. Wilson, eds., Vol. 1A, Elsevier, Amsterdam (and Van Nostrand, Princeton, N.J.), 1959, Ch. 2, Sec. 3.
4. *Methods for Chemical Analysis of Water and Wastes*, Environmental Research Center, Analytical Quality Control Laboratory, Cincinnati, Ohio 45268, 1971.
5. J. W. Mitchell, *Anal. Chem.*, **45**(6) : 492A–500A (1973).
6. T. T. Gorsuch, *The Destruction of Organic Matter*, International Series of Monographs in Analytical Chemistry, Vol. 39, Pergamon Press, 1970.
7. *Methods of Air Sampling and Analysis*, Intersociety Committee, Amer. Public Health Assoc., Washington, D.C., 1972.
8. M. L. Jackson, *Soil Chemical Analysis*, University of Wisconsin, 1956.
9. C. A. Black, ed., *Methods of Soil Analysis, Parts 1 and 2*, Amer. Soc. of Agronomy Inc., Wisconsin, 1965.
10. W. W. Meinke, "Characterization of Solids—Chemical Composition," in *Treatise on Solid State Chemistry*, Vol. 1, Plenum Press, 1973, Ch. 7.
11. W. F. Pickering, *Modern Analytical Chemistry*, Marcel Dekker Inc., New York, 1971.
12. Anal. Chem., *Fundamentals and Applications Review Issues*, each April in alternate years.
13. M. W. Skougstad, *Anal. Chem.*, **46** : 982A (1974).
14. R. F. Coleman, *Anal. Chem.*, **46** : 989A (1974).
15. J. H. Yoe, *J. Chem. Educ.*, **14** : 170 (1937).
16. T. S. West, *Analyst*, **87** : 630 (1962).
17. T. M. Florence, D. A. Johnson, and Y. J. Farrar, *Anal. Chem.*, **41** : 1652 (1969).

18. T. M. Florence and Y. J. Farrar, *Anal. Chem.*, **35** : 1613 (1963).
19. L. P. Hammett, *Physical Organic Chemistry*, McGraw Hill, N.Y., 1940.
20. H. H. Jaffe, *Chem. Rev.*, **53** : 191 (1953).
21. P. R. Haddad, P. W. Alexander, and L. E. Smythe, *Talanta*, **22** : 61 (1975).
22. C. H. Kim, C. M. Owens, and L. E. Smythe, *Talanta*, **21** : 445 (1974).
23. G. F. Kirkbright, *Talanta*, **13** : 1 (1960).
24. Environ. Protectn. Agency (U.S.A.), *Fed Reg.*, **36** (1971–present).
25. H. M. Englund and W. T. Berry, ed., *Proc. Intern. Clean Air Congress, 2nd.*, Academic Press, N.Y., 1971.
26. Intersoc. Committee, *Methods of Air Sampling and Analysis*, Amer. Pub. Hlth. Assn., Washington, D.C., 1972.
27. J. B. Homolya and J. P. Backmann, *Internat. Laboratory*, **12** : 37 (1971).
28. Analytical Instrument Development Inc., Metronics Associates, Polyscience Corp., all in U.S.A.
29. L. B. Kreuzer, *Anal. Chem.*, **46** : 239A (1974).
30. J. P. Matousek and K. G. Brodie, *Anal. Chem.*, **45** : 1606 (1973).
21. Various Methods Manuals and Bibliography, Millipore Corporation, Bedford, Massachusetts, 01730, 1974.
32. H. Weisz, *Microanalysis by the Ring Oven Technique*, Pergamon Press, N.Y., 1961.
33. R. M. Ross, *Proc. Inst. Environ. Sci.*, 18th Meeting, Prospect. Ill., 1972, p. 576.
34. R. L. Chuan, *Proc. Joint Conf. Sensing Environ. Pollut.*, Instrum. Soc. Am., New York, 1972.
35. M. Imada and P. K. Mueller, AIHL Report 114, Calif. Dept. Health, Berkeley, Calif., 1971.
36. K. Spurny and J. Oppelt, *Coll. Czech. Chem. Commun.*, **36** : 2683 (1971).
37. R. B. Husar and C. S. Heisler, *Proc. Conf. Methods in Air Pollt. Ind. Hyg. Studies*, State of Calif., Dept. Public Health, Berkeley, Calif., Oct., 1972.
38. L. Meites, *Handbook of Analytical Chemistry*, McGraw Hill, N.Y., 1963.
39. W. E. Ruch, *Quantitative Analysis of Gaseous Pollutants*, Ann Arbour, London, 1970.
40. R. Robinson, *Clinical Chemistry and Automation*, Griffin, 1971, Ch. 6.
41. A. B. Timms, *Amdel Bulletin (Aust.)*, 1967 (No. 3), 67.
42. J. D. Winefordner and R. C. Elser, *Anal. Chem.*, **43** : 24A (1971).
43. G. F. Kirkbright and T. S. West, *Chem. in Britain*, **8** : 428 (1972).
44. R. F. Browner, *Analyst*, **99** : 617 (1974).
45. L. E. Smythe, *Proc. Roy. Aust. Chem. Inst.*, **40** : 341 (1973).
46. S. R. Koirtyohann and J. W. Wen, *Anal. Chem.*, **45** : 1986 (1973).
47. J. Mika and T. Török, *Analytical Emission Spectroscopy*, London, Butterworths, 1974.
48. ASTM COMMITTEE E2, Methods for Emission Spectrochemical Analysis, Amer. Soc. Testing Mats., 1971.
49. B. F. Scribner and M. Margoshes, "Emission Spectroscopy," in *Treatise on Analytical Chemistry*, I. M. Kolthoff and P. J. Elving, eds., Part 1, Vol. 6, Interscience, New York, 1965, Ch. 64, pp. 3347–3461.
50. Model GX-3, Baird Atomic, Bedford, Mass.
51. Manufactured by Jarrell-Ash Co., Newtonville, Mass.
52. R. S. Juvet and S. P. Cram, *Anal. Chem.*, **46**(5) : 101R (1974).
53. B. I. Anvayer and Y. S. Drugov, *Zh. Anal. Khim.*, **26** : 1180 (1971).
54. W. A. Aue and H. H. Hill, *Anal. Chem.* **45** : 729 (1973).
55. L. E. Ettre and W. H. McFadden, eds., *Ancillary Techniques of Gas Chromatography*, Wiley–Interscience, 1969.
56. G. A. Junk, *Int. J. Mass Spectrom. Ion Phys.*, **8** : 1 (1972).
57. F. W. Karasek, *Anal. Chem.*, **44** : 32A (April, 1972).
58. R. N. Hager, *Anal. Chem.*, **45** : 1131A (1973).
59. R. K. Stevens and J. A. Hodgeson, *Anal. Chem.*, **45** : 443A (1973).
60. G. W. Nederbragt, A. Van der Horst, and J. Van Duijn, *Nature*, **206** (4949) : 87 (1965).
61. B. Draeger, W. Germ. Pat., 1,133,918 (July 26, 1962).
62. R. K. Stevens, J. D. Mulik, A. E. O'Keeffe, and K. J. Krost, *Anal. Chem.*, **43** : 827 (1971).

63. Oxides of Nitrogen Analyser 8440, Monitor Labs Incorp., San Diego, Calif.
64. C. E. White and R. J. Argauer, *Fluorescence Analysis. A Practical Approach*, Marcel Dekker, N.Y., 1970.
65. J. E. Stewart and H. R. Zulliger, *Indust. Research*, **16** : 71 (Oct., 1974).
66. K. F. J. Heinrich, *Scanning Electron Probe Microanalysis*, NBS Tech. Note 278, U.S. National Technical Information Service, Springfield, Virginia, 1968.
67. T. M. Florence, *Proceedings Royal Australian Chem. Inst.*, **39** : 211 (1972).
68. R. G. Clem, *Industrial Research*, **15** : 46 (Jan., 1973).
69. Model 715 Process Oxygen Monitor, Beckman Instruments Inc., Fullerton, Calif., U.S.A.
70. D. G. Davis, *Anal. Chem.*, **41**(5) : 21R (1974).
71. J. Koryta, *Anal. Chim. Acta*, **61** : 329 (1972).
72. G. J. Moody and J. D. R. Thomas, *Selective Ion-Sensitive Electrodes*, Merrow Technical Library, Merrow, Watford, England, 1971.
73. R. P. Buck, *Anal. Chem.*, **41**(5) : 28R (1974).
74. R. W. Cattrall and H. Freiser, *Anal. Chem.*, **43** : 1905 (1971).
75. G. J. Moody, R. B. Oke, and J. D. R. Thomas, *Analyst*, **95** : 910 (1970).
76. G. H. Griffiths, G. T. Moody, and J. D. R. Thomas, *Analyst*, **97** : 420 (1972).
77. M. Katz, *Measurement of Air Pollutants: Guide to the Selection of Methods*, World Health Organization, 1969.
78. D. Packham and L. Gibson, Report of unpublished work, *Instrument News*, No. 104 (1974), C.S.I.R.O., Australia.
79. MS-GC System—Mat 112, Varian Associates, Palo Alto, California, U.S.A.
80. P. Irving, *Industrial Research*, **16** : 78 (Oct., 1974).
81. L. R. Snyder, in: J. J. Kirkland, ed., *Modern Practice of Liquid Chromatography*, Wiley–Interscience, N.Y., 1971.
82. B. L. Karger, L. R. Snyder, and C. Horuath, *An Introduction to Separation Science*, Wiley, N.Y., 1973.
83. H. L. Kahn and Z. Bitterfield, *Industrial Research*, **15** : 32 (June, 1973).
84. W. P. Taggart, *Industrial Research*, **16** : 76 (Feb., 1974).
85. E. Grushka, *Anal. Chem.*, **46** : 510A (1974).
86. E. Stahl, *Thin-layer Chromatography*, Springer-Verlag, N.Y., 1969.
87. J. C. Touchstone, ed., *Quantitative Thin-layer Chromatography*, Wiley–Interscience, N.Y., 1973.
88. J. A. Perry, T. H. Jupille, and L. J. Glunz, *Industrial Research*, **16** : 55 (Feb., 1974).
89. J. P. Dixon, *Modern Methods of Organic Microanalysis*, Van Nostrand, Princeton, N.J., 1968.
90. G. Tölg, *Ultramicro Elemental Analysis*, Wiley–Interscience, N.Y., 1970.
91. T. S. Ma and M. Gutterson, *Anal. Chem.*, **46**(5) : 437R (1974).
92. H. A. C. McKay, *Principles of Radiochemistry*, Butterworths, London, 1971.
93. A. G. Maddock, ed., *Radiochemistry*, Vol. 8, Butterworths, London, 1974.
94. W. S. Lyon, E. Ricci, and H. Ross, *Anal. Chem.*, **46**(5) : 431R (1974).
95. *Rapid Methods for Measuring Radioactivity in the Environment*, Internat. Atomic Energy Agency, Vienna, Austria, 1971.
96. Various manuals issued by laboratories such as: Oak Ridge National Laboratory, Oak Ridge, Tenn., U.S.A., or Atomic Energy Research Establishment, Harwell, Berks., England (details in *Nuclear Science Abstracts*).
97. T. J. Burnett, *Sampling Methods and Refinements for Estimating Airborne Particulate Hazards*, ORNL Report **CF 52** : 1–11 (1952).
98. K. G. Vohra, M. C. Subbaramu, and A. M. Mohan Rao, *Nature*, **201** : 37 (1964).
99. K. G. Vohra, *Proc. 2nd Internat. Conf. Peaceful Uses At. Energy, Geneva*, **23** : 367 (1958).
100. G. D. Kelley, "Detection and Measurement of Nuclear Radiation," U.S.A.E.C. Report NAS-NS-3105, 1962.
101. E. P. Steinberg, "Counting Methods for the Assay of Radioactive Samples," in *Nuclear Instruments and Their Uses*, A. H. Snell, ed., Wiley, N.Y., 1962, p. 306.
102. B. Shleien, *J. Am. Ind. Hyg. Assoc.*, **24** : 180 (1963).

103. M. H. Wilkening and J. E. Hand, *J. Geophys. Res.*, **65** : 3367 (1960).
104. J. Sedlet, "Radon and Radium," in *Treatise on Analytical Chemistry*, I. M. Kolthoff and P. J. Elving, eds., Part II, Vol. 4, Wiley–Interscience, N.Y., 1966, p. 226.
105. "Measurement of Low-level Radioactivity," U.S. Report ICRU-22, Washington, D.C., 1972.
106. M. W. Skougstad, *Kirk-Othmer Encycl. Chem. Technol.*, 2nd Ed., **21** : 688 (1970).
107. *Instrumentation for Environmental Monitoring: Water, LBL-1*, Vol. 2, Technical Information Division, Lawrence Berkeley Laboratory, Berkeley, Calif., 94720.
108. *Standard Methods for the Examination of Water and Waste Water*, 13th Ed., Amer. Public Health Assoc., 1015 Eighteenth St., N.W., Washington, D.C. 20036, 1971.
109. E. Brown, M. W. Skougstad, and M. J. Fishman, *Methods for Collection and Analysis of Water Samples for Dissolved Minerals and Gases*, Book 5, U.S. Geol. Survey, 1970, Chap. A1–A4 (Available from: Supt. Documents, U.S. Govt. Printing Office, Washington, D.C. 20402).
110. *Annual Book of ASTM Standards*, Part 23. Water; Atmospheric Analysis, 1970. Amer. Soc. ASTM, 1916, Race Street, Philadelphia, Pa. 19103.
111. V. T. Stack, *Anal. Chem.*, **44**(8) : 32A (1972).
112. W. Lueck, *Messtechnik (Brunswick)*, **78** : 181 (1970); *Chem. Abstr.*, **74** : 45461v (1971).
113. B. K. Krylov, V. I. Kalmanovskii, and Ya. I. Yashin, *Zavod. Lab.*, **37** : 133 (1971); *Chem. Abstr.*, **74** : 134438k (1971).
114. H. C. Edgington and R. M. Roberts, U.S. Saline Water, Res. Develop. Progr. Rept., No. 625, 1971, 140 pp.
115. J. Hahn, *Anal. Chem.*, **44** : 1889 (1972).
116. J. Czerniec, Z. Gregorowicz, J. Fligier, and P. Czichon, *Z. Chem. Anal. (Warsaw)*, **16** : 1125 (1971); *Chem. Abstr.*, **76** : 89872a (1972).
117. T. Wierzbicki and H. Pieprzyk, *Chem. Anal. (Warsaw)*, **15** : 1041 (1970); *Chem. Abstr.*, **74** : 79394r (1971).
118. M. J. Fishman and D. E. Erdmann, "Water Analysis" (Review), *Anal. Chem.*, **45**(5) : 361R (1973).
119. J. E. Harwood and D. J. Huyser, *Water Res.*, **4** : 695 (1970).
120. P. C. Head, *Deep-Sea Res. Oceanogr. Abstr.*, **18** : 531 (1971).
121. I. Wudzinska, *Chem. Anal. (Warsaw)*, **16** : 1359 (1971); *Chem. Abstr.*, **77** : 9360q (1972).
122. A. A. Diggens and W. D. Meredith, *Meas. Contr.*, **4** : T48 (1971); *Chem. Abstr.*, **75** : 25108c (1971).
123. H. C. Marks, *Water Pollution Hanb.*, **3** : 1213 (1971).
124. J. F. Kopp and R. C. Kroner, *Trace Metals in Waters of the United States*, U. S. Dept. of Interior, F.W.P.C.A., Division of Pollution Surveillance, Cincinnatti, Ohio, 1967.
125. R. J. Gibbs, *Science*, **180** : 73 (1973).
126. C. R. Parker, *Water Analysis by Atomic Absorption Spectroscopy*, Varian Techtron Pty. Ltd., Springvale, Victoria, Australia, 1972.
127. K. J. Doolan and L. E. Smythe, *Talanta*, **20** : 241 (1973).
128. Y. Marcus, *Chem. Revs.*, **63** : 139 (1963).
129. I. M. Kolthoff, E. B. Sandell, E. J. Meehan, and S. Bruckenstein, *Quantitative Chemical Analysis*, Macmillan, 1969, Ch. 12–15.
130. H. F. Walton, *Anal. Chem.*, **46**(5) : 398R (1974).
131. C. B. Amphlett, *Inorganic Ion Exchangers*, Elsevier, 1964.
132. J. P. Riley and D. Taylor, *Anal. Chim. Acta*, **40** : 479 (1968).
133. R. L. Mitchell and R. O. Scott, *Spect. Acta*, **3** : 367 (1948).
134. W. P. Silvey and R. Brennan, *Anal. Chem.*, **34** : 784 (1962).
135. L. E. Smythe and T. M. Florence, "Recent Advances in the Analytical Chemistry of Beryllium," in *Progress in Nuclear Energy, Series IX*, Vol. 3, C. E. Crouthamel, ed., Pergamon Press, N.Y., 1963, Ch. 6.
136. D. C. Burrell, *Anal. Chim. Acta*, **38** : 447 (1967).
137. T. Joyner, M. L. Healy, D. Chakrabarti, and T. Koyanagi, Presented at Symposium on Trace Characterization, N.B.S., Washington, D.C., 1966.

138. R. L. Dem, W. G. Dunn, and E. G. Loder, *Anal. Chem.*, **33** : 607 (1961).
139. M. G. Lai, H. V. Weiss, *Anal. Chem.*, **34** : 1012 (1962).
140. S. L. Phillips, D. A. Mack, and W. D. McLeod, *Anal. Chem.*, **46**(3) : 345A (1974).
141. W. C. McCrone and J. G. Delly, *The Particle Atlas*, Vol. 1–4, 2nd ed., Ann Arbor Sci. Publ., Ann. Arbor, Mich., 1973.
142. L. D. Whittig, X-ray Diffraction Techniques for Mineral Identification and Mineralogical Composition," in *Methods of Soil Analysis*, C. A. Black, ed., American Society of Agronomy, Wisconsin, 1965, Ch. 49, pp. 671–696.
143. L. E. Smythe, *Talanta*, **15** : 1177 (1968).
144. E. E. Pickett and S. R. Koirtyohann, *Anal. Chem.*, **41**(14) : 28A (1969).
145. R. Mavrodineanu, ed., *Analytical Flame Spectroscopy*, Macmillan, N.Y., 1970.
146. B. V. L'vov, *Atomic Absorption Spectrochemical Analysis*, J. H. Dixon, transl., Adam Hilger, London, 1970.
147. W. Slavin, *Atomic Absorption Spectroscopy*, Wiley–Interscience, N.Y., 1968.
148. W. J. Price, *Analytical Atomic Absorption Spectroscopy*, Heyden, London, 1974.
149. G. D. Christian and F. J. Feldman, *Atomic Absorption Spectroscopy: Applications in Agriculture, Biology and Medicine*, Wiley–Interscience, N. Y., 1970.
150. J. D. Winefordner, ed., *Spectrochemical Methods of Analysis. Quantitative Analysis of Atoms and Molecules*, Vol. 9, Wiley–Interscience, N.Y., 1971.
151. R. Mavrodineanu, *Bibliography on Flame Spectroscopy, Analytical Applications, 1800–1966*, N.B.S. Miscell. Public., 281, 1967.
152. J. D. Winefordner and T. J. Vickers, "Flame Spectrometry" (Review), *Anal. Chem.*, **46**(5) : 192R (1974).
153. K. W. Busch and G. H. Morrison, *Anal. Chem.*, **45**(8) : 712A (1973).
154. D. G. Mitchell, K. W. Jackson, and K. M. Aldous, *Anal. Chem.*, **45**(14) : 1215A (1973).
155. V. A. Fassel and R. N. Kniseley, *Anal. Chem.*, **46**(13) : 1155A (1974).
156. A. Walsh, *Spectrochim. Acta*, **7** : 108 (1955).
157. R. R. Brooks and L. E. Smythe, *Anal. Chem. Acta*, **74** : 35 (1975).
158. R. J. Finlayson and L. E. Smythe, private communication, Dept. of Analyt. Chem., University of N.S.W., Sydney, 1975.
159. R. G. Clem, *Industrial Research*, **15** : 50 (Sept., 1973).
160. R. Brown, M. L. Jacobs, and H. E. Taylor, *Amer. Laboratory*, **4**(11) : 29 (1972).
161. R. A. Bingham and R. M. Elliott, *Anal. Chem.*, **43** : 43 (1971).
162. G. E. Batley, *Proc. Royal Australian Chem. Inst.*, **39** : 261 (1972).
163. F. Habashi and G. Thurston, *Anal. Chem.*, **39** : 243 (1967).
164. P. W. Alexander and G. L. Orth, *J. Electroanal. Chem.*, **31** : App. 3 (1971).
165. P. W. Alexander, *J. Electroanal. Chem.*, **36** : App. 21 (1972).
166. P. W. Alexander, private communication, Dept. of Analyt. Chem., University of N.S.W., Sydney, 1975.
167. H. B. Mark and G. A. Rechnitz, *Kinetics in Analytical Chemistry*, Interscience, N.Y., 1968.
168. R. A. Greinke and H. B. Mark, "Kinetic Aspects of Analytical Chemistry" (Review), *Anal. Chem.*, **46**(5) : 413R (1974).
169. G. A. Major, G. Dal Pont, J. Klye, and B. Newell, "Laboratory Techniques in Marine Chemistry", Report 51, Div. of Fisheries and Oceanography, Council Sci. and Indust. Res. Organization, Cronulla, Sydney, Australia, 1972.
170. D. W. Spencer and P. G. Brewer, *Crit. Rev. Solid State Sci.*, **1** : 409 (1970).
171. R. C. Weast, ed., *Handbook of Chemistry and Physics*, 54th Ed., The Chemical Rubber Co., Cleveland, Ohio, U.S.A., 1974.
172. A. A. Rosen, *Anal. Chem.*, **39**(10) : 26A (1967).
173. "Biological Analysis of Water and Wastewater," Application Manual AM 302, Millipore Corp., Bedford, MA., U.S.A., 1973.
174. J. Mitchell and J. Chiu, "Analysis of High Polymers" (Review), *Anal. Chem.*, **45**(5) : 273R (1973).
175. R. W. Perkins and L. A. Rancitelli, Report BNWL-SA-3993, 1971, 32pp.

176. W. S. Lyon, *Guide to Activation Analysis*, Van Nostrand, Princeton, N.J., 1964.
177. J. Hoste, J. Op De Beeck, R. Gijgels, R. Adams, P. Van Den Winkel, and D. De Soete, *Activation Analysis*, Chemical Rubber Co., Cleveland, Ohio, 1971.
178. J. M. A. Lenihan, S. J. Thompson, and V. P. Guinn, *Advances in Activation Analysis*, Vol. 11, Academic Press, London, 1972.
179. W. S. Lyon, Nucleonics (Review), *Anal. Chem.*, **46**(5) : 431R (1974).
180. R. R. Brooks, *Geobotany and Biogeochemistry in Mineral Exploration*, Harper and Row, N.Y., 1972.
181. D. Purves, *Environ. Pollut.*, **3** : 17 (1972).
182. F. Beavington, *Aust. J. Soil Res.*, **11** : 27 (1973).
183. *Trace Elements Today* (Review), *Ciba Review*, Ciba-Geigy, Basle, Switzerland, 1973.
184. W. J. Miller, K. Schwartz, and E. M. Carlisle, *Federation Proceedings*, **33**(6) : 1747 (1974).
185. H. A. Schroeder, *Med. Clinics of Nth. Amer.*, **58**(2) : 381 (1974).
186. A. M. Bond, *Anal. Chem.*, **43** : 134 (1971).
187. B. Bernas, *Anal. Chem.*, **40** : 1682 (1968).
188. F. T. Langmyhr and S. Sueen, *Anal. Chem. Acta*, **32** : 1 (1961).
189. N. H. Suhr and C. O. Igamells, *Anal. Chem.*, **38** : 730 (1966).
190. F. J. Langmyhr and P. E. Paus, Parts I–VIII of series of papers in *Anal. Chim. Acta*, 1968–70.
191. K. Eckschlager, *Errors Measurement and Results in Chemical Analysis*, R. A. Chalmers, transl., Van Nostrand, London, 1969.
192. G. H. Morrison and R. K. Skogerboe, "General Aspects of Trace Analysis," in *Trace Analysis Physical Methods*, Interscience, N.Y., 1965, Ch. 1.
193. J. D. Winefordner and T. J. Vickers, *Anal Chem.*, **36** : 1939 (1964).
194. P. A. S. John, W. J. McCarthy, and J. D. Winefordner, *Anal. Chem.*, **39** : 1945 (1967).
195. C. A. Bennett and N. L. Franklin, *Statistical Analysis in Chemistry and the Chemical Industry*, Chapman and Hall, London, 1954.
196. H. J. M. Bowen, *Analyst*, **92** : 124 (1967).
197. *Reports on the Panels on Analytical Chemistry of Nuclear Materials*, Nos. 18 and 62, International Atomic Energy Agency, Vienna, 1963 and 1966.
198. G. M. Morrison, *Anal. Chem.*, **43**(7) : 22A (1971).
199. L. B. Leopold, F. E. Clarke, B. B. Hanshaw, and J. R. Balsley, "A Procedure for Evaluating Environmental Impact," Circular 645, U.S. Geological Survey, National Center, Reston, Va. 22092, 1973.

"The System" and the Environment

In order to function effectively environmental chemists must certainly have some concept of economics—not only the economics of textbooks, but also the economics of real life in which the pulse of progress is fueled by the motive of profit. It is necessary to see how the often restrictive demands of environmental chemistry interact with the present expensive free enterprise economic system.

22

The Public Policy Issues Involved in Dealing with Environmental Degradation: A Dynamic Approach

Burton H. Klein

1. Strong versus Weak Uncertainties

From a public policy standpoint, the issues of environmental degradation are unique in that they involve making decisions under conditions of "strong uncertainties." No one can calculate in probabilistic terms either the risks to societies of environmental degradation or the costs that will be encountered to bring about a lower rate. With respect to the risk, there is no way of calculating to what extent the lives of people might be shortened by living in a densely populated and multifariously polluted city such as Los Angeles. Physicians

Burton H. Klein • Department of Economics, Division of Humanities and Social Sciences, California Institute of Technology, Pasadena, California, USA 91109

have only a very rough idea of how respiratory and heart diseases are affected by certain kinds of contaminants. However, the more that is learned about the risks of environmental degradation, the more serious they appear. This has been proven true with respect to radioactivity as well as with respect to cigarette smoking. And there is no reason to suppose it will not be true with respect to other pollutants.

This is not to suggest that because precise calculations cannot be made people behave irrationally if they take action to protect their own health. While some engineers and economists will deny that any effect can occur unless it can be measured, when it comes to health, people have no alternative but to be resourceful on the basis of highly ambiguous evidence. Although good estimates are not obtainable, I would guess that a very significant part of the outmigration during the past several years from Los Angeles and other cities having similar pollution problems was provoked by the effect of smog on health, and, more often than not, upon the advice of a physician. Physicians may not know a great deal about the effects of pollution on health, but they cannot very well advise their patients to wait until better knowledge is available. While there are economic risks involved in acting upon the basis of imperfect knowledge, for particular individuals there are more dangerous risks in waiting until good mortality data are available. A world of weak uncertainties can be defined as one having good mortality data, and, more generally speaking, as a world which can be described in probabilistic terms. In such a world it would be possible to obtain insurance against pollution. However, if such insurance is not available, people have no alternative but to balance the risks of continuing to live in a polluted area with those of resettling in a healthier environment.

Nor is it possible for public officials to be oblivious to the risks of environmental degradation. They may or may not be sensitive to the impact of pollution on health. But public officials are very sensitive to the impact upon the future tax base of a continued outmigration from a polluted area. In fact, I know firsthand that, on the basis of incomplete information about the reasons for the recent outmigration from the Los Angeles area, a number of political and business leaders saw in the threat of widespread pollution a real dilemma for the city.

To sum up this part of the discussion: When uncertainties are strong, risks cannot be calculated in probabilistic terms. Or, to put the proposition the other way around, if the uncertainties were weak, which indicates they could be calculated in probabilistic terms, people could insure themselves against living in high-pollution areas as they insure themselves against airplane accidents, and there would be no risk in living in high-pollution areas. People would simply have to decide whether or not living in a polluted area was worth the additional insurance premium. Governments might, of course, decide upon specific remedies to reduce the rate of environmental degradation. But if they did, it would be because it would cost less to save a polluted area than to allow it to die.

Nor is it possible to make good predictions of the longer-run costs likely to be involved in reducing environmental degradation. How rapidly the risks of environmental degradation are reduced will obviously depend on how imaginative business firms are in developing better menus of alternatives. To be sure, business firms may search for relatively minor advances that call for little imagination, in which case the costs involved in reducing environmental degradation would be highly predictable. But the rate of progress would probably be very slow. Conversely, advances calling for the greatest degree of imagination would be the least predictable. The reason for this is that imagination, if efficiently employed, leads to more significant ideological mutations, mutations which must be defined as highly unpredictable. For example, it would have been easier to predict the measures the American automobile industry has adopted to date than it would have been to predict the stratified charge engine developed by the Honda Corporation. And the reason is very clear: from the point of view of permitting better fuel consumption and better emissions performance the stratified charge engine represents a more impressive mutation.

What, then, determines whether business firms are likely to employ relatively little or a great deal of imagination in searching for better alternatives? The answer is: the degree of risk firms impose upon each other. As Fig. 1 shows, in an industry such as the computer industry, the advances represented by the stratified charge engine would not be looked upon as large. In that industry, engaging in a high degree of unpredictable behavior can be regarded as routine activity. In fact, as Fig. 1 shows, the advances follow upon each other so regularly the industry can be described as a "predictably unpredictable"

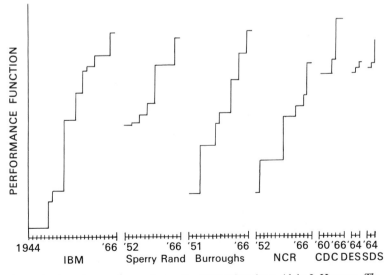

Figure 1. Actual computer performance. Data taken from Alvin J. Harman, *The International Computer Industry*, Harvard University Press, 1971.

industry engaged in the making of "fast history." It is no accident, however, that the computer industry happens to be one in which competitors impose large risks upon each other, as measured in terms of the markets which can be won or lost with the introduction of new computers. It is quite unlikely for an industry as a whole to behave in a predictably unpredictable manner while overcoming large discontinuities unless the firms impose large risks upon each other. After all, unless prodded by competitors, why else would firms continue to deal with large technological risks?

What, then, prodded Honda into taking a large technological risk? One obvious reason is that Honda is not a typical automobile firm. It originated as a bicycle firm and only recently entered the automobile business. Therefore, it is not likely to have the inflexible organizational structure associated with an American automobile firm, whose structure sharply limits the ability to engage in a higher degree of unpredictable behavior. For example, Honda engineers, working at Ford (Detroit) on the development of a stratified charge engine, argue that they are surprised that anything new is ever accomplished in the American automobile industry. They point out that, in Japan, they can obtain parts on the basis of rough drawings from general Japanese machinists in a matter of a few days, but at Ford, where specialized machinists can only work from highly detailed drawings, the same process can take months. Furthermore, Honda may not only have an organizational structure which is more consonant with risk taking, because it permits a higher degree of unpredictable behavior, but it also may have a greater incentive to engage in risk taking. If foreign manufacturers hope to compete in American markets they must have something special to sell. For only trivial differences in gasoline consumption the American consumer will not change his attachment from an American to a foreign product.

To conclude this part of the discussion: Viewed as isolated events, individual advances in technology are quite unpredictable. But a series of impressive advances can be made quite predictable when business firms impose a substantial degree of risk upon each other. Since the uncertainties are strong, they obviously cannot take out insurance against such competitive risks. They must, therefore, hire imaginative people to do to their competitors what their competitors can do to them! And it is the diversity of ideas generated, because competitors cannot predict each other's actions, which leads to smooth progress when plotted on a macroscale. Thus, microdiversity is the hidden hand of macrostability.

Why, then, are the issues of the environment to be considered as dynamic issues? As the terms are usually defined, a "static process" is one in which the initial conditions can be taken as given, and a "dynamic process" is one which involves unpredictable changes in the initial conditions. If this definition is accepted, then the environmental issues must be regarded as "dynamic issues" for all time to come. First, as was already pointed out, as man learns more about the risks of environmental degradation, it seems quite likely he will discover effects which cannot be predicted today. Moreover, as human evolution continues, new interactions with the physical and biological environments will result in the discovery of new risks (as they have since the Middle Ages). Second, growth will wear out technologies from an environmental point of view, and the more rapidly economies attempt to grow, the greater will be the need for dynamic behavior. Third, if effective action is to be taken to bring about smooth progress to reduce the risks of environmental degradation, a high degree of unpredictable behavior will be required to generate a diversity of ideas.

It is very unfortunate from a public policy point of view that the industries which contribute most to the problems of the environment, the automobile, petroleum, and electric power industries—industries for whom stable growth has become almost a religion—are those whose ability to engage in dynamic behavior is not very impressive. Furthermore, as far as the United States is concerned, the automobile and related industries are responsible for about one-sixth of its total employment.[1] Therefore, industrial spokesmen are in a good position to argue that "what is good for General Motors is good for the U.S.A." Or, as Keynes once put it: When you owe the bank one thousand pounds, the bank owns you. But when you owe the bank one million pounds, you own the bank!

What, then, are the policy options for dealing with environmental degradation? One is the internalization of costs: a prescription which has emerged from classical economic theory. Another is the policy which is being followed by most countries: a policy of direct regulation. And finally, there is a policy which can be described as "risk internalization." Its aim is to make business firms feel (because of their profits) the risk of not reducing environmental degradation at a more rapid rate. However, while the last proposal is based upon dynamic economic theory,[2] it certainly is not new. In fact, the concept of risk internalization was proposed in a somewhat different context in 1827 by an engineer who, as it happens, is also credited with the key ideas contained in the second law of thermodynamics: Sadi Carnot.

2. The Internalization of Costs

If it is assumed that an economic system is closed, in the sense that its knowledge can be taken as a given, and if it is further assumed that all people are more or less equally affected by environmental degradation, then the logic of the proposal which comes out of classical economic theory is unassailable. "Internalization of costs" means, in effect, that the polluters would be made to take into account the costs they are imposing upon society, and expenditures on improving the environment would be increased until the marginal benefits became equal to the marginal costs. One problem, as already was indicated, is that neither the costs nor the benefits are probabilistically known. Another is that, even if they were, they would affect people differently. And this, in turn, means that the amount to be spent on pollution abatement is not an economic but a political decision.

The advice economists are qualified to give concerns the most efficient means of bringing about reductions in the risks of environmental degradation. And by pretending that they can give advice on both the resources to be devoted to pollution abatement and the means for bringing about an appropriate reduction in pollution, economists end up by laying down pompous preconditions which have little or no operational content.

3. *Direct Regulation*

If it were possible to utilize relatively known technology to quickly clean up the environment, then direct regulation would probably be the most efficient way to get the job done. Providing that a centralized planning bureau could collect the same information as a decentralized planning process, then there would be no inherent advantage in decentralized decision making.

The problem with direct regulation is that the relevant information cannot be obtained from a centralized planning system. To be sure, in principle, at least, it should be possible to obtain good information about the impact of environmental degradation on health and about measures to lower the rate of environmental degradation. But, while responsiveness to feedback makes for a good cybernetic system, it is not the essence of dynamic behavior. The essence of dynamic behavior is the ability to generate new alternatives. And the only way a centralized planning agency could acquire a realistic knowledge of the alternatives is by promoting active competition between business firms. Unless business firms are confronted with real risks, how can a realistic knowledge of the alternatives be acquired?

Because regulation is not ordinarily defined as the promotion of competition, regulatory agencies have tended to represent deplorably inefficient instruments from a dynamic point of view. One reason is that the response of such organizations in adjusting to new circumstances inclines to be just above zero. This slow reaction occurs because in the process of promulgating a particular regulation the regulatory agency and the regulatory industry tend to become a coalition whose main objective in life is to prevent the regulation from being changed. In the United States, regulatory agencies are staffed mainly by lawyers. And, on the whole, lawyers tend to be strict constitutionalists who, once a regulation is promulgated, can be counted upon to resist all future efforts to change it. By contrast, an efficient dynamic system is one which possesses a significant degree of "openness": a real ability to interact with an environment to change the entire system: man's ideas plus his physical environment. Quite obviously, ideological change must go hand in hand with changes in the physical environment. But, because of the constraints they impose upon ideological change, regulatory agencies tend to have a low degree of dynamic efficiency.

The second problem with regulatory agencies is that the incentives they provide to industry are perverse. As was pointed out, to make smooth progress when reducing environmental risks requires the generation of a wide diversity of ideas. However, the incentives provided by regulation are inclined to promote consonance of behavior: the regulated industry obviously has an incentive to get the regulation postponed; consequently, from that point of view it is necessary to present a united front to the regulatory agency. And it is not entirely an accident that, despite its large research and development expenditures, the American automobile industry has featured a low diversity of ideas with respect to reducing pollution: even in the absence of regulation, it is not an industry known for risk taking. Unfortunately, instead of counteracting such a tendency regulation reinforces it.

While regulation does make possible some degree of progress, it is a terribly slow and inefficient way of bringing it about. It is slow because of the small degree of openness typically associated with regulatory agencies. It is expensive because it features the generation of a low diversity of ideas. The only way to make progress smoothly and inexpensively is by having a variety of options from which to choose. And the only way regulation can contribute to this result is by the promotion of competition.

4. The Internalization of Risk

As applied to business competition, the term "internalization of risk" means acting upon a technological risk to avoid a market risk. In a world in which new discoveries can be made, competitors face two risks: a competitive risk and a technological risk. And to act to avoid a competitive risk is to internalize a technological risk, in the sense that it makes the future rate of progress more predictable. Thus, risk internalization consists of competitors imposing a tax upon each other—the bigger the tax, the more rapid is likely to be the rate of progress, because the bigger the tax, the greater the incentive to engage in unpredictable behavior in overcoming discontinuities by generating new ideological mutations.

Once the logic of internalizing risks is understood, it quickly becomes apparent that competition provides only one way of internalizing risks. As another illustration, consider rate-making procedures in the field of long-distance and international telephone calls. Here, due to the long lag in reducing telephone rates in response to reduction in costs, it really paid AT&T to stay ahead of the ratemakers in discovering ways to reduce costs. And as Fig. 2 shows, the telephone system did a good job of staying ahead of the ratemakers.

In very general terms, risk internalization can be compared to a tax that contains both positive and negative incentives: rewards for achievers and punishments for slackers. An increase in the tax implies, therefore, a strengthening of both positive and negative incentives (i.e., a more differentiated reward system). Risk internalization provides a society with a dynamic insurance policy: assuming there are six to eight firms in an industry, progress will be smooth. And the rate will depend upon the degree to which risks are

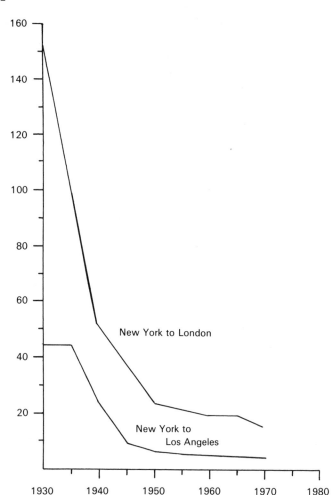

Figure 2. Long distance telephone rates (5 minutes, person to person).
Source: American Telephone and Telegraph, January, 1973.

internalized: the greater the "premiums"—the more differentiated the reward system—the more rapid the rate of progress.

More generally speaking, the objective in internalizing risks is to increase the degree of openness by making income after taxes depend on performance in improving productivity, however it is measured. And as the following quotation indicates, Sadi Carnot was familiar with the logic of the argument:

A tax on the rent of a farm would be much better than a tax on the land itself. Proprietors then could only avoid taxes by themselves improving their property. As it is, they merely collect the rents, and usually employ their surplus in unproductive expenditure, while the proprietary farmers devote theirs to the improvement of the land.[3]

Apparently, when Carnot wrote, a tax on land was on its raw unimproved value while the rent of land was on its improved value. If these assumptions are made, a tax on land would provide no incentive to improve it. But a tax on rent would provide an incentive to become a proprietary farmer, and to increase the value of the land. Thus, it is quite clear that Carnot had in mind a tax scheme aimed at increasing the degree of risk internalization.

Because the minds of most economists are to this day ruled by equilibrium economics, it is to be expected that they will not favor risk internalization schemes. However, it must be kept in mind that the issue is not one of regulation versus no regulation. It is, rather, an issue of what kind of regulation: between direct regulation which would provide perverse incentives for the generation of a diversity of ideas, and a risk internalization scheme which would provide positive incentives.

In one sense, a system of regulation does provide positive incentives. Legally speaking, the penalty of failing to meet the standards is a 100% excise tax. For example, an automobile company could be prevented from selling cars. But, practically speaking, the threat to escalate to full-scale antipollution war is not a credible threat, because in an industry in which millions of jobs might be affected no one would believe the threat.

Actually, it is no small wonder people in industry often proclaim, "We can live with regulation." What they mean is that they can engage in effective lobbying to get the regulations postponed. On the other hand, it is also easy to understand why direct regulations are very popular with politicians. The tough language contained in the laws makes good reading material for their constituents. In fact, if politicians want to be popular with both the public at large and business interests, the optimal policy is to carry a big stick but never use it. Another good example is to be found in recent measures taken in the United States to tighten the criminal penalties associated with the antitrust laws. Not only is the effect to make prosecution more difficult, but the politician can tell his constituents he favors a competitive society!

The advantage of incentive schemes over direct regulation is that they supply a more plausible threat when it comes to the internalization of risks. In order to internalize risks it is essential to adopt measures which will have a more-or-less calculable effect on business profits. Business firms must be left with no uncertainty that their profits after taxes will be larger if they display improved performance in reducing pollution.

Present regulation takes the form of the imposition of more and more stringent standards. There is nothing wrong with this general form of regulation. For the indefinite future there ought to be pressure on firms to discover ways to lower the rate of environmental degradation. Environmental taxes provide a more efficient method of regulation than direct controls because, whereas the first provide incentives to make technological discoveries to avoid the tax, the latter provide incentives to make political discoveries to escape from regulation.

This is not to say, of course, that with risk internalization measures the problem of creating perverse incentives would be completely obviated. For example, if an emissions tax were levied on automobiles only on the basis of emissions data obtained from newly manufactured cars, it could then be expected that automobile manufacturers would not put their best efforts into the development of emissions devices which are relatively easy to maintain. Indeed, this is already a problem. Therefore, to avoid this kind of perversity, an inspection scheme with a supplemental system of penalties would be required. Such measures would not, of course, please the consumer. But only by making the consumer feel the risk can automobile firms be motivated to develop emissions systems which would be easier to maintain. However, while such difficulties are inherent in all types of environmental taxes—whether on automobiles or newly constructed houses—such perverse effects are usually worse with direct regulation.

Nor do I want to imply there would be no difficulties in designing appropriate tax incentives. One difficulty is that of obtaining reasonably accurate measurements on the improvements in the performance of a technology from an environmental point of view. However, this is a difficulty which must be faced with any type of regulation. Another difficulty is that of deciding at what level to set the rates. It can be assumed that if the taxes were progressively raised, a level would be reached which elicited the appropriate degree of openness. However, given the inflexibility of tax systems in raising rates, it would obviously be desirable to be able to make predictions of the responsiveness of firms to increased degrees of risks to the environment. But to make predictions which might be accurate within a factor of two will require a good deal of empirical research.

The real constraint on the adoption of risk internalization measures is not, however, a lack of research. Generally speaking, policy-oriented research in the social sciences has not preceded the adoption of new policies; rather, it has followed. Thus, with one or two exceptions, economists in the United States did not become interested in the promotion of competition until long after the antitrust laws were adopted. And before Keynes argued that a mere reduction in wages would not restore the economy to a heavenly equilibrium, both the British and American governments were engaged in deficit financing. The real constraint is the attitude held in a number of countries that it is the responsibility of governments to spare their citizens from risk taking. The good life, politicians have come to believe, is life in which economic growth is so stable that no unpredictable event dares to occur. To be sure, biologists and psychologists now believe that people differ greatly in their propensities for security and adventure.[5] And, if this argument is correct, then there is nothing wrong with imposing a higher degree of risk on business firms and labor unions—the result would simply be that a different type of personality was favored. When economic policies favor microstability, the people who get to the top of business firms are lawyers and accountants, people who are trained to perform the same function in modern societies as genetic inbreeding performed in medieval societies. On the other hand, when they favor macrostability—smooth progress accompanied by a high degree of unpredictable microbehavior—people with a greater degree of openness are favored, as measured by their tolerance of ambiguity tests.

This distinction between the two types of people was once described by Thomas Jefferson as the "artificial" and the "natural" aristocracies. The principal difference, therefore, between a policy of direct regulation and one of risk internalization is that whereas the former favors the artificial aristocracy, the latter favors the natural aristocracy. In fact, the willingness of people to impose risk internalization measures on themselves can be regarded as a necessary condition for the preservation of democratic societies. The test of a democratic society, it may be assumed, is to survive in the face of trying circumstances by making good use of its natural aristocracy. But unless it is willing to impose tough risk internalization measures, how can a society meet this test? The basic reason, therefore, for adopting a dynamic approach is that it allows us to understand the logic of risk internalization.

References

1. U.S. Department of Commerce, Domestic and International Business Administration, *U.S. Industrial 1975 Outlook*, no. 0325–00020, Government Printing Office, Washington, D.C., p. 133.
2. B. M. Klein, *The Elements of Dynamic Economic Theory*, Harvard University Press, Cambridge, Mass., 1976.
3. R. H. Thurston, *Reflections on the Motive Power of Heat*, from the original French of N. L. S. Carnot, John Wiley and Sons, New York, 1890, p. 211.
4. Ibid.
5. C. D. Darlington, *The Evolution of Man and Society*, Simon & Schuster, New York, 1969; Erich Fromm, *The Anatomy of Human Destructiveness*, Holt, Rinehart & Winston, New York, 1973.

Aims of industrialists and technologists in grappling with problems of the environment often assume that all is known in the background fundamentals of chemistry which they hope to apply. In fact, there are startling gaps in basic chemistry which make many of the aims of the technologists—based upon general principles in chemistry—unattainable at this time. The next chapter points out some of the more basic research that is needed as a result of environmental chemical considerations.

23

A Few Lines in Research Relevant to Environmental Chemistry

J. O'M. Bockris

1. *General*

In earlier times, there was a sharp division between basic research, e.g., on nuclear fission, and applied research, e.g., on finding a brass polisher with improved chemical composition. The line of demarcation between pure and applied has grown more fuzzy in recent years, because there is no fundamental work for which applications cannot be foreseen,* and because most of the so-called applied work can only be done efficiently with a knowledge of the fundamental background.

However, another concept has entered into the research-planning situation in the last few years: mission-oriented research. One works toward a goal,

---

*However, Rutherford, the first research team leader to work on nuclear splitting, stated that he could not see how the process he had initiated could be of any practical significance.

J. O'M. Bockris • Department of Chemistry, School of Physical Sciences, Flinders University of South Australia, Bedford Park, South Australia 5042

and whether the work is defined as fundamental or applied is of no consequence. The concept of mission-oriented work was born during the NASA work on the moon mission in the United States during the 1960s. The mission was to make a manned landing on the moon. The research which had to be done to reach this objective was extremely varied in nature. It included fundamental work in mathematics, and indeed in most scientific fields. The main bulk of the work was in applied engineering.

It is this concept of mission-oriented work, a mixture of the applied with the pure, depending upon what is needed, which turns up in our present concepts of many research lines germane to environmental chemistry.

Another concept in research, which differs from the earlier concepts of pure and applied, is that of "fundamental applied research." This description does not involve a contradiction. One takes some applied objective: e.g., the direct conversion of coal to electricity, and researches those fundamental aspects of it which are preventing progress, e.g., the mechanism of the catalysis of hydrocarbons on metal electrodes, in their electrooxidation to CO_2.

It is these more recent concepts of the divisions of research—in which the puristic aspects of academic scientists are diminished in favor of the usefulness of the endeavor within a foreseeable time (decades!)—which permeate the spirit of the following remarks.

But let us recall being able to have too much of a good thing. It may well be that, in the next few decades, and perhaps for many decades, we will go on increasing our (more difficult) applied research at the expense of the easier, fundamental work. The new realizations which were introduced with relativistic physics in the early part of the century and with the quantum theory in the 1920s have not yet been completely assimilated into the general concepts of science, and certainly are little exploited. Thus, for the most part, we are still exploiting the electromagnetic theory of the nineteenth century: telecommunications, television, lighting, etc. It will take a minimum of another half-century or more to catch up on the exploitation of the quantum theory and relativity. But we must never go *too* far in changing the equilibrium constant between research which does have a conscious objective and the "blue sky" research with no conscious objective, except in the following up of curiosity, which indeed has given rise to most of the great advances in science, and therefore in the applied sciences, and, eventually, to the improvements in the standard of life. Just wondering, just asking blank questions contrary to the intellectual framework of the time, must forever be a *revered activity* in an evolving society. Research may be, in the future, a smaller light, but it will always be a match which lights the light which shows that society is living.

2. Surface Chemistry under Electric Fields in Biology

Some of the more basic thoughts about contraception will be concerned with the nature of the passage of spermatozoa across the surface of the ovum, a process about which we know distressingly little. The spermatazoa are colloidal, and alterations of the potential difference at the interface between the ovum and the surrounding fluid may be expected to affect not only the mobility of the sperm near the ovum, but also its ability to penetrate it.

This is an example of *surface chemistry* and its relation to electrical properties.

The chemistry of interfaces is involved with many of the problems associated with environmental chemical problems. However, most of the research work which has been done on the electrode–solution interface has involved metals. This chemistry must now orient itself away from the metal–solution interface, where the situation is understood, to the semiconductor–solution interface, and finally to the analogous interface involving an insulator, i.e., the semiconductor with a small electronic concentration in its conductance band in contact with an ionic solution. It is this model which comes nearest to a biological system, and which might be involved in a deeper look at the fertilization process.

The brain mechanism is, taken as a whole, above our ability to understand as yet at a molecular level. However, we are beginning to understand a few mechanisms, e.g., anesthetics. Similarly, we may be able to understand something of the action of pollutants on the brain by considering electron transfer reactions of adsorbed particles at the synapses. The mechanism may be analogized by the adsorption of organic materials on semiconductor–solution interfaces. The organic contains an electrical group, and the field exerted by this on the solid changes the electrochemical activity at the surface, and hence the potential difference at the interface. This changes the rate of the charge transfer reactions through the interfaces, and thus the overall rate of the chemical reaction.

There is a long way to go in this direction. An understanding, at a molecular level, of the biological mechanism of such matters as how the brain detects odors, and the effect of anesthetics, may lead to an understanding of the effect of trace metal particles and complex drugs on the bioelectrical properties of the brain.

3. *The Mechanism of Selectivity in Enzymes*

Enzyme reaction mechanisms are the central fundamental theme of the possibility of obtaining food from CO_2 and nitrogen. In many senses, enzymes still seem like magic in chemistry, for we have many gaps in our knowledge of how they work. The catalysis power of an enzyme is greater than that of most metal catalysts. It is the mechanism of *selectivity* in enzymes, and how this might be imitated artificially, which we must understand.

We must not forget a more down-to-earth aspect of enzyme chemistry: our ability to synthesize them. Since nature is so good at it, we now use bacteria as the source of enzymes, but in many cases envisaged for our future use of enzymes—e.g., the possible photoproduction of hydrogen from water—the enzymes will be needed in large quantities and would have to be made artificially.

4. *Electrocatalysis*

There are several problems awaiting solution—e.g., in recycling, or in dealing with sewage—where we want to take degraded molecules and convert them to a higher state. This requires further research on selectivity in electrode processes. The selectivity is better than in chemical catalysis, where one has only specific action of the catalyst, surface, pressure, and temperature, whereas in the electrode catalyst there is another variable, the charge at the interface and the potential difference across it. Selectivity in *electrocatalysis* is a subject of future import, but on the negative side is the tendency of electrocatalysts to fade in activity with time. We need to understand this mechanism and how to rejuvenate the catalyst.

One single electrochemical reaction must be underlined, and that is the solution phase reduction of oxygen to water. This is present in biological situations, but also in the reduction processes by which fuel cells operate and carry out chemical reactions with the by-product of electricity.

5. *Ionic Solution Process at High Pressures and Temperatures*

Resource exhaustion studies suggest not only recycling, but also increased opportunities for geochemical mining. There may be possibilities for *in situ* mining, i.e., mining which avoids the tearing apart of the materials of the earth, but which injects steam under high pressure for the dissolution of materials, after which they may be electrochemically extracted from the solution outside the mine. Underground gasification is another process which could help recover energy from coal speedily, although it does not seem to have a happy research history nor many prospects for the future.

Another technology which contains research opportunities and which is a part of environmental chemistry, because it is a part of resource chemistry, is the extraction of metals from the sea. Although the products of the sea provide no panacea in respect to food—most fish do not go very far from land because there is not much food for them in deep water—metals in the sea are dispersed through the world's oceans and are large indeed compared with those which are known to exist under minable conditions under the earth. Most of the apparent possibilities in sea mining fall down because of the energy needed to pump vast quantities of the sea through, for example, membranes. However, there are seas where the water flows at, for example, 4 mph under cost-free natural forces. Here, then, one could see the possibility of large-scale extractions of noble metals and uranium from water, and possibly some other metals, e.g., aluminum. The chemistry involved is that of the cheap synthesis of great quantities of chelating agent, which must be highly selective and must give up metal when introduced into a milieu of different pH.

6. *Chemical Reaction Kinetics in Solution*

Until the 1970s there was little stress upon recycling, and yet the exhaustion time of many of our metals is within the present century, and economic recycling processes are likely to take decades to develop. Thus, the ability to recycle to a high degree of completeness will determine whether we are able to continue to use many metals at all. It seems likely that the degree of completeness of recycling will be the determinant for the final viable population of the earth at a high living standard, rather than the supply of energy.

7. *Heterogeneous Catalysis*

Heterogeneous catalysis is a subject basic to the control of noxious exhausts. Selectivity, durability, reactivation, mechanisms of decay, all these are topics which have been investigated for many years, but the research has become asymptotic and needs rejuvenation and newer techniques. Nevertheless, we will never be completely able to eliminate noxious materials from certain gas streams, e.g., automotive exhausts. It will be a matter of degree. Just as the inevitability of the Carnot efficiency factor played little part in energy thinking until the 1960s, so the inevitability of leaving some pollutants behind in the atmosphere must be brought into account in our thinking with respect to our choice of fuels, whether it is acceptable to continue to use fossil fuels until the end of coal, or whether we should make an all-out effort to tap the inexhaustible sources of energy and to engineer the conversion and distribution equipment with all possible speed.

8. *New Processes in Cyclical Chemical Technology*

Another topic which comes out of the considerations of clean-up is what to do with that which has resulted from the clean-up. One example would be sulfur, which may well be produced from cleaner versions of many metallurgical processes where the metal comes from the sulfide. Here is the possibility of inventiveness in eliminating the present use-up of sulfur and in making cyclical processes of sulfur extraction and utilization.

9. *Large-Scale Chemical and Electrochemical Synthesis of Hydrogen*

The production of hydrogen is to be an important reaction in the near future: it will give us a clean fuel which may have advantages over the nearest competitor, electricity. The price of massively produced hydrogen will probably become as important as the prices of gasoline and electricity are now. The price of hydrogen will depend on that of the fundamental energy used to free H_2 from H_2O. But, within those limits, substantial differences can be determined by the method of synthesis. Should this be through *chemical* reactions, with the difficulty of the corrosive properties of hot aqueous solutions? Should it be electrochemical, while paying the full cost of the Carnot cycle? Or should it be photolytic? A choice among these questions will resolve a central issue in determining the future of the environment.

10. *The Stability of Aerosols*

Colloidal particles and their stability obtrudes into environmental chemistry. It is a subject which strangely became stuck in the 1940s, after which only sporadic work has been done, at least in Western countries. Yet our ability to deal with the destabilization of, for example, aerosols, may play an important role in our future.

11. *Photo Processes in the Atmosphere*

Several meteorological topics are vital ancillaries to environmental chemistry. Reflectivity from cloud will determine how much feedback is related to the greenhouse effect. As the temperature rises due to increase in the CO_2 in the atmosphere, more cloud will form because more water will evaporate. Will this cloud cause reflectance of the solar radiation away from the earth, thus leading to a cooling in world temperatures?

The CO_2 balance in the atmosphere is an important subject in environmental chemistry. The precise balance of CO_2 between its production in metabolism, its absorption in the sea, and its use in photosynthesis, all these need much more, and more exact, investigation; the amount of time we can use our fossil fuels after 2000 may depend upon the answer to the questions implied in considerations of these matters.

Investigations of CO_2 will play a part in another matter affecting our future, for we may have to have much to do with its reduction: it may form the basis of our fuel supply, if CH_3OH becomes the motor fuel of the future. We have much CO_2, for soon it will be one-half per cent of the atmosphere which we breathe. Its recovery, and catalytic and enzymatic reactions with hydrogen (from water and solar energy?), will give us the possibility of synthetic food production on a large scale. If the chemistry of the reduction of CO_2 can be coupled with knowledge of selectivity in enzymatic chemistry, prospects will be good.

12. *Inorganic Chemistry*

Inorganic chemistry needs a degree of rejuvenation in environmental chemical studies. Much knowledge which we need is lacking. For example, we do not know the potentials of many redox processes. What about the pressure of the sea and its effect upon such parameters as the partial molar volume of dissolved materials? We need to know this to find out effects at depth on, for example, the solubility of gases. There is a mountain of routine research and accurate new data gathering which must be done. Much attractive inorganic chemistry will have to be mined in respect to extraction processes from the sea, as well. For example, numerous shellfish have the ability to extract specific metals, e.g., vanadium, from the sea, and we do not yet know what the complexing agent is. Metalloorganic chemistry is an area relevant to seawater extraction.

13. *New Work on Solution Equilibria*

I refer to work which has had antecedents many years ago, e.g., the solubility of chlorine in solutions of various ionic strengths. What about its pressure coefficient? How fast does chlorine reequilibrate with the Cl^- in seawater? What is the pressure dependence of the reaction between chlorine and water to form oxygen and HCl? These questions must be answered if we are to have a satisfactory electrolysis of seawater, and yet also convert the Cl_2 coevolved with O_2 back to HCl and O_2.

14. *Solid-State Chemistry*

Solid-state chemistry is likely to play a great part in solar energy studies, e.g., in the study of new oxides which can be absorptive in the solar spectrum and which could contribute to thermoelectric junctions or photoelectric couples. It is remarkable how little we know about the absorptive spectra of many common oxides. Even simple parameters such as thermal expansivity are little known, although the substances are heated and cooled each day. Degradation processes, e.g., of warm cadmium sulfide by water vapor, are only poorly understood.

15. *Ion Exchange Processes*

Similar remarks apply to ion-exchange as did to extractions from the sea. Knowledge needed in all these areas has been obtained only in outline, only with examples.

16. *An Organic Chemistry Based on CO₂*

Organic chemistry is sometimes said to be at an asymptotic stage, but this need not be so in its application to environmental chemistry. Enzymatic processes have been stressed. A new organic chemistry could be based upon the reduction products of CO_2. New insecticides will be needed in the future; they must not poison us, as does DDT. And plastics? How can they be recycled without environmental damage?

17. *Electrochemical Energy Storage and Conversion*

Transportation problems provide plenty of research for the chemist, and not only the electrochemist. How can electric batteries be made light and yet contain more energy than does the lead battery? What reactions take place at the plates of the new cells? How do competitive reactions cause decay so that the batteries fail too early? Satisfactory batteries are never light enough, however, and this makes one turn to hydrogen and its use in internal combustion. Will the poisonous properties of nitric oxide spoil the use of hydrogen in internal combustion engines? What kind of supplementary reactions could be introduced to eliminate NO?

18. *Research into Economics and Political Processes*

It is not usual in a book on a scientific topic to carry statements about the topics which are still in a prescientific phase. However, it seems desirable to state that the applicability of chemistry to the environment depends enormously on the economic system. What seems very necessary is to *research* the systems in a scientific and objective way and to attempt to introduce into the study of politics and economics an increased range of sensors, and, therewith, an increase in scientific objectivity and range of prediction. Thus, at present, it is clear that economists are unable to adequately predict the near future. Statements of the type, "Inflation is not understood," are made. It is arguable that economics and politics cannot be exact sciences because they deal with the behavior of groups and crowds. This limitation clearly exists. However, the behavior of crowds, and perhaps of some smaller groups, is becoming more predictable.

Thus, research into politics and economics themselves is needed. To what extent, at this time, do these bodies of knowledge have the predictive power in assessing group reaction, and the interaction of people in masses toward economic measures arriving from new information?

Thus, the scientist sees that he has two opposed bodies to whom he offers his findings. The first of these are what might be called "groups," by which is meant the managers of capital, or, roughly, the directors of corporations. Their time unit is the year, and profits must be optimized within it. Persistent failure to succeed in this would lead to the termination of an admirable existence in affluence and power. Suggestions of expenditure—e.g., to recover, recycle, purify—are irritating heresy to these groups, hazards in the path of profitable progress to be avoided by astuteness.

The other body might be called the crowds, those not having to do with the management of capital. Although microfragments of such groups may resonate to the environmental scientist's findings, the vast majority has no understanding of them whatsoever, and only becomes exposed to the distorted anxiety-producing fragments which the media misleadingly selects from the whole on the basis of its sellability.

Thus, for the most part, the crowds remain in blissful ignorance of the environmental threats to survival (e.g., air-borne lead from automotive pollution), while occasionally reacting with a vigorous quaking to much lesser difficulties (e.g., mercury in fish).

Studies in economics and politics must grapple with gigantic problems arising from ecological considerations. The scientist brings the information. Should it be disseminated? How can it be disseminated without distortion? Understood? Absorbed? The relative significance understood in the face of so many clammerings for expenditure? How can the crowd be enlivened from the sodden mass of information-drunk blobs which it now is, directionless, hedonistic, influenced only by the groups who have money to pay the media to print the loaded message, making more offerings to the crowd's hedonism?

Economists and politicians must also be vitally and *urgently* concerned with the groups. The groups have the information and understand it. Their reaction is governed by self-interest. In theory, the system should work in this way: Group A will be interested in coal sales, say, and wish to suppress all information concerning lack of ability to remove sulfur. However, Group B is interested in electric car sales, and they will be able to purchase media space to represent, doubtless with equal one-sidedness, the advantages of the product they prefer.

However, of course, the system does not work, in this respect, because of the disparate size and power among the groups. Owners of the one product may be super-giants who can saturate the media with a certain message, whereas the vendors of the second product may be pygmies, hardly able to afford a straight advertisement. Hence, the blob, the sum of the crowds, remains ignorant and inactive in respect to product B, but Pavlovianly responsive to product A.

It is absolutely clear that in the study and practice of politics and economics lie many of the answers to the problems of environmental chemistry. Scientists and engineers can, by and large, solve these problems. Each solution, of course, comes with a price tag. Whether the price will be paid—and who decides that, and how, and when—is what needs to be investigated, brought out, discussed, and worked out to optimization by the community. Whether this is possible, how it works, how it can work better, and the implementing of that better working—that is the province of the study of economics and politics.

Thus, the last line of research to be suggested in this book is research into these fields, the fields of economics and politics and their relation to the lives of the crowds, their relation to long-term planning. Is the development adequate to grapple with the scientist–group–crowd relations to the advantage of the crowd? Can our system act and react in time to the benefit of the many? If not, what kind of system can?

Simplistic answers are not in order. Environmental problems are very great, and very rapid reactions (less than decades) are needed. It would be naive to state that such reactions necessitate a change in the political system, its metamorphic conversion to another system with a definite nature, an indefinite significance, and unknown consequences. It is correspondingly obvious that, for perhaps two or three generations, we do need considerable modification and evolution of our present political and economic systems to cope with the information given by the environmental scientist. Indeed, that is necessary if anything at all resembling our present way of life is to survive another generation.

Index

771